Cornerstones in Contemporary Inorganic Chemistry

Cornerstones in Contemporary Inorganic Chemistry

Editor

Duncan Gregory

MDPI • Basel • Beijing • Wuhan • Barcelona • Belgrade • Manchester • Tokyo • Cluj • Tianjin

Editor
Duncan Gregory
School of Chemistry
University of Glasgow
Glasgow
United Kingdom

Editorial Office
MDPI
St. Alban-Anlage 66
4052 Basel, Switzerland

This is a reprint of articles from the Special Issue published online in the open access journal *Inorganics* (ISSN 2304-6740) (available at: www.mdpi.com/journal/inorganics/special_issues/cornerstones_inorganic_chemistry).

For citation purposes, cite each article independently as indicated on the article page online and as indicated below:

LastName, A.A.; LastName, B.B.; LastName, C.C. Article Title. *Journal Name* **Year**, *Volume Number*, Page Range.

ISBN 978-3-0365-5706-9 (Hbk)
ISBN 978-3-0365-5705-2 (PDF)

© 2022 by the authors. Articles in this book are Open Access and distributed under the Creative Commons Attribution (CC BY) license, which allows users to download, copy and build upon published articles, as long as the author and publisher are properly credited, which ensures maximum dissemination and a wider impact of our publications.

The book as a whole is distributed by MDPI under the terms and conditions of the Creative Commons license CC BY-NC-ND.

Contents

Duncan H. Gregory
Cornerstones in Contemporary Inorganic Chemistry
Reprinted from: *Inorganics* 2022, 10, 108, doi:10.3390/inorganics10080108 1

Ines Wackerbarth, Ni Nyoman Agnes Tri Widhyadnyani, Simon Schmitz, Kathrin Stirnat, Katharina Butsch and Ingo Pantenburg et al.
Cu^{II} Complexes and Coordination Polymers with Pyridine or Pyrazine Amides and Amino Benzamides—Structures and EPR Patterns
Reprinted from: *Inorganics* 2020, 8, 65, doi:10.3390/inorganics8120065 7

Giacomo Manfroni, Simona S. Capomolla, Alessandro Prescimone, Edwin C. Constable and Catherine E. Housecroft
Isomeric 4,2′:6′,4- and 3,2′:6′,3-Terpyridines with Isomeric 4′-Trifluoromethylphenyl Substituents: Effects on the Assembly of Coordination Polymers with [Cu(hfacac)$_2$] (Hhfacac = Hexafluoropentane-2,4-dione)
Reprinted from: *Inorganics* 2021, 9, 54, doi:10.3390/inorganics9070054 27

Sourav Mondal, Wei-Xing Chen, Zhong-Ming Sun and John E. McGrady
Synthesis, Structure and Bonding in Pentagonal Bipyramidal Cluster Compounds Containing a *cyclo*-Sn5 Ring, [(CO)3MSn5M(CO)3]$^{4-}$-($M = Cr, Mo$)
Reprinted from: *Inorganics* 2022, 10, 75, doi: 10.3390/inorganics10060075 47

Dafydd D. L. Jones, Samuel Watts and Cameron Jones
Synthesis and Characterization of Super Bulky -Diketiminato Group 1 Metal Complexes
Reprinted from: *Inorganics* 2021, 9, 72, doi:10.3390/inorganics9090072 59

Xianji Qiao, Alex J. Corkett, Dongbao Luo and Richard Dronskowski
Silver Cyanoguanidine Nitrate Hydrate: Ag(C$_2$N$_4$H$_4$)NO$_3$·12 H$_2$O, a Cyanoguanidine Compound Coordinating by an Inner Nitrogen Atom
Reprinted from: *Inorganics* 2020, 8, 64, doi:10.3390/inorganics8120064 69

Christian Bäucker, Peter Becker, Keshia J. Morell and Rainer Niewa
Novel Fluoridoaluminates from Ammonothermal Synthesis: Two Modifications of K$_2$AlF$_5$ and the Elpasolite Rb$_2$KAlF$_6$
Reprinted from: *Inorganics* 2022, 10, 7, doi:10.3390/inorganics10010007 77

Kristen A. Pace, Vladislav V. Klepov, Mark D. Smith, Travis Williams, Gregory Morrison and Jochen A. Lauterbach et al.
Hydrothermal Synthesis and Structural Investigation of a Crystalline Uranyl Borosilicate
Reprinted from: *Inorganics* 2021, 9, 25, doi:10.3390/inorganics9040025 91

Chao Lin, Alexandre C. Foucher, Eric A. Stach and Raymond J. Gorte
A Thermodynamic Investigation of Ni on Thin-Film Titanates (ATiO$_3$)
Reprinted from: *Inorganics* 2020, 8, 69, doi:10.3390/inorganics8120069 105

Wei Zhao, Zhihao Liu, Yanan Zhao, Yi Luo and Shengbao He
Multivariate Linear Regression Models to Predict Monomer Poisoning Effect in Ethylene/Polar Monomer Copolymerization Catalyzed by Late Transition Metals
Reprinted from: *Inorganics* 2022, 10, 26, doi:10.3390/inorganics10020026 119

Joshua Northcote-Smith, Alice Johnson, Kuldip Singh, Fabrizio Ortu and Kogularamanan Suntharalingam
Breast Cancer Stem Cell Active Copper(II) Complexes with Naphthol Schiff Base and Polypyridyl Ligands
Reprinted from: *Inorganics* 2021, 9, 5, doi:10.3390/inorganics9010005 . . . . **129**

Giselle M. Vicatos, Ahmed N. Hammouda, Radwan Alnajjar, Raffaele P. Bonomo, Gabriele Valora and Susan A. Bourne et al.
Aqueous Solution Equilibria and Spectral Features of Copper Complexes with Tripeptides Containing Glycine or Sarcosine and Leucine or Phenylalanine
Reprinted from: *Inorganics* 2022, 10, 8, doi:10.3390/inorganics10010008 . . . . **143**

Federica Arrigoni, Giuseppe Zampella, Luca De Gioia, Claudio Greco and Luca Bertini
The Photochemistry of $Fe_2(S_2C_3H_6)(CO)_6$(-CO) and Its Oxidized Form, Two Simple [FeFe]-Hydrogenase CO-Inhibited Models. A DFT and TDDFT Investigation
Reprinted from: *Inorganics* 2021, 9, 16, doi:10.3390/inorganics9020016 . . . . **173**

Joseph P. A. Ostrowski, Ashley J. Wooles and Stephen T. Liddle
Synthesis and Characterisation of Molecular Polarised-Covalent Thorium-Rhenium and -Ruthenium Bonds
Reprinted from: *Inorganics* 2021, 9, 30, doi:10.3390/inorganics9050030 . . . . **191**

Alasdair Formanuik, Fabrizio Ortu, Iñigo J. Vitorica-Yrezabal, Floriana Tuna, Eric J. L. McInnes and Louise S. Natrajan et al.
Functionalized Tris(anilido)triazacyclononanes as Hexadentate Ligands for the Encapsulation of U(III), U(IV) and La(III) Cations
Reprinted from: *Inorganics* 2021, 9, 86, doi:10.3390/inorganics9120086 . . . . **201**

James W. Annis, Janet M. Fisher, David Thompsett and Richard I. Walton
Solvothermal Synthesis Routes to Substituted Cerium Dioxide Materials
Reprinted from: *Inorganics* 2021, 9, 40, doi:10.3390/inorganics9060040 . . . . **217**

Fernanda Paiva Franguelli, Kende Attila Béres and Laszló Kótai
Pyridinesilver Tetraoxometallate Complexes: Overview of the Synthesis, Structure, and Properties of Pyridine Complexed $AgXO_4$ (X = Cl, Mn, Re) Compounds
Reprinted from: *Inorganics* 2021, 9, 79, doi:10.3390/inorganics9110079 . . . . **251**

Mamo Gebrezgiabher, Yosef Bayeh, Tesfay Gebretsadik, Gebrehiwot Gebreslassie, Fikre Elemo and Madhu Thomas et al.
Lanthanide-Based Single-Molecule Magnets Derived [-25]from Schiff Base Ligands of Salicylaldehyde Derivatives
Reprinted from: *Inorganics* 2020, 8, 66, doi:10.3390/inorganics8120066 . . . . **265**

Alessia Tombesi and Claudio Pettinari
Metal Organic Frameworks as Heterogeneous Catalysts in Olefin Epoxidation and Carbon Dioxide Cycloaddition
Reprinted from: *Inorganics* 2021, 9, 81, doi:10.3390/inorganics9110081 . . . . **305**

Editorial
Cornerstones in Contemporary Inorganic Chemistry

Duncan H. Gregory

WestCHEM, School of Chemistry, University of Glasgow, Glasgow G12 8QQ, UK; duncan.gregory@glasgow.ac.uk

Citation: Gregory, D.H. Cornerstones in Contemporary Inorganic Chemistry. *Inorganics* **2022**, *10*, 108. https://doi.org/10.3390/inorganics10080108

Received: 24 June 2022
Accepted: 4 July 2022
Published: 28 July 2022

Publisher's Note: MDPI stays neutral with regard to jurisdictional claims in published maps and institutional affiliations.

Copyright: © 2022 by the author. Licensee MDPI, Basel, Switzerland. This article is an open access article distributed under the terms and conditions of the Creative Commons Attribution (CC BY) license (https://creativecommons.org/licenses/by/4.0/).

I am very happy to be able to present this Special Issue of *Inorganics*. The aim of this Special Issue, "Cornerstones in Contemporary Inorganic Chemistry", was to assemble a collection of seminal articles representing some of the most important aspects of inorganic chemistry in the early 21st century. I am delighted to say that the response to the call for papers was overwhelming but that it has been possible to select some of the finest works for publication in this issue. Inorganic chemistry remains a broad church, and the contributions to this Special Issue reflect just that. With a healthy balance of original papers and review articles, this Special Issue is able to capture the excitement at the cutting edge of some of the most topical of research areas and to address some of the key questions that occupy inorganic chemists today. This Special Issue also collates informative and thought-provoking works that can be referred to for a long time to come by researchers at all levels.

Altogether, this Special Issue compiles 14 original research articles and 4 review papers. The original research articles are considered first, and in no particular order, several different aspects of coordination chemistry are represented within this Special Issue. As is often the case in current times with many branches of chemistry (and science, more generally), the themes can be cross-cutting. Modern coordination chemistry has often taken the researcher beyond the bounds of the traditional coordination complex or molecule into the realms of supramolecular and inorganic polymeric chemistry. In this context, Axel Klein and co-workers examined the world of copper(II) complexes formed from pyridine/pyrazine amide and amino benzamide ligands [1]. Such ligands offer a variety of hard and soft donor atoms and functional groups, offering the potential to form isolated complexes or coordination polymers. These are both the types of ligands favoured as building blocks in metal organic frameworks (MOFs) and the ones that are featured in many biological systems of interest (for example, in metalloenzymes). In this work, the authors discovered that it was the donor O atom of the amide groups that bonds preferentially to Cu(II) (as opposed to the N atom of the amide NH_2 group), giving axially distorted octahedral (or square pyramidal) copper environments with the remainder of the coordination sphere typically made up from pyridine/pyrazine ring N donors. The crystallography is corroborated by EPR, UV-Vis and magnetic measurements.

The theme of copper coordination polymers is continued in the paper by Housecroft et al., which elegantly demonstrated how ligand conformation can dictate the way in which coordination polymers pack, even when such packing is governed by relatively weak intermolecular interactions [2]. The ligands in question are isomers of terpyridine (4,2′:6′,4″ and 3,2′:6′,3″), and it is the positions of the N donor atoms that are crucial in determining the type of 1D chain that is observed in the resulting coordination polymers formed by reactions with $Cu(hfacac)_2 \cdot H_2O$. The zig-zag chains formed with V-shaped 4,2′:6′,4″-terpyridine are held together by C–F···F–C contacts, whereas chains formed with 3,2′:6′,3″-terpyridine are packed together through π–π stacking. The latter interactions are rendered impossible by the conformation of functional groups on the former chains.

We turn from 1D coordination polymers to Zintl phases as McGrady et al. combined experiments and calculations (density functional theory; DFT) to examine two new heterometallic clusters [3]. The two clusters, $[(CO)_3CrSn_5Cr(CO)_3]^{4-}$ and $[(CO)_3MoSn_5Mo(CO)_3]^{4-}$, contain a similar pentagonal bipyramidal core, which can each be rationalised in terms of Sn_5^{4-} rings bridging two zero valent $M(CO)_3$ (M = Mo or Cr) fragments. Intriguingly, their calculations revealed two types of dominant bonding: the bonds within the Sn_5^{4-} ring and *trans*-annular bonds formed between the two apical M atoms. In the case of the two synthesised compounds, the LUMO is dominated by the Sn-Sn π-bonding in the rings, but the balance is subtle and other isoelectronic 28 electron clusters possess LUMOs localised on the M_2 apices, giving rise to the aforementioned *trans*-annular bond.

Of course, coordination complexes are not strictly the domain of d-block metals and Jones, Watts and Jones presented a study of alkali metal complexes formed with "super bulky" β-diketiminate (Nacnac) ligands [4]. Both lithium and potassium complexes were successfully produced, and each was shown to be highly reactive, offering the possibility of using the complexes as reagents themselves for the synthesis of new low-oxidation state metal complexes, which the ligands could stabilise kinetically. Finally, in terms of coordination chemistry, Richard Dronskowski and colleagues reported a new silver guanidine complex in their communication, which includes various aspects of its solid state chemistry [5]. In fact, the silver(I) cyanoguanidine nitrate hydrate, $Ag(C_2N_4H_4)NO_3 \cdot 1/2H_2O$, represents the first example of a monovalent silver complex coordinated through the inner (as opposed to terminal) nitrogen N atoms of the cyanoguanidine ligand. Given that the hydrate was an unexpected product in the intended synthesis of $Ag(C_2N_4H_4)NO_3$, the authors used first principles calculations to demonstrate that bonding through the inner N atom of the ligand is indeed energetically less favourable than coordinating via the terminal atoms.

This issue also contains several examples of investigations in solid-state chemistry that readily bridge into topics of materials chemistry and catalysis. Rainer Niewa and co-authors demonstrated the effectiveness of ammonothermal synthesis in the preparation of fluoridoaluminates, K_2AlF_5 and Rb_2KAlF_6, which contain infinite fluoroaluminate chains and isolated $[AlF_6]^{3-}$ octahedra, respectively [6]. The former is a known compound but is shown here to exist as two new polymorphs that both crystallise in the same space group under synthesis conditions only 20 K and 2 MPa apart. One modification contains a chain with double the number of unique "links" to the other, thus doubling the unit cell in one direction. Rb_2KAlF_6, meanwhile, is a new example of a complex fluoride with the elpasolite structure, essentially a distorted double perovskite with aluminium and potassium occupying the octahedral B sites in a completely ordered arrangement. Solvothermal synthesis was also employed by Zur Loye et al. to prepare new solid-state compounds, but on this occasion, more conventionally, the solvent was water [7]. Hydrothermal synthesis leads to $K_{1.8}Na_{1.2}[(UO_2)BSi_4O_{12}]$—the first reported example of a uranyl borosilicate. The borosilicate forms a 3D framework from the interlinking of two different structural motifs; otherwise, isolated UO_6 octahedra and infinite chains were constructed from unprecedented $[BSi_4O_{12}]^{5-}$ anions, themselves composed of vertex-sharing BO_4 and SiO_4 tetrahedra. The Na^+ and K^+ cations then inhabit the cavities within the pseudo-3D framework. Crucially, the complex crystal chemistry and the absence of peroxo groups were established using EXAFS, thermal analysis and calculations in addition to X-ray crystallography. Ultimately, fundamental studies such as this can help us understand how nuclear waste—and actinides such as U—can be captured in glasses.

Redispersed and exsolved catalysts (such as Rh and Ni) on perovskite substrates are attracting heightened interest for a range of chemical processes, particularly in fuel cells and for steam reforming. High temperature stability and coking resistance are particularly attractive qualities. In this context, Gorte and colleagues conducted a systematic study of Ni supported on thin film perovskite substrates deposited on $MgAl_2O_4$ by atomic layer deposition (ALD) [8]. Titanates, $ATiO_3$ (A = Ca, Sr, Ba), are considered, and the choice of alkaline earth metal is revealed to be crucial. Unfortunately, all the perovskite samples were less active than Ni on $MgAl_2O_4$ alone, and Ni is more easily oxidised on the perovskite substrates. Ni interacts most strongly with $CaTiO_3$ to the point where the catalyst is deactivated during CO_2 reforming of CH_4 at high CO_2 pressures. Additionally, on the topic of heterogeneous catalysis, Luo et al. used density functional theory (DFT) calculations and multivariate linear regression (MLR) to rationalise the poisoning effects in ethylene/polar monomer copolymerization catalysed by Brookhart-type catalysts [9]. Polar monomers with electron-rich functional groups such as the carbonyl, carboxyl and acyl groups demonstrated the strongest poisoning effects. Among several factors that mediate the poisoning effect, the metal-X distance in the σ-coordination structure was identified as important. The detailed models developed in this study should help improve copolymerization catalyst designs.

There follow several original papers that focus on topical bioinorganic and organometallic chemistry. The first is a communication by Ortu, Suntharalingam and co-workers, which describes new copper complex chemotherapeutics as a potential new treatment to remove breast cancer stem cells (CSCs) [10]. The copper complexes combine tridentate (O,N,S)-coordinated naphthol Schiff base ligands with 1,10-phenanthroline (and PF_6 counter ions) to kill breast CSCs in the micromolar range, with two of the complexes more than four times more effective in this respect than the established anti CSC agent, salinomycin. Copper(II) complexes have also demonstrated anti-inflammatory properties that can be exploited in the treatment of rheumatoid arthritis. Jackson et al. explored a series of Cu(II) tripeptide complexes, in which Cu(II) was square planar and first demonstrated that the complexes have sufficiently low stability constants to allow for their release in vivo [11]. Then, using techniques including EPR, NMR, absorption spectroscopy and DFT calculations, the authors were able to develop reasonable models to describe the coordination modes for each complex. Greco, Bertini and colleagues presented studies in which Fe^IFe^I $Fe_2(S_2C_3H_6)(CO)_6(\mu\text{-}CO)$ (^{1}a-CO) and its Fe^IFe^{II} cationic species (^{2}a$^+$-CO) were used as a basic model for the CO-inhibited [FeFe] hydrogenase active site [12]. The site was previously shown to undergo CO photolysis but the conditions and mechanism remained unclear. Time-dependent DFT (TDDFT) analyses of the ground state and low-lying excited-state potential energy surfaces (PESs) have delivered $Fe_2(S_2C_3H_6)(CO)_6$ (^{1}a) and $[Fe_2(S_2C_3H_6)(CO)_6]^+$ (^{2}a$^+$) as two simple models of the catalytic site. From a calculation of free energy barriers, it should be possible for ^{2}a$^+$-CO to be synthesized experimentally. Moreover, even when UV light is absorbed, during excited state processes, decay occurs through photo-dissociative channels involving a CO ligand, preventing irreversible enzyme photo-damage.

In one of the two more fundamental organometallic contributions, Stephen Liddle and colleagues described how alkane elimination reactions lead to two new exciting bimetallic thorium complexes [13]. These are especially notable since $[Th(Tren^{DMBS})Re(\eta^5\text{-}C_5H_5)_2]$ ($Tren^{DMBS} = \{N(CH_2CH_2NSiMe_2Bu^t)_3\}^{3-}$) is the first example of a complex with a Th-Re bond that has been structurally characterised, while the $[Th(Tren^{DMBS})Ru(\eta^5\text{-}C_5H_5)(CO)_2]$ complex represents only the second example of a structurally characterised analogous Th–Ru bond. The successful synthesis and isolation of these complexes has enabled a comparison to be made with uranium analogues and the Th-metal bonds were discovered to be more ionic. Finally, Louise Natrajan, David Mills and co-workers also considered f-block chemistry when they presented a study of the use of tripodal ligands to control the coordination sphere of uranium (and lanthanum) with the aim to explore the effects of the coordinatively unsaturated "steric pocket" on the activation of small molecules [14]. Two new two tris-anilido ligands with substituted aryl rings were considered for this purpose—{tacn(SiMe$_2$Nar)$_3$} (ar = $C_6H_3Me_2$-3,5 or C_6H_4Me-4)—which bring enhanced flexibility to the metal environment compared with previous examples.

In addition to the impressive collection of original papers above, this Special Issue contains four review articles covering solid-state materials chemistry, coordination chemistry, metal organic frameworks and catalysis. Richard Walton and colleagues provided a comprehensive overview of how solvothermal methods can be exploited to produce cerium dioxide, CeO_2 and a whole host of substituted ceria materials [15]. Substituents from transition metals through lanthanides to main group metals (both p- and s-block) can be introduced via this technique. Equally, nitrogen can be doped into the system and multiple cation substitutions can be performed. Perhaps almost as important, however, is the power of the technique in manipulating morphology and nanostructure, including allowing the synthetic chemist to deposit materials as structured layers. In an entirely different vein, Kótai et al. presented a timely review of pyridine complexes formed with silver perchlorate, permanganate and perrhenate (X = Cl, Mn, Rh, respectively) [16]. Given the excellent solubility of the AgXO$_4$ salts, it is not surprising that examples have been known for over a century and that their oxidative properties are well known and well utilised. Nevertheless, the literature contains some apparently contradictory information and a compilation of details on their synthesis, composition, structure and redox properties is overdue. Contrasts between reducing pyridine and oxidizing XO$_4^-$ can lead to quasi-intramolecular redox reactions, and the hydrogen bonds between the acidic C–H and polarized X–O bonds were discussed.

Single molecule magnets (SMMs) have been known now for several decades but are still generating considerable interest and excitement with prospects of high-density information storage and quantum computing. Lanthanide SMMs with Schiff base ligands form a most promising subset given their combination of magnetic ordering, anisotropy and relaxation pathways. Linnert, Thomas et al. presented an extensive review of these SMMs with a consideration of synthesis techniques through an evaluation of their structures to an in-depth exploration of their magnetic properties [17]. Among some of the highlights, the authors picked out dysprosium as a likely metal of choice; they considered salicylaldehyde derivatives in terms of useful Schiff base ligands and they concentrated on ways in which ligand and SMM design can be exploited to afford slow magnetic relaxation. Just as SMMs have been one of the revelations of inorganic chemistry over recent decades, so have MOFs. The review by Alessia Tombesi and Claudio Pettinari considered the role of MOFs specifically in the context of heterogeneous catalysts in olefin epoxidation and carbon dioxide cycloaddition [18]. The authors related the catalytic performance to specific building blocks such as organic linkers and metal nodes/clusters and considered the influence of mixed-metal species and nanocomposites. That article not only considered the relative merits of a large cross section of MOF materials but also delineated the major challenges that remain to be addressed. In common with many of the articles in this excellent Special Issue, that review should encourage researchers to devise new strategies to meet these future challenges.

Before I conclude, I thank all of the authors who have contributed to this Special Issue and fashioned it into the outstanding collection that it is. I also thank all my editorial and publishing colleagues who have enabled this issue to come into being.

Conflicts of Interest: The author declares no conflict of interest.

References

1. Wackerbarth, I.; Widhyadnyani, N.; Schmitz, S.; Stirnat, K.; Butsch, K.; Pantenburg, I.; Meyer, G.; Klein, A. Cu^{II} Complexes and Coordination Polymers with Pyridine or Pyrazine Amides and Amino Benzamides—Structures and EPR Patterns. *Inorganics* **2020**, *8*, 65. [CrossRef]
2. Manfroni, G.; Capomolla, S.; Prescimone, A.; Constable, E.; Housecroft, C. Isomeric 4,2′:6′,4″- and 3,2′:6′,3″-Terpyridines with Isomeric 4′-Trifluoromethylphenyl Substituents: Effects on the Assembly of Coordination Polymers with [Cu(hfacac)$_2$] (Hhfacac = Hexafluoropentane-2,4-dione). *Inorganics* **2021**, *9*, 54. [CrossRef]
3. Mondal, S.; Chen, W.; Sun, Z.; McGrady, J. Synthesis, Structure and Bonding in Pentagonal Bipyramidal Cluster Compounds Containing a cyclo-Sn5 Ring, [(CO)$_3$MSn$_5$M(CO)$_3$]$^{4-}$; (M = Cr, Mo). *Inorganics* **2022**, *10*, 75. [CrossRef]
4. Jones, D.; Watts, S.; Jones, C. Synthesis and Characterization of Super Bulky β-Diketiminato Group 1 Metal Complexes. *Inorganics* **2021**, *9*, 72. [CrossRef]
5. Qiao, X.; Corkett, A.; Luo, D.; Dronskowski, R. Silver Cyanoguanidine Nitrate Hydrate: Ag(C$_2$N$_4$H$_4$)NO$_3$·1/2 H$_2$O, a Cyanoguanidine Compound Coordinating by an Inner Nitrogen Atom. *Inorganics* **2020**, *8*, 64. [CrossRef]
6. Bäucker, C.; Becker, P.; Morell, K.; Niewa, R. Novel Fluoridoaluminates from Ammonothermal Synthesis: Two Modifications of K$_2$AlF$_5$ and the Elpasolite Rb$_2$KAlF$_6$. *Inorganics* **2022**, *10*, 7. [CrossRef]
7. Pace, K.; Klepov, V.; Smith, M.; Williams, T.; Morrison, G.; Lauterbach, J.; Misture, S.; zur Loye, H. Hydrothermal Synthesis and Structural Investigation of a Crystalline Uranyl Borosilicate. *Inorganics* **2021**, *9*, 25. [CrossRef]

8. Lin, C.; Foucher, A.; Stach, E.; Gorte, R. A Thermodynamic Investigation of Ni on Thin-Film Titanates (ATiO$_3$). *Inorganics* **2020**, *8*, 69. [CrossRef]
9. Zhao, W.; Liu, Z.; Zhao, Y.; Luo, Y.; He, S. Multivariate Linear Regression Models to Predict Monomer Poisoning Effect in Ethylene/Polar Monomer Copolymerization Catalyzed by Late Transition Metals. *Inorganics* **2022**, *10*, 26. [CrossRef]
10. Northcote-Smith, J.; Johnson, A.; Singh, K.; Ortu, F.; Suntharalingam, K. Breast Cancer Stem Cell Active Copper(II) Complexes with Naphthol Schiff Base and Polypyridyl Ligands. *Inorganics* **2021**, *9*, 5. [CrossRef]
11. Vicatos, G.; Hammouda, A.; Alnajjar, R.; Bonomo, R.; Valora, G.; Bourne, S.; Jackson, G. Aqueous Solution Equilibria and Spectral Features of Copper Complexes with Tripeptides Containing Glycine or Sarcosine and Leucine or Phenylalanine. *Inorganics* **2022**, *10*, 8. [CrossRef]
12. Arrigoni, F.; Zampella, G.; De Gioia, L.; Greco, C.; Bertini, L. The Photochemistry of Fe$_2$(S$_2$C$_3$H$_6$)(CO)$_6$(μ-CO) and Its Oxidized Form, Two Simple [FeFe]-Hydrogenase CO-Inhibited Models. A DFT and TDDFT Investigation. *Inorganics* **2021**, *9*, 16. [CrossRef]
13. Ostrowski, J.; Wooles, A.; Liddle, S. Synthesis and Characterisation of Molecular Polarised-Covalent Thorium-Rhenium and -Ruthenium Bonds. *Inorganics* **2021**, *9*, 30. [CrossRef]
14. Formanuik, A.; Ortu, F.; Vitorica-Yrezabal, I.; Tuna, F.; McInnes, E.; Natrajan, L.; Mills, D. Functionalized Tris (anilido)triazacyclononanes as Hexadentate Ligands for the Encapsulation of U(III), U(IV) and La(III) Cations. *Inorganics* **2021**, *9*, 86. [CrossRef]
15. Annis, J.; Fisher, J.; Thompsett, D.; Walton, R. Solvothermal Synthesis Routes to Substituted Cerium Dioxide Materials. *Inorganics* **2021**, *9*, 40. [CrossRef]
16. Franguelli, F.; Béres, K.; Kótai, L. Pyridinesilver Tetraoxometallate Complexes: Overview of the Synthesis, Structure, and Properties of Pyridine Complexed AgXO$_4$ (X = Cl, Mn, Re) Compounds. *Inorganics* **2021**, *9*, 79. [CrossRef]
17. Gebrezgiabher, M.; Bayeh, Y.; Gebretsadik, T.; Gebreslassie, G.; Elemo, F.; Thomas, M.; Linert, W. Lanthanide-Based Single-Molecule Magnets Derived from Schiff Base Ligands of Salicylaldehyde Derivatives. *Inorganics* **2020**, *8*, 66. [CrossRef]
18. Tombesi, A.; Pettinari, C. Metal Organic Frameworks as Heterogeneous Catalysts in Olefin Epoxidation and Carbon Dioxide Cycloaddition. *Inorganics* **2021**, *9*, 81. [CrossRef]

Article

Cu^{II} Complexes and Coordination Polymers with Pyridine or Pyrazine Amides and Amino Benzamides—Structures and EPR Patterns

Ines Wackerbarth [1], Ni Nyoman Agnes Tri Widhyadnyani [1], Simon Schmitz [1], Kathrin Stirnat [1], Katharina Butsch [1], Ingo Pantenburg [1], Gerd Meyer [1,2] and Axel Klein [1,*]

[1] Department für Chemie, Institut für Anorganische Chemie, Universität zu Köln, Greinstrasse 6, D-50939 Köln, Germany; ines.wackerbarth@gmail.com (I.W.); widhyadnyani.agnes@gmail.com (N.N.A.T.W.); s.schmitz@uni-koeln.de (S.S.); kstirnat@gmail.com (K.S.); Katharinabutsch@web.de (K.B.); ac118@uni-koeln.de (I.P.); gerdm@kth.se (G.M.)

[2] Department of Chemistry, KTH Royal Institute of Technology, Teknikringen 30, SE-100 44 Stockholm, Sweden

* Correspondence: axel.klein@uni-koeln.de; Tel.: +49-221-470-4006

Received: 9 November 2020; Accepted: 19 November 2020; Published: 1 December 2020

Abstract: Isonicotine amide, picoline amide, pyrazine 2-amide, 2- and 4-amino benzamides and various Cu^{II} salts were used to target Cu^{II} complexes of these ligands alongside with 1D and 2D coordination polymers. Under the criterion of obtaining crystalline and single phased materials a number of new compounds were reliably reproduced. Remarkably, for some of these compounds the ideal Cu:ligand ratio of the starting materials turned out to be very different from Cu:ligand ratio in the products. Crystal and molecular structures from single-crystal XRD were obtained for all new compounds; phase purity was checked using powder XRD. We observed exclusively the O_{amide} and not the NH_{2amide} function binding to Cu^{II}. In most of the cases; this occurred in chelates with the second pyridine, pyrazine or aminophenyl N function. μ-O,N ditopic bridging was frequently observed for the N = pyridine, pyrazine or aminophenyl functions, but not exclusively. The geometry around Cu^{II} in these compounds was very often axially elongated octahedral or square pyramidal. X-band EPR spectra of powder samples revealed various spectral symmetry patterns ranging from axial over rhombic to inverse axial. Although the EPR spectra cannot be unequivocally correlated to the observed geometry of Cu^{II} in the solid state structures, the EPR patterns can help to support assumed structures as shown for the compound [Cu(Ina)$_2$Br$_2$] (Ina = isonicotine amide). As UV-vis absorption spectroscopy and magnetic measurement in the solid can also be roughly correlated to the surrounding of Cu^{II}, we suggest the combination of EPR, UV-vis spectroscopy and magnetic measurements to elucidate possible structures of Cu^{II} compounds with such ligands.

Keywords: Cu^{II}; pyridine amides; pyrazine amide; amino benzamides; EPR spectroscopy

1. Introduction

Pyridine carboxamides, pyrazine carboxamides and amino-benzamides (Scheme 1) are interesting ditopic ligands for Cu^{II} coordination chemistry [1–32]. One reason is their use as versatile building blocks in coordination polymers and MOFs for various applications [1–3,8,9,11–13,16,20,21,26,27]. Another important reason is the biological relevance of some of the ligands used in this study (Scheme 1) [4–6,17,18,24,33]. An interesting general aspect of these ligands is the preference of Cu^{II} to the different (offered) donor atoms in the ligands. From the viewpoint of the HSAB principle the N atom of the heteroaromatic cores is softer than the amine function of 4-amino benzamide and both are softer than the O amide function of amides or corresponding carboxylates. From the viewpoint of Cu metalloenzymes the frequent binding of soft ligands such as histidine-N and cysteine-S confirms the tendency of Cu to softer ligands [33]. However, in various Cu enzymes also harder donor functions such as carboxylates were coordinated [34,35]. In metalloenzymes, carboxylate side chains of the peptide are very frequent ligands and were preferably coordinated by hard metal ions, as expected. In contrast to this, carboxamide side chains were less preferred as ligands in metalloenzymes [34–36], in agreement with the observation that the carboxamide function coordinates only effectively in chelate ligands as 2-pyridine or 2-pyrazine carboxamides [1–3,5,6,15–20,22–26,32,37–41].

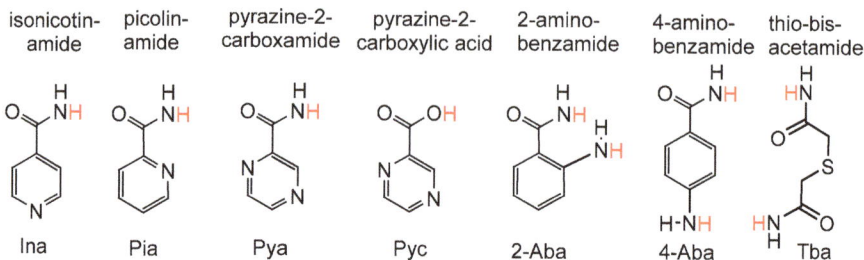

Scheme 1. Protoligands (ligand precursor prior to deprotonation) with abbreviations used in this study. In red: acidic protons.

Herein we report on a study aiming to explore the binding preferences of isonicotinamide (Ina), picolinamide (Pia), pyrazine-2-carboxamide (Pya) and 2- and 4-amino-benzamides (2- and 4-Aba; Scheme 1) towards Cu^{II}. We also added the versatile ligand thio-bisacetamide (Tba) to this study, as Tba represents the aliphatic amides and we expected that the ligand will form $S\hat{\,}O$ or $S\hat{\,}N$ chelates. The ligands were reacted with various Cu^{II} salts in EtOH without addition of bases to avoid deprotonation of the ligands. The compounds produced were studied in the solid state by single crystal and powder X-ray diffraction, EPR, magnetic measurements and UV-vis absorption spectroscopy. Being aware of the problem that our so-called "combinatorial approach"—which means dissolving metal salts and ligands and crystallising materials out of the mixture—might lead to a plethora of different structures, we aimed to work out reliable procedures for the preparation of defined structures which also allow reproducing the materials. We succeeded in doing so in many cases, and powder X-ray diffraction (PXRD) was used to evaluate the preparation methods and to prove the phase purity of the compounds. This also means that non-crystalline materials and materials which obviously contained mixtures of compounds as could be seen from their colours, which were obtained from further reactions of the seven ligands and the Cu salts, were not further analysed. Interestingly, this comprises all trials using nicotinamide as ligand.

Special focus was also laid on EPR spectroscopy of these solids. Having so many new structures and EPR data in hand we were able to correlate details in the EPR spectra such as spectral symmetry and g values to the local environment of the Cu^{II} centres. Magnetic and UV-vis absorption data of selected samples were also recorded.

2. Results and Discussion

2.1. Synthesis of the Cu Complexes

In a first series of experiments Cu^{II} salts were reacted with the ligands shown in Scheme 1 in a Cu:ligand ratio of 1:1 in EtOH. Single crystals of 18 compounds were obtained by slow evaporation of the solvent from the reaction mixtures (Table 1). The Cu:ligand ratios were found ranging from 1:1 to 1:4. Coligands to Cu^{II} were the anions from the Cu sources: Cl^-, Br^-, NO_3^-, BF_4^-, $Tfa\hat{\,}^-$, EtOH and H_2O. ClO_4^- was not coordinating but found as a counter ion. None of the ligands was found deprotonated in the structures. However, pyrazine-2-carboxamide was converted into pyrazine carboxylate forming the compound [Cu(Pyc)(Tfa)] (**11**) in neutral aqueous EtOH solution. In HCl acidic solutions isonicotinamide (Ina) was protonated yielding $(HIna)_2[CuCl_4]\cdot 2H_2O$ (**6**) and (HIna)[Cu(H$_2$O)Cl$_3$] (7) both containing 4-carboxamidopyridin-1-ium (HIna$^+$) cations. Under similar conditions 4-aminobenzamide was protonated to 4-benzamidyl-ammonium (4-AbaH$^+$) in [Cu(4-HAba)$_2$Cl$_4$] (17). A very interesting finding was that the complex [Cu(Ina)$_2$(NO$_3$)$_2$] (2) could be converted into [Cu(Ina)$_2$Br$_2$] (5) through grinding with KBr. Unfortunately, the obtained crystals of this compound were of poor quality preventing a structure solution. In a previous report of (5) space group $P2_1/c$ (No. 14) was concluded from 24 reflections by powder XRD and the authors assumed a polymeric structure with a distorted octahedral coordination around Cu^{II} [39]. From the new compound [Cu(Ina)$_4$(H$_2$O)$_2$](BF$_4$)$_2$ (1), the crystal structure of the corresponding ClO$_4^-$ salt was previously reported alongside with the structure of [Cu(Ina)$_2$(H$_2$O)$_2$(NO$_3$)$_2$] (3) [16]. All other crystal structures are new (see Figures S1–S40 and full data in Tables S1–S39 in the Supplementary Materials).

Table 1. Overview over the obtained crystalline compounds and structures.

Ligand	Cu^{II} Precursor	Compound	Formula (Weight)	Space Group
Ina	Cu(BF$_4$)$_2$·6H$_2$O	[Cu(Ina)$_4$(H$_2$O)$_2$](BF$_4$)$_2$ (1)	C$_{24}$H$_{28}$B$_2$Cu$_1$F$_8$N$_8$O$_6$ (761.69)	C2/c (No. 15)^d
Ina	Cu(NO$_3$)$_2$·3H$_2$O	$^1_\infty$[Cu(Ina)$_2$(NO$_3$)$_2$] (2)	C$_{12}$H$_{12}$Cu$_1$N$_6$O$_8$ (431.81)	P2$_1$2$_1$2$_1$ (No. 19)
Ina	Cu(NO$_3$)$_2$·3H$_2$O	[Cu(Ina)$_2$(H$_2$O)$_2$(NO$_3$)$_2$] (3)	C$_{12}$H$_{16}$Cu$_1$N$_6$O$_{10}$ (467.84)	P2$_1$/c (No. 14)^e
Ina	Cu(BF$_4$)$_2$·6H$_2$O	[Cu(Ina)$_2$(H$_2$O)$_2$(SiF$_6$)]·H$_2$O (4)	C$_{12}$H$_{20}$Cu$_1$F$_6$N$_4$O$_6$Si$_1$ (521.94)	P-1 (No. 2)
Ina	[Cu(Ina)$_2$(NO$_3$)$_2$]	[Cu(Ina)$_2$Br$_2$] (5)	C$_{12}$H$_{12}$Br$_2$Cu$_1$N$_4$O$_2$ (467.61)	P2$_1$ (No. 4)^f
HIna$^{+\,a}$	CuCl$_2$·2H$_2$O	(HIna)$_2$[CuCl$_4$]·2H$_2$O (6)	C$_{12}$H$_{18}$Cl$_4$CuN$_4$O$_4$ (487.64)	P-1 (No. 2)
HIna$^{+\,a}$	CuCl$_2$·2H$_2$O	(HIna)[Cu(H$_2$O)Cl$_3$] (7)	C$_6$H$_9$Cl$_3$Cu$_1$N$_2$O$_2$ (311.05)	P-1 (No. 2)
Pia	Cu(BF$_4$)$_2$·6H$_2$O	[Cu(Pia)$_2$(BF$_4$)$_2$] (8)	C$_{12}$H$_{12}$B$_2$Cu$_1$F$_8$N$_4$O$_2$ (481.41)	P-1 (No. 2)
Pya	Cu(BF$_4$)$_2$·6H$_2$O	$^2_\infty$[Cu(Pya)$_2$](BF$_4$)$_2$ (9)	C$_{10}$H$_{10}$B$_2$Cu$_1$F$_8$N$_6$O$_2$ (483.38)	P2$_1$/c (No. 14)
Pya	Cu(NO$_3$)$_2$·3H$_2$O	[Cu(Pya)$_2$(H$_2$O)(NO$_3$)](NO$_3$) (10)	C$_{10}$H$_{12}$Cu$_1$N$_8$O$_9$ (451.80)	P-1 (No. 2)
Pya	Cu(NO$_3$)$_2$·3H$_2$O	$^1_\infty$[Cu(Pya)(NO$_3$)$_2$] (11)	C$_5$H$_5$Cu$_1$N$_5$O$_7$ (310.67)	C2/c (No. 15)
Pyc$^{-\,b}$	Cu(Tfa)$_2$·H$_2$O	$^2_\infty$[Cu(Pyc)(Tfa)] (12)	C$_7$H$_3$Cu$_1$F$_3$N$_2$O$_4$ (299.65)	P2$_1$2$_1$2$_1$ (No. 19)
2-Aba	Cu(NO$_3$)$_2$·3H$_2$O	[Cu(2-Aba)$_2$(NO$_3$)$_2$] (13)	C$_{14}$H$_{16}$Cu$_1$N$_6$O$_8$ (459.86)	P2$_1$/n (No. 14)^g
4-Aba	Cu(NO$_3$)$_2$·3H$_2$O	$^1_\infty$[Cu$_2$(4-Aba)$_2$(H$_2$O)$_3$(NO$_3$)$_3$](NO$_3$) (14)	C$_{14}$H$_{22}$Cu$_2$N$_8$O$_{17}$ (701.46)	C2/c (No. 15)^h
4-Aba	Cu(ClO$_4$)$_2$·6H$_2$O	$^2_\infty$[Cu(4-Aba)$_2$(EtOH)$_2$](ClO$_4$)$_2$ (15)	C$_{18}$H$_{28}$Cl$_2$Cu$_1$N$_4$O$_{12}$ (626.88)	P2$_1$/c (No. 14)
4-Aba	Cu(BF$_4$)$_2$·6H$_2$O	$^2_\infty$[Cu(4-Aba)$_2$(EtOH)$_2$](BF$_4$)$_2$ (16)	C$_{18}$H$_{28}$B$_2$Cu$_1$F$_8$N$_4$O$_4$ (601.60)	P2$_1$/c (No. 14)
4-HAba$^{+\,c}$	Cu(ClO$_4$)$_2$·6H$_2$O + HCl	[Cu(4-HAba)$_2$Cl$_4$] (17)	C$_{14}$H$_{18}$Cl$_4$Cu$_1$N$_4$O$_2$ (479.67)	P2$_1$/c (No. 14)
Tba	Cu(ClO$_4$)$_2$·6H$_2$O	[Cu(Tba)$_2$](ClO$_4$)$_2$ (18)	C$_8$H$_{16}$Cl$_2$Cu$_1$N$_4$O$_{12}$S$_2$ (558.80)	C2/m (No. 12)

^{a} HIna$^+$ = 4-carboxamidopyridin-1-ium. ^{b} Pyc$^-$ = pyrazine-2-carboxylate. Tfa$^-$ = trifluoroacetate. ^{c} 4-AbaH$^+$ = 4-benzamidyl ammonium. ^{d} The ClO$_4$-salt was reported in C2/c previously, [16]. ^{e} Reported previously, [16]. ^{f} No valid structure refinement possible; previous assignment from powder XRD = P2$_1$/c (No. 14), from [22]. ^{g} A refinement in the standard space group P2$_1$/c was not successful. ^{h} The structure could not be fully refined: The nitrate anions around the N2 (coordinating) and N3 (non-coordinating) atoms show systematic disorder due to symmetric alternation. For the coordinating N2-nitrate there are two possible symmetric identical positions to bridge between the two Cu1 ions, both with an occupation of 0.5. This leads to two nitrate anions in very close proximity in the cell, but must be understood as alternating motif in consecutive cells. The non-coordinating nitrate N3 atom shows similar disorder (for details, see Tables S30 and S31).

In a further series of experiments, we tried to reproduce the obtained compounds and optimise their yields. Cl and Br turned out to be very strong ligands to Cu^{II} compared with the amides and we avoided them in these experiments. Similar observations that Cu–halide bonding dominates the structures of heteroleptic halide pyridine-amide Cu^{II} complexes were reported before [3,5,25–27,38,40]. Using the same reaction protocol but varying the Cu:ligand ratio, 10 out of these 18 compounds could be reliably reproduced in good to excellent yields and high phase purity as confirmed through elemental analyses and powder XRD (Scheme 2). In turn, this means that the compounds **2, 4, 11,** and **14** remained unreproducible.

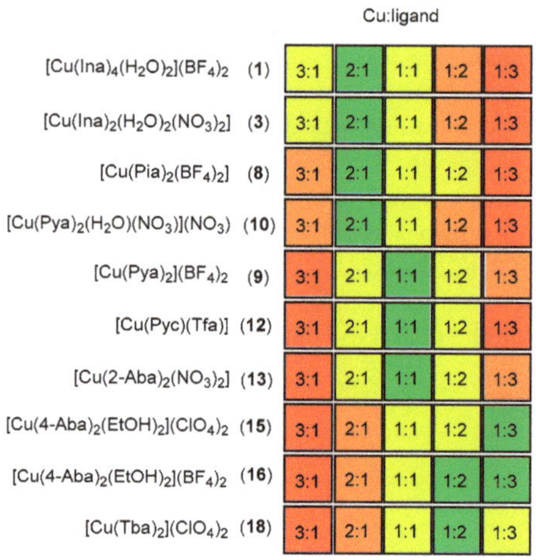

Compound	Cu:ligand				
	3:1	2:1	1:1	1:2	1:3
[Cu(Ina)$_4$(H$_2$O)$_2$](BF$_4$)$_2$ (1)	3:1	2:1	1:1	1:2	1:3
[Cu(Ina)$_2$(H$_2$O)$_2$(NO$_3$)$_2$] (3)	3:1	2:1	1:1	1:2	1:3
[Cu(Pia)$_2$(BF$_4$)$_2$] (8)	3:1	2:1	1:1	1:2	1:3
[Cu(Pya)$_2$(H$_2$O)(NO$_3$)](NO$_3$) (10)	3:1	2:1	1:1	1:2	1:3
[Cu(Pya)$_2$](BF$_4$)$_2$ (9)	3:1	2:1	1:1	1:2	1:3
[Cu(Pyc)(Tfa)] (12)	3:1	2:1	1:1	1:2	1:3
[Cu(2-Aba)$_2$(NO$_3$)$_2$] (13)	3:1	2:1	1:1	1:2	1:3
[Cu(4-Aba)$_2$(EtOH)$_2$](ClO$_4$)$_2$ (15)	3:1	2:1	1:1	1:2	1:3
[Cu(4-Aba)$_2$(EtOH)$_2$](BF$_4$)$_2$ (16)	3:1	2:1	1:1	1:2	1:3
[Cu(Tba)$_2$](ClO$_4$)$_2$ (18)	3:1	2:1	1:1	1:2	1:3

Scheme 2. Results of reproduction and optimisation reactions with yields and purities decreasing along the colour series: ▇▇▇▇. Yields were defined relative to the ligands (for details, see Experimental Section).

For some of the compounds such as [Cu(Tba)$_2$](ClO$_4$)$_2$ (**18**), [Cu(4-Aba)$_2$(EtOH)$_2$](BF$_4$)$_2$ (**16**), [Cu(Pya)$_2$](BF$_4$)$_2$ (**9**), and [Cu(2-Aba)$_2$(NO$_3$)$_2$] (**13**), the optimum Cu:ligand ratio was the same as the stoichiometry of the compound. Remarkably, this is also true for [Cu(Pyc)(Tfa)] (**12**) which was produced from a 1:1 mixture of Cu(Tfa)$_2$·H$_2$O and pyrazine-amide (Pya).

For other compounds, the optimum ratio lies more to the Cu side with the most remarkable example being [Cu(Ina)$_4$(H$_2$O)$_2$](BF$_4$)$_2$ (**1**) for which a 2:1 = Cu:ligand ratio was best to observe phase pure material, while a 1:4 ratio prevails in the compound. The most important parameter for the assessment of the synthesis was the phase purity. The second important parameter was the yield. However, it is important to note, that yields were either calculated based on Cu or based on the ligand. For example, for compound (**1**) a 2:1 ratio in the starting material meant a high yield relative to the ligand (88%) but, of course, a very low yield relative to the starting Cu^{II} precursor (22%).

IR spectroscopy allowed revealing some structural features of the compounds (data in the Experimental Section). The presence of BF$_4^-$, ClO$_4^-$, and NO$_3^-$ in the compounds as ligands or as counter anions is easily detectable through their characteristic vibrational modes. On the other hand, the data did not unequivocally allow determining whether these anions are ligands or counter ions. For example, in [Cu(Ina)$_4$(H$_2$O)$_2$](BF$_4$)$_2$ (**1**) the ν(B–F) vibration is found as a strong band at 1080 cm^{-1}, but as very strong broad resonance at 1050 cm^{-1} for [Cu(Pya)$_2$](BF$_4$)$_2$ (**9**). Both structures contain BF$_4^-$ as counter anion. The complex [Cu(Pia)$_2$(BF$_4$)$_2$] (**8**) contains two BF$_4$ ligands and the ν(B–F) is found at 1072 cm^{-1} as a strong band. Furthermore, for [Cu(Ina)$_2$(NO$_3$)$_2$] (**2**) the ν(NO$_3^-$) resonance is observed at 1383 cm^{-1}, but is shifted to 1283 cm^{-1} for [Cu(Pya)(NO$_3$)$_2$] (**11**) although both contain NO$_3^-$ ligands. Additionally, for [Cu(Pya)$_2$(H$_2$O)(NO$_3$)](NO$_3$) (**10**) only one ν(NO$_3^-$) vibration is found at 1350 cm^{-1} although one NO$_3^-$ is coordinated, while the other serves as counter ion.

Quite generally, the coordination to Cu leads to slight blue-shifts of characteristic ligand vibrations. However, there are exceptions like the compounds containing pyrazine 2-amide (Pya). The ν(C=O) resonances for the compounds were found markedly at lower energy compared with those of the uncoordinated ligand. Interestingly, the δ(NH$_2$) resonances of the amide ligands are usually red-shifted upon coordination and suffer from a marked loss of intensity. However, in none of our structures did the NH$_{2amide}$ function act as coordinating (see next paragraph). Thus, IR spectroscopy does not allow the unequivocal assignment of structural details.

2.2. Crystal Structures from Single Crystal XRD

The compounds listed in Table 1 could be obtained as single crystals suitable for single-crystal XRD directly from the reaction mixtures. Amongst the structures we found the 1D coordination polymers **2** (Figure 1), **11** (Figure 2), and **14** (Figure S29, Supplementary) and the 2D coordination polymers **9** (Figure 3), **12** (Figure S23), **15** (Figure S31), and **16** (Figure S34) in which the ditopic ligands isonicotine (Ina), pyrazine 2-amide (Pya), and 4-amino benzamide (4-Aba) bridge between Cu^{II} centres. Importantly, Ina, Pya and 4-Aba use the O_{amide} exclusively and not the NH_{2amide} function for the μ-O,N ditopic bridging with N = pyridine, pyrazine, or aminophenyl functions.

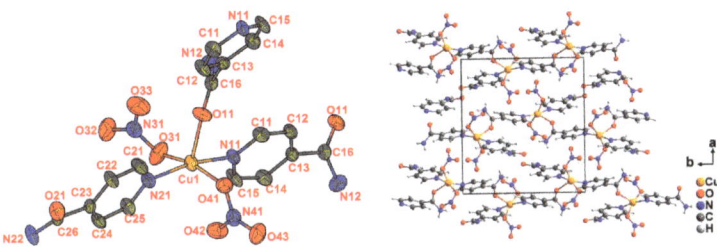

Figure 1. ORTEP representation (50% probability level) of the molecular entity within the structure of $^1_\infty[Cu(Ina)_2(NO_3)_2]$ (**2**) (**left**); packing along the c axis in the crystal (**right**); H atoms were omitted for clarity.

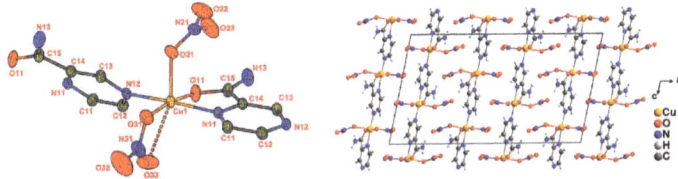

Figure 2. ORTEP representation (50% probability level) of the molecular entity within the structure of $^1_\infty[Cu(Pya)(NO_3)_2]$ (**11**) (**left**); packing along the b axis in the crystal (**right**); H atoms were omitted for clarity.

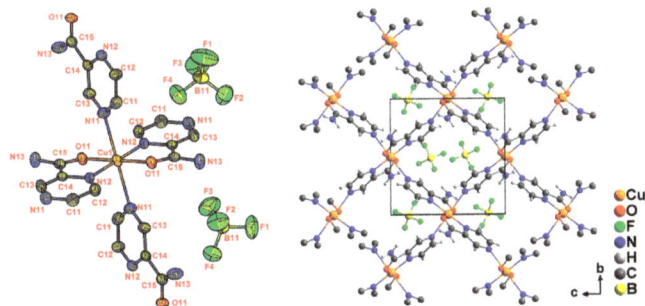

Figure 3. ORTEP representation (50% probability level) of the molecular entity within the structure of $^2_\infty[Cu(Pya)_2](BF_4)_2$ (**9**) (**left**); packing along the a axis in the crystal (**right**); H atoms were omitted for clarity.

At the same time, Pia, Pya, and 2-Aba form κ^2-N,O-chelates with Cu in **8** (Figure S1), **9** (Figure 3), **10** (Figure S18), **11** (Figure 2), and **13** (Figure S26), again exclusively using the O_{amide} instead of the NH_{2amide} function. Thus, in the Pya-containing compounds **9** and **11** both κ^2-N,O chelate binding and μ-O,N ditopic bridging was observed, while for Pia in compound **8**, the κ^2-N,O chelate binding prevented further μ-O,N ditopic bridging (Figure 4, middle). For compounds containing Ina or

4-Aba no chelate binding is possible and μ-*O*,*N* ditopic bridging was observed in some cases but not all. Out of the four compounds containing an 1,4-pyrazine unit three of them, **9**, **11**, and **12** show μ-*N*,*N*$_{pyrazine}$-bridging, while in **10** one pyrazine *N* position does not coordinate. In **12** the same pyrazine-2-carboxylate (Pyc⁻) ligand bridges two Cu^{II} centres in a μ-*N*,*N* fashion, bridges to another Cu^{II} using the carboxylate *O* function (μ-*O*,*O*), and shows κ²-*N*,*O* chelate binding at the same time, while the Tfa⁻ ligand binds terminally (Figure S23).

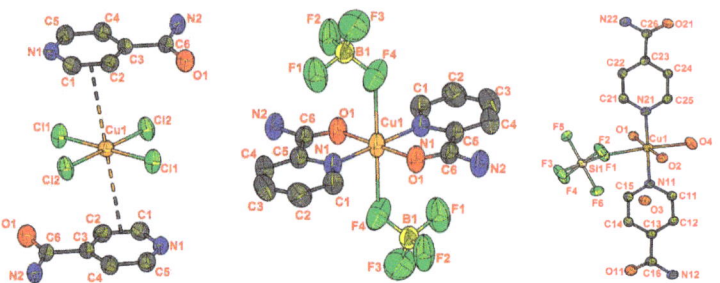

Figure 4. ORTEP representation (50% probability level) of the molecular entities within the structures of (HIna)$_2$[CuCl$_4$]·2H$_2$O (**6**) (**left**) of [Cu(Pia)$_2$(BF$_4$)$_2$] (**8**) (**middle**) and [Cu(Ina)$_2$(H$_2$O)$_3$(SiF$_6$)]·H$_2$O (**4**) (**right**); H atoms were omitted for clarity.

Nitrate (NO$_3$⁻) ligands are frequently found in terminal binding (**2** (Figure 1), **3**, **10**, **11**, **13**). However, in compound **11** one of the two NO$_3$⁻ ligands binds in a quasi-κ²-*O*,*O* fashion (Figure 2). In **14** both κ²-*O*,*O* and terminal binding were observed.

The thio-bisacetamide (Tba) ligand exhibits tridentate *N*^*S*^*N* binding in compound **18** as expected (Figure S39). In (HIna)[Cu(H$_2$O)Cl$_3$] (**7**) (Figure S13) the pyridine-*N* protonated HIna⁺ does not coordinate to Cu^{II}. A unique coordination mode is found in (HIna)$_2$[CuCl$_4$] (**6**) (Figure 4 and Figure S11) with two HIna⁺ cations occupying the axial positions with the C1=C2 bond of the pyridine ring (Figure 4, left). Although, the Cu···C=C$_{centroid}$ distance is rather long (3.360(1) Å) we consider this as a bonding contribution.

In the compounds **1**, **3**, **4**, and **7** containing the Ina ligand and **11** which contains the 4-NH$_3$ protonated 4-HAba⁺ ligand isolated Cu^{II} complexes were found. In their structures weak intermolecular forces as hydrogen bridges and π–π stacking are observed. Importantly, such forces were found in all compounds and we cannot neglect their role on the geometry around Cu^{II} which is decisive for the physical properties. Nevertheless, these non-covalent forces will not be further discussed, as we wanted to focus on the spectral consequences of the geometry around the Cu centre and not its origin.

The predominant coordination unit around Cu^{II} in most of the compounds reported herein is of the frequently observed tetragonally distorted octahedral type (OE in Table 2) [33,37,41–48]. The short equatorial positions are dominated through the *N* pyridine, pyrazine and aminophenyl functions. *O*$_{amide}$ occurs in this plane only in chelates with these *N* functions, NH$_{2amide}$–Cu binding was not observed. The Cu^{II}–ligand bonds in the equatorial plane lie in the range of 1.93 to 2.08 Å and are comparable with previously reported complexes of these ligands [13–32,38–46]. Remarkably, the shortest Cu–ligand bonds were of the Cu–*O*$_{amide}$ type and were found for the chelating Pia, Pya, and 2-Aba ligands, but also for terminating binding of the 4-Aba ligands including the 4-HAba⁺ cation. The Ina ligand forms exclusively short Cu–N bonds, while the *O*$_{amide}$ function does not contribute to the equatorial Cu–ligand bonding. For the compounds containing chlorido ligands, all four equatorial positions were Cl-occupied in (HIna)$_2$[CuCl$_4$] (**6**) but at the same time the *O*$_{amide}$ function coordinates tighter than one of the two chlorido ligands in [Cu(4-Haba)$_2$Cl$_4$] (**17**). The Cu–Cl distances range from 2.26 to 2.30 Å underlining that Cl⁻ is a reasonably strong ligand to Cu^{II}.

Table 2. Selected binding parameters (distances in Å, angles in °) around the Cu^{II} centres.

Complex	Geometry [a]	T (K)	Cu–L^{1} [b]	Cu–L^{2}	Cu–L^{3}	Cu–L^{4}	Cu–L^{5}	Cu–L^{6}
			L^{1}–Cu–L^{2}	L^{2}–Cu–L^{3}	L^{3}–Cu–L^{4}	L^{1}–Cu–L^{5}	L^{1}–Cu–L^{6}	L^{5}–Cu–L^{6}
[Cu(Ina)$_4$(H$_2$O)$_2$](BF$_4$)$_2$ (1)	OE	298(2)	2.024(3) N21	2.045(4) N11	=N11	=N21	2.392(5) O1	2.645(5) O2
			91.7(1)°	88.9(2)°	91.7(1)°	86.5(1)°	93.4(1)°	173.0(1)°	180°
$\frac{1}{\infty}$[Cu(Ina)$_2$(NO$_3$)$_2$] (2)	SPy+1	298(2)	1.987(5) O31	1.994(5) N21	2.002(5) N11	2.004(5) O41	2.314(5) O11	-
			90.1(2)°	87.7(2)°	89.1(2)°	78.9(2)°	-	165.5(2)°	-
[Cu(Ina)$_2$(H$_2$O)$_2$(NO$_3$)$_2$] (3) [c]	OE	296(2)	1.984(1) O2	1.992(1) N1			2.507(1) O3	
			90.0(1)°	90.0(1)°		95.6(1)°	84.4(1)°	180°	180°
[Cu(Ina)$_2$(H$_2$O)$_3$(SiF$_6$)]·H$_2$O (4)	OE	298(2)	1.969(2) O1	1.986(2) O2	1.999(2) N21	2.026(2) N11	2.409(2) F1	2.513(2) O4
			90.1(1)°	89.6(1)°	88.2(1)°	88.2(1)°	84.9(1)°	180°	170.9(1)°
(HIna)$_2$[CuCl$_4$]·2H$_2$O (6)	SP+2	298(2)	2.263(1) Cl1	2.270(1) Cl2	=Cl1	=Cl2	3.360(1) (py)	=L^5
(HIna)[Cu(H$_2$O)Cl$_3$] (7) [c]	SPy	298(2)	1.980(4) O1	2.251(1) Cl2	2.268(1) Cl1	2.279(1) Cl3	2.923(1) Cl2'	-
			87.4(1)°	93.2(1)°	92.6(1)°	-	-	172.5(1)°	-
[Cu(Pia)$_2$(BF$_4$)$_2$] (8)	OE	150(2)	1.935(3) O1	1.958(4) N1	=O1	=N1	2.589(3) F4	=F4
			83.3(2)°	96.6(1)°	83(2)°	89.7(1)°	90.3(1)°	180°	180°
$\frac{2}{\infty}$[Cu(Pya)$_2$](BF$_4$)$_2$ (9)	OE	298(2)	1.957(2) O11	1.984(3) N12	=O11	=N12	2.447(3) N11	=N11
			82.2(1)°	97.7(1)°	82.2(1)°	85.5(1)°	94.4(1)°	180°	180°
[Cu(Pya)$_2$(H$_2$O)(NO$_3$)](NO$_3$) (10)	Spy+1	298(2)	1.935(4) O11	1.938(4) O21	1.984(4) N21	1.989(4) N11	2.234(4) O5	2.818(4) O43
			82.7(1)°	97.6(1)°	82.7(1)°	94.3(2)°	89.7(1)°	172.1(1)°	169.8(1)°
$\frac{1}{\infty}$[Cu(Pya)(NO$_3$)$_2$] (11)	SPy	298(2)	1.953(2) O11	1.979(2) O31	1.998(2) N12	2.004(2) N11	2.238(2) O21	2.572(3) O33
			81.6(1)°	89.3(1)°	92.3(1)°	91.0(1)°	93.6(1)°	169.2(1)°	136.9(1)°
$\frac{2}{\infty}$[Cu(Pyc)(Tfa)] (12) [d]	SPy	298(2)	1.942(5) O1	1.944(7) O11	1.997(6) O12	2.065(8) N11	2.272(8) N12	-
			88.8(2)°	97.6(2)°	92.7(2)°	96.5(3)°	-	171.3(2)°	-
[Cu(2-Aba)$_2$(NO$_3$)$_2$] (13)	OE	298(2)	1.976(3) O1	2.010(4) N1	=O1	=N1	2.506(5) O3	=O3
			87.3(2)°	92.7(2)°	87.3(2)°	92.8(1)°	87.1(1)°	180°	180°
$\frac{1}{\infty}$[Cu$_2$(4-Aba)$_2$(H$_2$O)$_3$(NO$_3$)$_3$](NO$_3$) (14) [e]	OD	298(2)	1.942(4) O11	1.969(4) O21	2.016(4) N11	2.034(5) O21	2.262(4) O32	2.35(1) O23
			89.1(2)°	88.2(2)°	93.3(2)°	82.9(2)°	95.2(3)°	176.0(2)°	143.1(3)°
$\frac{2}{\infty}$[Cu(4-Aba)$_2$(EtOH)$_2$](ClO$_4$)$_2$ (15)	OE	150(2)	1.951(2) O11	2.077(4) N11	=N11	=O11	2.529(2) O21	=O21
			87.8(1)°	92.1(1)°	87.8(1)°	94.4(1)°	85.5(1)°	180°	180°
$\frac{2}{\infty}$[Cu(4-Aba)$_2$(EtOH)$_2$](BF$_4$)$_2$ (16)	OE	298(2)	1.951(2) O11	2.062(3) N11	=O11	=N11	2.545(2) O1	=O1
			87.9(1)°	92.1(1)°	87.9(1)°	87.3(1)°	92.6(1)°	180°	180°
[Cu(4-HAba)$_2$Cl$_4$] (17) [f]	OE	298(2)	1.957(6) O1	2.313(3) Cl1	=O1	=Cl	2.867(2) Cl2	=Cl2
			89.6(2)°	90.4(2)°	89.6(2)°	98.2(1)°	81.8(2)°	180°	180°
[Cu(Tba)$_2$](ClO$_4$)$_2$ (18)	OE	298(2)	1.976(2) O11	=O11	=O11	=O11	2.625(1) S11	=S11
			90.9(1)°	89.1(1)°	90.9(1)°	81.5(1)°	98.4(1)°	180°	180°

[a] OE = tetrahedrally elongated octahedral, OD = otherwise distorted octahedral, SPy = square pyramidal. [b] L_1 is defined by the shortest Cu–ligand bond, L_2 to L_4 lie in the equatorial plane, L_5 and L_6 in the elongated axial positions. [c] Data from [16]. [d] Pyc$^-$ = pyrazine-2-carboxylate. [e] Two species with slightly different bonding parameters. [f] 4-AbaH$^+$ = 4-benzamidyl ammonium.

For the labelling of the structures as OE in Table 2, two axial ligands must be present in a distance elongated to about 10–20% compared to the same ligand in an equatorial position [42]. Further observed cases were square pyramidal structures (Spy in Table 2), two structures containing one long axial and one even longer axial Cu–ligand bond (Spy+1) and the peculiar case of two very long distances to Cu^{II}

as in (HIna)$_2$[CuCl$_4$] (**6**) (Figure 4, left) called square planar +2 (SQ+2). In compound **10** (Figure S19) the sixth ligand is not only elongated but also displaced from the ideal position showing an "axial" angle of about 170°. Nevertheless, we assign this to Spy+1. In **14** (Figure S30) the coordination around Cu^{II} showed a severely and multiply distorted octahedron (OD).

Non-chelated O_{amide} binding to Cu^{II} was found in the axial (weak) positions only in one compound (**2**) and no NH_{2amide} coordination, in line with the observation that amide functions in amino acids do not represent an important class of ligands in metalloproteins [34–36] and the poor non-biologic coordination chemistry of amides and Cu [33,35,37]. Indeed, searching the CCDC database, gave only the Cu^{II} complex [Cu(L)$_2$(OH)$_2$] containing two chelating pyrazine-2,3-dicarboxamide ligands with Cu–NH$_{2amide}$ bonds [49]. Further frequent ligands in the axial positions of our structures were weak ligands as NO$_3^-$, EtOH, H$_2$O, BF$_4^-$, and the peculiar SiF$_6^{2-}$ (Figure 4). Nevertheless, the non-chelating $N_{4pyrazine}$ function of the Pyc$^-$ ligand (in **12**) was found in the axial position. Importantly, the axial non-chelating O_{amide} (in **2**) and $N_{4pyrazine}$ binding (in **12**) both represent structures with µ-O,N ditopic bridging. The axial position is occupied by the thio-bisacetamide (Tba) S atoms in compound **18** in line with the preferences of Cu^{II} to N over S coordination [33,34,50].

2.3. X-Band EPR Spectroscopy and Magnetic Measurements

X-Band EPR spectra of the microcrystalline materials were recorded at 298 and 110 K, Figures 5 and 6 show representative examples, Table 3 lists essential parameters.

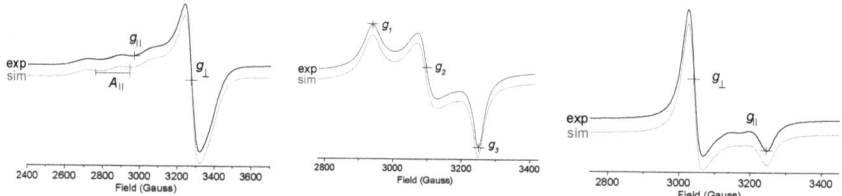

Figure 5. X-band EPR spectra of microcrystalline samples of [Cu(Ina)$_4$(H$_2$O)$_2$](BF$_4$)$_2$ (**1**) (**left**, frequency = 9.455063 GHz, simulation in red with parameters in Table 3), (Cu(H-Aba)$_2$Cl$_4$) (**17**) (**middle**, frequency = 9.454244 GHz, simulation in red), and [Cu$_2$(Aba)$_2$(H$_2$O)$_3$(NO$_3$)$_3$](NO$_3$) (**14**) (**right**, frequency = 9.448669 GHz, simulation in red), all measured at 298 K.

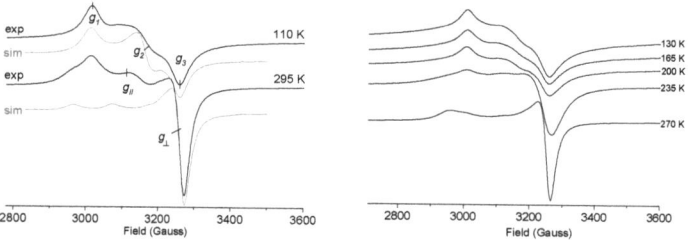

Figure 6. X-band EPR spectra of a microcrystalline sample of [Cu(4-Aba)$_2$(EtOH)$_2$](ClO$_4$)$_2$ (**15**) at 110 K (frequency = 9.459916 GHz) and 295 K (frequency = 9.448669 GHz) both with simulations in red (**left**) and at temperatures ranging from 130 to 270 K (**right**).

Table 3. Selected X-band EPR data of Cu^{II} compounds.[a]

Compound	T (K)	g_{av}	$g_∥/g_1$	g_2	$g_⊥/g_3$	$Δg$	Spectral Geom.	Cu^{II} Geom.
Group I								
[Cu(Ina)$_2$(NO$_3$)$_2$] (2)	298	2.143	2.280		2.074	0.206	axial	SPy+1
[Cu(Ina)$_4$(H$_2$O)$_2$](BF$_4$)$_2$ (1) [b]	298	2.118	2.260		2.048	0.212	axial	OE
(HIna)$_2$[CuCl$_4$]·2H$_2$O (6)	298	2.132	2.266	2.082	2.047	0.219	rhombic	SP+2
[Cu(Ina)$_2$(H$_2$O)$_3$(SiF$_6$)]·H$_2$O (4)	298	2.154	2.326		2.068	0.258	axial	OE
[Cu(Ina)$_2$(H$_2$O)$_2$(NO$_3$)$_2$] (3)	298	2.156	2.297		2.085	0.212	axial	OE
[Cu(4-HAba)$_2$Cl$_4$] (17)	298	2.175	2.286	2.169	2.066	0.219	rhombic	OE
[Cu(Pya)(NO$_3$)$_2$] (11)	298	2.179	2.372		2.083	0.289	axial	SPy
[Cu(Pya)$_2$(H$_2$O)(NO$_3$)](NO$_3$) (10)	298	2.183	2.400		2.075	0.325	axial	SPy+1
similar compounds								
[Cu(CCl$_3$COO)$_2$(MNA)$_2$]·2H$_2$O [c]	298	2.143	2.280		2.075	0.205	axial	OE
[Cu(Ina)$_2$(μ-N,S-SCN)$_2$] [d]	298	2.20	2.27		2.07	0.200	axial	OE
[Cu(meclof)$_2$(2-pyca)$_2$] [e]	298	2.142	2.290		2.068	0.222	axial	OE
[Cu(clof)$_2$(4-pymeth)$_2$(H$_2$O)] 2H$_2$O [f]	298	2.128	2.271		2.054	0.217	axial	SPy
[Cu(clof)$_2$(Et$_2$nia)$_2$] [f]	298	2.132	2.289		2.053	0.236	axial	[g]
[Cu(clof)$_2$(Ina)$_2$] [f]	298	2.153	2.285		2.087	0.198	axial	[g]
[Cu(tolf)$_2$(Et$_2$nia)$_2$)(H$_2$O)$_2$] [h]	298	2.130	2.294		2.048	0.246	axial	OE
[Cu(tolf)$_2$(Nia)$_2$] [h]	298	2.157	2.319		2.076	0.243	axial	[g]
Group II								
[Cu(Pya)$_2$](BF$_4$)$_2$ (9)	298	2.140	2.204	2.139	2.078	0.126	rhombic	OE
[Cu(4-Aba)$_2$(EtOH)$_2$](ClO$_4$)$_2$ (15)	110	2.137	2.231	2.123	2.059	0.172	rhombic	OE
[Cu(4-Aba)$_2$(EtOH)$_2$](ClO$_4$)$_2$ (15)	298	2.111	2.200		2.066	0.133	axial	OE
[Cu(4-Aba)$_2$(EtOH)$_2$](BF$_4$)$_2$ (16)	110	2.139	2.220	2.140	2.058	0.162	rhombic	OE
[Cu(4-Aba)$_2$(EtOH)$_2$](BF$_4$)$_2$ (16)	298	2.110	2.196		2.068	0.131	axial	OE
[Cu(TBA)$_2$](ClO$_4$)$_2$ (18)	298	2.130	2.228		2.080	0.147	axial	OE
similar compounds								
[Cu$_2$(Ina)$_2$(μ-1,1-N$_3$)$_2$(μ-1,3-N$_3$)$_2$] [d]	298	2.17	2.22		2.07	0.150	axial	SPy+1
[Cu(Ina)$_2$(μ-1,1-N$_3$)$_2$(μ-O,O-SO$_4$)]·2H$_2$O [d]	298	2.17	2.24	2.18	2.09	0.150	rhombic	SPy
Group III								
[Cu(Pyc)(Tfa)] (12)	298	2.138		2.138		0	isotropic	SPy
[Cu(Ina)$_2$Br$_2$] (5)	298	2.111		2.111		0	isotropic	SPy [i]
Group IV								
[Cu$_2$(Aba)$_2$(H$_2$O)$_3$(NO$_3$)$_3$](NO$_3$) (14)	298	2.141	2.190		2.050	0.112	inv. axial	OD

[a] Measured at 298 K or 110 K on microcrystalline powders; g_{av} = averaged g value = ($g_∥$+ $2g_⊥$)/3 or (g_1 + g_2 + g_3)/3; $Δg = g_∥ − g_⊥$ or $g_1 − g_3$. OE = tetragonal elongated octahedral, OD = otherwise distorted octahedral, SPy = square pyramidal. [b] An HFS $A_∥$ of 180 G was observed. [c] From [20]; MNA = N-methylnicotinamide; an HFS $A_∥$ of 165 G is observed. [d] From [15]. [e] From [18]; meclof = meclofenamate; 2-pyca = 2-pyridylcarbinol. [f] From [24]; clof = 2-(4-chlorophenoxy)-2-methylpropionic acid; 4-pymeth = 4-pyridylmethanol, Et$_2$nia = N,N-diethylnicotinamide, an HFS $A_∥$ of 165 G is observed for [Cu(clof)$_2$(Et$_2$nia)$_2$]. [g] No structure reported. [h] From [17], tolf = tolfenamic (N-(2-methyl-3-chlorophenyl)anthranilic) acid. [i] No unequivocal structure from XRD (see text); previous assignment OE from powder XRD and IR, [22].

EPR spectra of axial, rhombic and isotropic geometry were observed and, having full structural information from XRD, will be correlated to the local environment of the Cu^{II} centres in the following.

EPR spectra of axial symmetry with $g_∥ > g_⊥$ (Group I in Table 3, Figure 5 (left)) were observed for most of the tetragonally elongated octahedral species (OE) and square pyramidal systems with an extra-long distance to a sixth ligand (SPy+1) in line with previous reports [24,41–52]. Cu hyper fine structure (HFS, coupling to the ^{63}Cu (69.17%) and ^{65}Cu (30.83%) nuclei both with I = 3/2 [53]) was only observed in one case, [Cu(Ina)$_4$(H$_2$O)$_2$](BF$_4$)$_2$ (Figure 5 (left)), on the parallel g component ($A_\parallel$ = 180 G). The prevailing absence of HFS is in line with spectra of related species in the solid (see Table 3) [15–18,20,24]. The g anisotropy (Δg) in this group of complexes lies above values of 0.2 while the averaged g values (g_{av}) vary from 2.12 to 2.20. Exceptions are (HIna)$_2$[CuCl$_4$]·2H$_2$O (6) and [Cu(H–Aba)$_2$Cl$_4$] (17) for which a rhombic spectrum was observed (Figure 5 (middle) and Figure S43). On the other hand, the Δg values lie in the same range as for compounds of group I.

Rhombic spectra were recorded for [Cu(Pya)$_2$](BF$_4$)$_2$ (9), [Cu(4-Aba)$_2$(EtOH)$_2$](ClO$_4$)$_2$ (15) and [Cu(4-Aba)$_2$(EtOH)$_2$](BF$_4$)$_2$ (16) (group IV) when measured at 110 K (group II). Remarkably, the spectra of the two latter compounds are axial at 298 K and for [Cu(4-Aba)$_2$(EtOH)$_2$](ClO$_4$)$_2$ (15) a gradual transition between the rhombic and axial spectrum was observed between 110 and 298 K (Figure 6). The Δg values for this group of complexes lie markedly below 0.2 and their g_{av} values range from 2.1 to 2.17. A detailed inspection of the metrics around Cu^{II} showed very similar geometries for [Cu(4-Aba)$_2$(EtOH)$_2$](ClO$_4$)$_2$ (15) at 150 K and [Cu(4-Aba)$_2$(EtOH)$_2$](BF$_4$)$_2$ (16) at 298 K. We therefore concluded that in this group either very tiny changes in geometry and with this a change of the character of the electronic ground state from the predominant {dx^2-y^2}^{1} (axial) to a {dz^2}^{1} electronic ground state (rhombic or inverse axial) [42,47,48,54,55] were decisive and axial or rhombic spectra were observed. Importantly, this group is defined by a rather low g anisotropy Δg.

An isotropic broad spectrum with g values around 2.15 was observed for [Cu(Pyc)(Tfa)] (12) containing a square pyramidally coordinated Cu^{II} (group III). An isotropic spectrum was also found for the structurally ill-defined [Cu(Ina)$_2$Br$_2$] (5). Our findings for 5 are in line with the reported spectrum and g values for this compound by Atac et al. [22], while for the [Cu(Ina)$_2$Cl$_2$] derivative this group reports a rhombic spectrum. From our EPR results we conclude a square pyramidal coordination for Cu^{II} in [Cu(Ina)$_2$Br$_2$] (5).

A rather uncommon inverse axial EPR spectrum was observed for [Cu$_2$(Aba)$_2$(H$_2$O)$_3$(NO$_3$)$_3$](NO$_3$) (14). This type of symmetry is frequently observed for either axially compressed octahedral structures, {dz^2}^{1} ground states, or both, in contrast to the predominant {dx^2-y^2}^{1} and axially elongated octahedral coordination [38,48,54,55]. However, this structure represents several marked distortions from octahedral geometry, but not a compression. So, we can conclude a {dz^2}^{1} electronic ground state for the two Cu^{II} centres in this compound.

We selected the samples [Cu(Ina)$_2$(NO$_3$)$_2$] (2), [Cu(Pya)(NO$_3$)$_2$] (11), [Cu(Pyc)(Tfa)] (12), and [Cu(4-Aba)$_2$(EtOH)$_2$](ClO$_4$)$_2$ (15) for T-dependent magnetic measurements (Figure 7). 2 and 11 are 1D structures, while 12 and 15 represent 2D coordination polymers. Above 150 K a linear Curie-Weiss dependence is observed, in line with isolated Cu^{II} centres. At lower T, slight deviations were observed and the χ^{-1} graph tails down to zero, indicative for small anti-ferromagnetic coupling between the Cu centres in the solids. The shortest Cu···Cu contacts are 7.083 (1) Å (2), 5.822 (1) Å (11), 5.118 (2) and 7.079 (2) Å (12), and 8.463 (3) Å (15) and thus far too long for a super-exchange magnetic coupling. We assume that the alignment of the Cu^{II} centres in the 1D and 2D lattices is responsible for this effect. The magnetic moments (Table 4) are markedly below the spin-only value of 1.73 and positive Curie-Weiss constants are in line with this assumption. In contrast to this, for [Cu(Ina)$_2$Br$_2$] a μB of 1.90 has been reported alongside with [Cu(Ina)Cl$_2$] and [Cu(Ina)(NCS)$_2$] with 1.80 and 1.60 µB, respectively [39]. Higher values were due to spin-orbit coupling which is caused through heavy atoms such as Br or Cl. For our compounds only light N, O, or F atoms bind to Cu^{II} [3,5,10,15,19,52,56].

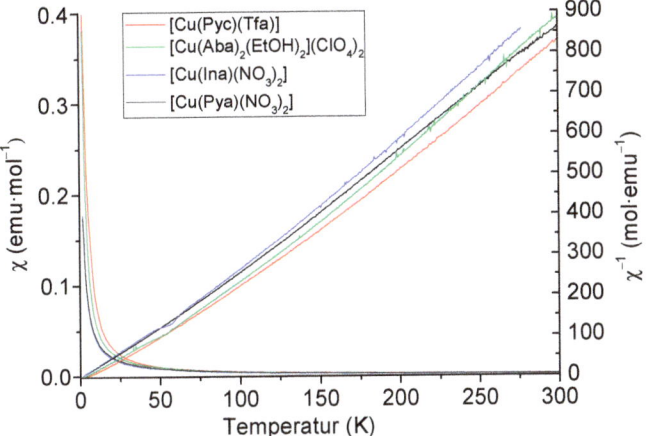

Figure 7. Thermal dependence of χ and χ^{-1} of selected Cu^{II} compounds.

Table 4. Magnetic moments and Curie-Weiss constants of selected Cu^{II} compounds.[a]

Compound	μ_B	θ_{CW} (K)	EPR Symmetry	Symmetry Around Cu^{II}
[Cu(Ina)$_2$(NO$_3$)$_2$] (2)	1.50	36.9	axial	SPy + 1
[Cu(Pya)(NO$_3$)$_2$] (11)	1.65	6.55	isotropic	SPy
[Cu(Pyc)(Tfa)] (12)	1.56	48.3	isotropic	SPy
[Cu(Aba)$_2$(EtOH)$_2$](ClO$_4$)$_2$ (15)	1.52	42.8	axial	OE

[a] Magnetic moments in Bohr magnetons μ_B and Curie-Weiss constants θ_{CW} in K.

2.4. Absorption Spectroscopy in the Solid and in Solution

All microcrystalline materials show colours corresponding to absorptions in the visible region. UV-vis absorption spectroscopy on selected samples (Table 5) reveals low-wavelengths absorptions ranging from 640 to 780 nm (15,600 to 12,820 cm^{-1}) which were assigned to the typical Cu^{II} d^9 system. Comparison of the three Pya-containing species reveals that the highest absorption energy is recorded for [Cu(Pya)$_2$](BF$_4$)$_2$ (9) with a tetrahedrally elongated (OE) structure, followed by [Cu(Pya)$_2$(H$_2$O)(NO$_3$)](NO$_3$) (10) having an SPy+1 configuration with the sixth ligand markedly elongated, while [Cu(Pya)(NO$_3$)$_2$] (11) exhibits a maximum red-shifted by 2810 cm^{-1} in line with its square pyramidal (SPy) coordination. The long-wavelength absorption observed for [Cu(Pyc)(Tfa)] (12) is rather high in energy in view of the only five coordinating atoms. This is probably due to the Tfa$^-$ ligand with a very short Cu–O distance indicative for its binding strength. For [Cu(Ina)$_2$Br$_2$] we found approximately the same absorption maximum as reported [22] in keeping with the assumption that we have obtained the same structure.

A view of a collection of similar structures (Table 5) confirmed that the absorption energy increased with increasing coordination number for very similar O and N coordinating pyridine-amides. For the unknown structures [Cu(clof)$_2$(Et$_2$nia)$_2$] [18] and [Cu(meclof)$_2$(Et$_2$nia)$_2$] [24], elongated octahedral coordination can be reasonably assumed from the UV-vis data, while the structure of [Cu(clof)$_2$(Ina)$_2$] probably represents a square pyramid.

Table 5. UV-vis absorption maxima of selected copper complexes.[a.]

Compound	λ (nm/cm^{-1}) of d-d Bands	Colour of the Crystals	Symmetry Around Cu^{II} [b]
[Cu(Pya)$_2$](BF$_4$)$_2$ (9)	613/16,310	blue	OE
[Cu(Pya)$_2$(H$_2$O)(NO$_3$)](NO$_3$) (10)	640/15,630	blue	SPy + 1
[Cu(Pya)(NO$_3$)$_2$] (11)	780/12,820	blue	SPy
[Cu(Pyc)(Tfa)] (12)	664/15,060	turquoise	SPy
[Cu(Ina)$_2$Br$_2$] (5)	691/14,470	dark green	_ [b]
similar compounds			
[Cu(tolf)$_2$(Et$_2$nia)$_2$(H$_2$O)$_2$] [c]	614/16,290	blue	OE
[Cu(meclof)$_2$(2-pyca)$_2$] [d]	615/16,260	blue	OE
[Cu(meclof)$_2$(Et$_2$nia)$_2$] [d]	605/16,530	blue	_ [e]
[Cu(clof)$_2$(4-pymeth)$_2$(H$_2$O)] 2H$_2$O [f]	620/16,130	blue	SPy
[Cu(clof)$_2$(Ina)$_2$] [f]	660/15,150	blue	_ [e]
[Cu(clof)$_2$(Et$_2$nia)$_2$] [f]	603/16,580	violet	_ [e]

[a] Measured as KBr pellets. [b] OE = tetrahedrally elongated octahedral, SPy = square pyramidal. [b] No unequivocal structure from XRD (see text); previous assignment OE from powder XRD and IR, [22], reported absorption at 675 nm/14,800 cm^{-1}. [c] From [17], tolf = tolfenamic (N-(2-methyl-3-chlorophenyl)anthranilic) acid. [d] From [18]; meclof = meclofenamate; 2-pyca = 2-pyridylcarbinol, Et$_2$nia = N,N-diethylnicotinamide. [e] No structure reported. [f] From [24]; clof = 2-(4-chlorophenoxy)-2-methylpropionic acid; 4-pymeth = 4-pyridylmethanol.

3. Experimental Section

3.1. Methods and Instrumentation

^{1}H and ^{13}C NMR spectra were recorded in DMSO-d_6 using a Bruker Avance II 300 spectrometer (Bruker, Rheinhausen, Germany). Elemental analyses were carried out using a HEKAtech CHNS EuroEA 3000 Analyzer (Hekatech, Wegberg, Germany). IR spectra were recorded using KBr or polyethylene pellets using an IFS/66v/S or an Alpha-T spectrometers (Bruker, Rheinhausen, Germany). UV-vis absorption spectra were measured on transparent KBr pellets using a Shimadzu UV-3600 photo spectrometer (Shimadzu Europe, Duisburg, Germany). EPR spectra were recorded in the X-band on a Bruker System ELEXSYS 500E equipped with a Bruker Variable Temperature Unit ER 4131VT (500 to 100 K) (Bruker, Rheinhausen, Germany). The g values were calibrated using a dpph sample. The magnetic measurements were carried out on finely ground samples using a MPMS XL7 (Quantum Design, Darmstadt, Germany) instrument measuring from 2 to 300 K at a magnetic field of 1 T.

3.2. Single Crystal Structure Determination

The measurements were performed at 293(2) K or 110(2) K using graphite monochromatised Mo-Kα radiation (λ = 0.71073 Å) on an IPDS II instrument (STOE & Cie., Darmstadt, Germany). The structures were solved by dual space methods using ShelXT-2015 [57] and refined by full-matrix least-squares techniques against F^2 (SHELXL-2017/1) [58,59]. The numerical absorption corrections (X-RED V1.31; STOE & Cie, 2005, Darmstadt, Germany) were performed after optimising the crystal shapes using the X-SHAPE V1.06 (STOE & Cie, 1999, Darmstadt, Germany) [60,61]. The non-hydrogen atoms were refined with anisotropic displacement parameters. H atoms were included by using appropriate riding models. More details on the crystal structures is provided in the Supplementary Materials. CCDC 1058949–1058951, 1058953–1058957, 1058960–1058963, and 2025467–2025471 contain the full crystallographic data. These data can be obtained free of charge at www.ccdc.cam.ac.uk/conts/retrieving.html or from the Cambridge Crystallographic Data Centre, 12 Union Road, Cambridge, CB2 1EZ UK. Fax: +44-1223-336-033; Email: deposit@ccdc.cam.ac.uk.

3.3. Powder X-ray Diffraction (PXRD)

Data collection was carried out with a Huber G670 diffractometer (Huber, Rimsting, Germany) equipped with a Ge(111) monochromator using Cu-Kα1 radiation with λ = 1.5405 Å and an image plate detector. The samples were measured as flat samples between two almost X-ray transparent foils. The foil gives rise to two broad reflections at 2θ ≈ 21.5° and 2θ ≈ 23.7°. PXRD data were visualised with the WinXPOW software package (STOE & Cie., 2012, Darmstadt, Germany) [62], which was also used to calculate line diagrams based on single crystal data. Gnuplot4.6 was used for the visualisation of PXRD patterns [63].

3.4. Syntheses

3.4.1. General

Picolinamide and nicotinamide were synthesised from their (commercially available) carboxylic acids (details in the Supplementary Materials), all other chemicals were used as supplied. Water-free reactions were carried out under inert gas conditions and performed using Schlenk techniques. Solvents were dried using an MBRAUN MB SPS-800 (MBRAUN, Garching, Germany) solvent purification system.

3.4.2. Synthesis of the Cu^{II} Compounds—General Method

The following compounds were synthesised through dissolving suitable Cu^{II} salts in EtOH, heating these solutions to 70 °C and then slowly adding the ligands to the mixture. Upon standing and slow evaporation of the solvent at ambient temperatures in the fume hood, the materials were obtained in yields from 39 to 75% referring to the Cu^{II} starting material and from 30 to 90% referring to the ligand.

[Cu(Ina)$_4$(H$_2$O)$_2$](BF$_4$)$_2$ (**1**). From 0.690 g (2 mmol) Cu(BF$_4$)$_2$·6H$_2$O and 0.122 g (1 mmol) Ina. Yield: 0.624 g (0.82 mmol, 41% referred to Cu and 82% to Ina) blue plates. Elemental analysis calculated for $C_{24}H_{28}B_2Cu_1F_8N_8O_6$ (761.69): C 37.85, H 3.71, N 14.71; found: C 37.83, H 3.70, N 14.71%. IR (KBr pellet): 3421 s; 3292 m, $ν_{as}$(NH$_2$); 3196 s, $ν_s$(NH$_2$); 3072 sh, ν(pyr–H); 1678 s, ν(C=O); 1606 s, δ(NH$_2$); 1554 s, 1549 s, ν(C=C), ν(C=N); 1425 sh, ν(C–C); 1415 s, ν(C–N); 1080 s, ν(B–F); 639 s, δ(N–C=O) cm^{-1}. PXRD is available in the Supplementary Materials.

[Cu(Ina)$_2$(NO$_3$)$_2$] (**2**). From 0.481 g (2 mmol) Cu(NO$_3$)$_2$·3H$_2$O and 0.122 g (1 mmol) Ina. Yield: 0.337 g (0.78 mmol, 39% referred to Cu and 78% to Ina), blue plates. Elemental analysis calculated for $C_{12}H_{12}Cu_1N_6O_8$ (431.81): C 33.38, H 2.80, N 19.46; found: C 33.36, H 2.81, N 19.44%. IR (PE pellet): 3426 s, 3311 m, $ν_{as}$(NH$_2$); 3181 s, $ν_s$(NH$_2$); 3082 m, ν(pyr–H); 1705 s, ν(C=O); 1623 s, δ(NH$_2$); 1614 s, 1546 s, ν(C=C), ν(C=N); 1424 s, ν(C–C); 1381 s, ν(C–N); 1383 s, ν(NO$_3^-$); 1118 m, ν(CN); 484 s, δ(N–C=O) cm^{-1}.

[Cu(Ina)$_2$(H$_2$O)$_2$(NO$_3$)$_2$] (**3**). From 0.482 g (2 mmol) Cu(NO$_3$)$_2$·3H$_2$O and 0.144 g (1 mmol) Ina. Yield: 0.383 g (0.82 mmol, 41% referred to Cu and 82% to Ina), blue blocks. Elemental analysis calculated for $C_{12}H_{16}Cu_1N_6O_{10}$ (467.84): C 30.81, H 3.45, N 17.96; found: C 30.83, H 3.46, N 17.94%. The compound was previously reported and obtained from a 1:2 mixture of Cu:ligand [16].

[Cu(Pia)$_2$(BF$_4$)$_2$] (**8**). From 0.690 g (2 mmol) Cu(BF$_4$)$_2$·6H$_2$O and 0.122 g (1 mmol) Pia. Yield: 0.385 g (0.8 mmol, 40% referred to Cu and 80% to Pia) light blue plates. Elemental analysis calculated for $C_{12}H_{12}B_2Cu_1F_8N_4O_2$ (481.41): C 29.94, H 2.51, N 11.64; found: C 29.89, H 2.50, N 11.63%. IR (KBr pellet): 3415 m, 3247 m, 3024 m, ν(NH$_2$); 1666 s, ν(C=O); 1560 s, 1444 s, ν(C=C), ν(C=N); 1311 w, ν(C=N); 1072 s, ν(B–F); 906 sh, ν(C–N) ring; 767 m, δ(C–N) ring; 656 m, δ(N–C=O) cm^{-1}. PXRD is available in the Supplementary Materials.

[Cu(Pya)$_2$](BF$_4$)$_2$ (**9**). From 0.345 g (1 mmol) Cu(BF$_4$)$_2$·6H$_2$O and 0.123 g (1 mmol) Pya. Yield: 0.208 g (0.43 mmol, 43% referred to Cu and 86% to Pya) light blue plates. Elemental analysis calculated for $C_{10}H_{10}B_2Cu_1F_8N_6O_2$ (483.38): C 24.85, H 2.09, N 17.39; found: C 24.88, H 2.10, N 17.36%. IR (KBr pellet): 3400 m, 3276 s, $ν_{as}$(N–H); 3239 s, $ν_s$(N–H); 1697 s, ν(C=O); 1619 m, ν(NH$_2$); 1578 m, 1542 w, 1496 m, ν(C=C), ν(N=C); 1414 m; 1459 m, amide; 1212 w, δ(CH); 1182 m, $ν_{14}$(ring); 1108-985 vs, ν(B–F); 871 m, γ(CH); 780 m, δ(OH)+$ν_1$(ring); 666 w, ω(NH$_2$); 602 m, δ(N–C=O); 456 m, $ν_{16}$(ring) cm^{-1}; δ(NH$_2$) and δ(CH) were probably obscured. PXRD is available in the Supplementary Materials.

[Cu(Pya)$_2$(H$_2$O)(NO$_3$)](NO$_3$) (**10**). From 0.482 g (2 mmol) Cu(NO$_3$)$_2$·3H$_2$O and 0.123 g (1 mmol) Pya. Yield: 0.406 g (0.9 mmol, 45% referred to Cu and 90% to Pya) blue plates. Elemental analysis calculated for C$_{10}$H$_{12}$Cu$_1$N$_8$O$_9$ (451.80): C 26.58, H 2.68, N 24.80; found: C 26.51, H 2.66, N 24.75%. IR (KBr pellet): 3600–2975 m, ν(O–H) H$_2$O; 3339 m, ν$_{as}$(N–H); 3150 s, ν$_s$(N–H); 1681 s, ν(C=O); 1605 s, ν(NH$_2$); 1567 m, 1530 w, 1492 m, ν(C=C), ν(N=C); 1449 s, 1397 m, amide; 1350 s, ν(NO$_3^-$); 1199 w, δ(CH); 1175 m, ν$_{14}$(ring); 1081 m, δ(NH$_2$); 1058 s, δ(CH); 1022 sh, ν$_{12}$(ring); 882 m, γ(CH); 803 m, δ(OH)+ν$_1$(ring); 690 w, ω(NH$_2$); 641–561 m, δ(N–C=O); 467 m, ν$_{16}$(ring) cm^{-1}. PXRD is available in the Supplementary Materials.

[Cu(Pya)(NO$_3$)$_2$] (**11**). From 0.241 g (1 mmol) Cu(NO$_3$)$_2$·3H$_2$O and 0.123 g (1 mmol) Pya. Yield: 0.233 g (0.245 mmol, 75% referred to Cu and Pya) blue plates. Elemental analysis calculated for C$_5$H$_5$Cu$_1$N$_5$O$_7$ (310.67): C 19.33, H 1.62, N 22.54; found: C 19.31, H 1.64, N 22.51%. IR (KBr pellet): 3386 m, 3287 s, ν$_{as}$(N–H); 3090 s, ν$_s$(N–H); 1693 s, ν(C=O); 1630 m, ν(NH$_2$); 1599 m, 1543 w, ν(C=C), ν(N=C); 1310 s, amide; 1212 w, δ(CH); 1488 s, 1421 s, 1282 s, ν(NO$_3^-$); 1116 m, ν$_{14}$ ring; 1077 m, δ(NH$_2$); 1058 s, δ(CH); 1021 s, ν$_{12}$ ring; 871 m, γ(CH); 803 m, δ(OH)+ν1(ring); 689 m, ω(NH$_2$); 483 w, δ(N–C=O) cm^{-1}.

[Cu(Pyc)(Tfa)](**12**). From 0.308 g (1 mmol) Cu(Tfa)$_2$·H$_2$O and 0.123 g (1 mmol) Pya. Yield: 0.263 g (0.88 mmol, 88% referred to Cu and Pya) blue blocks. Elemental analysis calculated for C$_7$H$_3$Cu$_1$F$_3$N$_2$O$_4$ (299.65): C 28.06, H 1.01, N 9.35; found: C 28.08, H 1.04, N 9.33%. IR (KBr pellet): 3236 s, ν(ring–H); 1696 vs, ν(C=O) of Tfa; 1728 s ν(C=O) of Pyc; 1587 m, 1520 m, 1440 m, ν(C=C), ν(N=C); 1207–1138 s, ν(C–F); 1065 w, δ(CH); 1047 m, ν$_{12}$(ring); 867 sh; γ(CH); 779 sh, δ(OH)+ν$_1$(ring); 468 m, ν$_{16}$(ring) cm^{-1}; Not observed (obscured): ν$_{as}$(N–H), ν$_s$(N–H), ν(NH$_2$), amide, δ(CH), δ(NH$_2$), ω(NH$_2$), and δ(N–C=O). FIR (KBr pellet): 614 m, ring; 580 m, ν$_1$(Cu–O); 516 m, ν$_2$(Cu–O); 475 m, ring; 420 m, ν$_3$(Cu–O); 317 m, ν$_1$(Cu–N11); 291 s, ν$_2$(Cu–N12) cm^{-1}. PXRD is available in the Supplementary Materials.

[Cu(2-Aba)$_2$(NO$_3$)$_2$] (**13**). From 0.241 g (1 mmol) Cu(NO$_3$)$_2$·3H$_2$O and 0.136 g (1 mmol) 2-Aba. Yield: 0.202 g (0.44 mmol, 44% referred to Cu and 88% to 2-Aba) green plates. Elemental analysis calculated for C$_{14}$H$_{16}$Cu$_1$N$_6$O$_8$ (459.86): C 36.57, H 3.51, N 18.28; found: C 36.55, H 3.54, N 18.30%. PXRD is available in the Supplementary Materials.

(Cu$_2$(4-Aba)$_2$(NO$_3$)$_3$(H$_2$O)$_3$)(NO$_3$) (**14**). From 0.241 g (1 mmol) Cu(NO$_3$)$_2$·3H$_2$O and 0.136 g (1 mmol) 4-Aba, Yield: 0.295 g (0.42 mmol, 84% referred to Cu and 4-Aba) green needles. Elemental analysis calculated for C$_{14}$H$_{22}$Cu$_2$N$_6$O$_8$ (701.46): C 23.97, H 3.16, N 15.97; found: C 23.98, H 3.13, N 15.97%.

[Cu(4-Aba)$_2$(EtOH)$_2$](ClO$_4$)$_2$ (**15**). From 0.370 g (1 mmol) Cu(ClO$_4$)$_2$·6H$_2$O and 0.408 g (3 mmol) 4-Aba.Yield: 0.564 g (0.90 mmol, 90% referred to Cu and 30% to 4-Aba) green blocks. Elemental analysis calculated for C$_{18}$H$_{28}$Cl$_2$Cu$_1$N$_4$O$_{12}$ (626.88): C 34.49, H 4.50, N 8.94; found: C 34.48, H 4.52, N 8.93%. IR (KBr pellet): 3441 s, ν(O–H) EtOH; 3362 ms; 3292 s, ν$_{as}$(N–H); 3252 m, ν$_s$(N–H); 3058 m, ν(aryl–H); 1650 s, 1611 s, ν(C=O), δ(NH$_2$); 1550 s, ν(C=C) ring; 1450 s, δ$_{s,as}$(CH$_2$,CH$_3$) EtOH; 1197-992 vs, ν(C–O) EtOH; 1046 s, 620 m, ClO$_4^-$ cm^{-1}; δ(N–C=O) probably obscured. PXRD is available in the Supplementary Materials.

[Cu(4-Aba)$_2$(EtOH)$_2$](BF$_4$)$_2$ (**16**). From 0.345 g (1 mmol) Cu(BF$_4$)$_2$·6H$_2$O and 0.408 g (3 mmol) 4-Aba. Yield: 0.553 g (0.92 mmol, 92% referred to Cu and 31% to 4-Aba) green blocks. Elemental analysis calculated for C$_{18}$H$_{28}$B$_2$Cu$_1$F$_8$N$_4$O$_4$ (601.60): C 35.94, H 4.69, N 9.31; found: C 35.93, H 4.66, N 9.33%. IR (KBr pellet): 3450 s, ν(O–H) EtOH; 3379 m, 3319 s, ν$_{as}$(N–H); 3277 s, ν$_s$(N–H); 3175 m, ν(aryl–H); 1652 s, 1610 s, ν(C=O), δ(NH$_2$); 1555 s, ν(C=C) ring; 1453 m, δ$_{s,as}$(CH$_2$,CH$_3$) EtOH; 1199-911 s, ν(C–O) EtOH, ν(B–F) cm^{-1}. PXRD is available in the Supplementary Materials.

[Cu(4-HAba)$_2$Cl$_4$] (**17**). From 0.370 g (1 mmol) Cu(ClO$_4$)$_2$·6H$_2$O and 0.136 g (1 mmol) 4-Aba and 5 mL concentrated aqueous HCl. Yield: 0.206 g (0.43 mmol, 43% referred to Cu and 86% to 4-Aba) brown platelets. Elemental analysis calculated for C$_{14}$H$_{18}$Cl$_4$Cu$_1$N$_4$O$_2$ (479.67): C 35.06, H 3.78, N 11.68; found: C 35.03, H 3.77, N 11.67%. IR (KBr pellet): 3378 s, ν$_{as}$(N–H); 3001-2650 s, ν$_s$(N–H); 2580–2560 m, ν(N–H) amine; 1650 s, 1611 m, ν(C=O), δ(NH$_2$); 1550 m, ν(C=C) ring; 579 m, δ(N–C=O) cm^{-1}. FIR (KBr pellet): 547 m, ν$_1$(Cu–O); 473 m, ring; 380 m, ν$_2$(Cu–O); 318 m, ν$_1$(Cu–Cl); 278 m, ν$_2$(Cu–Cl) cm^{-1}; ν(aryl–H) probably obscured.

[Cu(Tba)$_2$](ClO$_4$)$_2$ (**18**). From 0.370 g (1 mmol) Cu(ClO$_4$)$_2$·6H$_2$O and 0.272 g (2 mmol) Tba Yield: 0.508 g (0.91 mmol, 91% referred to Cu and Tba) brown platelets. Elemental analysis calculated for C$_8$H$_{16}$Cl$_2$Cu$_1$N$_4$O$_{12}$S$_2$ (558.80): C 17.20, H 2.89, N 10.03; found: C 17.23, H 2.88, N 10.03%. IR (KBr pellet): 3428 s, ν$_{as}$(N–H); 3329 s, ν$_s$(N–H); 2975 s, 2927 w, ν(C–H); 1653 s, 1600 m, ν(C=O), ν(NH$_2$); 1388 m, ν(C–N); 1086 vs, 580 s, ClO$_4^-$; 575 m, δ(N–C=O); 623 m, ν(S–C) cm^{-1}.

3.4.3. Further Syntheses

[Cu(Ina)$_2$(H$_2$O)$_3$(SiF$_6$)]·H$_2$O (**4**). Amounts of 173 mg (0.5 mmol) Cu(BF$_4$)$_2$·6H$_2$O and 62 mg (0.5 mmol) Ina were each dissolved in 15 mL EtOH, stirred at 70 °C for 30 min and then the two solutions were combined. Standing for about 2 weeks in a glass beaker while letting the solvent evaporate allowed collecting blue needles. Yield: 0.104 g (0.2 mmol, 40% referred to Cu and 80% to Ina). Elemental analysis calculated for C$_{12}$H$_{20}$Cu$_1$F$_6$N$_4$O$_6$Si$_1$ (521.94): C 27.61, H 3.86, N 10.73; found: C 27.63, H 3.86, N 10.74%. IR (KBr pellet): 3695-3270 m, ν(O–H) H$_2$O; 3423 s, 3316 s, ν$_{as}$(N–H); 3181 s, ν$_s$(N–H); 3079 m, ν(pyr–H); 1705 s, ν(C=O); 1626 m, δ(NH$_2$); 1612 s, 1552 s, ν(C=C), ν(C=N); 1421 s, ν(C–C); 1388 s, ν(C–N); 740 s, SiF$_6^{2-}$; 490 s, δ(N–C=O) cm^{-1}. FIR (KBr pellet) 286 m, ν(Cu–N); 255 m, ν(Cu–N); 186 m, ν(Cu–F) cm^{-1}.

[Cu(Ina)$_2$Br$_2$] (**5**) was synthesised by dissolving 216 mg (0.5 mmol) [Cu(Ina)$_2$(NO$_3$)$_2$] (**2**) and 71 mg (0.6 mmol) KBr each in 30 mL of EtOH at 70 °C. Mixing of the two solutions produced a green precipitate which was filtered off and dissolved in 20 mL of water. Slow evaporation gave 58 mg (0.125 mmol, 25%) green needles after about 3 weeks. Elemental analysis calculated for C$_{12}$H$_{12}$Cu$_1$Br$_2$N$_4$O$_2$ (467.61): C 30.82, H 2.59, N 11.98; found: C 30.83, H 2.56, N 11.94%. IR (KBr pellet): 3402 s, 3248 m, ν$_{as}$(NH$_2$); 3186 s, ν$_s$(NH$_2$); 3067 m, ν(pyr–H); 1714 s, ν(C=O); 1661 s, δ(NH$_2$); 1609 s, 1556 m, ν(C=C), ν(C=N); 1416 s, ν(C–C); 1118 w, ν(CN); 546 s, δ(N–C=O) cm^{-1}. Qualitatively, the compound can also be prepared by grinding [Cu(Ina)$_2$(NO$_3$)$_2$] (**2**) with KBr in a mortar. The compound was previously reported [39] but no details as to the Cu:ligand ratio were provided.

(HIna)$_2$[CuCl$_4$]·2H$_2$O (**6**). Amounts of 85.5 mg (0.5 mmol) CuCl$_2$·2H$_2$O and 61 mg (0.5 mmol) Ina were dissolved in EtOH and 1.3 mL concentrated HCl were added. The mixture was heated up to 70 °C and then left for slow evaporation of the solvent at ambient temperatures in the fume hood. After about 20 days, 126 mg (0.26 mmol, 52%) of blue plates were obtained. Elemental analysis calculated for C$_{12}$H$_{18}$Cl$_4$Cu$_1$N$_4$O$_4$ (487.64): C 29.56, H 3.72, N 11.49; found: C 29.55, H 3.70, N 11.50%.

(HIna)[Cu(H$_2$O)Cl$_3$] (**7**). Amounts of 85.5 mg (0.5 mmol) CuCl$_2$·2H$_2$O and 61 mg (0.5 mmol) Ina were suspended in 12 mL H$_2$O in a Teflon container and 3.5 mg (0.01 mmol) Gd$_2$O$_3$ and 1.3 mL concentrated HCl were added. The container was heated in an autoclave for 2 days at 140 °C. After cooling and opening the reaction mixture was immersed in a glass beaker. After 4 weeks evaporation, 53 mg (0.17 mmol, 34%) of light blue platelets were obtained. Elemental analysis calculated for C$_6$H$_9$Cl$_3$Cu$_1$N$_2$O$_2$ (311.05): C 23.17, H 2.92, N 9.01; found: C 23.13, H 2.96, N 9.00%. Remark: the role of Gd$_2$O$_3$ as catalyst was checked by a blind reaction without Gd$_2$O$_3$. The reaction occurred very slowly and only traces of the target product were obtained.

4. Conclusions

New Cu^{II} complexes and coordination polymers of isonicotineamide (Ina) and picolinamide (Pia), pyrazine 2-amide (Pya), 2- and 4-amino benzamides (2-Aba and 4-Aba) ligands were synthesised from various Cu^{II} sources. Under the criterion of crystallinity and phase purity we were able to reliably reproduce many of the materials in reasonable yields. Crystal and molecular structures from single-crystal XRD were obtained for all new compounds; phase purity was checked using powder XRD. For the optimisation of the reaction protocols, the Cu:ligand stoichiometry of the starting materials was varied and we observed marked deviations of the Cu:ligand ratio in products compared with the starting materials. For three compounds containing the pyridine amides Ina and Pia, high Cu loadings of 2:1 were necessary to obtain phase pure compounds with Cu:ligand ratios of 1:4 and 1:2. In contrast to this, for the amino benzamide 4-Aba the optimum ratio of starting materials was 1:3 or 1:2 to obtain compounds with a Cu:ligand ratio of 1:2. In the structures, we exclusively observed O_{amide} and not NH_{2amide} binding to Cu^{II}, in most of the cases supported by chelating of the second N binding function (pyridine, pyrazine, or aminophenyl) on the ligands. The ditopic ligands Ina, Pya, and 4-Aba frequently bridge µ-O,N between Cu^{II} centres forming 1D or 2D coordination polymers. Again, no NH_{2amide} binding occurred. However, the same ligands were also found in terminal non-bridging modes. Nitrate, NO_3^-, is often competing with amide ligand binding and might be one important reason for the saturation of Cu^{II} preventing the µ-O,N bridging and formation of polymers. The coordination surrounding Cu^{II} in these structures is dominated by the ubiquitous axial elongated octahedral geometry (OE). The second important coordination polyhedron is the square pyramid (SPy). The occurrence of these two polyhedra cannot be correlated with the polymeric or non-polymeric character of the compounds. The strong pyridine, pyrazine, or aminophenyl-N functions dominate the equatorial binding plane. O_{amide} occurs in this plane only in chelates with these N functions. Non-chelated O_{amide} binding to Cu^{II} was found in the axial (weak) position as expected, alongside with other weak ligands such as NO_3^-, EtOH, H_2O, BF_4^-, and the peculiar SiF_6^{2-}. X-band EPR spectra of powder samples revealed various spectral symmetry patterns ranging from axial over rhombic to inverse axial. Although the EPR spectra cannot be unequivocally correlated to the observed geometry of Cu^{II} in the solid state structures, the EPR patterns can help to support assumed structures as shown for the compound [Cu(Ina)$_2$Br$_2$] (Ina = isonicotine amide). As UV-vis absorption spectroscopy and magnetic measurements in the solid can also be reasonably correlated with the surrounding of Cu^{II}, we suggest the combination of EPR, UV-vis spectroscopy, and magnetic measurements to elucidate or support possible structures of Cu^{II} compounds with such ligands if no unequivocal structural information is available.

As we can reliably reproduce some of the presented materials here, we will study the biological properties, for example, the antiproliferative activity of selected compounds, in the near future.

Supplementary Materials: The following are available online at http://www.mdpi.com/2304-6740/8/12/65/s1, Figures S1–S40 showing crystal and molecular structures and powder XRD patterns of [Cu(Pia)$_2$(BF$_4$)$_2$] (**8**), [Cu(Ina)$_2$(NO$_3$)$_2$] (**2**), [Cu(Ina)$_4$(H$_2$O)$_2$](BF$_4$)$_2$ (**1**), [Cu(Ina)$_2$(H$_2$O)$_3$(SiF$_6$)]·H$_2$O (**4**), (HIna)$_2$[CuCl$_4$]·2H$_2$O (**6**), (HIna)[Cu(H$_2$O)Cl$_3$] (**7**), [Cu(Pya)$_2$](BF$_4$)$_2$ (**9**), [Cu(Pya)$_2$(H$_2$O)(NO$_3$)](NO$_3$) (**10**), [Cu(Pya)(NO$_3$)$_2$] (**11**), [Cu(Pyc)(Tfa)](**12**), [Cu(2-Aba)$_2$(NO$_3$)$_2$] (**13**), [Cu$_2$(4-Aba)$_2$(H$_2$O)$_3$(NO$_3$)$_3$](NO$_3$) (**14**), [Cu(4-Aba)$_2$(EtOH)$_2$](ClO$_4$)$_2$ (**15**), [Cu(4-Aba)$_2$(EtOH)$_2$](BF$_4$)$_2$ (**6**), [Cu(4-AbaH)$_2$Cl$_4$] (**17**), and [Cu(Tba)$_2$](ClO$_4$)$_2$ (**18**) and Figures S41–S48 showing X-Band EPR spectra of [Cu(Ina)$_2$(H$_2$O)$_2$(NO$_3$)$_2$] (**2**), [Cu(Tba)$_2$](ClO$_4$)$_2$ (**18**), [Cu(Pya)$_2$(H$_2$O)(NO$_3$)](NO$_3$) (**10**), [Cu(Ina)$_2$(H$_2$O)$_3$(SiF$_6$)]·H$_2$O (**4**), [Cu(Pya)(NO$_3$)$_2$] (**11**), [Cu(Pia)$_2$](BF$_4$)$_2$ (**8**), (HIna)$_2$[CuCl$_4$]·2H$_2$O (**6**), [Cu(Pyc)(Tfa)](**12**), and [Cu(Ina)$_2$Br$_2$] (**5**), together with Supplementary Tables S1–S39 containing crystal structure and refinement data of all compounds.

Author Contributions: I.W. and N.N.A.T.W. carried out the syntheses; S.S. and I.P. collected XRD data, solved the structures and provided figures and tables; K.B., K.S. and A.K. measured and analysed EPR spectra; G.M. and A.K. designed and supervised the project and provided the equipment; A.K. wrote the draft; G.M. and A.K. revised the manuscript. All authors have read and agreed to the published version of the manuscript.

Funding: General funding is gratefully acknowledged to the University of Cologne. A.K. acknowledges the German Academic Exchange Service (DAAD)—KD_0001052598-2 for a short-time guest lectureship and the Shiraz University, Iran, for support.

Acknowledgments: We thank Horst Schumacher, Department of Chemistry of the University of Cologne, for the PXRD measurements.

Conflicts of Interest: The authors declare no conflict of interest.

References

1. Riascos-Rodríguez, K.; Marks, S.; Evans, P.G.; Hernández-Rivera, S.P.; Ruiz-Caballero, J.L.; Pinñero, D.; Hernaández-Maldonado, A.J. Lithium Functionalization Promoted by Amide-Containing Ligands of a Cu(pzdc)(pia) Porous Coordination Polymer for CO_2 Adsorption Enhancement. *Cryst. Growth Des.* **2020**, *20*, 3898–3912. [CrossRef]
2. Li, F.-C.; Li, X.-L.; Tan, L.-K.; Wang, J.-T.; Yao, W.-Z. Evans–Showell-type polyoxometalate-based metal–organic complexes with novel 3D structures constructed from flexible bis-pyrazine-bis-amide ligands and copper metals: Syntheses, structures, and fluorescence and catalytic properties. *Dalton Trans.* **2019**, *48*, 2160–2169. [CrossRef]
3. Hearne, N.; Turnbull, M.M.; Landee, C.P.; van der Merwe, E.M.; Rademeyer, M. Halide-bi-bridged polymers of amide substituted pyridines and -pyrazines: Polymorphism, structures, thermal stability and magnetism. *CrystEngComm* **2019**, *21*, 1910–1927. [CrossRef]
4. Kumar, S.; Sharma, R.P.; Venugopalan, P.; Ferretti, V.; Tarpin, M.; Sayen, S.; Guillon, E. New copper(II) niflumate complexes with N-donor ligands: Synthesis, characterization and evaluation of anticancer potential against human cell lines. *Inorg. Chim. Acta* **2019**, *488*, 260–268. [CrossRef]
5. Kwiatek, D.; Kubicki, M.; Skokowski, P.; Gruszczyńska, J.; Lis, S.; Hnatejko, Z. Five subsequent new pyridine carboxamides and their complexes with d-electron ions. Synthesis, spectroscopic characterization and magnetic properties. *J. Mol. Struct.* **2019**, *1178*, 669–681. [CrossRef]
6. Kwiatek, D.; Kubicki, M.; Belter, J.; Jastrząb, R.; Wiśniewska, H.; Lis, S.; Hnatejko, Z. Synthesis, spectroscopic characterization and antifungal activity studies of five novel complexes with pyridine carboxamides. *Polyhedron* **2017**, *133*, 187–194. [CrossRef]
7. Nguyen, T.V.; Ong, T.D.; Lam, A.H.M.; Pham, V.T.; Phan, N.T.S.; Truong, T. Nucleophilic trifluormethylation of aryl boronic acid under heterogeneous Cu(INA)$_2$ catalysis at room temperature: The catalytic copper-based protocol. *Mol. Catal.* **2017**, *436*, 60–66. [CrossRef]
8. Chen, Y.; Li, L.; Li, J.; Ouyang, K.; Yang, J. Ammonia capture and flexible transformation of M-2(INA) (M = Cu, Co, Ni, Cd) series materials. *J. Hazard. Mater.* **2016**, *306*, 340–347. [CrossRef]
9. Chen, Y.; Li, L.; Yang, J.; Wang, S.; Li, J. Reversible flexible structural changes in multidimensional MOFs by guest molecules (I$_2$, NH$_3$) and thermal stimulation. *J. Solid Stat. Chem.* **2015**, *226*, 114–119. [CrossRef]
10. Zurowska, B. Structural and magnetic characterization of Cu–picolinate and Cu–quinaldinate and their mixed complexes with water or halides. *Inorg. Chim. Acta* **2014**, *418*, 136–152. [CrossRef]
11. Wang, P.; Li, G.; Chen, Y.; Chen, S.; James, S.L.; Yuan, W. Mechanochemical interconversion between discrete complexes and coordination networks—Formal hydration/dehydration by LAG. *CrystEngComm* **2012**, *14*, 1994–1997. [CrossRef]
12. He, Y.-C.; Yang, J.; Yang, G.-C.; Kan, W.-Q.; Ma, J.-F. Solid-state single-crystal-to-single-crystal transformation from a 2D layer to a 3D framework mediated by lattice iodine release. *Chem. Commun.* **2012**, *48*, 7859–7861. [CrossRef] [PubMed]
13. Lian, T.T.; Chen, S.-M. A new microporous Cu(II)-isonicotinate framework with 8-connected bcu topology. *Inorg. Chem. Commun.* **2012**, *18*, 8–10. [CrossRef]
14. Monfared, H.H.; Vahedpour, M.; Yeganeh, M.M.; Ghorbanloo, M.; Mayer, P.; Janiak, C. Concentration dependent tautomerism in green [Cu(HL^1)(L^2)] and brown [Cu(L^1)(HL^2)] with H$_2$L^1 = (E)-N'-(2-hydroxy-3-methoxybenzylidene)-benzoylhydrazone and HL^2 = pyridine-4-carboxylic (isonicotinic) acid. *Dalton Trans.* **2011**, *40*, 1286–1294. [CrossRef]
15. Dakovic, M.; Jaglicic, Z.; Kozlevcar, B.; Popovic, Z. Association of copper(II) isonicotinamide moieties via different anionic bridging ligands: Two paths of ferromagnetic interaction in the azide coordination compound. *Polyhedron* **2010**, *29*, 1910–1917. [CrossRef]
16. Dakovic, M.; Popovic, Z. Uncommon isonicotinamide supramolecular synthons in copper(II) complexes directed by nitrate and perchlorate anions. *Acta Cryst.* **2009**, *C65*, m361–m366.
17. Svorec, J.; Lörinc, S.; Moncol, J.; Melnik, M.; Koman, M. Structural and spectroscopic characterization of copper(II) tolfenamate complexes. *Trans. Met. Chem.* **2009**, *34*, 703–710. [CrossRef]
18. Lörinc, S.; Švorec, J.; Melnik, M.; Koman, M. Structure and spectral characterisation of copper(II) meclofenamate complexes. *Polyhedron* **2008**, *27*, 3545–3548. [CrossRef]

19. Kavalírova, J.; Korabik, M.; Stachová, P.; Moncol, J.; Sillanpää, R.; Lis, T.; Miklos, D.; Melník, M.; Mrozinski, J.; Valigura, D. Synthesis, spectral and magnetic properties of two different 2-nitrobenzoatocopper(II) complexes containing N,N-diethylnicotinamide. *Polyhedron* **2008**, *27*, 1333–1342. [CrossRef]
20. Moncol, J.; Mudra, M.; Lönnecke, P.; Hewitt, M.; Valko, M.; Morris, H.; Svorec, J.; Melnik, M.; Mazur, M.; Koman, M. Crystal structures and spectroscopic behavior of monomeric, dimeric and polymeric copper(II) chloroacetate adducts with isonicotinamide, N-methylnicotinamide and N,N-diethylnicotinamide. *Inorg. Chim. Acta* **2007**, *360*, 3213–3225. [CrossRef]
21. Pichon, A.; Lazuen-Garay, A.; James, S.L. Solvent-free synthesis of a microporous metal–organic framework. *CrystEngComm* **2006**, *8*, 211–214. [CrossRef]
22. Atac, A.; Yurdakul, S.; Ide, S. Synthesis and vibrational spectroscopic studies of isonicotinamide metal(II) halide complexes. *J. Mol. Struct.* **2006**, *783*, 79–87. [CrossRef]
23. Li, C.-B.; Liu, B.; Gao, G.-G.; Che, G.-B. Hydrogen bonding and π–π stacking in di-µ-isophthalato-bis[bis(isonicotinamide)copper(II)] trihydrate. *Acta Cryst.* **2005**, *E61*, m1705–m1707.
24. Moncol, J.; Kalinakova, B.; Svorec, J.; Kleinova, M.; Koman, M.; Hudecova, D.; Melnik, M.; Mazur, M.; Valko, M. Spectral properties and bio-activity of copper(II) clofibriates, part III: Crystal structure of $Cu(clofibriate)_2(2-pyridylmethanol)_2$, $Cu(clofibriate)_2(4-pyridylmethanol)_2(H_2O)$ dihydrate, and $Cu_2(clofibriate)_4(N,N-diethylnicotinamide)_2$. *Inorg. Chim. Acta* **2004**, *357*, 3211–3222. [CrossRef]
25. Goher, M.A.S.; Mautner, F.A. Spectral and structural characterization of copper(I) halide, nitrate and perchlorate complexes of pyrazine carboxamide (pyza) and X-ray crystal structure of polymeric $[Cu(pyza)_2I]_n$ complex. *Polyhedron* **2000**, *19*, 601–606. [CrossRef]
26. Aakeroy, C.B.; Beatty, A.M.; Leinen, D.S.; Lorimer, K.R. Deliberate combination of coordination polymers and hydrogen bonds in a supramolecular design strategy for inorganic/organic hybrid networks. *Chem. Commun.* **2000**, 935–936. [CrossRef]
27. Goher, M.A.S.; Abu-Youssef, M.A.M.; Mautner, F.A. ^{1}D polymeric copper(II) complexes containing bridging tridentate pyrazinato and terminal chloro or azido anions. Synthesis, spectral, structural and thermal study of $[CuCl(pyrazinato)(H_2O)]_n$ and $[Cu(N_3)(pyrazinato)(H_2O)]_n$ complexes. *Polyhedron* **1998**, *17*, 3305–3314. [CrossRef]
28. Goher, M.A.S.; Al-Salem, N.A.; Mautner, F.A.; Klepp, K.O. A copper(II) azide compound of pyrazinic acid containing a new dinuclear complex anion $[Cu_2(N_3)_6]^{2-}$. Synthesis, spectral and structural study of $KCu_2(pyrazinato)(N_3)_4$. *Polyhedron* **1997**, *16*, 825–831. [CrossRef]
29. Goher, M.A.S.; Mautner, F.A. New Unexpected Coordination Modes of Azide and Picolinato Anions Acting as Bridging Ligands Between Copper(II) and Sodium or Potassium Ions. Synthesis, Crystal Structures and Spectral Characterizations of $[MCu(picolinato)(N_3)_2]_n$ (M = Na or K) Complexes. *Polyhedron* **1995**, *14*, 1439–1446. [CrossRef]
30. Goher, M.A.S.; Mautner, F.A. Spectroscopic and Crystal Structure Study of $NaCu(Picolinato)_2(N_3)(H_2O)_2$. A Polymeric Structure Containing Simultaneous Bridging Pentadentate Picolinato Anion and µ(1,3) Azido Ligands Between Copper and Sodium Centred Polyhedra. *Polyhedron* **1994**, *13*, 2149–2155. [CrossRef]
31. Sileo, E.E.; Morando, P.J.; Della Vedova, C.; Blesa, M.A. The Thermal Decomposition of Copper(II) Nicotinate and Isonicotinate. *Thermochim. Acta* **1989**, *138*, 233–239. [CrossRef]
32. Sekizaki, M. The Crystal Structure of Bis(pyrazine-2-carboxamide)copper(II) Perchlorate. *Acta Cryst. B* **1973**, *B29*, 327–331. [CrossRef]
33. Mukherjee, R. *Comprehensive Coordination Chemistry II*; Chap. 6.6; Jon, A., McCleverty, J.A., Meyer, T.J., Eds.; Elsevier: Oxford, UK, 2004; Volume 6, p. 747910.
34. Kaim, W.; Schwederski, B.; Klein, A. *Bioinorganic Chemistry: Inorganic Elements in the Chemistry of Life*, 2nd ed.; Wiley: Chichester, UK, 2013.
35. Kaim, W.; Rall, J. Copper—A "Modern" Bioelement. *Angew. Chem. Int. Ed.* **1996**, *35*, 43–60. [CrossRef]
36. Erxleben, A. Interactions of copper complexes with nucleic acids. *Coord. Chem. Rev.* **2018**, *360*, 92–121. [CrossRef]
37. Girma, K.B.; Lorenz, V.; Blaurock, S.; Edelmann, F.T. Coordination chemistry of acrylamide. *Coord. Chem. Rev.* **2005**, *249*, 1283–1293. [CrossRef]
38. Allan, J.R.; McCloy, B.; Paton, A.D. Preparation, thermal, structural and electrical studies of dichlorohexa(anthranilamide) cobalt(II) and dichloro(anthranilamide) copper(II). *Thermochim. Acta* **1994**, *231*, 121–128. [CrossRef]

39. Ahuja, I.S.; Singh, R.; Rai, C.P. Complexes of Copper(II) with Nicotinic Acid and some Related Ligands. *Trans. Met. Chem.* **1977**, *2*, 257–260. [CrossRef]
40. Damous, M.; Denes, G.; Bouacida, S.; Hamlaoui, M.; Merazig, H.; Darand, J.-C. Di-μ-chlorido-bis[(2-aminobenzamide-$\kappa^2 N^2,O$)chloridocopper(II)]. *Acta Cryst. E* **2013**, *E69*, m488. [CrossRef]
41. Pérez, A.L.; Neuman, N.I.; Baggio, R.; Ramos, C.A.; Dalosto, S.D.; Rizzi, A.C.; Brondino, C.D. Exchange interaction between $S = 1/2$ centers bridged by multiple noncovalent interactions: Contribution of the individual chemical pathways to the magnetic coupling. *Polyhedron* **2017**, *123*, 404–410. [CrossRef]
42. Halcrow, M.A. Jahn-Teller Distortions in Transition Metal Compounds, and Their Importance in Functional Molecular and Inorganic Materials. *Chem. Soc. Rev.* **2013**, *42*, 1784–1795. [CrossRef]
43. Klein, A.; Butsch, K.; Elmas, S.; Biewer, C.; Heift, D.; Nitsche, S.; Schlipf, I.; Bertagnolli, H. Oxido-pincer complexes of copper(II)—An EXAFS and EPR study of mono- and binuclear [(pydotH$_2$)CuCl$_2$]$_n$ ($n = 1$ or 2). *Polyhedron* **2012**, *31*, 649–656. [CrossRef]
44. Thakurta, S.; Roy, P.; Rosair, G.; Gómez-García, C.J.; Garribba, E.; Mitra, S. Ferromagnetic exchange coupling in a new bis(μ-chloro)-bridged copper(II) Schiff base complex: Synthesis, structure, magnetic properties and catalytic oxidation of cycloalkanes. *Polyhedron* **2009**, *28*, 695–702. [CrossRef]
45. Klein, A.; Elmas, S.; Butsch, K. Oxido Pincer Ligands—Exploring the Coordination Chemistry of Bis(hydroxymethyl)pyridine Ligands for the Late Transition Metals. *Eur. J. Inorg. Chem.* **2009**, *2009*, 2271–2281. [CrossRef]
46. Yraola, F.; Albericio, F.; Corbella, M.; Royo, M. [{Cu(pzPh)(Opo)}$_2$(μ-Cl)$_2$]: A new dinuclear copper(II) complex with a chloride bridge and mixed blocking ligands. *Inorg. Chim. Acta* **2008**, *361*, 2455–2461. [CrossRef]
47. Koval, I.A.; Sgobba, M.; Huisman, M.; Lüken, M.; Saint-Aman, E.; Gamez, P.; Krebs, B.; Reedijk, J. A remarkable anion effect on the crystal packing of two analogous copper complexes from a thiophene-containing phenol-based ligand. *Inorg. Chim. Acta* **2006**, *359*, 4071–4078. [CrossRef]
48. Kilner, C.A.; McInnes, E.J.L.; Leech, M.A.; Beddard, G.S.; Howard, J.A.K.; Mabbs, F.E.; Collison, D.; Bridgeman, A.J.; Halcrow, M.A. A crystallographic, EPR and theoretical study of the Jahn–Teller distortion in [CuTp$_2$] (Tp$^-$ = tris{pyrazol-1-yl}hydridoborate). *Dalton Trans.* **2004**, 236–243. [CrossRef] [PubMed]
49. Klein, C.L.; Stevens, E.D.; O'Connor, C.I.; Majeste, R.J.; Trefonas, L.M. Magnetic Properties and Molecular Structure of Copper(II) Complexes of 2,3-Pyrazinedicarboxamide. *Inorg. Chim. Acta* **1983**, *70*, 151–158.
50. Laussmann, T.; Grzesiak, I.; Krest, A.; Stirnat, K.; Meier-Giebing, S.; Ruschewitz, U.; Klein, A. Copper Thiocyanate Complexes and Cocaine—A Case of Black Cocaine. *Drug Test. Anal.* **2015**, *7*, 56–64. [CrossRef]
51. Glowiak, T.; Podgórska, I. X-ray, Spectroscopic and Magnetic Studies of Hexaaquacopper(II) Di(diphenylphosphate) Diglycine. *Inorg. Chim. Acta* **1986**, *125*, 83–88. [CrossRef]
52. Hathaway, B.; Billing, D. The electronic properties and stereochemistry of mono-nuclear complexes of the copper(II) ion. *Coord. Chem. Rev.* **1970**, *5*, 143–207. [CrossRef]
53. Rieger, P.H. *Electron Spin Resonance—Analysis and Interpretation*; RCS Publishing: Cambridge, UK, 2007.
54. Halcrow, M.A.; Kilner, C.A.; Wolowska, J.; McInnes, E.J.L.; Bridgeman, A.J. Temperature dependence of the electronic ground states of two mononuclear, six-coordinate copper(II) centres. *New J. Chem.* **2004**, *28*, 228–233. [CrossRef]
55. Solanki, N.K.; Leech, M.A.; McInnes, E.J.L.; Mabbs, F.E.; Howard, J.A.K.; Kilner, C.A.; Rawson, J.M.; Halcrow, M.A. A crystallographic and EPR study of the fluxional Cu(II) ion in [CuL2][BF4]2 (L = 2,6-dipyrazol-1-ylpyridine). *J. Chem. Soc. Dalton Trans.* **2002**, 1295–1301. [CrossRef]
56. Mirzaei, M.; Lippolis, V.; Aragoni, M.C.; Ghanbari, M.; Shamsipur, M.; Meyer, F.; Demeshko, S.; Pourmortazavi, S.M. Extended structures in copper(II) complexes with 4-hydroxypyridine-2,6-dicarboxylate and pyrimidine derivative ligands: X-ray crystal structure, solution and magnetic studies. *Inorg. Chim. Acta* **2014**, *418*, 126–135. [CrossRef]
57. Sheldrick, G.M. ShelXT—Integrated space-group and crystal-structure determination. *Acta Crystallogr. Sect. A Found. Crystallogr.* **2015**, *71*, 3–8. [CrossRef] [PubMed]
58. Sheldrick, G.M. *SHELXL-2017/1, Program for the Solution of Crystal Structures*; University of Göttingen: Göttingen, Germany, 2017.
59. Sheldrick, G.M. Crystal structure refinement with SHELXL. *Acta Crystallogr. Sect. C Struct. Chem.* **2015**, *71*, 3–8. [CrossRef]
60. STOE X-RED. *Data Reduction Program*; Version 1.31/Windows; STOE & Cie: Darmstadt, Germany, 2005.

61. STOE X-SHAPE. *Crystal Optimisation for Numerical Absorption Correction*; Version 1.06/Windows; STOE & Cie: Darmstadt, Germany, 1999.
62. *STOE WinXPOW 1.07*; STOE & Cie GmbH: Darmstadt, Germany, 2000.
63. Williams, T.; Kelley, C.; Lang, R. Gnuplot 4.6—An Interactive Plotting Program. 2012. Available online: http://gnuplot.info/ (accessed on 23 November 2020).

Publisher's Note: MDPI stays neutral with regard to jurisdictional claims in published maps and institutional affiliations.

© 2020 by the authors. Licensee MDPI, Basel, Switzerland. This article is an open access article distributed under the terms and conditions of the Creative Commons Attribution (CC BY) license (http://creativecommons.org/licenses/by/4.0/).

Article

Isomeric 4,2':6',4"- and 3,2':6',3"-Terpyridines with Isomeric 4'-Trifluoromethylphenyl Substituents: Effects on the Assembly of Coordination Polymers with [Cu(hfacac)$_2$] (Hhfacac = Hexafluoropentane-2,4-dione)

Giacomo Manfroni, Simona S. Capomolla, Alessandro Prescimone, Edwin C. Constable and Catherine E. Housecroft *

Department of Chemistry, University of Basel, BPR 1096, Mattenstrasse 24a, CH-4058 Basel, Switzerland; giacomo.manfroni@unibas.ch (G.M.); simona.capomolla@unibas.ch (S.S.C.); alessandro.prescimone@unibas.ch (A.P.); edwin.constable@unibas.ch (E.C.C.)
* Correspondence: catherine.housecroft@unibas.ch

Citation: Manfroni, G.; Capomolla, S.S.; Prescimone, A.; Constable, E.C.; Housecroft, C.E. Isomeric 4,2':6',4"- and 3,2':6',3"-Terpyridines with Isomeric 4'-Trifluoromethylphenyl Substituents: Effects on the Assembly of Coordination Polymers with [Cu(hfacac)$_2$] (Hhfacac = Hexafluoropentane-2,4-dione). *Inorganics* 2021, 9, 54. https://doi.org/10.3390/inorganics9070054

Academic Editor: Duncan H. Gregory

Received: 1 June 2021
Accepted: 7 July 2021
Published: 10 July 2021

Publisher's Note: MDPI stays neutral with regard to jurisdictional claims in published maps and institutional affiliations.

Copyright: © 2021 by the authors. Licensee MDPI, Basel, Switzerland. This article is an open access article distributed under the terms and conditions of the Creative Commons Attribution (CC BY) license (https://creativecommons.org/licenses/by/4.0/).

Abstract: The isomers 4'-(4-(trifluoromethyl)phenyl)-4,2':6',4"-terpyridine (**1**), 4'-(3-(trifluoromethyl)phenyl)-4,2':6',4"-terpyridine (**2**), 4'-(4-(trifluoromethyl)phenyl)-3,2':6',3"-terpyridine (**3**), and 4'-(3-(trifluoromethyl)phenyl)-3,2':6',3"-terpyridine (**4**) have been prepared and characterized. The single crystal structures of **1** and **2** were determined. The 1D-polymers [Cu$_2$(hfacac)$_4$(**1**)$_2$]$_n$·2nC$_6$H$_4$Cl$_2$ (Hhfacac = 1,1,1,5,5,5-hexafluoropentane-2,4-dione), [Cu(hfacac)$_2$(**2**)]$_n$·2nC$_6$H$_5$Me, [Cu$_2$(hfacac)$_4$(**3**)$_2$]$_n$·nC$_6$H$_4$Cl$_2$, [Cu$_2$(hfacac)$_4$(**3**)$_2$]$_n$·nC$_6$H$_5$Cl, and [Cu(hfacac)$_2$(**4**)]$_n$·nC$_6$H$_5$Cl have been formed by reactions of **1**, **2**, **3** and **4** with [Cu(hfacac)$_2$]·H$_2$O under conditions of crystal growth by layering and four of these coordination polymers have been formed on a preparative scale. [Cu$_2$(hfacac)$_4$(**1**)$_2$]$_n$·2nC$_6$H$_4$Cl$_2$ and [Cu(hfacac)$_2$(**2**)]$_n$·2nC$_6$H$_5$Me are zig-zag chains and the different substitution position of the CF$_3$ group in **1** and **2** does not affect this motif. Packing of the polymer chains is governed mainly by C–F...F–C contacts, and there are no inter-polymer π-stacking interactions. The conformation of the 3,2':6',3"-tpy unit in [Cu$_2$(hfacac)$_4$(**3**)$_2$]$_n$·nC$_6$H$_4$Cl$_2$ and [Cu(hfacac)$_2$(**4**)]$_n$·nC$_6$H$_5$Cl differs, leading to different structural motifs in the 1D-polymer backbones. In [Cu(hfacac)$_2$(**4**)]$_n$·nC$_6$H$_5$Cl, the peripheral 3-CF$_3$C$_6$H$_4$ unit is accommodated in a pocket between two {Cu(hfacac)$_2$} units and engages in four C–H$_{phenyl}$...F–C$_{hfacac}$ contacts which lock the phenylpyridine unit in a near planar conformation. In [Cu$_2$(hfacac)$_4$(**3**)$_2$]$_n$·nC$_6$H$_4$Cl$_2$ and [Cu(hfacac)$_2$(**4**)]$_n$·nC$_6$H$_5$Cl, π-stacking interactions between 4'-trifluoromethylphenyl-3,2':6',3"-tpy domains are key packing interactions, and this contrasts with the packing of polymers incorporating **1** and **2**. We use powder X-ray diffraction to demonstrate that the assemblies of the coordination polymers are reproducible, and that a switch from a 4,2':6',4"- to 3,2':6',3"-tpy metal-binding unit is accompanied by a change from dominant C–F...F–C and C–F...H–C contacts to π-stacking of arene domains between ligands **3** or **4**.

Keywords: copper; 4,2':6',4"-terpyridine; 3,2':6',3"-terpyridine; coordination polymer; isomers

1. Introduction

The coordination chemistry of the 4,2':6',4"- and 3,2':6',3"-isomers of terpyridine (4,2':6',4"-tpy and 3,2':6',3"-tpy, Scheme 1) has attracted significant attention in the last decade because the vectorial properties of these isomers of tpy are suited to the assembly of coordination polymers and networks [1–6]. As Scheme 1 illustrates, 4,2':6',4"-tpy and 3,2':6',3"-tpy only coordinate through the outer pyridine donors, leaving the central nitrogen atom unbound. This provides a strategy for the design of coordination assemblies in which the surfaces of the solvent-accessible channels contain sites of Lewis basicity potentially leading to small molecule recognition through, e.g., C–H...N$_{pyridine}$ hydrogen bond formation [5], and sensing applications [5–9]. Moreover, 3,2':6',3"-tpy exhibits greater conformational flexibility than 4,2':6',4"-tpy (Scheme 1), leading to greater variation (or less predictability) in network assembly [4].

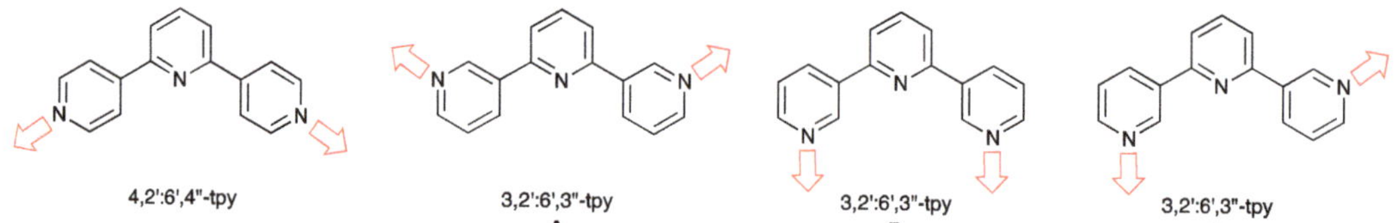

Scheme 1. 4,2':6',4"- and 3,2':6',3"-Terpyridines typically coordinate through the outer pyridine donors rendering them as divergent linkers. The three limiting, planar conformations of 3,2':6',3"-tpy are shown.

Although a wide range of 1D-, 2D- and 3D-assemblies incorporating 4,2':6',4"-tpy metal-binding domains is known, those with 3,2':6',3"-tpy ligands are less well explored [1–5]. Coordination assemblies involving Cu(II) or Cu(I) fall into several categories. A well-represented group involves $\{Cu_2(\mu\text{-OAc})_4\}$ paddle-wheel units connected into 1D-polymer chains by ditopic 4,2':6',4"-tpy or 3,2':6',3"-tpy linkers [10–16]. A number of architectures feature $\{Cu^{II}_2Cl_4\}$ [17] and $\{Cu^{I}_2I_2\}$ nodes [18–20], or $\{Cu^{I}_n(CN)_n\}$ building blocks [21–24]. We note, however, that when 1-(3,2':6',3"-terpyridin-4'-yl)ferrocene reacts with $CuCl_2$, $\{Cu^{II}_2Cl_4\}$ building blocks interconnect 3,2':6',3"-tpy units in conformation **C** (Scheme 1) to give a discrete molecular complex [25]. Introducing carboxylic acid substituents into the 4,2':6',4"- or 3,2':6',3"-tpy units is a strategy for increasing the donor capacity of the ligand, thereby increasing the dimensionality of the assembly. Typically, the carboxylic acid is deprotonated and the CO_2^- group supplements the N,N'-donor set of tpy. Examples include 4'-carboxylato-4,2':6',4"-terpyridine [26], 4'-(4-carboxylatophenyl)-4,2':6',4"-terpyridine [22,27], 3,5-dicyano-4'-(4-carboxylatophenyl)-4,2':6',4"-terpyridine [28], and 4'-(4-(3,5-dicarboxylatophenoxy)phenyl)-4,2':6',4"-terpyridine [29]. Assemblies combining Cu(II) and sulfonic acid-functionalized tpy ligands are less well represented [30]. In the case of 4'-(4-hydroxyphenyl)-4,2':6',4"-terpyridine (4'-(HOC_6H_4)-4,2':6',4"-tpy), the ligand coordinates only through the 4,2':6',4"-tpy unit in the 2D-networks [{Cu(OH$_2$)(4'-(HOC$_6$H$_4$)-4,2':6',4"-tpy)$_2$Cl}$_n$][NO$_3$]$_n$ and [{Cu(OH$_2$)$_2$(4'-(HOC$_6$H$_4$)-4,2':6',4"-tpy)$_2$}$_n$][NO$_3$]$_{2n}$ [31]. 3-Dimensional architectures directed by Cu(II) nodes in which the 4,2':6',4"-tpy ligands are merely linkers are exemplified by the chiral [{Cu$_2$(DMSO)$_3$(4'-(MeOC$_6$H$_4$)-4,2':6',4"-tpy)$_4$}$_n$][BF$_4$]$_{4n}$ (4'-(MeOC$_6$H$_4$)-4,2':6',4"-tpy = 4'-(4-methoxyphenyl)-4,2':6',4"-terpyridine) [32]. A further way of increasing dimensionality of an assembly is to introduce a 4'-pyridinyl (usually 4'-pyridin-4-yl) substituents into the tpy metal-binding domain. Zaworotko and coworkers have reported a beautiful series of 3D-networks featuring 8-connected [Cu$_2$(py-4,2':6',4"-tpy)$_8$(μ-MF$_6$)]$^{2+}$ building blocks in which py-4,2':6',4"-tpy is 4'-(pyridin-4-yl)-4,2':6',4"-terpyridine, and M = Si, Ge, Sn, Ti or Zr [33].

Both the {Cu(acac)$_2$} and {Cu(hfacac)$_2$} units (Hacac = pentane-2,4-dione, Hhfacac = 1,1,1,5,5,5-hexafluoropentane-2,4-dione) are ubiquitous in coordination chemistry, although it is interesting that the latter is far better represented in the Cambridge Structural Database (CSD) [34] than the former. A search of the CSD, version 2020.3.1 [35] using ConQuest version 2020.3.1 [35] revealed 1039 hits for compounds containing a {Cu(hfacac)$_2$} unit compared to 172 containing {Cu(acac)$_2$}; of these 172 hits, 62 are different determinations of polymorphs of the structure of [Cu(acac)$_2$] (CSD refcode ACACCU). One reason for the dominance of [hfacac]$^-$ containing compounds may be that the presence of the CF$_3$ substituents improves the solubility of the Cu(II) salt in a wider range of solvents with respect to [Cu(acac)$_2$]. Although coordination polymers containing {Cu(hfacac)$_2$} nodes are well established (557 hits in the CSD version 2020.3.1), examples incorporating divergent terpyridine ligands are rare. Moreno and coworkers described the syntheses and structural characterization of [Cu(hfacac)$_2$(L$_1$)]$_n$, [Cu(hfacac)$_2$(L$_2$)]$_n$·nCHCl$_3$, [Cu(hfacac)$_2$(L$_3$)]$_n$·nCHCl$_3$, and [Cu(hfacac)$_2$(L$_4$)]$_n$ (L$_1$–L$_4$ are defined in Scheme 2) [36]. All four compounds are

1D-coordination polymers with 4,2′:6′,4″- or 3,2′:6′,3″-tpy domains linking octahedral Cu(II) centers. However, whereas [Cu(hfacac)$_2$(L$_1$)]$_n$ and [Cu(hfacac)$_2$(L$_3$)]$_n$·CHCl$_3$ contain a *cis*-arrangement of pyridine N-donors, the latter are in a *trans*-arrangement in [Cu(hfacac)$_2$(L$_2$)]$_n$·CHCl$_3$ and [Cu(hfacac)$_2$(L$_4$)]$_n$. In [Cu(hfacac)$_2$(L$_3$)]$_n$·CHCl$_3$, the 3,2′:6′,3″-tpy unit adopts conformation **A** shown in Scheme 1. Moreno has also reported that the reaction of L$_1$ with [Cu(ttfacac)$_2$] (Httfacac = 4,4,4-trifluoro-1-(thiophen-2-yl)butane-1,3-dione) yielded the discrete, trinuclear complex [Cu$_3$(ttfacac)$_6$(L$_1$)$_2$] [37]. Other relevant 1D-coordination polymers containing octahedral {Cu(hfacac)$_2$(N$_{py}$)$_2$} building blocks include [Cu(hfacac)$_2$(4,4′-bpy)]$_n$ (4,4′-bpy = 4,4′-bipyridine) [38] and [Cu(hfacac)$_2$(dpss)]$_n$ (dpss = di(pyridin-2-yl disulfide) [39] in which the N$_{py}$ donors are *trans*, and [Cu(hfacac)$_2$(bpyb)]$_n$ (bpyb = 1,4-bis(pyridin-2-yl)buta-1,3-diyne in which the N$_{py}$ donors are mutually *cis* [39]. A combination of 1,3,5-tris(pyridin-4-ylethynyl)benzene (L$_5$, Scheme 2) with [Cu(hfacac)$_2$] gives a 2D-network directed by the 3-connecting L$_5$ ligand; *trans*-{Cu(hfacac)$_2$(N$_{py}$)$_2$} units are present [40].

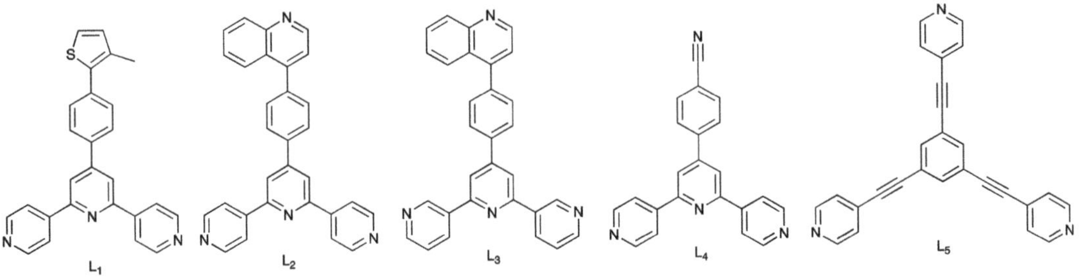

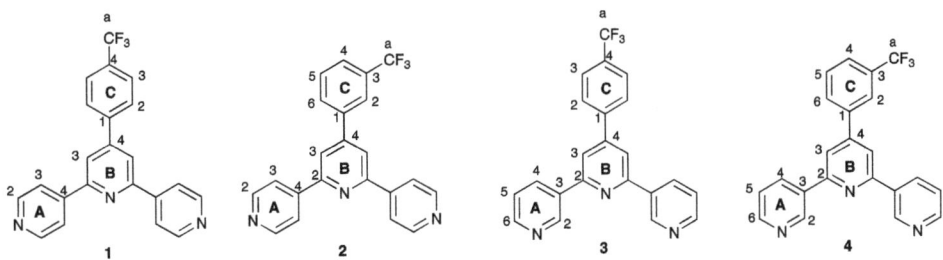

Scheme 2. The structures of ligands L$_1$, L$_2$, L$_3$, L$_4$ [36] and L$_5$ [40].

In order to complement both our own investigations of the coordination chemistry of 4,2′:6′,4″-tpy and 3,2′:6′,3″-tpy ligands and copper(II) salts [11–16], and the structural diversity of the known coordination polymers featuring {Cu(hfacac)$_2$} nodes, we decided to explore the assemblies formed between [Cu(hfacac)$_2$] and ligands **1–4** shown in Scheme 3. The ligand series was selected to combine isomers of the tpy metal-binding domain with isomers of the 4′-trifluoromethylphenyl substituent. The trifluoromethyl substituents were incorporated to give a complementarity to the [hfacac]$^-$ ligands, and potentially introduce additional supramolecular packing interactions within the lattice. The availability of the different trifluoromethylphenyl isomers as substituents allows the subtlety of the supramolecular interactions to be probed.

Scheme 3. Structures of ligands **1–4** with the numbering system used for NMR spectroscopic assignments.

2. Results and Discussion

2.1. Syntheses and Characterization of Ligands 1–4

Compounds **1–4** were prepared using the one-pot strategy of Wang and Hanan [41], (Scheme 4 for **1**). After purification, compounds **1–4** were isolated as colorless, microcrystalline solids in yields of between 31.3 and 49.3%. The four compounds are isomers and in the MALDI-TOF mass spectrum of each, the base peak corresponded to the [M+H]$^+$ ion (Figures S1–S4 in the Supporting Material). For **3** and **4**, which contain the 3,2′:6′,3″-tpy unit, the isotope pattern is as expected (Figures S3 and S4). However, for the derivatives of 4,2′:6′,4″-tpy (compounds **1** and **2**), the relative intensities of the peaks at m/z 378.1 and 379.1 (Figures S1 and S2) are consistent with both [M+H]$^+$ and [M+2H]$^{2+}$ ions. This is consistent with the greater basicity of 4,2′:6′,4″-tpy versus 3,2′:6′,3″-tpy. To support this, we considered the model compounds 3-phenylpyridine (3-Phpy) and 4-phenylpyridine (4-Phpy). The pK_a values of the conjugate acids [H(4-Phpy)]$^+$ and [H(3-Phpy)]$^+$ are 5.38 and 4.81, respectively [42], confirming that [H(3-Phpy)]$^+$ is a stronger acid than [H(4-Phpy)]$^+$ and, therefore, 4-Phpy is a stronger base then 3-Phpy. The solid-state IR spectra of **1–4** are shown in Figures S5–S8, and exhibit similar fingerprint regions. The absorption spectra of MeCN solutions of the terpyridine ligands are all similar (Figure S9) and the absorptions at λ_{max} = 248–250 nm and λ_{max} = 294–306 nm (see Sections 3.2–3.5) are assigned to $\pi^* \leftarrow \pi$ transitions.

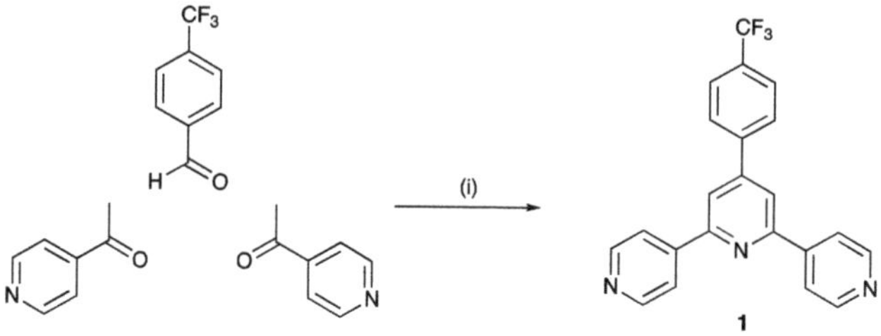

Scheme 4. Synthetic route to ligand **1**. Analogous routes were used for the preparations of **2**, **3**, and **4**. Conditions: (i) KOH, EtOH; NH$_3$ (aqueous), room temperature, overnight (ca. 21 h).

Compounds **1** and **2** were significantly less soluble in common organic solvents than **3** and **4**, and DMSO-d_6 was used for recording NMR spectra. While **3** and **4** dissolved easily in DMSO-d_6 under ambient conditions, complete dissolution of **1** and **2** was achieved only with heating. Precipitation of **1** in the NMR tube resulted in broadened signals in the ^{1}H NMR spectrum. The ^{1}H and ^{31}C{^{1}H} NMR spectra of compounds **1–4** were assigned using 2D-methods, and were consistent with the structures shown in Scheme 3. A singlet at around δ –61 ppm (see Sections 3.2–3.5) was observed in the ^{19}F{^{1}H} NMR spectrum of each compound, consistent with one CF$_3$ environment. The ^{1}H NMR spectra are compared in Figure S10, while Figure S11 displays the ^{31}C{^{1}H} NMR spectra. While compound **1** is C_2 symmetric on the NMR timescale, the symmetry is lowered on moving the CF$_3$ substituent from the 4- to 3-position of the 4′-phenyl ring (Figure S10, **1** to **2**). A comparison of the spectra of **1** and **3**, and of **2** and **4** in Figure S10 shows the effects of going from the 4,2′:6′,4″- to 3,2′:6′,3″-tpy domain while retaining the CF$_3$ group in a common position. Similar effects are seen by comparing the ^{31}C{^{1}H} NMR spectra in Figure S11. The characteristic quartets for C^a (see Scheme 3) with J_{CF} = 272 Hz, and for C^{C3} or C^{C4} (J_{CF} = 31 Hz) are highlighted in Figure S11.

2.2. Single Crystal Structures of 1 and 2

Single crystals of **1** were grown by diffusion of Et$_2$O into a CHCl$_3$ solution of the compound, and X-ray quality crystals of **2** grew as a hot DMSO solution of **2** was allowed to cool to room temperature. Compounds **1** and **2** crystallize in the monoclinic space group $P2_1/c$, and triclinic space group $P–1$, respectively. The molecular structures of **1** and one of the two crystallographically independent molecules of **2** are shown in Figure 1a,b, respectively. The conformations of the 4,2′:6′,4″-tpy units differ slightly. In **1**, the angles between the planes of the rings containing N1/N2 and N2/N3 are 31.1 and 2.8°, respectively. In the two independent molecules of **2**, the corresponding angles are 23.2 and 2.4°, and 26.5 and 21.7°. In **1** and molecule 1 of compound **2**, the angle between the planes of the phenyl ring and the central pyridine ring are 33.4 and 35.7°, respectively. These are typical of a 4′-substituted arene ring and minimize unfavorable H...H repulsions. In contrast, the phenyl and central pyridine rings are almost coplanar in the second molecule of **2** (angle between the ring planes = 5.7°). This is associated with face-to-face π-stacking between the independent molecules of **2** which extends across the whole molecular framework shown in Figure 2a. The centroid...centroid separations for pairs of rings containing N3/C33, N2/N5, N1/N4 and N6/C9 are 3.70, 3.67, 3.76, and 3.78 Å, and the corresponding angles between the planes of the stacked rings are 4.9, 3.3, 0.8, and 6.1°. Stacking of molecules continues to assemble columns along the crystallographic *a*-axis (Figure S12). The head-to-tail arrangement of the molecules (Figure 2a) facilitates C–H...F hydrogen bond formation between adjacent molecules which are augmented by C–H...N hydrogen bonds leading to an extended array (Figure 2b). Contact parameters are given in the caption to Figure 2b. Packing of molecules of **1** is also dominated by face-to-face π-stacking, which may contribute to the low solubilities of the compounds. As in **2**, the stacking interaction in **1** extends across the whole molecule, and extended columnar assemblies are formed (Figure 2c). The centroid...centroid distances between pairs of stacked rings containing N3/N3^i, N2/N2^i, N1/C10^i and C10/N1^i (symmetry code i = *x*, $^3/_2$–*y*, –$^1/_2$+*z*) are 4.0, 4.2, 3.8, 3.7 Å, and the corresponding angles between the ring planes are 5.1, 0.6, 3.1, and 3.1°. Thus, despite the change in the position of the CF$_3$ substituent on going from **1** to **2**, the structural motifs and packing interactions bear a striking resemblance to one another. These observations complement a recent study by Yi et al. which highlights the scarcity of investigations of π-stacking interactions between trifluoromethylated aromatics [43].

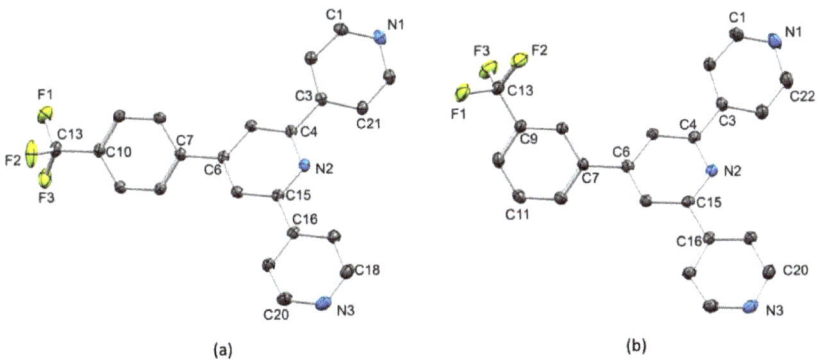

Figure 1. The molecular structures of (**a**) **1** and (**b**) one of the two crystallographically independent molecules of **2**; the structure of the second molecule of **2** is similar to that shown. H atoms are omitted for clarity, and ellipsoids are drawn at 40% probability level. Selected bond lengths for **1**: F1–C13 = 1.3292(16), F2–C13 = 1.3357(16), F3–C13 = 1.3372(16), C10–C13 = 1.4958(17), C6–C7 = 1.4859(16), C4–C3 = 1.4889(16), C15–C16 = 1.4930(17) Å. Selected bond lengths for **2** molecule 1: F1–C13 = 1.339(3), F2–C13 = 1.330(4), F3–C13 = 1.345(4), C9–C13 = 1.496(4), C7–C6 = 1.483(4), C3–C4 = 1.489(4), C15–C16 = 1.494(4) Å; the bond lengths in **2** molecule 2 are similar.

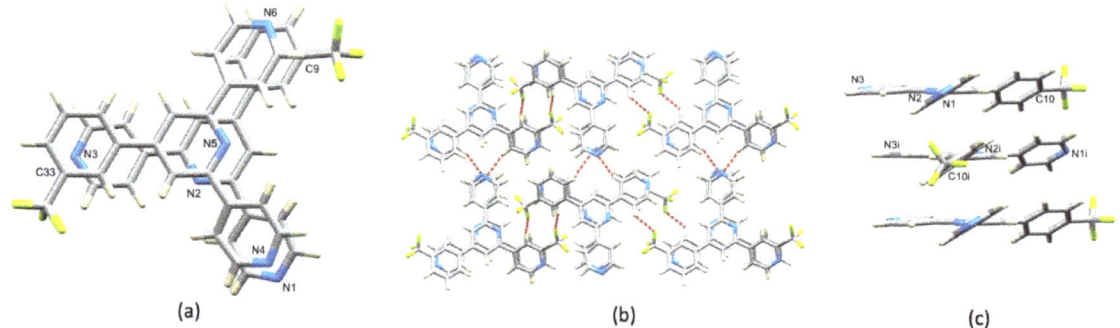

Figure 2. (a) Face-to-face π-stacking between the two independent molecules of **2**. (b) The π-stacked pairs of molecules of **2** are interconnected by C–H...F and C–H...N interactions: N1...H17^i–C17^i = 2.59 Å; C19–H19...F4^{ii} = 2.65 Å; F2...H39^{iii}–C39^{iii} = 2.62 Å; N4...H41^{iv}–C41^{iv} = 2.71 Å (symmetry codes i = x, 1+y, z; ii = 1−x, 1−y, −z; iii = 1−x, 1−y, 1−z; iv = x, 1+y, z). (c) Face-to-face π-stacking between molecules of **1** leads to columnar assemblies.

2.3. Single-Crystal Structures of the Coordination Polymers [Cu$_2$(hfacac)$_4$(**1**)$_2$]$_n$·2nC$_6$H$_4$Cl$_2$, [Cu(hfacac)$_2$(**2**)]$_n$·2nC$_6$H$_5$Me, [Cu$_2$(hfacac)$_4$(**3**)$_2$]$_n$·nC$_6$H$_4$Cl$_2$, [Cu$_2$(hfacac)$_4$(**3**)$_2$]$_n$·nC$_6$H$_5$Cl and [Cu(hfacac)$_2$(**4**)]$_n$·nC$_6$H$_5$Cl

Single crystals were grown under ambient conditions by layering a solution of [Cu(hfacac)$_2$]·H$_2$O in either toluene, chlorobenzene or 1,2-dichlorobenzene over a chloroform solution of **1**, **2**, **3** or **4**. For each terpyridine ligand, X-ray quality crystals were obtained only for one or two of the solvent combinations, and structural analysis of [Cu$_2$(hfacac)$_4$(**1**)$_2$]$_n$·2nC$_6$H$_4$Cl$_2$, [Cu(hfacac)$_2$(**2**)]$_n$·2nC$_6$H$_5$Me, [Cu$_2$(hfacac)$_4$(**3**)$_2$]$_n$·nC$_6$H$_4$Cl$_2$, [Cu$_2$(hfacac)$_4$(**3**)$_2$]$_n$·nC$_6$H$_5$Cl and [Cu(hfacac)$_2$(**4**)]$_n$·nC$_6$H$_5$Cl revealed the assembly of a 1D-coordination polymer in each case. [Cu$_2$(hfacac)$_4$(**3**)$_2$]$_n$·nC$_6$H$_4$Cl$_2$, [Cu$_2$(hfacac)$_4$(**3**)$_2$]$_n$·nC$_6$H$_5$Cl both crystallize in the triclinic space group $P\bar{1}$ with similar cell dimensions (a = 11.9939(3), b = 12.1658(3), c = 12.9674(3) Å, α = 102.257(2), β = 103.145(2), γ = 91.214(2)° for [Cu$_2$(hfacac)$_4$(**3**)$_2$]$_n$·nC$_6$H$_4$Cl$_2$, and a = 11.9906(3), b = 11.9911(3), c = 13.0617(3) Å, α = 103.144(2), β = 102.547(2), γ = 91.491(2)° for [Cu$_2$(hfacac)$_4$(**3**)$_2$]$_n$·nC$_6$H$_5$Cl). Since the polymers are essentially isostructural, we discuss only the structure of [Cu$_2$(hfacac)$_4$(**3**)$_2$]$_n$·nC$_6$H$_4$Cl$_2$.

Figures S13–S16 show the molecular structures of the asymmetric units in [Cu$_2$(hfacac)$_4$(**1**)$_2$]$_n$·2nC$_6$H$_4$Cl$_2$, [Cu(hfacac)$_2$(**2**)]$_n$·2nC$_6$H$_5$Me, [Cu$_2$(hfacac)$_4$(**3**)$_2$]$_n$·nC$_6$H$_4$Cl$_2$, and [Cu(hfacac)$_2$(**4**)]$_n$·nC$_6$H$_5$Cl with symmetry generated atoms. In all four compounds, each copper(II) center is octahedrally sited with a *trans*-arrangement of pyridine donors. Each of the ligands **1**, **2**, **3**, and **4** coordinates through the outer pyridine rings and links two Cu(II) centers. The bond lengths and angles in the compounds are unexceptional and selected values are given in Table 1. Table 2 presents the angles between the planes of adjacent aromatic rings in each of the coordinated ligands **1**–**4**. The most striking difference is in the angle between the central pyridine ring (with N2) and phenyl ring for the polymer containing **4**. For the compounds containing **1**, **2** and **3**, the twist angles (28.5–34.9°) are typical for minimizing steric interactions between the H atoms on adjacent rings. The near coplanarity of the rings in [Cu(hfacac)$_2$(**4**)]$_n$·nC$_6$H$_5$Cl appears to be associated with a combination of effects which are connected to the conformation of the 3,2′:6′,3″-tpy unit (see later). The four structures are discussed below in a comparative way with a focus on the effects of changing the substitution position of the CF$_3$ group while retaining the same terpyridine isomer, and the effects of going from the 4,2′:6′,4″-tpy to 3,2′:6′,3″-tpy metal-binding domain.

Table 1. Space groups and selected bond lengths and angles in the copper(II) coordination polymers.

Compound	Space Group	Cu–N/Å	Cu–O/Å	N–Cu–O/°
[Cu$_2$(hfacac)$_4$(**1**)$_2$]$_n$·2nC$_6$H$_4$Cl$_2$	Pbca	2.014(2), 2.016(2)	1.967(2), 1.982(2), 2.298(2), 2.302(2)	91.38(9), 89.24(9), 91.35(9), 88.05(9), 88.83(9), 88.02(9), 91.74(9), 91.37(9)
[Cu(hfacac)$_2$(**2**)]$_n$·2nC$_6$H$_5$Me	Cc	2.002(6), 2.012(6)	2.046(6), 2.197(6), 2.031(6), 2.234(6)	90.6(2), 90.0(2), 89.5(2), 89.1(2), 90.1(2), 90.6(2), 89.8(2), 90.3(2)
[Cu$_2$(hfacac)$_4$(**3**)$_2$]$_n$·nC$_6$H$_4$Cl$_2$	P-1	2.064(3), 2.023(4)	1.955(3), 2.339(3), 1.990(3), 2.275(3)	88.70(12), 91.30(12), 98.90(12), 81.10(12), 92.38(13), 87.62(13), 88.12(13), 91.88(13)
[Cu(hfacac)$_2$(**4**)]$_n$·nC$_6$H$_5$Cl	Pnma	2.051(3)	1.971(2), 2.291(3)	91.19(10), 88.81(10), 95.60(10), 84.40(10)

Table 2. Angles between the planes of pairs of connected rings in coordinated ligands **1–4**.

Compound	Angle between Planes of Adjacent Pyridine Rings/°	Angle between Ring with N2 and Phenyl Ring/°
[Cu$_2$(hfacac)$_4$(**1**)$_2$]$_n$·2nC$_6$H$_4$Cl$_2$	12.9, 23.6	28.5
[Cu(hfacac)$_2$(**2**)]$_n$·2nC$_6$H$_5$Me	7.8, 26.7	34.9
[Cu$_2$(hfacac)$_4$(**3**)$_2$]$_n$·nC$_6$H$_4$Cl$_2$	15.7, 21.7	30.0
[Cu(hfacac)$_2$(**4**)]$_n$·nC$_6$H$_5$Cl	16.6, 16.6	0.9

Ligand **1** presents a V-shaped building block and, combined with the *trans*-arrangement of the pyridine donors in the Cu(II) coordination sphere, this leads to a zigzag 1D-polymer chain in [Cu$_2$(hfacac)$_4$(**1**)$_2$]$_n$·2nC$_6$H$_4$Cl$_2$ (Figure 3a). The chains associate through short C–F...F–C interactions (Figure 3b) with F...F distances of 2.92, 2.97, 2.76 and 2.93 Å, which are less than or similar to the sum of the van der Waals radii (2.92–2.94 Å) [44,45]. These contacts involve the ordered CF$_3$ groups containing C13, C27 and C32 (Figure S13). Although distinct from halogen bonds [46], weak F...F contacts are recognized as contributing towards crystal packing interactions [43,47,48]. At first glance, the packing shown in Figure 3b appears to be reminiscent of the characteristic nesting of zigzag chains in [Cu$_2$(µ-OAc)$_4$(4'-X-4,2':6',4''-tpy)]$_n$ to form 2D-sheets [11,13,14]. However, the chains in [Cu$_2$(hfacac)$_4$(**1**)$_2$]$_n$·2nC$_6$H$_4$Cl$_2$ are offset (highlighted in red in Figure 3c) and a second set of chains slices obliquely through the first as shown in Figure 3c. Interestingly, π-stacking interactions between ligands **1** do not contribute to the packing interactions, although the 1,2-dichlorobenzene solvate molecule does form face-to-face π-stacking contacts with the central pyridine ring of **1** (centroid ... centroid = 3.78 Å, angle between the ring planes = 3.0°).

A zigzag polymer is also present in [Cu(hfacac)$_2$(**2**)]$_n$·2nC$_6$H$_5$Me (Figure 4a, and as in [Cu$_2$(hfacac)$_4$(**1**)$_2$]$_n$·2nC$_6$H$_4$Cl$_2$, the dominant packing interactions in [Cu(hfacac)$_2$(**2**)]$_n$·2nC$_6$H$_5$Me are weak C–F...F–C contacts. Four of the five crystallographically independent CF$_3$ groups are involved in such interactions, and these CF$_3$ units are ordered. The C–F...F–C network in [Cu(hfacac)$_2$(**2**)]$_n$·2nC$_6$H$_5$Me is more complex than in the polymer containing ligand **1**, with each F atom of the CF$_3$ group in **2** forming a C–F...F–C contact with an {Cu(hfacac)$_2$} unit in a different polymer chain (Figure 4b,c). The F...F distances for these interactions are 2.94, 2.92 and 2.82 Å. The 1D-polymers are arranged parallel to one another (Figure 4c, and Figure S17 in the Supporting Material). As in [Cu$_2$(hfacac)$_4$(**1**)$_2$]$_n$·2nC$_6$H$_4$Cl$_2$, there are no π-stacking interactions between arene rings in adjacent chains in [Cu(hfacac)$_2$(**2**)]$_n$·2nC$_6$H$_5$Me. It is tempting to suggest that this is due to the steric hindrance of the {Cu(hfacac)$_2$} domains. We note that there are also no π-stacking interactions between 4,2′:6′,4″-tpy domains in the 1D-polymers [Cu(hfacac)$_2$(L$_2$)]$_n$·CHCl$_3$ and [Cu(hfacac)$_2$(L$_4$)]$_n$ (see Scheme 2 for L$_2$ and L$_4$) [36], although Moreno and coworkers did observe π-stacking of 4,2′:6′,4″-tpy units in the molecular complex [Cu$_3$(ttfacac)$_6$(L$_1$)$_2$] (L$_1$, see Scheme 2) [37]. Another similarity between [Cu$_2$(hfacac)$_4$(**1**)$_2$]$_n$·2nC$_6$H$_4$Cl$_2$ and [Cu(hfacac)$_2$(**2**)]$_n$·2nC$_6$H$_5$Me is the role of the solvent molecules. In the latter, one toluene molecule engages in a face-of-face π-stacking interaction with one pyridine ring of **2** (centroid . . . centroid = 3.71 Å, angle between the ring planes = 2.7°). Additionally, the same pyridine ring (with N1) exhibits a CH... π contact with the second toluene molecule (C–H...centroid = 2.95 Å, angle C–H...centroid = 149.3°).

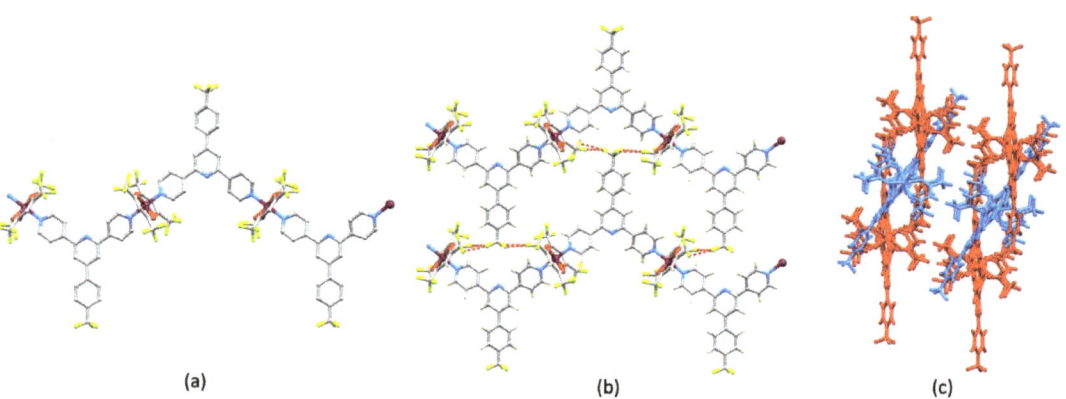

Figure 3. The structure of [Cu$_2$(hfacac)$_4$(**1**)$_2$]$_n$·2nC$_6$H$_4$Cl$_2$. (**a**) Part of one 1D-polymer with H atoms omitted. (**b**) Chains associate through short C–F...F–C contacts (hashed red lines). (**c**) The chains shown in (**b**) are offset (two pairs are shown in red) and a second set of chains (in blue) slices obliquely through the first.

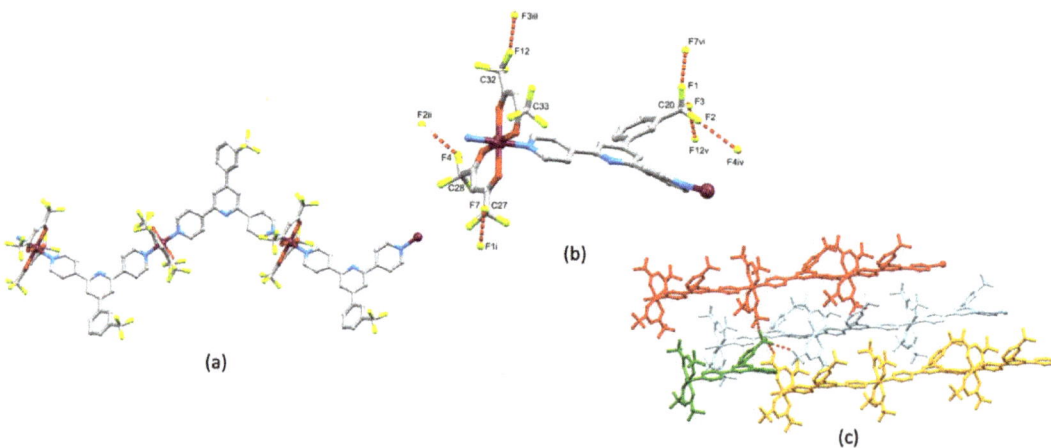

Figure 4. The structure of [Cu(hfacac)$_2$(**2**)]$_n$·2nC$_6$H$_5$Me. (**a**) Part of one 1D-polymer with H atoms omitted. (**b**) Four of the five CF$_3$ groups are involved in C–F...F–C interactions and this leads to (**c**) the CF$_3$ group in **2** being linked to three different polymer chains; the asymmetric unit is shown in green. Symmetry codes: i = $^3/_2$+x, $^1/_2$+y, z; ii = $^3/_2$+x, $-^3/_2$−y, $^1/_2$+z; iii = $^1/_2$+x, $-^3/_2$−y, $^1/_2$+z; iv = $-^3/_2$+x, $-^3/_2$−y, $-^1/_2$+z; v = $-^1/_2$+x, $-^3/_2$−y, $-^1/_2$+z; vi = $-^3/_2$+x, $-^1/_2$+y, z.

Having investigated the effects of moving the substitution position of the CF$_3$ group while retaining a 4,2′:6′,4″-tpy metal-binding unit on going from **1** to **2**, we turned our attention to ligands **3** and **4** with 3,2′:6′,3″-tpy domains. [Cu$_2$(hfacac)$_4$(**3**)$_2$]$_n$·nC$_6$H$_4$Cl$_2$

crystallizes in the triclinic space group $P{-}1$ with one independent ligand **3** and two half-{Cu(hfacac)$_2$} units, with each of Cu1 and Cu2 lying on an inversion center. Figure 5a displays part of the 1D-polymer chain present in the structure, and shows the octahedral Cu(II) coordination geometry. The 3,2′:6′,3″-tpy unit adopts conformation **C** in Scheme 1, and the alternating arrangement of these units along the chain is dictated by symmetry. As we have previously discussed [15], for a *trans*-arrangement of pyridine donors at a metal center, ligand conformation **C** can, in principle, lead to assembly algorithms **I**, **II** or **III** (Scheme 5) of which two are represented in [Cu$_2$(hfacac)$_4$(**3**)$_2$]$_n$·nC$_6$H$_4$Cl$_2$ (Figure 5a).

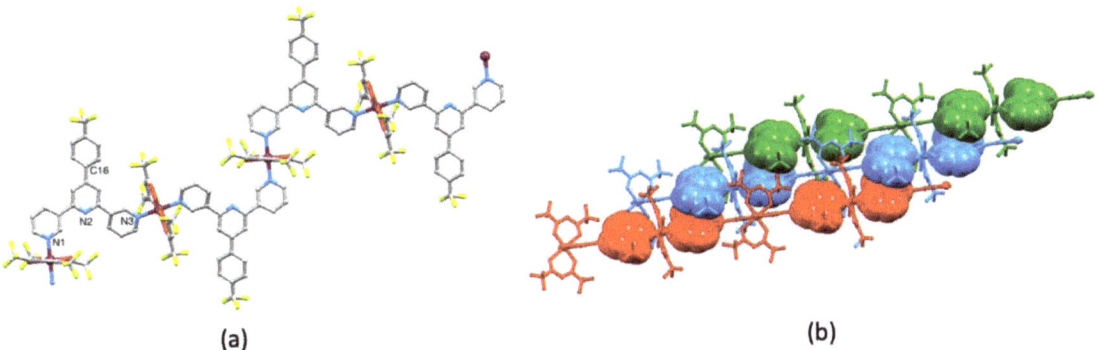

(a) (b)

Figure 5. The structure of [Cu$_2$(hfacac)$_4$(**3**)$_2$]$_n$·nC$_6$H$_4$Cl$_2$. (**a**) Part of the 1D-coordination polymer showing that the 3,2′:6′,3″-tpy unit exhibits conformation **C** (see Scheme 1) and the two crystallographically different Cu centers (compare with Scheme 5). (**b**) Stacking interactions between adjacent 1D-polymers.

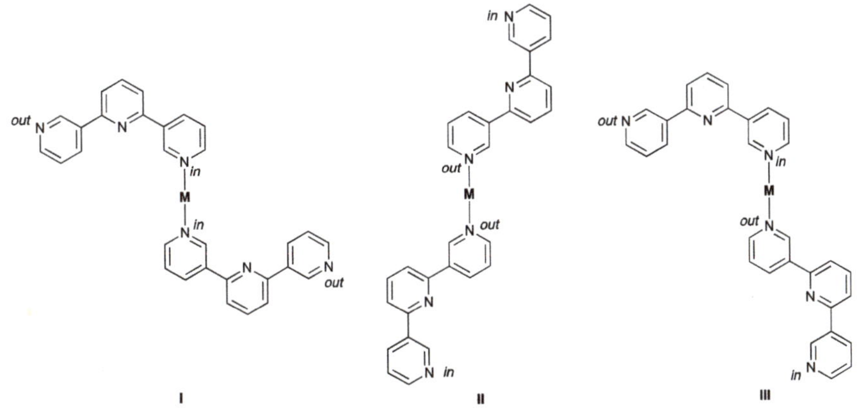

Scheme 5. With 3,2′:6′,3″-tpy in conformation **C** (Scheme 1), three coordination patterns are possible for a *trans*-arrangement of ligands at a metal center. We have previously [15] used the labels in and out to describe the orientation of the lone pair of each coordinating N atom with respect to the central pyridine ring.

Compared to the assemblies with ligands **1** and **2**, a major difference in the packing polymer chains in [Cu$_2$(hfacac)$_4$(**3**)$_2$]$_n$·nC$_6$H$_4$Cl$_2$ (and also in the polymer with **4**, see below) is the role of face-to-face π-stacking interactions. The change in the relative positions of the {Cu(hfacac)$_2$} units caused by a change in the positions of the N-donors on going from **1** and **2** to **3**, may alleviate steric congestion, allowing a closer approach of the arene units. Figure 5b depicts the centrosymmetric pairing of 3,2′:6′,3″-tpy units in adjacent chains with the pyridine ring containing N1 engaging in a π-π stack with the phenyl ring with atom C16 (symmetry code = 2−x, 1−y, 1−z). The centroid...centroid distance is 3.82 Å, and the angle between the ring planes is 18.7°. As Figure 5b shows, the stacking interactions interconnect 1D-polymers through the lattice. Additional π-stacking interactions are provided by the 1,2-$C_6H_4Cl_2$ molecule which resides over the pyridine rings containing N2 and N3. This mirrors the role of the aromatic solvent in the assemblies with ligands **1** and **2**. These interactions are supplemented by extensive C–F...F–C and C–F...H–C contacts (Figure S18 in the Supplementary Materials). The C–F...F–C distances are 2.85 and 2.87 Å, which are within the sum of the van der Waals radii (2.92–2.94 Å) [44,45]. The shortest H...F contacts lie in the range 2.50–2.57 Å, which compare with 2.67 Å using Bondi's van der Waals radii [44] or 2.57 Å using radii recommended by Rowland and Taylor [45]. Longer H...F contacts (>2.64 Å) have not been included in Figure S18. We note that disordering of some CF_3 groups (see Figure S15) precludes a detailed discussion of interactions involving these units.

[Cu(hfacac)$_2$(**4**)]$_n$·nC$_6$H$_5$Cl crystallizes in the orthorhombic space group *Pmna* and the asymmetric unit contains half of a molecule of **4**, the second half being generated by a mirror plane. Consequently, the CF_3 group in **4** is disordered over two sites (see Section 3.11) and in the figures and discussion below, only one of these sites is considered. Figure 6a shows part of the 1D-polymer chain. The 3,2′:6′,3″-tpy adopts conformation **A** in Scheme 1 and, as noted earlier, the near coplanarity of the rings containing N2 and C6 (Figure S16 and Table 2) is striking. This can be traced back to the accommodation of the 3-$CF_3C_6H_4$ ring within a pocket between two {Cu(hfacac)$_2$} units which follows from the conformation of the 3,2′:6′,3″-tpy. This leads to the presence of short intramolecular C–H...F–C interactions between the *ortho*-H atoms of the phenyl ring and CF_3 groups (which are ordered) of the two {Cu(hfacac)$_2$} units (Figure 6b). The H4...F6 and H7...F6 distances are 2.68 and 2.45 Å, respectively, with C–H...F angles of 154.1 and 152.3°, respectively. The shorter contact is well within the lower (see above) estimate of the sum of the van der Waals radii (2.57 Å [45]) and is towards the shorter end of the range of contacts seen in a survey of the CSD reported in 2005 [48]. The 1D-polymers associate through face-to-face π-stacking of phenyl and central-pyridine rings (angle between ring planes = 0.9°, cenroid...centroid = 3.72 Å) as depicted in Figure 6c, and the interactions extend infinitely through the lattice (Figure S19). The arrangement of neighboring stacks allows association through C–H...π contacts (Figure 6d) with the C–H$_{phenyl}$...centroid$_{phenyl}$ distance being 3.15 Å. The role of the chlorobenzene solvent could not be assessed because of disordering (see Section 3.11).

2.4. PXRD Analysis

After single crystals had been selected for single-crystal X-ray structure determination, the remaining crystals in each crystallization tube were collected and were washed with $CHCl_3$ and the aromatic solvent used in the crystallization experiment (toluene, chlorobenzene or 1,2-dichlorobenzene). The bulk samples were analyzed by IR spectroscopy and PXRD. The IR spectra are shown in Figures S20–S23 in the Supporting Materials. When compared to the IR spectra of ligands **1**–**4**, a strong absorption is observed in the spectra of the coordination polymers containing **1**, **2**, **3**, and **4**, respectively, at 1653, 1650, 1650 and 1646 cm^{-1} which is absent in the spectra of the ligands. This is assigned to one of the C=O stretching modes which appear at 1644 and 1614 cm^{-1} in [Cu(hfacac)$_2$] [49].

Confirmation that the single crystals selected were representative of the bulk crystalline materials came from a comparison of the experimental PXRD patterns (shown in red in Figure 7a–d) with the patterns predicted from the single crystal structures (black traces in Figure 7). For each coordination polymer, all peaks in the predicted pattern had a matching partner in the experimental PXRD pattern, and no additional peaks were observed. The differences in intensities (blue traces in Figure 7) can be justified in terms of differences in the preferred orientations of the crystallites in the bulk powder samples.

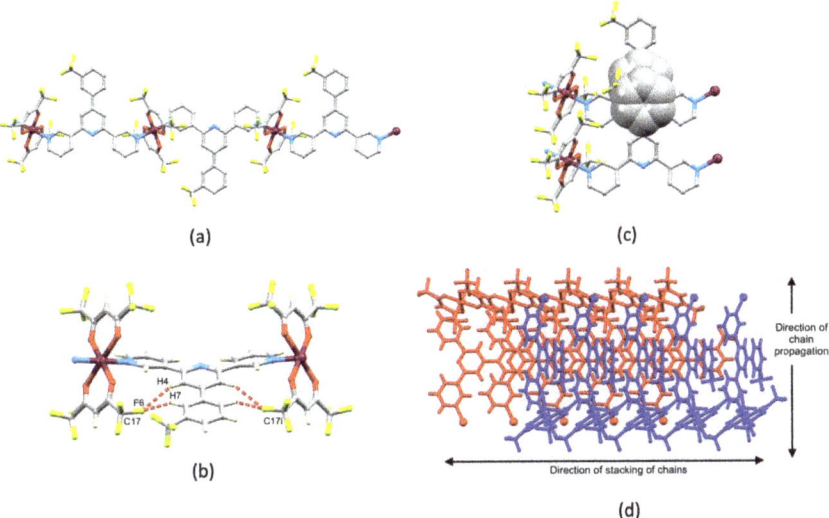

Figure 6. The structure of [Cu(hfacac)$_2$(**4**)]$_n$·nC$_6$H$_5$Cl. (**a**) Part of the 1D-polymer. The CF$_3$ group in **4** is disordered over two sites related by a mirror plane and only one position is shown. (**b**) Intramolecular C–H...F–C contacts; symmetry code i = x, $^3/_2-y$, z. (**c**) π-Stacking between the phenyl ring and the pyridine ring containing N2^{ii} (symmetry code ii = 1+x, y, z). (**d**) The arrangement of adjacent stacks of polymer chains.

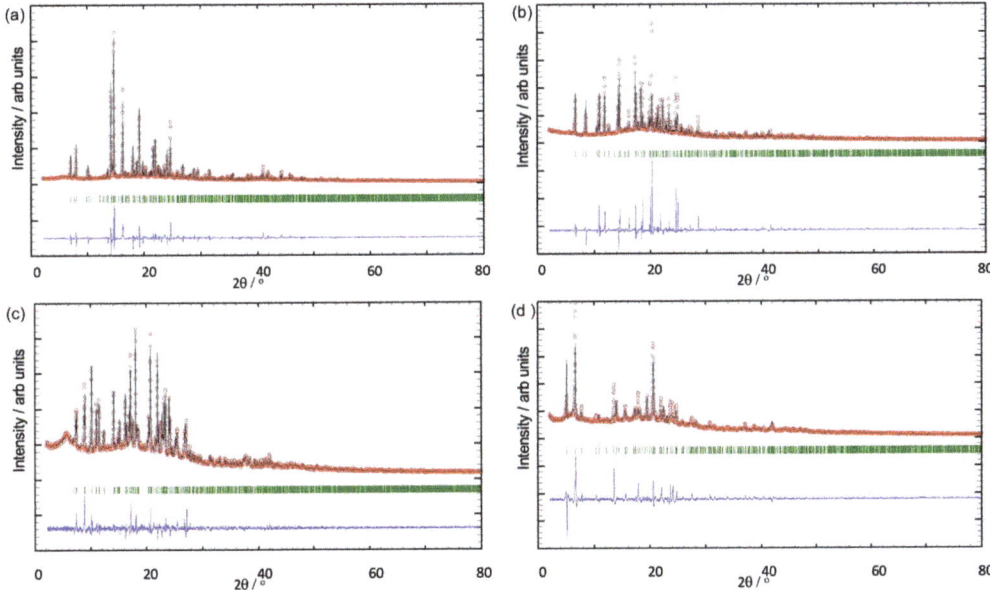

Figure 7. X-Ray diffraction (CuKα1 radiation) patterns (red circles) of the bulk crystalline materials of (**a**) [Cu$_2$(hfacac)4(**1**)2]$_n$·2nC6H4Cl$_2$, (**b**) [Cu(hfacac)$_2$(**2**)]$_n$·2nC$_6$H$_5$Me, (**c**) [Cu$_2$(hfacac)$_4$(**3**)$_2$]$_n$·nC$_6$H$_4$Cl$_2$, and (**d**) [Cu$_2$(hfacac)$_4$(**4**)]$_n$·nC$_6$H$_5$Cl, fitting to the predicted patterns from the single-crystal structures. The black lines are the best fits from the Rietveld refinements, and green lines display the Bragg peak positions. Each blue plot gives the difference between calculated and experimental points (see text).

2.5. Preparative Scale Reactions

To complete the investigation, we performed preparative scale syntheses of the copper(II) complexes using a 1:1 ratio of [Cu(hfacac)$_2$]·H$_2$O to ligand **1**, **2**, **3** and **4**. A solution of [Cu(hfacac)$_2$]·H$_2$O in toluene, chlorobenzene or 1,2-dichlorobenzene was added to a chloroform solution of each ligand and the green precipitates that formed were isolated and dried under vacuum. Satisfactory elemental analyses were obtained for [Cu$_2$(hfacac)$_4$(**1**)$_2$]$_n$·nC$_6$H$_4$Cl$_2$, [Cu(hfacac)$_2$(**2**)]$_n$, [Cu(hfac)$_2$(**3**)]$_n$, and [Cu(hfacac)$_2$(**4**)]$_n$. A PXRD pattern was measured for each compound, and comparisons of these experimental data with the patterns from the bulk crystalline materials from single-crystal growth are displayed in Figure 8 and Figures S24–S26. Good matches are seen for all compounds, providing support that the same coordination polymers are produced on a preparative scale as in single-crystal growth under conditions of layering.

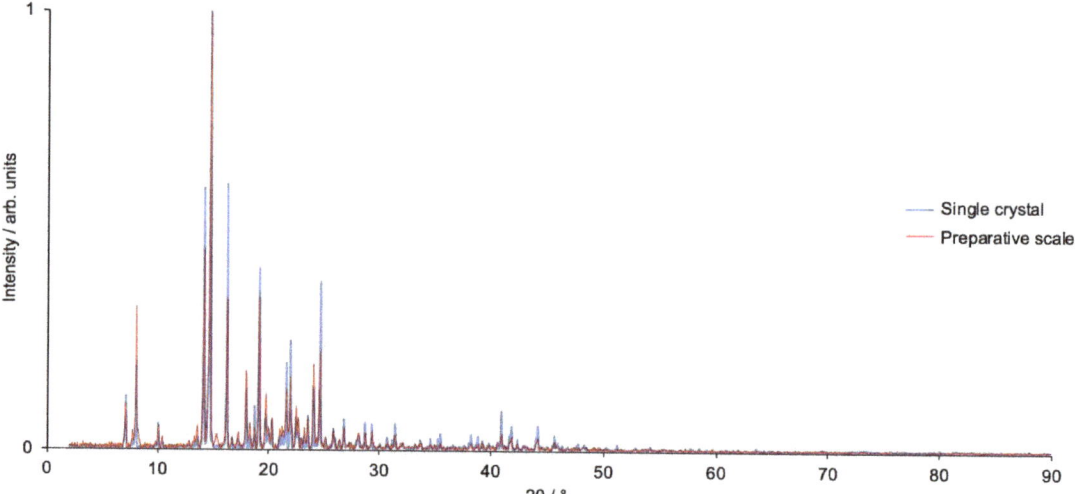

Figure 8. A comparison of the PXRD pattern of [Cu$_2$(hfacac)$_4$(**1**)$_2$]$_n$·nC$_6$H$_4$Cl$_2$ prepared on a preparative scale, and that of the bulk single crystals of [Cu$_2$(hfacac)$_4$(**1**)$_2$]$_n$·2nC$_6$H$_4$Cl$_2$. The difference in solvent arises from drying the synthesized material.

3. Materials and Methods

3.1. General

3-Acetylpyridine and 4-trifluoromethylbenzaldehyde were purchased from Acros Organics (Fisher Scientific AG, 4153 Reinach, Switzerland). 4-Acetylpyridine was bought from Sigma Aldrich (Riedstr. 2, 89555 Steinheim, Germany), 4-trifluoromethylbenzaldehyde from Fluorochem Ltd. (Glossop, UK) and Cu(hfacac)$_2$ monohydrate was bought from abcr GmbH (Im Schlehert 10, 76187 Karlsruhe, Germany). All chemicals were used as received. Analytical thin-layer chromatography was conducted with pre-coated silica gel 60 F$_{254}$ aluminium sheets (Merck KGaA, 64293 Darmstadt, Germany) and visualised using ultraviolet (UV) light (254 nm). Flash column chromatography was performed on a Biotage Selekt system (Biotage, 75103 Uppsala, Sweden) with self-packed silica gel columns (SiliaFlash® P60, 40–63 μm, 230-400 mesh from SiliCycle Inc., Québec, QC, Canada) using ethyl acetate in cyclohexane (gradient) as eluent and monitoring and collecting at 254 nm.

^{1}H and ^{13}C{^{1}H} NMR spectra were recorded on a Bruker Avance III-500 spectrometer (Bruker BioSpin AG, 8117 Fällanden, Switzerland) at 298 K. The ^{1}H and ^{13}C NMR chemical shifts were referenced with respect to the residual solvent peak (δ 2.50 and δ 39.52 respectively for DMSO-d6). ^{19}F{^{1}H} NMR spectra were recorded at 298 K on a Bruker Avance III-600 spectrometer (Bruker BioSpin AG, 8117 Fällanden, Switzerland). MALDI-

TOF mass spectra were recorded on a Shimadzu MALDI 8020 (Shimadzu Schweiz GmbH, 4153 Reinach, Switzerland) using α-cyano-4-hydroxycinnamic acid as matrix. PerkinElmer UATR Two (Perkin Elmer, 8603 Schwerzenbach, Switzerland) and Cary-5000 (Agilent Technologies Inc., Santa Clara, CA 95051, US) instruments were used to record FT-infrared (IR) and UV-VIS absorption spectra, respectively. Melting temperatures were determined using a Stuart melting point SMP 30 device (Cole-Parmer, Stone, UK).

3.2. 4'-(4-(Trifluoromethyl)Phenyl)-4,2':6',4''-Terpyridine (1)

4-Trifluoromethylbenzaldehyde (1.74 g, 10.0 mmol, 1.0 eq) was dissolved at room temperature in EtOH (50 mL). 4-Acetylpyridine (2.8 mL, 25.0 mmol, 2.5 eq) and crushed KOH (1.12 g, 20.0 mmol, 2.0 eq) were then added to the colorless solution. Immediate color change upon the addition of KOH from colorless to orange observed. Then slow addition of aqueous NH_3 (32%, 38.5 mL) followed. The reaction mixture was stirred at room temperature overnight (21 h). The precipitate was collected by filtration and washed with H_2O (3 × 10 mL) followed by EtOH (3 × 10 mL). The light red solid was reprecipitated from a MeOH (40 mL)/CH_2Cl_2 (1 mL)/ and chloroform (1 mL) mixture and dried *in vacuo* overnight yielding **1** (1.39 g, 3.68 mmol, 36.8%) as a colorless solid. M.p. 266.7–268.7 °C. ^{1}H NMR (500 MHz, DMSO-d6): δ/ppm 8.79 (m, 4H, H^{A2}), 8.56 (s, 2H, H^{B3}), 8.34 (m, 6H, H^{A3+C2}), 7.96 (d, J = 8.0 Hz, 2H, H^{C3}). $^{13}C\{^1H\}$ NMR (126 MHz, DMSO-d6): δ/ppm 154.5 (C^{B2}), 150.4 (C^{A2}), 148.8 (C^{B4}), 145.1 (C^{A4}), 141.0 (C^{C1}), 129.8 (q, J_{CF} = 31 Hz, C^{C4}), 128.5 (C^{C2}), 125.9 (q, J_{CF} = 4 Hz, C^{C3}), 124.1 (q, J_{CF} = 272 Hz, C^a), 121.2 (C^{A3}), 119.4 (C^{B3}). $^{19}F\{^1H\}$ NMR (565 MHz, DMSO-d6): δ/ppm −61.1. UV-VIS (MeCN, 2.0 × 10^{-5} mol dm^{-3}) λ/nm 250 (ε/dm^{-3} mol^{-1} cm^{-1} 42,420), 306 (7600). MALDI-TOF-MS *m/z* 378.10 $[M+H]^+$ (calc. 378.12). Found C 69.98, H 3.81, N 11.17; required for $C_{22}H_{14}F_3N_3$ C 70.02, H 3.74, N 11.14. See Figure S5 for the IR spectrum of **1**.

Single crystals of **1** were grown as follows. Ligand **1** (ca. 10 mg) was added to $CHCl_3$ (2 mL) in a small vial to give a clear solution. The open vial was then placed in a larger vial containing Et_2O. Slow diffusion of the non-solvent led to colorless plate-shaped crystals after 7 days.

3.3. 4'-(3-(Trifluoromethyl)Phenyl)-4,2':6',4''-Terpyridine (2)

3-Trifluoromethylbenzaldehyde (1.74 g, 10.0 mmol, 1.0 eq) was dissolved at room temperature in EtOH (50 mL). 4-Acetylpyridine (2.8 mL, 25.0 mmol, 2.5 eq) and crushed KOH (1.12 g, 20.0 mmol, 2.0 eq) were then added to the colorless solution. Immediate color change upon the addition of KOH from colorless to orange was observed. Then slow addition of aqueous NH_3 (32%, 38.5 mL) followed. The reaction mixture was stirred at room temperature overnight (21 h). The formed precipitate was collected by filtration and washed with H_2O (3 x 10 mL) followed by EtOH (3 × 10 mL). The light brown solid was reprecipitated from a MeOH (40 mL)/ CH_2Cl_2 (1 mL) and chloroform (1 mL) mixture and dried in vacuo overnight affording **2** (1.86 g, 4.93 mmol, 49.3%) as a colorless solid. M.p. 237.0–239.4 °C. ^{1}H NMR (500 MHz, DMSO-d6): δ/ppm 8.78 (m, 4H, H^{A2}), 8.57 (s, 2H, H^{B3}), 8.47 (s, 1H, H^{C2}), 8.42 (d, J = 8.4 Hz, 1H, H^{C6}), 8.36 (m, 4H, H^{A3}), 7.91 (d, J = 7.7 Hz, 1H, H^{C4}), 7.83 (t, J = 7.8 Hz, 1H, H^{C5}). $^{13}C\{^1H\}$ NMR (126 MHz, DMSO-d6): δ/ppm 154.6 (C^{B2}), 150.4 (C^{A2}), 148.8 (C^{B4}), 145.1 (C^{A4}), 138.0 (C^{C1}), 132.6 (C^{C6}), 130.1 (C^{C5}), 130.0 (q, J_{CF} = 31 Hz, C^{C3}), 126.2 (q, J_{CF} = 4 Hz, C^{C4}), 124.3 (q, J_{CF} = 4 Hz, C^{C2}), 124.1 (q, J_{CF} = 272 Hz, C^a), 121.2 (C^{A3}), 119.3 (C^{B3}). $^{19}F\{^1H\}$ NMR (565 MHz, DMSO-d6): δ/ppm −60.7. UV-VIS (MeCN, 2.0 × 10^{-5} mol dm^{-3}) λ/nm 250 (ε/dm^{-3} mol^{-1} cm^{-1} 42,270), 294 (7300). MALDI-TOF-MS *m/z* 378.15 $[M+H]^+$ (calc. 378.12). Found C 70.06, H 3.95, N 11.31; required for $C_{22}H_{14}F_3N_3$ C 70.02, H 3.74, N 11.14. See Figure S6 for the IR spectrum of **2**.

Single crystals of **2** were grown as follows. Compound **2** (ca. 10 mg) was added to DMSO (0.7 mL) in an NMR tube to give a white suspension. The NMR tube was then heated using a heat gun to give a clear solution, and as the solution was allowed to cool to room temperature, colorless plate-shaped crystals grew within an hour.

3.4. 4′-(4-(Trifluoromethyl)Phenyl)-3,2′:6′,3″-Terpyridine (3)

4-Trifluoromethylbenzaldehyde (1.74 g, 10.0 mmol, 1.0 eq) was dissolved at room temperature in EtOH (50 mL). 3-Acetylpyridine (2.8 mL, 25.0 mmol, 2.5 eq) and crushed KOH (1.12 g, 20.0 mmol, 2.0 eq) were then added to the light yellow solution. Immediate color change upon the addition of KOH from yellow to orange was observed. Then slow addition of aqueous NH_3 (32%, 38.5 mL) followed. The reaction mixture was stirred at room temperature overnight (24 h). The formed precipitate was collected by filtration and washed with H_2O (3 × 10 mL) followed by EtOH (3 × 10 mL). Purification by column chromatography (380 g self-packed silica gel column, Biotage Select, eluent: EtOAc in cyclohexane 20–100%) gave **3** (1.56 g, 4.12 mmol, 41.2%) as a colorless crystalline solid. M.p. 209.4–211.3 °C. ^1H NMR (500 MHz, DMSO-d6): δ/ppm 9.53 (dd, J = 2.3, 0.7 Hz, 2H, H^{A2}), 8.73–8.70 (m, 4H, H^{A4+A6}), 8.44 (s, 2H, H^{B3}), 8.33 (d, J = 8.1 Hz, 2H, H^{C2}), 7.94 (d, J = 8.3 Hz, 2H, H^{C3}), 7.59 (m, 2H, H^{A5}). $^{13}C\{^1H\}$ NMR (126 MHz, DMSO-d6): δ/ppm 154.8 (C^{B2}), 150.2 (C^{A6}), 148.4 (C^{B4}), 148.3 (C^{A2}), 141.2 (q, J_{CF} = 1 Hz, C^{C1}), 134.5 (C^{A4}), 133.8 (C^{A3}), 129.7 (q, J_{CF} = 31 Hz, C^{C4}), 128.5 (C^{C2}), 125.8 (q, J_{CF} = 4 Hz, C^{C3}), 124.1 (q, J_{CF} = 272 Hz, C^a), 123.8 (C^{A5}), 117.9 (C^{B3}). $^{19}F\{^1H\}$ NMR (565 MHz, DMSO-d6): δ/ppm −61.0. UV-VIS (MeCN, 2.0 × 10^{-5} mol dm^{-3}) λ/nm 249 (ε/dm^{-3} mol^{-1} cm^{-1} 43,620), 297 (6550). MALDI-TOF-MS m/z 378.09 [M+H]$^+$ (calc. 378.12). Found C 69.89, H 3.82, N 11.50; required for $C_{22}H_{14}F_3N_3$ C 70.02, H 3.74, N 11.14. See Figure S7 for the IR spectrum of **3**.

3.5. 4′-(3-(Trifluoromethyl)Phenyl)-3,2′:6′,3″-Terpyridine (4)

3-Trifluoromethylbenzaldehyde (1.74 g, 10.0 mmol, 1.0 eq) was dissolved at room temperature in EtOH (50 mL). 3-Acetylpyridine (2.8 mL, 25.0 mmol, 2.5 eq) and crushed KOH (1.12 g, 20.0 mmol, 2.0 eq) were then added to the light yellow solution. Immediate color change upon the addition of KOH from yellow to orange was observed. Then slow addition of aqueous NH_3 (32%, 38.5 mL) followed. The reaction mixture was stirred at room temperature overnight (24 h). The formed precipitate was collected by filtration and washed with H_2O (3 × 10 mL) and EtOH (3 × 10 mL). The product was recrystallized from MeOH and dried in vacuo to yield **4** (1.18 g, 3.13 mmol, 31.3%) as a colorless solid. M.p. 165.3–167.0 °C. ^1H NMR (500 MHz, DMSO-d6): δ/ppm 9.55 (dd, J = 2.3, 0.8 Hz, 2H, H^{A2}), 8.73 (dt, J = 7.9, 2.0 Hz, 2H, H^{A4}), 8.70 (dd, J = 4.8, 1.6 Hz, 2H, H^{A6}), 8.48 (m, 1H, H^{C2}), 8.47 (s, 2H, H^{B3}), 8.42 (d, J = 7.8 Hz, 1H, H^{C6}), 7.89 (d, J = 7.8 Hz, 1H, H^{C4}), 7.82 (t, J = 7.8 Hz, 1H, H^{C5}), 7.59 (m, 2H, H^{A5}). $^{13}C\{^1H\}$ NMR (126 MHz, DMSO-d6): δ/ppm 154.8 (C^{B2}), 150.2 (C^{A6}), 148.4 (overlapping C^{A2+B4}), 138.3 (C^{C1}), 134.5 (C^{A4}), 133.8 (C^{A3}), 131.7 (C^{C6}), 130.1 (C^{C5}), 130.0 (q, J_{CF} = 31 Hz, C^{C3}), 126.0 (q, J_{CF} = 4 Hz, C^{C4}), 124.3 (q, J_{CF} = 4 Hz, C^{C2}), 124.1 (q, J_{CF} = 272 Hz, C^a), 123.8 (C^{A5}), 117.8 (C^{B3}). $^{19}F\{^1H\}$ NMR (565 MHz, DMSO-d6): δ/ppm −60.7. UV-VIS (MeCN, 2.0 × 10^{-5} mol dm^{-3}) λ/nm 248 (ε/dm^{-3} mol^{-1} cm^{-1} 40,870), 296 (6350). MALDI-TOF-MS m/z 378.10 [M+H]$^+$ (calc. 378.12). Found C 69.81, H 3.71, N 11.31; required for $C_{22}H_{14}F_3N_3$ C 70.02, H 3.74, N 11.14. See Figure S8 for the IR spectrum of **4**.

3.6. $[Cu_2(hfacac)_4(1)_2]_n \cdot 2nC_6H_4Cl_2$

A 1,2-dichlorobenzene (5 mL) solution of [Cu(hfacac)$_2$]·H$_2$O (29.7 mg, 0.060 mmol) was layered over a CHCl$_3$ solution (4 mL) of ligand **1** (11.3 mg, 0.030 mmol). Green block-like crystals grew after 1 day. A single crystal was selected for X-ray diffraction. The remaining crystals were washed with chloroform and 1,2-dichlorobenzene and analyzed by PXRD and IR spectroscopy.

For a preparative scale reaction, **1** (28.7 mg, 0.076 mmol) was dissolved in CHCl$_3$ (4 mL). Then a solution of [Cu(hfacac)$_2$]·H$_2$O (37.7 mg, 0.076 mmol) in 1,2-dichlorobenzene (5 mL) was added, and the green solution was stirred at room temperature. Immediate formation of a pale green precipitate was observed and stirring of the suspension at room temperature was continued for 44 h. The suspension was then centrifuged, the solid collected and dried in vacuo. [Cu$_2$(hfacac)$_4$(**1**)$_2$]$_n$·$nC_6H_4Cl_2$ (64.4 mg, 0.035 mmol, 91.2%) was isolated as a pale green powder. Elemental analysis: found C 45.34, H 2.24, N 4.24;

required for $C_{70}H_{36}Cl_2Cu_2F_{30}N_6O_8$ C 45.27, H 1.95, N 4.53. PXRD analysis was performed (see text).

3.7. $[Cu(hfacac)_2(2)]_n \cdot 2nC_6H_5Me$

A toluene (5 mL) solution of $[Cu(hfacac)_2] \cdot H_2O$ (29.7 mg, 0.060 mmol) was layered over a CHCl$_3$ solution (4 mL) of compound **2** (11.3 mg, 0.030 mmol). Blue plate-like crystals grew after 4 days, and one X-ray quality crystal was chosen. The remaining crystals were washed with CHCl$_3$ and toluene and this bulk sample was analyzed by IR spectroscopy and PXRD.

A preparative scale reaction was also carried out. Ligand **2** (28.7 mg, 0.076 mmol) was dissolved in CHCl$_3$ (4 mL), and then a solution of $[Cu(hfacac)_2] \cdot H_2O$ (37.7 mg, 0.076 mmol) in toluene (5 mL) was added. The green solution was stirred at room temperature and immediate formation of a pale green precipitate was observed. After 44 h, the precipitate that formed was collected by centrifugation and dried *in vacuo*. $[Cu(hfacac)_2(2)]_n$ (43.7 mg, 0.051 mmol, 67.1%) was isolated as a pale green powder. Elemental analysis: found C 44.92, H 2.23, N 5.17; required for $C_{32}H_{16}CuF_{15}N_3O_4$ C 44.95, H 1.89, N 4.91. PXRD analysis was performed (see text).

3.8. $[Cu_2(hfacac)_4(3)_2]_n \cdot nC_6H_4Cl_2$

A solution of $[Cu(hfacac)_2] \cdot H_2O$ (29.7 mg, 0.060 mmol) in 1,2-dichlorobenzene (5 mL) was layered over a CHCl$_3$ solution (4 mL) of compound **3** (11.3 mg, 0.030 mmol). Green plate-like crystals grew after 11 days. An X-ray quality single crystal was selected and the residual crystals were washed with CHCl$_3$ and 1,2-dichlorobenzene and analyzed by PXRD and IR spectroscopy.

On a preparative scale, compound **3** (28.7 mg, 0.076 mmol) was dissolved in chloroform (4 mL), and then a solution of $[Cu(hfacac)_2] \cdot H_2O$ (37.7 mg, 0.076 mmol) in 1,2-dichlorobenzene (5 mL) was added. The green solution was stirred at room temperature and the formation of a pale green precipitate was immediately observed. After 44 h, the suspension was then centrifuged, and the solid was collected and dried *in vacuo*. $[Cu(hfac)_2(3)]_n$ (57.4 mg, 0.034 mmol, 88.3%) was isolated as a pale green powder. Elemental analysis: found C 45.13, H 2.08, N 4.83; required for $C_{32}H_{16}CuF_{15}N_3O_4$ C 44.95, H 1.89, N 4.91. PXRD analysis was performed (see text).

3.9. $[Cu_2(hfacac)_4(3)_2]_n \cdot nC_6H_5Cl$

A solution of $[Cu(hfacac)_2] \cdot H_2O$ (29.7 mg, 0.060 mmol) in chlorobenzene (5 mL) was layered over a CHCl$_3$ solution (4 mL) of ligand **3** (11.3 mg, 0.030 mmol). Green plate-like crystals grew after 11 days. Single-crystal X-ray crystallography confirmed that the structure of $[Cu_2(hfacac)_4(3)_2]_n \cdot nC_6H_5Cl$ was essentially isostructural with $[Cu_2(hfacac)_4(3)_2]_n \cdot nC_6H_4Cl_2$, and no bulk sample characterization was carried out.

3.10. $[Cu(hfacac)_2(4)]_n \cdot nC_6H_5Cl$

A chlorobenzene (5 mL) solution of $[Cu(hfacac)_2] \cdot H_2O$ (29.7 mg, 0.060 mmol) was layered over a CHCl$_3$ solution (4 mL) of **4** (11.3 mg, 0.030 mmol). Green block-like crystals grew after 10 days. After the selection of a single crystal, the residual crystals were washed with CHCl$_3$ and chlorobenzene and this bulk sample was analyzed by PXRD and IR spectroscopy.

On a preparative scale, compound **4** (28.7 mg, 0.076 mmol) was dissolved in CHCl$_3$ (4 mL). Then a solution of $[Cu(hfacac)_2] \cdot H_2O$ (37.7 mg, 0.076 mmol) in chlorobenzene (5 mL) was added, and the green solution was stirred at room temperature. Immediate formation of a pale green precipitate was observed. After 44 h, the formed precipitate was collected by centrifugation and dried *in vacuo*. $[Cu(hfacac)_2(4)]_n$ (49.1 mg, 0.057 mmol, 75.0%) was obtained as a pale green powder. Elemental analysis: found C 44.78, H 2.03, N 5.22; required for $C_{32}H_{16}CuF_{15}N_3O_4$ C 44.95, H 1.89, N 4.91. PXRD analysis was performed (see text).

3.11. Crystallography

Single crystal data were collected either on a STOE StadiVari diffractometer (STOE & Cie GmbH, 64295 Darmstadt, Germany) equipped with a Metaljet D2 source (GaKα radiation) and a Pilatus300K detector, or on a Bruker APEX-II diffractometer (Bruker BioSpin AG, 8117 Fällanden, Switzerland) with CuKα radiation. For the former, data processing used STOE software (X-Area 1.90, STOE, 2020), and structures were solved using Superflip [50,51] and Olex2 [52], and the model was refined with ShelXL v. 2014/7 [53]. For the latter, data reduction, solution, and refinement used the programs APEX [54], ShelXT [55], Olex2 [52], and ShelXL v. 2014/7 [53]. All H atoms were included at geometrically calculated positions and refined using a riding model with U_{iso} = 1.2 of the parent atom. Structure analysis used CSD Mercury 2020.1 [56]. In the four coordination polymers, some CF_3 groups were disordered and these F atoms were refined isotropically. In $[Cu_2(hfacac)4(\mathbf{1})2]_n \cdot 2nC_6H_4Cl_2$, one CF_3 group on each $[hfacac]^-$ ligand was rotationally disordered and was modeled over three positions with 0.5, 0.3 and 0.2 fractional occupancies for one CF_2 group, and 0.4, 0.4 and 0.2 fractional occupancies for the second. In $[Cu(hfacac)_2(\mathbf{2})]_n \cdot 2nC_6H_5Me$, one CF_3 group of an $[hfacac]^-$ ligand was disordered and was modeled over two positions with fractional occupancies of 0.6 and 0.4; one toluene molecule was also disordered, and the methyl group was modeled over two sites of equal occupancies. In $[Cu_2(\mathbf{3})_2(hfacac)_4]_n \cdot nC_6H_4Cl_2$, one $[hfacac]^-$ ligand contains one rotationally disordered CF_3 which was modeled over two equal occupancy sites; the CF_3 group in 3 was also disordered, and again was modeled over two sites with fractional occupancies of 0.5. In $[Cu_2(hfacac)_4(\mathbf{4})]_n \cdot nC_6H_5Cl$, the asymmetric unit contains half of one molecule of 4, and the CF_3 group is disordered over two sites related by a mirror, and the whole group was refined isotropically; additionally, the CF_3 is rotationally disordered and was modeled over sites of fractional occupancies 0.3 and 0.2. A mask was used to treat the solvent region in $[Cu_2(hfacac)_4(\mathbf{4})]_n \cdot nC_6H_5Cl$ and the electron density removed equated to one C_6H_5Cl molecule per 2 Cu atoms; this was added to the formulae and appropriate numbers. In the structural discussions, only the major (or one of the equal) occupancy sites are considered in each disordered entity.

PXRD data were collected at room temperature in transmission mode using a Stoe Stadi P diffractometer (STOE & Cie GmbH, 64295 Darmstadt, Germany), equipped with CuKα1 radiation (Ge(111) monochromator and a DECTRIS MYTHEN 1K detector. Whole-pattern decomposition (profile matching) analysis [57–59] of the diffraction patterns was done using the package FULLPROF SUITE (v. September 2020) [59,60] using a previously determined instrument resolution function based on a NIST640d standard. The structural models were derived from the single crystal X-ray diffraction data. Refined parameters in Rietveld were scale factor, zero shift, lattice parameters, Cu and halogen atomic positions, background points, and peaks shapes as a Thompson-Cox-Hastings pseudo-Voigt function. Preferred orientations as a March–Dollase multi-axial phenomenological model were incorporated into the analysis.

1: $C_{22}H_{14}F_3N_3$, M_r = 377.36, colorless plate, monoclinic, space group $P2_1/c$, a = 10.5418(13), b = 21.653(3), c = 7.4248(9) Å, β = 94.146(4)°, V = 1690.3(4) Å^3, T = 150 K, Z = 4, μ(CuK$_\alpha$) = 0.947. Total 22080 reflections, 3110 unique (R_{int} = 0.0257). Refinement of 3056 reflections (253 parameters) with $I > 2\sigma(I)$ converged at final R_1 = 0.0347 (R_1 all data = 0.0350), wR_2 = 0.0916 ($wR2$ all data = 0.0919), F(000) = 776, gof = 1.028. CCDC 2077591.

2: $C_{22}H_{14}F_3N_3$, M_r = 377.36, colorless plate, triclinic, space group P–1, a = 7.5432(5), b = 10.8309(8), c = 21.3995(15) Å, α = 91.518(3), β = 98.791(2), γ = 98.418(2)°, V = 1707.0(2) Å^3, T = 150 K, Z = 4, μ(CuK$_\alpha$) = 0.937. Total 16588 reflections, 5996 unique (R_{int} = 0.0235). Refinement of 5541 reflections (505 parameters) with $I > 2\sigma(I)$ converged at final R_1 = 0.1767 (R_1 all data = 0.1784), wR_2 = 0.0601 ($wR2$ all data = 0.0629), F(000) = 776, gof = 1.109. CCDC 2077593.

$[Cu_2(hfacac)_4(\mathbf{1})_2]_n \cdot 2nC_6H_4Cl_2$: $C_{76}H_{40}Cl_4Cu_2F_{30}N_6O_8$, M_r = 2004.02, green block, orthorhombic, space group $Pbca$, a = 22.4688(13), b = 14.3450(8), c = 24.8918(14) Å, V = 8023.0(8) Å^3, T = 150 K, Z = 4, μ(CuK$_\alpha$) = 3.050. Total 65061 reflections, 7433 unique (R_{int} = 0.0438).

Refinement of 6996 reflections (570 parameters) with $I > 2\sigma(I)$ converged at final $R_1 = 0.1509$ (R_1 all data = 0.1534), $wR_2 = 0.0589$ ($wR2$ all data = 0.0616), F(000) = 3992, gof = 1.029. CCDC 2077592.

[Cu$_2$(hfacac)$_4$(**2**)]$_n$·2nC$_6$H$_5$Me: C$_{46}$H$_{32}$CuF$_{15}$N$_3$O$_4$, M_r = 1039.28, blue plate, monoclinic, space group Cc, a = 8.8604(2), b = 25.3335(7), c = 20.7490(6) Å, β = 97.537(2)°, V = 4617.2(2) Å^3, T = 150 K, Z = 4, μ(GaK$_\alpha$) = 3.168. Total 13164 reflections, 5535 unique (R_{int} = 0.0508). Refinement of 5261 reflections (575 parameters) with $I > 2\sigma(I)$ converged at final $R_1 = 0.0802$ (R_1 all data = 0.0832), $wR_2 = 0.2115$ ($wR2$ all data = 0.2173), F(000) = 2100, gof = 1.019. CCDC 2077595.

[Cu$_2$(hfacac)$_4$(**3**)$_2$]$_n$·nC$_6$H$_4$Cl$_2$: C$_{70}$H$_{36}$Cl$_2$Cu$_2$F$_{30}$N$_6$O$_8$, M_r = 1857.03, green plate, triclinic, space group P–1, a = 11.9939(3), b = 12.1658(3), c = 12.9674(3) Å, α = 102.257(2), β = 103.145(2), γ = 91.214(2)°, V = 1795.76(8) Å^3, T = 150 K, Z = 1, μ(GaK$_\alpha$) = 4.466. Total 40977 reflections, 7061 unique (R_{int} = 0.0352). Refinement of 6430 reflections (510 parameters) with $I > 2\sigma(I)$ converged at final $R_1 = 0.0829$ (R_1 all data = 0.0884), $wR_2 = 0.2286$ ($wR2$ all data = 0.2355), F(000) = 924, gof = 1.068. CCDC 2077596.

[Cu$_2$(hfacac)$_4$(**3**)$_2$]$_n$·nC$_6$H$_5$Cl: C$_{70}$H$_{37}$ClCu$_2$F$_{30}$N$_6$O$_8$, M_r = 1822.58, green plate, triclinic, space group P–1, a = 11.9906(3), b = 11.9911(3), c = 13.0617(3) Å, α = 103.144(2), β = 102.547(2), γ = 91.491(2)°, V = 1779.51(8) Å^3, T = 150 K, Z = 1, μ(GaK$_\alpha$) = 4.273. Total 53503 reflections, 6936 unique (R_{int} = 0.0598). Refinement of 6607 reflections (510 parameters) with $I > 2\sigma(I)$ converged at final $R_1 = 0.0940$ (R_1 all data = 0.0965), $wR_2 = 0.2580$ ($wR2$ all data = 0.2613), F(000) = 908, gof = 1.090. CCDC 2077594.

[Cu$_2$(hfacac)$_4$(**4**)]$_n$·nC$_6$H$_5$Cl: C$_{38}$H$_{21}$ClCuF$_{15}$N$_3$O$_4$, M_r = 967.57, green block, orthorhombic, space group $Pnma$, a = 6.5155(4), b = 26.2371(17), c = 22.3188(15) Å, V = 3815.4(4) Å^3, T = 200.0 K, Z = 4, μ(CuK$_\alpha$) = 0.845. Total 24327 reflections, 3512 unique (R_{int} = 0.0333). Refinement of 3376 reflections (266 parameters) with $I > 2\sigma(I)$ converged at final $R_1 = 0.0640$ (R_1 all data = 0.0654), $wR_2 = 0.1707$ ($wR2$ all data = 0.1717), F(000) = 1932, gof = 1.128. CCDC 2077597.

4. Conclusions

We have prepared and characterized four new ligands **1–4** which comprise pairs with either 4,2′:6′,4″- or 3,2′:6′,3″-tpy metal-binding domains and with isomeric 4′-trifluoromethylphenyl substituents. The single crystal structures of **1** and **2** were determined. Despite the change in the substitution position of the CF$_3$ group upon going from **1** to **2**, the packing interactions in the two compounds are similar and are dominated by face-to-face π-stacking, with the stacking interaction extending across the whole molecular framework. Reactions of **1**, **2**, **3** and **4** with [Cu(hfacac)$_2$]·H$_2$O under conditions of crystal growth by layering using a combination of CHCl$_3$ and an aromatic solvent resulted in the formation of [Cu$_2$(hfacac)$_4$(**1**)$_2$]$_n$·2nC$_6$H$_4$Cl$_2$, [Cu(hfacac)$_2$(**2**)]$_n$·2nC$_6$H$_5$Me, [Cu$_2$(hfacac)$_4$(**3**)$_2$]$_n$·nC$_6$H$_4$Cl$_2$, [Cu$_2$(hfacac)$_4$(**3**)$_2$]$_n$·nC$_6$H$_5$Cl, and [Cu(hfacac)$_2$(**4**)]$_n$·nC$_6$H$_5$Cl. All are 1D-coordination polymers, and the two polymers containing **3** are essentially isostructural. PXRD analysis of the bulk crystalline products confirmed that the single crystals used for structure determination were representative of the bulk materials. PXRD was used to confirm that the same coordination compounds could be prepared on a preparative scale.

The 1D-polymers [Cu$_2$(hfacac)$_4$(**1**)$_2$]$_n$·2nC$_6$H$_4$Cl$_2$ and [Cu(hfacac)$_2$(**2**)]$_n$·2nC$_6$H$_5$Me are zig-zag chains which follows from the V-shaped 4,2′:6′,4″-tpy building block. This structural motif is unaffected by changing the substitution position of the CF$_3$ group in **1** versus **2**. In both structures, packing interactions are dominated by C–F...F–C contacts, but the arrangement of the 1D-chains is significantly altered as a consequence of the directionalities of the C–CF$_3$ domains in **1** and **2**. There are no inter-polymer face-of-face π-stacking interactions, but instead, aromatic solvent molecules are incorporated into the lattice and engage in π-stacking contacts with the arene-backbone of both polymers.

In [Cu$_2$(hfacac)$_4$(**3**)$_2$]$_n$·nC$_6$H$_4$Cl$_2$, the 3,2′:6′,3″-tpy adopts conformation **C** with an out/in arrangement of N-donors. A combination of this with Cu atoms on inversion centers leads to an alternating arrangement of 3,2′:6′,3″-tpy units in the 1D-polymer. In contrast, the 3,2′:6′,3″-tpy unit in $[Cu(hfacac)_2(4)]_n \cdot nC_6H_5Cl$ exhibits conformation **A** (Scheme 1). The near coplanarity of the phenyl and central pyridine rings in **4** is notable and arises from the phenyl ring being locked in position by four C–H$_{phenyl}$...F–C$_{hfacac}$ contacts. In both $[Cu_2(hfacac)_4(3)_2]_n \cdot nC_6H_4Cl_2$ and $[Cu(hfacac)_2(4)]_n \cdot nC_6H_5Cl$, π-stacking interactions between 4′-trifluoromethylphenyl-3,2′:6′,3″-tpy domains are key packing interactions, and this contrasts with the packing of polymers incorporating **1** and **2**.

We have demonstrated that the assemblies of the coordination polymers in this work are reproducible, and that a switch from a 4,2′:6′,4″- to 3,2′:6′,3″-tpy metal-binding unit is accompanied by a change from dominant C–F...F–C and C–F...H–C contacts to π-stacking of arene domains between ligands **3** or **4**. The switch from a 3-CF$_3$ to 4-CF$_3$ substituent in the 4′-phenyl group has less significant consequences.

Supplementary Materials: The following are available online at https://www.mdpi.com/article/10.3390/inorganics9070054/s1, Figures S1–S4: Mass spectra of **1–4**; Figures S5–S8: IR spectra of **1–4**; Figure S9: Solution absorption spectra of **1–4**; Figures S10 and S11: ^{1}H and ^{13}C{^{1}H} NMR spectra of **1–4**; Figure S12: Packing of molecules of **2**; Figures S13–S16: Molecular structures of the asymmetric units in the coordination polymers; Figure S17: Packing in $[Cu(hfacac)_2(2)]_n \cdot 2nC_6H_5Me$; Figure S18: C–F...F–C and C–F...H–C contacts present in $[Cu_2(hfacac)_4(3)_2]_n \cdot nC_6H_4Cl_2$; Figure S19: Packing in $[Cu(hfacac)_2(4)]_n \cdot nC_6H_5Cl$; Figure S20–S23: IR spectra of the coordination polymers; Figures S24–S26: Additional PXRD data.

Author Contributions: G.M. and S.S.C. contributed equally to the research and writing. Project conceptualization, administration, supervision, and funding acquisition: C.E.H. and E.C.C.; investigation and data analysis: G.M. and S.S.C.; single-crystal X-ray diffraction and PXRD: A.P., G.M. and S.S.C.; manuscript writing: C.E.H., G.M. and S.S.C.; manuscript editing and review: all authors. All authors have read and agreed to the published version of the manuscript.

Funding: This research was funded in part by the Swiss National Science Foundation, grant number 200020_182559.

Data Availability Statement: The data presented in this study are available on request from the corresponding author. The data are not publicly accessible at present.

Acknowledgments: We thank the University of Basel for support of our research.

Conflicts of Interest: The authors declare no conflict of interest.

References

1. Housecroft, C.E. 4,2′:6′,4″-Terpyridines: Diverging and Diverse Building Blocks in Coordination Polymers and Metallomacrocycles. *Dalton Trans.* **2014**, *43*, 6594–6604. [CrossRef]
2. Housecroft, C.E. Divergent 4,2′:6′,4″- and 3,2′:6′,3″-Terpyridines as Linkers in 2- and 3-Dimensional Architectures. *CrystEngComm* **2015**, *17*, 7461–7468. [CrossRef]
3. Housecroft, C.E.; Constable, E.C. Ditopic and tetratopic 4,2′:6′,4″-Terpyridines as Structural Motifs in 2D- and 3D-Coordination Assemblies. *Chimia* **2019**, *73*, 462–467. [CrossRef]
4. Housecroft, C.E.; Constable, E.C. The Terpyridine Isomer Game: From Chelate to Coordination Network Building Block. *Chem. Commun.* **2020**, *56*, 10786–10794. [CrossRef] [PubMed]
5. Elahi, S.M.; Raizada, M.; Sahu, P.K.; Konar, S. Terpyridine-Based 3D Metal–Organic-Frameworks: A Structure–Property Correlation. *Chem. Eur. J.* **2021**, *27*, 5858–5870. [CrossRef] [PubMed]
6. Housecroft, C.E.; Constable, E.C. Isomers of terpyridine as ligands in coordination polymers and networks containing zinc(II) and cadmium(II). *Molecules* **2021**, *26*, 3110. [CrossRef]
7. Liu, B.; Hou, L.; Wu, W.-P.; Dou, A.-N.; Wang, Y.-Y. Highly selective luminescence sensing for Cu^{2+} ions and selective CO$_2$ capture in a doubly interpenetrated MOF with Lewis basic pyridyl sites. *Dalton Trans.* **2015**, *44*, 4423–4427. [CrossRef]
8. Wu, Y.; Wu, J.; Luo, Z.; Wang, J.; Li, Y.; Han, Y.; Liu, J. Fluorescence detection of Mn^{2+}, Cr$_2$O$_7^{2-}$ and nitroexplosives and photocatalytic degradation of methyl violet and rhodamine B based on two stable metal–organic frameworks. *RSC Adv.* **2017**, *7*, 10415–10423. [CrossRef]
9. Yuan, F.; Yuan, C.-M.; Hu, H.-M.; Wang, T.-T.; Zhou, C.-S. Structural diversity of a series of terpyridyl carboxylate coordination polymers: Luminescent sensor and magnetic properties. *J. Solid State Chem.* **2018**, *258*, 588–601. [CrossRef]

10. Dorofeeva, V.N.; Mishura, A.M.; Lytvynenko, A.S.; Grabovaya, N.V.; Kiskin, M.A.; Kolotilov, S.V.; Eremenko, I.L.; Novotortsev, V.M. Structure and Electrochemical Properties of Copper(II) Coordination Polymers with Ligands Containing Naphthyl and Anthracyl Fragments. *Theor. Exper. Chem.* **2016**, *52*, 111–118. [CrossRef]
11. Nijs, T.; Klein, Y.M.; Mousavi, S.F.; Ahsan, A.; Nowakowska, S.; Constable, E.C.; Housecroft, C.E.; Jung, T.A. The Different Faces of 4′-Pyrimidinyl-Functionalized 4,2′:6′,4″-Terpyridines: Metal–Organic Assemblies from Solution and on Au(111) and Cu(111) Surface Platforms. *J. Am. Chem. Soc.* **2018**, *140*, 2933–2939. [CrossRef] [PubMed]
12. Constable, E.C.; Housecroft, C.E.; Vujovic, S.; Zampese, J.A.; Crochet, A.; Batten, S.R. Do perfluoroarene···arene and C–H···F interactions make a difference to the structures of 4,2′:6′,4″-terpyridine-based coordination polymers? *CrystEngComm* **2013**, *15*, 10068–10078. [CrossRef]
13. Li, L.; Zhang, Y.Z.; Yang, C.; Liu, E.; Golen, J.A.; Zhang, G. One-dimensional copper(II) coordination polymers built on 4′-substituted 4,2′:6′,4″- and 3,2′:6′,3″-terpyridines: Syntheses, structures and catalytic properties. *Polyhedron* **2016**, *105*, 115–122. [CrossRef]
14. Constable, E.C.; Housecroft, C.E.; Neuburger, M.; Vujovic, S.; Zampese, J.A. Molecular recognition between 4′-(4-biphenylyl)-4,2′:6′,4″-terpyridine domains in the assembly of d^9 and d^{10} metal ion-containing one-dimensional coordination polymers. *Polyhedron* **2013**, *60*, 120–129. [CrossRef]
15. Rocco, D.; Novak, S.; Prescimone, A.; Constable, E.C.; Housecroft, C.E. Manipulating the conformation of 3,2′:6′,3″-terpyridine in [Cu$_2$(μ-OAc)$_4$(3,2′:6′,3″-tpy)]$_n$ 1D-polymers. *Chemistry* **2021**, *3*, 15. [CrossRef]
16. Rocco, D.; Manfroni, G.; Prescimone, A.; Klein, Y.M.; Gawryluk, D.J.; Constable, E.C.; Housecroft, C.E. Single and double-stranded 1D-coordination polymers with 4′-(4-alkyloxyphenyl)-3,2′:6′,3″-terpyridines and {Cu$_2$(μ-OAc)$_4$} or {Cu$_4$(μ$_3$-OH)$_2$(μ-OAc)$_2$(μ$_3$-OAc)$_2$(AcO-κO)$_2$} motifs. *Polymers* **2020**, *12*, 318. [CrossRef]
17. Klein, Y.M.; Prescimone, A.; Constable, E.C.; Housecroft, C.E. Coordination behaviour of 1-(4,2′:6′,4″-terpyridin-4′-yl)ferrocene and 1-(3,2′:6′,3″-terpyridin-4′-yl)ferrocene: Predictable and unpredictable assembly algorithms. *Aust. J. Chem.* **2017**, *70*, 468–477. [CrossRef]
18. Khatua, S.; Goswami, S.; Biswas, S.; Tomar, K.; Jena, H.S.; Konar, S. Stable Multiresponsive Luminescent MOF for Colorimetric Detection of Small Molecules in Selective and Reversible Manner. *Chem. Mater.* **2015**, *27*, 5349–5360. [CrossRef]
19. Khatua, S.; Biswas, P. Flexible Luminescent MOF: Trapping of Less Stable Conformation of Rotational Isomers, In Situ Guest-Responsive Turn-Off and Turn-On Luminescence and Mechanistic Study. *ACS Appl. Mater. Interfaces* **2020**, *12*, 22335–22346. [CrossRef]
20. Khatua, S.; Bar, A.K.; Konar, S. Tuning Proton Conductivity by Interstitial Guest Change in Size-Adjustable Nanopores of a Cu^I-MOF: A Potential Platform for Versatile Proton Carriers. *Chem. Eur. J.* **2016**, *22*, 16277–16285. [CrossRef]
21. Li, X.-Z.; Zhou, X.-P.; Li, D.; Yin, Y.-G. Controlling interpenetration in CuCN coordination polymers by size of the pendant substituents of terpyridine ligands. *CrystEngComm* **2011**, *13*, 6759–6765. [CrossRef]
22. Xi, Y.; Wei, W.; Xu, Y.; Huang, X.; Zhang, F.; Hu, C. Coordination Polymers Based on Substituted Terpyridine Ligands: Synthesis, Structural Diversity, and Highly Efficient and Selective Catalytic Oxidation of Benzylic C–H Bonds. *Cryst. Growth Des.* **2015**, *15*, 2695–2702. [CrossRef]
23. Liu, C.; Ding, Y.-B.; Shi, X.-H.; Zhang, D.; Hu, M.-H.; Yin, Y.-G.; Li, D. Interpenetrating Metal–Organic Frameworks Assembled from Polypyridine Ligands and Cyanocuprate Catenations. *Cryst. Growth Des.* **2009**, *9*, 1275–1277. [CrossRef]
24. Wang, J.; Yuan, F.; Hu, H.-M.; Bai, C.; Xue, G.-L. Nitro explosive and cation sensing by a luminescent 2D Cu(I) coordination polymer with multiple Lewis basic sites. *Inorg. Chem. Commun.* **2016**, *73*, 37–40. [CrossRef]
25. Klein, Y.M.; Lanzilotto, A.; Prescimone, A.; Krämer, K.W.; Decurtins, S.; Liu, S.-X.; Constable, E.C.; Housecroft, C.E. Coordination behaviour of 1-(3,2′:6′,3″-terpyridin-4′-yl)ferrocene: Structure and magnetic and electrochemical properties of a tetracopper dimetallomacrocycle. *Polyhedron* **2017**, *129*, 71–76. [CrossRef]
26. Yuan, F.; Xie, J.; Hu, H.-M.; Yuan, C.-M.; Xu, B.; Yang, M.-L.; Dong, F.-X.; Xue, G.-L. Effect of pH/metal ion on the structure of metal–organic frameworks based on novel bifunctionalized ligand 4′-carboxy-4,2′:6′,4″-terpyridine. *CrystEngComm* **2013**, *15*, 1460–1467. [CrossRef]
27. Gong, Y.; Zhang, M.M.; Zhang, P.; Shi, H.F.; Jiang, P.G.; Lin, J.H. Metal–organic frameworks based on 4-(4-carboxyphenyl)-2,2′,4,4′-terpyridine: Structures, topologies and electrocatalytic behaviors in sodium laurylsulfonate aqueous solution. *CrystEngComm* **2014**, *16*, 9882–9890. [CrossRef]
28. Yang, W.; Lin, X.; Jia, J.; Blake, A.J.; Wilson, C.; Hubberstey, P.; Champness, N.R.; Schröder, M. A biporous coordination framework with high H$_2$ storage density. *Chem. Commun.* **2008**, 359–361. [CrossRef]
29. Wu, Y.; Lu, L.; Feng, J.; Li, Y.; Sun, Y.; Ma, A. Design and construction of diverse structures of coordination polymers: Photocatalytic properties. *J. Solid State Chem.* **2017**, *245*, 213–218. [CrossRef]
30. Zhang, L.; Li, C.-J.; He, J.-E.; Chen, Y.-Y.; Zheng, S.-R.; Fan, J.; Zhang, W.-G. Construction of New Coordination Polymers from 4′-(2,4-disulfophenyl)- 3,2′:6′3″-terpyridine: Polymorphism, pH-dependent- syntheses, structures, and properties. *J. Solid State Chem.* **2016**, *233*, 444–454. [CrossRef]
31. Zhu, S.; Dai, X.-J.; Wang, X.-G.; Cao, Y.-Y.; Zhao, X.-J.; Yang, E.-C. Two Bulky Conjugated 4′-(4-Hydroxyphenyl)-4,2′:6′,4″-terpyridine-based Layered Complexes: Synthesis, Structure, and Photocatalytic Hydrogen Evolution Activity. *Z. Anorg. Allg. Chem.* **2019**, *645*, 516–522. [CrossRef]

32. Zuo, T.; Luo, D.; Huang, Y.-L.; Li, Y.Y.; Zhou, X.-P.; Li, D. Chiral 3D coordination polymers consisting of achiral terpyridyl precursors: From spontaneous resolution to enantioenriched induction. *Chem. Eur. J.* **2020**, *26*, 1936–1940. [CrossRef]
33. Lusi, M.; Fechine, P.B.A.; Chen, K.-J.; Perry, J.J.; Zaworotko, M.J. A rare cationic building block that generates a new type of polyhedral network with "cross-linked" pto topology. *Chem. Commun.* **2016**, *52*, 4160–4162. [CrossRef] [PubMed]
34. Groom, C.R.; Bruno, I.J.; Lightfoot, M.P.; Ward, S.C. The Cambridge Structural Database. *Acta Cryst.* **2016**, *B72*, 171–179. [CrossRef] [PubMed]
35. Bruno, I.J.; Cole, J.C.; Edgington, P.R.; Kessler, M.; Macrae, C.F.; McCabe, P.; Pearson, J.; Taylor, R. New software for searching the Cambridge Structural Database and visualizing crystal structures. *Acta Cryst.* **2002**, *B58*, 389–397. [CrossRef]
36. Toledo, D.; Vega, A.; Pizarro, N.; Baggio, R.; Pena, O.; Roisnel, T.; Pivan, J.-Y.; Moreno, Y. Comparitive study on structural, magnetic and spectroscopic properties of four new copper(II) coordination polymers with 4′-substituted terpyridine ligands. *J. Solid State Chem.* **2017**, *253*, 78–88. [CrossRef]
37. Toledo, D.; Ahumada, G.; Manzur, C.; Roisnel, T.; Pena, O.; Hamon, J.-R.; Pivan, J.-Y.; Moreno, Y. Unusual trinuclear complex of copper(II) containing a 4′-(3-methyl-2-thienyl)-4,2′:6′,4″-terpyridine ligand. Structural, spectroscopic, electrochemical and magnetic properties. *J. Mol. Struct.* **2017**, *1146*, 213–221. [CrossRef]
38. Lopez-Periago, A.; Vallcorba, O.; Frontera, C.; Domingo, C.; Ayllón, J.A. Exploring a novel preparation method of 1D metal organic frameworks based on supercritical CO_2. *Dalton Trans.* **2015**, *44*, 7548–7553. [CrossRef]
39. Delgado, S.; Barrilero, A.; Molina-Ontoria, A.; Medina, M.-E.; Pastor, C.J.; Jiménez-Aparicio, R.; Priego, J.L. Novel Coordination Polymers Generated from Angular 2,2′-Dipyridyl Ligands and Bis(hexafluoroacetylacetonate) Copper(II): Crystal Structures and Magnetic Properties. *Eur. J. Inorg. Chem.* **2006**, 2745–2759. [CrossRef]
40. Winter, S.; Weber, E.; Eriksson, L.; Csoregh, I. New coordination polymer networks based on copper(ii) hexafluoroacetylacetonate and pyridine containing building blocks: Synthesis and structural study. *New J. Chem.* **2006**, *30*, 1808–1819. [CrossRef]
41. Wang, J.; Hanan, G.S. A facile route to sterically hindered and non-hindered 4′-aryl-2,2′:6′,2″-terpyridines. *Synlett* **2005**, *2005*, 1251–1254. [CrossRef]
42. Smith, R.M.; Martell, A.E. *Critical Stability Constants*; Plenum Press: New York, NY, USA, 1975; Volume 2.
43. Yi, H.; Albrecht, M.; Pan, F.; Valkonen, A.; Rissanen, K. Stacking of Sterically Congested Trifluoromethylated Aromatics in their Crystals—The Role of Weak F···π or F···F Contacts. *Eur. J. Org. Chem.* **2020**, 6073–6077. [CrossRef]
44. Bondi, A. van der Waals volumes and radii. *J. Phys. Chem.* **1964**, *68*, 441–451. [CrossRef]
45. Rowland, R.S.; Taylor, R. Intermolecular Nonbonded Contact Distances in Organic Crystal Structures: Comparison with Distances Expected from van der Waals Radii. *J. Phys. Chem.* **1996**, *100*, 7384–7391. [CrossRef]
46. Spilfogel, T.S.; Titi, H.M.; Friščić, T. Database Investigation of Halogen Bonding and Halogen...Halogen Interactions between Porphyrins: Emergence of Robust Supamolecular Motifs and Frameworks. *Cryst. Growth Des.* **2021**, *21*, 1810–1832. [CrossRef]
47. Levina, E.O.; Chernyshov, I.Y.; Voronin, A.P.; Alekseiko, L.N.; Stash, A.I.; Vener, M.V. Solving the enigma of weak fluorine contacts in the solid state: A periodic DFT study of fluorinated organic crystals. *RSC Adv.* **2019**, *9*, 12520–12537. [CrossRef]
48. Reichenbächer, K.; Süss, H.I.; Hulliger, J. Fluorine in crystal engineering—"The little atom that could". *Chem. Soc. Rev.* **2005**, *34*, 22–30. [CrossRef]
49. Nakamoto, K. *Infrared and Raman Spectra of Inorganic Compounds, Part B*, 6th ed.; Wiley: Hoboken, NJ, USA, 2009; p. 99.
50. Palatinus, L.; Chapuis, G. Superflip—A Computer Program for the Solution of Crystal Structures by Charge Flipping in Arbitrary Dimensions. *J. Appl. Cryst.* **2007**, *40*, 786–790. [CrossRef]
51. Palatinus, L.; Prathapa, S.J.; Van Smaalen, S. EDMA: A Computer Program for Topological Analysis of Discrete Electron Densities. *J. Appl. Cryst.* **2012**, *45*, 575–580. [CrossRef]
52. Dolomanov, O.V.; Bourhis, L.J.; Gildea, R.J.; Howard, J.A.K.; Puschmann, H. Olex2: A Complete Structure Solution, Refinement and Analysis Program. *J. Appl. Cryst.* **2009**, *42*, 339–341. [CrossRef]
53. Sheldrick, G.M. Crystal Structure Refinement with ShelXL. *Acta Cryst.* **2015**, *C27*, 3–8. [CrossRef]
54. *Software for the Integration of CCD Detector System Bruker Analytical X-Ray Systems*; Bruker axs: Madison, WI, USA, 2001.
55. Sheldrick, G.M. ShelXT-Integrated space-group and crystal-structure determination. *Acta Cryst.* **2015**, *A71*, 3–8. [CrossRef] [PubMed]
56. Macrae, C.F.; Sovago, I.; Cottrell, S.J.; Galek, P.T.A.; McCabe, P.; Pidcock, E.; Platings, M.; Shields, G.P.; Stevens, J.S.; Towler, M.; et al. Mercury 4.0: From visualization to analysis, design and prediction. *J. Appl. Cryst.* **2020**, *53*, 226–235. [CrossRef] [PubMed]
57. LeBail, A.; Duroy, H.; Fourquet, J.L. Ab-initio structure determination of $LiSbWO_6$ by X-ray powder diffraction. *Mat. Res. Bull.* **1988**, *23*, 447–452. [CrossRef]
58. Pawley, G.S. Unit-cell refinement from powder diffraction scans. *J. Appl. Cryst.* **1981**, *14*, 357–361. [CrossRef]
59. Rodríguez-Carvajal, J. Recent Advances in Magnetic Structure Determination by Neutron Powder Diffraction. *Physica B* **1993**, *192*, 55–69. [CrossRef]
60. Roisnel, T.; Rodríguez-Carvajal, J. WinPLOTR: A Windows tool for powder diffraction patterns analysis Materials Science Forum. In Proceedings of the Seventh European Powder Diffraction Conference (EPDIC 7), Barcelona, Spain, 20–23 May 2000; pp. 118–123.

Article

Synthesis, Structure and Bonding in Pentagonal Bipyramidal Cluster Compounds Containing a *cyclo*-Sn$_5$ Ring, [(CO)$_3$MSn$_5$M(CO)$_3$]$^{4-}$ (M = Cr, Mo)

Sourav Mondal [1,†], Wei-Xing Chen [2,†], Zhong-Ming Sun [2,*] and John E. McGrady [1,*]

1. Department of Chemistry, University of Oxford, South Parks Road, Oxford OX1 3QZ, UK; sourav.mondal@chem.ox.ac.uk
2. State Key Laboratory of Elemento-Organic Chemistry, Tianjin Key Lab of Rare Earth Materials and Applications, School of Material Science and Engineering, Nankai University, Tianjin 300350, China; 1120200450@mail.nankai.edu.cn
* Correspondence: sunlab@nankai.edu.cn (Z.-M.S.); john.mcgrady@chem.ox.ac.uk (J.E.M.)
† These authors contributed equally to this work.

Abstract: In this paper, we report the synthesis and structural characterisation of two hetero-metallic clusters, [(CO)$_3$CrSn$_5$Cr(CO)$_3$]$^{4-}$ and [(CO)$_3$MoSn$_5$Mo(CO)$_3$]$^{4-}$, both of which have a pentagonal bipyramidal core. The structures are similar to that of previously reported [(CO)$_3$MoPb$_5$Mo(CO)$_3$]$^{4-}$ and our analysis of the bonding suggests that they are best formulated as containing Sn$_5^{4-}$ rings bridging two zerovalent M(CO)$_3$ fragments. The electronic structure is compared to two isolobal M$_2$E$_5$ clusters, [CpCrP$_5$CrCp]$^-$ and Tl$_7^{7-}$, both of which show clear evidence for *trans*-annular bonds between the apical atoms that is not immediately obvious in the title clusters. Our analysis shows that the balance between E-E and M-M bonding is a delicate one, and shifts in the relative energies of the orbitals on the E$_5$ and M$_2$ fragments generate a continuum of bonding situations linked by the degree of localisation of the cluster LUMO.

Keywords: Zintl clusters; X-ray crystallography; Density Functional Theory

Citation: Mondal, S.; Chen, W.-X.; Sun, Z.-M.; McGrady, J.E. Synthesis, Structure and Bonding in Pentagonal Bipyramidal Cluster Compounds Containing a *cyclo*-Sn$_5$ Ring, [(CO)$_3$MSn$_5$M(CO)$_3$]$^{4-}$ (M = Cr, Mo). *Inorganics* 2022, 10, 75. https://doi.org/10.3390/inorganics10060075

Academic Editor: Duncan H. Gregory

Received: 11 May 2022
Accepted: 26 May 2022
Published: 30 May 2022

Publisher's Note: MDPI stays neutral with regard to jurisdictional claims in published maps and institutional affiliations.

Copyright: © 2022 by the authors. Licensee MDPI, Basel, Switzerland. This article is an open access article distributed under the terms and conditions of the Creative Commons Attribution (CC BY) license (https://creativecommons.org/licenses/by/4.0/).

1. Introduction

The chemistry of *cyclo*-Pn$_5$ rings of the pnictogens is now well established, and many coordination compounds of P$_5^-$ and its heavier congeners, As$_5^-$ and Sb$_5^-$, have been reported in the literature [1–3]. These are of interest from a synthetic perspective, but also in the context of the isolobal relationship between Pn and CH [4,5] which can be used, for example, to rationalise the very similar structural chemistry of complexes of P$_5^-$ and C$_5$H$_5^-$ [6]. Rather less common are *cyclo*-Tt$_5$ rings containing the tetrel elements Si, Ge, Sn and Pb. An isoelectronic relationship to P$_5^-$ would require a formal charge of −6, and indeed Todorov and Sevov have reported both Sn$_5^{6-}$ and Pb$_5^{6-}$ in a family of Zintl phases, Na$_8$BaPb$_6$, Na$_8$BaSn$_6$ and Na$_8$EuSn$_6$, Figure 1, (where charge balance is maintained by the presence of an isolated Tt^{4-} ion in the unit cell in addition to a Tt$_5^{6-}$ ring) [7]. Lighter homologues such as Si$_5^{6-}$ and Ge$_5^{6-}$ have an even longer history, having been identified in binary phases such as Li$_{12}$Si$_7$ and Li$_{11}$Ge$_6$ as far back as the late 1970s [8–10]. The stability of the planar Tt$_5^{6-}$ rings can be rationalised either on the grounds of aromaticity (the rings have 6 π electrons), or in terms of Wade's rules (the clusters have a 2n + 6 = 16 skeletal electron count or, equivalently, a total valence electron count of 4n + 6 = 26, consistent with a 5-vertex *arachno* cluster based on a pentagonal bipyramid with two missing *trans* vertices) [11]. These two perspectives, both equally valid, highlight the important point that the bonding in ring and cluster compounds of the heavier main group elements is complex, and different interpretations are often possible.

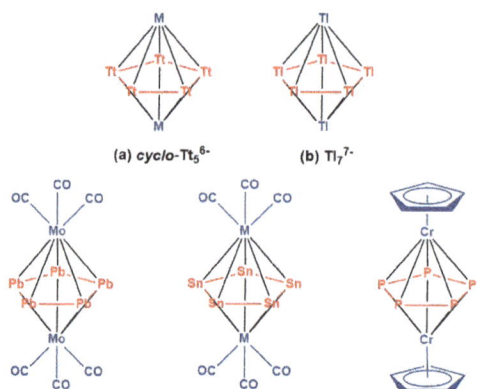

Figure 1. Examples of pentagonal bipyramidal cluster architectures from the transition metal series and main-group: (**a**) *cyclo*-Tt$_5^{6-}$ (Tt = Si, Sn, Pb; M = alkali, alkaline- or rare-earth metals); (**b**) Tl$_7^{7-}$ in the binary phase K$_{10}$Tl$_7$ [12]; (**c**) [(CO)$_3$MoPb$_5$Mo(CO)$_3$]$^{4-}$ [13], (**d**) [(CO)$_3$MSn$_5$M(CO)$_3$]$^{4-}$ (M = Cr, Mo), this work and (**e**) [CpCrP$_5$CrCp]$^{0/-}$ [14].

Soon after Sevov's report of the Sn$_5^{6-}$ and Pb$_5^{6-}$ rings [7], Fässler, Kaupp and co-workers described the synthesis of a molecular analogue, [(CO)$_3$MoPb$_5$Mo(CO)$_3$]$^{4-}$, which they formulated as a Pb$_5^{4-}$ ring bridging two neutral Mo(CO)$_3$ fragments [13]. The 4π electron count is now consistent with anti-aromatic, rather than aromatic, character, but the Pb-Pb bond lengths in this cluster (3.0138–3.0647 Å) are nevertheless marginally shorter than those in the aromatic (6π) Pb$_5^{6-}$ unit in Na$_8$BaPb$_6$ (3.047–3.117 Å). An analysis of the electronic structure reveals very substantial transfer of charge from the Pb$_5^{4-}$ ring to the Mo(CO)$_3$ fragments, such that the most intense C-O stretching frequency (1737 cm^{-1}) is substantially reduced from the value of 1960 cm^{-1} in the precursor, (mesitylene)Mo(CO)$_3$ [15], and this covalency clearly precludes a simple correlation between bond lengths and formal charge on the five-membered ring. Gholiee et al. have, more recently, published a detailed computational survey of the family of pentagonal bipyramidal clusters with general formula [(CO)$_3$ME$_5$M(CO)$_3$]$^{4-}$, M = Cr, Mo, W, E = Si, Ge, Sn and Pb [16]. Their analysis, using the M06-2X functional in combination with a def2-TZVPP basis, concurs with the conclusion from Fässler and Kaupp, that there is substantial charge transfer from the formally Tt$_5^{4-}$ ring to the M(CO)$_3$ fragments, such that the former actually carries a positive natural charge: clearly the formal charge is at best a first approximation, and the balance between M-M, M-Tt and Tt-Tt bonding is a delicate one.

In this paper, we report the synthesis and structural characterisation of two tin analogues of Fässler's lead cluster, [(CO)$_3$MSn$_5$M(CO)$_3$]$^{4-}$, M = Cr (**1**) and Mo (**2**), both of which were included in Gholiee's computational survey [16]. Both compounds are formed from the reaction of K$_4$Sn$_9$ with the organometallic precursors M(MeCN)$_3$(CO)$_3$ (M = Cr, Mo), and both feature a planar *cyclo*-Sn$_5$ motif capped by two M(CO)$_3$ fragments. This new structural data provides a platform to consider the general features of bonding in the family of pentagonal bipyramids. In the final section of this paper, we take the opportunity to compare these two new clusters to closely related species that share the pentagonal bipyramidal architecture, and to compare the different electron-counting models that have been applied in this context.

2. Materials and Methods

All manipulations and reactions were performed under a dry nitrogen atmosphere in the glove box. Ethylenediamine (en) (99%, Sigma-Aldrich, Beijing, China), and N, N-Dimethylformamide (Sigma-Aldrich, 99.8%) were freshly distilled and stored under nitrogen before use. Toluene (Sigma-Aldrich, 99.8%) was distilled from sodium/benzophenone under nitrogen and stored under nitrogen. The [2.2.2]crypt (4,7,13,16,21,24-Hexaoxa-1,10-diazabicyclo (8.8.8) hexacosane) (98%), and 18-Crown-6 (1,4,7,10,13,16-Hexaoxacyclooctadecane) (99%), both purchased from Sigma-Aldrich, were dried in vacuum for 12 h prior to use. K_4Sn_9 was synthesised by heating a mixture of K and Sn (K: 99%, Sn: 99.99%, both from Aladdin (Shanghai, China)) at 850 °C for 48 h in a tubular tantalum container and then cooling to room temperature at a rate of 10 °C/h. Finally, black powders were obtained. The $M(MeCN)_3(CO)_3$ (M = Cr, Mo) starting materials were synthesised according to literature procedures [17].

2.1. Synthesis of $K_2(en)_3[K([2.2.2]-crypt)]_2[Cr(CO)_3Sn_5Cr(CO)_3]\cdot 2en$, (1)

K_4Sn_9 (100 mg, 0.817 mmol) and [2.2.2]crypt (150 mg, 0.399 mmol) were dissolved in 3 mL en solution in a 10 mL vial, and the mixture was stirred vigorously for 30 min at room temperature. The organometallic precursor $Cr(MeCN)_3(CO)_3$ (30 mg 0.116 mmol) was then added and stirring continued for 3 h at room temperature. The resulting red-brown solution was centrifuged and filtered with standard glass wool, then carefully layered with 3 mL toluene. After 5 days, small dark-brown needle-like crystals of **1** (Figure S1, left) were observed in the bottom of the test tube in approximately 26% yield (based on precursor $Cr(MeCN)_3(CO)_3$ used).

2.2. Synthesis of $K[K(18-crown-6)]_3[Mo(CO)_3Sn_5Mo(CO)_3]$ (2)

K_4Sn_9 (100 mg, 0.817 mmol) and 18-Crown-6 (105 mg, 0.397 mmol) were dissolved in 3 mL en solution in a 10 mL vial, and the mixture was stirred vigorously for 30 min at room temperature. The organometallic precursor $Mo(MeCN)_3(CO)_3$ (35 mg 0.116 mmol) was then added and stirring continued for 4 h at room temperature. The resulting red-brown solution was centrifuged and filtered with standard glass wool, then carefully layered with 3 mL toluene. After 7 days, dark-brown needle-like crystals of **2** (Figure S1, right) were observed in the bottom of the test tube in approximately 19% yield (based on precursor $Mo(MeCN)_3(CO)_3$ used).

2.3. X-ray Diffraction

Suitable crystals of **1** and **2** were selected for X-ray diffraction analyses. Crystallographic data were collected on Rigaku XtalAB Pro MM007 DW diffractometer with graphite monochromated Cu Kα radiation (λ = 1.54184 Å). The structures of crystals **1** (Figures S2 and S3) and **2** (Figures S4 and S5) were solved using direct methods and then refined using SHELXL-2014 and OLEX2 [18,19]. All the non-hydrogen atoms were refined anisotropically, except for those in the split positions. The uncoordinated solvent molecules in **1** and **2** could not be modeled properly, so the solvent molecules are removed using SQUEEZE in PLATON [20]. The cluster anion in **1** has a disordered structure which was solved by the Split SAME process [19]. The unique bond angles and distances were constrained by Dfix order. We use the SIMU order to limit the anisotropic displacement parameters of the interconnected atoms to be similar. A summary of the crystallographic data for the title compounds is listed in Table S1, and selected bond distances and bond angles are given in Table S2. CCDC entries CCDC-2122111 and CCDC-2122112 for compounds **1** and **2** contain the supplementary crystallographic data for this paper. These data can be obtained free of charge via www.ccdc.cam.ac.uk/data_requested/cif (accessed on 15 November 2021).

2.4. Electrospray Ionisation Mass Spectrometry (ESI-MS)

Negative ion mode ESI-MS of a MeCN solution made up from crystals of **1** were measured on an LTQ linear ion trap spectrometer by Agilent Technologies ESI-TOF-MS (6230). The spray voltage was 5.48 kV and the capillary temperature was maintained at 300 °C. The capillary voltage was 30 V. The samples were prepared inside a glove box and very rapidly transferred to the spectrometer in an airtight syringe by direct infusion with a Harvard syringe pump at 0.2 mL/min.

2.5. Energy Dispersive X-ray (EDX) Analysis

EDX analysis was performed to support the elemental composition proposed in the XRD experiment. These were carried out using a scanning electron microscope (Hitachi S-4800) equipped with a Bruker AXS XFlash detector 4010. Data acquisition was performed with an acceleration voltage of 20 kV and an accumulation time of 150 s (Figures S8 and S9).

2.6. Computational Details

All DFT calculations were performed using the ORCA 5.0.1 software [21,22]. A range of functionals was considered, including the generalised gradient approximation (GGA) functional proposed by Perdew, Burke and Ernzerhof (PBE) [23], the hybrid B3LYP [24–26], the meta-hybrid M06-2X [27] and the double-hybrid, B2PLYP [28,29]. Scalar relativistic effects were included using the zeroth-order relativistic approximation (ZORA) [30–32]. A valence triple-zeta polarised relativistically reconstracted Karlsruhe basis set (ZORA-def2-TZVP) [33–35] was employed for H, C and O and segmented all-electron relativistically contracted (SARC) basis sets were used for heavier elements. The RI-J approximation to the Coulomb integrals (J) was made using the (SARC/J) [34–36] auxiliary basis set appropriate for ZORA calculations. The conductor-like polarisable continuum mode (CPCM) was implemented taking water as a solvent with dielectric constant (ϵ) 80.4 to model the confining potential of the cation lattice [37–40]. The multiwfn package was used to analyse the topology of the electron density and to compute bond orders [41]. Graphics containing molecular structures were generated using Chemcraft [42].

3. Results

3.1. Synthesis and Structural Characterisation

The synthesis of both **1** and **2** was achieved by reacting a transition metal carbonyl precursor, $M(MeCN)_3(CO)_3$, with K_4Sn_9 in the presence of a chelating ligand, 18-Crown-6 or [2.2.2]-crypt, that binds K^+ cations in the final product. In initial tests, both chelating ligands were used with both transition metals, but in the case of 18-Crown-6 with Cr or [2.2.2]-crypt with Mo, the major products were the known clusters $[M(CO)_3Sn_9]^{4-}$ [43], rather than the Sn_5-containing species. The influence of the chelating ligand most likely reflects subtle differences in the solubility of the two competing anions with cations of different size. X-ray diffraction analysis reveals that **1** crystallises in the triclinic space group P$\bar{1}$ and contains a $[(CO)_3CrSn_5Cr(CO)_3]^{4-}$ anion with two isolated K^+ ions and two $[K([2.2.2]\text{-crypt})]^+$ cations per cluster (Figures S2 and S3). The anionic clusters are arranged in approximately linear chains, with the K^+ ions bridging adjacent units via four K^+-O (CO) contacts of ~2.85 Å. The coordination sphere of the K^+ ions is completed by a single chelating en ligand. The K^+ ions chelated by $[K([2.2.2]\text{-crypt})]^+$ are separated from the centroid of the cluster by distances in excess of 7 Å. **2** (Figures S4 and S5) crystallises in the monoclinic space group $P2_1/c$ with three $[K(18\text{-Crown-6})]^+$ cations and one further, isolated, K^+ in the unit cell in addition to the $[(CO)_3MoSn_5Mo(CO)_3]^{4-}$ anion. The isolated K^+ cations form bridges between adjacent cluster units, in this case via three K^+-O (CO) contacts between 2.70 Å and 3.45 Å as well as three short K^+-Sn contacts to the equatorial rings. These short K^+-Sn contacts are responsible for the small distortion of the Sn_5 ring away from a perfect pentagon. The structures of the two anions are shown in Figure 2 and further details are give in Supporting Information, Figures S2–S5. The $[(CO)_3MSn_5M(CO)_3]^{4-}$ anions (M = Cr, **1**; Mo, **2**) are, like the Pb analogue [13], pentagonal bipyramids with a Sn_5 ring capped by two $M(CO)_3$ fragments at the vertices. The M−M distances along the principal axis are 2.948(9) Å and 3.1393(5) Å in **1** and **2**, respectively. Fässler and co-workers have emphasised the point that the Mo-Mo distance of 3.2156(8) Å in $[(CO)_3MPb_5M(CO)_3]^{4-}$ is longer than that in unambiguously Mo-Mo bonded species such as $[Mo_2(CO)_{10}]^{2-}$ (Mo-Mo = 3.123 Å) [44]. The corresponding Mo-Mo distance of 3.1393(5) Å in the Sn_5 analogue, **2**, is, however, almost identical to that in [Mo$_2$(CO)$_{10}$]$^{2-}$ while the Cr-Cr distance of 2.948(9) Å in **1** is shorter than that in [Cr$_2$(CO)$_{10}$]$^{2-}$, 2.970 Å [44]. The unambiguous identification of metal-metal bonds in systems with bridging ligands is notoriously difficult, as illustrated by the spirited debate over the existence or otherwise of an Fe-Fe bond in Fe$_2$(CO)$_9$ [45–47], but in this case, it seems clear, at least, that direct metal-metal bonding cannot be excluded a priori based on the structural data alone. The Sn$_5$ rings are almost perfectly planar, with the sums of the interior angles close to the ideal value of 540° (**1**: 539.850°; **2**: 539.928°). The Sn–Sn bond lengths in **1** (av. 2.878 Å) and **2** (av. 2.934 Å) (Figure 2) bracket the values in the Sn$_5^{6-}$ clusters: 2.883 Å in Na$_8$EuSn$_6$, 2.921 Å in Na$_8$BaSn$_6$, again highlighting the absence of a simple relationship between bond length and formal charge [7].

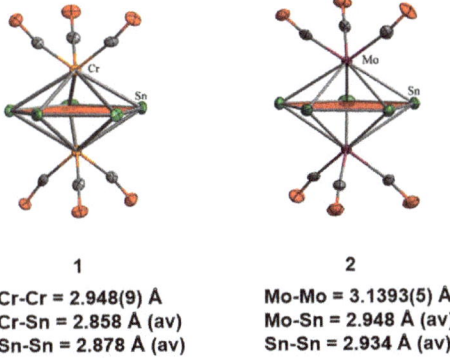

1
Cr-Cr = 2.948(9) Å
Cr-Sn = 2.858 Å (av)
Sn-Sn = 2.878 Å (av)

2
Mo-Mo = 3.1393(5) Å
Mo-Sn = 2.948 Å (av)
Sn-Sn = 2.934 Å (av)

Figure 2. Molecular structure of the anions [(CO)$_3$CrSn$_5$Cr(CO)$_3$]$^{4-}$ and [(CO)$_3$MoSn$_5$Mo(CO)$_3$]$^{4-}$ in compounds **1** and **2**, respectively. Thermal ellipsoids are drawn at the 50% probability level.

3.2. ESI-MS

The ESI-MS of an acetonitrile solution of **1** shown in Figure 3 shows peaks at m/z = 865.3735 and 1281.5919, identified as [Cr$_2$Sn$_5$(CO)$_6$]$^-$ and [HK(2.2.2-crypt)Cr$_2$Sn$_5$(CO)$_6$]$^-$, respectively. No other small fragment peaks were found, indicating that compound **1** is relatively stable in acetonitrile solution under mass spectrometry conditions, albeit with a much reduced overall charge compared to the anion present in the crystal. Very similar features are apparent in the spectrum of **2** (Figures S6 and S7), where a prominent peak assigned to [K(18−Crown−6)Mo$_2$Sn$_5$(CO)$_6$]$^-$ is observed at m/z~1256.

3.3. Electronic Structure Analysis

Optimised structural parameters for the two Sn$_5$ clusters [(CO)$_3$MSn$_5$M(CO)$_3$]$^{4-}$, M = Cr, Mo, are compared in Table 1, along with corresponding values for Fässler's Pb$_5$ cluster [13], using a range of exchange-correlation functionals. The gradient corrected functional (PBE) give somewhat shorter M-M bonds than their hybrid counterparts (B3LYP) [26], which is a relatively common observation in systems with weak metal–metal bonds [48]. In contrast, the double hybrid function B2PLYP [28,29] predicts M-M bond lengths that are shorter than any of the other functionals, and indeed even shorter than the experimental values. Despite the functional dependence of the absolute bond lengths, the calculations replicate the important trends in the crystallographic data, irrespective of functional. Within the pair of Sn$_5$ clusters, the optimised Cr-Cr bond length is ~0.2 Å shorter than the Mo-Mo bond and the Sn-Sn bonds in the Sn$_5$ ring are ~0.05 Å shorter in the Cr complex than its Mo counterpart. Likewise, the elongation of the Mo-Mo distance in [Mo(CO)$_3$Pb$_5$Mo(CO)$_3$]$^{4-}$ vs. [(CO)$_3$MoSn$_5$Mo(CO)$_3$]$^{4-}$ seen in the X-ray data is reproduced across all functionals.

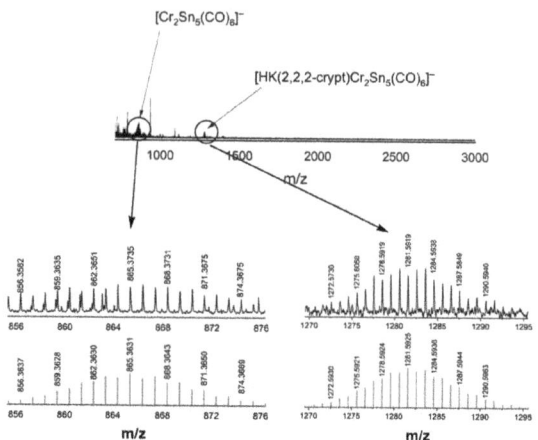

Figure 3. Negative-ion mode ESI-MS spectra of a freshly prepared solution of **1** in MeCN. The region of the spectrum corresponding to $[Cr_2Sn_5(CO)_6]^-$ is expanded (Top: measured spectrum. Bottom: calculated spectrum).

Table 1. Optimised bond lengths for $[(CO)_3MSn_5M(CO)_3]^{4-}$, M = Cr, Mo and $[(CO)_3MoPb_5Mo(CO)_3]^{4-}$ for PBE, B3LYP, M062X and B2PLYP functionals.

		M-M	M-Tt	Tt-Tt
$[(CO)_3CrSn_5Cr(CO)_3]^{4-}$	X-ray	2.95	2.86	2.88
	PBE	2.97	2.88	2.91
	B3LYP	3.07	2.91	2.91
	M062X	2.97	2.86	2.88
	B2PLYP	2.85	2.87	2.93
$[(CO)_3MoSn_5Mo(CO)_3]^{4-}$	X-ray	3.14	2.95	2.93
	PBE	3.19	2.99	2.98
	B3LYP	3.27	3.02	2.98
	M062X	3.20	2.97	2.95
	B2PLYP	3.15	2.98	2.97
$[(CO)_3MoPb_5Mo(CO)_3]^{4-}$	X-ray	3.21	3.05	3.04
	PBE	3.22	3.08	3.09
	B3LYP	3.31	3.11	3.10
	B2PLYP	3.17	3.06	3.09

A schematic molecular orbital diagram showing the interaction between a Sn_5^{4-} fragment and a $(CO)_3M-M(CO)_3$ unit is summarised in Figure 4. Quantitative versions for both clusters are shown in the Supporting Information, Figure S10. Note that the orbitals of the Sn_5^{4-} ring are labelled according to D_{5h} point symmetry while those of $M(CO)_3$ and $(CO)_3M-M(CO)_3$ are labelled according to C_{3v} and D_{3h}, respectively. The complete cluster has only C_s point symmetry. The bonding interaction between the Sn_5^{4-} ring and the transition metal center is dominated by charge transfer from the highest occupied orbitals on Sn_5^{4-} with Sn-Sn σ (e_1') and Sn-Sn π (e_2'') character, which overlap with in- and out-of-phase combinations of the degenerate LUMO of $M(CO)_3$ ($2e'$ and $2e''$, respectively) to generate near-degenerate pairs with a' and a'' symmetry (($5a'$,$3a''$), ($6a'$,$4a''$). The air- and moisture-sensitivity of both **1** and **2** has precluded the measurement of infra-red spectra for either, but our computed C-O stretching frequencies (1712 cm^{-1} and 1744 cm^{-1} for **1** and **2**, respectively) indicate a similar red-shift to that observed in the Pb_5 analogue (measured, 1737 cm^{-1} [13], calculated 1744 cm^{-1}). In addition, there is a further significant interaction between the out-of-phase combination of d_{z^2} orbitals and the Sn_5 π a_2'' orbital, which destabilises the antibonding combination to the extent that it constitutes the LUMO of the complex (8a'). This LUMO, an iso-surface plot which is shown in Figure 4, is clearly localised primarily on the Sn$_5$ ring, with relatively minor contributions from the out-of-phase combination of Cr d_{z^2} orbitals.

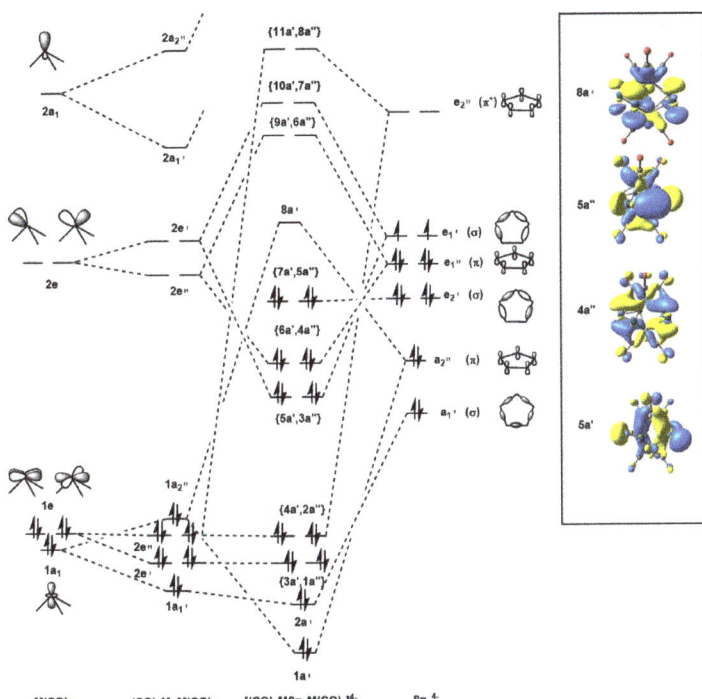

Figure 4. Schematic molecular orbital diagram for [(CO)$_3$MSn$_5$M(CO)$_3$]$^{4-}$. Note that the orbitals of the M(CO)$_3$, M$_2$(CO)$_6$ and Sn$_5$ fragments are labelled according to C_{3v}, D_{3h} and D_{5h} point symmetry, respectively. The cluster itself has only C_S point symmetry.

4. Discussion

Electron-Counting and Metal–Metal Bonding in Pentagonal Bipyramidal Clusters

In order to place these new Sn$_5$ clusters into an appropriate context, it is useful to compare them to other 7-vertex cluster compounds with the same or similar formal electron counts. A useful point of comparison is with the 7-vertex main-group clusters Tl$_4$Bi$_3^{3-}$ and Tl$_7^{7-}$, the former synthesised recently by Dehnen [49] and co-workers and the latter by Corbett and co-workers in 2000 [12]. The [Tl$_4$Bi$_3$]$^{3-}$ cluster has a total valence-electron count of 4n + 2 = 30, a skeletal-electron count of 2n + 2 = 16, and is a classically *closo* cluster with a distance of 4.086(3) Å between the two apical Tl atoms. Corbett's Tl$_7^{7-}$ cluster, in contrast, has two electrons fewer (4n = 28) and is compressed along the principal axis, generating a short *trans*-annular Tl-Tl bond of 3.4622(9) Å. Its a_2''-symmetric LUMO (the counterpart of which is occupied in [Tl$_4$Bi$_3$]$^{3-}$), shown in Figure 5, has Tl-Tl π^* character around the equator, just like the LUMOs of [(CO)$_3$MSn$_5$M(CO)$_3$]$^{4-}$, but also substantial Tl 6s character on the apical Tl atoms, and it is this feature that gives rise to the direct Tl-Tl bond [12,50].

How can we reconcile the well-established Wade's rules electron-counting model used in these main-group clusters with the electronic structure of $[(CO)_3MSn_5M(CO)_3]^{4-}$? Fässler and Kaupp formulated the skeletal electron count as 2n = 14, treating the Mo(CO)$_3$ fragment as a zero-electron donor. The assumption here is that the occupied 3d orbitals, including the $3d_{z^2}$ orbitals aligned along the principal axis, are not involved in the cluster bonding. The linear combinations of $3d_{z^2}$ are, however, the direct analogues of the 6s radial lone pairs in $[Tl_4Bi_3]^{3-}$, and if we do include these we reach a total-electron count of 28 (=4n), highlighting an isolobal analogy to Tl_7^{7-}. So why, then, do we consider a *trans*-annular Tl-Tl bond to be present in the main-group cluster, but not in $[(CO)_3MSn_5Cr(CO)_3]^{4-}$? The resolution to this question lies in the nature of the LUMO, and in particular its distribution over the five atoms of the equatorial ring and the two along the principal axis. To understand the significance of this orbital, we note that the hypothetical 30-electron (*closo*) clusters, $[(CO)_3MSn_5Cr(CO)_3]^{6-}$, can be formulated without ambiguity as a Sn_5^{6-} ring bridging two $M^0(CO)_3$ fragments, with no *trans*-annular bond. If the LUMO of the 28-electron cluster is an out-of-phase combination of orbitals on the apical atoms, a *trans*-annular bond is fully developed while the equatorial five-membered ring retains the full complement of six π electrons. At the opposite extreme, if the LUMO is localised entirely in the equatorial plane, the electron deficiency is accommodated in an anti-aromatic 4π ring while both in- and out-of-phase combinations of $3d_{z^2}$ are occupied, annihilating the *trans*-annular bond. The LUMOs shown in Figure 5 suggest that the amplitude of the LUMO is indeed shifted towards the apical atoms in Tl_7^{7-} compared to $[(CO)_3CrSn_5Cr(CO)_3]^{4-}$, justifying the rather different descriptions of bonding applied in the two cases. An even more extreme example of a *trans*-annular bond can be found in the P$_5$-bridged clusters, CpCrP$_5$CrCp and Cp*CrP$_5$CrCp* and their one-electron reduced analogues (Figure 1e) [14,51]. The neutral Cp cluster has a Cr-Cr separation is 2.69(1) Å, considerably shorter than that in **1**, and Hoffmann's analysis of the bonding [52] concludes that there is indeed Cr-Cr bonding present. The one-electron reduced analogue [CpCrP$_5$CrCp]$^-$ noted in Scherer's original report of the synthesis of the Cp* compound [14] is particularly interesting in the present context because its total valence electron count of 28 establishes a further isolobal relationship to both Tl_7^{7-} and $[(CO)_3CrSn_5Cr(CO)_3]^{4-}$. The optimised structure of [CpCrP$_5$CrCp]$^-$ is summarised in Table 2, with full cartesian coordinates given in the Supplementary Materials: the Cr-Cr bond length of 2.61 Å is indicative of an even stronger Cr-Cr interaction than in the neutral species. The 28 available electrons in [CpCrP$_5$CrCp]$^-$ can either be partitioned to give a 6π-electron P$_5^-$ ring and two d^5 CpCr fragments (and hence a Cr-Cr bond) or, alternatively, as a 4π-electron P$_5^+$ ring with two d^6 CpCr$^-$ fragments (and hence no Cr-Cr bond). The LUMO, shown in Figure 5c, is clearly strongly localised on the apical Cr atoms with relatively minor contributions on the equatorial P$_5$ ring, confirming that the electron-deficiency is accommodated on the apical atoms.

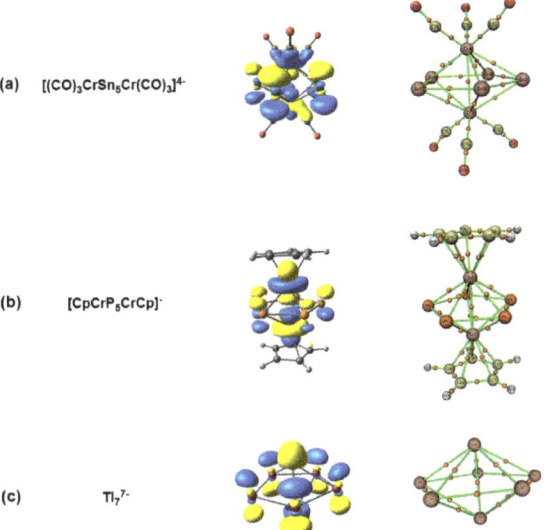

Figure 5. Contour plots of the LUMOs of (a) $[(CO)_3CrSn_5Cr(CO)_3]^{4-}$, (b) $[CpCrP_5CrCp]^-$ and (c) Tl_7^{7-} and the corresponding molecular graphs. Bond Critical Points (BCPs) are shown in the molecular graphs as orange dots.

An important point to take from this analysis is that there is no black-and-white distinction between the electronic structure models for $[(CO)_3CrSn_5Cr(CO)_3]^{4-}$, $[CpCrP_5CrCp]^-$ and Tl_7^{7-}. All three share a common skeletal-electron count of 28, two fewer that than the *closo* count of 30, and they constitute a continuum of situations defined by the shape of the LUMO, and hence by the way that the electron deficiency (relative to the *closo* form) is accommodated. In terms of the molecular orbital diagram shown in Figure 4, the limiting cases defined by $[(CO)_3CrSn_5Cr(CO)_3]^{4-}$ and $[CpCrP_5CrCp]^-$ are connected by a shift in the relative energies of the metal- and E_5-based fragments. A downward shift of the orbitals on the E_5 ring will increase the metal d_{z^2} character in the LUMO, increasing the importance of the *trans*-annular bonded resonance form while an upward shift increases the contribution of the anti-aromatic 4π configuration. The Quantum Theory of Atoms in Molecules (QTAIM) offers an alternative perspective on the nature of the bonding in these systems, although we note that the separation of metal–metal and metal–ligand bonding remains far from simple [53]. The delocalisation index, δ, offers a direct measure of Cr-Cr bond strength, and the values of 0.40 and 0.93 for $[(CO)_3CrSn_5Cr(CO)_3]^{4-}$ and $[CpCrP_5CrCp]^-$, respectively, correlate directly with the computed values for the Mayer bond order (0.39 and 0.72) extracted from the wavefunction itself (Table 2). A survey of the topology of the electron density identifies bond critical points (BCPs) midway between the Cr centres for both $[(CO)_3CrSn_5Cr(CO)_3]^{4-}$ and $[CpCrP_5CrCp]^-$ (see molecular graphs in Figure 5), but the electron density at the BCP (ρ_{BCP}) is considerably larger for the latter (0.055) than the former (0.028). Following Macchi's classification of the local indicators of bond type [54], the balance between kinetic (G_b) and potential (V_b) energies is consistent with an open-shell-type interaction in both cases, although the local dominance of V_b over G_b in $[CpCrP_5CrCp]^-$ is indicative of enhanced covalent Cr-Cr character. All of the indicators, structural and electronic, are therefore consistent with the conclusion that Cr-Cr bonding is considerably stronger in $[CpCrP_5CrCp]^-$ than it is in $[(CO)_3CrSn_5Cr(CO)_3]^{4-}$. The Mayer bond order, delocalisation index and properties of the *trans*-annular BCP for the Tl_7^{7-} cluster suggest that it sits closer to $[(CO)_3CrSn_5Cr(CO)_3]^{4-}$ than to $[CpCrP_5CrCp]^-$ on this spectrum of bond types.

Table 2. Comparison of Mayer bond orders (BO) and QTAIM parameters for the *trans*-Annular M-M Bond Critical Points (BCPs) for 28-electron M_2E_5 clusters, $[(CO)_3CrSn_5Cr(CO)_3]^{4-}$, $[CpCrP_5CrCp]^-$, and Tl_7^{7-}.

	$[(CO)_3CrSn_5Cr(CO)_3]^{4-}$	$[CpCrP_5CrCp]^-$	Tl_7^{7-}
$r_{M-M}/\text{Å}$	2.97	2.62	3.32
BO_{M-M}	0.39	0.72	0.31
δ_{M-M}	0.40	0.93	0.52
ρ_{BCP}/au	0.028	0.055	0.020
G_b/au	0.010	0.026	0.009
V_b/au	−0.015	−0.045	−0.010

5. Conclusions

In this paper we have reported the synthesis and structural characterisation of two new cluster compounds, $[(CO)_3MSn_5M(CO)_3]^{4-}$, M = Cr, Mo, which feature a pair of zerovalent $M(CO)_3$ fragments bridged by a *cyclo*-Sn_5 unit. The Sn clusters are isostructural with the Pb analogue, $[(CO)_3MoPb_5Mo(CO)_3]^{4-}$, reported previously by Fässler and co-workers. The LUMOs of both clusters have dominant Sn-Sn π character, consistent with their formulation as Sn_5^{4-} rings coordinated to two zero-valent $M(CO)_3$ fragments. The electronic structure makes a striking contrast with the isoelectronic $[CpCrP_5CrCp]^-$ anion, where the relative energies of the orbitals on the E_5 and M_2 fragments are reversed, causing the LUMO to have dominant Cr-Cr σ^* rather than Sn-Sn π character. The Tl_7^{7-} cluster is isolobal with both clusters, and the axial compression in this *hypo*-electronic cluster is strikingly reminiscent of the short Cr-Cr bond in $[CpCrP_5CrCp]^-$. These three 28-electron clusters therefore mark three distinct points on a continuum between two limits where (i) the LUMO is localised in the equatorial E_5 ring (as in $[(CO)_3MSn_5M(CO)_3]^{4-}$), and (ii) the LUMO is localised on the M_2 caps, forming a *trans*-annular bond (as in $[CpCrP_5CrCp]^-$). This continuum perspective establishes a link between apparently quite different electron-counting models.

Supplementary Materials: The following Supplementary Materials can be downloaded at: https://www.mdpi.com/article/10.3390/inorganics10060075/s1, Details of the crystallographic analysis (Figures S1–S5 and Tables S1 and S2), the ESI for **2** (Figures S6 and S7) and EDX (Figures S8 and S9) experiments. Cartesian coordinates of the optimized structures are given in Supporting Information, and quantitative MO diagrams for **1** and **2** are presented in Figure S10.

Author Contributions: Synthesis and characterization, W.-X.C. and Z.-M.S.; electronic structure analysis, S.M. and J.E.M. All authors have read and agreed to the published version of the manuscript.

Funding: This work was supported by the National Natural Science Foundation of China (92161102 and 21971118) and the Natural Science Foundation of Tianjin City (No. 20JCYBJC01560 and B2021202077). Z.M.S. thanks the 111 project (B18030) from China.

Data Availability Statement: Crystallographic data are available free of charge via www.ccdc.cam.ac.uk/data_requested/cif, entries CCDC-2122111 and CCDC-2122112 for compounds **1** and **2**, respectively (accessed on 15 November 2021).

Acknowledgments: S.M. acknowledges the Indian Government for a scholarship (National Overseas Scholarship).

Conflicts of Interest: The authors declare no conflict of interest.

Abbreviations

The following abbreviations are used in this manuscript:

DFT   Density Functional Theory
QTAIM   Quantum Theory of Atoms in Molecules

References

1. Scherer, O.J. Complexes with Substituent-free Acyclic and Cyclic Phosphorus, Arsenic, Antimony, and Bismuth Ligands. *Angew. Chem. Int. Ed. Engl.* **1990**, *29*, 1104–1122. [CrossRef]
2. Turbervill, R.S.P.; Goicoechea, J.M. From Clusters to Unorthodox Pnictogen Sources: Solution-Phase Reactivity of $[E_7]^{3-}$ (E = P–Sb) Anions. *Chem. Rev.* **2014**, *114*, 10807–10828. [CrossRef]
3. Zhang, M.; Wang, W.; Sun, Z.; Meng, L.; Li, X. Construction of $Pn_{10}M$ Sandwich Compounds from Pn_5^- and Pn_5M (Pn = N-Bi; M = Li, Na, K, Be, Mg, Ca, Fe, Co and Ni): A Theoretical Assessment. *Comp. Theor. Chem.* **2016**, *1098*, 50–55. [CrossRef]
4. Dillon, K.B.; Mathey, F.; Nixon, J.F. *Phosphorus: The Carbon Copy*, 1st ed.; Wiley: New York, NY, USA, 1998.
5. Simpson, M.C.; Protasiewicz, J.D. Phosphorus as a Carbon Copy and as a Photocopy: New Conjugated Materials Featuring Multiply Bonded Phosphorus. *Pure App. Chem.* **2013**, *85*, 801–815. [CrossRef]
6. Heinl, C.; Peresypkina, E.; Balázs, G.; Mädl, E.; Virovets, A.V.; Scheer, M. The Missing Parent Compound $[(C_5H_5)Fe(\eta^5-P_5)]$: Synthesis, Characterization, Coordination Behavior and Encapsulation. *Chem. Eur. J.* **2021**, *27*, 7542–7548. [CrossRef] [PubMed]
7. Todorov, I.; Sevov, S.C. Heavy-Metal Aromatic Rings: Cyclopentadienyl Anion Analogues Sn_5^{6-} and Pb_5^{6-} in the Zintl Phases Na_8BaPb_6, Na_8BaSn_6, and Na_8EuSn_6. *Inorg. Chem.* **2004**, *43*, 6490–6494. [CrossRef] [PubMed]
8. von Schnering, H.G.; Nesper, R.; Curda, J.; Tebbe, K.F. $Li_{12}Si_7$, a Compound Having a Trigonal Planar Si_4 Cluster and Planar Si_5 Rings. *Angew. Chem. Int. Ed. Engl.* **1980**, *19*, 1033–1034. [CrossRef]
9. Nesper, R.; Curda, J.; Von Schnering, H. Li_8MgSi_6, a Novel Zintl Compound Containing Quasi-Aromatic Si_5 rings. *J. Solid State Chem.* **1986**, *62*, 199–206. [CrossRef]
10. Frank, U.; Müller, W. $Li_{11}Ge_6$—A Phase with Isolated, Plane, Five-Membered Ge-Rings. *Z. Naturf. B* **1975**, *30*, 313–315. [CrossRef]
11. Wade, K. Structural Significance of Number of Skeletal Bonding Electron-Pairs in Carboranes, Higher Boranes and Borane Anions, and Various Transition-Metal Carbonyl Cluster Compounds. *J. Chem. Soc. D Chem. Comm.* **1971**, *15*, 792–793. [CrossRef]
12. Kaskel, S.; Corbett, J.D. Synthesis and Structure of $K_{10}Tl_7$: The First Binary Trielide Containing Naked Pentagonal Bipyramidal Tl_7 Clusters. *Inorg. Chem.* **2000**, *39*, 778–782. [CrossRef]
13. Yong, L.; Hoffmann, S.D.; Fässler, T.F.; Riedel, S.; Kaupp, M. $[Pb_5\{Mo(CO)_3\}_2]^{4-}$: A complex containing a planar Pb_5 unit. *Ang. Chem. Int. Ed. Engl.* **2005**, *44*, 2092–2096. [CrossRef]
14. Scherer, O.J.; Schwalb, J.; Wolmershäuser, G.; Kaim, W.; Gross, R. $cyclo-P_5$ as Complex Ligand—The Phosphorus Analogue of the Cyclopentadienyl Ligand. *Angew. Chem. Int. Ed. Engl.* **1986**, *25*, 363–364. [CrossRef]
15. Armstrong, R.S.; Aroney, M.J.; Barnes, C.M.; Nugent, K.W. Infrared and Raman Spectra of $(\eta^6-mesitylene)M(CO)_3$ Complexes (M = Cr, Mo or W): An Insight Into Metal-Arene Bonding. *App. Organomet. Chem.* **1990**, *4*, 569–580. [CrossRef]
16. Gholiee, Y.; Salehzadeh, S.; Khodaveisi, S. Significant Geometry and Charge Difference Between the E_5^{4-} Bare Clusters of Group 14 Zintl anions and Their Coordinated Form in $[E_5\{M(CO)_3\}_2]^{4-}$ (E = Si, Ge, Sn, Pb; M = Cr, Mo, W) complexes. *New J. Chem.* **2019**, *43*, 7797–7805. [CrossRef]
17. Tate, D.P.; Knipple, W.R.; Augl, J.M. Nitrile Derivatives of Chromium Group Metal Carbonyls. *Inorg. Chem.* **1962**, *1*, 433–434. [CrossRef]
18. Sheldrick, G.M. *SHELXT*—Integrated Space-Group and Crystal-Structure Determination. *Acta Cryst. Section A* **2015**, *71*, 3–8. [CrossRef]
19. Dolomanov, O.V.; Bourhis, L.J.; Gildea, R.J.; Howard, J.A.K.; Puschmann, H. *OLEX2*: A Complete Structure Solution, Refinement and Analysis Program. *J. App. Cryst.* **2009**, *42*, 339–341. [CrossRef]
20. Spek, A.L. *PLATON SQUEEZE*: A Tool for the Calculation of the Disordered Solvent Contribution to the Calculated Structure Factors. *Acta Cryst. Section C* **2015**, *71*, 9–18. [CrossRef]
21. Neese, F. The ORCA Program System. *Wiley Interdiscip. Rev. Comput. Mol. Sci.* **2012**, *2*, 73–78. [CrossRef]
22. Neese, F. Software Update: The ORCA Program System, Version 4.0. *Wiley Interdiscip. Rev. Comput. Mol. Sci.* **2018**, *8*, e1327. [CrossRef]
23. Perdew, J.P.; Burke, K.; Ernzerhof, M. Generalized Gradient Approximation Made Simple. *Phys. Rev. Lett.* **1996**, *77*, 3865–3868. [CrossRef] [PubMed]
24. Becke, A.D. Density-Functional Thermochemistry. III. The Role of Exact Exchange. *J. Chem. Phys.* **1993**, *98*, 5648–5652. [CrossRef]
25. Lee, C.; Yang, W.; Parr, R.G. Development of the Colle-Salvetti Correlation-Energy Formula into a Functional of the Electron Density. *Phys. Rev. B* **1988**, *37*, 785. [CrossRef]
26. Hertwig, R.H.; Koch, W. On the Parameterization of the Local Correlation Functional. What is Becke-3-LYP? *Chem. Phys. Lett.* **1997**, *268*, 345–351. [CrossRef]
27. Zhao, Y.; Truhlar, D.G. The M06 Suite of Density Functionals for Main Group Thermochemistry, Thermochemical Kinetics, Noncovalent Interactions, Excited States, and Transition Elements: Two New Functionals and Systematic Testing of Four M06-Class Functionals and 12 Other Functionals. *Theor. Chem. Acc.* **2008**, *120*, 215–241.
28. Grimme, S. Semiempirical Hybrid Density Functional with Perturbative Second-Order Correlation. *J. Chem. Phys.* **2006**, *124*, 034108. [CrossRef]
29. Neese, F.; Schwabe, T.; Grimme, S. Analytic Derivatives for Perturbatively Corrected "Double Hybrid" Density Functionals: Theory, Implementation, and Applications. *J. Chem. Phys.* **2007**, *126*, 124115. [CrossRef]
30. van Lenthe, E.; Baerends, E.J.; Snijders, J.G. Relativistic Regular Two-Component Hamiltonians. *J. Chem. Phys.* **1993**, *99*, 4597–4610. [CrossRef]

31. van Lenthe, E.; Baerends, E.J.; Snijders, J.G. Relativistic Total Energy Using Regular Approximations. *J. Chem. Phys.* **1994**, *101*, 9783–9792. [CrossRef]
32. van Lenthe, E.; Ehlers, A.; Baerends, E.J. Geometry Optimizations in the Zero Order Regular Approximation for Relativistic Effects. *J. Chem. Phys.* **1999**, *110*, 8943–8953. [CrossRef]
33. Weigend, F.; Ahlrichs, R. Balanced Basis Sets of Split Valence, Triple Zeta Valence and Quadruple Zeta Valence Quality for H to Rn: Design and Assessment of Accuracy. *Phys. Chem. Chem. Phys.* **2005**, *7*, 3297–3305. [CrossRef]
34. Rolfes, J.D.; Neese, F.; Pantazis, D.A. All-Electron Scalar Relativistic Basis Sets for the Elements Rb–Xe. *J. Comp. Chem.* **2020**, *41*, 1842–1849. [CrossRef]
35. Pantazis, D.A.; Neese, F. All-Electron Scalar Relativistic Basis Sets for the 6p Elements. *Theor. Chem. Acc.* **2012**, *131*, 1292. [CrossRef]
36. Weigend, F. Accurate Coulomb-fitting Basis Sets for H to Rn. *Phys. Chem. Chem. Phys.* **2006**, *8*, 1057–1065. [CrossRef]
37. Klamt, A.; Schüürmann, G. COSMO: A New Approach to Dielectric Screening in Solvents with Explicit Expressions for the Screening Energy and its Gradient. *J. Chem. Soc. Perkin Trans.* **1993**, *2*, 799–805. [CrossRef]
38. Andzelm, J.; Kölmel, C.; Klamt, A. Incorporation of Solvent Effects into Density Functional Calculations of Molecular Energies and Geometries. *J. Chem. Phys.* **1995**, *103*, 9312–9320. [CrossRef]
39. Barone, V.; Cossi, M. Quantum Calculation of Molecular Energies and Energy Gradients in Solution by a Conductor Solvent Model. *J. Phys. Chem. A* **1998**, *102*, 1995–2001. [CrossRef]
40. Cossi, M.; Rega, N.; Scalmani, G.; Barone, V. Energies, Structures, and Electronic Properties of Molecules in Solution with the C-PCM Solvation Model. *J. Comp. Chem.* **2003**, *24*, 669–681. [CrossRef]
41. Lu, T.; Chen, F. Multiwfn: A Multifunctional Wavefunction Analyzer. *J. Comp. Chem.* **2012**, *33*, 580–592. [CrossRef]
42. Zhurko G. A. Chemcraft—Graphical Program for Visualization of Quantum Chemistry Computations. Ivanovo, Russia, 2005. Available online: https://chemcraftprog.com (accessed on 15 November 2021).
43. Kesanli, B.; Fettinger, J.; Eichhorn, B. The *closo*-$[Sn_9M(CO)_3]^{4-}$ Zintl Ion Clusters where M = Cr, Mo, W: Two Structural Isomers and Their Dynamic Behavior. *Chem. Eur. J.* **2001**, *7*, 5277–5285. [CrossRef]
44. Handy, L.B.; Ruff, J.K.; Dahl, L.F. Structural Characterization of the Dinuclear Metal Carbonyl Anions $[M_2(CO)_{10}]^{2-}$ (M= chromium, molybdenum) and $[Cr_2(CO)_{10}H]^-$. Marked Stereochemical Effect of a Linearly Protonated Metal-Metal Bond. *J. Am. Chem. Soc.* **1970**, *92*, 7312–7326. [CrossRef]
45. Green, J.C.; Green, M.L.H.; Parkin, G. The Occurrence and Representation of Three-Centre Two-Electron Bonds in Covalent Inorganic Compounds. *Chem. Commun.* **2012**, *48*, 11481–11503. [CrossRef]
46. Labinger, J.A. Does Cyclopentadienyl Iron Dicarbonyl Dimer Have a Metal–Metal bond? Who's Asking? *Inorg. Chim. Acta* **2015**, *424*, 14–19. [CrossRef]
47. McGrady, J.E. *Molecular Metal-Metal Bonds: Compounds, Synthesis, Properties*; Wiley: New York, NY, USA, 1999.
48. Spivak, M.; Arcisauskaite, V.; López, X.; McGrady, J.E.; de Graaf, C. A Multi-Configurational Approach to the Electronic Structure of Trichromium Extended Metal Atom Chains. *Dalton Trans.* **2017**, *46*, 6202–6211. [CrossRef]
49. Lichtenberger, N.; Franzke, Y.J.; Massa, W.; Weigend, F.; Dehnen, S. The Identity of "Ternary" A/Tl/Pb or K/Tl/Bi Solid Mixtures and Binary Zintl Anions Isolated From Their Solutions. *Chem. Eur. J.* **2018**, *24*, 12022–12030. [CrossRef]
50. McGrady, J.E. A Unified Approach to Electron Counting in Main-Group Clusters. *J. Chem. Ed.* **2004**, *81*, 733. [CrossRef]
51. Goh, L.Y.; Wong, R.C.S.; Chu, C.K.; Hambley, T.W. Reaction of $[(Cr(cp)(CO)_3)_2]$ (cp = η^5-C_5H_5) with Elemental Phosphorus. Isolation of $[Cr_2(cp)_2(P_5)]$ as a Thermolysis Product and its X-ray Crystal Structure. *J. Chem. Soc. Dalton Trans.* **1990**, *3*, 977–982
52. Tremel, W.; Hoffmann, R.; Kertesz, M. Inorganic Rings, Intact and Cleaved, Between Two Metal Fragments. *J. Am. Chem. Soc.* **1989**, *111*, 2030–2039. [CrossRef]
53. Lepetit, C.; Fau, P.; Fajerwerg, K.; Kahn, M.L.; Silvi, B. Topological Analysis of the Metal-Metal Bond: A Tutorial Review. *Coord. Chem. Rev.* **2017**, *345*, 150–181. [CrossRef]
54. Macchi, P.; Proserpio, D.M.; Sironi, A. Experimental Electron Density in a Transition Metal Dimer: Metal-Metal and Metal-Ligand Bonds. *J. Am. Chem. Soc.* **1998**, *120*, 13429–13435. [CrossRef]

Article

Synthesis and Characterization of Super Bulky β-Diketiminato Group 1 Metal Complexes

Dafydd D. L. Jones, Samuel Watts and Cameron Jones *

School of Chemistry, Monash University, P.O. Box 23, Melbourne, VIC 3800, Australia; dafydd.jones@monash.edu (D.D.L.J.); samwatts4@outlook.com (S.W.)
* Correspondence: cameron.jones@monash.edu; Tel.: +61-(0)3-9902-0391

Abstract: Sterically bulky β-diketiminate (or Nacnac) ligand systems have recently shown the ability to kinetically stabilize highly reactive low-oxidation state main group complexes. Metal halide precursors to such systems can be formed via salt metathesis reactions involving alkali metal complexes of these large ligand frameworks. Herein, we report the synthesis and characterization of lithium and potassium complexes of the super bulky anionic β-diketiminate ligands, known [^{TCHP}Nacnac]$^-$ and new [$^{TCHP/Dip}$Nacnac]$^-$ (^{Ar}Nacnac = [(ArNCMe)$_2$CH]$^-$) (Ar = 2,4,6-tricyclohexylphenyl (TCHP) or 2,6-diisopropylphenyl (Dip)). The reaction of the proteo-ligands, ^{Ar}NacnacH, with nBuLi give the lithium etherate compounds, [(^{TCHP}Nacnac)Li(OEt$_2$)] and [($^{TCHP/Dip}$Nacnac)Li(OEt$_2$)], which were isolated and characterized by multinuclear NMR spectroscopy and X-ray crystallography. The unsolvated potassium salts, [{K(^{TCHP}Nacnac)}$_2$] and [{K($^{TCHP/Dip}$Nacnac)}$_\infty$], were also synthesized and characterized in solution by NMR spectroscopy. In the solid state, these highly reactive potassium complexes exhibit differing alkali metal coordination modes, depending on the ligand involved. These group 1 complexes have potential as reagents for the transfer of the bulky ligand fragments to metal halides, and for the subsequent stabilization of low-oxidation state metal complexes.

Keywords: lithium; potassium; Nacnac; β-diketiminate; tricyclohexylphenyl; steric bulk

1. Introduction

The use of sterically demanding β-diketiminate ligand systems for the stabilization of reactive low-oxidation state metal compounds has become commonplace in organometallic chemistry [1–8]. One of the most common β-diketiminate ligands used is [^{Dip}Nacnac]$^-$ ([(DipNCMe)$_2$CH]$^-$, Dip = 2,6-diisopropylphenyl). This has seen widespread use due to its ease of synthesis from commercially available starting materials and the high kinetic stabilization it imparts to the coordinated metal. A particularly successful use of β-diketiminates has been the stabilization of magnesium(I) dimers, e.g., LMg–MgL (L = β-diketiminate), which have subsequently been widely applied as soluble, selective reducing reagents in organic and inorganic synthesis [9–12]. Attempts to increase the steric profile of β-diketiminate ligands, in the hope that compounds elusive to stabilization by the ^{Dip}Nacnac system can be isolated, are worthwhile pursuits, with notable successes coming from the groups of Piers [13], Holland [14] and Hill [15]. With respect to the stabilization of group 2 compounds, bulky β-diketiminate ligands based on 2,6-bis(diphenylmethyl)-aryl N-substituents, [^{Ar}Nacnac]$^-$ ([(ArNCMe)$_2$CH]$^-$, Ar = 2,6-[C(H)Ph$_2$]$_2$-4-MeC$_6$H$_2$ (Ar*); 2,6-[C(H)Ph$_2$]$_2$-4-iPrC$_6$H$_2$ (Ar†)) were used to isolate the first monomeric magnesium hydrides [15]. More recently, it has been shown that increasing the steric bulk of the Nacnac ligand, by the replacement of the N-bound Dip groups with either DIPeP (DIPeP = 2,6-diisopentylphenyl) or TCHP (TCHP = 2,4,6-tricyclohexylphenyl) groups, allows for the isolation of the magnesium(I) compounds, [{(^{Ar}Nacnac)Mg}$_2$] (Ar = DIPeP, TCHP), both of which show elongated Mg–Mg bonds compared to previously reported examples [16–18].

The activation of β-diketiminate coordinated Mg–Mg bonded fragments by the addition of a sub-stoichiometric amount of a Lewis base is known to generate unsymmetrical magnesium(I) compounds with elongated Mg–Mg bonds. These have been shown to increase the reactivity of the magnesium(I) compound towards inert molecules such as CO and ethylene [19,20]. Recently, we showed that a TCHP-substituted magnesium(I) dimer, upon irradiation, was able to reduce benzene to generate "Birch-like" cyclohexadienyl bridged products, e.g., [{(^{TCHP}Nacnac)Mg}$_2$(μ-C$_6$H$_6$)] [21]. Further to this, Harder and co-workers have shown that Nacnac ligands incorporating the DIPeP aryl moiety are able to stabilize molecular magnesium(0) compounds [22,23], as well as calcium complexes, which activate dinitrogen, as in [{(^{DIPeP}Nacnac)Ca}$_2$(μ-N$_2$)] [24].

Routes to low-oxidation state metal complexes bearing Nacnac ligands typically proceed via the reduction of β-diketiminato metal halide precursors, which can be synthesized by the reaction of an alkali metal salt of the β-diketiminate ligand with a metal halide (Scheme 1). As such, the synthesis of alkali metal complexes of new very bulky ligands is desirable, with the view to using such complexes as ligand transfer reagents in the preparation of metal halide precursors to highly reactive low-valent metal complexes. Potentially, compounds previously inaccessible using the archetypal ^{Dip}Nacnac ligand system can be realized using this approach. Herein, we describe the synthesis and characterization of a new unsymmetrical, bulky β-diketimine pro-ligand incorporating the TCHP aryl moiety, as well as four alkali metal complexes derived from this, and a previously reported TCHP-substituted β-diketimine.

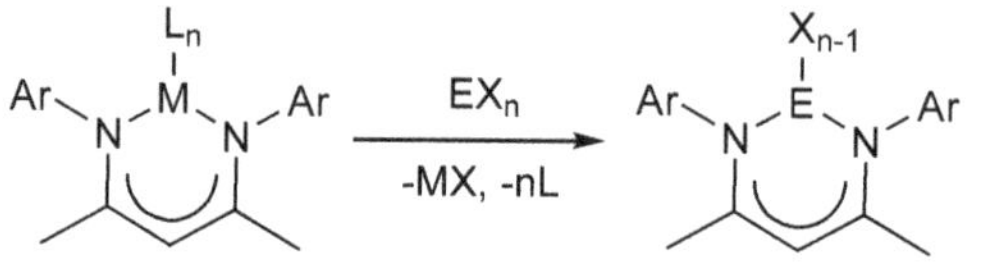

Scheme 1. Generic reaction of a β-diketiminato alkali metal salt with a metal/metalloid halide.

Ar = bulky aryl
M = Li, Na, K
L = donor ligand
E = metal or metalloid element
X = halide

2. Results and Discussion

The β-diketimine, ^{TCHP}NacnacH, **1**, was recently reported to be prepared by a condensation reaction between acacH, H$_2$C(MeC=O)$_2$ and two equivalents of (TCHP)NH$_2$ [17]. Here, the new, unsymmetrical β-diketimine, $^{TCHP/Dip}$NacnacH **2**, was similarly synthesized by the condensation of one equivalent of (TCHP)NH$_2$ with ^{Dip}NacacH, {DipN(H)CMe}CHC(O)Me, which gave **2** in a 43% isolated yield [25]. It is of note that, in the current study, during one preparation of **1**, a low yield of the β-ketimine, ^{TCHP}NacacH, {(TCHP)N(H)CMe}CHC(O)Me **3**, was obtained as a by-product. This likely results from the use of slightly less than two equivalents of (TCHP)NH$_2$ in the synthesis. A rational synthesis of **3** was not attempted, but it was spectroscopically characterized. The molecular structure of **2** is shown in Figure 1a, which reveals it to exist as its ene-imine conjugated tautomer in the solid state, as is typical for β-diketimines. The proton on N2 is hydrogen bonded to N1, which leads to the two double bonds within the C$_3$N$_2$ backbone (viz. N1-C1 and C3-C4) adopting a cis-configuration, relative to each other. A similar structure is adopted by β-ketimine, **3** (Figure 1b), in which the double bonds within the C$_3$NO backbone are C1-O1 and C3-C4.

The bulky β-diketimines, **1** and **2**, were deprotonated by treating diethyl ether solutions of the compounds with a slight excess of nBuLi at −78 °C. Upon warming to room temperature and stirring the reaction solutions overnight, they were concentrated and stored at −30 °C to yield colorless crystals of lithium β-diketiminate complexes [(^{TCHP}Nacnac)Li(OEt$_2$)], **4**, and [($^{TCHP/Dip}$Nacnac)Li(OEt$_2$)], **5**, in moderate isolated yields. The solution state NMR spectroscopic data for the compounds are consistent with them retaining their solid-state structures (see below) in solution. However, it is noteworthy that the ^{13}C{^{1}H} NMR spectrum of compound **4** exhibits 10 cyclohexyl carbon signals (δ = 25–45 ppm), which suggests restricted rotation around one of the sterically more encumbered ortho-cyclohexyl rings of the TCHP substituents.

Compounds **4** and **5** have very similar solid-state structures, as is evident from Figure 2. The molecular structure of these compounds are typical of previously reported β-diketiminate lithium etherate complexes, e.g., [(^{Dip}Nacnac)Li(OEt$_2$)] [26]. That is, they are monomeric and their lithium centers possess trigonal planar coordination geometries with one coordinated ether ligand. The β-diketiminate ligand binds as an N,N-chelate to the lithium, with Li-N bond distances between 1.921 and 1.935 Å, within the normal range for such complexes [26]. The Li-O distances are slightly longer when compared to the previously reported [(^{Dip}Nacnac)Li(OEt$_2$)] complex (1.953(4) Å for **4** and 1.975(4) Å for **5**; cf. 1.911(4) Å for [(^{Dip}Nacnac)Li(OEt$_2$)]), which is likely caused by steric repulsion from its bulky TCHP aryl groups. This also leads to a slight narrowing of the N-Li-N angle for **4** and **5** (97.67(18)° and 97.79(17)°, respectively) compared to the angle in [(^{Dip}Nacnac)Li(OEt$_2$)] (99.9(2)°). The bond lengths within the C$_3$N$_2$ backbones of both compounds imply significant electronic delocalizations over those fragments.

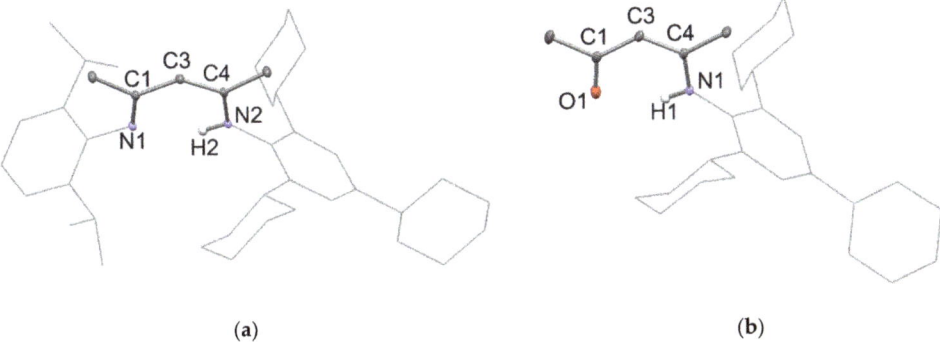

(a)   (b)

Figure 1. (**a**) Partial thermal ellipsoid plot (25% probability surface for displayed ellipsoids) of the molecular structure of **2** (hydrogen atoms, except H2, omitted; Dip and TCHP groups shown as wireframe for clarity). Selected bond lengths (Å) and angles (°): N(1)-C(1) 1.317(2), C(1)-C(3) 1.419(2), C(3)-C(4) 1.391(2), N(2)-C(4) 1.337(2), N(1)-C(1)-C(3) 121.09(14), C(1)-C(3)-C(4) 126.12(14); (**b**) thermal ellipsoid plot (25% probability surface) of the molecular structure of **3** (hydrogen atoms, except H1, omitted; TCHP group shown as wireframe for clarity). Selected bond lengths (Å) and angles (°): O(1)-C(1) 1.239(2), C(1)-C(3) 1.420(3), C(3)-C(4) 1.381(2), N(1)-C(4) 1.342(2), O(1)-C(1)-C(3) 123.14(17), C(1)-C(3)-C(4) 123.61(17).

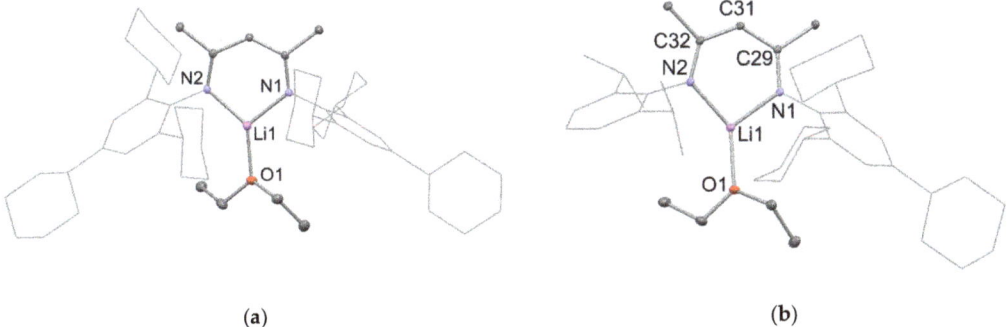

(a)   (b)

Figure 2. (**a**) Partial thermal ellipsoid plot (25% probability surface for displayed ellipsoids) of the molecular structure of **4** (hydrogen atoms omitted; TCHP groups shown as wireframe for clarity). Selected bond lengths (Å) and angles (°): Li(1)-O(1) 1.953(4), N(1)-Li(1) 1.921(4), N(2)-Li(1) 1.924(4), N(1)-Li(1)-N(2) 97.67(18); (**b**) thermal ellipsoid plot (25% probability surface) of the molecular structure of **5** (hydrogen atoms omitted; Dip and TCHP groups shown as wireframe for clarity). Selected bond lengths (Å) and angles (°): Li(1)-O(1) 1.975(4), N(1)-Li(1) 1.927(4), N(2)-Li(1) 1.935(4), N(1)-Li(1)-N(2) 97.79(17).

Unsolvated potassium salts of the TCHP-substituted β-diketiminates could also be synthesized using a similar methodology to that previously reported for the synthesis of [K(^{Dip}Nacnac)] [27]. That is, β-diketimine, **1**, was combined with benzyl potassium in toluene, pro-ligand **2** was combined with potassium bis(trimethylsilyl)amide (KHMDS) in toluene and both reactions were monitored by NMR spectroscopy. No reaction could be observed after 16 h at room temperature for either pro-ligand, contrary to the synthesis of [K(^{Dip}Nacnac)], which showed good conversion under these conditions. This can be attributed to the increased steric bulk from the TCHP aryl group, perturbing deprotonation of **1** and **2** by the base. However, heating these solutions at 60 °C overnight led to clean conversion to the potassium salts [{K(^{TCHP}Nacnac)}$_2$], **6**, and [{K($^{TCHP/Dip}$Nacnac)}$_\infty$], **7**, as determined by ^{1}H NMR spectroscopy. It should be noted that compound **6** could also be formed using KHMDS as the base, though this route gave variable yields of product. Attempts to isolate compound **6** repeatedly led to a degree of decomposition of the complex to **1** upon manipulation of the reaction mixture. Moreover, in the presence of silicone grease, compound **6** decomposed to unknown products in solution over time. This suggests that compound **6** is highly reactive, and as such could only be characterized by NMR spectroscopy after formation in situ. As monitoring the formation of **6** from **1** by NMR spectroscopy showed complete conversion, we believe this compound could be used as a ligand transfer reagent after its formation in situ, and in the absence of silicone grease.

Crystals of **6** suitable for an X-ray diffraction study were formed by the addition of *n*-hexane to a concentrated toluene solution of **6** at room temperature. The molecular structure (Figure 3) shows the compound crystallizing as a dimer, with the potassium centers having trigonal planar geometries, being N,N-chelated by a β-diketiminate, as well as possessing an intermolecular C···K interaction with a *meta*-carbon of an opposing TCHP group of the second [K(^{TCHP}Nacnac)] monomer unit. The N-K distances (2.633(3) and 2.619(3) Å) are in the expected range and correspond well to those previously reported for [K(^{Dip}Nacnac)] (2.681 Å mean) [27]. The C···K distance of 3.309(3) Å is outside the sum of the covalent radii for the two elements (2.71 Å [28]), but well within the sum of their van der Waals radii (4.45 Å [29]).

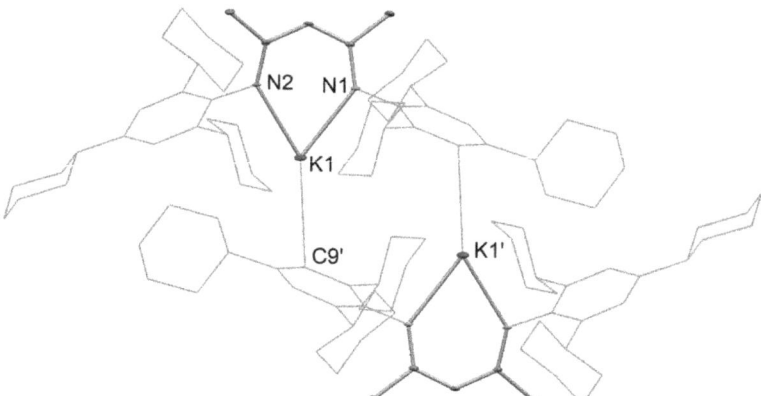

Figure 3. Partial thermal ellipsoid plot (25% probability surface for displayed ellipsoids) of the molecular structure of **6** (hydrogen atoms omitted; TCHP groups shown as wireframe for clarity). Selected bond lengths (Å) and angles (°): K(1)-N(1) 2.633(3), K(1)-N(2) 2.619(3), K(1)-C(9)' 3.309(3), N(1)-K(1)-N(2) 68.23(10).

Compound **7** crystallized as a colorless material from the toluene reaction solution, upon cooling it to room temperature from 60 °C. However, it should be noted that, like **6**, compound **7** undergoes decomposition to the proteo-ligand **2** in the presence of silicone grease. Crystals of **7** suitable for an X-ray diffraction were obtained from a saturated benzene solution, and its molecular structure is shown in Figure 4. It is polymeric, similar to previously reported [K(^{Dip}Nacnac)] [27]. The polymer is propagated via intermolecular C···K interactions between the potassium centers and *para*- and *meta*-carbons of a Dip group on an adjacent monomer unit. The K···C bond distances are 3.178(4) Å for K–C$_{para}$

and 3.254(3) Å for K–C$_{meta}$. These separations are slightly shorter than the equivalent C···K interactions in [K(^{Dip}Nacnac)], which are 3.214(3) Å for K–C$_{para}$, and 3.351(3) Å and 3.276(3) Å for K–C$_{meta}$ interactions. The K-N distances in the compound are comparable to those in **6**.

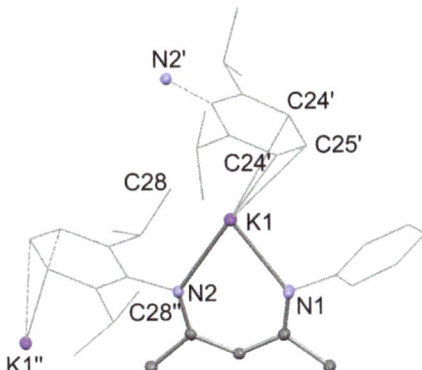

Figure 4. Partial thermal ellipsoid plot (25% probability surface for displayed ellipsoids) of the molecular structure of **7** (hydrogen atoms omitted; Dip groups shown as wireframe, and TCHP group truncated for clarity). Selected bond lengths (Å) and angles (°): K(1)-N(1) 2.634(3), K(1)-N(2) 2.713(3), K(1)-C(24)′ 3.254(3), K(1)-C(25)′ 3.178(4), N(1)-K(1)-N(2) 68.82(9).

3. Materials and Methods

3.1. General Considerations

All manipulations were carried out using standard Schlenk and glove box techniques under an atmosphere of high purity dinitrogen. Toluene and hexane were distilled over molten potassium, while diethyl ether was distilled over 1:1 Na/K alloy. Benzene-d$_6$ was stored over a mirror of sodium and degassed three times via freeze-pump-thawing before use. The ^{1}H, ^{13}C{^{1}H} and ^{7}Li NMR spectra were recorded on a Bruker AvanceIII 400 spectrometer and were referenced to the residual resonances of the solvent used or external 1 M LiCl. Mass spectra were collected using an Agilent Technologies 5975D inert MSD with a solid-state probe. FTIR spectra were collected as Nujol mulls on an Agilent Cary 630 attenuated total reflectance (ATR) spectrometer. Microanalyses were carried out at the Science Centre, London Metropolitan University or using a PerkinElmer- 2400 CHNS/O Series II System. Melting points were determined in sealed glass capillaries under dinitrogen and are uncorrected. The starting materials, ^{TCHP}NacnacH **1** [17], ^{Dip}NacacH [30] (TCHP)NH$_2$ [31] and benzyl potassium [32], were prepared by literature procedures. All other reagents were used as received.

3.2. Syntehsis of $^{TCHP/Dip}$NacnacH **2** and Data for ^{TCHP}NacacH **3**

Preparation of $^{TCHP/Dip}$NacnacH, 2: 2,4,6-Tricyclohexylaniline (4.57 g, 13.5 mmol), ^{Dip}NacacH (3.50 g, 13.5 mmol) and p-toluenesulfonic acid (2.32 g, 13.5 mmol) were dissolved in toluene (100 mL), and the mixture was heated at reflux for 3 days in a Dean–Stark apparatus. Volatiles were then removed in vacuo, the resultant oil was dissolved in dichloromethane (100 mL), the extract washed with a saturated Na$_2$CO$_3$ solution (100 mL, 1 M) and the organic layer was separated. The organic layer was dried over MgSO$_4$ and volatiles were removed in vacuo. Methanol (5 mL) was added to the oily residue, which was then sonicated and filtered to yield compound **2** as a crude white powder. This was dried and recrystallized from hot ethyl acetate to yield pure $^{TCHP/Dip}$NacnacH (2.25 g). A further crop of the product could be obtained by concentration of the mother liquor, and letting it stand at room temperature (1.00 g). Total yield: 3.25 g, 42%. X-ray quality crystals could be grown from the slow evaporation of a diethyl ether solution of $^{TCHP/Dip}$NacnacH.

M.p: 224–227 °C; ^{1}H NMR (400 MHz, C$_6$D$_6$, 298 K): δ = 1.24 (d, $^3J_{HH}$ = 8 Hz, 6H, CH(CH$_3$)$_2$), 1.28 (d, $^3J_{HH}$ = 8 Hz, 6H, CH(CH$_3$)$_2$), 1.31–1.68 (m, 17H, Cyc-CH$_2$), 1.70 (s, 3H, NCCH$_3$), 1.72 (s, 3H, NCCH$_3$), 1.74–2.06 (m, 13H, Cyc-CH$_2$), 2.52–2.60 (m, 1H, p-Cyc-H), 2.95–3.03 (m, 2H, o-Cyc-H), 3.29–3.38 (sept, $^3J_{HH}$ = 8 Hz, 2H, CH(CH$_3$)$_2$), 4.89 (s, 1H, NCCH), 7.18 (s, 2H, ArH), 7.20 (m, 3H, ArH), 12.29 (s, 1H, NH). ^{13}C{^{1}H} NMR (101 MHz, C$_6$D$_6$, 298 K): δ = 20.6 (NCCH$_3$), 21.4 (NCCH$_3$), 23.8, 24.3 (CH(CH$_3$)$_2$), 26.6, 26.8, 27.4, 2 × 27.7 (Cyc-C), 28.6 (CH(CH$_3$)$_2$), 34.7, 35.3 (Cyc-C), 39.6 (o-Cyc-CH), 45.3 (p-Cyc-CH), 94.1 (NCCH), 123.0, 123.5, 125.0, 137.3, 141.2, 143.3, 143.7, 145.8 (ArC), 159.7 (NCCH$_3$), 163.8 (NCCH$_3$). IR ν/cm^{-1} (ATR): 2920 (s), 2847 (s), 1618 (s), 1545 (s), 1442 (s), 1359 (w), 1270 (m), 1175 (m). Acc. Mass/ESI m/z: calc. 581.4835 found: 581.4841 [M+H]$^+$.

N.B. During one synthesis of **1**, β-ketimine **3** was obtained as a low yield by-product.

Data for ^{TCHP}NacacH 3. M.p 183–185 °C. ^{1}H NMR (400 MHz, C$_6$D$_6$, 298 K): N.B. integration of resonances for cyclohexyl groups are estimated due to complex overlapping signals, δ = 1.12–1.44 (m, 10H, Cyc-CH$_2$), 1.46 (s, 3H, NCCH$_3$), 1.48–1.98 (m, 20H, Cyc-CH$_2$), 2.06 (s, 3H, OCCH$_3$), 2.46–2.52 (m, 1H, p-Cyc-CH), 2.78–2.86 (m, 2H, o-Cyc-CH), 5.14 (s, 1H, NCCH), 7.17 (s, 2H, ArH), 12.67 (br, 1H, NH). ^{13}C{^{1}H} NMR (101 MHz, C$_6$D$_6$, 298 K): δ = 19.1 (NCCH$_3$), 2 × 26.6, 27.3, 2 × 27.4 (Cyc-C), 29.1 (OCCH$_3$), 33.5, 35.1, 35.6, 40.0, 45.2 (Cyc-C), 95.9 (NCCH), 122.9, 132.5, 145.5, 147.8 (ArC), 162.9 (NCCH$_3$), 195.6 (OCCH$_3$). IR ν/cm^{-1} (ATR): 2921 (s), 2848 (s), 1611 (s), 1567 (s), 1492 (w), 1445(s), 1351 (m), 1274 (s), 1235 (w), 1185 (w), 1119 (w), 1014 (w), 998 (w), 951 (w), 919 (w), 893 (w), 865 (m), 774 (w), 741 (m), 674 (w). Acc. Mass/ESI m/z: calc. for [M+H]$^+$ 422.3417 found: 422.3422.

3.3. Syntehsis of Complexes 4–7

Preparation of [(^{TCHP}Nacnac)Li(OEt$_2$)], 4. ^{TCHP}NacnacH (1.15 g, 1.5 mmol) was dissolved in diethyl ether (20 mL) and the solution cooled to −78 °C. nBuLi (1.05 mL, 1.6 M in hexanes, 1.7 mmol) was added dropwise via syringe and the resultant solution was allowed to warm to room temperature and stirred for 2 h. The solution was concentrated in vacuo and stored at −30 °C overnight, yielding large colorless blocks of **4** (0.62 g) suitable for X-ray diffraction. Further concentration of the mother liquor and storage at −30 °C gave a second crop of crystals (0.14 g). Total yield: 0.76 g, 60%. M.p 112–114 °C (decomp): ^{1}H NMR (400 MHz, C$_6$D$_6$, 298 K) N.B. integration for cyclohexyl groups are estimated due to complex overlapping signals: δ = 0.67 (br, 6H, (CH$_3$CH$_2$)$_2$O), 1.27–1.84 (m, 50H, Cyc-CH$_2$), 1.90 (s, 6H, NCCH$_3$), 1.98–2.03 (m, 10H, Cyc-CH$_2$), 2.52–2.58 (m, 2H, p-Cyc-CH), 2.96–3.04 (br m, 8H, overlapping o-Cyc-CH & (CH$_3$CH$_2$)$_2$O), 4.88 (s, 1H, NCCH), 7.13 (s, 4H, ArH). ^{13}C{^{1}H} NMR (101 MHz, C$_6$D$_6$, 298 K): δ = 14.3 ((CH$_3$CH$_2$)$_2$O), 23.6 (NCCH$_3$), 26.8, 27.0, 27.6, 27.9, 28.1, 34.8, 35.1, 35.6, 39.4, 45.2 (Cyc-C), 65.2 ((CH$_3$CH$_2$)$_2$O), 92.5 (NCCH), 122.4, 139.8, 141.8, 147.9 (ArC), 164.3 (NCCH$_3$). ^{7}Li NMR (155 MHz, C$_6$D$_6$, 298 K): δ = 1.91. IR ν/cm^{-1} (ATR): 2920 (s), 2849 (m), 1601 (m), 1550 (m), 1438 (m), 1401 (s), 1348 (m), 1282 (m), 1250 (w), 1187 (w), 1116 (m), 1070 (m), 1024 (w), 994 (w), 950 (w), 921 (w), 890 (w), 862 (m), 835 (w), 779 (w), 724 (w), 696 (m). EI/MS (70 eV) m/z (%): 83.1 (Cyc$^+$, 26), 364.3 (Cyc)$_3$C$_6$H$_2$NCMe$^+$, 100), 659.6 (^{TCHP}NacnacH–Cyc$^+$, 10), 727.8 (^{TCHP}NacnacH–CH$_3^+$, 17), 742.8 (^{TCHP}NacnacH$^+$, 13), 748.6 (^{TCHP}NacnacLi$^+$, 8). Elemental analysis calculated for C$_{57}$H$_{87}$LiN$_2$O: C 83.16%, H 10.65%, N 3.40%, found: C 83.02%, H 10.40%, N 3.32%.

Preparation of [($^{TCHP/Dip}$Nacnac)Li(OEt$_2$)], 5. $^{TCHP/Dip}$NacnacH (500 mg, 0.86 mmol) was dissolved in diethyl ether (20 mL) and the solution cooled to −78 °C. nBuLi (0.56 mL, 1.6 M in hexanes, 0.90 mmol) was added dropwise via syringe and the resultant solution was allowed to warm to room temperature and stirred overnight. The solution was concentrated in vacuo and stored at −30 °C overnight, yielding large colorless blocks of **5** (258 mg, 45%) suitable for X-ray diffraction. M.p 134–136 °C (decomp): ^{1}H NMR (400 MHz, C$_6$D$_6$, 298 K) N.B. integration for cyclohexyl groups are estimated due to complex overlapping signals: δ = 0.47–0.50 (t, $^3J_{HH}$ = 7.0 Hz, 6H, (CH$_3$CH$_2$)$_2$O), 1.26–1.27 (d, $^3J_{HH}$ = 6.9 Hz, 6H, CH(CH$_3$)$_2$), 1.32–1.33 (d, $^3J_{HH}$ = 6.9 Hz, 6H, CH(CH$_3$)$_2$), 1.37–1.83 (m, 25H, Cyc-CH$_2$), 1.88 (s, 3H, NCCH$_3$), 1.93 (s, 3H, NCCH$_3$), 1.95–1.98 (m, 5H, Cyc-CH$_2$), 2.50–2.58 (m, 1H, p-Cyc-CH), 2.75–2.81 (q, $^3J_{HH}$ = 7.0 Hz, 4H, (CH$_3$CH$_2$)$_2$O), 2.94–3.00 (m,

2H, o-Cyc-CH), 3.40 (sept, $^3J_{HH}$ = 6.9 Hz, 2H, CH(CH$_3$)$_2$) 4.96 (s, 1H, NCCH), 7.08–7.20 (m, 5H, overlapping TCHP-ArH & Dip-ArH). ^{13}C{^{1}H} NMR (101 MHz, C$_6$D$_6$, 298 K): δ = 13.7 ((CH$_3$CH$_2$)$_2$O), 23.4, 23.5 (NCCH$_3$), 24.0, 24.3 (CH(CH$_3$)$_2$), 26.7, 27.0, 27.6, 27.7, 27.9, (Cyc-C), 28.3 (CH(CH$_3$)$_2$), 34.2, 35.2, 35.6, 39.5, 45.3 (Cyc-C), 63.7 ((CH$_3$CH$_2$)$_2$O), 92.8 (NCCH), 122.3, 123.1, 123.4,'139.5, 140.9, 141.9, 147.9, 150.1 (ArC), 163.7, 164.0 (NCCH$_3$). ^{7}Li NMR (155 MHz, C$_6$D$_6$, 298 K): δ = 1.88. IR ν/cm^{-1} (ATR): 621 (w), 1549 (s), 1518 (m), 1148 (w), 1093 (w), 1058 (w), 1016 (w), 953 (w), 926 (w), 863 (m), 831 (w), 788 (w), 758 (s), 731 (s). A reproducible elemental analysis could not be obtained for this compound as it consistently co-crystallized with small amounts of $^{TCHP/Dip}$NacnacH, which could not be separated by repeated recrystallizations.

Preparation of [{K(^{TCHP}Nacnac)}$_2$], 6: ^{TCHP}NacnacH (400 mg, 0.54 mmol) and benzyl potassium (72 mg, 0.55 mmol) were dissolved in toluene (10 mL) and the mixture stirred overnight at 60 °C. The solution was then cooled to room temperature and volatiles were removed in vacuo. The subsequent solid was dissolved in n-hexane (5 mL) and the extract left to stand at room temperature, yielding colorless crystals of **6** suitable for X-ray diffraction studies. Analysis of the crystalline material by NMR spectroscopy consistently showed contamination of the product with ^{TCHP}NacnacH, which could not be removed, despite repeated recrystallizations.

In situ preparation of 6: ^{TCHP}NacnacH (10 mg, 0.013 mmol) and benzyl potassium (3 mg, 0.023 mmol) were dissolved in C$_6$D$_6$ (0.6 mL) in a J. Young's NMR tube equipped with a Teflon screw cap and the mixture heated overnight at 60 °C. The subsequent reaction solution was analyzed by NMR spectroscopy, showing near complete conversion to compound **6**. ^{1}H NMR (400 MHz, C$_6$D$_6$, 298 K) N.B. integration for cyclohexyl groups are estimated due to complex overlapping signals: δ = 1.36–1.87 (m, 55H, Cyc-CH$_2$), 1.89 (s, 6H, NCCH$_3$), 2.02–2.06 (m, 5H, Cyc-CH$_2$), 2.54–2.62 (m, 2H, p-Cyc-CH), 3.03–3.10 (m, 4H, o-Cyc-CH), 4.74 (s, 1H, NCCH), 7.16 (s, 4H, TCHP-ArH). ^{13}C{^{1}H} NMR (101 MHz, C$_6$D$_6$, 298 K): δ = 23.8 (NCCH$_3$), 26.8, 27.0, 27.6, 27.8, 28.2, 35.0, 2 × 35.6, 38.8, 45.2 (Cyc-C), 90.1 (NCCH), 122.8, 138.6, 139.6, 140.0, 149.4, (ArC), 161.2, (NCCH$_3$).

Preparation of [{K($^{TCHP/Dip}$Nacnac)}$_∞$], 7. $^{TCHP/Dip}$NacnacH (516 mg, 0.89 mmol) and KHMDS (195 mg, 0.98 mmol) were dissolved in toluene (8 mL) and the mixture heated overnight at 60 °C. The solution was cooled to room temperature, after which time micro-crystalline **7** was deposited. The suspension was filtered, and the solid was dried in vacuo giving compound **7** as a spectroscopically near pure colorless solid (500 mg, 91%). Crystals suitable for X-ray diffraction were grown by slow cooling a saturated solution of **7** in C$_6$D$_6$ from 60 °C to room temperature. M.p. > 260 °C: ^{1}H NMR (400 MHz, C$_6$D$_6$, 298 K) N.B. integration for cyclohexyl groups are estimated due to complex overlapping signals: δ = 1.14–1.17 (d, $^3J_{HH}$ = 6.9 Hz, 6H, CH(CH$_3$)$_2$), 1.24–1.30 (m, 2H, Cyc-CH$_2$), 1.32–1.34 (d, $^3J_{HH}$ = 6.9 Hz, 6H, CH(CH$_3$)$_2$), 1.37–1.84 (m, 24H, Cyc-CH$_2$), 1.89 (s, 3H, NCCH$_3$), 1.91 (s, 3H, NCCH$_3$), 1.97–2.04 (m, 4H, Cyc-CH$_2$), 2.53–2.59 (m, 1H, p-Cyc-CH), 2.96–3.04 (m, 2H, o-Cyc-CH), 3.39 (sept, $^3J_{HH}$ = 6.9 Hz, 2H, CH(CH$_3$)$_2$), 4.80 (s, 1H, NCCH), 7.01–7.19 (m, 5H, overlapping TCHP-ArH & Dip-ArH). ^{13}C{^{1}H} NMR (101 MHz, C$_6$D$_6$, 298 K): δ = 23.6, 23.7 (NCCH$_3$), 24.0, 24.4 (CH(CH$_3$)$_2$), 26.8, 26.9, 27.7, 27.9 (Cyc-C), 28.0 (CH(CH$_3$)$_2$), 34.5, 35.6, 35.8, 39.1, 45.2 (Cyc-C), 91.2 (NCCH), 121.2, 122.8, 123.7, 138.4, 139.6, 140.0, 149.1, 151.2 (ArC), 160.1, 160.9 (NCCH$_3$). IR ν/cm^{-1} (ATR): 1566 (m), 1518 (m), 1143 (m), 1016 (w), 923 (w), 890 (w), 862 (m), 819 (w), 792 (m), 770 (m), 727 (s). A reproducible elemental analysis could not be obtained for this compound as it consistently co-crystallized with small amounts of $^{TCHP/Dip}$NacnacH, which could not be separated by repeated recrystallizations.

3.4. Crystallographic Details

Crystals of **2–7** suitable for X-ray structural determination were mounted in silicone oil. Crystallographic measurements were carried out at 123(2) K, and were made using a Rigaku Synergy diffractometer using a graphite monochromator with Cu Kα radiation (λ = 1.54184 Å). The structures were solved by direct methods and refined on F^2 by full matrix least squares (SHELX16) [33] using all unique data. All non-hydrogen atoms were anisotropic with hydrogen atoms typically included in calculated positions (riding model). Crystal data, details of the data collection and refinement are given in Table S1 (in Supplementary Materials). Crystallographic data for the structures have been deposited with the Cambridge Crystallographic Data Centre (CCDC no. 2105059-2105064). Copies of this information may be obtained free of charge from The Director, CCDC, 12 Union Road, Cambridge, CB2 1EZ, UK (fax: +44-1223-336033; email: deposit@ccdc.cam.ac.uk or www: http://www.ccdc.cam.ac.uk).

4. Conclusions

In summary, the syntheses of four alkali metal complexes of two super bulky β-diketiminate ligand systems are described. These complexes were characterized by X-ray crystallography and multinuclear NMR spectroscopy. The solid-state structures of the lithium etherate complexes are similar to previously reported examples. The molecular structures of the potassium salts show varied coordination environments in potassium, depending on the accompanying ligand framework. That is, [{K(^{TCHP}Nacnac)}$_2$] 6 is dimeric with C···K interactions involving a *meta*-aryl carbon of the opposing monomer unit. In contrast, [{K{($^{TCHP/Dip}$Nacnac)}$_\infty$] 7 is polymeric, and shows comparable connectivity between monomer units to that seen in the previously reported complex [K(^{Dip}Nacnac)]. These alkali metal compounds have significant potential for use as ligand transfer reagents in the synthesis of metal complexes incorporating large TCHP-substituted β-diketiminates. We are currently exploring the use of these ligands for the kinetic stabilization of low-oxidation state metal complexes.

Supplementary Materials: The following are available online at https://www.mdpi.com/article/10.3390/inorganics9090072/s1, Table S1: Summary of crystallographic data for 2–7; Figures S1–S14: ^{1}H, ^{13}C{^{1}H}, and ^{7}Li{^{1}H} NMR spectra of 2–7; CIF and CheckCIF files of 2–7.

Author Contributions: Conceptualization, D.D.L.J. and C.J.; methodology, D.D.L.J. and S.W.; validation, D.D.L.J. and S.W.; formal analysis, D.D.L.J., S.W. and C.J.; investigation, D.D.L.J. and S.W.; writing—original draft preparation, D.D.L.J.; writing—review and editing, C.J.; visualization, D.D.L.J. and C.J.; supervision, C.J.; project administration, C.J.; funding acquisition, C.J. All authors have read and agreed to the published version of the manuscript.

Funding: CJ is grateful to the Australian Research Council for funding part of this work. Moreover, this material is based upon work supported by the Air Force Office of Scientific Research under award number FA2386-21-1-4048.

Institutional Review Board Statement: Not applicable.

Informed Consent Statement: Not applicable.

Data Availability Statement: NMR spectra and crystal data are given in the supporting information. Crystal data, details of the data collection and refinement are given in Table S1. Crystallographic data for the structures have been deposited with the Cambridge Crystallographic Data Centre (CCDC no. 2105059-2105064). Copies of this information may be obtained free of charge from The Director, CCDC, 12 Union Road, Cambridge, CB2 1EZ, UK (fax: +44-1223-336033; email: deposit@ccdc.cam.ac.uk or http://www.ccdc.cam.ac.uk).

Conflicts of Interest: The authors declare no conflict of interest. The funders had no role in the design of the study; in the collection, analyses, or interpretation of data; in the writing of the manuscript, or in the decision to publish the results.

References

1. Bourget-Merle, L.; Lappert, M.F.; Severn, J.R. The Chemistry of β-Diketiminatometal Complexes. *Chem. Rev.* **2002**, *102*, 3031–3066. [CrossRef] [PubMed]
2. Sarish, S.P.; Nembenna, S.; Nagendran, S.; Roesky, H.W. Chemistry of Soluble β-Diketiminatoalkaline-Earth Metal Complexes with M−X Bonds (M = Mg, Ca, Sr; X = OH, Halides, H). *Acc. Chem. Res.* **2011**, *44*, 157–170. [CrossRef] [PubMed]
3. Yao, S.; Driess, M. Lessons from Isolable Nickel(I) Precursor Complexes for Small Molecule Activation. *Acc. Chem. Res.* **2012**, *45*, 276–287. [CrossRef]
4. Tsai, Y. The chemistry of univalent metal β-diketiminates. *Coord. Chem. Rev.* **2012**, *256*, 722–758. [CrossRef]

5. Zhu, D.; Budzelaar, P.H.M. N-Aryl β-diiminate complexes of the platinum metals. *Dalton Trans.* **2013**, *42*, 11343–11354. [CrossRef]
6. Chen, C.; Bellows, S.M.; Holland, P.L. Tuning steric and electronic effects in transition-metal β-diketiminate complexes. *Dalton Trans.* **2015**, *44*, 16654–16670. [CrossRef] [PubMed]
7. Hohloch, S.; Kriegel, B.M.; Bergman, R.G.; Arnold, J. Group 5 chemistry supported by β-diketiminate ligands. *Dalton Trans.* **2016**, *45*, 15725–15745. [CrossRef]
8. Camp, C.; Arnold, J. On the non-innocence of "Nacnacs": Ligand-based reactivity in β-diketiminate supported coordination compounds. *Dalton Trans.* **2016**, *45*, 14462–14498. [CrossRef]
9. Green, S.P.; Jones, C.; Stasch, A. Stable Magnesium(I) Compounds with Mg-Mg Bonds. *Science* **2007**, *318*, 1754–1757. [CrossRef]
10. Stasch, A.; Jones, C. Stable dimeric magnesium(i) compounds: From chemical landmarks to versatile reagents. *Dalton Trans.* **2011**, *40*, 5659–5672. [CrossRef]
11. Jones, C. Dimeric magnesium(I) β-diketiminates: A new class of quasi-universal reducing agent. *Nat. Rev. Chem.* **2017**, *1*, 0059. [CrossRef]
12. Bonyhady, S.; Jones, C.; Nembenna, S.; Stasch, A.; Edwards, A.; McIntyre, G. β-Diketiminate-Stabilized Magnesium(I) Dimers and Magnesium(II) Hydride Complexes: Synthesis, Characterization, Adduct Formation, and Reactivity Studies. *Chem. Eur. J.* **2010**, *16*, 938–955. [CrossRef]
13. Kenward, A.L.; Ross, J.A.; Piers, W.E.; Parvez, M. Metalation-Resistant β-Diketiminato Ligands for Thermally Robust Organoscandium Complexes. *Organometallics* **2009**, *28*, 3625–3628. [CrossRef]
14. Cowley, R.E.; Holland, P.L. Ligand Effects on Hydrogen Atom Transfer from Hydrocarbons to Three-Coordinate Iron Imides. *Inorg. Chem.* **2012**, *51*, 8352–8361. [CrossRef]
15. Arrowsmith, M.; Maitland, B.; Kociok-Köhn, G.; Stasch, A.; Jones, C.; Hill, M.S. Mononuclear Three-Coordinate Magnesium Complexes of a Highly Sterically Encumbered β-Diketiminate Ligand. *Inorg. Chem.* **2014**, *53*, 10543–10552. [CrossRef]
16. Gentner, T.X.; Rösch, B.; Ballmann, G.; Langer, J.; Elsen, H.; Harder, S. Low Valent Magnesium Chemistry with a Super Bulky β-Diketiminate Ligand. *Angew. Chem. Int. Ed.* **2019**, *58*, 607–611. [CrossRef]
17. Yuvaraj, K.; Douair, I.; Jones, D.D.L.; Maron, L.; Jones, C. Sterically controlled reductive oligomerisations of CO by activated magnesium(i) compounds: Deltate vs. ethenediolate formation. *Chem. Sci.* **2020**, *11*, 3516–3522. [CrossRef]
18. Boutland, A.J.; Dange, D.; Stasch, A.; Maron, L.; Jones, C. Two-Coordinate Magnesium(I) Dimers Stabilized by Super Bulky Amido Ligands. *Angew. Chem. Int. Ed.* **2016**, *55*, 9239–9243. [CrossRef] [PubMed]
19. Yuvaraj, K.; Douair, I.; Paparo, A.; Maron, L.; Jones, C. Reductive Trimerization of CO to the Deltate Dianion Using Activated Magnesium(I) Compounds. *J. Am. Chem. Soc.* **2019**, *141*, 8764–8768. [CrossRef] [PubMed]
20. Yuvaraj, K.; Douair, I.; Maron, L.; Jones, C. Activation of Ethylene by N-Heterocyclic Carbene Coordinated Magnesium(I) Compounds. *Chem. Eur. J.* **2020**, *26*, 14665–14670. [CrossRef] [PubMed]
21. Jones, D.D.L.; Douair, I.; Maron, L.; Jones, C. Photochemically Activated Dimagnesium(I) Compounds: Reagents for the Reduction and Selective C−H Bond Activation of Inert Arenes. *Angew. Chem. Int. Ed.* **2021**, *60*, 7087–7092. [CrossRef]
22. Rösch, B.; Gentner, T.X.; Eyselein, J.; Langer, J.; Elsen, H.; Harder, S. Strongly reducing magnesium(0) complexes. *Nature* **2021**, *592*, 717–721. [CrossRef] [PubMed]
23. Jones, C. Highly reactive form of magnesium stabilized by bulky ligands. *Nature* **2021**, *592*, 687–688. [CrossRef] [PubMed]
24. Rösch, B.; Gentner, T.X.; Langer, J.; Färber, C.; Eyselein, J.; Zhao, L.; Frenking, G.; Harder, S. Dinitrogen complexation and reduction at low-valent calcium. *Science* **2021**, *371*, 1125–1128. [CrossRef]
25. Weerawardhana, E.A.; Pena, A.; Zeller, M.; Lee, W.-T. Synthesis and characterization of iron and cobalt complexes with an asymmetric N-alkyl,N'-aryl-β-diketiminate ligand. *Inorg. Chim. Acta* **2017**, *460*, 29–34. [CrossRef]
26. Stender, M.; Wright, R.J.; Eichler, B.E.; Prust, J.; Olmstead, M.M.; Roesky, H.W.; Power, P.P. The synthesis and structure of lithium derivatives of the sterically encumbered β-diketiminate ligand [{(2,6-Pri$_2$H$_3$C$_6$)N(CH$_3$)C}$_2$CH]$^-$, and a modified synthesis of the aminoimine precursor. *J. Chem. Soc. Dalton Trans.* **2001**, 3465–3469. [CrossRef]
27. Clegg, W.; Cope, E.K.; Edwards, A.J.; Mair, F.S. Structural Characterization of [(2,6-Pri$_2$C$_6$H$_3$)NC(Me)C(H)C(Me)N(2,6-Pri$_2$C$_6$H$_3$)K·PhCH$_3$]$_\infty$: A Heavy Alkali Metal Diazapentadienyl Complex. *Inorg. Chem.* **1998**, *37*, 2317–2319. [CrossRef]
28. Pyykö, P.; Atsumi, M. Molecular Single-Bond Covalent Radii for Elements 1–118. *Chem. Eur. J.* **2009**, *15*, 186–197. [CrossRef]
29. Mantina, M.; Chamberlain, A.C.; Valero, R.; Cramer, C.J.; Truhlar, D.G. Consistent van der Waals Radii for the Whole Main Group. *J. Phys. Chem. A* **2009**, *113*, 5806–5812. [CrossRef] [PubMed]
30. Dove, A.P.; Gibson, V.C.; Marshall, E.L.; White, A.J.; Williams, D.J. Magnesium and zinc complexes of a potentially tridentate β-diketiminate ligand. *Dalton Trans.* **2004**, 570–578. [CrossRef] [PubMed]
31. Savka, R.; Plenio, H. Metal Complexes of Very Bulky N,N'-Diarylimidazolylidene N-Heterocyclic Carbene (NHC) Ligands with 2,4,6-Cycloalkyl Substituents. *Eur. J. Inorg. Chem.* **2014**, *2014*, 6246–6253. [CrossRef]
32. Kundu, S.; Sinhababu, S.; Siddiqui, S.S.; Luebben, A.V.; Dittrich, B.; Yang, T.; Frenking, G.; Roesky, H.W. Comparison of Two Phosphinidenes Binding to Silicon(IV)dichloride as well as to Silylene. *J. Am. Chem. Soc.* **2018**, *140*, 9409–9412. [CrossRef] [PubMed]
33. Sheldrick, G.M. SHELX-16. University of Göttingen, 2016.

Communication

Silver Cyanoguanidine Nitrate Hydrate: Ag(C$_2$N$_4$H$_4$)NO$_3$·½ H$_2$O, a Cyanoguanidine Compound Coordinating by an Inner Nitrogen Atom

Xianji Qiao [1], Alex J. Corkett [1], Dongbao Luo [1,2] and Richard Dronskowski [1,2,*]

1. Chair of Solid-State and Quantum Chemistry, Institute of Inorganic Chemistry, RWTH Aachen University, D-52056 Aachen, Germany; xianji.qiao@ac.rwth-aachen.de (X.Q.); alexander.corkett@ac.rwth-aachen.de (A.J.C.); dongbao.luo@ac.rwth-aachen.de (D.L.)
2. Hoffmann Institute of Advanced Materials, Shenzhen Polytechnic, 7098 Liuxian Blvd, Nanshan District, Shenzhen 518055, China
* Correspondence: drons@HAL9000.ac.rwth-aachen.de; Tel.: +49-241-80-93642; Fax: +49-241-80-92642

Received: 5 November 2020; Accepted: 17 November 2020; Published: 24 November 2020

Abstract: Silver(I) cyanoguanidine nitrate hydrate, Ag(C$_2$N$_4$H$_4$)NO$_3$·½H$_2$O, was synthesized as the first cyanoguanidine solid-state complex in which monovalent Ag is coordinated through inner nitrogen N atoms. Its structure was characterized by single-crystal X-ray diffraction, crystallizing in the acentric orthorhombic space group $P2_12_12$ with a = 10.670(3) Å, b = 18.236(5) Å, and c = 3.5078(9) Å. The differing chemical bondings of Ag(C$_2$N$_4$H$_4$)NO$_3$·½H$_2$O and Ag(C$_2$N$_4$H$_4$)$_3$NO$_3$ were compared on the basis of first-principle calculations.

Keywords: cyanoguanidine; silver; crystal structure; chemical bonding

1. Introduction

Cyanoguanidine, also called dicyandiamide, is not only the dimer of molecular cyanamide (or carbodiimide) but also a commercially important compound that has found medicinal and industrial applications [1–4] and has long been known to crystallographers. The molecule has also been widely used as a precursor for synthesizing organonitrogen compounds [5]. It exists in two tautomeric forms, differing in the protonation and bonding of its nitrogen atoms (Figure 1). The structural conformations of both forms are planar with one tautomer displaying two terminal amine groups (Figure 1a) and the other expressing only one terminal amine group but two imine groups in the middle of the molecule (Figure 1b). The coordination chemistry of cyanoguanidine has earned significant attention owing to these differing functional groups [6–11], which readily coordinate late transition metals, forming complexes with monovalent [6] and divalent copper [7], silver [8,11], and divalent zinc [9,10], all of which have been structurally characterized. As alluded to in Figure 1, the inner N atoms as well as the terminal N atoms are able, at least in principle, to coordinate to metal atoms due to the existence of lone pairs at those N atoms. However, to the best of our knowledge, all reported metal cyanoguanidine compounds involve a coordination only with the terminal cyano N atoms of the cyanoguanidine molecules.

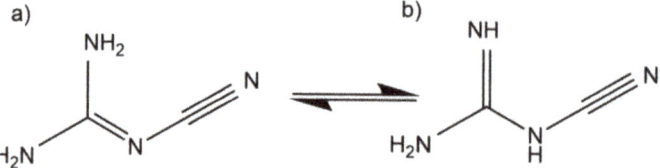

Figure 1. The two tautomeric forms of cyanoguanidine $H_4C_2N_4$ with (**a**) one tautomer displaying two terminal amine groups and (**b**) only one terminal amine group.

In an attempt to grow high-quality $Ag(C_2N_4H_4)_3NO_3$ single crystals by slow diffusion, we unexpectedly made the discovery of the title compound $Ag(C_2N_4H_4)NO_3 \cdot \frac{1}{2}H_2O$. Here, we report the synthesis and crystal structure of this silver(I) cyanoguanidine nitrate hydrate, the first example of a metal cyanoguanidine compound coordinating with inner nitrogen atoms.

2. Results and Discussion

2.1. Structure Description

The crystal structure was analyzed by single-crystal XRD. $Ag(C_2N_4H_4)NO_3 \cdot \frac{1}{2}H_2O$ crystallizes in the orthorhombic space group $P2_12_12$ (No. 18) with $a = 10.670(3)$ Å, $b = 18.236(5)$ Å, and $c = 3.5078(9)$ Å, as well as four formula units per cell (Figure 2a). The Ag atom is coordinated by two terminal N3 atoms, one inner N2 atom, and two O atoms of the nitrate group, O1 and O2, in a distorted square-pyramidal environment (Figure 2b). One of the Ag–N3 bonds is 2.283(8) Å (symmetry code: $1-x, 1-y, 1+z$), which is lengthwise consistent with those given in the literature [8], while the other Ag–N3 bond of 2.543(8) Å (symmetry code: $1-x, 1-y, z$) is somewhat longer and also weaker than the preceding one. The Ag–N2, Ag–O1, and Ag–O2 distances are 2.277(7) Å, 2.444(7) Å, and 2.733(8) Å, respectively, and within the expected distance range of Ag–N and Ag–O single bonds [8,12,13]. The elongated Ag–N3 bond probably results from further coordinative interactions of the type Ag···N and Ag···O (Ag–N2, Ag–N3, and Ag–O1) [8]. As we have discussed in the previous section, the cyanoguanidine molecule exists in two tautomeric forms (Figure 1), and here the C2–N4 and C2–N5 bond lengths come very close to 1.32 Å, a result strongly suggestive of the first form with two terminal amine groups (Figure 1a), the N4–C2–N5 angle also being almost 120°. Additional selected bond lengths and angles are given in Table 1.

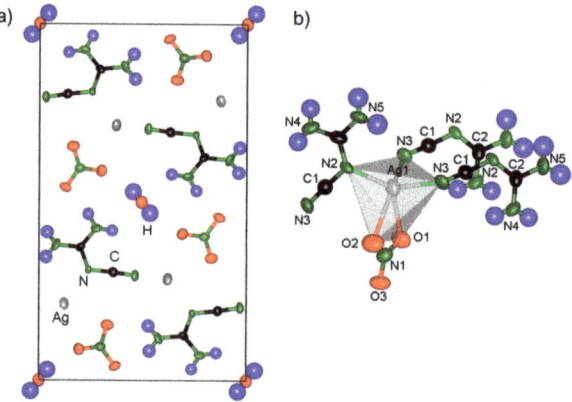

Figure 2. (**a**) Crystal structure of $Ag(C_2N_4H_4)NO_3 \cdot \frac{1}{2}H_2O$ and (**b**) coordination environment of Ag.

Table 1. Selected bond lengths and angles in Ag($C_2N_4H_4$)$NO_3 \cdot \frac{1}{2}H_2O$.

Bond Length (Å)	Bond Angle (°)
Ag1–N2 2.277(7)	O2–N1–O3 120.9(8)
Ag1–N3 (−½+x, ½−y, −z) 2.283(8)	O2–N1–O1 119.1(7)
Ag1–O1 2.444(7)	O3–N1–O1 119.9(8)
Ag1–N3 (−½+x, ½−y, 1−z) 2.543(8)	C1–N2–C2 116.2(7)
O1–N1 1.257(10)	C2–N4–H4A 120(4)
O2–N1 1.244(9)	C2–N4–H4B 114(4)
O3–N1 1.242(9)	H4A–N4–H4B 120(4)
N2–C1 1.316(10)	C2–N5–H5A 127(5)
N2–C2 1.357(12)	C2–N5–H5B 117(5)
N3–C1 1.152(10)	H5A–N5–H5B 115(4)
N4–C2 1.324(12)	N3–C1–N2 176.5(11)
N4–H4A 1.004(13)	N5–C2–N4 120.4(9)
N4–H4B 1.002(13)	N5–C2–N2 116.6(9)
N5–C2 1.324(10)	N4–C2–N2 122.9(7)
N5–H5A 1.008(13)	H1–O1W–H1 105(2)
N5–H5B 1.009(13)	
O1W–H1 (x, y, z) 0.976(13)	
O1W–H1(1−x, 1−y, 1−z) 0.976(13)	

The cyanoguanidine unit may act both as an acceptor (cyanimine, =N–C≡N) and as a donor (amine, –NH$_2$), resulting in four different hydrogen bonding modes. The first is a contact between the amine group and nitrate (N4–H4B⋯O1, N4–H4B⋯O3, N4–H4B⋯N1), the second an interaction between the amine group and neighboring nitrate (N5–H5A⋯O3, N5–H5B⋯O2), and the third an interaction between a nitrate O atom and a water O atom (O1W–H1⋯O1). Finally, there is hydrogen bonding between the amine group and a water O atom (N4–H4A⋯O1W). The entire complex is linked by hydrogen bonds forming a three-dimensional structure (Figure 3). The hydrogen bonds are listed in Table 2.

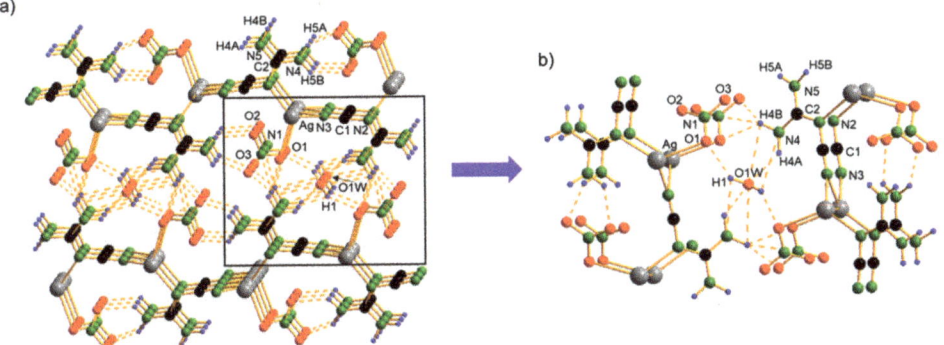

Figure 3. (a) The three-dimensional hydrogen-bond network in Ag($C_2N_4H_4$)$NO_3 \cdot \frac{1}{2}H_2O$ along c and (b) a selection of hydrogen bonds along a.

2.2. Chemical Bonding Analysis

As the discovery of Ag($C_2N_4H_4$)$NO_3 \cdot \frac{1}{2}H_2O$ was a consequence of preparing yet another silver cyanoguanidine nitrate, Ag($C_2N_4H_4$)$_3NO_3$ [8], questions of chemical equilibrium as regards the energetic competition between the two phases pop up, so their structural and bonding character was analyzed in more detail. As shown in Figure 2b, the Ag atom in Ag($C_2N_4H_4$)$NO_3 \cdot \frac{1}{2}H_2O$ is surrounded by two amine and one imine N atoms of the $H_4C_2N_4$ molecule and also two O atoms of the nitrate anion, while the Ag atom in Ag($C_2N_4H_4$)$_3NO_3$ is surrounded by five amine N atoms.

Additionally, a comparison of the bond lengths (Figure 4) indicates that the Ag–N bonds are shorter in Ag(C$_2$N$_4$H$_4$)$_3$NO$_3$ than the Ag–N and Ag–O bonds in Ag(C$_2$N$_4$H$_4$)NO$_3$·½H$_2$O.

Table 2. Hydrogen-bonding geometry (Å, °).

D–H	d(D–H)	d(H·A)	<DHA	d(D·A)	A	Symmetry Code
N4–H4A	1.004(13)	2.13(6)	136(6)	2.939(11)	O1W	
N4–H4B	1.002(13)	2.29(5)	135(6)	3.075(10)	O1	$-x+½, y+½, -z+1$
N4–H4B	1.002(13)	2.13(5)	148(8)	3.028(10)	O3	$-x+½, y+½, -z+1$
N4–H4B	1.002(13)	2.54(4)	156(8)	3.477(11)	N1	$-x+½, y+½, -z+1$
N5–H5A	1.008(13)	2.20(5)	149(7)	3.107(10)	O3	$x-½, -y+½, -z+1$
N5–H5B	1.009(13)	2.32(6)	137(5)	3.132(11)	O2	$x-½, -y+½, -z$
O1W–H1	0.976(13)	2.15(6)	146(9)	3.007(10)	O1	$-x+½, y+½, -z$

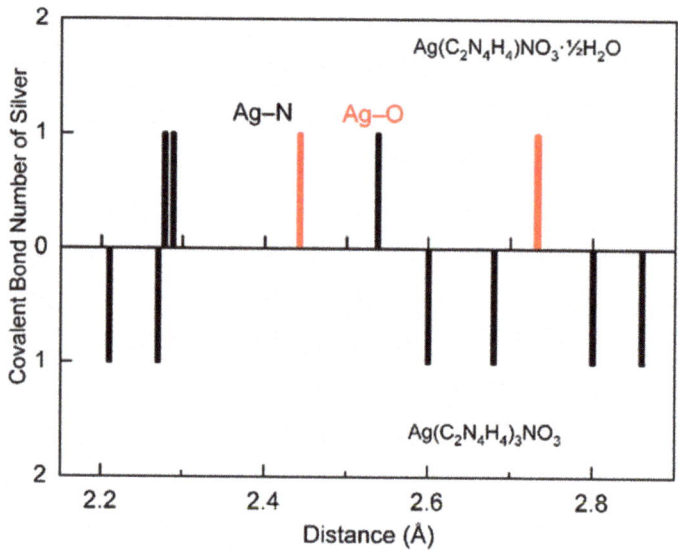

Figure 4. Bond-length histogram of Ag(H$_4$C$_2$N$_4$)NO$_3$·½H$_2$O (top) and Ag(H$_4$C$_2$N$_4$)$_3$NO$_3$ (bottom).

In order to quantitatively probe the relative stability of these phases, crystal orbital Hamilton population (COHP) analyses were performed to study the bonding mechanisms, as shown in Figure 5. Note that the integrated COHP (ICOHP) up to the Fermi level measures covalency, with more negative ICOHP values corresponding to stronger bonding interactions. Figure 5 reveals that the enhanced stability of Ag(C$_2$N$_4$H$_4$)$_3$NO$_3$, compared with Ag(C$_2$N$_4$H$_4$)NO$_3$·½H$_2$O, appears to arise from the increased strength of the two short Ag–N bonds in Ag(C$_2$N$_4$H$_4$)$_3$NO$_3$ (ICOHP = −1.58 eV for Ag–N1 and −1.31 eV for Ag–N5), which are significantly stronger than the two short Ag–N bonds in Ag(C$_2$N$_4$H$_4$)NO$_3$·½H$_2$O (−1.08 eV for Ag–N2 and −1.14 eV for Ag–N3). Furthermore, we note that the inner Ag–N2 mode in Ag(C$_2$N$_4$H$_4$)NO$_3$·½H$_2$O is weaker than the terminal Ag–N3 interaction despite being slightly shorter. It seems, therefore, that the inner N coordination mode is less strongly bonding than equivalent terminal N bonds, thus explaining the preferential formation of Ag(C$_2$N$_4$H$_4$)$_3$NO$_3$ over Ag(C$_2$N$_4$H$_4$)NO$_3$·½H$_2$O and the absence of other metal cyanoguanidines that coordinate with inner N atoms.

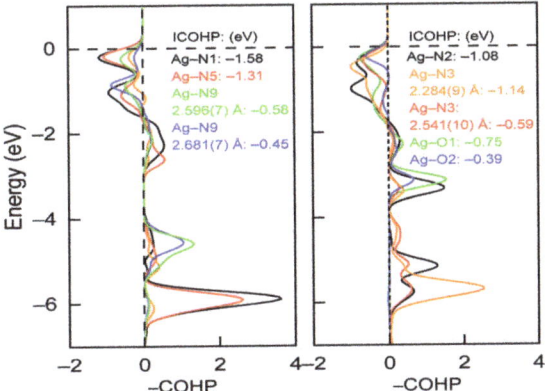

Figure 5. COHP analysis of various bonds in Ag(C$_2$N$_4$H$_4$)$_3$NO$_3$ (left) and Ag(C$_2$N$_4$H$_4$)NO$_3$·½H$_2$O (right).

3. Materials and Methods

3.1. Synthesis

Ag(C$_2$N$_4$H$_4$)NO$_3$·½H$_2$O was prepared by mixing aqueous solutions of AgNO$_3$ (339.7 mg, 2 mmol in 20 mL; Alfa, Kandel, Rheinland-Pfalz, Germany) and H$_4$C$_2$N$_4$ (504.0 mg, 2 mmol in 20 mL; Alfa, Kandel, Rheinland-Pfalz, Germany). This reaction yielded a large amount of white precipitate that was isolated by filtration and washed with water before being dried under vacuum. The product was then recrystallized by dissolving 0.5 g in 20 mL water and stirring for 1 h. Colorless, transparent needle-like crystals were obtained after water evaporation. It should be noted that attempts to grow high-quality single crystals by evaporation were hindered by the preferential formation of yet another silver cyanoguanidine compound, Ag(C$_2$N$_4$H$_4$)$_3$NO$_3$ [8], whose crystal structure was clarified many years ago. In order to help understand this preferential formation, electronic structure calculations were performed as detailed below.

3.2. Single-Crystal Diffraction

A selected crystal was mounted on a glass fiber. Intensity data were collected with a Bruker SMART APEX CCD detector (Bruker AXS GmbH, Karlsruhe, Germany) equipped with an Incoatec microsource (Mo-Kα radiation, λ = 0.71073 Å, multilayer optics). The software package SHELXL-2014 (Version 2014/7, George M. Sheldrick, Göttingen, Germany) [14,15] was used to solve and refine the crystal structure. Full details concerning the structure determination are available in CIF format and have been deposited under the CCDC entry number 1973094. Further crystallographic details can be found in the Supplementary Materials. Copies of the data can be obtained free of charge via https://www.ccdc.cam.ac.uk/Community/csd-communications/ or from the CCDC, 12 Union Road, Cambridge CB2 1EZ, U.K.; Fax: + 44-1223-336033; E-mail: deposit@ccdc.cam.ac.uk.

3.3. Chemical Bonding Analysis

The chemical bonding analysis was performed using the LOBSTER package (Version 3.2.0, Richard Dronskowski, Aachen, Germany) [16–19], starting from a static self-consistency calculation using the entire k-mesh with the symmetry switched off. For bonding analysis, the "pbevaspfit2015" basis set including 1s orbitals for hydrogen, 2s and 2p orbitals for nitrogen, carbon, and oxygen, and the 5s, 4p, and 4d orbitals for silver was employed.

4. Conclusions

The first silver(I) cyanoguanidine nitrate in which silver is coordinated by inner imine nitrogen atoms was prepared by mixing aqueous solutions of $AgNO_3$ and $H_4C_2N_4$. LOBSTER calculations were used to compare the chemical bonding interactions of $Ag(C_2N_4H_4)NO_3 \cdot \frac{1}{2}H_2O$ and $Ag(C_2N_4H_4)_3NO_3$ such as to explain the limited stability of $Ag(C_2N_4H_4)NO_3 \cdot \frac{1}{2}H_2O$. ICOHP analysis reveals that the inner N interaction in $Ag(C_2N_4H_4)_3NO_3 \cdot \frac{1}{2}H_2O$ is somewhat weaker than equivalent terminal N bonding modes, thereby explaining the propensity for terminal N bonding in metal cyanoguanidines.

Supplementary Materials: The following are available online at http://www.mdpi.com/2304-6740/8/12/64/s1: The CIF and the checkCIF output files.

Author Contributions: The manuscript was written through contributions of all authors. Conceptualization, X.Q.; Methodology, X.Q.; Software, X.Q. and D.L.; Validation, X.Q.; Formal Analysis, X.Q., A.J.C. and D.L.; Investigation, X.Q.; Resources, R.D.; Data Curation, R.D., X.Q. and A.J.C.; Writing-Original Draft Preparation, X.Q.; Writing-Review & Editing, R.D., A.J.C. and X.Q.; Visualization, X.Q., A.J.C.; Supervision, R.D.; Project Administration, R.D.; Funding Acquisition, R.D. All authors have read and agreed to the published version of the manuscript.

Funding: This research received no external funding.

Acknowledgments: We thank Markus Mann and George Ogutu for synthetic assistance. We also thank Tobias Storp for assistance with the single-crystal X-ray diffraction measurement. Furthermore, we thank the IT Center of RWTH Aachen University for the provision of computing time.

Conflicts of Interest: The authors declare no conflict of interest.

References

1. Barrio, J.; Volokh, M.; Shalom, M. Polymeric carbon nitrides and related metal-free materials for energy and environmental applications. *J. Mater. Chem. A* **2020**, *8*, 11075–11116.
2. Williams, P.; Ferrer, E.G.; Baeza, N.; Piro, O.E.; Castellano, E.E.; Baran, E.J. Transition metal promoted addition of methanol to cyanoguanidine. Molecular structure and properties of the generated copper(II) and nickel(II) complexes. *Z. Anorg. Allg. Chem.* **2005**, *631*, 1502–1506.
3. Kose, M.; Duman, S.E.; McKee, V.; Akyol, I.; Kurtoglu, M. Hydrogen bond directed 1D to 3D structures of square-planar Ni(II) complexes and their antimicrobial studies. *Inorg. Chim. Acta* **2017**, *462*, 281–288.
4. Gungor, O.; Kurtar, S.K.; Koçer, F.; Kose, M. A guanylurea ligand and its Cu(II), Ni(II) and Zn(II) complexes: Antibacterial activities and DNA binding properties. *Inorg. Nano Met. Chem.* **2020**, 1–9. [CrossRef]
5. Albertin, G.; Antoniutti, S.; Castro, J. Preparation of diethycyanamide and cyanoguanidine complexes of iridium. *ChemistrySelect.* **2018**, *3*, 11054–11058.
6. Begley, M.J.; Hubberstey, P.; Russell, C.E.; Walton, P.H. Bis(μ-pyridazine)-bis[(2-cyanoguanidine)copper(I)] cation: A molecule containing two co-ordinatively unsaturated copper(I) centres. *J. Chem. Soc. Dalton Trans.* **1994**, *17*, 2483–2488.
7. Begley, M.J.; Hubberstey, P.; Stroud, J. Copper(II)-diethylenetriamine complexes with monodentate N-donor ligands: Crystal and molecular structure of diethylenetriamine 2-cyanoguanidine copper(II) nitrate. *Polyhedron* **1997**, *16*, 805–813.
8. Dronskowski, R.; Liu, X.H. Bis(cyanoguanidine) silver(I) nitrate-cyanoguanidine(1/1). *Acta Cryst. C* **2003**, *59*, m243–m245.
9. Fowkes, A.; Harrison, W.T.A. A monoclinic polymorph of dichlorobis(cyanoguanidine) zinc(II). *Acta Crystallogr. Sect. E* **2005**, *61*, m2021–m2022.
10. Ritche, L.K.; Harrison, W.T. A *catena*-Poly[[dichlorozinc(II)]-μ-cyanoguanidine]. *Acta Crystallogr. Sect. E* **2007**, *63*, m617–m618.
11. Bessler, K.E.; Sousa, A.T.; Deflon, V.M.; Niquet, E. Silver complexes with cyanoguanidine: Preparation and crystal structures of $(Ag(cgn)_2)F$ and $(Ag(cgn)_2)(BF_4)$. *Z. Anorg. Allg. Chem.* **2003**, *629*, 1091–1095.
12. Meyer, P.; Rimsky, A.; Chevalier, R. Structure d'une phase du nitrate d'argent instable à température et pression ordinaires. *Acta Cryst. B* **1976**, *32*, 1143–1146.
13. Meyer, P.; Rimsky, A.; Chevalier, R. Structure du nitrate d'argent à pression et température ordinaires. Exemple de cristal parfait. *Acta Cryst. B* **1978**, *34*, 1457–1462.

14. Sheldrick, G.M. A short history of SHELX. *Acta Crystallogr. Sect. A* **2008**, *64*, 112–122.
15. Sheldrick, G.M. Crystal structure refinement with SHELXL. *Acta Crystallogr. Sect. C* **2015**, *71*, 3–8.
16. Deringer, V.L.; Tchougréeff, A.L.; Dronskowski, R. Crystal orbital Hamilton population (COHP) analysis as projected from plane-wave basis sets. *J. Phys. Chem. A* **2011**, *115*, 5461–5466. [PubMed]
17. Maintz, S.; Deringer, V.L.; Tchougréeff, A.L.; Dronskowski, R. Analytic projection from plane-wave and PAW wavefunctions and application to chemical-bonding analysis in solids. *J. Comput. Chem.* **2013**, *34*, 2557–2567. [PubMed]
18. Dronskowski, R.; Blöchl, P.E. Crystal orbital Hamilton populations (COHP): Energy-resolved visualization of chemical bonding in solids based on density-functional calculations. *J. Phys. Chem.* **1993**, *97*, 8617–8824.
19. Nelson, R.; Ertural, C.; George, J.; Deringer, V.L.; Hautier, G.; Dronskowski, R. LOBSTER: Local orbital projections, atomic charges, and chemical-bonding analysis from projector-augmented-wave-based density-functional theory. *J. Comput. Chem.* **2020**, *41*, 1931–1940. [PubMed]

Publisher's Note: MDPI stays neutral with regard to jurisdictional claims in published maps and institutional affiliations.

 © 2020 by the authors. Licensee MDPI, Basel, Switzerland. This article is an open access article distributed under the terms and conditions of the Creative Commons Attribution (CC BY) license (http://creativecommons.org/licenses/by/4.0/).

Article

Novel Fluoridoaluminates from Ammonothermal Synthesis: Two Modifications of K_2AlF_5 and the Elpasolite Rb_2KAlF_6

Christian Bäucker, Peter Becker, Keshia J. Morell and Rainer Niewa *

Institute of Inorganic Chemistry, University of Stuttgart, Pfaffenwaldring 55, 70569 Stuttgart, Germany; c.baeucker@gmail.com (C.B.); peter.heuberg@t-online.de (P.B.); keshia.morell@gmx.de (K.J.M.)
* Correspondence: rainer.niewa@iac.uni-stuttgart.de

Abstract: Two new modifications of the pentafluoridoaluminate K_2AlF_5 were obtained from ammonothermal synthesis at 753 K, 224 MPa and 773 K, 220 MPa, respectively. Both crystallize in the orthorhombic space group type $Pbcn$, with close metric relations and feature kinked chains of cis-vertex-connected AlF_6 octahedra resulting in the Niggli formula $^1_\infty\{[AlF_{2/2}^eF_{4/1}^t]^{2-}\}$. The differences lie in the number of octahedra necessary for repetition within the chains, which for K_2AlF_5-2 is realized after four and for K_2AlF_5-3 after eight octahedra. As a result, the orthorhombic unit cell for K_2AlF_5-3 is doubled in chain prolongation direction [001] as compared to K_2AlF_5-2 (1971.18(4) pm versus 988.45(3) pm, respectively), while the unit cell parameters within the other two directions are virtually identical. Moreover, the new elpasolite Rb_2KAlF_6 is reported, crystallizing in the cubic space group $Fm\bar{3}m$ with a = 868.9(1) pm and obtained under ammonothermal conditions at 723 K and 152 MPa.

Keywords: ammonothermal synthesis; fluoride; aluminum

Citation: Bäucker, C.; Becker, P.; Morell, K.J.; Niewa, R. Novel Fluoridoaluminates from Ammonothermal Synthesis: Two Modifications of K_2AlF_5 and the Elpasolite Rb_2KAlF_6. *Inorganics* **2022**, *10*, 7. https://doi.org/10.3390/inorganics10010007

Academic Editor: Hans-Conrad zur Loye

Received: 19 December 2021
Accepted: 8 January 2022
Published: 10 January 2022

Publisher's Note: MDPI stays neutral with regard to jurisdictional claims in published maps and institutional affiliations.

Copyright: © 2022 by the authors. Licensee MDPI, Basel, Switzerland. This article is an open access article distributed under the terms and conditions of the Creative Commons Attribution (CC BY) license (https://creativecommons.org/licenses/by/4.0/).

1. Introduction

Ammonothermal synthesis, which uses supercritical ammonia as a solvent medium, is a pathway to the production and crystal growth of a variety of materials, particularly nitrides, amides, and even halides, depending on the administered mineralizer [1,2]. Especially due to the available nitrides from ammonothermal reactions, such as AlN [3], GaN [4,5], and InN [6], this technique has gained considerable interest [7]. Exploratory work in ammonothermal synthesis recently led to solid intermediate amides, which may give valuable information about the dissolved species during the process as well as the condensation process eventually producing nitride materials [8–10]. Similar to the hydrothermal technique, we frequently observe the formation of several modifications of the same compound under rather similar pressure–temperature conditions during synthesis [11].

For ammonothermal synthesis, typically supercritical conditions are applied (critical point of pure ammonia at 405.2 K and 11.3 MPa [12]), which are most often realized by applying elevated temperatures to a sealed reaction vessel, therefore reaching a high pressure from the expanding ammonia contained within. In this experimental set-up, pressure and thus ammonia density fundamentally depend on the temperature and filling degree of the autoclave with ammonia. The supercritical state of the solvent in combination with suitable mineralizers is intended to provide sufficient solubility and supersaturation for solid product formation and ideally crystal growth. The mineralizer can provide ammonobasic or ammonoacidic conditions within a wide range of pH and serves for the formation of dissolved complex species.

K_2AlF_5 was first reported by de Kozak et al. as a dehydrate of $K_2AlF_5·H_2O$, which was obtained by hydrothermal synthesis [13]. During thermogravimetric investigations, the water-free K_2AlF_5 was observed to form upon heating. In the crystal structure of the monohydrate, very slightly kinked infinite trans-vertex-sharing octahedra with Niggli formula $^1_\infty\{[AlF_{2/2}{}^eF_{4/1}{}^t]^{2-}\}$ (e = corner-sharing, t = terminal) are prearranged to result in straight chains in the dehydrate. This compound, in the following denoted K_2AlF_5-1, crystallizes in $P4/mmm$ and spontaneously transforms back to its monohydrate after days in air. While these fluoridoaluminates are characterized by infinite chains of trans-vertex-sharing octahedra around Al, isolated octahedra $[AlF_6]^{3-}$ are well known, for example from the mineral elpasolite, K_2NaAlF_6 [14,15], which might be described as a double perovskite. Fluorides with the elpasolite structure are numerous [14,16,17]. Here we report two new modifications of K_2AlF_5 featuring infinite chains of cis-vertex-sharing octahedra around Al with different conformations and the novel elpasolite Rb_2KAlF_6, obtained from ammonothermal synthesis using a mineralizer system containing alkali metal amides and fluoride ions in a near-ammononeutral regime.

2. Results

Two new modifications of K_2AlF_5 were obtained from supercritical ammonia under very similar conditions of 753 K, 224 MPa versus 773 K, 220 MPa. While we monitor the pressure in the autoclaves during the process, the temperatures given represent the furnace temperatures outside the autoclave. Average temperatures within the autoclave are about 100 K lower, and an intended temperature gradient of about 50 K naturally develops due to heat loss at the autoclave installations for filling, pressure monitoring, and safety purposes sticking out the furnace [18]. Thus from experimental conditions, we cannot derive any information on the relative stabilities of these different modifications. Upon providing both potassium and rubidium ions in solution at slightly reduced pressures (723 K and 152 MPa) the novel elpasolite Rb_2KAlF_6 was formed.

All three fluoridoaluminates were obtained as colorless crystals in the hot zone of the autoclave in a temperature gradient, thus exhibiting lower solubility at higher temperatures for the given conditions (so-called retrograde solubility). The temperature dependence of the solubility for a given mineralizer system typically depends on the pressure and temperature regime applied [2,19]. The mineralizer system itself has a fundamental impact on solubility and temperature dependence via the formation of dissolved species and intermediates. Here we used a combination of metal fluorides and alkali metal amides in molar ratios such that near-ammononeutral conditions are expected if completely reacted. While the dissolved intermediates are difficult to experimentally study in situ due to the extreme conditions administered, it is likely that they are represented in the present case by complex fluoridoaluminate ions.

2.1. Crystal Structures

Potassium Pentafluoridoaluminates K_2AlF_5

We present two novel modifications of K_2AlF_5, which both crystallize in the orthorhombic space group type $Pbcn$ and are labeled K_2AlF_5-2 and K_2AlF_5-3 in the following. According to structure determination from single-crystal X-ray diffraction, K_2AlF_5-3 (Figure 1, bottom) features a doubled c-axis in comparison to K_2AlF_5-2 (Figure 1, top), while the a and b axes are nearly identical. This fact can readily be understood from the crystal structures discussed in the following. Selected crystallographic data and refinement parameters are gathered in Table 1. Fractional atomic coordinates and isotropic displacement parameters of K_2AlF_5-2 and K_2AlF_5-3 can be found in Tables 2 and 3, respectively.

The crystal structure of K_2AlF_5-2 (Figure 1, top) features cis-vertex-sharing AlF_6-octahedra, which form infinite zig-zag chains running parallel [001]. These chains obey the Niggli formula $^1_\infty\{[AlF_{2/2}{}^eF_{4/1}{}^t]^{2-}\}$ and are aligned in the motif of a hexagonal rod packing (Figure 2, left). Repetition of the zig-zag chain is realized after four octahedra. The terminal fluoride ligands F(1–4) feature Al–F distances in the range of 176 pm to 179 pm, which is noticeably shorter than the distances to the bridging fluorides atoms (F(5) with 189 pm and F(6) with 192 pm). Selected interatomic distances are gathered in Table 4. Despite these slightly different distances, the octahedra are close to ideal as indicated by the internal angles near 90° and 180°. The angles at the bridging fluoride ligands ∢ (Al–F–Al) vary from linear (180° due to symmetry restrictions) to slightly kinked with 168° at F(5) and F(6), respectively.

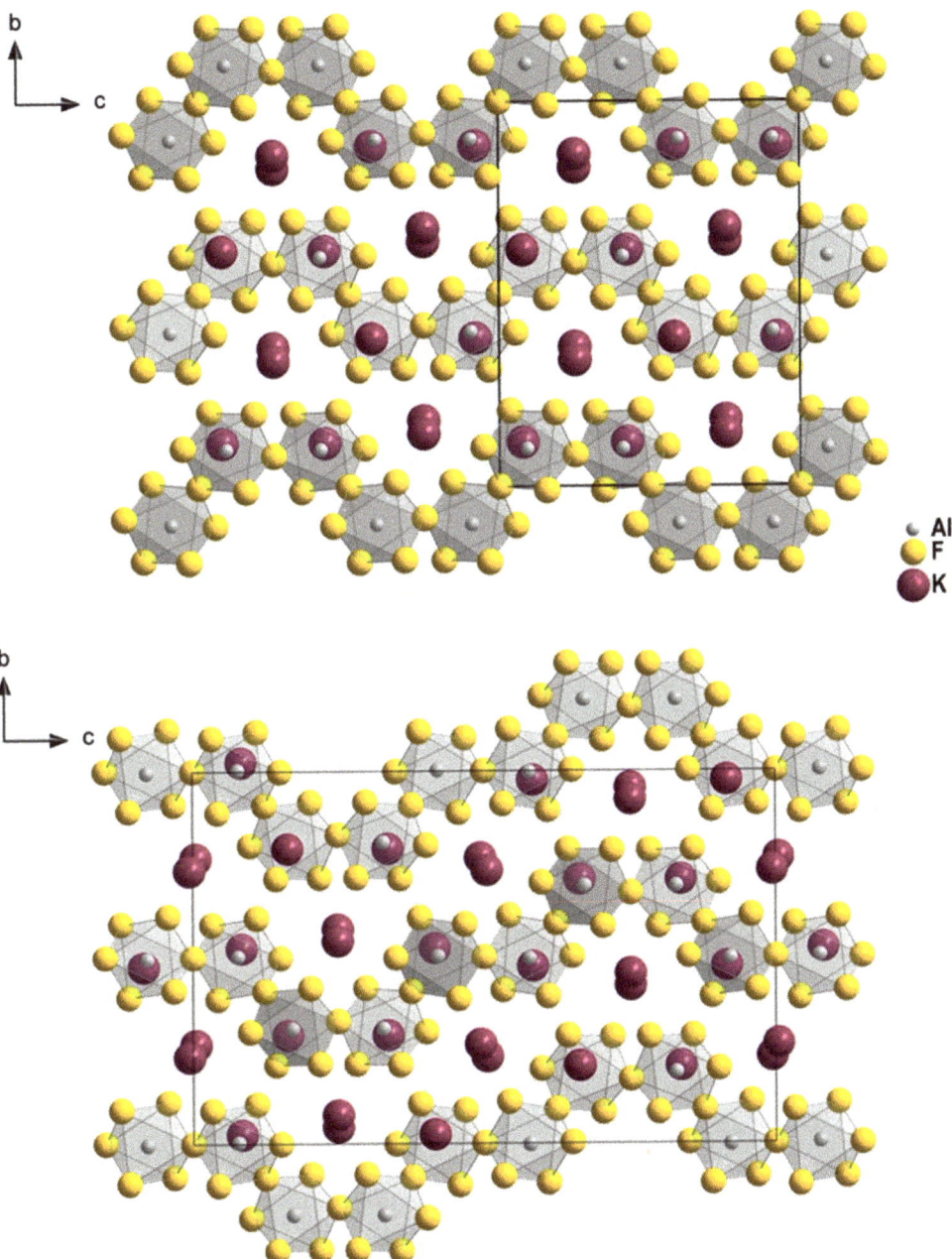

Figure 1. Sections of the crystal structures of K_2AlF_5-2 (**top**) and K_2AlF_5-3 (**bottom**), viewing direction [100].

Table 1. Selected crystallographic parameters and refinement data of K_2AlF_5-2, K_2AlF_5-3, and Rb_2KAlF_6.

Compound	K_2AlF_5-2	K_2AlF_5-3	Rb_2KAlF_6
Crystal system	Orthorhombic	Orthorhombic	Cubic
Space group	$Pbcn$	$Pbcn$	$Fm\bar{3}m$
a/pm	718.12(3)	718.370(10)	868.88(11)
b/pm	1265.94(5)	1264.49(2)	-
c/pm	988.45(3)	1971.18(4)	-
Z	8	16	4
Density (calculated)/g·cm^{-3}	2.959	2.970	3.554
Volume/10^6 pm^3	898.60(6)	1790.56(5)	656.0(2)
Index ranges hkl	$-9 \leq h \leq 9$ $-16 \leq k \leq 16$ $-12 \leq l \leq 11$	$-9 \leq h \leq 8$ $-16 \leq k \leq 16$ $-25 \leq l \leq 25$	$-10 \leq h \leq 9$ $-10 \leq k \leq 10$ $-11 \leq l \leq 11$
$2\theta_{max}/°$	54.94	54.96	54.96
$F(000)$	768	1536	640
T/K	293(2)	293(2)	293(2)
μ(Mo-K$_\alpha$)/mm^{-1}	2.30	2.31	15.73
Measured reflections/sym. independent	14490/1034	27919/2061	1072/58
R_{int}/R_σ	0.0857/0.0357	0.0557/0.0201	0.1148/0.0319
R_1 with $|F_o| > 4\sigma(F_o)$	0.0308	0.0219	0.0318
$R_1/wR_2/GooF$	0.0517/0.0732/1.107	0.0287/0.0554/1.112	0.0392/0.0949/1.189
Largest peak/hole in the difference electron density map/10^6 pm^{-3}	0.64/−0.43	0.31/−0.32	0.85/−0.41

Table 2. Fractional atomic coordinates and equivalent isotropic displacement parameters (in 10^4 pm^2) for K_2AlF_5-2.

Atom	Site	x/a	y/b	z/c	U_{eq}
Al	8d	0.1544(1)	0.09746(6)	0.40752(7)	0.0148(2)
F(1)	8d	0.0028(2)	0.1963(1)	0.4657(2)	0.0237(4)
F(2)	8d	0.2039(2)	0.3988(1)	0.0533(2)	0.0248(4)
F(3)	8d	0.2149(2)	0.4889(1)	0.3435(2)	0.0242(4)
F(4)	8d	0.2904(2)	0.1911(1)	0.3155(2)	0.0234(4)
F(5)	4a	0	0	0	0.0224(5)
F(6)	4c	0	0.0811(2)	1/4	0.0215(5)
K(1)	4c	0	0.32802(6)	1/4	0.0216(2)
K(2)	8d	0.35685(9)	0.11414(5)	0.07843(6)	0.0288(2)
K(3)	4c	1/2	0.36182(6)	1/4	0.0238(2)

Table 3. Fractional atomic coordinates and equivalent isotropic displacement parameters (in 10^4 pm^2) for K_2AlF_5-3.

Atom	Site	x/a	y/b	z/c	U_{eq}
Al(1)	8d	0.34006(6)	0.00306(4)	0.07804(2)	0.01285(12)
Al(2)	8d	0.15454(6)	0.30892(4)	0.32833(2)	0.01289(12)
F(1)	8d	0.00118(13)	0.21130(8)	0.14313(5)	0.02101(22)
F(2)	8d	0.00529(13)	0.40872(8)	0.37647(5)	0.02002(22)
F(3)	4b	1/2	1/2	0	0.01986(30)
F(4)	4c	0	0.32874(11)	1/4	0.01899(29)
F(5)	8d	0.28728(14)	0.21407(8)	0.28163(5)	0.02057(22)

Table 3. Cont.

Atom	Site	x/a	y/b	z/c	U_{eq}
F(6)	8d	0.27138(14)	0.38826(7)	0.04254(5)	0.01934(21)
F(7)	8d	0.47905(14)	0.10612(8)	0.10966(5)	0.02259(23)
F(8)	8d	0.30459(14)	0.49862(8)	0.15048(5)	0.02174(23)
F(9)	8d	0.19137(14)	0.08868(8)	0.03231(5)	0.02165(23)
F(10)	8d	0.28759(14)	0.41739(8)	0.29745(5)	0.02110(22)
F(11)	8d	0.29631(14)	0.30022(8)	0.40163(5)	0.02154(22)
K(1)	4c	1/2	0.04458(4)	1/4	0.02156(13)
K(2)	4c	0	0.07646(4)	1/4	0.01799(13)
K(3)	8d	0.15370(5)	0.02462(3)	0.41609(2)	0.02395(11)
K(4)	8d	0.36071(5)	0.29171(3)	0.16355(2)	0.02299(11)
K(5)	8d	0.01092(5)	0.26658(3)	0.49444(2)	0.01940(11)

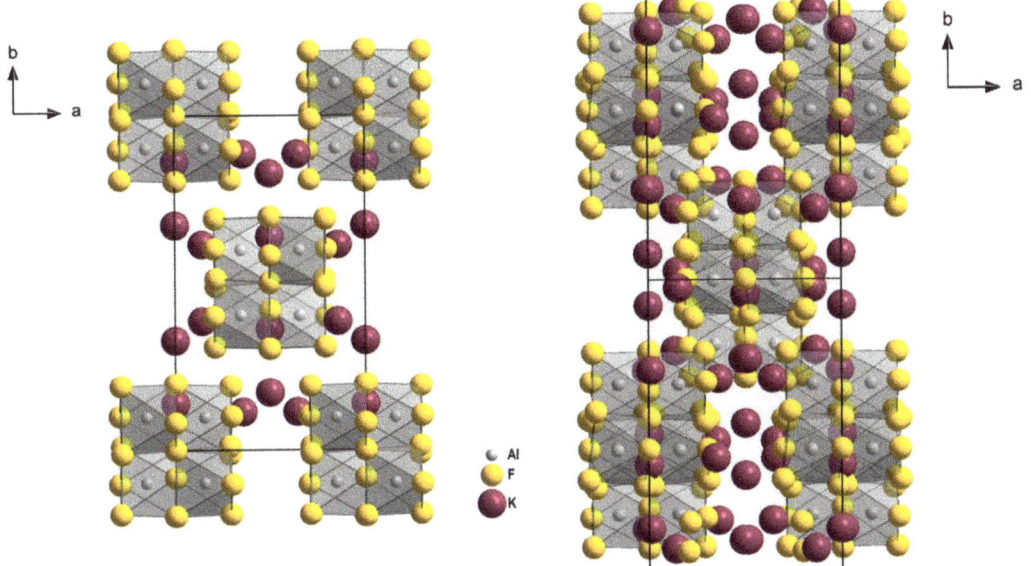

Figure 2. Section of the crystal structures of K_2AlF_5-2 (**left**) and K_2AlF_5-3 (**right**), shown along [001] emphasizing the motif of hexagonal rod packings by the infinite fluoroaluminate chains. The crystallographic unit cells are indicated by black lines.

Table 4. Selected interatomic distances (in pm) in K_2AlF_5-2.

	Distance/pm		Distance/pm		Distance/pm		Distance/pm
Al–F(1)	175.5(2)	K(1)–F(2)	259.4(2)	K(2)–F(4)	258.3(2)	K(3)–F(4)	271.2(2)
Al–F(2)	176.5(2)	K(1)–F(2)	259.4(2)	K(2)–F(3)	271.3(2)	K(3)–F(4)	271.2(2)
Al–F(3)	178.1(2)	K(1)–F(1)	270.6(2)	K(2)–F(2)	277.2(2)	K(3)–F(3)	276.3(2)
Al–F(4)	178.5(2)	K(1)–F(1)	270.7(2)	K(2)–F(2)	281.7(2)	K(3)–F(3)	276.3(2)
Al–F(5)	189.37(7)	K(1)–F(3)	271.7(2)	K(2)–F(1)	281.8(2)	K(1)–F(6)	277.6(2)
Al–F(6)	192.24(8)	K(1)–F(3)	271.7(2)	K(2)–F(1)	283.1(2)	K(3)–F(1)	290.5(2)
		K(1)–F(4)	278.8(2)	K(2)–F(4)	291.0(2)	K(3)–F(1)	290.5(2)
		K(1)–F(4)	278.8(2)	K(2)–F(5)	304.23(7)	K(3)–F(2)	291.9(2)
		K(1)–F(6)	312.6(2)	K(2)–F(6)	310.13(7)	K(3)–F(2)	291.9(2)
				K(2)–F(3)	310.6(2)	K(3)–F(5)	302.76(5)
				K(2)–F(3)	311.8(2)	K(3)–F(5)	302.76(5)

The potassium ions interconnect three chains each (Figure 3). As it can be taken from Figure 4, K(1) is coordinated by nine fluoride ions, K(2) and K(3) by eleven fluoride ligands. Coordination by the octahedra of the chains is realized mono-, bi-, and three-dentate. Distances with about 258 to 312 pm are in the expected range, given those found for K_2AlF_5-1, K_2AlF_5-3, and Rb_2KAlF_6 discussed below.

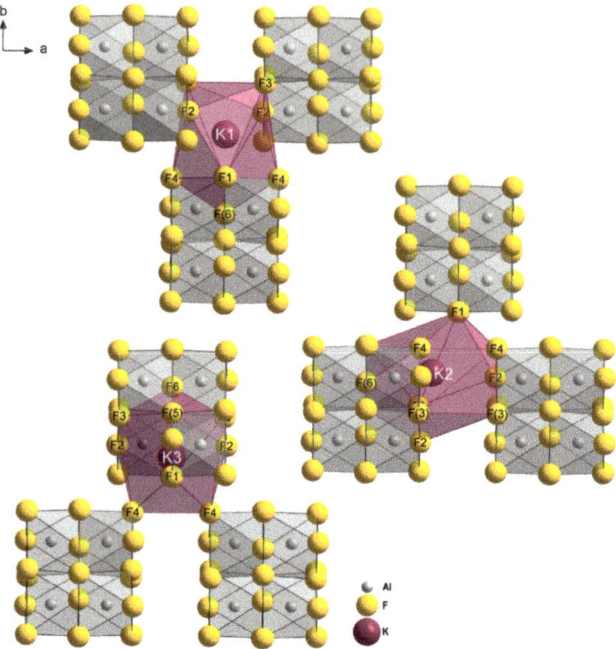

Figure 3. Interconnection of the infinite fluoridoaluminate zig-zag chains by potassium ions in K_2AlF_5-2.

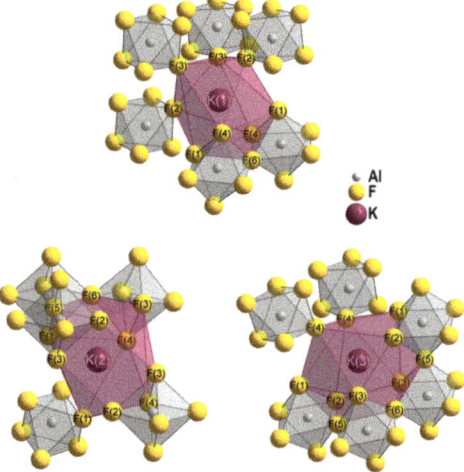

Figure 4. Potassium site coordination by AlF_6-octahedra in K_2AlF_5-2. Each potassium site connects three fluoridoaluminate chains.

Similar to K_2AlF_5-2, K_2AlF_5-3 crystallizes in the orthorhombic space group type *Pbcn*, but with a doubled *c*-axis (Figure 1, bottom). It crystallizes as an isotype of K_2FeF_5, first described by Vlasse et al. in the noncentrosymmetric space group type $Pn2_1a$, which was later corrected by Le Bail et al. [20,21]. The crystal structure of K_2AlF_5-3 shows close structural similarity to K_2AlF_5-2, while apparently no direct group–subgroup relation exists. The crystal structure of K_2AlF_5-3 also contains infinite zig-zag chains of *cis*-vertex-sharing AlF_6-octahedra, which are aligned parallel [001] in the motif of a hexagonal rod packing (Figure 2, right) and consistent with the Niggli formula $^1_\infty\{[AlF_{2/2}{}^eF_{4/1}{}^t]^{2-}\}$. However, in K_2AlF_5-3 the chains exhibit a longer repetition length, completed only after eight octahedra. Figure 1 gives a comparison of the arrangements of the chains in K_2AlF_5-2 and K_2AlF_5-3. The already known modification K_2AlF_5-1 (Cs_2MnF_5 structure type in *P4/mmm* with $a \approx 597$ pm, $c \approx 370$ pm), in contrast, is built from straight chains of *trans*-vertex-sharing octahedra with a 180° angle at the bridging fluoride ligands (Figure 5) [13]. The coordination polyhedron of potassium can be described as a bicapped square prism.

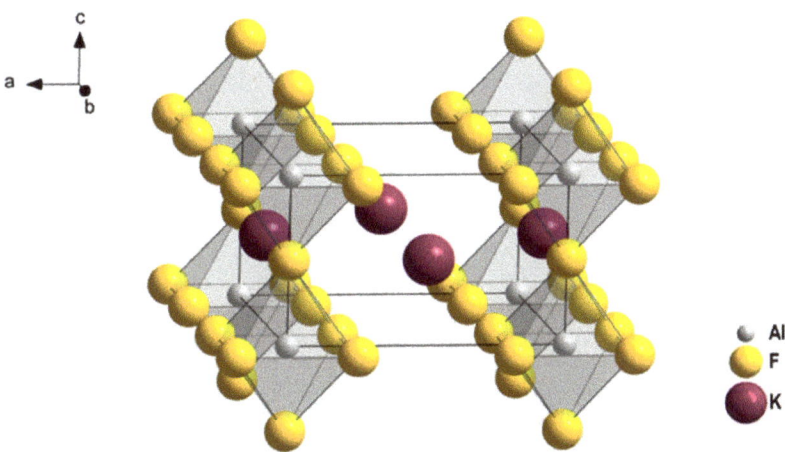

Figure 5. Section of the crystal structure of K_2AlF_5-1 according to [13].

The five different potassium sites in K_2AlF_5-3 are coordinated by nine, ten, or eleven fluoride ions (see Figure 6). In contrast to K_2AlF_5-2, the coordination by the octahedra of the chains is only realized mono- or three-dentate, but no bi-dentate coordination appears. Each site connects three chains of AlF_6-octahedra, similar to the situation in K_2AlF_5-2 (see Figure 7). Still, internal distances in both modifications are rather similar: d(Al–F) in K_2AlF_5-3 to terminal fluoride ligands range from 176 pm to 180 pm, while those to bridging fluoride ions are longer with 191 and 192 pm (Table 5). Again the internal angles within the octahedra are close to ideal the of 90° and 180°, while the angles at the bridging fluoride ions ∡ (Al–F–Al) diversify between 180° for F(3) (due to symmetry restrictions) and 165° at F(4). Distances between potassium and fluoride ions in both modifications are also rather similar, although the furthest distances in the coordination are increased to 334 pm in K_2AlF_5-3.

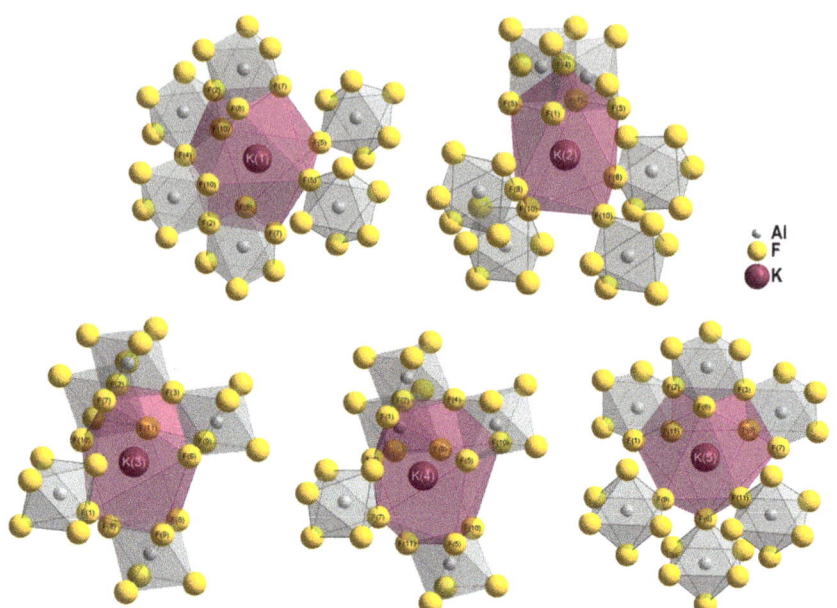

Figure 6. Potassium site coordination by AlF_6-octahedra in K_2AlF_5-3. Each site connects three fluoridoaluminate chains.

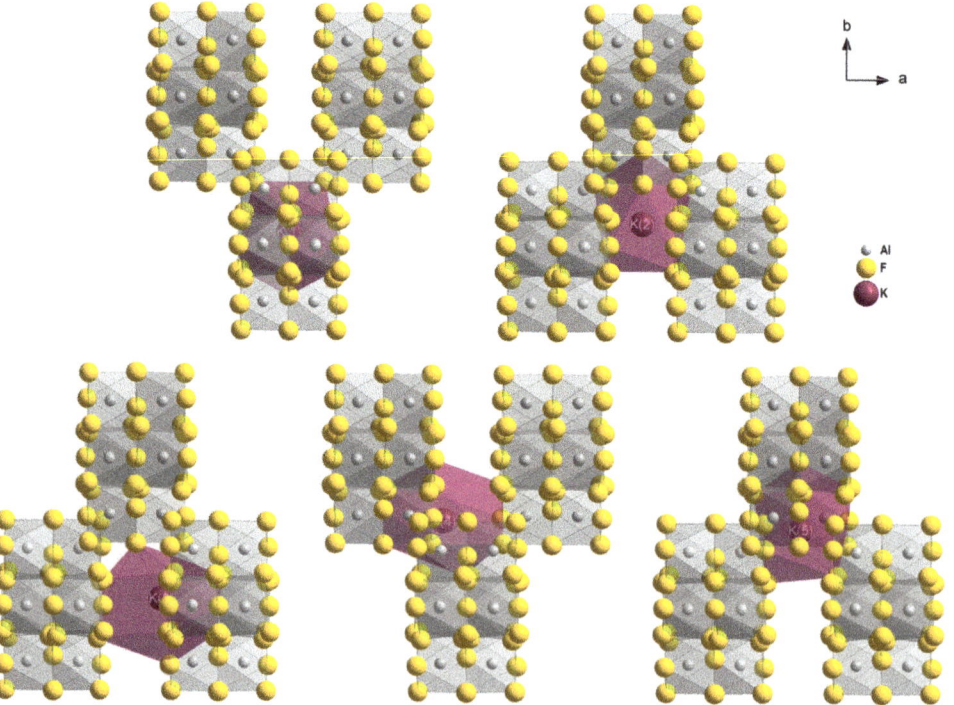

Figure 7. Interconnection of the infinite fluoridoaluminate zig-zag chains by potassium ions in K_2AlF_5-3.

Table 5. Selected interatomic distances (in pm) in K_2AlF_5-3.

	Distance/pm		Distance/pm		Distance/pm		Distance/pm
Al(1)–F(7)	175.60(11)	Al(2)–F(1)	175.82(10)	K(1)–F(5)	2 × 270.50(10)	K(2)–F(8)	2 × 260.54(9)
Al(1)–F(8)	176.68(10)	Al(2)–F(11)	177.12(10)	K(1)–F(4)	1 × 272.93(15)	K(2)–F(10)	2 × 269.24(10)
Al(1)–F(9)	176.79(10)	Al(2)–F(10)	177.92(10)	K(1)–F(10)	2 × 278.02(11)	K(2)–F(1)	2 × 271.02(10)
Al(1)–F(6)	179.95(10)	Al(2)–F(5)	178.75(10)	K(1)–F(7)	2 × 287.76(10)	K(2)–F(5)	2 × 277.05(11)
Al(1)–F(2)	190.68(10)	Al(2)–F(2)	190.85(10)	K(1)–F(8)	2 × 299.57(10)	K(2)–F(4)	1 × 319.00(15)
Al(1)–F(3)	192.04(5)	Al(2)–F(4)	191.82(5)	K(1)–F(2)	2 × 302.77(10)		
K(3)–F(9)	271.56(10)	K(4)–F(5)	258.05(10)	K(5)–F(6)	264.78(11)		
K(3)–F(10)	273.60(10)	K(4)–F(8)	265.97(10)	K(5)–F(6)	267.91(10)		
K(3)–F(6)	277.78(10)	K(4)–F(7)	271.27(11)	K(5)–F(11)	269.94(10)		
K(3)–F(9)	279.92(11)	K(4)–F(6)	275.55(10)	K(5)–F(9)	272.95(10)		
K(3)–F(8)	284.93(11)	K(4)–F(11)	278.09(11)	K(5)–F(11)	278.04(11)		
K(3)–F(1)	285.88(11)	K(4)–F(1)	280.47(11)	K(5)–F(7)	278.49(11)		
K(3)–F(11)	287.44(11)	K(4)–F(5)	291.99(11)	K(5)–F(1)	280.16(10)		
K(3)–F(7)	287.75(11)	K(4)–F(10)	308.22(11)	K(5)–F(9)	291.23(11)		
K(3)–F(2)	295.96(10)	K(4)–F(2)	311.84(10)	K(5)–F(2)	293.93(10)		
K(3)–F(3)	300.36(4)	K(4)–F(10)	312.53(11)	K(5)–F(3)	295.47(4)		
K(3)–F(6)	334.39(11)	K(4)–F(4)	313.65(4)				

2.2. Raman Spectroscopy

The Raman signals obtained from single crystals of K_2AlF_5-2 and K_2AlF_5-3 are assigned in accordance with spectra reported for K_3AlF_6 [22], Na_3AlF_6 [23], and KF:AlF_3 melts [24]. The Raman spectra resulted in three signals between 700 and 300 cm^{-1} for both K_2AlF_5-2 and K_2AlF_5-3 (Figure 8). The two strong and sharp signals at approximately 600 and 456 cm^{-1} can be assigned to the Al–F stretching modes ν_1 and ν_2. The third signal at around 377 cm^{-1} is rather weak and belongs to the F–Al–F bending mode. The remaining signals below 300 cm^{-1} relate to lattice vibration modes with signals at 282, 190, 160, and 71 cm^{-1} for K_2AlF_5-2 and 240, 178, and 65 cm^{-1} for K_2AlF_5-3, respectively.

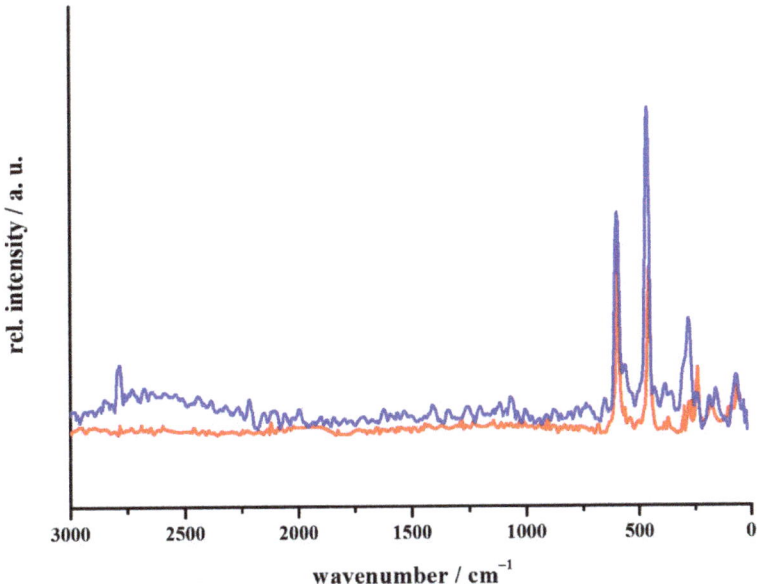

Figure 8. Raman spectra of K_2AlF_5-2 (blue) and K_2AlF_5-3 (red).

Compared with the literature, the observed signals are shifted to higher wavenumbers. However, literature data are from compounds containing isolated complex $[AlF_6]^{3-}$ ions and are mostly concerned with salt melts at higher temperatures, for which a shift of signals towards lower wavenumbers with rising temperature is reported [25].

Potassium Dirubidium Hexafluoridoaluminate, Rb_2KAlF_6

In an experiment, providing rubidium next to potassium ions within the reaction vessel, a new cubic hexafluoridoaluminate with elpasolite structure, Rb_2KAlF_6, was obtained. General crystallographic parameters are gathered in Table 1, the fractional atomic coordinates, as well as isotropic displacement parameters, are collected in Table 6. The structure features one crystallographic site each for aluminum, fluoride, potassium, and rubidium arranged in an ordered double perovskite motif with an alternating filling of octahedra formed from fluoride ions by aluminum and potassium ions. Consequently, the structure contains mutually isolated $[AlF_6]^{3-}$ octahedra arranged in the motif of a cubic closed packing. In this hierarchical view, potassium ions occupy octahedral voids, while rubidium ions are located in tetrahedral voids in the packing of $[AlF_6]^{3-}$ octahedra. This leads to twelvefold coordination of Rb by F in a distorted cuboctahedron (4 × triangular face of $[AlF_6]^{3-}$ octahedra, Figure 9). A slight antisite disorder by mutual substitution of potassium and rubidium ions appears possible, while refinements resulted in around 5% for both alkali metal sites. However, since for both sites the resulting figures are in the order of the standard deviations and the number of 58 unique reflections is already low in comparison to the number of refined parameters (7 for the fully ordered structure) we have discarded to further consider this minor disorder. Isostructural fluoridoaluminates comprise K_2LiAlF_6 [16], K_2NaAlF_6 [14], and Rb_2NaAlF_6 [17].

Table 6. Fractional atomic coordinates and equivalent isotropic displacement parameters (in 10^4 pm^2) of Rb_2KAlF_6.

Atom	Site	x/a	y/b	z/c	U_{eq}
Al	4a	0	0	0	0.0166(17)
K	4b	1/2	0	0	0.0224(14)
Rb	8c	1/4	1/4	1/4	0.0310(8)
F	24e	0	0.2073(8)	0	0.0386(18)

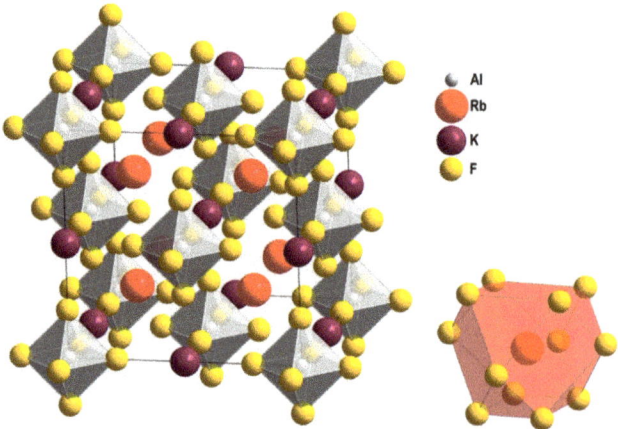

Figure 9. Extended unit cell of Rb_2KAlF_6 (**left**) and distorted cuboctahedral coordination of rubidium ions (**right**).

Distances within Rb_2KAlF_6 are within the expected ranges, with d(Al–F) = 180.1(7) pm, d(K–F) = 254.3(7) pm, and d(Rb–F) = 309.4(1) pm. Compared to the above-discussed modifications of K_2AlF_5, the slightly longer distance of the terminal fluoride ligands to aluminum reflects the isolated complex ions $[AlF_6]^{3-}$ including the increased negative charge. The distance between potassium and fluoride ions, on the other hand, presents as shorter, due to the reduced coordination number of six.

3. Materials and Methods

The entire handling of all compounds was conducted under argon with $p(O_2) < 0.1$ ppm (glovebox: MBRAUN Inertgas-Systeme, Garching, Germany). The synthesis of the reported compounds was performed under similar conditions as discussed for the synthesis of intermediates in ammonothermal InN synthesis [6,26]. A custom-made autoclave from nickel base alloy was used, as well as a Si_3N_4 liner (air-pressure sintered silicon nitride, Ingenieurkeramik, a QSIL company, Frankenblick, Germany), which protects the autoclave from corrosion [27,28]. The autoclaves were assembled and disassembled in the glovebox. Ammonia (Linde, Pullach, Germany, purity $\geq$ 99.999) was filled into the autoclave by condensation (cooling in an ethanol/dry ice bath), using a self-made tensi-eudiometer according to Hüttig [29]. The synthesis was performed in a one-zone tubular furnace (LOBA 1200-60-400-1 OW, HTM Reetz, Berlin, Germany) set up in a vertical position, generating a temperature gradient from the heated lower part of the reaction vessel (warmer temperature zone) to the unheated upper part (colder temperature zone) [30,31]. The pressure was monitored with a pressure transmitter and a digital analyzer (HBM P2VA2 and DA 2510, Hottinger Brüel and Kjaer, Darmstadt, Germany).

For the synthesis of K_2AlF_5-2, equimolar amounts of InF_3 (125.4 mg, 0.73 mmol) and AlF_3 (61.3 mg, 0.73 mmol) were used as aluminum and fluoride sources together with a sixfold amount of KNH_2 (241.3 mg, 4.38 mmol), which served as mineralizer and provided the potassium (equimolar amounts of F^- and K^+). At the beginning of the experiment, KNH_2 was spatially separated from the metal fluorides by placing in a Si_3N_4 crucible, equipped with a cap with a hole. The crucible prevents the reactants and mineralizer from a premature solid-state reaction, the cap reduces the diffusion rate of dissolved KNH_2 into the solution. A total of 19.0 g ammonia were condensed into the autoclave, which corresponds to a filling degree of 100% regarding the free volume in the liner and crucible. The autoclave was heated up to 753 K within five hours and maintained at this temperature for 60 h, reaching a maximum pressure of 224 MPa. It was subsequently cooled within 15 h to room temperature. K_2AlF_5-3 was obtained from a synthesis using Al (50.0 mg, 1.85 mmol), Cr (96.4 mg, 1.85 mmol), NH_4F (205.7, 5.56 mmol), and KNH_2 (306.4 mg, 5.56 mmol) in 17.0 g of ammonia (90% filling degree), again providing equimolar amounts of the ammonoacid and the ammonobase. The autoclave was heated to 773 K within five hours and kept at that temperature for 24 h. At maximum, a pressure of 220 bar was reached. The autoclave was subsequently cooled to room temperature over a time of 48 h. Both products were obtained as colorless crystals from the hot zone of the autoclave. Chromium and indium were not found to take part in the reactions.

The synthesis of Rb_2KAlF_6 included AlF_3 (90.7 mg, 1.08 mmol), Ga (75.3 mg, 1.08 mmol), NH_4F (119.5 mg, 3.23 mmol), and $RbNH_2$ (355.4 mg, 6.45 mmol), residue KNH_2 from earlier reactions, and a total of 17.5 g ammonia (93% filling degree). Loaded with the starting chemicals, the autoclave was heated to 723 K within 4.5 h and then kept at that temperature for 72 h, reaching a maximum pressure of 152 MPa. Subsequently, the autoclave was cooled to room temperature within 72 h, after which the product was found in the hot zone of the autoclave. Elemental gallium was recovered unchanged after the reaction.

InF_3 (Alfa Aesar, Thermo Fisher, Kandel, Germany, 99.95% metal basis, anhydrous), AlF_3 (abcr, Karlsruhe, Germany, 99.99% metal basis), Al (Alfa Aesar, Thermo Fischer Kandel GmbH, Germany, 99.97% metal basis), Cr (Sigma Aldrich, Taufkirchen, Germany, 99.0%), NH_4F (Sigma Aldrich, Taufkirchen, Germany, 99.99%), and self-made KNH_2 and $RbNH_2$, synthesized from potassium (Sigma Aldrich, Merck, Darmstadt, Germany, 98%) and rubidium (donation) reacting with ammonia at 373 K for 24 h were used for synthesis.

Single-crystal X-ray diffraction data collection was performed on a κ-CCD (Bruker Cooperation, Billerica, MA, USA) with Mo-K_α radiation. Solving and refinement of all crystal structures was done with the SHELX-2013 software package [32,33].

Single-crystal Raman spectroscopy was done on an XploRa Raman spectrometer (Horiba Europe, Oberursel, Germany) equipped with a confocal polarization microscope (Olympus BX51, Olympus Europa, Hamburg, Germany). For single-crystal X-ray diffraction and Raman spectroscopy, the crystals were measured in a sealed glass capillary.

4. Conclusions

Ammonothermal synthesis in presence of fluoride ions is a promising technique for the production of complex fluorides including new modifications, which possibly are difficult to assess by other methods. We have obtained two new modifications of K_2AlF_5, build up by zig-zag chains of *cis*-vertex-sharing AlF_6-octahedra with different conformations. Upon addition of rubidium next to potassium ions within the system, formation of the quaternary elpasolite, Rb_2KAlF_6, is preferred, which contains mutually isolated $[AlF_6]^{3-}$ octahedra.

Author Contributions: Conceptualization, C.B., P.B. and R.N.; methodology, C.B., P.B. and K.J.M.; formal analysis, C.B., P.B. and K.J.M.; investigation, C.B., P.B. and K.J.M.; resources, R.N.; data curation, C.B., P.B. and K.J.M.; writing—original draft preparation, C.B., P.B. and R.N.; writing—review and editing, C.B., P.B. and R.N.; visualization, C.B. and P.B.; supervision, R.N.; project administration, R.N.; funding acquisition, R.N. All authors have read and agreed to the published version of the manuscript.

Funding: This research was funded by the Deutsche Forschungsgemeinschaft (DFG), grant number NI489/16-1.

Institutional Review Board Statement: Not applicable.

Informed Consent Statement: Not applicable.

Data Availability Statement: Supplementary crystallographic data can be obtained online free of charge via http://www.ccdc.cam.ac.uk/conts/retrieving.html (accessed on 17 December 2021), deposition numbers CSD 2129567 (Rb_2KAlF_6), 2129568 (K_2AlF_5-2), and 2129569 (K_2AlF_5-3).

Acknowledgments: We thank Sebastian Kunkel for performing the Raman measurements and Falk Lissner for the collection of X-ray intensity data.

Conflicts of Interest: The authors declare no conflict of interest.

References

1. Richter, T.M.M.; Niewa, R. Chemistry of Ammonothermal Synthesis. *Inorganics* **2014**, *2*, 29–78. [CrossRef]
2. Meissner, E.; Niewa, R. *Ammonothermal Synthesis and Crystal Growth of Nitrides: Chemistry and Technology*; Meissner, E., Niewa, R., Eds.; Springer: Cham, Switzerland, 2021; ISBN 9783030563042.
3. Peters, D. Ammonothermal synthesis of aluminum nitride. *J. Cryst. Growth* **1990**, *104*, 411–418. [CrossRef]
4. Purdy, A.P. Ammonothermal Synthesis of Cubic Gallium Nitride. *Chem. Mater.* **1999**, *11*, 1648–1651. [CrossRef]
5. Dwiliński, R.; Baranowski, J.M.; Kamińska, M.; Doradziński, R.; Garczyński, J.; Sierzputowski, L. On GaN Crystallization by Ammonothermal Method. *Acta Phys. Pol. A* **1996**, *90*, 763–766. [CrossRef]
6. Hertrampf, J.; Becker, P.; Widenmeyer, M.; Weidenkaff, A.; Schlücker, E.; Niewa, R. Ammonothermal Crystal Growth of Indium Nitride. *Cryst. Growth Des.* **2018**, *18*, 2365–2369. [CrossRef]
7. Häusler, J.; Schnick, W. Ammonothermal Synthesis of Nitrides: Recent Developments and Future Perspectives. *Chem. Eur. J.* **2018**, *24*, 11864–11879. [CrossRef]
8. Bäucker, C.; Niewa, R. A New Modification of $Rb[Al(NH_2)_4]$ and Condensation in Solid State. *Crystals* **2020**, *10*, 1018. [CrossRef]
9. Bäucker, C.; Bauch, S.; Niewa, R. Synthesis and Characterization of the Amidomanganates $Rb_2[Mn(NH_2)_4]$ and $Cs_2[Mn(NH_2)_4]$. *Crystals* **2021**, *11*, 676. [CrossRef]
10. Becker, P.; Cekovski, T.B.; Niewa, R. Indium Ammoniates from Ammonothermal Synthesis: $InAlF_6(NH_3)_2$, $[In(NH_3)_6][AlF_6]$, and $[In_2F(NH_3)_{10}]_2[SiF_6]_5 \cdot 2 NH_3$. *Crystals* **2021**, *11*, 679. [CrossRef]
11. Hertrampf, J.; Alt, N.S.A.; Schlücker, E.; Niewa, R. Three Solid Modifications of $Ba[Ga(NH_2)_4]_2$: A Soluble Intermediate in Ammonothermal GaN Crystal Growth. *Eur. J. Inorg. Chem.* **2017**, *2017*, 902–909. [CrossRef]
12. Xiang, H.W. Vapor Pressures, Critical Parameters, Boiling Points, and Triple Points of Ammonia and Trideuteroammonia. *J. Phys. Chem. Ref. Data* **2004**, *33*, 1005–1011. [CrossRef]

13. de Kozak, A.; Gredin, P.; Pierrard, A.; Renaudin, J. The crystal structure of a new form of the dipotassium pentafluoroaluminate hydrate, $K_2AlF_5 \cdot H_2O$, and of its dehydrate, K_2AlF_5. *J. Fluorine Chem.* **1996**, *77*, 39–44. [CrossRef]
14. Schneider, S.; Hoppe, R. Über neue Verbindungen Cs_2NaMF_6 und K_2NaMF_6 sowie über Cs_2KMnF_6. *Z. Anorg. Allg. Chem.* **1970**, *376*, 268–276. [CrossRef]
15. Moras, L.R. Crystal structure of dipotassium sodium fluoroaluminate (elpasolite). *J. Inorg. Nucl. Chem.* **1974**, *36*, 3876–3878. [CrossRef]
16. Graulich, J.; Drüeke, S.; Babel, D. Röntgenstrukturuntersuchungen an den polymorphen Elpasolithen K_2LiAlF_6 und Rb_2LiGaF_6. *Z. Anorg. Allg. Chem.* **1998**, *624*, 1460–1464. [CrossRef]
17. Yakubovich, O.V.; Kiryukhina, G.V.; Dimitrova, O.V. Crystal structure of Rb-elpasolite Rb_2NaAlF_6. *Crystallogr. Rep.* **2013**, *58*, 412–415. [CrossRef]
18. Zhang, S. Intermediates during the Formation of GaN under Ammonothermal Conditions. Ph.D. Thesis, University of Stuttgart, Stuttgart, Germany, 2014.
19. Ehrentraut, D.; Meissner, E.; Bockowski, M. *Technology of Gallium Nitride Crystal Growth*; Ehrentraut, D., Bockowski, M., Meissner, E., Eds.; Springer: Berlin, Germany, 2010; ISBN 9783642048302.
20. Vlasse, M.; Matejka, G.; Tressaud, A.; Wanklyn, B.M. The crystal structure of K_2FeF_5. *Acta Crystallogr. B* **1977**, *33*, 3377–3380. [CrossRef]
21. Le Bail, A.; Desert, A.; Fourquet, J.L. Reinvestigation of the structure of K_2FeF_5. *J. Solid State Chem.* **1990**, *84*, 408–412. [CrossRef]
22. Ma, N.; You, J.; Lu, L.; Wang, J.; Wang, M.; Wan, S. Micro-structure studies of the molten binary K_3AlF_6–Al_2O_3 system by in situ high temperature Raman spectroscopy and theoretical simulation. *Inorg. Chem. Front.* **2018**, *5*, 1861–1868. [CrossRef]
23. Parker, S.F.; Ramirez-Cuesta, A.J.; Daemen, L.L. The structure and vibrational spectroscopy of cryolite, Na_3AlF_6. *RSC Adv.* **2020**, *10*, 25856–25863. [CrossRef]
24. Gilbert, B.; Materne, T. Reinvestigation of Molten Fluoroaluminate Raman Spectra: The Question of the Existence of AlF_5^{2-} Ions. *Appl. Spectrosc. AS* **1990**, *44*, 299–305. [CrossRef]
25. Daniel, P.; Bulou, A.; Rousseau, M.; Nouet, J.; Fourquet, J.L.; Leblanc, M.; Burriel, R. A study of the structural phase transitions in AlF_3: X-ray powder diffraction, differential scanning calorimetry (DSC) and Raman scattering investigations of the lattice dynamics and phonon spectrum. *J. Phys. Condens. Matter* **1990**, *2*, 5663–5677. [CrossRef]
26. Becker, P.; Cekovski, T.B.; Niewa, R. Two Intermediates in Ammonothermal InN Crystal Growth: $[In(NH_3)_5Cl]Cl_2$ and $InF_2(NH_2)$. *Z. Anorg. Allg. Chem.* **2021**, in press. [CrossRef]
27. Alt, N.S.A.; Meissner, E.; Schluecker, E. Development of a novel in situ monitoring technology for ammonothermal reactors. *J. Cryst. Growth* **2012**, *350*, 2–4. [CrossRef]
28. Hertweck, B.; Schimmel, S.; Steigerwald, T.G.; Alt, N.S.A.; Wellmann, P.J.; Schluecker, E. Ceramic liner technology for ammonoacidic synthesis. *J. Supercrit. Fluids* **2015**, *99*, 76–87. [CrossRef]
29. Hüttig, G.F. Apparat zur gleichzeitigen Druck- und Raummessung von Gasen. (Tensi-Eudiometer.). *Z. Anorg. Allg. Chem.* **1920**, *114*, 161–173. [CrossRef]
30. Schimmel, S.; Tomida, D.; Ishiguro, T.; Honda, Y.; Chichibu, S.; Amano, H. Numerical Simulation of Ammonothermal Crystal Growth of GaN—Current State, Challenges, and Prospects. *Crystals* **2021**, *11*, 356. [CrossRef]
31. Erlekampf, J.; Seebeck, J.; Savva, P.; Meissner, E.; Friedrich, J.; Alt, N.S.A.; Schlücker, E.; Frey, L. Numerical time-dependent 3D simulation of flow pattern and heat distribution in an ammonothermal system with various baffle shapes. *J. Cryst. Growth* **2014**, *403*, 96–104. [CrossRef]
32. Sheldrick, G.M. A short history of SHELX. *Acta Crystallogr. A* **2008**, *64*, 112–122. [CrossRef] [PubMed]
33. Sheldrick, G.M. Crystal structure refinement with SHELXL. *Acta Crystallogr. C* **2015**, *71*, 3–8. [CrossRef]

Article

Hydrothermal Synthesis and Structural Investigation of a Crystalline Uranyl Borosilicate

Kristen A. Pace [1,2,†], Vladislav V. Klepov [2,‡], Mark D. Smith [2], Travis Williams [3], Gregory Morrison [1,2], Jochen A. Lauterbach [3], Scott T. Misture [1,4] and Hans-Conrad zur Loye [1,2,*]

1. Center for Hierarchical Waste form Materials, Columbia, SC 29208, USA; kpace@lanl.gov (K.A.P.); MORRI383@mailbox.sc.edu (G.M.); misture@alfred.edu (S.T.M.)
2. Department of Chemistry and Biochemistry, University of South Carolina, Columbia, SC 29208, USA; klepov@northwestern.edu (V.V.K.); MDSMITH3@mailbox.sc.edu (M.D.S.)
3. Department of Chemical Engineering, University of South Carolina, Columbia, SC 29208, USA; traswill@gmail.com (T.W.); LAUTERAJ@cec.sc.edu (J.A.L.)
4. Kazuo Inamori School of Engineering, Alfred University, Alfred, NY 14802, USA
* Correspondence: zurloye@mailbox.sc.edu
† Present Address: Pit Technologies Division, Los Alamos National Laboratory, Los Alamos, NM 87545, USA.
‡ Present Address: Department of Chemistry, Northwestern University, Evanston, IL 60208, USA.

Citation: Pace, K.A.; Klepov, V.V.; Smith, M.D.; Williams, T.; Morrison, G.; Lauterbach, J.A.; Misture, S.T.; zur Loye, H.-C. Hydrothermal Synthesis and Structural Investigation of a Crystalline Uranyl Borosilicate. *Inorganics* 2021, 9, 25. https://doi.org/10.3390/inorganics9040025

Academic Editor: Duncan H. Gregory

Received: 5 March 2021
Accepted: 29 March 2021
Published: 6 April 2021

Publisher's Note: MDPI stays neutral with regard to jurisdictional claims in published maps and institutional affiliations.

Copyright: © 2021 by the authors. Licensee MDPI, Basel, Switzerland. This article is an open access article distributed under the terms and conditions of the Creative Commons Attribution (CC BY) license (https://creativecommons.org/licenses/by/4.0/).

Abstract: The relevance of multidimensional and porous crystalline materials to nuclear waste remediation and storage applications has motivated exploratory research focused on materials discovery of compounds, such as actinide mixed-oxoanion phases, which exhibit rich structural chemistry. The novel phase $K_{1.8}Na_{1.2}[(UO_2)BSi_4O_{12}]$ has been synthesized using hydrothermal methods, representing the first example of a uranyl borosilicate. The three-dimensional structure crystallizes in the orthorhombic space group $Cmce$ with lattice parameters a = 15.5471(19) Å, b = 14.3403(17) Å, c = 11.7315(15) Å, and V = 2615.5(6) Å^3, and is composed of UO_6 octahedra linked by $[BSi_4O_{12}]^{5-}$ chains to form a $[(UO_2)BSi_4O_{12}]^{3-}$ framework. The synthesis method, structure, results of Raman, IR, and X-ray absorption spectroscopy, and thermal stability are discussed.

Keywords: crystalline borosilicate; actinides; supercritical hydrothermal synthesis; waste forms

1. Introduction

In the U.S., alkali borosilicate glasses are used as a primary nuclear waste form material for the immobilization of high-level waste. The glasses are durable and can host a wide range of elements in the disordered structures, which are composed of network forming SiO_4 and BO_4 tetrahedra as well as trigonal planar BO_3 anions that support compositional variety [1]. However, the low solubility of actinides in glasses with high actinide loadings has motivated researchers to study the crystal chemistry of actinide compounds, both to identify phases that may be forming under simulated conditions and to develop an understanding of the fundamental chemistry of waste form relevant phases [2,3]. Moreover, it has been recognized that for some waste streams, crystalline waste forms may offer an advantage over glasses such as higher waste loading capacities [4,5]. The structural complexity exhibited by multidimensional and porous crystalline materials can also be exploited, enabling the incorporation of specific radionuclides and, in some cases, ion exchange capabilities [6–8].

Among the many compound classes that have been studied for waste form applications, uranyl silicates and borates have received considerable attention. The variable arrangement of SiO_4 tetrahedra in uranyl silicate compounds has resulted in a multitude of different structure types, many of which are three-dimensional [9–17]. Likewise, the ability of BO_3 and BO_4 units to polymerize to form polyborate units has given rise to a rich structural chemistry in uranyl borates [3,18–24]. It is therefore unsurprising that incorporation of mixed-oxoanions into the structures of uranyl compounds often facilitates further structural diversity. For example, a few uranyl borophosphate [25,26], borogermanate [27], aluminoborate [28], and aluminophosphate [29] compounds are known, most of which exhibit complex open-framework structures. No such uranyl compounds containing borosilicate units have been reported, although there are several existing non-uranyl compounds exhibiting a variety of different borosilicate units. While uranyl borosilicates are relevant to waste form applications, in the case of non-actinide borosilicates, the structural diversity afforded by the presence of the borosilicate units often promotes the formation of noncentrosymmetric structures, making some of these compounds potentially promising candidates as nonlinear optical materials [30]. Some of the known examples within these compound classes contain isolated BO_3 and SiO_4 structural units [31,32], condensed BO_4 and SiO_4 tetrahedra [33,34], trigonal planar BO_3 units condensed with tetrahedral BO_4 and SiO_4 [35], as well as HBO_4 and SiO_4 hydroxyborosilicate units [36].

Herein, we present the first example of a uranyl borosilicate compound, $K_{1.8}Na_{1.2}[(UO_2)BSi_4O_{12}]$, which was synthesized using hydrothermal synthesis methods with supercritical water as the reaction medium. The structure of the compound was investigated, revealing residual electron density near the uranyl oxygen that was attributed to either the presence of a light atom in an unusual coordination or to an unidentified crystal defect, prompting further investigation of the compound using a combination of X-ray absorption and vibrational spectroscopy techniques, characterization of its thermal properties, as well as the employment of first principles calculations to determine the optimized geometry of the uranium coordination environment [37,38]. Using single crystal X-ray diffraction, the compound was found to crystallize in a centrosymmetric structure type with a complex three-dimensional crystal structure containing unique $[BSi_4O_{12}]^{5-}$ chains composed of BO_4 and SiO_4 tetrahedra. Characterizations of the thermal and spectroscopic properties were performed using thermogravimetric analysis in addition to Raman, IR, and extended X-ray absorption fine structure spectroscopy. These data, in addition to the results of first principles calculations, suggest that the residual electron density in the structure is likely due to a crystal defect. The successful preparation of $K_{1.8}Na_{1.2}[(UO_2)BSi_4O_{12}]$ serves to further expand on the diverse structural chemistry of uranium and moreover, highlights an under-explored set of synthetic conditions that may be exploited to obtain other novel actinide borosilicate compounds.

2. Materials and Methods

2.1. Reagents

$UO_3 \cdot xH_2O$ (International Bio-Analytical Industries, ACS grade), $NaBO_2 \cdot 4H_2O$ (Acros Organics, 98.5%), SiO_2 (Alfa Aesar, 99.9%), KOH pellets (Alfa Aesar, 99.99%), and KOH solution (Fisher Scientific, 45% w/w) were used as received. Caution: although the uranium precursors used contain depleted uranium, standard safety procedures for handling radioactive materials must be followed.

2.2. Synthesis

$UO_3 \cdot xH_2O$ (150.3 mg), $NaBO_2 \cdot 4H_2O$ (689.3 mg), and SiO_2 (300.4 mg) were added to a mineralizer solution of aqueous KOH (224.4 mg in 2 mL deionized water) in an approximate molar ratio of 1:10:10:8 of U:Na,B:Si:K. A 5% by mass excess of $UO_3 \cdot xH_2O$ was added to the reaction mixture to compensate for the unknown degree of hydration of the uranium starting material. The mixture was sealed in a 12.7 cm silver tube that was loaded into a high-pressure vessel containing approximately 20 mL of water to serve as counterpressure. The vessel was placed in a programmable oven and heated to 400 °C for 48 h, followed by a period of slow cooling to 350 °C at a rate of 3 °C per hour. The vessel was then allowed to naturally cool to room temperature. An estimated 30 MPa of pressure was generated during the reaction based on the pressure-temperature diagram of water.

Upon cooling, the silver tube was cut open and the mother liquor was decanted. The contents of the tube were sonicated in deionized water prior to vacuum filtering and thoroughly washing the products in water and acetone. The product mixture consisted of brown/black polycrystalline powders of unknown composition in addition to clusters of large yellow prismatic crystals (Figure S1) in approximately 70% yield. The large yellow prismatic crystals were manually separated from the product mixture and ground to a powder to obtain a phase pure sample for bulk property measurements.

To obtain high quality single crystals suitable for single crystal X-ray diffraction measurements, efforts were made to optimize the reaction conditions by varying the hydroxide concentration. In order to have more precise control over the KOH concentration, subsequent attempts employed identical reaction conditions with substitution of KOH pellets for KOH solution diluted from a 45% w/w stock solution. Reactions containing 2 mL of 2M KOH solution (1:10:10:8 = U:Na,B:Si:K) resulted only in the formation of brown polycrystalline material, identified as a mixture of $K(UO)Si_2O_6$ [17] and $Na_7UO_2(UO)_2(UO_2)_2Si_4O_{16}$ [39] via powder X-ray diffraction. In comparison, reactions containing 2 mL of 1.5M KOH solution (1:10:10:6 = U:Na,B:Si:K) resulted in polycrystalline $Na_7UO_2(UO)_2(UO_2)_2Si_4O_{16}$ along with an approximately 40% yield of high quality single crystals of $K_{1.8}Na_{1.2}[(UO_2)BSi_4O_{12}]$ that were used for single crystal X-ray analysis.

2.3. Single Crystal X-ray Diffraction

Single crystal X-ray diffraction data were collected at 301(2) K using a Bruker D8 QUEST diffractometer equipped with an Incoatec microfocus source (Mo Kα radiation, λ = 0.71073 Å). Data were integrated and corrected for absorption effects using SAINT+ and SADABS programs [40]. An initial structure model was obtained with SHELXT and subsequently refined with SHELXL-2018 using the ShelXle interface [41,42]. The ADDSYM program implemented into PLATON software was used to check for possible missing symmetry, and no alternative symmetry was found [43]. Crystallographic data for $K_{1.8}Na_{1.2}[(UO_2)BSi_4O_{12}]$ are provided in Table 1. A residual electron density peak of 4.17 e/Å^3 is located 1.34 Å from the uranyl oxygen and 1.82 Å from the uranium atom, along with a second peak of 2.25 e/Å^3 located trans to the 4.17 e/Å^3 peak across the uranium atom, suggesting the oxygen atoms of the uranyl group form a disordered crystallographic domain. No twinning was identified, and several different crystals were screened but the residual electron density was present in each structure solution.

Table 1. Crystal Data and Structure Refinement.

Formula	$K_{1.78}Na_{1.22}[(UO_2)BSi_4O_{12}]$
space group	Cmce
formula weight (g/mol)	682.77
temperature (K)	301(2)
a (Å)	15.5471(19)
b (Å)	14.3403(17)
c (Å)	11.7315(15)
volume (Å^3)	2615.5(6)
Z	8
density (g cm^{-3})	3.468
crystal dimensions (mm^3)	0.04 × 0.03 × 0.02
absorption coefficient (mm^{-1})	13.460
data collection and refinement	
collected reflections	72,412
independent reflections	2963
R_{int}	0.0341
refined restraints/parameters	1/120
goodness-of-fit on F^2	1.251
final R indices [$I > 2\sigma(I)$]	R_1 = 0.0356, wR_2 = 0.0678
final R indices (all data)	R_1 = 0.0376, wR_2 = 0.0686
largest diff. peak and hole (e$^-$/Å^3)	4.154 and −2.158

Single crystal X-ray data were also collected on the large crystals obtained from initial reactions utilizing KOH pellets rather than KOH stock solution, and refinement of the structure suggested a marginal difference in composition, $K_{1.7}Na_{1.3}[(UO_2)BSi_4O_{12}]$, although with poor refinement statistics (Table S1). Similar to $K_{1.8}Na_{1.2}[(UO_2)BSi_4O_{12}]$, residual electron density was also observed in the structure solution of $K_{1.7}Na_{1.3}[(UO_2)BSi_4O_{12}]$; however, only a single large peak of 10.35 e/Å^3 located 1.33 Å from the uranyl oxygen and 1.94 Å from the uranium atom was observed. Several different crystals were screened, and the same result was obtained in each case. The residual electron density persisted in data collected at low temperature as well as in structure solutions obtained in lower symmetry space groups, and no twinning was identified.

2.4. Topological Analysis

The ToposPro software package (Blatov, Shevchenko, Samara, Russia) was used to perform crystal structure analysis [44,45]. The ADS program was used to obtain underlying nets using the standard structure simplification procedure [46].

2.5. Powder X-ray Diffraction

Powder X-ray diffraction (PXRD) measurements were performed on a Bruker D2 Phaser diffractometer (Cu Kα radiation, λ = 1.54184 Å) equipped with a LYNXEYE XE-T detector. Data were collected over a 2θ range of 5–65° with a step size of 0.04°. PXRD data confirmed the phase purity of the obtained samples.

2.6. Spectroscopy

Raman spectroscopy measurements were performed on a Horiba XploraPLUS Raman microscope (Horiba, Kyoto, Japan) that was used to acquire spectra over the range of 525 to 950 cm^{-1} with a 638 nm laser as the excitation source. Scans were performed at 25% laser power and were collected on approximately 15 mg of powder sample that was deposited between two microscope slides and sealed on all sides with tape as containment during the measurement. Fourier-transform infrared spectroscopy data were collected on a powder sample using a PerkinElmer Spectrum 100 FT-IR spectrometer (PerkinElmer, Waltham, MA, United States) in the range of 650 to 4000 cm^{-1}.

Extended X-ray absorption fine structure (EXAFS) measurements were made using the MRCAT beamline, sector 10-BM-A,B at the Advanced Photon Source [47]. Measurements were made using the uranium L$_{III}$ edge at 17.16 keV in transmission. Measurements were made for the new uranium borosilicate phase of interest and also for γ-UO_3 and α-U_3O_8 as standards. Data analysis was performed using the Athena and Artemis programs from the Demeter package [48] with a k weight of 3 and usable data range to 12.5 Å^{-1}.

2.7. Thermal Properties

Thermogravimetric and differential thermal analysis (TGA/DTA) measurements were performed on a polycrystalline powder sample using a TA SDT Q600 TGA (TA Instruments, New Castle, DE, Unites States). The sample was heated from room temperature to 800 °C under a flow of nitrogen gas at a purge rate of 100 mL per minute, and the resulting powder was analyzed by PXRD for phase identification after heating.

2.8. First Principles Calculations

We performed first principles calculations to carry out geometry optimizations in the form of density functional theory (DFT), using the Vienna Ab-initio Package (VASP) planewave code (VASP Software GmbH, Vienna, Austria) [49,50] generalized gradient approximation of Perdew, Burke and Ernzerhof (PBE) [51] and projector augmented wave (PAW) method [52,53]. Spin-polarized calculations were performed with 520 eV cut-off energy for the plane wave basis set, 10^{-4} eV energy convergence criteria and $3 \times 3 \times 3$ k-point mesh. The ground state geometries at 0 K were optimized by relaxing the cell volume, atomic positions, and cell symmetry until the maximum force on each atom was less than 0.01 eV/Å. Implementation of Hubbard U correction of 4.0 eV for uranium f-orbitals resulted in similar optimized geometries.

3. Results and Discussion

3.1. Synthesis

Many factors are known to influence crystal growth in supercritical hydrothermal synthesis, including pH, temperature and pressure, mineralizer type, reactant solubilities, solution redox potential, reaction container type, etc. [54,55]. While there is valuable information to be gleaned from analysis of the reaction processes occurring in these conditions, such studies are often precluded by the presence of soluble intermediate species [56]. In the absence of the ability to perform mechanistic studies, part of our investigations of this system aimed to identify key factors dictating phase formation. Our findings indicated that the two predominating factors influencing the formation of $K_{1.8}Na_{1.2}[(UO_2)BSi_4O_{12}]$ are reaction dwell time and KOH concentration.

The optimal reaction time was determined to be 48 h, where shorter reaction times resulted in lower yields of $K_{1.8}Na_{1.2}[(UO_2)BSi_4O_{12}]$ and instead favored the formation of the co-crystallizing phases $Na_7UO_2(UO)_2(UO_2)_2Si_4O_{16}$ [39] and $K(UO)Si_2O_6$ [17]. Longer reaction times did not result in appreciable increase in yield or larger single crystals of $K_{1.8}Na_{1.2}[(UO_2)BSi_4O_{12}]$. The discrepancy between the products obtained from reactions beginning with KOH pellets versus KOH stock solution served as an indication of the relatively narrow range of conditions in which $K_{1.8}Na_{1.2}[(UO_2)BSi_4O_{12}]$ will form, where phase formation did not occur in reactions with KOH concentrations greater than 1.5 M. The optimal concentration range was determined to be approximately 0.5–1.5 M.

3.2. Structure Description

$K_{1.8}Na_{1.2}[(UO_2)BSi_4O_{12}]$ crystallizes in the centrosymmetric orthorhombic space group $Cmce$ with lattice parameters a = 15.5471(19) Å, b = 14.3403(17) Å, c = 11.7315(15) Å, and V = 2615.5(6) Å^3. The asymmetric unit consists of one uranium atom, four potassium atoms, one sodium atom, two silicon atoms, one boron atom, and nine oxygen atoms. The structure is composed of isolated UO_6 octahedra that are coordinated through vertex-sharing of the equatorial oxygen atoms to $[BSi_4O_{12}]^{5-}$ chains, forming a framework with channels in which disordered Na^+ and K^+ cations reside (Figure 1). The borosilicate chains are made up of BO_4 and SiO_4 tetrahedra that run parallel to the c axis, where each BO_4 tetrahedron shares all four vertices with SiO_4 tetrahedra within the chain and each SiO_4 tetrahedron shares vertices with two adjacent SiO_4 tetrahedra, one BO_4 tetrahedron, and a distorted UO_6 octahedron. A table of selected interatomic distances is provided in the supporting information in Table S2.

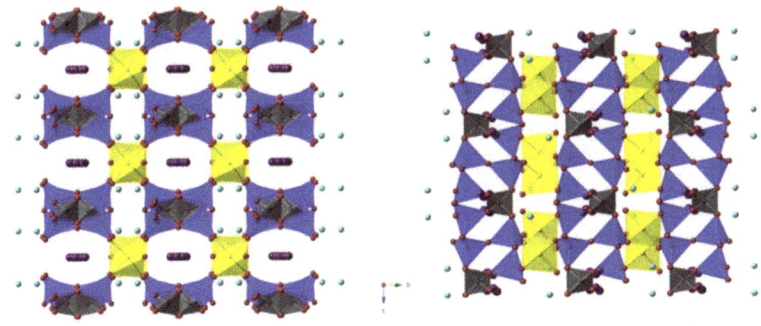

Figure 1. The three-dimensional $[(UO_2)BSi_4O_{12}]^{3-}$ framework is shown, where yellow octahedra represent UO_6 groups, gray tetrahedra are BO_4 groups, blue tetrahedra are SiO_4 groups, and purple and light blue spheres are K^+ cations and disordered Na^+/K^+, respectively.

The fundamental building block (FBB) of the borosilicate chain, $[BSi_4O_{12}]^{5-}$, is composed of four SiO_4 tetrahedra and one BO_4 tetrahedron that are connected through vertex-sharing, as shown in Figure 2a. The borosilicate FBB is an open-branched dreier-single ring that may be described as 5☐:<4☐>☐ according to the notation proposed by Burns, Grice, and Hawthorne [57,58]. The simplified underlying net of the borosilicate chain topology is shown in Figure 2b. A variety of borosilicate units have been observed in non-uranyl compounds [31–36]; to the best of our knowledge, $K_{1.8}Na_{1.2}[(UO_2)BSi_4O_{12}]$ represents the first example of a compound containing $[BSi_4O_{12}]^{5-}$ units.

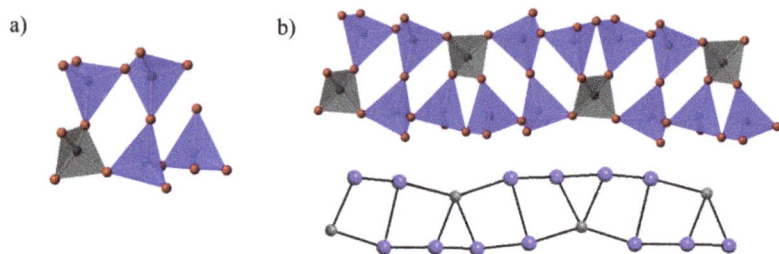

Figure 2. (a) The $[BSi_4O_{12}]^{5-}$ FBB of the borosilicate chain is shown and (b) a simplified net with BO_4 and SiO_4 tetrahedra shown as nodes within the chain.

Within the larger voids of the $[(UO_2)BSi_4O_{12}]^{3-}$ framework, K^+ cations are disordered over three sites that have a combined total occupancy of one, whereas the smaller channels contain disordered Na^+/K^+ cations located on a single site staggered between UO_6 octahedra, with occupancies of Na^+ and K^+ refining to 0.61 and 0.39, respectively. It was expected that larger alkali cations could be incorporated into the structure considering the extensive disorder of the K^+ cations in the large channels of the framework; however, attempts to synthesize the Rb and Cs analogs were unsuccessful.

The U-O bond lengths within the distorted UO_6 octahedra are typical, with the equatorial bond distances ranging from 2.233(2) to 2.245(3) Å and axial uranyl U(1)–O(1) and U(1)–O(2) bond distances of 1.819(4) and 1.813(4) Å, respectively. The bond valence sum (BVS) at the uranium site, using the parameters $r_0 = 2.051$ and $B = 0.519$, was calculated to be 5.93, which is consistent with U(VI). In the structure of $K_{1.7}Na_{1.3}[(UO_2)BSi_4O_{12}]$, the UO_6 equatorial and uranyl U(1)–O(2) bond distances are equivalent to those of $K_{1.8}Na_{1.2}[(UO_2)BSi_4O_{12}]$ within estimated standard deviations. The presence of the single large residual electron density peak (10.35 e/Å^3) located 1.33 Å from the uranyl oxygen O(1) and 1.94 Å from U(1), prompted us to consider the residual electron density as a possible oxygen atom. Assigning the peak as an oxygen atom and freely refining the occupancies of both the assigned oxygen site and the uranyl O(1) atom resulted in a partial occupancy of approximately 0.40 for the assigned oxygen site and full occupancy of O(1), constituting a possible O_2^{2-} peroxo group in lieu of a single uranyl oxygen (Figure 3). Using this model that takes into account the coordination of a peroxo group to the uranium atom, the BVS at the uranium site was calculated to be 6.67. Selected interatomic distances for $K_{1.7}Na_{1.3}[(UO_2)BSi_4O_{12}]$ are provided in Table S3.

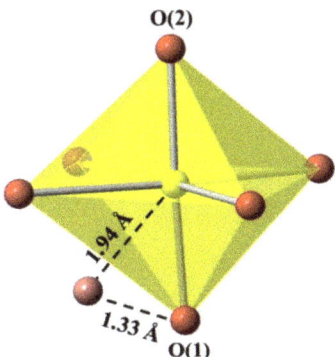

Figure 3. Residual electron density near the uranium atom and uranyl oxygen O(1) can be refined as a partially occupied oxygen (shown in light pink), resulting in a highly unusual coordination environment in which a peroxo group is coordinated to uranium along the axial direction.

While many uranyl peroxo compounds are known [59–64], reported examples are characterized by peroxo interactions existing within the equatorial plane; none that exhibit peroxo interactions along the axial direction have been reported. Furthermore, O-O bond lengths in uranyl peroxo compounds typically range between about 1.45 to 1.52 Å [65], and thus, the significantly shorter peroxo bond in this model (1.33 Å) warrants some skepticism. As an alternative to a peroxo group, the observed residual electron density might simply be an artifact from an unknown crystal defect, possibly relating to the incorporation of slightly more Na^+ on the disordered Na^+/K^+ site in $K_{1.7}Na_{1.3}[(UO_2)BSi_4O_{12}]$ than in $K_{1.8}Na_{1.2}[(UO_2)BSi_4O_{12}]$. Given the novelty of such a coordination environment for uranium, as well as this possibility of a crystal defect, further analysis was needed to either confirm or reject the peroxo structure model.

3.3. Spectroscopy

In order to probe for a peroxo group in the structure of $K_{1.7}Na_{1.3}[(UO_2)BSi_4O_{12}]$, Raman spectroscopy was performed. Shown in Figure 4, the predominating feature in the Raman spectrum is a band at 776 cm^{-1}, corresponding to the symmetric stretching of the uranyl moiety. This is consistent with the uranyl bond lengths determined from the single crystal X-ray data [66]. A few weak bands are observed in the spectrum between 548 to 744 cm^{-1} and are attributable to BO_4 and SiO_4 bending modes [34,67]. For η^2-peroxo complexes, the O-O symmetric stretching is normally observed as a strong band near 870 cm^{-1} [68], although this band has been observed to occur anywhere in the range of approximately 830 to 880 cm^{-1} in the spectra of uranyl peroxo compounds [69,70]. This region is bare in the Raman spectrum of $K_{1.7}Na_{1.3}[(UO_2)BSi_4O_{12}]$, with the exception of a weak band centered at approximately 830 cm^{-1}. Even in the case of a partially occupied peroxo group, it is expected that a band corresponding to the symmetric stretching mode would be fairly intense, thus we attribute the weak band at 830 cm^{-1} to the antisymmetric stretching of the uranyl moiety [16,71].

The IR spectrum of $K_{1.7}Na_{1.3}[(UO_2)BSi_4O_{12}]$ is complex, with significant overlap of the absorption bands of BO_4, SiO_4, and UO_2^{2+} observed. The spectrum, shown in Figure 5, exhibits two bands at 686 and 750 cm^{-1} that are attributed to the BO_4 and SiO_4 bending modes, in addition to antisymmetric and symmetric stretching absorptions of BO_4 and SiO_4 in the region of 750 to 1134 cm^{-1} [35,72]. The vibrational band at 901 cm^{-1} is assigned to the uranyl antisymmetric stretching [72,73].

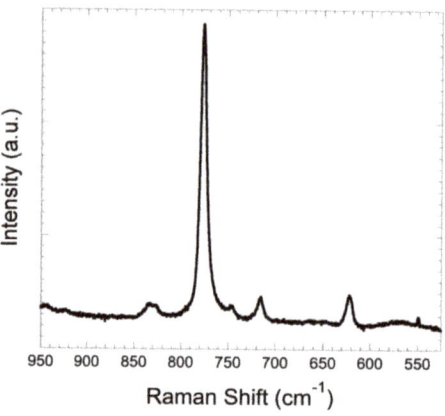

Figure 4. The Raman spectrum of $K_{1.7}Na_{1.3}[(UO_2)BSi_4O_{12}]$, which exhibits no bands in the range of 820 to 880 cm^{-1} where symmetric stretching of peroxo is expected to occur.

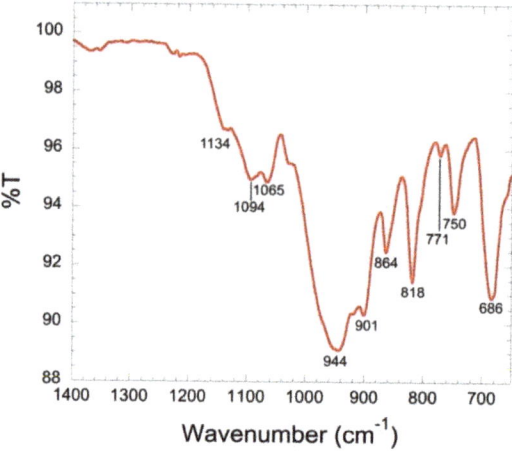

Figure 5. IR spectrum of $K_{1.7}Na_{1.3}[(UO_2)BSi_4O_{12}]$.

EXAFS was used to further probe the presence or absence of the peroxo oxygen in a 'side-on' configuration. The first coordination shell if the peroxo group is present has two oxygen ions at a distance of ~1.8 Å from the central uranium, with four additional oxygen ions at ~2.25 Å. The peroxo group oxygen is present at 1.95 Å, approximately halfway between the other oxygen distances. The structure model without the peroxo group has the same first-shell configuration but without the peroxo oxygen. Both structure models have nearly identical atom locations, with the primary difference being the presence of the peroxo oxygen, so it was only necessary to fit the first shell of the EXAFS data to distinguish between them.

First-shell fits were successful for the γ-UO$_3$ and α-U$_3$O$_8$ materials, and the fit to the new uranium borosilicate was good with or without the peroxo group, with R-values of 0.026 with the peroxo group and 0.022 without (Figure 6). For the model without the peroxo ion, the 4-parameter fit yielded sensible values with shifts of the oxygen ions of less than 0.01 Å and mean square relative displacement of 0.0020(7). However, the model for the peroxo group moved the peroxo oxygen to 2.13 Å while the initial model placed the oxygen ion at 1.95 Å. This result indicates that the peroxo oxygen is not a feature of the first coordination shell and that the model with only 6-fold coordination of the uranium ion is correct.

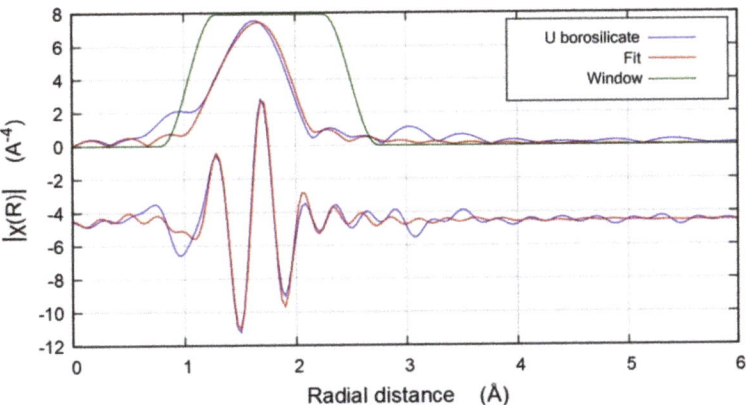

Figure 6. EXAFS first shell fit for uranium borosilicate. The real space magnitude and phase are shown with the first shell fit window.

3.4. Thermal Properties

TGA measurements were performed to assess the thermal behavior of $K_{1.7}Na_{1.3}[(UO_2)BSi_4O_{12}]$. It was expected that if the peroxo oxygen in question were present in the structure, a distinct weight loss step associated with loss of an oxygen would be observed. Based on a 40% occupancy of the peroxo oxygen as determined from structure refinements, a weight loss of approximately 0.9% was calculated. As shown in Figure 7, a total weight loss of only 0.45% was observed over the measurement range of 25 °C to 800 °C. PXRD patterns collected on TGA samples after heating revealed the material to be thermally stable up to 800 °C, with no significant changes observed in the patterns before and after heating (Figure 8). To verify the weight loss observed was not attributable to the loss of oxygen from a peroxo group, single crystal X-ray data were collected on a single crystal sample that was synthesized and subsequently heated to 800 °C. The residual electron density was still present in the structure model, and no other differences were observed in the structure model obtained for the sample heated post-synthesis.

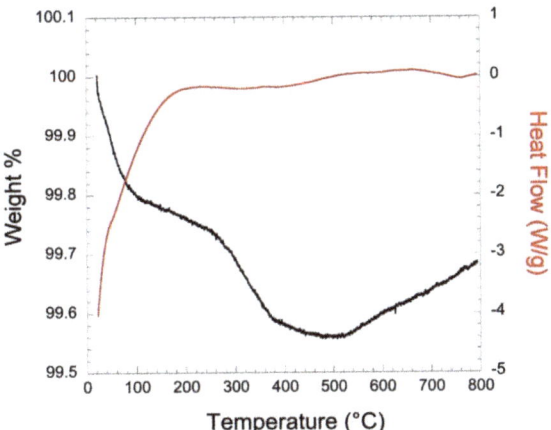

Figure 7. TGA/DTA data for $K_{1.7}Na_{1.3}[(UO_2)BSi_4O_{12}]$ collected from room temperature to 800 °C under a flow of nitrogen gas.

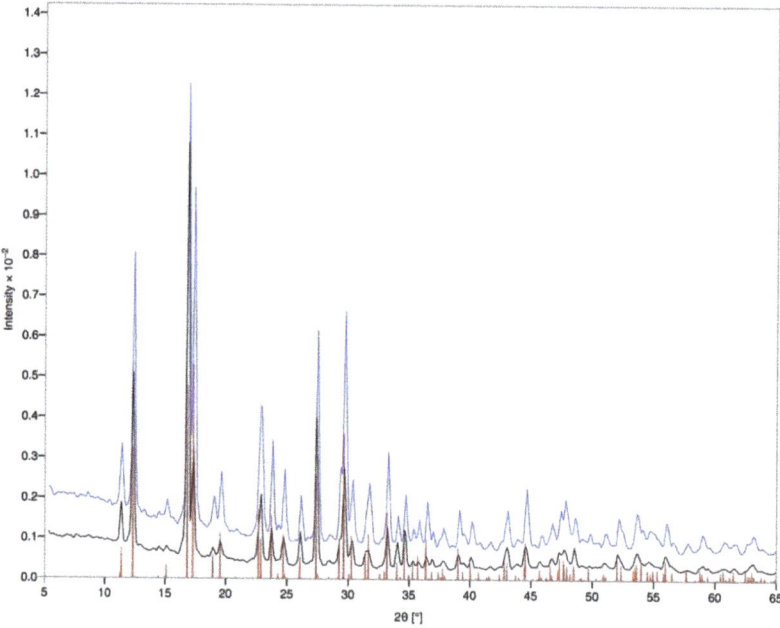

Figure 8. PXRD patterns of $K_{1.7}Na_{1.3}[(UO_2)BSi_4O_{12}]$ for TGA before (black) and after (blue) heating to 800 °C. The calculated pattern from the CIF is shown in red.

3.5. Energy Optimization Calculations

First principles calculations were used to optimize the geometry of the structure in order to determine the optimized U-O bond lengths of the uranyl oxygens and the peroxo oxygen. The results of the calculations are provided in Table 2. Three sets of calculations were performed in which the structure contained a fully occupied peroxo oxygen, a half-occupied peroxo oxygen, and no peroxo oxygen, allowing for a comparison of uranyl bond distances as a function of peroxo oxygen occupancy. The results of the calculations revealed the optimal $U(1)-O_{peroxo}$ and $U(1)-O(1)$ bond distances for the structure containing a half-occupied peroxo oxygen should be significantly elongated at approximately 2.17 and 2.16 Å, respectively. Only nominal differences in calculated bond lengths between the structure containing half- and fully occupied peroxo oxygen were determined, suggesting that any peroxo oxygen occupancy would result in elongated U-O bonds.

Table 2. DFT-Optimized Uranyl Bond Lengths in $K_{1.7}Na_{1.3}[(UO_2)BSi_4O_{12}]$ (Å).

U-O	Experimental	Calculated		
		Fully Occupied Peroxo	Half-Occupied Peroxo	No Peroxo
U(1)-O(1)	1.83	2.15	2.16	1.87
U(1)-O(2)	1.81	1.86	1.88	1.86
U(1)-O_{peroxo}	1.94	2.16	2.17	-

In comparison to the experimentally determined bond lengths, the calculated bond lengths for the structure containing no peroxo oxygen are still slightly elongated but are more consistent with the structure model obtained from single crystal X-ray data. Thus, in combined consideration of the structure, thermal, and spectroscopy data, the calculations better support the structure model exhibiting no axially coordinated peroxo.

4. Conclusions

In summary, hydrothermal synthesis methods were employed for the preparation of the first example of a uranyl borosilicate phase, $K_{1.8}Na_{1.2}[(UO_2)BSi_4O_{12}]$. The framework structure is made up of UO_6 octahedra connected by $[BSi_4O_{12}]^{5-}$ borosilicate chains containing BO_4 and SiO_4 tetrahedra, which constitutes a new borosilicate FBB. This highlights the structural diversity that is accessible in this new class of compounds, warranting further exploration of crystalline uranyl borosilicate compounds.

Successful synthesis of the phase was determined to occur under a relatively narrow set of conditions, where slight deviations in the KOH concentration of the reaction was initially thought to produce a related but distinct secondary phase with a possible partially occupied peroxo group in place of one of the uranyl oxygen atoms. The composition of this phase–with no peroxo accounted for–was determined to be $K_{1.7}Na_{1.3}[(UO_2)BSi_4O_{12}]$ from single crystal X-ray data. No evidence of O_2^{2-} symmetric stretching was observed in the Raman spectrum and first-shell fits of the EXAFS spectrum were consistent with a UO_6 coordination lacking a peroxo group. No loss of peroxo was observed upon heating the sample in TGA measurements, and in fact, no thermal decomposition was found to occur over the measurement range up to 800 °C. Furthermore, first principles calculations indicated the presence of a peroxo group should result in significantly longer U-O bonds than those determined experimentally. We therefore conclude the residual electron density in the structure models are attributable to a crystal defect and not a peroxo group.

Supplementary Materials: The following are available online at https://www.mdpi.com/article/10.3390/inorganics9040025/s1, Figure S1: Optical image of single crystals of $K_{1.7}Na_{1.3}[(UO_2)BSi_4O_{12}]$, Table S1: $K_{1.70}Na_{1.30}[(UO_2)BSi_4O_{12}]$ Crystal Data and Structure Refinement, Table S2: Selected Interatomic Distances for $K_{1.8}Na_{1.2}[(UO_2)BSi_4O_{12}]$ (Å), Table S3: Selected Interatomic Distances for $K_{1.7}Na_{1.3}[(UO_2)BSi_4O_{12}]$ (Å), checkCIF output file.

Author Contributions: Synthesis, data collection, and topological analysis were performed by K.A.P.; X-ray diffraction data collection and interpretation was performed by K.A.P., V.V.K., M.D.S., G.M. and H.-C.z.L.; EXAFS data interpretation and fittings were performed by S.T.M.; Raman data collection and interpretation was performed by K.A.P., T.W. and J.A.L.; first principles calculations were performed by V.V.K. The manuscript was written by K.A.P. with contributions from V.V.K. and S.T.M. and feedback from all the authors. All authors have read and agreed to the published version of the manuscript.

Funding: This work was supported as part of the Center for Hierarchical Waste Form Materials, an Energy Frontier Research Center funded by the U.S. Department of Energy (DOE), Office of Science, Basic Energy Sciences under Award No. DE-SC0016574 and by the South Carolina Smartstate Center for Strategic Approaches to the Generation of Electricity (SAGE). The EXAFS study was performed at MRCAT. MRCAT operations are supported by the DOE and the MRCAT member institutions. This research used resources of the Advanced Photon Source, a U.S. DOE Office of Science User Facility operated for the DOE Office of Science by Argonne National Laboratory under Contract No. DE-AC02-06CH11357.

Institutional Review Board Statement: Not applicable.

Informed Consent Statement: Not applicable.

Data Availability Statement: Accession code: CCDC 2035885. These data can be obtained free of charge via www.ccdc.cam.ac.uk/data_request/cif, or by emailing data_request@ccdc.cam.ac.uk, or by contacting The Cambridge Crystallographic Data Centre, 12 Union Road, Cambridge CB2 1EZ, UK; Fax +44 1223 336033.

Conflicts of Interest: The authors declare no conflict of interest.

References

1. National Research Council. *Waste Forms Technology and Performance*; Final Report by National Research Council of the National Academies; The National Academies Press: Washington, DC, USA, 2011.

2. Burns, P.C.; Olson, R.A.; Finch, R.J.; Hanchar, J.M.; Thibault, Y. KNa$_3$(UO$_2$)$_2$(Si$_4$O$_{10}$)$_2$(H$_2$O)$_4$, a new compound formed during vapor hydration of an actinide-bearing borosilicate waste glass. *J. Nucl. Mater.* **2000**, *278*, 290–300. [CrossRef]
3. Wang, S.; Alekseev, E.V.; Ling, J.; Skanthakumar, S.; Soderholm, L.; Depmeier, W.; Albrecht-Schmitt, T.E. Neptunium diverges sharply from uranium and plutonium in crystalline borate matrixes: Insights into the complex behavior of the early actinides relevant to nuclear waste storage. *Angew. Chem. Int. Ed. Engl.* **2010**, *49*, 1263–1266. [CrossRef]
4. Weber, W.J.; Navrotsky, A.; Stefanovsky, S.V.; Vance, E.R.; Vernaz, E.Y. Materials science of high-level nuclear waste immobilization. *MRS Bull.* **2009**, *34*, 46–53. [CrossRef]
5. Zur Loye, H.-C.; Besmann, T.; Amoroso, J.; Brinkman, K.; Grandjean, A.; Henager, C.H.; Hu, S.; Misture, S.T.; Phillpot, S.R.; Shustova, N.B.; et al. Hierarchical Materials as Tailored Nuclear Waste Forms: A Perspective. *Chem. Mater.* **2018**, *30*, 4475–4488. [CrossRef]
6. Xu, Y.; Wen, Y.; Grote, R.; Amoroso, J.; Shuller Nickles, L.; Brinkman, K.S. A-site compositional effects in Ga-doped hollandite materials of the form Ba$_x$Cs$_y$Ga$_{2x+y}$Ti$_{8-2x-y}$O$_{16}$: Implications for Cs immobilization in crystalline ceramic waste forms. *Sci. Rep.* **2016**, *6*, 27412. [CrossRef]
7. Wang, S.; Alekseev, E.; Diwu, J.; Casey, W.; Phillips, B.; Depmeier, W.; Albrecht-Schmitt, T. NDTB-1: A Supertetrahedral Cationic Framework That Removes TcO$_4^-$ from Solution. *Angew. Chem. Int. Ed.* **2010**, *49*, 1057–1060. [CrossRef] [PubMed]
8. Halasyamani, P.S.; Walker, S.M.; O'Hare, D. The First Open Framework Actinide Material (C$_4$N$_2$H$_{12}$)U$_2$O$_4$F$_6$ (MUF-1). *J. Am. Chem. Soc.* **1999**, *121*, 7415–7416. [CrossRef]
9. Jackson, J.M.; Burns, P.C. A re-evaluation of the structure of weeksite, a uranyl silicate framework mineral. *Can. Mineral.* **2001**, *39*, 187–195. [CrossRef]
10. Wang, X.; Huang, J.; Liu, L.; Jacobson, A.J. The novel open-framework uranium silicates Na$_2$(UO$_2$)(Si$_4$O$_{10}$)·2.1H$_2$O (USH-1) and RbNa(UO$_2$)(Si$_2$O$_6$)·H$_2$O (USH-3). *J. Mater. Chem.* **2002**, *12*, 406–410. [CrossRef]
11. Chen, C.-S.; Kao, H.-M.; Lii, K.-H. K$_5$(UO$_2$)$_2$[Si$_4$O$_{12}$(OH)]: A uranyl silicate containing chains of four silicate tetrahedra linked by SiO···HOSi hydrogen bonds. *Inorg. Chem.* **2005**, *44*, 935–940. [CrossRef]
12. Liu, H.-K.; Chang, W.-J.; Lii, K.-H. High-Temperature, High-Pressure Hydrothermal Synthesis and Characterization of an Open-Framework Uranyl Silicate with Nine-Ring Channels: Cs$_2$UO$_2$Si$_{10}$O$_{22}$. *Inorg. Chem.* **2011**, *50*, 11773–11776. [CrossRef] [PubMed]
13. Liu, H.-K.; Peng, C.-C.; Chang, W.-J.; Lii, K.-H. Tubular Chains, Single Layers, and Multiple Chains in Uranyl Silicates: A$_2$[(UO$_2$)Si$_4$O$_{10}$] (A = Na, K, Rb, Cs). *Cryst. Growth Des.* **2016**, *16*, 5268–5272. [CrossRef]
14. Morrison, G.; Tran, T.T.; Halasyamani, P.S.; zur Loye, H.-C. K$_8$(K$_5$F)U$_6$Si$_8$O$_{40}$: An Intergrowth Uranyl Silicate. *Inorg. Chem.* **2016**, *55*, 3215–3217. [CrossRef]
15. Morrison, G.; Smith, M.D.; zur Loye, H.-C. Understanding the Formation of Salt-Inclusion Phases: An Enhanced Flux Growth Method for the Targeted Synthesis of Salt-Inclusion Cesium Halide Uranyl Silicates. *J. Am. Chem. Soc.* **2016**, *138*, 7121–7129. [CrossRef]
16. Li, H.; Kegler, P.; Bosbach, D.; Alekseev, E.V. Hydrothermal Synthesis, Study, and Classification of Microporous Uranium Silicates and Germanates. *Inorg. Chem.* **2018**, *57*, 4745–4756. [CrossRef]
17. Chen, C.-S.; Lee, S.-F.; Lii, K.-H. K(UO)Si$_2$O$_6$: A pentavalent-uranium silicate. *J. Am. Chem. Soc.* **2005**, *127*, 12208–12209. [CrossRef]
18. Wang, S.; Alekseev, E.V.; Stritzinger, J.T.; Depmeier, W.; Albrecht-Schmitt, T.E. Crystal chemistry of the potassium and rubidium uranyl borate families derived from boric acid fluxes. *Inorg. Chem.* **2010**, *49*, 6690–6696. [CrossRef] [PubMed]
19. Wang, S.; Alekseev, E.V.; Stritzinger, J.T.; Depmeier, W.; Albrecht-Schmitt, T.E. How are centrosymmetric and noncentrosymmetric structures achieved in uranyl borates? *Inorg. Chem.* **2010**, *49*, 2948–2953. [CrossRef]
20. Wang, S.; Alekseev, E.V.; Ling, J.; Liu, G.; Depmeier, W.; Albrecht-Schmitt, T.E. Polarity and Chirality in Uranyl Borates: Insights into Understanding the Vitrification of Nuclear Waste and the Development of Nonlinear Optical Materials. *Chem. Mater.* **2010**, *22*, 2155–2163. [CrossRef]
21. Wang, S.; Alekseev, E.V.; Stritzinger, J.T.; Liu, G.; Depmeier, W.; Albrecht-Schmitt, T.E. Structure–Property Relationships in Lithium, Silver, and Cesium Uranyl Borates. *Chem. Mater.* **2010**, *22*, 5983–5991. [CrossRef]
22. Wu, S.; Wang, S.; Polinski, M.; Beermann, O.; Kegler, P.; Malcherek, T.; Holzheid, A.; Depmeier, W.; Bosbach, D.; Albrecht-Schmitt, T.E.; et al. High structural complexity of potassium uranyl borates derived from high-temperature/high-pressure reactions. *Inorg. Chem.* **2013**, *52*, 5110–5118. [CrossRef]
23. Hao, Y.; Klepov, V.V.; Murphy, G.L.; Modolo, G.; Bosbach, D.; Albrecht-Schmitt, T.E.; Kennedy, B.J.; Wang, S.; Alekseev, E.V. Influence of Synthetic Conditions on Chemistry and Structural Properties of Alkaline Earth Uranyl Borates. *Cryst. Growth Des.* **2016**, *16*, 5923–5931. [CrossRef]
24. Hao, Y.; Klepov, V.V.; Kegler, P.; Modolo, G.; Bosbach, D.; Albrecht-Schmitt, T.E.; Wang, S.; Alekseev, E.V. Synthesis and Study of the First Zeolitic Uranium Borate. *Cryst. Growth Des.* **2018**, *18*, 498–505. [CrossRef]
25. Wu, S.; Polinski, M.J.; Malcherek, T.; Bismayer, U.; Klinkenberg, M.; Modolo, G.; Bosbach, D.; Depmeier, W.; Albrecht-Schmitt, T.E.; Alekseev, E.V. Novel fundamental building blocks and site dependent isomorphism in the first actinide borophosphates. *Inorg. Chem.* **2013**, *52*, 7881–7888. [CrossRef]
26. Hao, Y.; Murphy, G.L.; Bosbach, D.; Modolo, G.; Albrecht-Schmitt, T.E.; Alekseev, E.V. Porous Uranyl Borophosphates with Unique Three-Dimensional Open-Framework Structures. *Inorg. Chem.* **2017**, *56*, 9311–9320. [CrossRef] [PubMed]

27. Hao, Y.C.; Hu, C.L.; Xu, X.; Kong, F.; Mao, J.G. SrGe$_2$B$_2$O$_8$ and Sr$_3$Ge$_2$B$_6$O$_{16}$: Novel strontium borogermanates with three-dimensional and layered anionic architectures. *Inorg. Chem.* **2013**, *52*, 13644–13650. [CrossRef]
28. Wu, S.; Beermann, O.; Wang, S.; Holzheid, A.; Depmeier, W.; Malcherek, T.; Modolo, G.; Alekseev, E.V.; Albrecht-Schmitt, T.E. Synthesis of uranium materials under extreme conditions: UO$_2$[B$_3$Al$_4$O$_{11}$(OH)], a complex 3D aluminoborate. *Chem. Eur. J.* **2012**, *18*, 4166–4169. [CrossRef]
29. Klepov, V.V.; Juillerat, C.A.; Alekseev, E.V.; zur Loye, H.-C. Overstepping Löwenstein's Rule—A Route to Unique Aluminophosphate Frameworks with Three-Dimensional Salt-Inclusion and Ion Exchange Properties. *Inorg. Chem.* **2019**, *58*, 724–736. [CrossRef]
30. Mutailipu, M.; Poeppelmeier, K.R.; Pan, S. Borates: A Rich Source for Optical Materials. *Chem. Rev.* **2021**, *121*, 1130–1202. [CrossRef]
31. Lin, X.; Zhang, F.; Pan, S.; Yu, H.; Zhang, F.; Dong, X.; Han, S.; Dong, L.; Bai, C.; Wang, Z. Ba$_4$(BO$_3$)$_3$(SiO$_4$)·Ba$_3$X (X = Cl, Br): New salt-inclusion borosilicate halides as potential deep UV nonlinear optical materials. *J. Mater. Chem. C* **2014**, *2*, 4257. [CrossRef]
32. Miao, Z.; Yang, Y.; Wei, Z.; Yang, Z.; Yu, S.; Pan, S. NaCa$_5$BO$_3$(SiO$_4$)$_2$ with Interesting Isolated [BO$_3$] and [SiO$_4$] Units in Alkali- and Alkaline-Earth-Metal Borosilicates. *Inorg. Chem.* **2019**, *58*, 3937–3943. [CrossRef]
33. Heyward, C.C.; McMillen, C.D.; Kolis, J.W. Hydrothermal Growth of Lanthanide Borosilicates: A Useful Approach to New Acentric Crystals Including a Derivative of Cappelenite. *Inorg. Chem.* **2014**, *54*, 905–913. [CrossRef] [PubMed]
34. Wu, H.; Yu, H.; Pan, S.; Huang, Z.; Yang, Z.; Su, X.; Poeppelmeier, K.R. Cs$_2$B$_4$SiO$_9$: A Deep-Ultraviolet Nonlinear Optical Crystal. *Angew. Chem. Int. Ed.* **2013**, *52*, 3406–3410. [CrossRef] [PubMed]
35. Heyward, C.; McMillen, C.D.; Kolis, J. Hydrothermal synthesis and structural analysis of new mixed oxyanion borates: Ba$_{11}$B$_{26}$O$_{44}$(PO$_4$)$_2$(OH)$_6$, Li$_9$BaB$_{15}$O$_{27}$(CO$_3$) and Ba$_3$Si$_2$B$_6$O$_{16}$. *J. Solid State Chem.* **2013**, *203*, 166–173. [CrossRef]
36. Chang, L.-W.; Liu, H.-K.; Lii, K.-H. Flux synthesis and characterization of two barium hydroxyborosilicates with triple and single tetrahedral layers: Ba$_2$[Si$_3$B$_3$O$_{12}$(OH)] and Ba[Si$_2$BO$_6$(OH)]. *J. Solid State Chem.* **2020**, *287*, 121331. [CrossRef]
37. Medvedev, A.G.; Mikhaylov, A.A.; Churakov, A.V.; Vener, M.V.; Tripol'skaya, T.A.; Cohen, S.; Lev, O.; Prikhodchenko, P.V. Potassium, Cesium, and Ammonium Peroxogermanates with Inorganic Hexanuclear Peroxo Bridged Germanium Anion Isolated from Aqueous Solution. *Inorg. Chem.* **2015**, *54*, 8058–8065. [CrossRef] [PubMed]
38. Manceau, A.; Gorshkov, A.I.; Drits, V.A. Structural chemistry of Mn, Fe, Co, and Ni in manganese hydrous oxides: Part II. Information from EXAFS spectroscopy and electron and X-ray diffraction. *Am. Mineral.* **1992**, *77*, 1144–1157.
39. Lee, C.-S.; Lin, C.-H.; Wang, S.-L.; Lii, K.-H. [Na$_7$U^{IV}O$_2$(U^{V}O)$_2$(U$^{V/VI}$O$_2$)$_2$Si$_4$O$_{16}$]: A mixed-valence uranium silicate. *Angew. Chem. Int. Ed.* **2010**, *49*, 4254–4256. [CrossRef]
40. *APEXIII Version 2016. 5-0, SAINT Version 7.60A, SADABS Version 2016/2*; Bruker Analytical X-ray Systems: Madison, WI, USA, 2016.
41. Sheldrick, G.M. SHELXT—Integrated space-group and crystal-structure determination. *Acta Cryst.* **2015**, *A71*, 3–8. [CrossRef]
42. Sheldrick, G.M. A short history of SHELX. *Acta Cryst.* **2008**, *A64*, 112–122. [CrossRef]
43. Spek, A.L. Single-crystal structure validation with the program PLATON. *J. Appl. Crystallogr.* **2003**, *36*, 7–13. [CrossRef]
44. Blatov, V.A.; Shevchenko, A.P.; Proserpio, D.M. Applied Topological Analysis of Crystal Structures with the Program Package ToposPro. *Cryst. Growth Des.* **2014**, *14*, 3576–3586. [CrossRef]
45. Blatov, V.A.; Shevchenko, A.P.; Serezhkin, V.N. Multipurpose Crystallochemical Analysis with the Program Package TOPOS. *IUCr CompComm Newslett.* **2006**, *7*, 4–38.
46. Alexandrov, E.V.; Blatov, V.A.; Kochetkov, A.V.; Proserpio, D.M. Underlying nets in three-periodic coordination polymers: Topology, taxonomy and prediction from a computer-aided analysis of the Cambridge Structural Database. *CrystEngComm* **2011**, *13*, 3947–3958. [CrossRef]
47. Kropf, A.J.; Katsoudas, J.; Chattopadhyay, S.; Shibata, T.; Lang, E.A.; Zyryanov, V.N.; Ravel, B.; McIvor, K.; Kemner, K.M.; Scheckel, K.G.; et al. The New MRCAT (Sector 10) Bending Magnet Beamline at the Advanced Photon Source. *AIP Conf. Proc.* **2010**, *1234*, 299–302. [CrossRef]
48. Ravel, B.; Newville, M. ATHENA, ARTEMIS, HEPHAESTUS: Data analysis for X-ray absorption spectroscopy using IFEFFIT. *J. Synchrotron Radiat.* **2005**, *12*, 537–541. [CrossRef]
49. Kresse, G.; Furthmüller, J. Efficiency of ab-initio total energy calculations for metals and semiconductors using a plane-wave basis set. *Comput. Mater. Sci.* **1996**, *6*, 15–50. [CrossRef]
50. Kresse, G. Efficient iterative schemes for ab initio total-energy calculations using a plane-wave basis set. *Phys. Rev. B* **1996**, *54*, 11169–11186. [CrossRef]
51. Perdew, J.P.; Burke, K.; Ernzerhof, M. Generalized gradient approximation made simple. *Phys. Rev. Lett.* **1997**, *78*, 1396–1399. [CrossRef]
52. Blöchl, P.E. Projector augmented-wave method. *Phys. Rev. B* **1994**, *50*, 17953–17979. [CrossRef]
53. Kresse, G. From ultrasoft pseudopotentials to the projector augmented-wave method. *Phys. Rev. B* **1999**, *59*, 1758–1775. [CrossRef]
54. Rabenau, A. The role of hydrothermal synthesis in materials science. *J. Mater. Educ.* **1988**, *10*, 543–591.
55. Stritzinger, J.T.; Alekseev, E.V.; Polinski, M.J.; Cross, J.N.; Eaton, T.M.; Albrecht-Schmitt, T.E. Further evidence for the stabilization of U(V) within a tetraoxo core. *Inorg. Chem.* **2014**, *53*, 5294–5299. [CrossRef]
56. McMillen, C.D.; Kolis, J.W. Hydrothermal synthesis as a route to mineralogically-inspired structures. *Dalton Trans.* **2016**, *45*, 2772–2784. [CrossRef]

57. Burns, P.C.; Grice, J.D.; Hawthorne, F.C. Borate Minerals. I. Polyhedral Clusters and Fundamental Building Blocks. *Can. Mineral.* **1995**, *33*, 1131–1151.
58. Ewald, B.; Huang, Y.-X.; Kniep, R. Structural Chemistry of Borophosphates, Metalloborophosphates, and Related Compounds. *Z. Anorg. Allg. Chem.* **2007**, *633*, 1517–1540. [CrossRef]
59. Qiu, J.; Ling, J.; Sieradzki, C.; Nguyen, K.; Wylie, E.M.; Szymanowski, J.E.S.; Burns, P.C. Expanding the Crystal Chemistry of Uranyl Peroxides: Four Hybrid Uranyl-Peroxide Structures Containing EDTA. *Inorg. Chem.* **2014**, *53*, 12084–12091. [CrossRef] [PubMed]
60. Qiu, J.; Ling, J.; Jouffret, L.; Thomas, R.; Szymanowski, J.E.S.; Burns, P.C. Water-soluble multi-cage super tetrahedral uranyl peroxide phosphate clusters. *Chem. Sci.* **2014**, *5*, 303–310. [CrossRef]
61. Zhang, Y.; Bhadbhade, M.; Price, J.R.; Karatchevtseva, I.; Collison, D.; Lumpkin, G.R. Kinetics vs. thermodynamics: A unique crystal transformation from a uranyl peroxo-nanocluster to a nanoclustered uranyl polyborate. *RSC Adv.* **2014**, *4*, 34244–34247. [CrossRef]
62. Thangavelu, S.G.; Cahill, C.L. Uranyl-Promoted Peroxide Generation: Synthesis and Characterization of Three Uranyl Peroxo [$(UO_2)_2(O_2)$] Complexes. *Inorg. Chem.* **2015**, *54*, 4208–4221. [CrossRef]
63. Qiu, J.; Dembowski, M.; Szymanowski, J.E.S.; Toh, W.C.; Burns, P.C. Time-Resolved X-ray Scattering and Raman Spectroscopic Studies of Formation of a Uranium-Vanadium-Phosphorus-Peroxide Cage Cluster. *Inorg. Chem.* **2016**, *55*, 7061–7067. [CrossRef] [PubMed]
64. Dembowski, M.; Colla, C.A.; Hickam, S.; Oliveri, A.F.; Szymanowski, J.E.S.; Oliver, A.G.; Casey, W.H.; Burns, P.C. Hierarchy of Pyrophosphate-Functionalized Uranyl Peroxide Nanocluster Synthesis. *Inorg. Chem.* **2017**, *56*, 5478–5487. [CrossRef]
65. Qiu, J.; Spano, T.L.; Dembowski, M.; Kokot, A.M.; Szymanowski, J.E.; Burns, P.C. Sulfate-Centered Sodium-Icosahedron-Templated Uranyl Peroxide Phosphate Cages with Uranyl Bridged by $\mu\text{-}\eta^1{:}\eta^2$ Peroxide. *Inorg. Chem.* **2017**, *56*, 1874–1880. [CrossRef] [PubMed]
66. Bartlett, J.R. On the determination of uranium-oxygen bond lengths in dioxouranium(VI) compounds by Raman spectroscopy. *J. Mol. Struct.* **1989**, *193*, 295–300. [CrossRef]
67. Frost, R.L.; Bouzaid, J.M.; Martens, W.N.; Reddy, B.J. Raman spectroscopy of the borosilicate mineral ferroaxinite. *J. Raman Spectrosc.* **2007**, *38*, 135–141. [CrossRef]
68. Bastians, S.; Crump, G.; Griffith, W.P.; Withnall, R. Raspite and studtite: Raman spectra of two unique minerals. *J. Raman Spectrosc.* **2004**, *35*, 726–731. [CrossRef]
69. Dembowski, M.; Bernales, V.; Qiu, J.; Hickam, S.; Gaspar, G.; Gagliardi, L.; Burns, P.C. Computationally-Guided Assignment of Unexpected Signals in the Raman Spectra of Uranyl Triperoxide Complexes. *Inorg. Chem.* **2017**, *56*, 1574–1580. [CrossRef]
70. McGrail, B.T.; Sigmon, G.E.; Jouffret, L.J.; Andrews, C.R.; Burns, P.C. Raman Spectroscopic and ESI-MS Characterization of Uranyl Peroxide Cage Clusters. *Inorg. Chem.* **2014**, *53*, 1562–1569. [CrossRef] [PubMed]
71. Frost, R.L.; Čejka, J.; Weier, M.L.; Martens, W. Molecular structure of the uranyl silicates—A Raman spectroscopic study. *J. Raman Spectrosc.* **2006**, *37*, 538–551. [CrossRef]
72. Frost, R.L.; Čejka, J.; Weier, M.L.; Martens, W.; Kloprogge, J.T. A Raman and infrared spectroscopic study of the uranyl silicates-weeksite, soddyite and haiweeite. *Spectrochim. Acta A* **2006**, *64*, 308–315. [CrossRef]
73. Faulques, E.; Kalashnyk, N.; Massuyeau, F.; Perry, D.L. Spectroscopic markers for uranium phosphates: A vibronic study. *RSC Adv.* **2015**, *5*, 71219–71227. [CrossRef]

Article

A Thermodynamic Investigation of Ni on Thin-Film Titanates (ATiO$_3$)

Chao Lin [1], Alexandre C. Foucher [2], Eric A. Stach [2] and Raymond J. Gorte [1,*]

[1] Department of Chemical and Biomolecular Engineering, University of Pennsylvania, 34th Street, Philadelphia, PA 19104, USA; linchao@seas.upenn.edu
[2] Department of Materials Science and Engineering, University of Pennsylvania, Philadelphia, PA 19104, USA; afoucher@seas.upenn.edu (A.C.F.); stach@seas.upenn.edu (E.A.S.)
* Correspondence: gorte@seas.upenn.edu

Received: 25 November 2020; Accepted: 9 December 2020; Published: 11 December 2020

Abstract: Thin, ~1-nm films of CaTiO$_3$, SrTiO$_3$, and BaTiO$_3$ were deposited onto MgAl$_2$O$_4$ by Atomic Layer Deposition (ALD) and then studied as catalyst supports for ~5 wt % of Ni that was added to the perovskite thin films by Atomic Layer Deposition. Scanning Transmission Electron Microscopy demonstrated that both the Ni and the perovskites uniformly covered the surface of the support following oxidation at 1073 K, even after redox cycling, but large Ni particles formed following a reduction at 1073 K. When compared to Ni/MgAl$_2$O$_4$, the perovskite-containing catalysts required significantly higher temperatures for Ni reduction. Equilibrium constants for Ni oxidation, as determined from Coulometric Titration, indicated that the oxidation of Ni shifted to lower P_{O_2} on the perovskite-containing materials. Based on Ni equilibrium constants, Ni interactions are strongest with CaTiO$_3$, followed by SrTiO$_3$ and BaTiO$_3$. The shift in the equilibrium constant was shown to cause reversible deactivation of the Ni/CaTiO$_3$/MgAl$_2$O$_4$ catalyst for CO$_2$ reforming of CH$_4$ at high CO$_2$ pressures, due to the oxidation of the Ni.

Keywords: Atomic Layer Deposition; Ni catalyst; reforming catalyst; thermodynamics

1. Introduction

Perovskite-supported catalysts have received a great deal of attention recently because they offer the possibility of redispersing the metal if the crystallites become large due to sintering. Redispersion in these so-call "intelligent" catalysts occurs when the metal cations become part of the bulk lattice under oxidizing conditions, and then ex-solve back to the surface upon high-temperature reduction [1–8]. Originally proposed for use in automotive-emission-control catalysis, the idea has also shown promise for Solid Oxide Fuel Cell (SOFC) anodes [9,10] and steam-reforming catalysts [11–13], due, in part, to the improved thermal stabilities, but also due to improved tolerance against coking [14,15].

The application of perovskite supports has been limited due to the low specific surface areas of the perovskites, the sluggish kinetics for ingress and egress of metal particles, and the fact that some ex-solved metal remains embedded in the bulk, as discussed in more detail elsewhere [16]. Our group has prepared thin films of different perovskites on high-surface-area MgAl$_2$O$_4$ using Atomic Layer Deposition (ALD) in order to circumvent these limitations. For the materials to maintain a high surface area, the film thickness is limited to approximately 1 nm. This is demonstrated by the fact that a 1-nm film of SrTiO$_3$ on a 200-m^2/g support would have 0.96 g SrTiO$_3$/g of support, which results in a significantly lower specific surface area due to both the increased sample mass and decreased pore sizes of the sample. Although metals that are supported on these thin-film supports are not catalytically identical to bulk, ex-solution metals [17], many of their properties are similar. For example, most of these catalysts are only active after high-temperature reduction [18]. Perovskite-supported,

Ni catalysts also exhibit a high tolerance against coking [19,20]. Finally, supported Pt catalysts that are based on both thin-film and bulk perovskites showed similar differences from conventional Pt catalysts in their relative inactivity for hydrogenation reactions when compared to oxidation reactions [18].

There are indications that some of the support effects that are associated with perovskites are due to the formation of chemical bonds with the metal. First, the nature of interactions is metal and perovskite specific, as expected when chemical bonding is important. While Pt and Rh can enter the bulk $CaTiO_3$ lattice and both metals exhibit strong support interactions in their thin-film variants, Pd cannot be doped into $CaTiO_3$, and Pd supported on $CaTiO_3$ films shows normal behavior [17,18]. Metal-perovskite interactions were also shown to depend on the perovskite composition in a comparison of Rh catalysts that are supported on thin films of $CaTiO_3$, $SrTiO_3$, and $BaTiO_3$ [17]. Rh interacted very strongly with $CaTiO_3$, but only weakly with $BaTiO_3$. A second piece of evidence for bonding interactions is that the perovskite support can change the thermodynamic properties for oxidation of the supported metal. This was demonstrated for Ni on $LaFeO_3$ films, where the equilibrium constant for the reaction Ni to NiO was observed to shift by four orders of magnitude to lower PO_2 in the presence of the perovskite [20].

A better understanding of the nature of metal-perovskite interactions is clearly required before these materials can find wider application. In the present study, the equilibrium properties for Ni oxidation were examined on thin films of $CaTiO_3$, $SrTiO_3$, and $BaTiO_3$ that were supported on $MgAl_2O_4$. Ni-based catalysts are convenient for this comparison, because previous work has shown that the support can change the equilibrium constant for Ni oxidation to NiO by a significant amount. For the reaction, $Ni + O_2 = NiO$, the equilibrium constant, K_{NiO}, is equal to $P_{O_2}^{-}$ and is given by $\exp(-\Delta G/RT)$, where ΔG is the free energy of reaction. K_{NiO}, and therefore ΔG, can be obtained by measuring the P_{O_2} at which both Ni and NiO exist in equilibrium. We will show that, similar to what was observed for Rh on these perovskite supports, Ni interacts most strongly with $CaTiO_3$, followed by $SrTiO_3$ and $BaTiO_3$.

2. Results

Table 1 provides a list of the samples that were used in this study, together with some of their key properties. The $CaTiO_3/MgAl_2O_4$, $SrTiO_3/MgAl_2O_4$, and $BaTiO_3/MgAl_2O_4$ supports were prepared while using procedures that are identical to those used in an earlier study of the support effects on Rh [17]. As in that case, the targeted perovskite film thicknesses were 1 nm, assuming that the perovskite film densities were the same as that of the corresponding bulk perovskites and that the 120-m^2/g $MgAl_2O_4$ was uniformly covered. The Ni loadings were achieved using five ALD cycles of the Ni precursor. The properties that are shown in the Table were measured after the sample had been oxidized and reduced five times at 1073 K in order to ensure that the samples had reached an equilibrium state.

Table 1. Key properties of the samples used in the study.

	Specific Surface Area (m^2/g) after 5 Redox Cycles at 1073 K	ALD Thin Film Loading (wt %)	Ni Loading (wt %)
Ni/$MgAl_2O_4$	115	N/A	4.8
Ni/$CaTiO_3$/$MgAl_2O_4$	73	29	4.6
Ni/$SrTiO_3$/$MgAl_2O_4$	67	33	3.7
Ni/$BaTiO_3$/$MgAl_2O_4$	59	39	4.1

The perovskite-containing samples that are listed in Table 1 were analyzed by Scanning Transmission Electron Microscopy/Energy Dispersive X-ray Spectroscopy (STEM/EDS) and XRD. In all cases, the samples were pretreated while using five redox cycles at 1073 K before the measurements were performed. The results are shown after oxidation at 1073 K and after reduction at 1073 K. Figures 1–3 show the STEM/EDS results for the oxidized and reduced samples. For the oxidized samples, the images for each of the three samples were indistinguishable from that of the $MgAl_2O_4$ support; and, the EDS maps showed a spatial distribution of A-site cations, Ti, and Ni that matched well with the Mg and Al variations. All of this implies that each of the species deposited by ALD are uniformly distributed over the surface of the $MgAl_2O_4$ support and remain uniformly distributed, even after redox cycling. The $CaTiO_3$, $SrTiO_3$, and $BaTiO_3$ films remained largely unchanged after reduction at 1073 K, but Ni particles were formed on each of the three samples after this treatment. The Ni particles appeared to be slightly smaller for $Ni/CaTiO_3/MgAl_2O_4$, ~10 to 20 nm, and largest on $Ni/BaTiO_3/MgAl_2O_4$, ~30 to 50 nm, with $Ni/SrTiO_3/MgAl_2O_4$ in between. It is noteworthy that the formation of Ni particles was reversible on each of the samples, since the uniform Ni films were restored by oxidation.

Figure 4 reports the XRD patterns for the oxidized and reduced samples. For oxidized $Ni/CaTiO_3/MgAl_2O_4$ and $Ni/SrTiO_3/MgAl_2O_4$, all of the peaks can be assigned to either the $MgAl_2O_4$ support or the corresponding perovskite phase. Upon reduction, peaks that are associated with metallic Ni were also observed. For $BaTiO_3/MgAl_2O_4$, there is significant overlap in the peak positions for $BaTiO_3$ and $MgAl_2O_4$; however, the relative ratio of peaks, particularly those that are centered at 32 and 37 degrees 2θ, make it clear that the major peaks are due to a mixture of $MgAl_2O_4$ and $BaTiO_3$. Small peaks that could be indexed to $BaCO_3$ were also observed on the $Ni/BaTiO_3/MgAl_2O_4$ sample. Again, the XRD pattern of the reduced sample was unchanged, except for the additional features that are associated with metallic Ni.

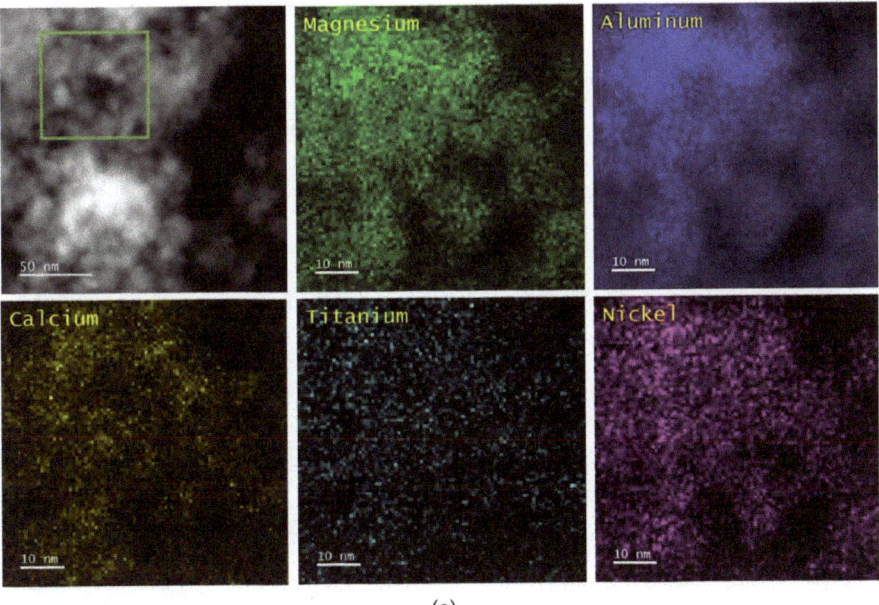

(a)

Figure 1. Cont.

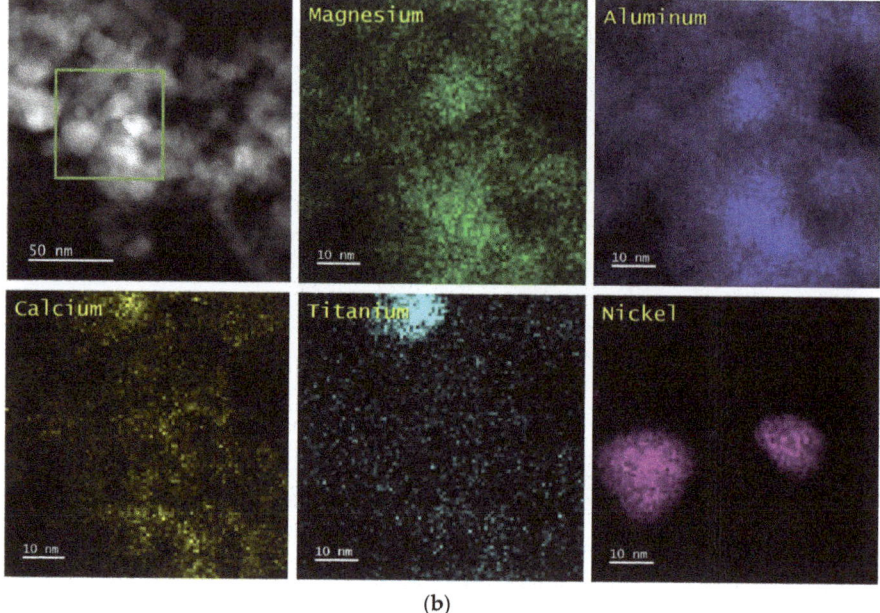

(b)

Figure 1. (a) Representative Scanning Transmission Electron Microscopy (STEM) results for Ni/CaTiO$_3$/MgAl$_2$O$_4$ after 1073 K oxidation followed by 773 K reduction with Energy Dispersive X-ray Spectroscopy (EDS) maps of corresponding elements. (b) Representative STEM results for Ni/CaTiO$_3$/MgAl$_2$O$_4$ after 1073 K oxidation followed by 1073 K reduction with EDS maps of corresponding elements.

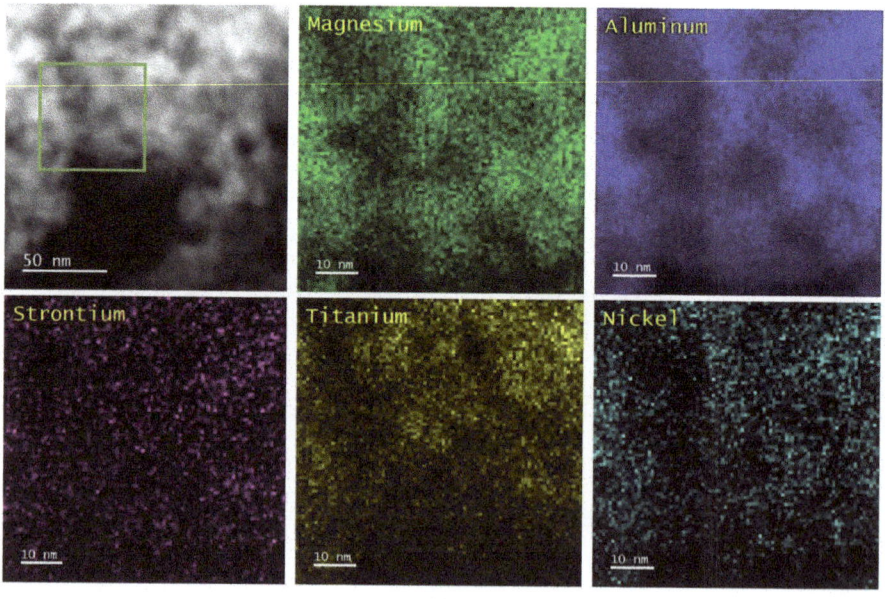

(a)

Figure 2. Cont.

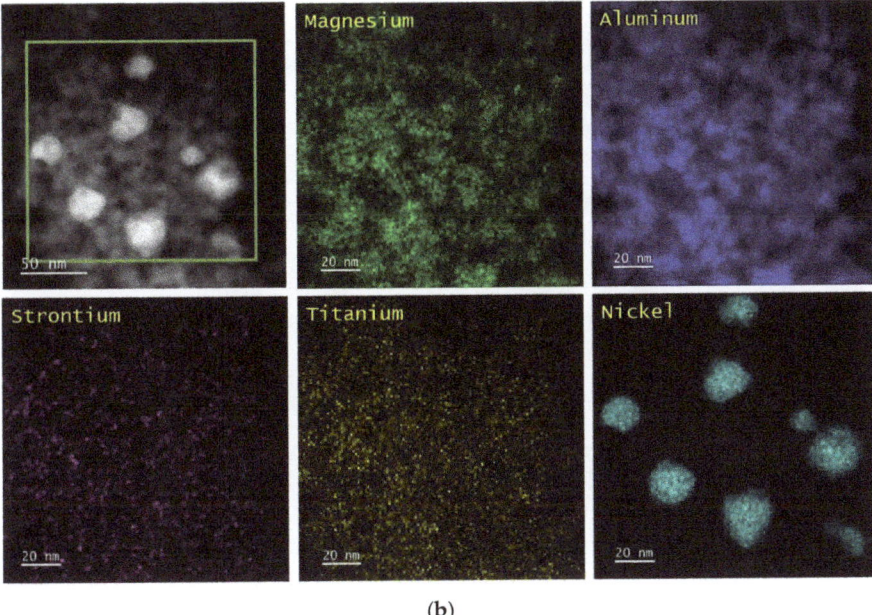

(b)

Figure 2. (a) Representative STEM results for Ni/SrTiO$_3$/MgAl$_2$O$_4$ after 1073 K oxidation followed by 773 K reduction with EDS maps of corresponding elements. (b) Representative STEM results for Ni/SrTiO$_3$/MgAl$_2$O$_4$ after 1073 K oxidation followed by 1073 K reduction with EDS maps of corresponding elements.

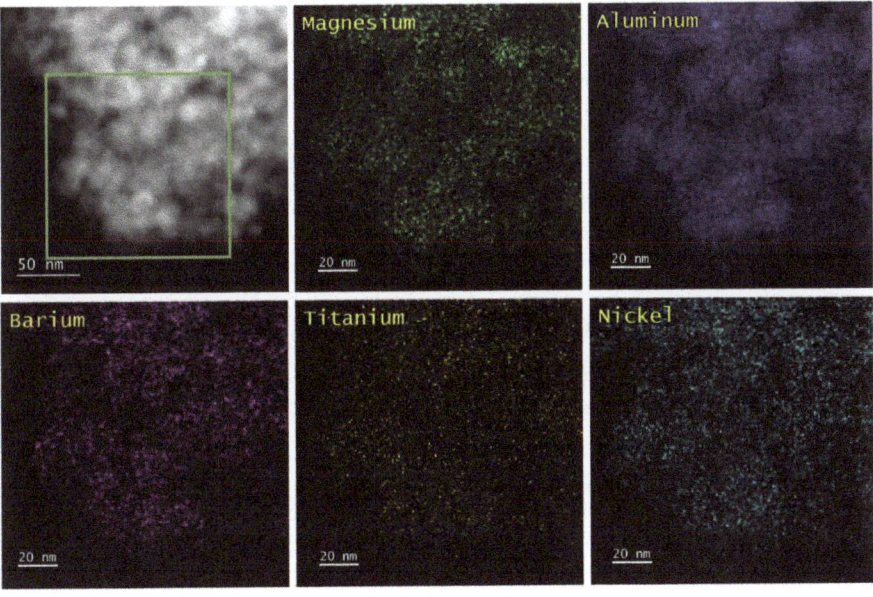

(a)

Figure 3. *Cont.*

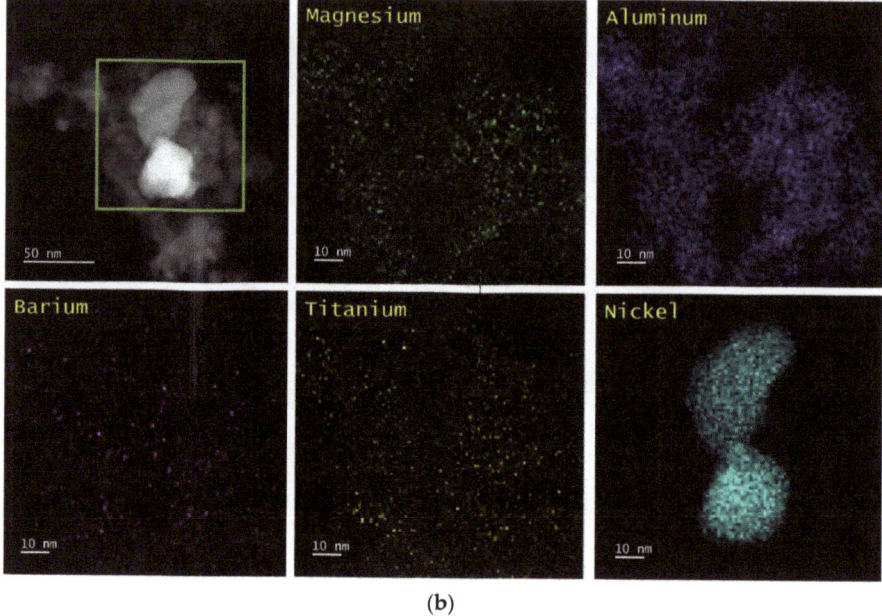

(b)

Figure 3. (a) Representative STEM results for Ni/BaTiO$_3$/MgAl$_2$O$_4$ after 1073 K oxidation followed by 773 K reduction with EDS maps of corresponding elements. (b) Representative STEM results for Ni/BaTiO$_3$/MgAl$_2$O$_4$ after 1073 K oxidation, followed by 1073 K reduction with EDS maps of corresponding elements.

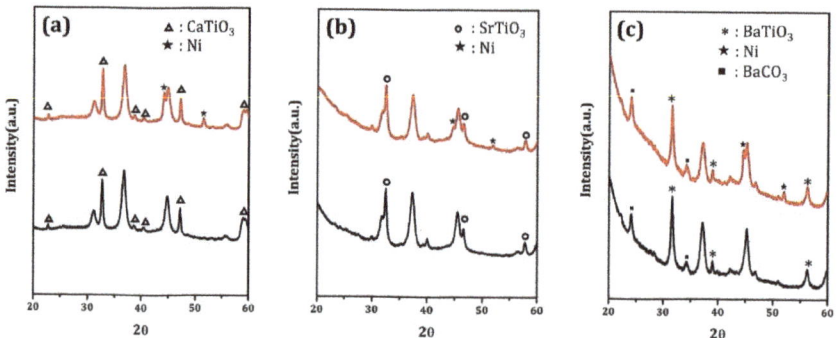

Figure 4. X-ray diffraction (XRD) patterns for (a) Ni/CaTiO$_3$/MgAl$_2$O$_4$, (b) Ni/SrTiO$_3$/MgAl$_2$O$_4$, (c) Ni/BaTiO$_3$/MgAl$_2$O$_4$ after (black) 1073 K oxidation followed by 773 K reduction and (red) 1073 K oxidation followed by 1073 K reduction.

As discussed elsewhere [16], the presence of intense perovskite diffraction peaks for uniformly-deposited films that are nominally 1-nm thick suggests the formation of relatively large, two-dimensional crystals, randomly oriented with respect to the support. Based on the width of the peaks, the nominal sizes of the perovskite crystallites are 22 nm for CaTiO$_3$, 18 nm for SrTiO$_3$, and 18 nm for BaTiO$_3$. The particle sizes were also estimated from XRD peak widths for the Ni particles in the reduced samples. These gave reasonable agreement with the values that were obtained from STEM, with values of 21 nm for Ni/CaTiO$_3$/MgAl$_2$O$_4$, 25 nm for Ni/SrTiO$_3$/MgAl$_2$O$_4$, and 31 nm for Ni/BaTiO$_3$/MgAl$_2$O$_4$.

The extent of reduction was measured as a function of temperature for each of the samples in Table 1, with results being shown in Table 2. The samples were first oxidized at 1073 K in flowing air and then reduced in 10%H_2-He mixture at the indicated temperature for 30 min. The amount of oxygen required to completely oxidize the sample at 1073 K was then measured using flow titration. First, there was no measurable reduction of the $CaTiO_3/MgAl_2O_4$ and $SrTiO_3/MgAl_2O_4$ supports, even at 1073 K; but, there was some reduction of the $BaTiO_3/MgAl_2O_4$ sample at the highest temperature. Therefore, with the exception of Ni/$BaTiO_3/MgAl_2O_4$ at 1073 K, the results that are presented in Table 2 indicate the extent of reduction of the Ni in each of the samples. For Ni/$MgAl_2O_4$, only 38% of the Ni was reduced at 773 K and 87% at 1073 K. That such harsh conditions are required for reducing some of the Ni may indicate that a fraction of Ni has reacted with the $MgAl_2O_4$ support after the five redox cycles at 1073 K, with Ni replacing Mg and forming $NiAl_2O_4$. For Ni/$CaTiO_3/MgAl_2O_4$ and Ni/$SrTiO_3/MgAl_2O_4$, a reduction of Ni was insignificant at temperatures below 1073 K. At that temperature, most of the Ni in these samples was reduced. There was some reduction of Ni at lower temperatures on Ni/$BaTiO_3/MgAl_2O_4$, but 1073 K was again required for a majority of the Ni to be reduced.

Table 2. Oxygen uptakes from flow titration for various samples after different reduction pretreatments.

Samples	Oxygen Uptakes (µmol O/g) and Corresponding Ni Reduction Extents * (%) after Dry H_2 Reduction at Various Temperatures			
	773-K Reduction	873-K Reduction	973-K Reduction	1073-K Reduction
Ni/$MgAl_2O_4$	310 (38%)	410 (50%)	490 (59%)	720 (87%)
Ni/$CaTiO_3/MgAl_2O_4$	0 (0%)	0 (0%)	30 (4%)	920 (116%)
Ni/$SrTiO_3/MgAl_2O_4$	0 (0%)	0 (0%)	90 (14%)	700 (109%)
Ni/$BaTiO_3/MgAl_2O_4$	60 (9%)	100 (14%)	230 (33%)	1100 (158%)
$CaTiO_3/MgAl_2O_4$	-	-	-	0
$SrTiO_3/MgAl_2O_4$	-	-	-	0
$BaTiO_3/MgAl_2O_4$	-	-	-	230

* Values in parenthesis indicate % reduction of Ni, assuming oxygen uptake is only for Ni.

Coulometric Titration (CT) measurements were performed at 1073 K on each of the samples in order to better understand the effect of the perovskite supports on Ni reducibility, and the results are shown in Figure 5. At this temperature, tabulated values of ΔG for bulk Ni would indicate that the equilibrium P_{O_2} should be 10^{-14} atm. Data for Ni/$MgAl_2O_4$, reproduced here from a previous publication [20], found that the amount of oxygen that was taken up by the sample was very close to the value obtained from the flow titration reported in Table 2 and that oxidation of the Ni component occurred between 10^{-15} and 10^{-16} atm. The relatively small change in the equilibrium P_{O_2} from that reported for bulk Ni may be due to the Ni particle size or to interactions with the $MgAl_2O_4$ support. The equilibrium P_{O_2} for each of the perovskite-containing samples were shifted to lower values. The shift was the largest for Ni/$CaTiO_3/MgAl_2O_4$, with oxidation occurring in the range of 10^{-18} to 10^{-19} atm, and the smallest for Ni/$BaTiO_3/MgAl_2O_4$, between 10^{-16} and 10^{-17} atm. The Ni/$BaTiO_3/MgAl_2O_4$ sample also took up slightly more oxygen in the CT experiments, which was in agreement with the flow-titration results. Even though the equilibrium P_{O_2} for Ni were shifted to lower values by interactions with the titanates, the changes were less than that previously observed with Ni/$LaFeO_3/MgAl_2O_4$ [20], for which the equilibrium P_{O_2} was 10^{-20} atm at 1073 K.

In order to determine whether these equilibrium properties affect catalytic activities, methane dry reforming (MDR) was investigated in all four Ni catalysts presented in Table 1. In order to determine the effect of reduction temperature on each of the catalysts, methane conversions were measured at 873 K, while using 4.2% CH_4, 8.3% CO_2, and a 72,000 mL·g^{-1}·h^{-1} Gas Hourly Space Velocity (GHSV), and the results are shown in Figure 6a. In each case, the catalysts were oxidized at 1073 K and then reduced in dry 10% H_2 for 30 min at the indicated temperatures before measuring the conversion. Ni/MgAl$_2$O$_4$ showed significant conversion after reduction at 773 K, and the conversions increased with the reduction temperature. This corresponds well to the flow-titration results, which indicated that Ni was somewhat reduced on this sample at 773 K, but that higher reduction temperatures increased the fraction of Ni reduction. Each of the perovskite-containing samples was completely inactive for reduction at temperatures below 1073 K, but showed reasonable activity after reduction at that temperature. The differential reaction rates are shown for each of the samples in the Arrhenius plots presented in Figure 6b. The rates were about a factor of two lower on the perovskite-containing samples, but all of the catalysts showed a similar temperature dependence.

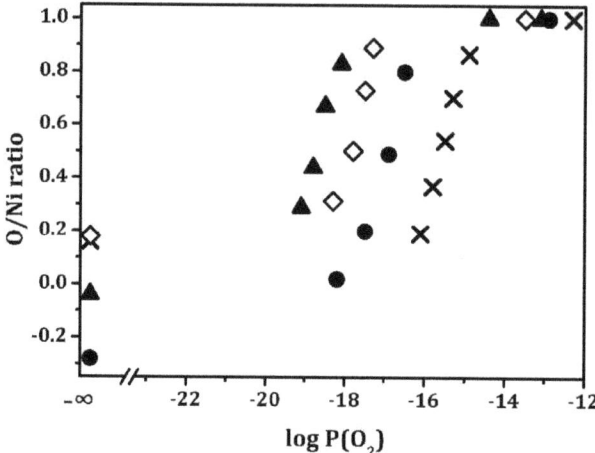

Figure 5. Oxidation isotherm obtained using coulometric titration at 1073 K for (×) Ni/MgAl$_2$O$_4$; (▲) Ni/CaTiO$_3$/MgAl$_2$O$_4$; (◊) Ni/SrTiO$_3$/MgAl$_2$O$_4$; and, (•) Ni/BaTiO$_3$/MgAl$_2$O$_4$.

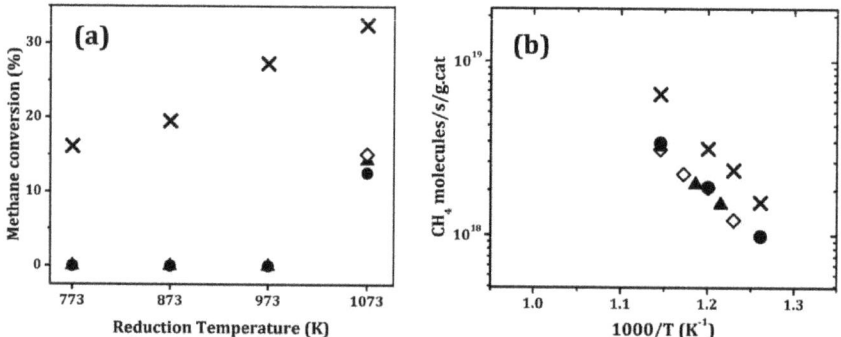

Figure 6. (a) Methane conversion for methane dry reforming (MDR) reaction at 873 K as a function of reduction temperature for (×) Ni/MgAl$_2$O$_4$; (▲) Ni/CaTiO$_3$/MgAl$_2$O$_4$; (◊) Ni/SrTiO$_3$/MgAl$_2$O$_4$; and, (•) Ni/BaTiO$_3$/MgAl$_2$O$_4$. Total flow rate was controlled at 120 mL min^{-1} with CH$_4$ and CO$_2$ flow being 5 mL min^{-1} and 10 mL min^{-1}, respectively. (b) Arrhenius plots for methane dry reforming (MDR) reaction for (×) Ni/MgAl$_2$O$_4$; (▲) Ni/CaTiO$_3$/MgAl$_2$O$_4$; (◊) Ni/SrTiO$_3$/MgAl$_2$O$_4$; and, (•) Ni/BaTiO$_3$/MgAl$_2$O$_4$ after each catalyst was reduced at 1073 K. The total flow rate was controlled at 120 mL/min with CH$_4$ and CO$_2$ flow being 5 mL min^{-1} and 10 mL min^{-1}, respectively.

With Ni on LaFeO$_3$ films, the shift to lower P_{O_2} in the equilibrium constants for Ni oxidation is sufficient for Ni to be oxidized under some MDR reaction conditions [20]. Although equilibrium constants for the titanates are significantly closer to that of bulk Ni, we investigated the MDR reaction on Ni/MgAl$_2$O$_4$ and Ni/CaTiO$_3$/MgAl$_2$O$_4$ as a function of the CH$_4$:CO$_2$ ratio to see whether oxidation of Ni could also occur in these catalysts. The reaction was studied at integral conversions to produce CO and H$_2$; an equilibrium P_{O_2} is then established by the CO$_2$:CO ratio. For these measurements, the CH$_4$ flow rate was kept at 5 mL min^{-1} and the CO$_2$ and He flows were tuned in order to achieve a desired CO$_2$:CH$_4$ ratio while keeping the space velocity the same. All of the conversions were measured at 873 K, and the results are shown in Figure 7.

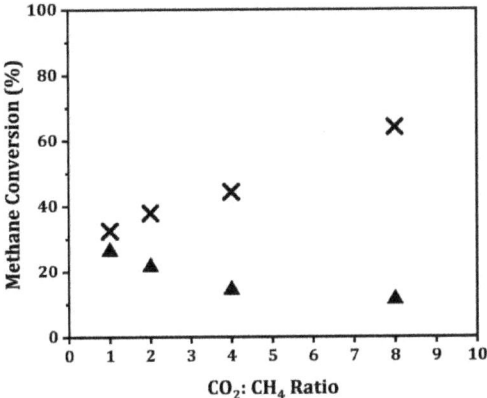

Figure 7. Methane conversion for MDR reaction at 873K for (×) Ni/MgAl$_2$O$_4$ and (▲) Ni/CaTiO$_3$/MgAl$_2$O$_4$ as a function of CO$_2$:CH$_4$ ratio. The total flow rate and the methane flow rate were controlled at 120 mL/min and 5 mL/min, respectively, while He and CO$_2$ flows were adjusted accordingly to achieve the desired ratios. The loading is 100 mg for Ni/MgAl$_2$O$_4$ and 200 mg for Ni/CaTiO$_3$/MgAl$_2$O$_4$ in order to ensure similar starting point for the methane conversion. Both of the samples were reduced at 1073K prior to catalytic measurements.

For the Ni/MgAl$_2$O$_4$ sample, we observed an increase in CH$_4$ conversion with increasing CO$_2$ concentration. This is the expected result for reactions with a positive reaction order. For Ni/CaTiO$_3$/MgAl$_2$O$_4$, the CH$_4$ conversion decreased with increasing CO$_2$ concentration, similar to what was previously observed for Ni/LaFeO$_3$/MgAl$_2$O$_4$ [20]. This deactivation was not due to coking, since the conversions shown in Figure 7 were stable and did not depend on whether the CO$_2$ concentration was increasing or decreasing. With the LaFeO$_3$-supported Ni, the deactivation was easier to observe because of the larger change in the equilibrium constant.

3. Discussion

Much of the work on perovskite-supported catalysts has focused on their exsolution properties and resistance to sintering. Thin-film perovskite supports are not identical to bulk perovskites, since the films are thinner than the nominal particle size of the metals, as discussed in more detail elsewhere [16]. However, there are many similarities between the bulk and thin-film supports, which suggests that bonding interactions at the metal-perovskite interface are similar in these two cases.

In agreement with a previous study of Ni on LaFeO$_3$ films [20], the work presented here demonstrates that the support can affect the oxidation thermodynamics of the Ni. Because the electronic effects of the support on a metal can only extend to a few atomic distances from the interface [21], most of the Ni must be in direct contact with the support. Indeed, the STEM/EDS data indicate that is the case, at least for the oxidized samples. While thermodynamic properties do not indicate how large the kinetic barriers will be for reduction, the fact that the perovskite-supported catalysts require high temperatures for Ni reduction is additional evidence that there must be strong bonding interactions between the support and Ni.

Changes in Ni geometry upon different pretreatments are also worth noting. While previous studies suggested that interactions between Ni and some supports can lead to the redispersion of Ni, the conditions used were found to be harsher and the redispersion less complete than the results that are reported here [22]. Our results would indicate that Ni atoms can migrate relatively long distances, along the surface of the perovskite thin films. It seems likely that Ni is incorporated as a strongly interacting phase, different from normal NiO, when supported on perovskite thin films, thus leading to the changes in equilibrium P_{O_2}.

Similar to what we observed in an earlier study of Rh on $CaTiO_3$, $SrTiO_3$, and $BaTiO_3$ films, support interactions depend on the A-site cations in this series. With Rh, the differences were much stronger than what we observed here with Ni. Rh that was supported on $CaTiO_3/MgAl_2O_4$ was almost unreducible and catalytically inactive, while the support effects for Rh on $BaTiO_3/MgAl_2O_4$ were negligible. The ordering of support interactions for Ni were the same as with Rh, in that $CaTiO_3$ interacts most strongly and $BaTiO_3$ least. How the A-site cations affect these interactions is uncertain. One possibility is that differences in the surface energies of these titanates change the surface terminations. For example, we have previously established by Low Energy Ion Scattering that $CaTiO_3$ films are terminated by Ca [18] and it is possible that $BaTiO_3$ prefers to be B-site terminated due to the relative sizes of the ions. The free volumes for these three perovskite materials also change with the size of the A-site cations, and this may affect how easily catalytic metal cations can enter the lattice.

It is also interesting to ask why the differences for Ni on the various titanates were less than what was observed for Rh. One possible reason for this could be the nominal charge on the cations. It was previously shown that Pd, which, like Ni, is expected to exist in the +2 state when oxidized, does not interact with $CaTiO_3$ and it cannot enter the $CaTiO_3$ lattice [18]. The ability of Rh to have multiple, stable oxidation states may also play a role in allowing Rh to interact more strongly with the perovskite lattices.

There is obviously much that we still need to learn about both ex-solution catalysts and the differences between bulk and thin-film perovskites. The fact that support interactions with these perovskite-supported metals are so specific to the compositions of the metal and the support represents both a challenge and an opportunity to tailor the catalytic properties for specific applications.

4. Materials and Methods

$MgAl_2O_4$ was prepared in our laboratory and had a surface area of 120 m^2/g after being stabilized by calcination to 1173 K, as described elsewhere [17,23]. The equipment and procedures used for ALD have also been reported previously [24,25]. The system is essentially a high-temperature adsorption apparatus that is attached to a mechanical vacuum pump. The evacuated samples were exposed to vapors from the precursors for approximately 10 min before purging excess precursor by evacuation. The precursors choices for the A-site cations were the same as one of the previous publications [17], bis(2,2,6,6-tetramethyl-3,5-heptanedionato) calcium (Ca(TMHD)$_2$, Strem, Newburyport, MA, USA), bis(2,2,6,6-tetramethyl-3,5-heptanedionato) strontium hydrate (Sr(TMHD)$_2$, Strem, USA), and bis(2,2,6,6-tetramethyl-3,5-heptanedionato) barium hydrate (Ba(TMHD)$_2$, Strem, USA) were chosen for Ca, Sr, and Ba ALD processes, respectively, and the exposure was carried out at 573 K. Because the TMHD ligands can only be removed at elevated temperatures by air [26], the oxidation step of the ALD cycle for these precursors was performed by transferring the samples to a muffle furnace at 773 K for 5 min. For the TiO_2-ALD process, Titanium chloride (TiCl$_4$, Sigma-Aldrich, St. Louis, MO, USA) was chosen as the precursor. TiCl$_4$ was kept at 363 K to ensure the generation of a sufficient amount of the precursor vapor. The deposition and the oxidation were both performed at 423 K. For the deposition half cycle, the samples were exposed to TiCl$_4$ vapor for three times for 1 min. After that, humidified air (10% water content) was introduced to the sample to ensure complete oxidation. Ni was also deposited by ALD while using Ni(TMHD)$_2$ (Strem, USA) as the precursor and a deposition temperature of 523 K.

The ALD growth rates were measured gravimetrically and checked by Inductively Coupled Plasma Optical Emission Spectrometry (ICP-OES, Spectro Genesis). The dissolution processes were performed at 333 K overnight while using 10 mL of *aqua regia* for the samples. After that, the solutions were diluted to suitable ranges for ICP measurements. The X-ray diffraction (XRD) patterns were recorded on a Rigaku MiniFlex diffractometer that was equipped with a Cu Kα source (λ = 0.15416 nm) and then patterns for 2 theta range from 20 to 60 degrees were recorded. BET surface areas were measured while using homebuilt equipment at 78 K [24]. *Ex-situ* Scanning Transmission Electron Microscopy and Energy Dispersive X-ray Spectroscopy (STEM-EDS) measurements were performed using a JEOL NEOARM operated at 200 kV.

The rates for methane dry reforming were determined from differential conversions in a 0.25-inch, quartz, tubular-flow reactor at atmospheric pressure, with products analyzed while using a gas chromatograph (SRI 8610C) that was equipped with a TCD detector. The catalyst loadings were 100 mg for all of the samples. The total gas flow rate was maintained at 120 mL min^{-1}. Methane flow was kept at 5 mL min^{-1} throughout the measurements. CO_2 and He flows were adjusted to achieve different CO_2:CH_4 ratios while maintaining a constant flow rate.

The reducibilities of the samples were quantified by flow titration and Coulometric Titration (CT), as described in detail elsewhere [25]. The flow-titration experiments were performed in a tubular reactor at 1 atm with 200-mg samples. After reducing the samples in flowing H_2 at desired temperatures, followed by purging with He, the samples were exposed to dry air at 1073 K at a flow rate of 5 mL min^{-1} while monitoring the reactor effluent with a mass spectrometer. The O_2 uptake was then determined by integrating the difference between the N_2 and O_2 signals in the effluent. Equilibrium constants for oxidation of Ni were probed by CT at 1073 K. In a CT experiment, 500-mg samples were placed in the center of a YSZ (Yttria-Stabilized Zirconia) tube that had Pt paste on the inner wall and Ag paste on the outer wall. After flowing a mixture of 10% H_2, 10% H_2O, and 80% He over the sample at 1073 K, the tube was sealed with cajon fittings. The known quantities of oxygen were then electrochemically pumped into or out of the YSZ tube by applying a known charge across the metal electrodes using a Gamry instruments potentiostat. The equilibrium P_{O_2} was calculated from the electrode open-circuit potential while using the Nernst Equation after allowing the system to come to equilibrium with the electrodes at open circuit.

5. Conclusions

Ni catalysts that were supported on 1 nm films of $CaTiO_3$, $SrTiO_3$, and $BaTiO_3$ on $MgAl_2O_4$ were found to have different properties as compared to the Ni/$MgAl_2O_4$ catalyst. 1073-K reduction is required to activate the Ni/$ATiO_3$/$MgAl_2O_4$ catalysts. By contrast, the Ni/$MgAl_2O_4$ catalyst was found to be active for methane dry reforming (MDR) upon 773-K reduction, and the catalytic performance increased with reduction temperature due to a more complete reduction of Ni. Equilibrium oxidation constants for Ni, which were obtained from Coulometric Titration (CT), suggested that the equilibrium P_{O_2} for Ni oxidation is shifted to lower values for the Ni/$ATiO_3$/$MgAl_2O_4$ samples. The equilibrium P_{O_2} for Ni oxidation was found to decrease in the order of Ni/$CaTiO_3$/$MgAl_2O_4$ < Ni/$SrTiO_3$/$MgAl_2O_4$ < Ni/$BaTiO_3$/$MgAl_2O_4$ < Ni/$MgAl_2O_4$. Consequently, catalyst deactivation was observed on the Ni/$CaTiO_3$/$MgAl_2O_4$ sample when increasing the feed CO_2:CH_4 ratio, which was potentially due to Ni oxidation or changes in Ni reaction order.

Author Contributions: Conceptualization, C.L. and R.J.G.; formal analysis, C.L., R.J.G., A.C.F. and E.A.S.; data curation, C.L. and A.C.F.; writing—original draft preparation, C.L.; writing—review and editing, R.J.G., A.C.F. and E.A.S.; supervision, R.J.G. and E.A.S.; funding acquisition, R.J.G. and E.A.S. All authors have read and agreed to the published version of the manuscript.

Funding: This research was funded by Department of Energy, Office of Basic Energy Sciences, Chemical Sciences, Geosciences and Biosciences Division, Grant No. DE-FG02-13ER16380. This work was performed in part at the Singh Center for Nanotechnology at the University of Pennsylvania, a member of the National Nanotechnology Coordinated Infrastructure (NNCI) network, which is supported by the National Science Foundation (Grant NNCI-1542153).

Acknowledgments: C.L. and R.J.G. are grateful to the Department of Energy, Office of Basic Energy Sciences, Chemical Sciences, Geosciences and Biosciences Division, Grant No. DE-FG02-13ER16380 for support of this work. A.C.F and E.A.S. acknowledge support for a Fellowship to A.C.F. from the Vagelos Institute for Energy Science and Technology at the University of Pennsylvania. The authors gratefully acknowledge use of facilities and instrumentation supported by NSF through the University of Pennsylvania Materials Research Science and Engineering Center (MRSEC) (DMR-1720530).

Conflicts of Interest: The authors declare no conflict of interest.

References

1. Tanaka, H.; Taniguchi, M.; Uenishi, M.; Kajita, N.; Tan, I.; Nishihata, Y.; Mizuki, J.I.; Narita, K.; Kimura, M.; Kaneko, K. Self-regenerating Rh-and Pt-based perovskite catalysts for automotive-emissions control. *Angew. Chem. Int. Ed.* **2006**, *45*, 5998–6002. [CrossRef] [PubMed]
2. Nishihata, Y.; Mizuki, J.; Akao, T.; Tanaka, H.; Uenishi, M.; Kimura, M.; Okamoto, T.; Hamada, N. Self-regeneration of a Pd-perovskite catalyst for automotive emissions control. *Nature* **2002**, *418*, 164. [CrossRef] [PubMed]
3. Tanaka, H.; Tan, I.; Uenishi, M.; Kimura, M.; Dohmae, K. Regeneration of palladium subsequent to solid solution and segregation in a perovskite catalyst: An intelligent catalyst. *Top. Catal.* **2001**, *16*, 63–70. [CrossRef]
4. Uenishi, M.; Taniguchi, M.; Tanaka, H.; Kimura, M.; Nishihata, Y.; Mizuki, J.; Kobayashi, T. Redox behavior of palladium at start-up in the Perovskite-type LaFePdO$_x$ automotive catalysts showing a self-regenerative function. *Appl. Catal. B Environ.* **2005**, *57*, 267–273. [CrossRef]
5. Tanaka, H.; Fujikawa, H. Perovskite-Pd three-way catalysts for automotive applications. In *SAE Technical Paper*; SAE International: Warrendale, PA, USA, 1993.
6. Tanaka, H.; Uenishi, M.; Taniguchi, M.; Tan, I.; Narita, K.; Kimura, M.; Kaneko, K.; Nishihata, Y.; Mizuki, J.i. The intelligent catalyst having the self-regenerative function of Pd, Rh and Pt for automotive emissions control. *Catal. Today* **2006**, *117*, 321–328. [CrossRef]
7. Tanaka, H.; Fujikawa, H.; Takahashi, I. Excellent oxygen storage capacity of perovskite-Pd three-way catalysts. *Sae Trans.* **1995**, 289–301.
8. Tanaka, H.; Taniguchi, M.; Kajita, N.; Uenishi, M.; Tan, I.; Sato, N.; Narita, K.; Kimura, M. Design of the intelligent catalyst for Japan ULEV standard. *Top. Catal.* **2004**, *30*, 389–396. [CrossRef]
9. Marcucci, A.; Zurlo, F.; Sora, I.N.; Placidi, E.; Casciardi, S.; Licoccia, S.; Di Bartolomeo, E. A redox stable Pd-doped perovskite for SOFC applications. *J. Mater. Chem. A* **2019**, *7*, 5344–5352. [CrossRef]
10. Zhu, T.; Troiani, H.E.; Mogni, L.V.; Han, M.; Barnett, S.A. Ni-substituted Sr(Ti, Fe)O$_3$ SOFC anodes: Achieving high performance via metal alloy nanoparticle exsolution. *Joule* **2018**, *2*, 478–496. [CrossRef]
11. Oemar, U.; Ang, M.; Hee, W.; Hidajat, K.; Kawi, S. Perovskite La$_x$M$_{1-x}$Ni$_{0.8}$Fe$_{0.2}$O$_3$ catalyst for steam reforming of toluene: Crucial role of alkaline earth metal at low steam condition. *Appl. Catal. B Environ.* **2014**, *148*, 231–242. [CrossRef]
12. Deng, J.; Cai, M.; Sun, W.; Liao, X.; Chu, W.; Zhao, X.S. Oxidative Methane Reforming with an Intelligent Catalyst: Sintering-Tolerant Supported Nickel Nanoparticles. *ChemSusChem* **2013**, *6*, 2061–2065. [CrossRef] [PubMed]
13. Chai, Y.; Fu, Y.; Feng, H.; Kong, W.; Yuan, C.; Pan, B.; Zhang, J.; Sun, Y. A Nickel-Based Perovskite Catalyst with a Bimodal Size Distribution of Nickel Particles for Dry Reforming of Methane. *ChemCatChem* **2018**, *10*, 2078–2086. [CrossRef]
14. Neagu, D.; Oh, T.-S.; Miller, D.N.; Ménard, H.; Bukhari, S.M.; Gamble, S.R.; Gorte, R.J.; Vohs, J.M.; Irvine, J.T. Nano-socketed nickel particles with enhanced coking resistance grown in situ by redox exsolution. *Nat. Commun.* **2015**, *6*, 1–8. [CrossRef] [PubMed]
15. Oh, T.-S.; Rahani, E.K.; Neagu, D.; Irvine, J.T.; Shenoy, V.B.; Gorte, R.J.; Vohs, J.M. Evidence and model for strain-driven release of metal nanocatalysts from perovskites during exsolution. *J. Phys. Chem. Lett.* **2015**, *6*, 5106–5110. [CrossRef] [PubMed]
16. Mao, X.; Lin, C.; Graham, G.W.; Gorte, R.J. A Perspective on Thin-Film Perovskites as Supports for Metal Catalysts. *ACS Catal.* **2020**, *10*, 8840–8849. [CrossRef]

17. Lin, C.; Foucher, A.C.; Ji, Y.; Stach, E.A.; Gorte, R.J. Investigation of Rh–Titanate (ATiO$_3$) Interactions on High-Surface-Area Perovskite Thin Films Prepared by Atomic Layer Deposition. *J. Mater. Chem. A* **2020**, *8*, 16973–16984. [CrossRef]
18. Lin, C.; Foucher, A.C.; Ji, Y.; Curran, C.D.; Stach, E.A.; McIntosh, S.; Gorte, R.J. "Intelligent" Pt catalysts studied on high-surface-area CaTiO$_3$ films. *ACS Catal.* **2019**, *9*, 7318–7327. [CrossRef]
19. Lin, C.; Jang, J.B.; Zhang, L.; Stach, E.A.; Gorte, R.J. Improved Coking Resistance of "Intelligent" Ni Catalysts Prepared by Atomic Layer Deposition. *ACS Catal.* **2018**, *8*, 7679–7687. [CrossRef]
20. Mao, X.; Foucher, A.C.; Stach, E.A.; Gorte, R.J. Changes in Ni-NiO equilibrium due to LaFeO$_3$ and the effect on dry reforming of CH$_4$. *J. Catal.* **2020**, *381*, 561–569. [CrossRef]
21. Ro, I.; Resasco, J.; Christopher, P. Approaches for understanding and controlling interfacial effects in oxide-supported metal catalysts. *ACS Catal.* **2018**, *8*, 7368–7387. [CrossRef]
22. Remiro, A.; Arandia, A.; Oar-Arteta, L.; Bilbao, J.; Gayubo, A.G. Regeneration of NiAl$_2$O$_4$ spinel type catalysts used in the reforming of raw bio-oil. *Appl. Catal. B Environ.* **2018**, *237*, 353–365. [CrossRef]
23. Onn, T.M.; Monai, M.; Dai, S.; Fonda, E.; Montini, T.; Pan, X.; Graham, G.W.; Fornasiero, P.; Gorte, R.J. Smart Pd catalyst with improved thermal stability supported on high-surface-area LaFeO$_3$ prepared by atomic layer deposition. *J. Am. Chem. Soc.* **2018**, *140*, 4841–4848. [CrossRef] [PubMed]
24. Onn, T.M.; Zhang, S.; Arroyo-Ramirez, L.; Chung, Y.-C.; Graham, G.W.; Pan, X.; Gorte, R.J. Improved thermal stability and methane-oxidation activity of Pd/Al$_2$O$_3$ catalysts by atomic layer deposition of ZrO$_2$. *ACS Catal.* **2015**, *5*, 5696–5701. [CrossRef]
25. Onn, T.M.; Mao, X.; Lin, C.; Wang, C.; Gorte, R.J. Investigation of the thermodynamic properties of surface ceria and ceria–zirconia solid solution films prepared by atomic layer deposition on Al$_2$O$_3$. *Inorganics* **2017**, *5*, 69.
26. Lin, C.; Mao, X.; Onn, T.M.; Jang, J.; Gorte, R.J. Stabilization of ZrO$_2$ powders via ALD of CeO$_2$ and ZrO$_2$. *Inorganics* **2017**, *5*, 65. [CrossRef]

Publisher's Note: MDPI stays neutral with regard to jurisdictional claims in published maps and institutional affiliations.

© 2020 by the authors. Licensee MDPI, Basel, Switzerland. This article is an open access article distributed under the terms and conditions of the Creative Commons Attribution (CC BY) license (http://creativecommons.org/licenses/by/4.0/).

Article

Multivariate Linear Regression Models to Predict Monomer Poisoning Effect in Ethylene/Polar Monomer Copolymerization Catalyzed by Late Transition Metals

Wei Zhao [1], Zhihao Liu [1], Yanan Zhao [1,*], Yi Luo [1,2,*] and Shengbao He [2]

1. State Key Laboratory of Fine Chemicals, School of Chemical Engineering, Dalian University of Technology, Dalian 116024, China; weizhao3494@163.com (W.Z.); zhihao-liu@outlook.com (Z.L.)
2. PetroChina Petrochemical Research Institute, Beijing 102206, China; hsb@petrochina.com.cn
* Correspondence: yananzhao@dlut.edu.cn (Y.Z.); luoyi@dlut.edu.cn (Y.L.)

Abstract: This study combined density functional theory (DFT) calculations and multivariate linear regression (MLR) to analyze the monomer poisoning effect in ethylene/polar monomer copolymerization catalyzed by the Brookhart-type catalysts. The calculation results showed that the poisoning effect of polar monomers with relatively electron-deficient functional groups is weaker, such as ethers, and halogens. On the contrary, polar monomers with electron-rich functional groups (carbonyl, carboxyl, and acyl groups) exert a stronger poisoning effect. In addition, three descriptors that significantly affect the poisoning effect have been proposed on the basis of the multiple linear regression model, viz., the chemical shift of the vinyl carbon atom and heteroatom of polar monomer as well as the metal-X distance in the σ-coordination structure. It is expected that these models could guide the development of efficient catalytic copolymerization system in this field.

Keywords: density functional theory; multivariate linear regression; poisoning effect of polar monomers; Brookhart-type catalysts

1. Introduction

Compared with the polyolefins, the incorporation of polar monomers into nonfunctionalized polyolefin backbones can significantly improve various properties of polymers, such as flexibility, adhesion, protective properties, surface properties, solvent resistance, which leads to expanding the range of applications [1–6]. It is well known that metal-catalyzed coordination-insertion copolymerization of olefinic hydrocarbons with polar monomers is the most convenient and economical synthetic strategy. Generally, the early-transition-metal complexes with high oxygen affinity are easily poisoned by polar functional groups [7]. It is, therefore, necessary to use late-transition-metal complexes (Ni or Pd) with low oxygen affinity to catalyze the coordination copolymerization of polar monomers.

In this context, various late-transition-metal catalysts have been developed (Figure 1a) [8,9]. In the mid-1990s, a groundbreaking work was achieved by using cationic Ni/Pd catalysts based on α-diimine ligands by Brookhart (II in Figure 1a,) [10]. Since then, a series of complexes [11–14] have been developed on the basis of the Brookhart-type catalysts, which are only suitable for a small part of simple monomers such as acrylates [15], vinyl ketones [10], and silyl vinyl ethers [16–19]. In 2002, Drent-type catalysts were reported by Drent and co-workers (III in Figure 1a,) [20]. In addition, they expanded the scope of substrate for copolymerization, such as vinyl fluoride [21], vinyl ethers [22], and some important methylene-spaced polar monomers (with a spacer between the polar group and the double bond) [23–28]. However, there are still some obvious disadvantages in these catalytic systems: low copolymerization activity, low insertion rate and low molecular weight [29], etc. Among them, the low copolymerization activity is a common challenge for these catalysts. It is noteworthy that the main reason for the low activity of copolymerization is the occurrence of the poisoning effect. The tolerance of the same catalyst to different functional groups is obviously different, but there is a lack of systematic study on this difference. Therefore, systematically exploring the poisoning effect of polar monomers is helpful for improving the copolymerization activity. Herein, multivariate linear regression analyses on the poisoning effect of different polar monomers have been conducted based on DFT calculations at the molecular level, taking the conventional Brookhart-type catalysts **II** as a model. Various polar monomers [30–32] and the copolymerization mechanism are shown in Figure 1 and Figure S1. When $\Delta\Delta E(\pi\text{-}\sigma) < 0$, the π-complex (double bond coordination) is more stable than the heteroatom coordination complex (σ-complex), and vice versa. As shown in Figure 1c, if the σ-complex (**B2**) is more stable than the vinyl-coordination π-complex (**B3**), the catalyst could be poisoned and inactive. Through multiple linear regression analysis, it has been demonstrated that the chemical shifts of a vinyl carbon atom ($^{mon}NMR_C{}^\beta$) and the coordinating heteroatom ($^{mon}NMR_X$) of polar monomers, as well as the metal-heteroatom distance ($^{B2}bond_{Pd/Ni\text{-}X}$), are the key factors governing the poisoning effect. The current prediction models are expected to be useful for the prediction of the poisoning effect of other polar monomers in the polymerization catalyzed by Brookhart-type catalysts.

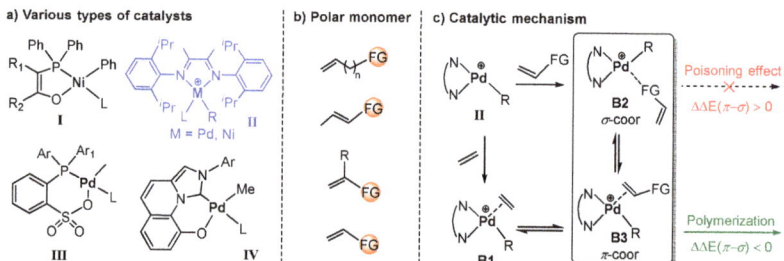

Figure 1. (**a**) Four conventional catalyst structures [8–10,20]. (**b**) Several different polar monomer structures. (**c**) The copolymerization mechanism of ethylene/polar monomer meditated by Brookhart-type catalysts **II**.

2. Computational Details

All DFT calculations were performed with Gaussian 16 program [33]. The D3 [34] dispersion-corrected density functional method B3LYP functional [35–37], together with the 6-311G(d) basis set for nonmetal atoms (C, H, O, N, P, S, F, Si and Cl) and the LANL2DZ [38–40] basis set, as well as the associated poseupotential for metal atoms (Pd and Ni) was used for geometry optimizations. All optimizations were carried out in the gas-phase. The noncovalent interaction (NCI) analysis [41] by Multiwfn [42] and VMD [43] softwares was carried out for important structures. The optimized tridimensional geometrical structures were represented by CYLView [44]. Color map and multiple linear regression analysis were performed with Matlab program. Taking diimide palladium as an example, single-point calculations were further performed at the higher level by using the density functional method M06 [45]; 6-311 + G (d, p) was used for the nonmetal atoms; the basis set LANL2DZ [38–40], as well as the associated pseudopotential, were applied for the Pd atom. In these single-point calculations, the solvation effect of toluene ($\varepsilon = 2.37$) was considered through the CPCM model [46,47]. It can be seen from comparison of results by solvated and gas-phase that the effect of solvation exerted a minor effect on the trend of poisoning effects (see Figure S1).

3. Results and Discussion

The poisoning effect of several different types of polar monomers (see Figure S2) were explored by DFT calculation and multiple linear regression analysis. Generally, the energy difference $\Delta\Delta E(\pi\text{-}\sigma)$ (Figure 1c) between double bond coordination and heteroatom coordination is directly related to the poisoning effect. Firstly, the coordination process of a series of polar monomers catalyzed by complex II_{Pd} and II_{Ni} were calculated. Through theoretical calculation, it was found that the energy difference of some polar monomers was too high ($\Delta\Delta E(\pi\text{-}\sigma) > 0$), such as acrylonitrile, methylene spacer vinyl and internal olefin monomers; thus, there are very obvious poisoning effects. On the contrary, the poisoning effect of ether olefins and halogenated olefins is relatively weaker theoretically ($\Delta\Delta E(\pi\text{-}\sigma) < 0$, Figure 2). However, although the poisoning effect of halogenated olefins is weak (Figure 2a), the β-halide or or β-OR elimination reaction can easily occur in the polymerization, resulting in low polymerization activity [48,49]. For polar monomers 42, 56, 57, this is a kind of styrenic monomers, which are easy to coordinate metal to produce stable and inactive η^3-complexes [50]. These results are consistent with the experimental phenomena [51]. Moreover, the number of polar monomers that cannot poison Ni complexes (Figure 2a) is significantly less than that of Pd complexes (Figure 2a,b).

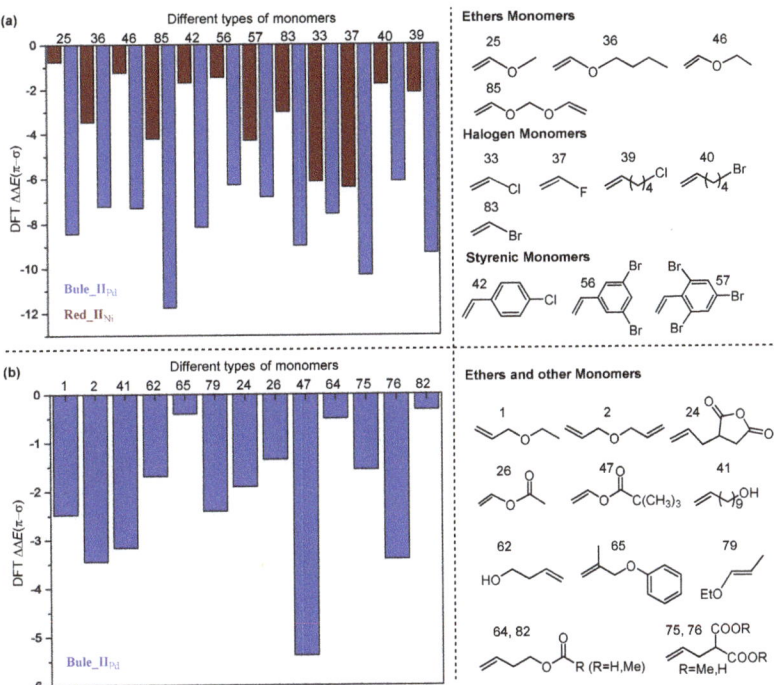

Figure 2. (a) The polar monomers with weak poisoning effect catalyzed by II_{Ni} and II_{Pd} complexes ($\Delta\Delta E(\pi\text{-}\sigma) < 0$). (b) Other monomers with weak poisoning effect catalyzed by II_{Pd} complex.

Before the MLR analysis, it is necessary to calculate the electronic and stereoscopic descriptors of different polar monomers by DFT calculation. At present, as hundreds of molecular descriptors are available in the literature [42–54], we selected a number of them based on our previous experience on olefin polymerization [55–58] and the literature about MLR analysis of transition-metal-based reactivity. For the predictive effect and the convenience of the MLR model, only one of heteroatom coordination and double bond coordination should be selected for descriptor calculation. Firstly, the descriptors of heteroatom coordination structure **B2** and polar monomers are calculated for multivariate linear regression. A total of 21 descriptors were calculated, including Sterimol values [59] ($^{B2}B1_{Ni/Pd-X}$, $^{B2}B5_{Ni/Pd-X}$ and $^{B2}L_{Ni/Pd-X}$), steric hindrance of metal center ($^{B2}Steric_{Ni/Pd}$), bond length ($^{B2}bond_{Ni/Pd-X}$), LUMO ($^{B2}LUMO$), dihedral angle ($^{B2}\angle XNi/PdN_1C_1$ and $^{B2}\angle C_3Ni/PdN_2C_2$), Infrared freq ($^{mon}IR_{C=C}$) and Freq Intensities ($^{mon}v_{C=C}$), NMR ($^{mon}NMR_C{}^\alpha$, $^{mon}NMR_C{}^\beta$ and $^{mon}NMR_X$),

NBO (^{B2}NBO$_{Ni/Pd}$, ^{mon}NBO$_X$, ^{mon}NBO$_C{}^\alpha$ and ^{mon}NBO$_C{}^\beta$), Polarizability ($^{mon}\alpha$), HOMO (^{mon}HOMO), volume (^{mon}V) and Dipole Moment ($^{mon}\mu$). Having both computed $\Delta\Delta E(\pi\text{-}\sigma)$ and descriptors in hand, we performed univariate correlation analysis for the whole data set (87 reactions) to investigate the relationship between descriptors and $\Delta\Delta E(\pi\text{-}\sigma)$, and to see the variation trends in poisoning effect (by complex $\mathbf{II_{Ni}}$). For this purpose, a correlation matrix for the selected parameters was generated, as represented by a color map [60] (Figure 3). As shown in Figure 3, relatively strong correlations were found between $\Delta\Delta E(\pi\text{-}\sigma)$ and some electronic parameters involving $\mathbf{D_2}$ (^{mon}NMR$_C{}^\beta$, |R| = 0.76), $\mathbf{D_3}$ (^{mon}NMR$_X$, |R| = 0.59), $\mathbf{D_5}$ (^{mon}NBO$_X$, |R| = 0.60) and $\mathbf{D_{15}}$ (^{B2}bond$_{Ni\text{-}X}$, |R| = 0.70). These descriptors with high correlation coefficient |R| can presumably exert a major influence on $\Delta\Delta E(\pi\text{-}\sigma)$. Conversely, steric hindrance descriptors exert little effect on $\Delta\Delta E(\pi\text{-}\sigma)$, such as $\mathbf{D_{17}}$–$\mathbf{D_{21}}$.

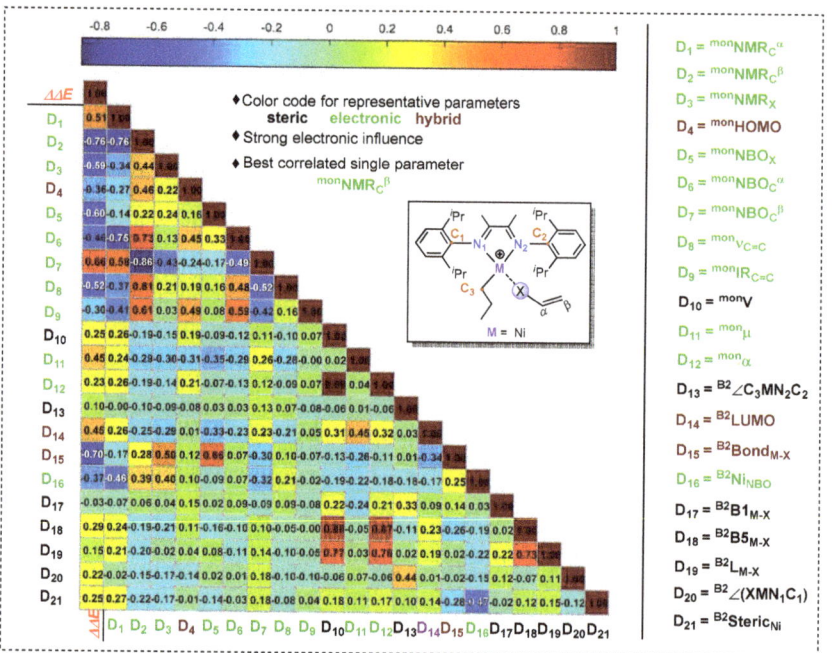

Figure 3. Correlation color map. The first column corresponds to the single-parameter correlations of the $\Delta\Delta E(\pi\text{-}\sigma)$, while the others represent the interparameter correlations [60].

According to the color map in Figure 3, we selected four single parameters $\mathbf{D_2}$ (^{mon}NMR$_C{}^\beta$, |R| = 0.76), $\mathbf{D_3}$ (^{mon}NMR$_X$, |R| = 0.59), $\mathbf{D_5}$ (^{mon}NBO$_X$, |R| = 0.60) and $\mathbf{D_{15}}$ (^{B2}bond$_{Ni\text{-}X}$, |R| = 0.70) with relatively higher absolute values of R to performed univariate correlation analysis to investigate the relationship between descriptors and $\Delta\Delta E(\pi\text{-}\sigma)$ and to see the variation trends of poisoning effect. However, when we evaluated the full set of data (87 reactions) in Figure S1, the single parameters cannot describe well the trend of the $\Delta\Delta E(\pi\text{-}\sigma)$ value ($R^2 < 0.60$, Figure 4), suggesting that multivariate linear regression analysis was required to describe the combined multivariate influences of polar monomers on poisoning effect.

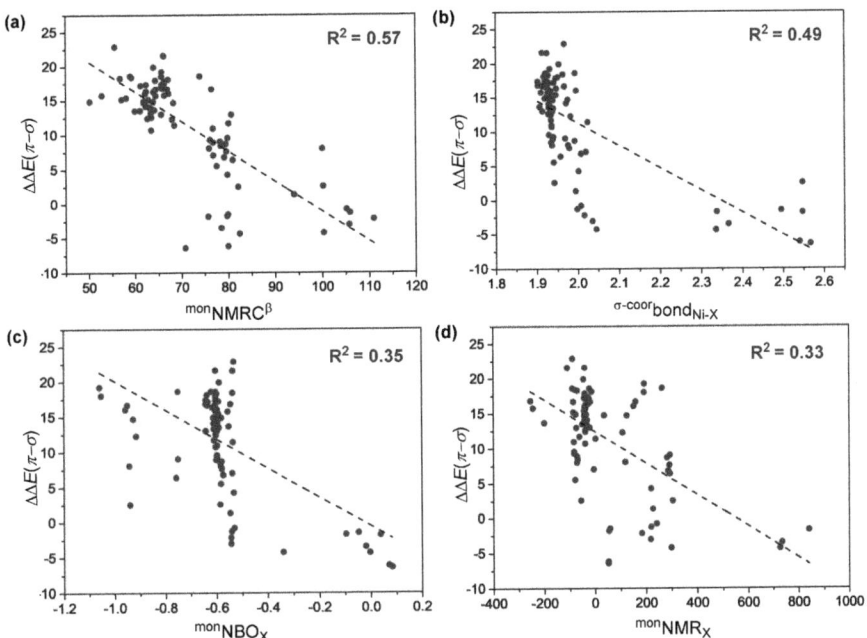

Figure 4. Representative univariate trends. Different polar monomers catalyzed by **II$_{Ni}$** complex.

Next, Linear regression modeling was applied to correlate the poisoning effect (expressed as $\Delta\Delta E(\pi\text{-}\sigma)$) to the above calculated descriptors. A total of 87 reactions were used in this paper, of which 60 reactions were randomly selected as the training set and 27 reactions as the test set. Considering that the accuracy of the simplified regression model can be expressed more clearly via a graphical manner, a plot of the calculated $\Delta\Delta E(\pi\text{-}\sigma)$ values and that predicted by this model is depicted in Figure 5. When complex **II$_{Ni}$** was used, the results demonstrated a high correlation ($R^2 = 0.85$, $Q^2 = 0.83$; Q^2 represents the value of leave-one-out cross validation and showed that there was no overfitting) between predicted $\Delta\Delta E(\pi\text{-}\sigma)$ values and calculated ones (Figure 5a). Moreover, when complex **II$_{Pd}$** is used, the result shows a relatively lower correlation between predicted $\Delta\Delta E(\pi\text{-}\sigma)$ values and calculated ones ($R^2 = 0.79$, $Q^2 = 0.78$, Figure 5b). Notably, as the models were acquired from normalized descriptors, the resulting coefficients can indicate the significance of the represented descriptor. Therefore, it can be seen from the model that, for complex **II$_{Ni}$** and **II$_{Pd}$** systems, the descriptors $^{mon}NMR_C^{\beta}$ and $^{B2}Bond_{Ni/Pd\text{-}X}$ exert a significant impact on the poisoning effect. Nevertheless, $^{mono}NMR_X$ holds a relatively weaker effect on the poisoning effect of **II$_{Ni}$** systems and almost no effect on **II$_{Pd}$** systems. Besides, the descriptors of π-coordination structure **B3** (in complex **II$_{Ni}$** systems) and polar monomers are also calculated (Figure S3), but the prediction results ($R^2 = 0.82$) based on these descriptors are not as satisfactory as the above results ($R^2 = 0.85$, Figure 5a).

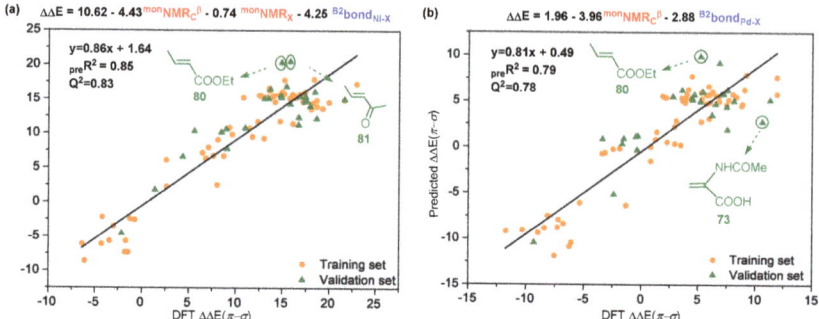

Figure 5. Plot of computed vs. predicted $\Delta\Delta E(\pi\text{-}\sigma)$ (kcal/mol) for complexes $\mathbf{II_{Ni}}$ (a) and $\mathbf{II_{Pd}}$ (b) using the multivariate linear regression models.

It is worth noting that there were several deviation points in both complexes $\mathbf{II_{Ni}}$ (a) and $\mathbf{II_{Pd}}$ (b) systems (points in green circle, Figure 5a,b), and these deviation points may reduce the prediction accuracy of the MLR model; thus, it is necessary to analyze the origins of these deviation points. Through structural analysis, it was found that these deviation points share something in common, that is, there are obvious non-covalent weak interactions in the σ-coordination structure (Figure 6), including hydrogen bond, H···π interactions and heteroatom···π interactions. In order to confirm the effect of non-covalent weak interaction on energy of σ-coordination structure, the aromatic ring on the ligand was substituted with methyl, and the $\Delta\Delta E(\pi\text{-}\sigma)$ was calculated (Figure S4). By comparing the calculation results, it can be seen that the non-covalent weak interaction between polar monomers and aromatic ring on ligands exerted an obvious effect on the energy of $\Delta\Delta E(\pi\text{-}\sigma)$, but the non-covalent weak interaction is not considered in above descriptors. Therefore, there is a certain deviation in the prediction results of the system with obvious non-covalent weak interaction. Furthermore, external validation of the MLR model for the complex $\mathbf{II_{Ni}}$ (a) and $\mathbf{II_{Pd}}$ were performed separately; results are shown in Figure 7. These results indicate that these MLR models have certain extrapolation and prediction ability. It is noted that some monomers containing an active hydrogen or a strong electrophilic group could undergo side reactions in the polymerization system rather than polymerize. However, these monomers were considered for expending the scope of data set and were helpful for constructing the prediction models [61–63]. It is also noteworthy that there may be secondary interactions between polar monomer molecules in the polymerization system [4,64]. Our work mainly focuses on the influence of the electronic or stereoscopic effect of the monomer itself on the poisoning effect of the catalyst. The actual experimental reaction system was very complex; there will be a variety of interactions. In order to explore the main reasons affecting the poisoning effect, this work simplifies the experimental conditions and provides a valuable reference for experimenters to screen suitable polymerized monomers.

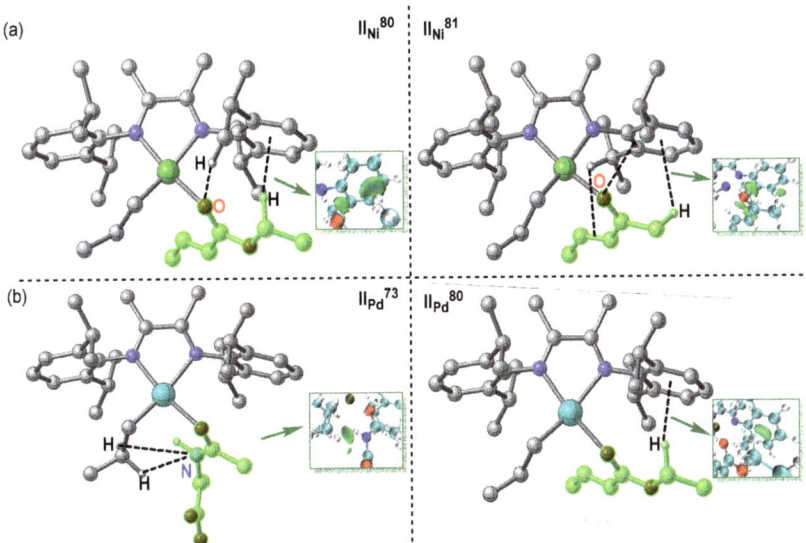

Figure 6. Structure and non-covalent interaction of deviation points in complexes II_{Ni} (a) and II_{Pd} (b) systems.

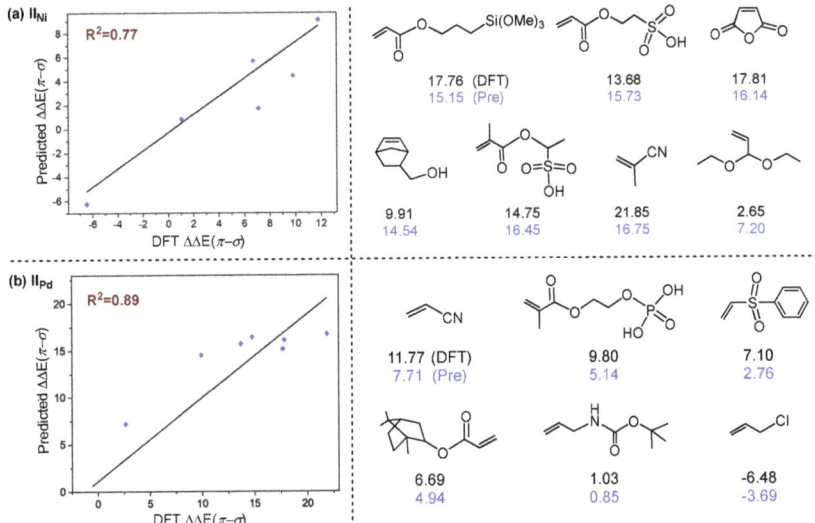

Figure 7. External verification of the multivariate linear regression models for complexes II_{Ni} (a) and II_{Pd} (b). Blue represents predicted $\Delta\Delta E(\pi\text{-}\sigma)$ (kcal/mol), black represents computed $\Delta\Delta E(\pi\text{-}\sigma)$ (kcal/mol).

4. Conclusions

In summary, the poisoning effect of different polar monomers by Brookhart-type catalyst were explored through the combination of DFT calculations and multiple linear regression analyses. The results show that the combination of the structure of heteroatom-coordination complex and the descriptors of polar monomer can establish a relationship between the structure and the poisoning effect represented by $\Delta\Delta E(\pi\text{-}\sigma)$. It is found that the descriptors of $^{mon}NMR_C{}^\beta$ and $^{B2}Bond_{Ni/Pd-X}$ are the key factors affecting the poisoning effect. In addition, in the case of the Ni system, the $^{mon}NMR_X$ descriptor also plays a certain role in the poisoning effect. Besides, by analyzing the σ-coordination structure of the deviation point, it was found that the non-covalent interaction between polar monomers and the catalyst may be the main reason for the deviation of the predicted value by the multiple linear regression model. Moreover, the result of external verification shows that such prediction models possess a certain ability to predict and extrapolate the poisoning effect of other polar monomers. Such a combination of DFT-derived energy difference and multidimensional quantitative analysis is expected to be effective in assessing the other polymerization performance and can provide new and efficient monomer screening strategies for experimentalists.

Supplementary Materials: The following supporting information can be downloaded at: https://www.mdpi.com/article/10.3390/inorganics10020026/s1, Figure S1: Gas-phase vs. solvation effect $\Delta\Delta E(\pi\text{-}\sigma)$ (kcal/mol) for complex IIPd; Figure S2: structures of polar monomers; Figure S3: plot of computed vs. predicted $\Delta\Delta E(\pi\text{-}\sigma)$ (kcal/mol) for complex IINi (base on B3 structure) using the multivariate linear regression models; Figure S4: changes of structure and $\Delta\Delta E(\pi\text{-}\sigma)$ after substitution of aromatic ring on catalyst (complex IINi) with methyl. Electronic energy data of polar monomers coordination, polar monomers and other structures; and electronic and stereoscopic descriptors of polar monomers, complexes B2 and B3 (Excel) (XLSX). Optimized stationary points (ZIP).

Author Contributions: Conceptualization, Y.L., Y.Z. and S.H.; methodology, W.Z., Y.Z. and Z.L.; investigation, W.Z. and Y.Z.; data curation, W.Z., Z.L. and Y.Z.; writing—original draft preparation, W.Z. and Y.Z.; writing—review and editing, Y.L., Y.Z. and W.Z.; funding acquisition, Y.L., Y.Z. All authors have read and agreed to the published version of the manuscript.

Funding: This research was funded by the NSFC (Nos. 22071015, U1862115, and 22101041. China Postdoctoral Science Foundation (2021M700664)).

Institutional Review Board Statement: Not applicable.

Informed Consent Statement: Not applicable.

Data Availability Statement: This research did not report any data.

Acknowledgments: This work was supported by the NSFC (Nos. 22071015, U1862115, and 22101041, China Postdoctoral Science Foundation (2021M700664)). The authors also thank the Network and Information Center of Dalian University of Technology for part of computational resources.

Conflicts of Interest: The authors declare no conflict of interest. The funders had absolutely no role in the design of the study; in the collection, analyses, or interpretation of data; in the writing of the manuscript, or in the decision to publish the results.

References

1. Dong, J.Y.; Hu, Y. Design and synthesis of structurally well-defined functional polyolefins via transition metal-mediated olefin polymerization chemistry. *Coord. Chem. Rev.* **2006**, *250*, 47–65. [CrossRef]
2. Chung, T.M. *Functionalization of Polyolefins*; Elsevier: Amsterdam, The Netherlands, 2002.
3. Rünzi, T.; Mecking, S. Saturated polar-substituted polyethylene elastomers from insertion polymerization. *Adv. Funct. Mater.* **2014**, *24*, 387–395. [CrossRef]
4. Dai, S.; Chen, C. Palladium-catalyzed direct synthesis of various branched, carboxylic acid-functionalized polyolefins: Characterization, derivatization, and properties. *Macromolecules* **2018**, *51*, 6818–6824. [CrossRef]
5. Na, Y.; Dai, S.; Chen, C. Direct synthesis of polar-functionalized linear low-density polyethylene (LLDPE) and low-density polyethylene (LDPE). *Macromolecules* **2018**, *51*, 4040–4048. [CrossRef]
6. Sui, X.; Hong, C.; Pang, W.; Chen, C. Unsymmetrical α-diimine palladium catalysts and their properties in olefin (co) polymerization. *Mater. Chem. Front.* **2017**, *1*, 967–972. [CrossRef]
7. Ittel, S.D.; Johnson, L.K.; Brookhart, M. Late-metal catalysts for ethylene homo-and copolymerization. *Chem. Rev.* **2000**, *100*, 1169–1204. [CrossRef]
8. Klabunde, U.; Itten, S.D. Nickel catalysis for ethylene homo-and co-polymerization. *J. Mol. Catal.* **1987**, *41*, 123–134. [CrossRef]
9. Yasuda, H.; Nakano, R.; Ito, S.; Nozaki, K. Palladium/IzQO-catalyzed coordination-insertion copolymerization of ethylene and 1,1-disubstituted ethylenes bearing a polar functional group. *J. Am. Chem. Soc.* **2018**, *140*, 1876–1883. [CrossRef]

10. Johnson, L.K.; Mecking, S.; Brookhart, M. Copolymerization of ethylene and propylene with functionalized vinyl monomers by palladium (II) catalysts. *J. Am. Chem. Soc.* **1996**, *118*, 267–268. [CrossRef]
11. Chen, Z.; Brookhart, M. Exploring ethylene/polar vinyl monomer copolymerizations using Ni and Pd α-diimine catalysts. *Acc. Chem. Res.* **2018**, *51*, 1831–1839. [CrossRef]
12. Wang, F.; Chen, C. A continuing legend: The Brookhart-type α-diimine nickel and palladium catalysts. *Polym. Chem.* **2019**, *10*, 2354–2369. [CrossRef]
13. Culver, D.B.; Tafazolian, H.; Conley, M.P. A bulky Pd (II) α-Diimine catalyst supported on sulfated zirconia for the polymerization of ethylene and copolymerization of ethylene and methyl acrylate. *Organometallics* **2018**, *37*, 1001–1006. [CrossRef]
14. Hyatt, M.G.; Guironnet, D. Silane as chain transfer agent for the polymerization of ethylene catalyzed by a palladium (II) diimine catalyst. *ACS Catal.* **2017**, *7*, 5717–5720. [CrossRef]
15. Mecking, S.; Johnson, L.K.; Wang, L.; Brookhart, M. Mechanistic studies of the palladium-catalyzed copolymerization of ethylene and α-olefins with methyl acrylate. *J. Am. Chem. Soc.* **1998**, *120*, 888–899. [CrossRef]
16. Luo, S.; Jordan, R.F. Copolymerization of silyl vinyl ethers with olefins by (α-diimine) PdR⁺. *J. Am. Chem. Soc.* **2006**, *128*, 12072–12073. [CrossRef]
17. Chen, Z.; Liu, W.; Daugulis, O.; Brookhart, M. Mechanistic studies of Pd (II)-catalyzed copolymerization of ethylene and vinylalkoxysilanes: Evidence for a β-silyl elimination chain transfer mechanism. *J. Am. Chem. Soc.* **2016**, *138*, 16120–16129. [CrossRef]
18. Chen, Z.; Leatherman, M.D.; Daugulis, O.; Brookhart, M. Nickel-catalyzed copolymerization of ethylene and vinyltrialkoxysilanes: Catalytic production of cross-linkable polyethylene and elucidation of the chain-growth mechanism. *J. Am. Chem. Soc.* **2017**, *139*, 16013–16022. [CrossRef]
19. Zhou, S.; Chen, C. Synthesis of silicon-functionalized polyolefins by subsequent cobalt-catalyzed dehydrogenative silylation and nickel-catalyzed copolymerization. *Sci. Bull.* **2018**, *63*, 441–445. [CrossRef]
20. Drent, E.; van Dijk, R.; van Ginkel, R.; van Oort, B.; Pugh, R.I. Palladium catalysed copolymerisation of ethene with alkylacrylates: Polar comonomer built into the linear polymer chain. *Chem. Commun.* **2002**, *7*, 744–745. [CrossRef]
21. Weng, W.; Shen, Z.; Jordan, R.F. Copolymerization of ethylene and vinyl fluoride by (phosphine-sulfonate) Pd (Me)(py) catalysts. *J. Am. Chem. Soc.* **2007**, *129*, 15450–15451. [CrossRef] [PubMed]
22. Luo, S.; Vela, J.; Life, G.R.; Jordan, R.F. Copolymerization of ethylene and alkyl vinyl ethers by a (phosphine-sulfonate) PdMe catalyst. *J. Am. Chem. Soc.* **2007**, *129*, 8946–8947. [CrossRef]
23. Nakamura, A.; Anselment, T.M.J.; Claverie, J.; Goodall, B.; Jordan, R.F.; Mecking, S.; Rieger, B.; Sen, A.; van Leeuwen, P.W.N.M.; Nozaki, K. Ortho-phosphinobenzenesulfonate: A superb ligand for palladium-catalyzed coordination-insertion copolymerization of polar vinyl monomers. *Acc. Chem. Res.* **2013**, *46*, 1438–1449. [CrossRef]
24. Ito, S. Palladium-catalyzed homo-and copolymerization of polar monomers: Synthesis of aliphatic and aromatic polymers. *Bull. Chem. Soc. Jpn.* **2018**, *91*, 251–261. [CrossRef]
25. Schuster, N.; Rünzi, T.; Mecking, S. Reactivity of functionalized vinyl monomers in insertion copolymerization. *Macromolecules* **2016**, *49*, 1172–1179. [CrossRef]
26. Jian, Z.; Falivene, L.; Boffa, G.; Sánchez, S.O.; Caporaso, L.; Grassi, A.; Mecking, S. Direct synthesis of telechelic polyethylene by selective insertion polymerization. *Angew. Chem.* **2016**, *128*, 14590–14595. [CrossRef]
27. Zhang, W.; Waddell, P.M.; Tiedemann, M.A.; Padilla, C.E.; Mei, J.; Chen, L.; Carrow, B.P. Electron-rich metal cations enable synthesis of high molecular weight, linear functional polyethylenes. *J. Am. Chem. Soc.* **2018**, *140*, 8841–8850. [CrossRef]
28. Zhang, D.; Chen, C. Influence of polyethylene glycol unit on palladium-and nickel-catalyzed ethylene polymerization and copolymerization. *Angew. Chem. Int. Ed.* **2017**, *56*, 14672–14676. [CrossRef] [PubMed]
29. Tan, C.; Chen, C. Emerging palladium and nickel catalysts for copolymerization of olefins with polar monomers. *Angew. Chem. Int. Ed.* **2019**, *58*, 7192–7200. [CrossRef] [PubMed]
30. Liu, Z.; Li, G.; Liang, G.; Gao, H.; Wu, Q. Enhancing thermal stability and living fashion in α-diimine–nickel-catalyzed (co) polymerization of ethylene and polar monomer by increasing the steric bulk of ligand backbone. *Macromolecules* **2017**, *50*, 2675–2682.
31. Zhong, S.; Tan, Y.; Zhong, L.; Gao, J.; Liao, H.; Jiang, L.; Gao, H.; Wu, Q. Precision synthesis of ethylene and polar monomer copolymers by palladium-catalyzed living coordination copolymerization. *Macromolecules* **2017**, *50*, 5661–5669. [CrossRef]
32. Hong, C.; Wang, X.; Chen, C. Palladium-catalyzed dimerization of vinyl ethers: Mechanism, catalyst optimization, and polymerization applications. *Macromolecules* **2019**, *52*, 7123–7129. [CrossRef]
33. Frisch, M.J.; Trucks, G.W.; Schlegel, H.B.; Scuseria, G.E.; Robb, M.A.; Cheeseman, J.R.; Scalmani, G.; Barone, V.; Petersson, G.A.; Nakatsuji, H.; et al. *Gaussian 16, Revision A.03*; Gaussian, Inc.: Wallingford, CT, USA, 2016.
34. Grimme, S.; Antony, J.; Ehrlich, S.; Krieg, H. A consistent and accurate ab initio parametrization of density functional dispersion correction (DFT-D) for the 94 elements H-Pu. *J. Chem. Phys.* **2010**, *132*, 154104. [CrossRef] [PubMed]
35. Lee, C.; Yang, W.; Parr, R.G. Development of the colic-salvetti correlation-energy formula into a functional of the electron density. *Phys. Rev. B* **1988**, *37*, 785–789. [CrossRef] [PubMed]
36. Becke, A.D. Density-functional exchange-energy approximation with correct asymptotic behavior. *Phys. Rev. A* **1988**, *38*, 3098–3100. [CrossRef]
37. Becke, A.D. Density-functional thermochemistry. III. The Role of Exact Exchange. *J. Chem. Phys.* **1993**, *98*, 5648–5652. [CrossRef]

38. Wadt, W.R.; Hay, P.J. Ab initio effective core potentials for molecular calculations. potentials for main group elements Na to Bi. *J. Chem. Phys.* **1985**, *82*, 284–298. [CrossRef]
39. Hay, P.J.; Wadt, W.R. Ab initio effective core potentials for molecular calculations. Potentials for K to Au including the outermost core orbitals. *J. Chem. Phys.* **1985**, *82*, 299–310. [CrossRef]
40. Hay, P.J.; Wadt, W.R. Ab initio effective core potentials for molecular calculations. Potentials for the transition metal atoms Sc to Hg. *J. Chem. Phys.* **1985**, *82*, 270–283. [CrossRef]
41. Johnson, E.R.; Keinan, S.; Mori-Sanchez, P.; Contreras-García, J.; Cohen, A.J.; Yang, W.T. Revealing noncovalent interactions. *J. Am. Chem. Soc.* **2010**, *132*, 6498–6506. [CrossRef]
42. Lu, T.; Chen, F. Multiwfn: A multifunctional wavefunction analyzer. *J. Comput. Chem.* **2012**, *33*, 580–592. [CrossRef]
43. Humphrey, W.; Dalke, A.; Schulten, K. VMD: Visual molecular dynamics. *J. Mol. Graphics* **1996**, *14*, 33–38. [CrossRef]
44. Legault, C.Y. Cylview, Version 1.0 b. 2009. Available online: http://www.cylview.org/ (accessed on 17 January 2022).
45. Zhao, Y.; Truhlar, D.G. Density functionals with broad applicability in chemistry. *Acc. Chem. Res.* **2008**, *41*, 157–167. [CrossRef] [PubMed]
46. Barone, V.; Cossi, M. Quantum calculation of molecular energies and energy gradients in solution by a conductor solvent model. *J. Phys. Chem. A* **1998**, *102*, 1995–2001. [CrossRef]
47. Cossi, M.; Rega, N.; Scalmani, G.; Barone, V. Energies, structures, and electronic properties of molecules in solution with the C-PCM solvation model. *J. Comput. Chem.* **2003**, *24*, 669–681. [CrossRef] [PubMed]
48. Kang, M.; Sen, A.; Zakharov, L.; Rheingold, A.L. Diametrically opposite trends in alkene insertion in late and early transition metal compounds: Relevance to transition-metal-catalyzed polymerization of polar vinyl monomers. *J. Am. Chem. Soc.* **2002**, *124*, 12080–12081. [CrossRef]
49. Foley, S.R.; Stockland, R.A.; Shen, H.; Jordan, R.F. Reaction of vinyl chloride with late transition metal olefin polymerization catalysts. *J. Am. Chem. Soc.* **2003**, *125*, 4350–4361. [CrossRef]
50. Liao, G.; Xiao, Z.; Chen, X.; Du, C.; Zhong, L.; Cheung, C.S.; Gao, H. Fast and regioselective polymerization of para-alkoxystyrene by palladium catalysts for precision production of high-molecular-weight polystyrene derivative. *Macromolecules* **2020**, *53*, 256–266. [CrossRef]
51. Chen, M.; Chen, C. Direct and tandem routes for the copolymerization of ethylene with polar functionalized internal olefins. *Angew. Chem. Int. Ed.* **2020**, *59*, 1206–1210. [CrossRef]
52. Maity, B.; Cao, Z.; Kumawat, J.; Gupta, V.; Cavallo, L. A multivariate linear regression approach to predict ethene/1-Olefin copolymerization statistics promoted by group 4 catalysts. *ACS Catal.* **2021**, *11*, 4061–4070. [CrossRef]
53. Santiago, C.B.; Guo, J.Y.; Sigman, M.S. Predictive and mechanistic multivariate linear regression models for reaction development. *Chem. Sci.* **2018**, *9*, 2398–2412. [CrossRef]
54. Todeschini, R.; Consonni, V. *Handbook of Molecular Descriptors*; John Wiley & Sons: Hoboken, NJ, USA, 2008; Volume 11.
55. Sun, J.; Chen, M.; Luo, G.; Chen, C.; Luo, Y. Diphosphazane-monoxide and phosphine-sulfonate palladium catalyzed ethylene copolymerization with polar monomers: A computational study. *Organometallics* **2019**, *38*, 638–646. [CrossRef]
56. Mehmood, A.; Xu, X.; Raza, W.; Kim, K.H.; Luo, Y. Mechanistic studies for palladium catalyzed copolymerization of ethylene with vinyl ethers. *Polymers* **2020**, *12*, 2401. [CrossRef] [PubMed]
57. Mehmood, A.; Xu, X.; Kang, K.; Luo, Y. Origin of different chain-end microstructures in ethylene/vinyl halide copolymerization catalysed by phosphine-sulfonate palladium complexes. *New J. Chem.* **2020**, *44*, 16941–16947. [CrossRef]
58. Mehmood, A.; Xu, X.; Raza, W.; Kukkar, D.; Kim, K.H.; Luo, Y. Computational study of the copolymerization mechanism of ethylene with methyl 2-acetamidoacrylate catalyzed by phosphine-sulfonate palladium complexes. *New J. Chem.* **2021**, *45*, 16670–16678. [CrossRef]
59. Cramer III, R.D. Quantitative drug design. *Annu. Rep. Med. Chem.* **1976**, *11*, 301–310.
60. Guo, J.Y.; Minko, Y.; Santiago, C.B.; Sigman, M.S. Developing comprehensive computational parameter sets to describe the performance of pyridine-oxazoline and related ligands. *ACS Catal.* **2017**, *7*, 4144–4151. [CrossRef]
61. Imuta, J.I.; Kashiwa, N.; Toda, Y. Catalytic regioselective introduction of allyl alcohol into the nonpolar polyolefins: Development of one-pot synthesis of hydroxyl-capped polyolefins mediated by a new metallocene IF catalyst. *J. Am. Chem. Soc.* **2002**, *124*, 1176–1177. [CrossRef]
62. Ji, G.; Chen, Z.; Wang, X.Y.; Ning, X.S.; Xu, C.J.; Zhang, X.M.; Tao, W.J.; Li, J.F.; Gao, Y.; Shen, Q.; et al. Direct copolymerization of ethylene with protic comonomers enabled by multinuclear Ni catalysts. *Nat. Commun.* **2021**, *12*, 6283. [CrossRef]
63. Marques, M.M.; Correia, S.G.; Ascenso, J.R.; Ribeiro, A.F.G.; Gomes, P.T.; Dias, A.R.; Foster, P.; Raush, M.D.; Chien, J.C.W. Polymerization with TMA-protected polar vinyl comonomers. I. Catalyzed by group 4 metal complexes with η^5-type ligands. *J. Polym. Sci. Polym. Chem.* **1999**, *37*, 2457–2469. [CrossRef]
64. Chen, T.; Chen, Z.; Chen, C. An ionic cluster strategy for performance improvements and product morphology control in metal-catalyzed olefin–polar monomer copolymerization. *J. Am. Chem. Soc.* **2022**, *144*, 2245–2254.

Communication

Breast Cancer Stem Cell Active Copper(II) Complexes with Naphthol Schiff Base and Polypyridyl Ligands

Joshua Northcote-Smith, Alice Johnson, Kuldip Singh, Fabrizio Ortu * and Kogularamanan Suntharalingam *

School of Chemistry, University of Leicester, Leicester LE1 7RH, UK; jns7@leicester.ac.uk (J.N.-S.); alice.johnson@leicester.ac.uk (A.J.); ks42@le.ac.uk (K.S.)
* Correspondence: fabrizio.ortu@leicester.ac.uk (F.O.); k.suntharalingam@leicester.ac.uk (K.S.); Tel.: +44-(0)116-294-4670 (F.O.); +44-(0)116-294-4562 (K.S.)

Abstract: Breast cancer stem cells (CSCs) are a sub-population of tumour cells that can promote breast cancer relapse and metastasis. Current treatments are unable to completely remove breast CSCs, therefore it is essential to develop new chemotherapeutics that can remove breast CSCs at clinically compatible doses. Here we present the synthesis, characterisation, and anti-breast CSC properties of copper(II) complexes, [Cu(L^2)(1,10-phenanthroline)]PF_6 (**2**) and [Cu(L^3)(1,10-phenanthroline)]PF_6 (**3**) comprising of a tridentate (O,N,S) coordinated naphthol Schiff base ligand (L^2 = (E)-1-(((2-(methylthio)ethyl)imino)methyl)naphthalen-2-ol or L^3 = (E)-1-(((2-(ethylthio)ethyl)imino)methyl) naphthalen-2-ol and 1,10-phenanthroline. The copper(II) complexes (**2** and **3**) kill breast CSCs, cultured in monolayer and three-dimensional systems, in the micromolar range. Notably, **2** and **3** are more potent towards breast CSC mammospheres than salinomycin (up to 4.5-fold), an established anti-breast CSC agent. Further, cell-based studies indicate that **2** and **3** are readily taken up by breast CSCs and elevate intracellular reactive oxygen species (ROS) levels upon short exposure times (0.5–1 h). The latter is likely to be the underlying mechanism by which **2** and **3** induces breast CSC death.

Keywords: metallopharmaceuticals; cancer stem cells; copper; bioinorganic chemistry; medicinal inorganic chemistry; reactive oxygen species

1. Introduction

Life-threatening diseases such as cancer and pathogenic infections impose a huge socio-economic burden on humankind. Despite significant advances in cancer treatments annual cancer-related deaths within the European Union (EU) remain extremely high. Latest statistics indicate that cancer caused over 1.25 million deaths (accounting for 26% of all deaths) within EU28 countries in 2013 [1]. Amongst the various cancer types breast cancer caused the third most fatalities within EU28 countries (around 93,500 deaths). The economic impact of breast cancer across the EU, in terms of healthcare resources, productivity losses because of early death, lost working days, and informal care costs, amounts to around €15.0 billion [2]. Given these harrowing statistics, continuing to conduct basic research on new breast cancer medicines and therapies that could ultimately reduce mortality rates and lessen financial burden is of significant importance. Breast cancer relapse and metastasis, the leading cause of breast cancer associated deaths, is strongly linked to the existence of breast cancer stem cells (CSCs) [3,4]. Breast CSCs are a sub-set of breast cancer cells that have the intrinsic ability to differentiate, self-renew, and form secondary tumours [5,6]. Breast CSCs are known to evade current breast cancer therapy (including surgery, chemotherapy, and radiation), promote re-population of primary cancer sites, and potentially lead to fatal incidences of cancer relapse and metastasis [7,8]. Therefore it is vital that anti-breast chemotherapeutics possess the ability to remove breast CSCs at therapeutically significant doses.

We and others have developed several breast CSC potent and selective metal complexes over the last six to seven years [9]. One of the most promising classes of anti-breast CSC agents reported thus far, are copper(II) complexes bearing polypyridyl and/or Schiff base ligands [10–13]. It should be noted that copper(II) complexes have been widely studied as anticancer agents (with limited studies focused on their anti-CSC potential), but none have been approved for clinic use [14,15]. The most advanced copper(II) complexes, called Casiopeinas, are currently in Phase I clinical trials [16–19]. The very latest studies suggest that Casiopeinas may be limited by dose-dependent cardiotoxicity [20], probably due to speciation. The breast CSC active copper(II) complexes developed by our group induce breast CSC death by generating intracellular reactive oxygen species (ROS), often activating the p38 and JNK stress pathways and caspase-dependent apoptosis [10–13]. The success of these copper(II) complexes against breast CSCs is thought to arise partly from the vulnerability of breast CSCs to subtle changes in their redox state [21,22]. Very recently we reported a copper(II) complex **1**, containing a O,N,S-Schiff base ligand **L¹**, and 4,7-diphenyl-1,10-phenanthroline (see Supplementary Materials Figure S1 for chemical structures of **L¹** and **1**), capable of killing breast CSCs via cytotoxic and immunogenic mechanisms [23]. This was the first metal complex to induce immunogenic cell death (ICD) of breast CSCs and promote their engulfment by immune cells. This was an important step in our ongoing efforts to develop clinically viable anti-breast cancer drug candidates, as the removal of CSCs by immunological activation could serve as an effective method to eliminate residual CSCs after conventional CSC inactive treatments. In this study, we explore the breast CSC activity of structurally related copper(II) complexes (**2** and **3**) where the Schiff base ligand **L¹** in **1** has been modified. Specifically a naphthalene moiety was incorporated into the Schiff base ligand scaffold to yield (E)-1-(((2-(methylthio)ethyl)imino)methyl)naphthalen-2-ol (**L²**) which was used to prepare the corresponding copper(II) complex with 1,10-phenanthroline, [Cu(**L²**)(1,10-phenanthroline)]PF$_6$ (**2**) (Scheme 1). Furthermore, the methyl group on the sulphur atom in **L²** was replaced with an ethyl group to yield (E)-1-(((2-(ethylthio)ethyl)imino)methyl)naphthalen-2-ol (**L³**), which was subsequently used to prepare the corresponding copper(II) complex with 1,10-phenanthroline, [Cu(**L³**)(1,10-phenanthroline)]PF$_6$ (**3**) (Scheme 1). These structural modifications were expected to modulate breast CSC uptake and intracellular ROS generation. Herein we report the synthesis, characterisation, and anti-breast CSC properties of the copper(II) complexes, [Cu(**L²**)(1,10-phenanthroline)]PF$_6$ (**2**) and [Cu(**L³**)(1,10-phenanthroline)]PF$_6$ (**3**).

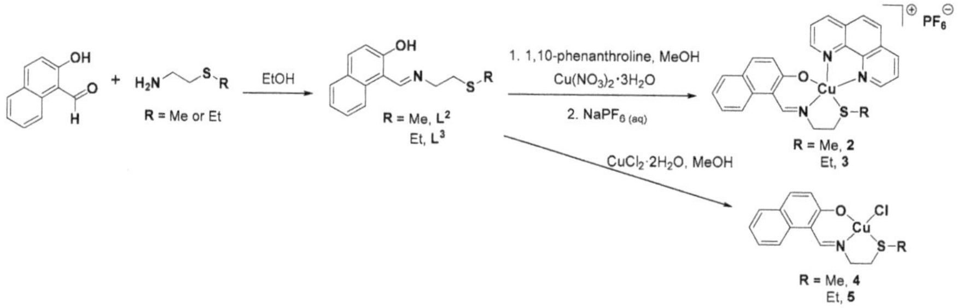

Scheme 1. The reaction scheme for the preparation of the naphthol Schiff base ligands, **L²** and **L³** and the corresponding copper(II) complexes **2–5**.

2. Results and Discussion

2.1. Synthesis and Characterisation of the Ligands and Copper(II) Complexes

The copper(II) complexes [Cu(**L²**)(1,10-phenanthroline)]PF$_6$ (**2**) and [Cu(**L³**)(1,10-phenanthroline)]PF$_6$ (**3**) were prepared as outlined in Scheme 1. The naphthol Schiff base ligands, (E)-1-(((2-(methylthio)ethyl)imino)methyl)naphthalen-2-ol (L^2) and (E)-1-(((2-(ethylthio)ethyl)imino)methyl)naphthalen-2-ol (L^3) were prepared by reacting equimolar amounts of 2-(methylthio)ethylamine or 2-(ethylthio)ethylamine with 2-hydroxy-1-naphthaldehyde in ethanol for 16 h. The ligands, L^2 and L^3 were isolated in excellent yields (96–100%) as orange solids and fully characterised by ^{1}H and ^{13}C NMR, infrared (IR) spectroscopy, high-resolution ESI mass spectrometry, and elemental analysis (see Section 3 and Figures S2–S8). Characteristic signals at 9.10–9.11 ppm and 177.84–177.85 ppm in the ^{1}H and ^{13}C NMR spectra for L^2 and L^3 respectively confirmed formation of the imine functionality. Disappearance for the aldehyde peak at 10.82 ppm and 193.33 ppm in the ^{1}H and ^{13}C NMR spectra corresponding to the 2-hydroxy-1-naphthaldehyde starting material established full conversion to the imine product (Figures S9 and S10). Further, the IR spectra of L^2 and L^3 exhibited bands at 1610–1616 cm^{-1} corresponding to the C=N$_{imine}$ stretch. The purity of L^2 and L^3 was confirmed by elemental analysis (see Section 3). The copper(II) complexes, **2** and **3** were prepared by reacting equimolar amounts of 1,10-phenanthroline with copper(II) nitrate hydrate in methanol, followed by the addition of the Schiff base ligand, L^2 or L^3 and excess sodium hexafluorophosphate. The complexes were isolated in reasonable yields (45–69%) as (dark) green solids and fully characterised by high-resolution ESI mass spectrometry, IR spectroscopy, and elemental analysis (see Section 3 and Figures S8, S11 and S12). Distinctive molecular ion peaks corresponding to **2** and **3** with the appropriate isotopic pattern were observed in the HR ESI-MS (m/z = 487.0775 [**2**-PF$_6$]$^+$; 501.0936 [**3**-PF$_6$]$^+$; Figures S11 and S12). The IR spectra for **2** and **3** displayed C=N$_{imine}$ bands between 1616–1618 cm^{-1} indicating the presence of the imine functionality associated to L^2 and L^3 (Figure S8). Furthermore, the IR spectra for **2** and **3** did not display a broad O–H stretch (Figure S8), supporting the tridentate complexation of L^2 and L^3 to the copper centre as depicted in Scheme 1. The purity of **2** and **3** was established by elemental analysis (see Section 3). [Cu(L^2)Cl] (**4**) and [Cu(L^3)Cl] (**5**) were also prepared to serve as control compounds (as outlined in Scheme 1)—copper(II) complexes without 1,10-phenanthroline. The synthetic protocol and full characterisation of **4** and **5** (high-resolution ESI mass spectrometry, IR spectroscopy, and elemental analysis) is reported in the Section 3 (Figures S8, S13 and S14).

Single (dark green needle-like) crystals of **2** and **3** suitable for X-ray diffraction studies were obtained by slow evaporation of an acetone solution of **2** and by vapour diffusion of diethyl ether into an acetonitrile solution of **3** (CCDC 2046679–2046680, Figure 1, Table S1). Selected bond distances and angles are presented in Tables S2 and S3. The cationic component of **2** exhibits a distorted trigonal bipyramidal geometry with the copper(II) centre coordinated to L^2 in a tridentate manner (via the O, N, and S atoms) and 1,10-phenanthroline in a bidentate manner. Within the CuNOS equatorial plane, the O(1)–Cu–S(1), O(1)–Cu–N(2), and N(2)–Cu–S(1) bond angles average to 119.90(15)°, and the axial N(1)–Cu–N(3) angle is 177.78(12). This is consistent with a distorted trigonal bipyramidal geometry (τ_5 = 0.96) (Figure 1). The cationic component of **3** also adopts a distorted trigonal bipyramidal geometry (τ_5 = 0.94), however, the structure is disordered due to the flexibility of the thioethyl moiety in L^3 (Figure 1). Alike **2**, the copper(II) centre in **3** binds to L^3 in a tridentate manner (via the O, N, and S atoms) and 1,10-phenanthroline in a bidentate manner. Overall, the Cu–N$_{imine}$, Cu–N$_{polypyridyl}$, Cu–S, and Cu–O bond lengths observed for **2** and **3** are consistent with bond parameters reported for related 5-coordinate copper(II) complexes (Tables S2 and S3) [23–25].

2.2. Lipophilicity and Solution Stability of the Copper(II) Complexes

It is important to understand the lipophilicity of compounds intended for medical applications as this directly affects aqueous solubility, absorption, distribution, and toxicity [26]. The ability of a given compound to partition between octanol and water (P) is a good indicator of their inherent lipophilicity. The lipophilicity of **2–5** was gauged in this manner. As shown in Table S4, the experimentally determined LogP values for **2–5** were between -0.63 ± 0.03 and 0.34 ± 0.03. These values suggest that **2–5** are amphiphilic and thus should be internalised easily by cells and be partially soluble in water. To probe the stability of **2** and **4**, taken as demonstrative compounds within the copper(II) complex series, UV-Vis spectroscopy studies were performed in physiologically relevant solutions. The absorbance of **2** and **4** (50 µM) in PBS:DMSO (200:1) altered significantly over the course of 24 h at 37 °C upon addition of ascorbic acid (10 equivalents), a cellular reductant (Figures S15 and S16). For complex **2**, the absorbance of the π–π* and MLCT bands decreased by up to 54% after 24 h whereas for complex **4** the bands completely disappeared in the same timeframe, suggesting that the 1,10-phenanthroline ligand in **2** provides some protection against ascorbic acid reduction-mediated instability. It is worth pointing out that the reaction between **2** or **4** and ascorbic acid (mostly likely a reduction reaction) is time-dependent and may be proceeding via many reaction pathways (Figures S15 and S16). The addition of bathocuproine disulfonate (BCS, 2 equivalents), which is a strong and selective copper(I) chelator, to **2** and **4** (50 µM) and ascorbic acid (10 equivalents) in PBS:DMSO (200:1) produced a distinct absorbance band at 480 nm corresponding to $[Cu^{I}(BCS)_2]^{3-}$ (Figures S17 and S18). This clearly shows that the copper(II) centre in **2** and **4** is reduced to copper(I) in the presence of ascorbic acid. Collectively, analysis of the UV-Vis spectroscopy data indicates that **2** and **4** are reduced from the copper(II) form to the copper(I) form under biologically reducing conditions. Before beginning cell-based studies, UV-Vis spectroscopy was used to probe the stability of **2** and **4** in mammary epithelial cell growth medium (MEGM) over the course of 24 h at 37 °C (Figures S19 and S20). Both complexes were deemed to be reasonably stable under these condition, with the 1,10-phenanthroline-containing complex **2** displaying greater stability than **4**.

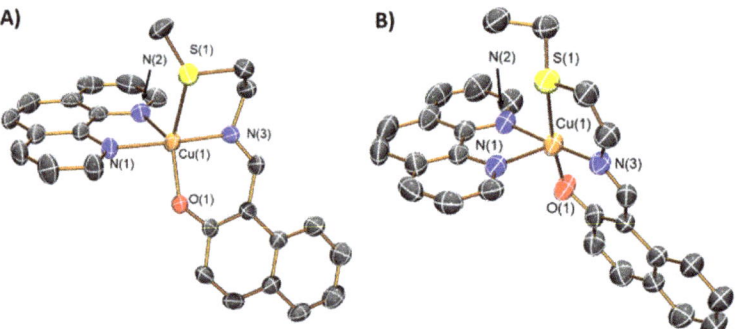

Figure 1. X-ray structures of the copper(II) complexes **2** (**A**) and **3** (**B**) comprising of L^2 or L^3 and 1,10-phenanthroline. Ellipsoids are shown at 50% probability. C in grey, N in dark blue, S in yellow, O in red, and Cu in orange. H atoms, disorder components, and the hexafluorophosphate counter anion have been omitted for clarity.

2.3. In Vitro Cytotoxicity of the Copper(II) Complexes

The monolayer cytotoxicity of **2–5** towards bulk breast cancer cells (HMLER) and breast CSCs (HMLER-shEcad) was assessed using the colorimetric MTT assay. IC_{50} values were determined from dose–response curves (Figures S21–S24) and are summarized in Table 1. The copper(II) complexes, **2–5** all displayed micromolar potency towards HMLER and HMLER-shEcad cells. The mono-cationic 1,10-phenanthroline-containing complexes **2** and **3** exhibited significantly ($p < 0.05$) higher potency towards HMLER and HMLER-shEcad cells than the corresponding neutral chloride-containing complexes **4** and **5** (Table 1). This suggests that the presence of the 1,10-phenanthroline ligand in **2** and **3** enhances their breast CSC toxicity. Complexes **2** and **3** displayed similar potency for breast CSCs and bulk breast cancer cells, whereas **4** and **5** displayed greater potency for bulk breast cancer cells over breast CSCs. The former suggests that **2** and **3** have the potential to kill entire heterogeneous breast cancer cell populations (consisting of bulk breast cancer cells and breast CSCs) with a single micromolar dose. The similarity in IC$_{50}$ values of **2** and **3** (towards HMLER and HMLER-shEcad cells) indicates that modification of the thioalkyl moiety within the Schiff base ligand, from a methyl group to an ethyl group, had little effect on potency. Notably, the potency of **2** and **3** for CSC-enriched HMLER-shEcad cells was slightly greater than that of salinomycin, an established clinically-tested anti-breast CSC agent [10]. Control cytotoxicity studies showed that the potency of the free naphthol Schiff base ligands **L^2** and **L^3**, CuCl$_2$·2H$_2$O [27], and dichloro(1,10-phenanthroline)copper(II) [10] towards HMLER and HMLER-shEcad cells were all significantly lower ($p < 0.05$) than **2** and **3** (Table 1, Figures S25 and S26). This suggests that the cytotoxicity of **2** and **3** towards bulk breast cancer cells and breast CSCs is likely to result from the intact cellular entry of the copper(II) complexes, which allows for the synergistic co-delivery of all the complex components.

Table 1. IC$_{50}$ values of the copper(II) complexes **2–5**, the naphthol Schiff base ligands **L^2** and **L^3**, salinomycin, CuCl$_2$·2H$_2$O, and dichloro(1,10-phenanthroline)copper(II) against HMLER and HMLER-shEcad cells and HMLER-shEcad mammospheres determined after 72 h or 120 h incubation (mean of three independent experiments ± SD).

Compound	HMLER IC$_{50}$ [μM]	HMLER-ShEcad IC$_{50}$ [μM]	Mammosphere IC$_{50}$ [μM]
2	3.19 ± 0.11	3.40 ± 0.05	4.14 ± 0.20
3	3.25 ± 0.18	3.43 ± 0.10	6.08 ± 0.88
4	7.57 ± 0.23	12.70 ± 0.14	42.55 ± 8.13
5	8.87 ± 0.54	12.90 ± 0.14	49.10 ± 4.38
salinomycin [1]	11.40 ± 0.40	4.20 ± 0.30	18.50 ± 1.50
L^2	12.91 ± 0.71	32.41 ± 0.95	64.29 ± 0.20
L^3	11.20 ± 0.10	20.41 ± 0.35	52.06 ± 0.31
CuCl$_2$·2H$_2$O [1]	47.55 ± 0.18	47.92 ± 1.80	n.d.
dichloro(1,10-phenanthroline)copper(II) [1]	4.90 ± 0.17	7.86 ± 0.26	9.49 ± 0.20

[1] Taken from references [10,13,27]. n.d. not determined.

Breast CSCs are known to form three-dimensional, tumour-like structures, known as mammospheres when grown under low-attachment, serum-free conditions [28]. The ability of a given compound to inhibit mammosphere formation from single cell suspensions (with respect to number and size) is often used as a gauge for CSC potency. The ability of **2–5** to inhibit HMLER-shEcad mammosphere formation was assessed using an inverted microscope. The addition of **2** and **3** (at 0.5 μM, a non-lethal dose after 5 days incubation) appreciably reduced the number of mammospheres formed, whereas equal dosage with **4** and **5** did not significantly decrease the number of mammospheres formed (Figure 2). Although **2** and **3** inhibited mammosphere formation to a significant level ($p < 0.05$), their inhibitory effect was lower than salinomycin (Figure 2). Specifically, **2** and **3** reduced the number of mammospheres formed by 32–35% compared to the untreated control, whereas salinomycin induced a 60% reduction in the number of mammospheres formed compared to the untreated control. All of the copper(II) complexes **2–5** decreased the size of mammospheres formed (upon treatment at the IC$_{20}$ value for 5 days) to some extent, with **2** and **3** reducing the size more than **4** and **5** (Figure 3). Overall this shows that the presence of the 1,10-phenanthroline moiety in **2** and **3** enhances their mammosphere inhibitory effect. Salinomycin (upon treatment at the IC$_{20}$ value for 5 days) markedly reduced mammosphere size (Figure S27). As expected, addition of the naphthol Schiff base ligands **L^2** and **L^3** (at 0.5 μM or IC$_{20}$ value, both non-lethal doses, after 5 days incubation) had little effect on the number and size of mammospheres formed (Figures 2 and S28).

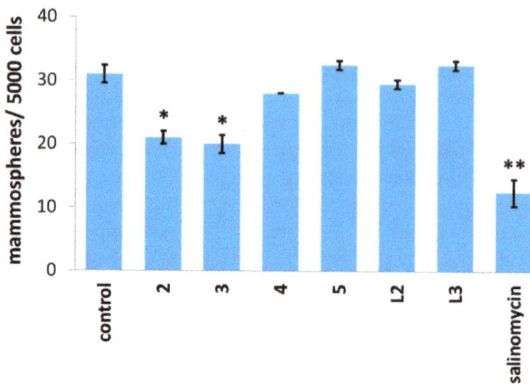

Figure 2. Representation of the number of mammospheres formed from HMLER-shEcad cell suspensions treated with **2–5**, L^2, L^3 or salinomycin at 0.5 μM, a non-lethal dose after 5 days incubation. Standard deviation was used to signify the associated errors. Student t-test, * = $p < 0.05$, ** = $p < 0.01$.

Figure 3. Bright-field images (taken using an inverted microscope) representative of untreated HMLER-shEcad mammospheres and those treated with **2–5** for 5 days at their respective IC_{20} values. Scale bar = 100 μm.

To determine the ability of **2–5** to reduce mammosphere viability, TOX8 a resazurin-based reagent was used. The IC_{50} values (concentration required to reduce mammosphere viability by 50%) were extrapolated from dose-response curves (Figures S29–S32) and are summarised in Table 1. The IC_{50} values for **2–5** were in the micromolar range, with the most effective complexes **2** and **3** displaying 4.5- and 3-fold greater mammosphere potency than salinomycin, respectively (Table 1) [13]. The mammosphere potency of **2** and **3** were significantly ($p < 0.05$) higher than **4** and **5**. This highlights the importance of the 1,10-phenanthroline moiety in **2** and **3** to their mammosphere toxicity, and is consistent with the monolayer cytotoxicity results. Furthermore control studies showed that the potency of the naphthol Schiff base ligands L^2 and L^3, and dichloro(1,10-phenanthroline)copper(II) [11] towards HMLER-shEcad mammospheres was significantly lower ($p < 0.05$) than **2** and **3** (Figures S33 and S34, Table 1). Therefore, the mammosphere potency observed for **2** and **3** is likely to be due to the intact mammosphere uptake of the copper(II) complexes, which facilitates the concerted co-delivery of all the complex components.

2.4. Mechanism of Action of the Copper(II) Complexes

In order to better understand the discrepancy in breast CSC monolayer and mammosphere potency of the mono-cationic 1,10-phenanthroline-containing complexes **2** and **3**, and the corresponding neutral chloride-containing complexes **4** and **5**, cellular uptake studies were performed. HMLER-shEcad cells were dosed with **2–5** (5 μM for 24 h) and the copper content was determined by inductively coupled plasma mass spectrometry (ICP-MS). As depicted in Figure 4, the 1,10-phenanthroline-containing complexes **2** (124 ng of Cu/10^6 cells) and **3** (188 ng of Cu/10^6 cells) were internalized by HMLER-shEcad cells to a significantly ($p < 0.05$) greater extent than the corresponding neutral chloride-containing complexes **4** (78 ng of Cu/10^6 cells) and **5** (95 ng of Cu/10^6 cells). The higher breast CSC uptake of **2** relative to **4**, and **3** relative to **5** could explain the variance in breast CSC potency of the complexes (with and without 1,10-phenanthroline).

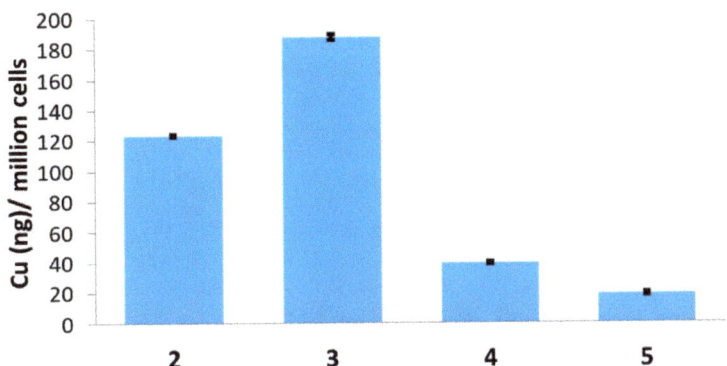

Figure 4. Copper content (ng of Cu/10^6 cells) in HMLER-shEcad cells treated with **2–5** (5 µM for 24 h).

Copper(II) complexes bearing Schiff base and/or phenanthroline ligands are known to induce cell death by elevating intracellular ROS levels [10–13]. To determine if **2**- and **3**-induced breast CSC toxicity is related to ROS production, intracellular ROS levels were quantified at different exposure times (0.5–24 h) using 6-carboxy-2′,7′-dichlorodihydrofluorescein diacetate (DCFH-DA), a well-established ROS indicator. HMLER-shEcad cells treated with **2** (2 × IC_{50} value) displayed a significant increase in ROS levels after 1 h exposure (92.5% increase, $p < 0.05$) relative to untreated control cells (Figure S35). Shorter (0.5 h) or prolonged (3–24 h) exposure of **2** did not increase intracellular ROS levels compared to untreated cells. HMLER-shEcad cells dosed with **3** (2 × IC_{50} value) displayed a significant increase in ROS levels after 0.5 h exposure (75.1% increase, $p < 0.05$) relative to untreated control cells (Figure S36). Prolonged (1–24 h) exposure of **3** did not increase intracellular ROS levels compared to untreated cells, although a statistically insignificant increase ($p > 0.05$) in ROS levels was observed after 3 h exposure. Therefore, for HMLER-shEcad cells treated with **2** and **3**, a sharp ROS burst occurs within the first hour of exposure and then ROS levels attenuate to levels observed in untreated control cells. Similar time-dependent ROS generation properties have been previously reported for other CSC-potent copper(II) complexes [12,13]. H_2O_2-treated (positive control, 150 µM for 0.5–24 h) HMLER-shEcad cells exhibited a significant increase ($p < 0.05$) in ROS levels after short (0.5–3 h, up to 14-fold) and long (6–24 h, up to 2-fold) exposure times, compared to untreated control cells (Figure S37). Collectively, the ROS studies suggest that **2**- and **3**-induced breast CSC death could be associated to intracellular ROS generation.

3. Materials and Methods

3.1. General Procedures

All synthetic procedures were performed under normal atmospheric conditions. Fourier transform infrared (FTIR) spectra were recorded with an IRAffinity-1S Shimadzu spectrophotometer. Electron spray ionisation mass spectra were recorded on a Micromass Quattro spectrometer. UV-Vis absorption spectra were recorded on a Cary 3500 UV-Vis spectrophotometer. ^1H and ^{13}C NMR spectra were recorded on a BrukerAvance 400 MHz Ultrashield NMR spectrometer. ^1H NMR spectra were referenced internally to residual solvent peaks, and chemical shifts are expressed relative to tetramethylsilane, $SiMe_4$ ($\delta = 0$ ppm). Elemental analysis of the compounds prepared was performed commercially by London Metropolitan University or the University of Cambridge. 2-hydroxy-1-naphthaldehyde, 2-(methylthio)ethylamine, 2-(ethylthio)ethylamine, 1,10-phenanthroline, $Cu(NO_3)_2 \cdot 3H_2O$,

CuCl$_2$·2H$_2$O, and NaPF$_6$ were purchased from Sigma Aldrich or Alfa Aesar and used as received.

3.2. Synthesis of (E)-1-(((2-(Methylthio)ethyl)imino)methyl)naphthalen-2-ol, L^2

A mixture of 2-hydroxy-1-naphthaldehyde (344 mg, 2.0 mmol) and 2-(methylthio) ethylamine (191 mg, 2.1 mmol) were refluxed in ethanol (20 mL) for 16 h. The reaction mixture was then evaporated under vacuum to afford **L^2** as an orange solid (472 mg, 96%); ^{1}H NMR (400 MHz, DMSO-d_6): δ 13.93 (s, 1H, O*H*), 9.11 (d, 1H, N=C*H*), 8.06 (d, 1H, Ar–*H*), 7.73 (d, 1H, Ar–*H*), 7.64 (d, 1H, Ar–*H*), 7.43 (ddd, 1H, Ar–*H*), 7.19 (ddd, 1H, Ar–*H*), 6.71 (d, 1H, Ar–*H*), 3.84 (dd, 2H, C*H$_2$*), 2.83 (t, 2H, C*H$_2$*), 2.13 (s, 3H, C*H$_3$*); ^{13}C NMR (162 MHz, DMSO-d_6): δ 177.85 (N=*C*H), 159.74 (*Ar*), 137.60 (*Ar*), 134.85 (*Ar*), 129.34 (*Ar*), 128.34 (*Ar*), 126.05 (*Ar*), 125.66 (*Ar*), 122.62 (*Ar*), 118.95 (*Ar*), 106.13 (*Ar*), 50.33 (*C*H$_2$), 34.47 (*C*H$_2$), 15.05 (*C*H$_3$); IR (solid, ATR, cm^{-1}): 3021, 2971, 2917, 1616, 1541, 1529, 1491, 1439, 1399, 1354, 1314, 1282, 1254, 1193, 1177, 1137, 1040, 996, 955, 924, 879, 855, 827, 743, 723, 686, 654, 541, 513, 497, 481, 437, 413; HR ESI-MS Calcd. for C$_{14}$H$_{16}$NOS [M+H]$^+$ 246.0953 a.m.u. Found [M+H]$^+$: 246.0954 a.m.u.; Anal. Calcd. for C$_{14}$H$_{15}$NOS·0.25H$_2$O (%): C, 67.30; H, 6.25; N, 5.61. Found: C, 67.62; H, 6.06; N, 5.99.

3.3. Synthesis of (E)-1-(((2-(Ethylthio)ethyl)imino)methyl)naphthalen-2-ol, L^3

A mixture of 2-hydroxy-1-naphthaldehyde (172 mg, 1.0 mmol) and 2-(ethylthio) ethylamine (116 mg, 1.1 mmol) were refluxed in ethanol (20 mL) for 16 h. The reaction mixture was evaporated under vacuum to afford **L^3** as an orange solid (259 mg, 100%); ^{1}H NMR (400 MHz, DMSO-d_6): δ 13.94 (s, 1H, O*H*), 9.10 (d, 1H, N=C*H*), 8.06 (d, 1H, Ar–*H*), 7.73 (d, 1H, Ar–*H*), 7.63 (dd, 1H, Ar–*H*), 7.43 (ddd, 1H, Ar–*H*), 7.19 (ddd, 1H, Ar–*H*), 6.72 (d, 1H, Ar–*H*), 3.82 (dd, 2H, C*H$_2$*), 2.86 (t, 2H, C*H$_2$*), 2.60 (q, 2H, C*H$_2$*), 1.20 (t, 3H, C*H$_3$*); ^{13}C NMR (162 MHz, DMSO-d_6): δ 177.84 (N=*C*H), 159.70 (*Ar*), 137.60 (*Ar*), 134.85 (*Ar*), 129.34 (*Ar*), 128.34 (*Ar*), 126.04 (*Ar*), 125.66 (*Ar*), 122.61 (*Ar*), 118.95 (*Ar*), 106.12 (*Ar*), 50.97 (*C*H$_2$), 31.96 (*C*H$_2$), 25.33 (*C*H$_2$), 15.28 (*C*H$_3$); IR (solid, ATR, cm^{-1}): 3049, 3025, 2969, 2920, 1610, 1540, 1493, 1440, 1402, 1344, 1256, 1206, 1183, 1137, 1070, 1034, 995, 961, 867, 832, 740, 638, 542, 515, 504, 480, 434, 414; HR ESI-MS Calcd. for C$_{15}$H$_{18}$NOS [M+H]$^+$ 260.1109 a.m.u. Found [M+H]$^+$: 260.1112 a.m.u.; Anal. Calcd. for C$_{15}$H$_{17}$NOS·0.2H$_2$O (%): C, 68.51; H, 6.67; N, 5.33. Found: C, 68.62; H, 6.38; N, 5.55.

3.4. Synthesis of [Cu(L^2)1,10-Phenanthroline][PF$_6$], 2

1,10-phenanthroline (74 mg, 0.41 mmol) and Cu(NO$_3$)$_2$·3H$_2$O (99 mg, 0.41 mmol) dissolved in methanol (10 mL) were stirred at room temperature for 0.5 h. The colour of the solution changed from blue to light green. **L^2** (100 mg, 0.41 mmol) in methanol (10 mL) was added dropwise. The dark green mixture was stirred at room temperature for 72 h. The mixture was then filtered to remove the precipitate. The filtrate was reduced to ~10 mL. An excess of NaPF$_6$ (250 mg, 1.5 mmol) in water (50 mL) was added and the mixture stirred for 0.5 h. The resultant precipitate was collected and washed thoroughly with water and diethyl ether to give **2** as a dark green solid (116 mg, 45%); IR (solid, ATR, cm^{-1}): 1618, 1602, 1585, 1540, 1518, 1430, 1413, 1392, 1366, 1342, 1254, 1223, 1194, 1144, 1106, 1029, 979, 828, 752, 722, 647, 557, 523, 476, 416; HR ESI-MS Calcd. for C$_{26}$H$_{22}$CuN$_3$OS [M-PF$_6$]$^+$ 487.0780 a.m.u. Found [M-PF$_6$]$^+$ 487.0775 a.m.u.; Anal. Calcd. for C$_{26}$H$_{22}$CuN$_3$OSPF$_6$ (%): C, 49.33; H, 3.50; N, 6.64. Found: C, 49.14; H, 3.34; N, 6.46.

3.5. Synthesis of [Cu(L^3)1,10-Phenanthroline][PF$_6$], 3

1,10-phenanthroline (110 mg, 0.61 mmol) and Cu(NO$_3$)$_2$·3H$_2$O (147 mg, 0.61 mmol) dissolved in methanol (10 mL) were stirred at room temperature for 0.5 h. The colour of the solution changed from blue to light green. **L^3** (160 mg, 0.62 mmol) in methanol (10 mL) was added dropwise. The dark green mixture was stirred at room temperature for 72 h. The mixture was then filtered to remove the precipitate. The filtrate was reduced to ~10 mL. An excess of NaPF$_6$ (400 mg, 2.4 mmol) in water (~30 mL) was added and the mixture stirred for 0.5 h. The resultant precipitate was collected and washed thoroughly with water and diethyl ether to give **3** as a green solid (272 mg, 69%); IR (solid, ATR, cm^{-1}): 1616, 1604, 1583, 1539, 1509, 1460, 1437, 1428, 1416, 1393, 1363, 1345, 1254, 1224, 1187, 1145, 1104, 1094, 1036, 1006, 979, 828, 751, 725, 645, 555, 521, 490, 476, 450, 421, 387; HR ESI-MS Calcd. for $C_{27}H_{24}CuN_3OS$ [M-PF$_6$]$^+$ 501.0936 a.m.u. Found [M-PF$_6$]$^+$ 501.0936 a.m.u.; Anal. Calcd. for $C_{27}H_{24}CuN_3OSPF_6$ (%): C, 50.12; H, 3.74; N, 6.49. Found: C, 49.79; H, 3.45; N, 6.18.

3.6. Synthesis of [Cu(L^2)Cl], 4

L^2 (96.4 mg, 0.39 mmol) in methanol (20 mL) was added to CuCl$_2$·2H$_2$O (66.5 mg, 0.39 mmol) in methanol (5 mL). The dark green mixture was stirred overnight. The resultant precipitate was collected and washed with cold methanol and diethyl ether to yield **4** as a dark green solid (20 mg). The filtrate was reduced and left overnight in the freezer ($-20\,^\circ$C) to yield a second crop of **4** as a dark green solid (16 mg) (total 36 mg, 27%); IR (solid, ATR, cm^{-1}): 1615, 1604, 1589, 1537, 1503, 1452, 1431, 1409, 1392, 1359, 1339, 1306, 1250, 1210, 1183, 1164, 1140, 1089, 1029, 976, 957, 940, 859, 825, 778, 765, 746, 646, 583, 553, 516, 474, 448, 417, 387; HR ESI-MS Calcd. for $C_{14}H_{14}CuNOS$ [M-Cl]$^+$ 307.0092 a.m.u. Found [M-Cl]$^+$ 307.0096 a.m.u.; Anal. Calcd. for $C_{14}H_{14}CuNOSCl$ (%): C, 48.98; H, 4.11; N, 4.08. Found: C, 48.77; H, 3.98; N, 4.02.

3.7. Synthesis of [Cu(L^3)Cl], 5

L^3 (119 mg, 0.46 mmol) in methanol (5 mL) was added to CuCl$_2$·2H$_2$O (78 mg, 0.46 mmol) in methanol (5 mL). The dark green mixture was refluxed for 2 h. The mixture was then put in the freezer ($-20\,^\circ$C) overnight. The resultant precipitate was collected and washed with cold methanol and diethyl ether to yield **5** as a dark green solid (52 mg, 32%); IR (solid, ATR, cm^{-1}): 1614, 1604, 1537, 1507, 1454, 1433, 1412, 1391, 1361, 1340, 1308, 1253, 1206, 1183, 1172, 1162, 1141, 1093, 1077, 1033, 1000, 977, 949, 869, 830, 752, 644, 584, 558, 549, 513, 473, 449, 421, 384; HR ESI-MS Calcd. for $C_{15}H_{16}CuNOS$ [M-Cl]$^+$ 321.0249 a.m.u. Found [M-Cl]$^+$ 321.0257 a.m.u.; Anal. Calcd. for $C_{15}H_{16}CuNOSCl$ (%): C, 50.42; H, 4.51; N, 3.92. Found: C, 50.75; H, 4.25; N, 3.83.

3.8. X-ray Single Crystal Diffraction Analysis

Single crystals of complexes **2** and **3** were obtained by slow evaporation of an acetone solution of **2** and by vapour diffusion of diethyl ether into an acetonitrile solution of **3**. Crystals suitable for X-ray diffraction analysis were selected and mounted on a Bruker Apex 2000 CCD area detector diffractometer using standard procedures. Data was collected using graphite-monochromated Mo-Kα radiation (λ = 0.71073) at 150(2) K. Absorption corrections were applied using a multiscan method (SADABS) [29]. The structures were solved using SHELXS [30]; the datasets were refined by full-matrix least-squares on all unique F^2 values, with anisotropic displacement parameters for all non-hydrogen atoms, and with constrained riding hydrogen geometries [31]; U_{iso}(H) was set at 1.2 (1.5 for methyl groups) times U_{eq} of the parent atom. The largest features in final difference syntheses were close to heavy atoms and were of no chemical significance. SHELX was employed through OLEX2 for structure solution and refinement [29,30,32]. ORTEP-3 and POV-Ray were employed for molecular graphics [33]. The structures have been deposited with the Cambridge Crystallographic Data Centre (CCDC 2046679 and 2046680). This information can be obtained free of charge from www.ccdc.cam.ac.uk/data_request/cif.

3.9. Measurement of Water-Octanol Partition Coefficient (LogP)

The LogP values for **2–5** were determined using the shake-flask method and UV-Vis spectroscopy. The 1-octanol used in this experiment was pre-saturated with water. An aqueous solution of **2–5** (500 µL, 100 µM) was incubated with 1-octanol (500 µL) in a 1.5 mL tube. The tube was shaken at room temperature for 24 h. The two phases were separated by centrifugation and the **2–5** content in each phase was determined by UV-Vis spectroscopy.

3.10. Cell Lines and Cell Culture Conditions

R. A. Weinberg (Whitehead Institute, MIT) (Cambridge, USA) generously gave the human mammary epithelial cell lines, HMLER and HMLER-shEcad, used in this study. The cells were grown in Mammary Epithelial Cell Growth Medium (MEGM) with supplements and growth factors (BPE, hydrocortisone, hEGF, insulin, and gentamicin/amphotericin-B). Standard cell culture conditions were used to grow the cells (310 K, 5% CO_2).

3.11. Cytotoxicity MTT Assay

The cytotoxicity of **2–5**, L^2, and L^3 was determined using the colorimetric MTT assay. Five thousand HMLER and HMLER-shEcad cells were added to each well of a 96-well plate. Upon allowing the cells to attach to the bottom of each well of a 96-well plate overnight, different concentrations of the test compounds (0.2–100 µM) were added, and incubated for 72 h (total volume 200 µL). 10 mM stock solutions of the compounds in DMSO were prepared and appropriately diluted using Mammary Epithelial Cell Growth Medium (MEGM). The untreated control wells contained 0.5% DMSO, which was the final concentration of DMSO in each treated well. Upon 72 h incubation, a 4 mg mL^{-1} solution of MTT dissolved in PBS was added to each well (20 µL). The 96-well plate was then incubated for an additional 4 h. After aspiration of the MEGM/MTT solution in each well, 200 µL of DMSO was added to each well to dissolve any purple formazan crystals. The absorbance of the resultant solutions in each well was read at 550 nm. The absorbance values corresponding to each well were normalised to DMSO-containing control wells and plotted as concentration of test compound versus % cell viability. IC_{50} values were calculated from the resulting dose dependent curves. The reported IC_{50} values are the average of three independent experiments (n = 18).

3.12. Tumorsphere Formation and Viability Assay

Five thousand HMLER-shEcad cells were added to each well of an ultralow-attachment 96-well plate (Corning) in MEGM containing B27 (Invitrogen), 20 ng/mL EGF, and 4 µg/mL heparin (Sigma), and incubated for 120 h. The HMLER-shEcad cells were also treated with **2–5**, L^2, L^3, and salinomycin (0–133 µM). Wells where HMLER-shEcad mammospheres were incubated with **2–5**, L^2, L^3, and salinomycin (at 0.5 µM and their respective IC_{20} values for 5 days) were manually counted and imaged using an inverted microscope. The resazurin-based dye, TOX8 (Sigma) was used to determine mammosphere viability. Upon incubation for 16 h, the fluorescence of the solutions in each well was read at 590 nm (λ_{ex} = 560 nm). The fluorescence values corresponding to each well were normalised to DMSO-containing controls and plotted as concentration of test compound versus % mammospheres viability. IC_{50} values were calculated from the resulting dose dependent curves. The reported IC_{50} values are the average of two independent experiments, each consisting of two replicates per concentration level.

3.13. Cellular Uptake

To determine the internalisation of **2–5** by HMLER-shEcad cells, about one million HMLER-shEcad cells were seeded in separate 60 mm Petri dishes overnight. The HMLER-shEcad cells were then dosed with **2–5** (at 5 µM) and incubated for 24 h. After the incubation period, the cells were harvested using standard procedures and the number of cells was counted using a haemocytometer. The resultant cellular pellets were dissolved in 65% HNO_3 (250 µL) overnight. The solutions were then appropriately diluted using UltraPure water and analysed using inductively coupled plasma mass spectrometry (ICP-MS, ThermoScientific ICAP-Qc quadrupole ICP mass spectrometer). The copper levels found in the cellular pellets are given as Cu (ng) per million cells. The data presented corresponds to the mean of four determinations for each data point.

3.14. Intracellular ROS Assay

HMLER-shEcad cells (5×10^3) were seeded in each well of a 96-well plate. After incubating the cells overnight, they were treated with **2** and **3** ($2 \times IC_{50}$ value for 0.5–24 h) and incubated with 6-carboxy-2′,7′-dichlorodihydrofluorescein diacetate (20 µM) for 90 min. The intracellular ROS level was determined by measuring the fluorescence of the solutions in each well at 529 nm (λ_{ex} = 504 nm).

4. Conclusions

In summary, we report the preparation, spectroscopic and analytical characterisation, and in vitro anti-breast CSC properties of two copper(II) complexes (**2** and **3**) bearing a naphthol Schiff base ligand (**L^2** or **L^3**) and 1,10-phenanthroline. According to the X-ray crystal structures, the cationic components of **2** and **3** exhibit five-coordinate distorted trigonal bipyramid geometries. Notably, **2** and **3** simultaneously kill bulk breast cancer and breast CSCs in the micromolar range. As **2** and **3** are equipotent towards bulk breast cancer cells and breast CSCs, they have the potential to remove heterogenous breast tumour populations (containing bulk breast cancer cells and breast CSCs) with a single (micromolar) dose. The 1,10-phenanthroline moiety is a contributing factor in the breast CSC potency of **2** and **3**, and this is shown by the fact that the corresponding chloride-containing complexes without 1,10-phenanthroline **4** and **5** are 3.7–3.8-fold (monolayer) and 8.6–15.5-fold (mammosphere) less toxic towards breast CSCs than **2** and **3**. Remarkably, **2** and **3** reduced breast CSC mammosphere viability to a greater extent (up to 4.5-fold lower IC_{50} value) than salinomycin, an established anti-breast CSC agent. Cellular uptake studies show that **2** and **3** were internalised by breast CSCs more readily than the corresponding chloride-containing complexes **4** and **5**, which could explain their differences in breast CSC potency. Furthermore, **2** and **3** are able to generate significant levels of intracellular ROS (75–93% increase relative to untreated control breast CSCs) after short exposure times (0.5–1 h), and this may be the mechanism by which they induce breast CSC death. Overall our studies highlight the promising anti-CSC properties of redox-modulating copper(II) complexes containing Schiff base and 1,10-phenanthroline ligands. These findings support the continued development of this class of redox-active metal complexes as potential CSC-directed chemotherapeutics. Future studies will focus on understanding the full cytotoxic and immunological mechanism of action of this family of copper(II) complexes and determining their translatable scope.

Supplementary Materials: The following are available online at https://www.mdpi.com/2304-6740/9/1/5/s1, Figures S1–S37, Tables S1–S4, and the CIF and corresponding checkCIF output files.

Author Contributions: K.S. (Kogularamanan Suntharalingam) and J.N.-S. conceived and designed the experiments; J.N.-S., A.J., K.S. (Kuldip Singh), F.O. and K.S. (Kogularamanan Suntharalingam) performed the experiments; J.N.-S., A.J., K.S. (Kuldip Singh), F.O. and K.S. (Kogularamanan Suntharalingam) analyzed the data; K.S. (Kogularamanan Suntharalingam), and J.N.-S. wrote the manuscript; J.N.-S., A.J., K.S. (Kuldip Singh), F.O. and K.S. (Kogularamanan Suntharalingam) reviewed and edited the manuscript. All authors have read and agreed to the published version of the manuscript.

Funding: K.S. is supported by an EPSRC New Investigator Award (EP/S005544/1).

Institutional Review Board Statement: Not applicable.

Informed Consent Statement: Not applicable.

Data Availability Statement: Not applicable.

Acknowledgments: We are grateful to Robert Weinberg for providing the HMLER and HMLER-shEcad cell lines used in this study.

Conflicts of Interest: The authors declare no conflict of interest.

References

1. Still 1 in 4 Deaths Caused by Cancer in the EU. Available online: https://ec.europa.eu/eurostat/documents/2995521/7149996/3-03022016-BP-EN.pdf/0bbc3389-8c0d-44a0-9b0c-2a0bff49f466 (accessed on 3 February 2016).
2. Luengo-Fernandez, R.; Leal, J.; Gray, A.; Sullivan, R. Economic burden of cancer across the European Union: A population-based cost analysis. *Lancet Oncol.* **2013**, *14*, 1165–1174. [CrossRef]
3. Al-Hajj, M.; Wicha, M.S.; Benito-Hernandez, A.; Morrison, S.J.; Clarke, M.F. Prospective identification of tumorigenic breast cancer cells. *Proc. Natl. Acad. Sci. USA* **2003**, *100*, 3983–3988. [CrossRef] [PubMed]
4. Gupta, P.B.; Chaffer, C.L.; Weinberg, R.A. Cancer stem cells: Mirage or reality? *Nat. Med.* **2009**, *15*, 1010–1012. [CrossRef] [PubMed]
5. Balic, M.; Lin, H.; Young, L.; Hawes, D.; Giuliano, A.; McNamara, G.; Datar, R.H.; Cote, R.J. Most early disseminated cancer cells detected in bone marrow of breast cancer patients have a putative breast cancer stem cell phenotype. *Clin. Cancer Res.* **2006**, *12*, 5615–5621. [CrossRef] [PubMed]
6. Owens, T.W.; Naylor, M.J. Breast cancer stem cells. *Front. Physiol.* **2013**, *4*, 225. [CrossRef] [PubMed]
7. Calcagno, A.M.; Salcido, C.D.; Gillet, J.P.; Wu, C.P.; Fostel, J.M.; Mumau, M.D.; Gottesman, M.M.; Varticovski, L.; Ambudkar, S.V. Prolonged drug selection of breast cancer cells and enrichment of cancer stem cell characteristics. *J. Natl. Cancer Inst.* **2010**, *102*, 1637–1652. [CrossRef] [PubMed]
8. Phillips, T.M.; McBride, W.H.; Pajonk, F. The response of CD24$^{-/low}$/CD44$^+$ breast cancer-initiating cells to radiation. *J. Natl. Cancer Inst.* **2006**, *98*, 1777–1785. [CrossRef]
9. Laws, K.; Suntharalingam, K. The Next Generation of Anticancer Metallopharmaceuticals: Cancer Stem Cell-Active Inorganics. *ChemBioChem* **2018**, *19*, 2246–2253. [CrossRef]
10. Boodram, J.N.; McGregor, I.J.; Bruno, P.M.; Cressey, P.B.; Hemann, M.T.; Suntharalingam, K. Breast Cancer Stem Cell Potent Copper(II)-Non-Steroidal Anti-Inflammatory Drug Complexes. *Angew. Chem. Int. Ed.* **2016**, *55*, 2845–2850. [CrossRef]
11. Eskandari, A.; Boodram, J.N.; Cressey, P.B.; Lu, C.; Bruno, P.M.; Hemann, M.T.; Suntharalingam, K. The breast cancer stem cell potency of copper(II) complexes bearing nonsteroidal anti-inflammatory drugs and their encapsulation using polymeric nanoparticles. *Dalton Trans.* **2016**, *45*, 17867–17873. [CrossRef]
12. Lu, C.; Eskandari, A.; Cressey, P.B.; Suntharalingam, K. Cancer Stem Cell and Bulk Cancer Cell Active Copper(II) Complexes with Vanillin Schiff Base Derivatives and Naproxen. *Chem. Eur. J.* **2017**, *23*, 11366–11374. [CrossRef] [PubMed]
13. Lu, C.; Laws, K.; Eskandari, A.; Suntharalingam, K. A reactive oxygen species-generating, cyclooxygenase-2 inhibiting, cancer stem cell-potent tetranuclear copper(II) cluster. *Dalton Trans.* **2017**, *46*, 12785–12789. [CrossRef] [PubMed]
14. Marzano, C.; Pellei, M.; Tisato, F.; Santini, C. Copper complexes as anticancer agents. *Anticancer Agents Med. Chem.* **2009**, *9*, 185–211. [CrossRef] [PubMed]
15. Santini, C.; Pellei, M.; Gandin, V.; Porchia, M.; Tisato, F.; Marzano, C. Advances in Copper Complexes as Anticancer Agents. *Chem. Rev.* **2014**, *114*, 815–862. [CrossRef]
16. Galindo-Murillo, R.; Garcia-Ramos, J.C.; Ruiz-Azuara, L.; Cheatham, T.E., 3rd; Cortes-Guzman, F. Intercalation processes of copper complexes in DNA. *Nucleic Acids Res.* **2015**, *43*, 5364–5376. [CrossRef]
17. Ruiz-Azuara, L.; Bravo-Gomez, M.E. Copper compounds in cancer chemotherapy. *Curr. Med. Chem.* **2010**, *17*, 3606–3615. [CrossRef]
18. Vertiz, G.; Garcia-Ortuno, L.E.; Bernal, J.P.; Bravo-Gomez, M.E.; Lounejeva, E.; Huerta, A.; Ruiz-Azuara, L. Pharmacokinetics and hematotoxicity of a novel copper-based anticancer agent: Casiopeina III-Ea, after a single intravenous dose in rats. *Fundam. Clin. Pharmacol.* **2014**, *28*, 78–87. [CrossRef]
19. Wehbe, M.; Leung, A.W.Y.; Abrams, M.J.; Orvig, C.; Bally, M.B. A Perspective—Can copper complexes be developed as a novel class of therapeutics? *Dalton Trans.* **2017**, *46*, 10758–10773. [CrossRef]
20. Silva-Platas, C.; Villegas, C.A.; Oropeza-Almazan, Y.; Carranca, M.; Torres-Quintanilla, A.; Lozano, O.; Valero-Elizondo, J.; Castillo, E.C.; Bernal-Ramirez, J.; Fernandez-Sada, E.; et al. Ex Vivo Cardiotoxicity of Antineoplastic Casiopeinas Is Mediated through Energetic Dysfunction and Triggered Mitochondrial-Dependent Apoptosis. *Oxid. Med. Cell Longev.* **2018**, *2018*, 8949450. [CrossRef]
21. Diehn, M.; Cho, R.W.; Lobo, N.A.; Kalisky, T.; Dorie, M.J.; Kulp, A.N.; Qian, D.; Lam, J.S.; Ailles, L.E.; Wong, M.; et al. Association of reactive oxygen species levels and radioresistance in cancer stem cells. *Nature* **2009**, *458*, 780–783. [CrossRef]
22. Shi, X.; Zhang, Y.; Zheng, J.; Pan, J. Reactive oxygen species in cancer stem cells. *Antioxid. Redox Signal.* **2012**, *16*, 1215–1228. [CrossRef]
23. Kaur, P.; Johnson, A.; Northcote-Smith, J.; Lu, C.; Suntharalingam, K. Immunogenic Cell Death of Breast Cancer Stem Cells Induced by an Endoplasmic Reticulum-Targeting Copper(II) Complex. *ChemBioChem* **2020**, *21*, 3618–3624. [CrossRef] [PubMed]
24. Dhar, S.; Chakravarty, A.R. Efficient visible light induced nuclease activity of a ternary mono-1,10-phenanthroline copper(II) complex containing 2-(methylthio)ethylsalicylaldimine. *Inorg. Chem.* **2003**, *42*, 2483–2485. [CrossRef] [PubMed]
25. Reddy, P.A.; Santra, B.K.; Nethaji, M.; Chakravarty, A.R. Metal-assisted light-induced DNA cleavage activity of 2-(methylthio)phenylsalicylaldimine Schiff base copper(II) complexes having planar heterocyclic bases. *J. Inorg. Biochem.* **2004**, *98*, 377–386. [CrossRef] [PubMed]
26. Waring, M.J. Lipophilicity in drug discovery. *Expert Opin. Drug Discov.* **2010**, *5*, 235–248. [CrossRef] [PubMed]

27. Eskandari, A.; Suntharalingam, K. A reactive oxygen species-generating, cancer stem cell-potent manganese(II) complex and its encapsulation into polymeric nanoparticles. *Chem. Sci.* **2019**, *10*, 7792–7800. [CrossRef] [PubMed]
28. Dontu, G.; Abdallah, W.M.; Foley, J.M.; Jackson, K.W.; Clarke, M.F.; Kawamura, M.J.; Wicha, M.S. In vitro propagation and transcriptional profiling of human mammary stem/progenitor cells. *Genes Dev.* **2003**, *17*. [CrossRef]
29. Sheldrick, G. *SADABS: Program for Absorption Correction Using Area Detector Data*; University of Göttingen: Göttingen, Germany, 1996.
30. Sheldrick, G. A short history of SHELX. *Acta Cryst. Sect. A* **2008**, *64*, 112–122. [CrossRef] [PubMed]
31. Sheldrick, G.M. Crystal structure refinement with SHELXL. *Acta Crystallogr. Sect. C* **2015**, *71*, 3–8. [CrossRef]
32. Dolomanov, O.V.; Bourhis, L.J.; Gildea, R.J.; Howard, J.A.K.; Puschmann, H. OLEX2: A complete structure solution, refinement and analysis program. *J. Appl. Crystallogr.* **2009**, *42*, 339–341. [CrossRef]
33. Farrugia, L.J. WinGX and ORTEP for Windows: An update. *J. Appl. Cryst.* **2012**, *45*, 849–854. [CrossRef]

Article

Aqueous Solution Equilibria and Spectral Features of Copper Complexes with Tripeptides Containing Glycine or Sarcosine and Leucine or Phenylalanine

Giselle M. Vicatos [1], Ahmed N. Hammouda [1,2], Radwan Alnajjar [1,2], Raffaele P. Bonomo [3], Gabriele Valora [3], Susan A. Bourne [1] and Graham E. Jackson [1,*]

[1] Department of Chemistry, University of Cape Town, Cape Town 7701, South Africa; VCTGIS001@myuct.ac.za (G.M.V.); ahmed.hammouda@uct.ac.za (A.N.H.); ALNRAD001@myuct.ac.za (R.A.); susan.bourne@uct.ac.za (S.A.B.)
[2] Department of Chemistry, Faculty of Science, University of Benghazi, Benghazi 16063, Libya
[3] Dipartimento di Scienze Chimiche, Università degli Studi di Catania, 95131 Catania, Italy; rbonomo@unict.it (R.P.B.); gabriele.valora@unict.it (G.V.)
* Correspondence: graham.jackson@uct.ac.za; Tel.: +27-216-502-531

Abstract: Copper(II) complexes of glycyl-L-leucyl-L-histidine (GLH), sarcosyl-L-leucyl-L-histidine (Sar-LH), glycyl-L-phenylalanyl-L-histidine (GFH) and sarcosyl-L-phenylalanyl-L-histidine (Sar-FH) have potential anti-inflammatory activity, which can help to alleviate the symptoms associated with rheumatoid arthritis (RA). From pH 2–11, the MLH, ML, MLH$_{-1}$ and MLH$_{-2}$ species formed. The combination of species for each ligand was different, except at the physiological pH, where CuLH$_{-2}$ predominated for all ligands. The prevalence of this species was supported by EPR, ultraviolet-visible spectrophotometry, and mass spectrometry, which suggested a square planar CuN$_4$ coordination. All ligands have the same basicity for the amine and imidazole-N, but the methyl group of sarcosine decreased the stability of MLH and MLH$_{-2}$ by 0.1–0.34 and 0.46–0.48 log units, respectively. Phenylalanine increased the stability of MLH and MLH$_{-2}$ by 0.05–0.29 and 1.19–1.21 log units, respectively. For all ligands, ^{1}H NMR identified two coordination modes for MLH, where copper(II) coordinates via the amine-N and neighboring carbonyl-O, as well as via the imidazole-N and carboxyl-O. EPR spectroscopy identified the MLH, ML and MLH$_{-2}$ species for Cu-Sar-LH and suggested a CuN$_2$O$_2$ chromophore for ML. DFT calculations with water as a solvent confirmed the proposed coordination modes of each species at the B3LYP level combined with 6-31++G**.

Keywords: copper; speciation; equilibria; thermodynamic stability; peptides

1. Introduction

Rheumatoid arthritis (RA) is a chronic, inflammatory, systemic and debilitating disease characterized by the destruction of diarthrodial joints. It is considered to be an autoimmune disease, and the exact cause is still unknown [1–3]. Often the symptoms are treated with steroidal and non-steroidal anti-inflammatory drugs, and sometimes surgical intervention is necessary in cases of severe joint deformation [4,5]. It is well known that copper(II) complexes have anti-inflammatory activity and cause a reduction in inflammation in a dose-dependent manner when administered subcutaneously [6–10]. Copper(II) exists in the plasma by binding non-reversibly to ceruloplasmin, reversibly to serum albumin, as well as by being distributed among low molecular mass ligands. These low molecular mass ligands have anti-inflammatory activity and are thought to be involved in transportation between cells in the body [11–13]. Therefore, in principle, an increase in the fraction of copper(II) complexes with low molecular mass ligands would lead to augmenting their anti-inflammatory activity when administered either orally or by subcutaneous injection [14].

Sorenson [15] and Jackson et al. [5,16] have investigated copper(II) complexes with low molecular mass ligands and have shown that they are effective in reducing the inflammation associated with RA and have reduced toxicity. This suggests that the anti-inflammatory effect of endogenous copper(II) can be enhanced by exogenous sources. The preferred route of administration is dermal absorption as it is both harmless and convenient for patients.

In normal plasma, serum albumin is a major reversible, metal binding protein, accounting for 40 μg of copper(II) per ml of plasma [17,18]. The copper(II) binds to the amine, the amide and the imidazole of Asp-Ala-His in the C-terminus of the protein [19,20]. Using this sequence as a design template, Zvimba [5,18] and Odisitse [1,6,21,22] synthesized and tested a series of amide ligands. Zvimba [18] found that replacing an amine with an amide significantly reduced the in vivo copper(II) mobilizing ability of the ligand, but increased the lipophilicity of the complex. On the other hand, in murine biodistribution studies, Odisitse [22] found significant trans-dermal absorption and retention of copper(II) using amide ligands with a terminal pyridine.

Since peptides are a readily available source of diverse amides, in recent years research has centered around developing tripeptide ligands that would form a complex with copper(II) and studying their anti-inflammatory activity. For example, Grunchlik et al. [23] studied the role that the two tripeptides, Gly-His-Lys and Gly-Gly-His, had on skin inflammation, and concluded that copper(II) peptides could be used on skin as an alternative to corticosteroids or nonsteroidal anti-inflammatory drugs. Hostynek et al. [24] studied the tripeptide glycyl-L-histidyl-L-lysine cuprate diacetate and found that, through trans-dermal absorption, a potentially effective therapeutic amount of copper(II) complexed to the tripeptide was delivered to treat the inflammatory disease. Elmagbari [25] studied a series of tripeptides and found that the copper(II) complexes were hydrophilic. Specifically, the results from Vicatos [26] and Hammouda [27] were used to base the design of the tripeptides for this study.

Vicatos [26] studied two tripeptides, sarcosyl-L-leucyl-phenylalanine (Sar-Leu-Phe) and glycyl-L-leucyl-phenylalanine (Gly-Leu-Phe), and found that they were poor at mobilizing copper(II) in vivo, as they preferred to bind to zinc(II). Hammouda [27] studied a series of tripeptides, finding that the copper(II) mobilizing capacities were higher when histidine was in the third position, as opposed to having histidine in the second position, or if histidine is not present in the tripeptide. However, partition coefficient and membrane permeability studies showed that the copper(II) complexes of Vicatos' peptides were more lipophilic and were 2.1–8.8 times more permeable than the ligand complexes that Hammouda reported [27]. From the two studies of Hammouda [27] and Vicatos [26], it was concluded that an imidazole group in the third position should be included in the ligand design to ensure that the ligands would be selective for copper(II). The permeability coefficient measurements of Vicatos [26] suggest that the amino acid in position 2 of the tripeptide should be non-polar leucine or phenylalanine, and the first position of the ligand can either be glycine or sarcosine. N-methylated ligands have been shown to increase the lipophilicity and biological half-life of complexes compared to non-N-methylated ligands [28,29]. Thus, sarcosine was included in this study. The resultant ligands are therefore GLH, Sar-LH, GFH and Sar-FH.

2. Results and Discussions

2.1. Potentiometry

In an aqueous solution, GLH, Sar-LH, GFH and Sar-FH are zwitterions, which have three available sites for protonation. Glycine and sarcosine have a primary and secondary amine, respectively, while histidine has both a carboxyl group and an imidazole nitrogen. With decreasing pH, the amine group ($pK_a \approx 9.8$) will become protonated first, then the imidazole ring ($pK_a \approx 6.0$), and, finally, the carboxyl group ($pK_a \approx 1.8$). Their protonation constants ($\log \beta_{pqr}$) are reported in Table 1.

Table 1. Protonation and stability constants of GLH, Sar-LH, GFH and Sar-FH with copper(II). $\beta_{pqr} = [M_pL_qH_r]/[M]^p[L]^q[H]^r$, I = in 0.15 mol.dm^{-3} (NaCl), T = 25 °C.

Ligand	p q r	log β_{pqr}	Complex	p q r	log β_{pqr}
GLH	0 1 1 0 1 2 0 1 3	8.21 15.10 17.88	Cu-GLH	1 1 1 1 1 −1 1 1 −2	12.71 2.77 −2.24
Sar-LH	0 1 1 0 1 2 0 1 3	8.45 15.32 18.05	Cu-Sar-LH	1 1 1 1 1 0 1 1 −2	12.37 7.38 −2.70
GFH	0 1 1 0 1 2 0 1 3	7.95 14.82 17.65	Cu-GFH	1 1 1 1 1 −2	12.76 −1.03
Sar-FH	0 1 1 0 1 2 0 1 3	8.22 15.09 17.96	Cu-Sar-FH	1 1 1 1 1 −2	12.66 −1.51

With the introduction of copper(II), these four ligands formed complexes, but each ligand formed a different set of species over a pH range from 2–11. The stability constants for the different species are seen in Table 1, and their distribution diagrams are seen in Figure 1. For these copper(II) complexes to satisfy the aim of undergoing transdermal absorption and releasing copper(II) ions into the blood plasma, it is essential to analyze their stability. It is noted that the complexes must be stable enough to form, but not so stable that copper(II) cannot be released once it is in the blood plasma.

In particular, the stability between the N-methylated group on the complexes with sarcosine versus the non-N-methylated group on the complexes with glycine was analyzed. To do this, constants from GLH were compared with constants from Sar-LH and constants from GFH were compared with constants from Sar-FH. For the protonation constants, the species from Sar-LH or Sar-FH, i.e., with the methyl group, are 0.17–0.31 log units bigger and more stable than their glycine counterparts. For the copper(II) complexation stability constants, the MLH and MLH$_{-2}$ species without the methyl group (GLH and GFH) increased the stability by 0.1–0.34 log units and 0.46–0.48 log units, respectively, and therefore are more stable than the methylated species (Sar-LH and Sar-FH). The stability constants of the ML and MLH$_{-1}$ species could not be compared.

The methyl group has an electron-donating inductive effect, and so it was expected that the ligands/complexes with sarcosine would have larger stability coefficients and thus be more stable than the complexes with glycine. This was seen for the protonation constants, and therefore it is confirmed that the methyl group does affect the stability. However, it was mostly found that the complexes without the methyl group were slightly more stable, which leads to the suggestion that the methyl group also has steric effects. A steric effect makes the inductive effect less prominent. Another reason could be due to the ammonium ions or charged amine groups preferentially forming hydrogen bonds with water and thereby decreasing the available charge. This phenomenon was also reported for solvated alkylamines, where the order of base strength was rearranged. The true base strength in a vacuum should have been $NH_3 < RNH_2 < R_2NH < R_3N$, since the methyl groups increase, which subsequently increases the electron density on the nitrogen atom. However, the order for aqueous solutions was found to be $NH_3 < RNH_2, R_2NH > R_3N$, which was explained by the preference of the amine to form hydrogen bonds with water [30].

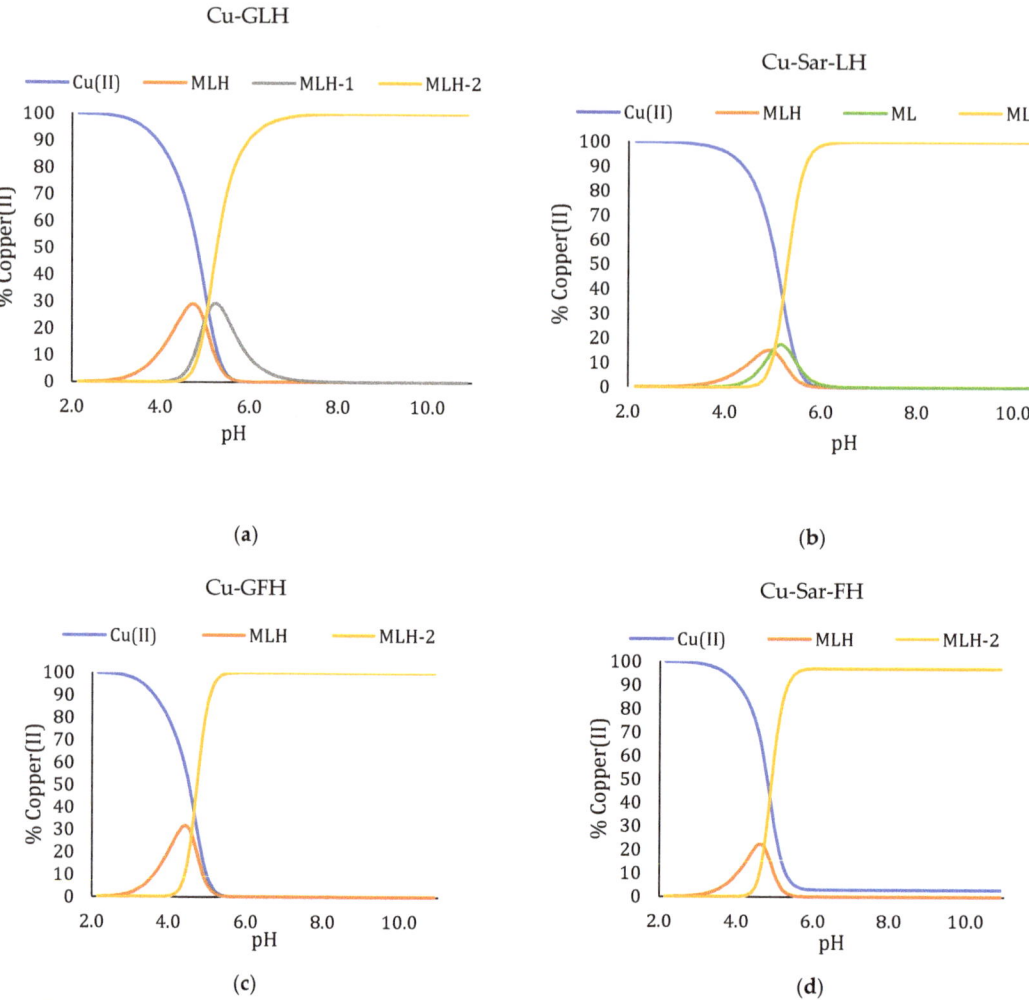

Figure 1. Protonation species distribution curve for copper(II) and (**a**) GLH (1:1 ratio), (**b**) Sar-LH (1:1 ratio), (**c**) GFH (1:1 ratio) and (**d**) Sar-FH (1:1 ratio) at 25 °C in 0.15 mol dm^{-3} (NaCl).

An analysis of the leucine and phenylalanine amino acids can also be conducted to determine how they affect the stability constants. Thus, GLH and GFH are compared, and Sar-LH and Sar-FH are compared. Starting with the protonation constants, the species with leucine (GLH and Sar-LH) are 0.09–0.28 log units bigger and more stable than the species with phenylalanine (GFH and Sar-FH). For the copper(II) complexes, the MLH and MLH$_{-2}$ species with phenylalanine (GFH and Sar-FH) increased the stability by 0.05–0.29 log units and 1.19–1.21 log units, respectively. The stability constants for the ML and MLH$_{-1}$ species could not be compared.

The isobutyl group of leucine is electron donating, which should increase the basicity of the ligand. On the other hand, the benzyl group of phenylalanine is electron withdrawing, which should decrease the basicity of the ligand. This was seen in the protonation constants. However, when analyzing the metal stability constants, a similar scenario is seen as with the comparison between the metal complexes of glycine and sarcosine, where steric or entropy effects influence the stability of the metal complexes.

An indication of how the MLH species coordinates was found by comparing the stability constants of this species with the stability constants of other ligands found in the literature. The MLH species most likely has the copper(II) either coordinated to the amine and neighboring carbonyl-O, or to the imidazole-N and the carboxyl-O. For the former comparison, the literature compounds glycylglycine (GGOMe) and glycylsarcosine (GSOMe) were selected. They coordinate to copper(II) through the amine-N and carbonyl-O. Their log K values are 4.11 and 5.18, respectively [31]. To compare these values with the MLH species of Cu-GLH, Cu-Sar-LH, Cu-GFH and Cu-Sar-FH, the corresponding protonation of the imidazole group had to be subtracted from the complex stability constant. This gives log K values of 5.82, 5.5, 5.89 and 5.71, respectively. These values correspond closely to the literature values above, suggesting that this is the mode of coordination of these ligands. Unfortunately, the comparison could not be made for the imidazole-N and carboxyl-O, since literature for this comparison could not be found. However, after an ^{1}H NMR analysis (Section 3.4), there is reason to believe that copper(II) also coordinates to the imidazole-N and to the carboxyl-O. This "double-sided" coordination was also proposed for the MLH species of copper(II) complexes containing peptides with a histidyl residue [32]. Figure 2 shows the two coordination modes of MLH for GLH, which can also be transferred to Sar-LH, GFH and Sar-FH.

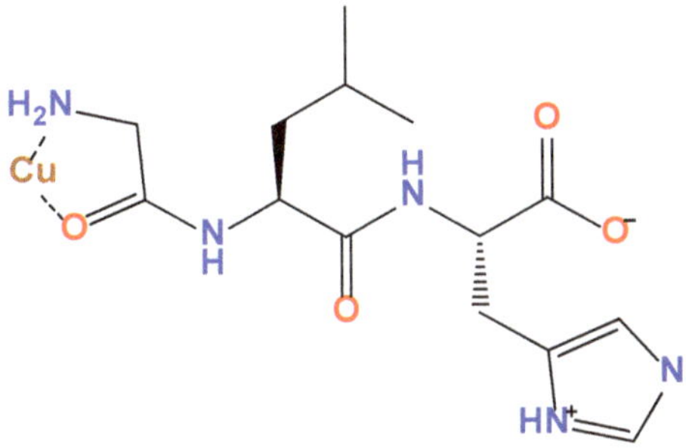

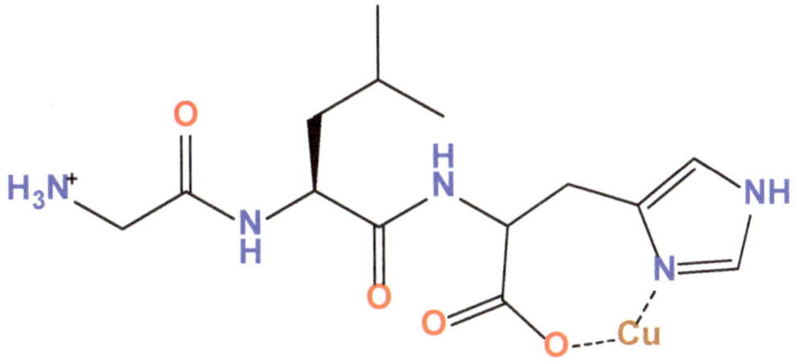

Figure 2. The two coordination modes (**a**) with an amine-N and carbonyl-O coordination and (**b**) with an imidazole-N and carboxyl-O coordination for the MLH species of GLH.

The ML species of Cu-Sar-LH occurs at a higher pH than MLH, and so it is most likely that copper(II) only coordinates via the amine route and not via the imidazole ring, since the charge of the carboxylic acid moiety does not need to be neutralized at high pH values. Therefore, a possible coordination for the ML species is to the amine and neighboring amide-N with a protonated imidazole-N. The most likely mechanism in going from an MLH to ML species is for the sarcosine amide to switch from a carbonyl-O to an amide-N coordination, while the imidazole-N remains protonated. The transition from MLH to ML gives a pK_a value of 4.99, which is a value typical of a metal-assisted amide deprotonation. The proposed switch has been seen in the ML to MLH$_{-1}$ species of Cu-di, tri and tetraglycine [33,34]. These ligands have pK_a values of 4.23, 5.41 and 5.56, respectively, and represent the first amide deprotonation. Additionally, a Cu-GGG complex [34] showed that the neighboring carbonyl-O could also be involved in the coordination. Therefore, the two possible coordination modes for the ML species (Figure 3) could be a bidentate coordination consisting of the amine and neighboring amide-N with a protonated imidazole-N, or a tridentate coordination consisting of the amine, neighboring amide-N and the carbonyl-O, with a protonated imidazole-N.

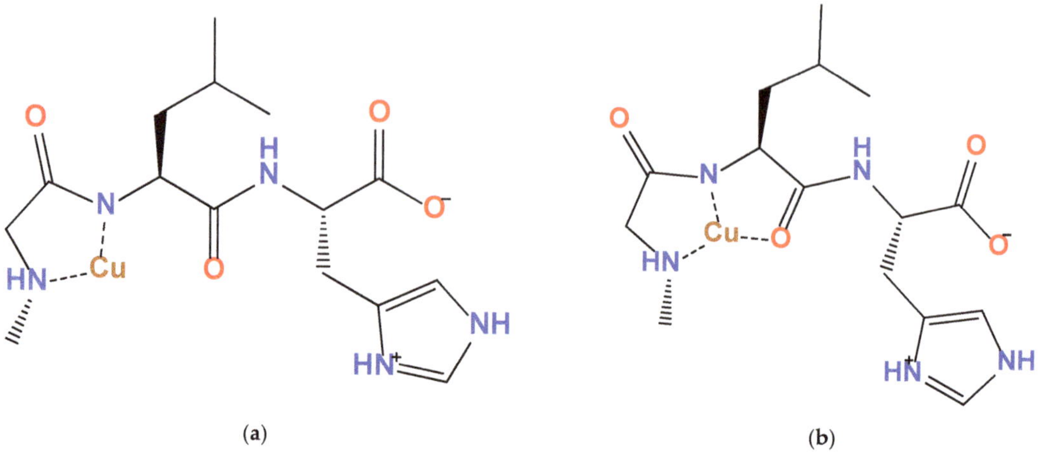

Figure 3. The two coordination modes (**a**) with an amine-N and neighboring amide-N coordination and (**b**) with an amine-N, neighboring amide-N and carbonyl-O coordination for the ML species of Sar-LH.

The coordination mode of the MLH$_{-2}$ species cannot be determined using pK_a values, but EPR and UV-vis have proposed the coordination to the amine-N, both amide-Ns and to the imidazole-N. This 4N-complex has been seen in other copper(II) peptide complexes [33,35]. The pK_a value of the ML to MLH$_{-2}$ species for Cu-Sar-LH is not attainable, but it can be speculated that this transition happens in one step where another coordination switch occurs between the carbonyl-O of leucine and its neighboring amide-N, as well as the coordination to the imidazole-N. Likewise, the pK_a values of the MLH to MLH$_{-2}$ species for Cu-GFH and Cu-Sar-FH are not attainable, but the coordination to two amide-Ns and to the imidazole-N probably occur in one step.

Similar to the ML species of Cu-Sar-LH, for the MLH$_{-1}$ species of Cu-GLH, copper(II) is also thought to coordinate only via the amine route. The pK_a value for the MLH$_{-1}$ to MLH$_{-2}$ transition is 5.01 and was originally thought to suggest a second amide deprotonation. This would produce an MLH$_{-1}$ coordination mode where copper(II) is coordinated to the amine, neighboring amide-N and imidazole-N. Transitioning to the MLH$_{-2}$ species would then just include the coordination to the second amide-N. However, since the value of 5.01 represents the second amide deprotonation, when comparing this value with Cu-GGG (pK_a = 6.86) [33,34], it is not close enough to confirm a second amide deprotonation. Two other MLH$_{-1}$ coordination possibilities are firstly to the amine and two amide-Ns with a protonated imidazole-N, or secondly to the amine, two amide-Ns and the carboxyl-O with a protonated imidazole-N. Transitioning to the MLH$_{-2}$ species would then include a coordination to the imidazole-N. However, the protonation of the imidazole-N has a log K of 6.89, which is also not close enough to the 5.01 pK_a value to confirm a coordination mode. All three coordination modes are reasonable and proposed as possible structures (Figure 4).

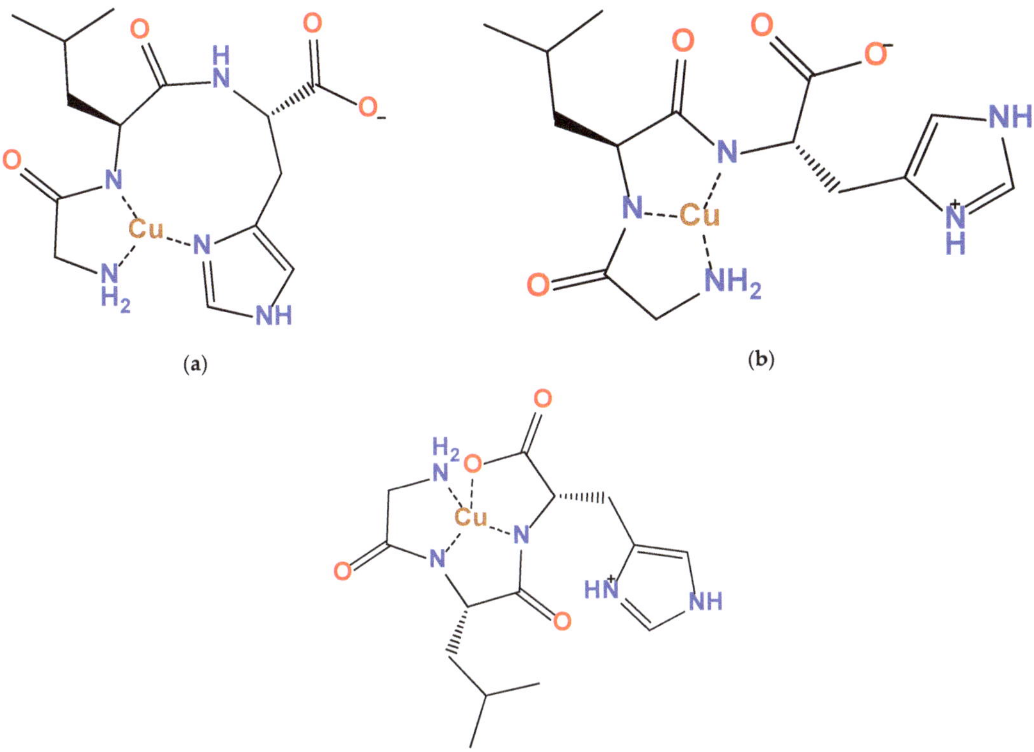

Figure 4. The three coordination modes (**a**) with an amine-N, neighboring amide-N and imidazole-N coordination, (**b**) with an amine-N and two amide-Ns coordination and (**c**) with an amine-N, two amide-Ns and carboxyl-O coordination for the MLH$_{-1}$ species of GLH.

2.2. Ultraviolet-Visible Spectrophotometry (UV-Vis)

Color changes occurred during the potentiometric titrations for the copper(II) complexes. As the pH of the copper(II) solution increased, its color changed from a clear to a violet-pink color. This color change showed the formation of species over the pH range 2–11 and aided in the identification of the species coordination modes [6]. The spectra of the four ligands (Figure 5) appear to be similar, since one absorption band with similar λ_{max} values (Table 2) can be seen in all spectra.

Table 2. The maximum wavelengths and their corresponding molar extinction coefficients of the MLH$_{-2}$ species from Cu-GLH, Cu-Sar-LH, Cu-GFH and Cu-Sar-FH.

Complex	λ_{max} (nm)	ε (dm^3 mol^{-1} cm^{-1})
Cu-GLH	518	88
Cu-Sar-LH	523	105
Cu-GFH	517	85
Cu-Sar-FH	521	98

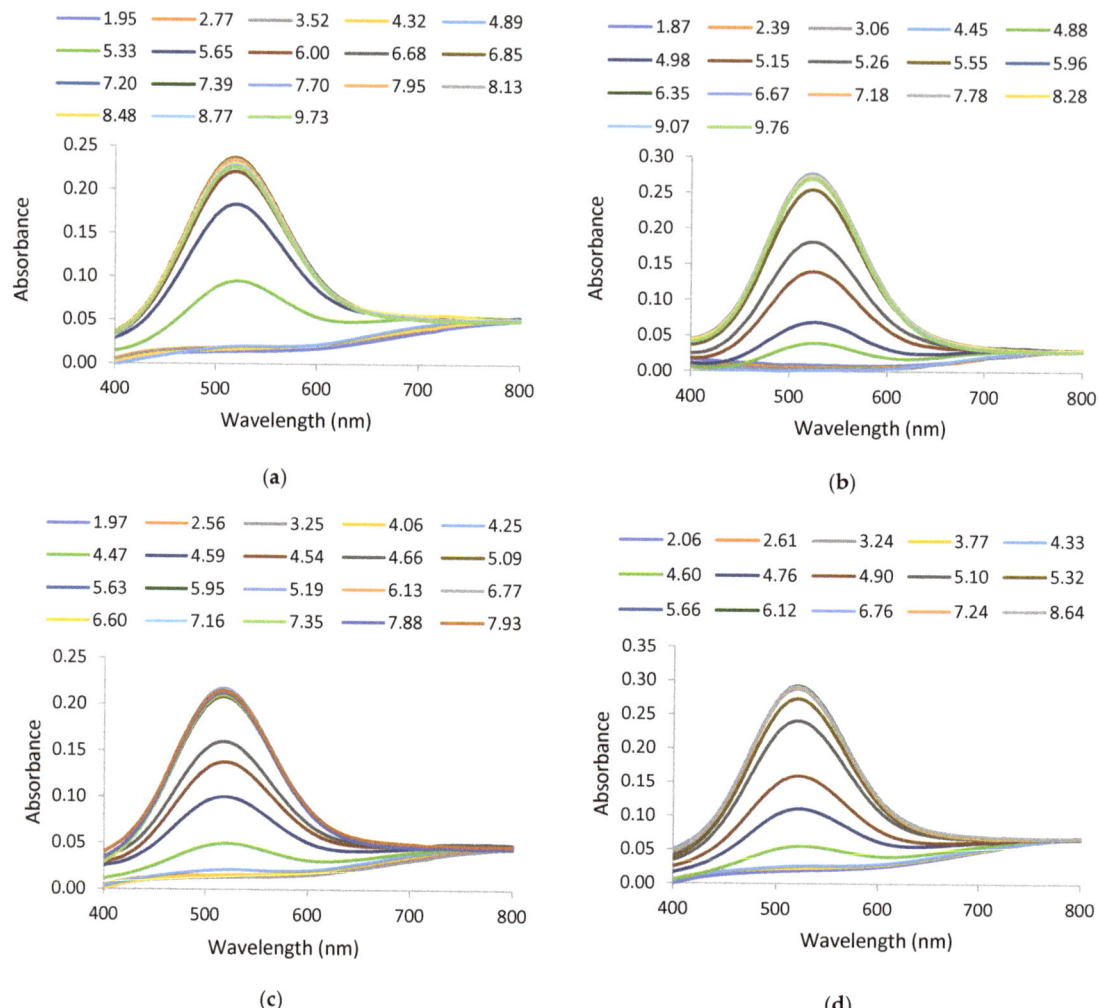

Figure 5. Electronic spectra of solutions containing (**a**) Cu-GLH (3.04 × 10^{-3} M of GLH and 2.69 × 10^{-3} M of copper(II)), (**b**) Cu-Sar-LH (2.93 × 10^{-3} M of Sar-LH and 2.58 × 10^{-3} M of copper(II)), (**c**) Cu-GFH (3.09 × 10^{-3} M of GFH and 2.55 × 10^{-3} M of copper(II)) and (**d**) Cu-Sar-FH (3.66 × 10^{-3} M of Sar-FH and 2.97 × 10^{-3} M of copper(II)).

Copper(II) ions have a d^9 configuration, which gives rise to Jahn–Teller distortion. This causes a lack of symmetry in an octahedrally coordinated copper(II) complex, which allows electron d–d transitions to occur and form an absorption band [36]. The four copper(II) complexes are expected to form tetragonally distorted octahedral complexes, which is expected to produce only a single absorption band in the visible region. Three spin-allowed transitions would occur, which are $^2A_{1g} \leftarrow\ ^2B_{1g}$, $^2B_{2g} \leftarrow\ ^2B_{1g}$ and $^2E_g \leftarrow\ ^2B_{1g}$ [5,37], but because these bands are broad and cannot be distinguished, they appear as a single absorption band between 555 and 769 nm [37].

The absorption spectra in Figure 5 shows only one complex species (λ_{max} = 517–523 nm). For all the ligands, this complex species is the MLH$_{-2}$ species, since the λ_{max} peaks follow the MLH$_{-2}$ formation trend. None of the other supposed species can be identified. Molar extinction coefficients for MLH$_{-2}$ (Table 2) were calculated by taking the absorbance band produced at the highest pH and dividing it by the copper(II) concentration. Octahedral environments typically have extinction coefficients of about 10 dm^3 mol^{-1} cm^{-1} [36,38]. The higher molar extinction coefficients for the MLH$_{-2}$ species verify that the coordination of the complex is unsymmetrical. When using Sigel and Martin's [39,40] empirical equation and parameters, the equatorial coordination to the amine-N, two amide-Ns and an imidazole-N gives a calculated λ_{max} of 531 nm. This is close to the observed λ_{max} range of 517–523 nm [41,42] and is thus reasonable to suggest a square planar geometry with a CuN$_4$ coordination.

2.3. Electron Paramagnetic Resonance (EPR) Measurements

The four copper (II) complexes, Cu-GLH, Cu-Sar-LH, Cu-GFH and Cu-Sar-FH, were examined at pH 7, and following the species distribution diagrams, only one copper(II) species should be found at this pH. At first, the RT spectra were run in a more acidic pH to check the presence of prominent complex species in the system, which took into account that all the EPR spectra were run in the absence of added ionic strength.

The isotropic EPR spectrum is obtained at RT conditions, which is characterized by four lines (copper nuclear spin equal to 3/2, and then 2I + 1 lines). In contrast, an anisotropic EPR spectrum is obtained at LT conditions, which generally has an axial symmetry, characterized by two sets of four transitions occurring at different values of the magnetic field. These two sets are either parallel or perpendicular transitions. The parallel transitions of the LT EPR spectrum are well separated by the hyperfine coupling constant, whose values, together with the g parallel values, give information on the geometry of the copper(II) complex. By contrast, the perpendicular hyperfine coupling constant has a much lower numeric value than the parallel one, which is not resolved and presented as a single transition at higher magnetic fields. The two sets of transitions can overlap, and because the parallel lines are easily recognizable due to their low intensity, two or three of the parallel lines are generally visible.

The RT spectra of the four copper(II) complexes at pH 7 are reported in Figure 6. Looking at these spectra, it is evident that they all look very similar to each other. This can be a result of the MLH$_{-2}$ species, which is the predominant species in the pH range 4.5–11. Due to the presence of donor nitrogen atoms on the ligand at a higher magnetic field, it is possible to see that all the RT spectra contain the fourth line. This shows a superhyperfine (shf) structure that is coming from the delocalization of the spin density of the copper(II) free electron on the nitrogen atoms. Since the nitrogen nuclear spin is 1, from the multiplicity of the 2ΣI + 1 lines, it is possible to count the number of nitrogen atoms bound to copper(II) in the complex equatorial plane. Nine shf lines indicate that copper(II) is bound to four quasi-equivalent nitrogen atoms, and so the chromophore of this species can be considered as CuN$_4$.

The magnetic parameters from the LT EPR spectrum can be seen in Table 3. All the LT frozen EPR spectra show $g_{||} > g_{\perp} \geq 2.04$ [43]. This suggests that the copper(II) ground state could reasonably be assigned to the $d_{x^2-y^2}$ orbital in an octahedral, square planar or square-based pyramidal geometry. The low value of the $g_{||}$ and the relatively high absolute value of the parallel hyperfine constant suggest that these copper(II) complex species have a square planar geometry in a probable macrochelate complex. Since only subtle and negligible differences are observed among the copper(II) complexes with these ligands, they can be considered to have the same stereochemistry.

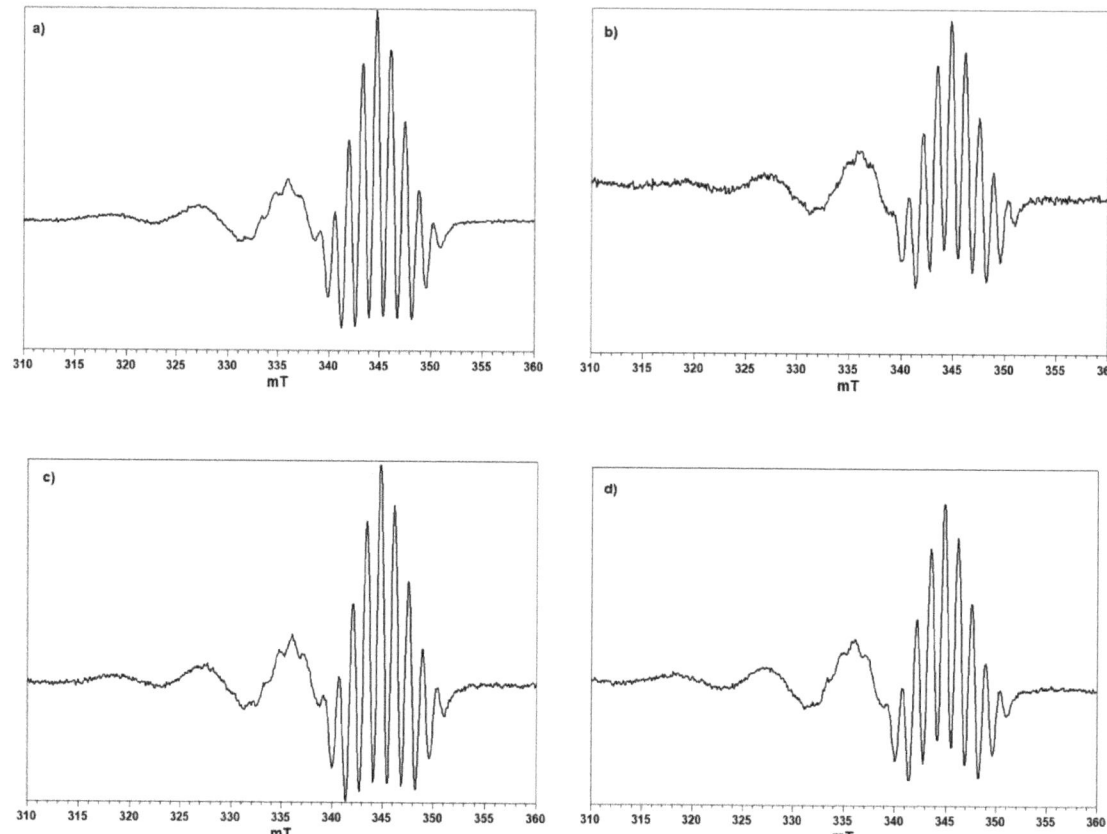

Figure 6. RT EPR 2nd derivative spectra recorded in aqueous solution for the copper(II) complexes, (**a**) Cu-GLH, (**b**) Cu-Sar-LH, (**c**) Cu-GFH and (**d**) Cu-Sar-FH.

Table 3. Spin Hamiltonian parameters of copper(II) complexes with Cu-GLH, Cu-Sar-LH, Cu-GFH and Cu-Sar-FH at pH 7.0, which have been drawn out from RT EPR spectra and LT frozen aqueous solution EPR spectra. All the hyperfine coupling constants are expressed in 10^4 cm^{-1} units. Presumed errors in the last decimal figure are reported between brackets.

Complex	g_{iso} (3)	a_{iso} (3)	$g_{\|\|}$ (4)	$A_{\|\|}$ (4)	$g_\perp$ (7)	$A_\perp$ (7)	a_{iso}^N(1)	$A_\perp^N$ (1)	$A_{\|\|}^N$ (1)
Cu-GLH	2.092	86	2.174	210	2.046	23	14	-	15
Cu-Sar-LH	2.094	88	2.171	211	2.046	24	14	-	15
Cu-GFH	2.091	84	2.175	208	2.040	27	14	11	16
Cu-Sar-FH	2.091	86	2.172	208	2.040	28	14	-	15

When lowering the pH of the aqueous solutions to pH 5, only Cu-Sar-LH displays signals from other species that are present in the system. The RT spectrum showing three different complex species of Cu-Sar-LH at pH 5.1 can be seen in Figure 7.

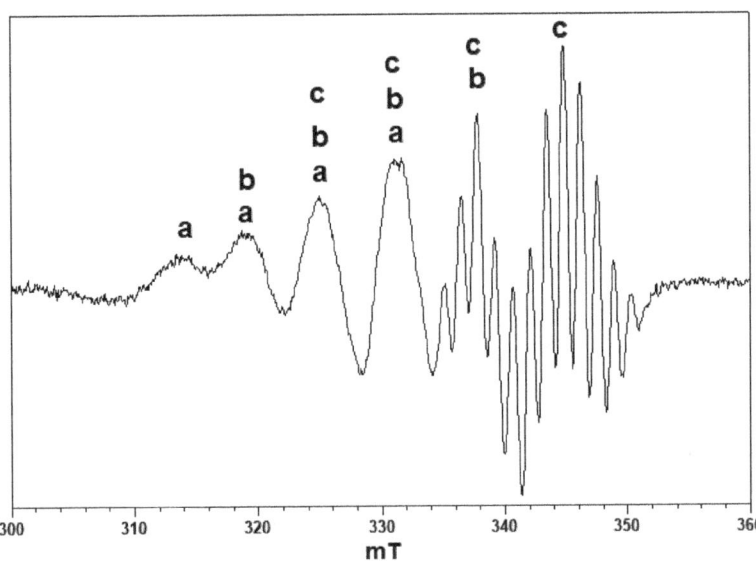

Figure 7. RT EPR 2nd derivative spectrum in an aqueous solution of Cu-Sar-LH at pH 5.1. Three copper(II) species are found and are designated as **a**, **b** and **c**.

Species **c** is the same species that was found at higher pH values and is thus the MLH_{-2} species. It can also be noted that it is evident that the MLH_{-2} species has already formed at pH 5.1. Species **a** and **b** both occur simultaneously with species **c** at this pH and give the following isotropic magnetic values: where **a** gives g_{iso} = 2.151 ± 0.007 and a_{iso} = 59 ± 0.005 × 10^{-4} cm^{-1} and **b** gives g_{iso} = 2.111 ± 0.007, a_{iso} = 62 ± 0.005 × 10^{-4} cm^{-1} and a_{iso}^N = 13 ± 0.002 × 10^{-4} cm^{-1}. Species **b** shows an shf interaction with a pattern of five lines and an approximate intensity distribution of 1:2:3:2:1, and thus this indicates a CuN_2O_2 chromophore. The last two lines of species **b** overlap with the shf lines of species **c**. Looking at the isotropic magnetic parameters, which show a low value of g_{iso} and a high value of a_{iso}, it is probable that the pattern (species **b**) is due to the nitrogen atoms coming from the amine and neighboring amide-N. At pH 5.1, the imidazole-N is still protonated, and therefore the species would be the ML species. Species **a** shows isotropic magnetic parameters, which are compatible with the formation of MLH.

Unfortunately, the RT EPR spectra of Cu-GLH, Cu-GFH and Cu-Sar-FH were not well resolved in the pH range of 4.8–5.2. This meant that it was not possible to determine isotropic magnetic parameters for other species that could be present.

For all four complexes, the shf structure at LT is present in all the EPR spectra of the frozen solutions (pH 7–8). Unfortunately, it is only resolved in the parallel part of the spectrum for Cu-GFH (Figure 8). This LT spectrum has the nine-line shf structure that was already seen in the RT EPR spectra in Figure 6.

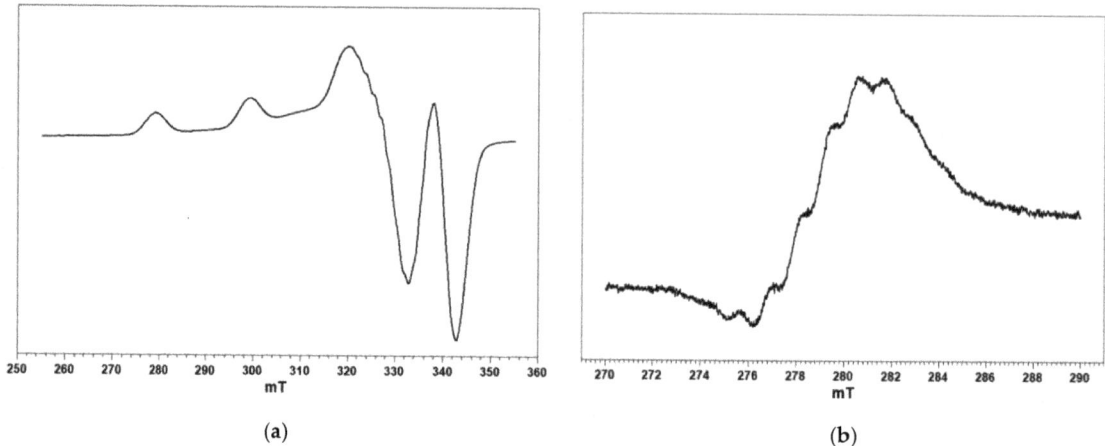

Figure 8. (a) LT EPR spectra of Cu-GFH in frozen aqueous solution at pH 7–8 and (b) LT EPR 2nd derivative spectrum of the lowest magnetic field feature of Cu-GFH in frozen aqueous solution at pH 7–8.

2.4. H NMR Spectroscopy

Copper(II) is a paramagnetic metal ion that affects both the chemical shifts and relaxation rates of the ligand nuclei. This is due to the strong interactions between the unpaired electron of copper(II) and the magnetic dipoles of the nuclei of the ligand [44,45]. These interactions can be described as a through-bond effect (Fermi contact interaction) and a through-space effect (dipolar interaction). As a result, ^{1}H NMR can be used to determine the binding sites of copper(II) to the ligands. As copper(II) coordinates to the ligand, the dipolar interaction causes the relaxation rates to increase, and that produces broadening in the ^{1}H NMR signals. The increase in the relaxation rates are dependent on the distance between the nuclei of the ligand and the copper(II) ion, and so the closer the signals are to the binding sites, the more they broaden [46–51]. To be exact, the transverse relaxation time (T_2) determines the line width of the peaks, which is defined as:

$$T_2 = \frac{1}{\pi W_{1/2}} \quad (1)$$

where $W_{1/2}$ is the line width at half height [52].

The technique to view the broadening of the ^{1}H NMR signals is to titrate the ligands with copper(II) at a predetermined pH. This will cause the ^{1}H NMR signals to differentially broaden, which is visually convenient for structural analysis. This is possible because NMR is a relatively slow spectroscopic technique, where the relaxation rates are in minutes, while the reactions for copper(II) exchange are fast. This means that an average is seen between the free and bound ligand spectra, which consequently causes the signals to broaden only gradually [48].

Figure 9 shows the ^{1}H NMR peak to proton assignments of the four ligands, as well as their change in chemical shifts for selected protons, as a function of pH. From the inflection points of these curves, the protonation stepwise formation constants can be estimated. The estimated pK_a for L→LH is 8.8 for GLH, Sar-LH and Sar-FH, and 8.6 for GFH, and the estimated pK_a for LH→LH$_2$ is 6.8 for GLH and Sar-LH, and 6.9 for GFH and Sar-FH. These estimated stepwise formation constants agree with the potentiometric results. The full ^{1}H NMR spectra for the four ligands are given in the Supplementary Information (Figure S1).

For all the ligands, the amide proton peaks **d** and **g/i**, as well as the proton signals with the assignments, **c** and **e**, were indistinguishable from each other, and so proton–proton correlation spectra (1D-TOCSY) (Figure S2) were collected to differentiate between the signals. The results for GLH are given in Figure S2. The irradiating peak **d** (Figure S2a) affects peaks **b** and **b′**, which means peak **d** belongs to the amide-NH of the histidine amino acid. Peak **i** should then belong to the leucine amino acid, and as expected, by irradiating peak **i** (Figure S2b), peaks **b**, **b′** and **j** disappear, while peaks **h**, **f**, **f′** and **g** are affected. It is noted in both spectra that peaks **a** and **a′** are also affected; this is because they overlap the amide peaks and therefore have the same frequency as the amide groups. Figure S2c shows the expanded region of Figure S2b that contains peaks **c** and **e**; since peak **i** was irradiated and the second peak is affected, this means that this is peak **e**.

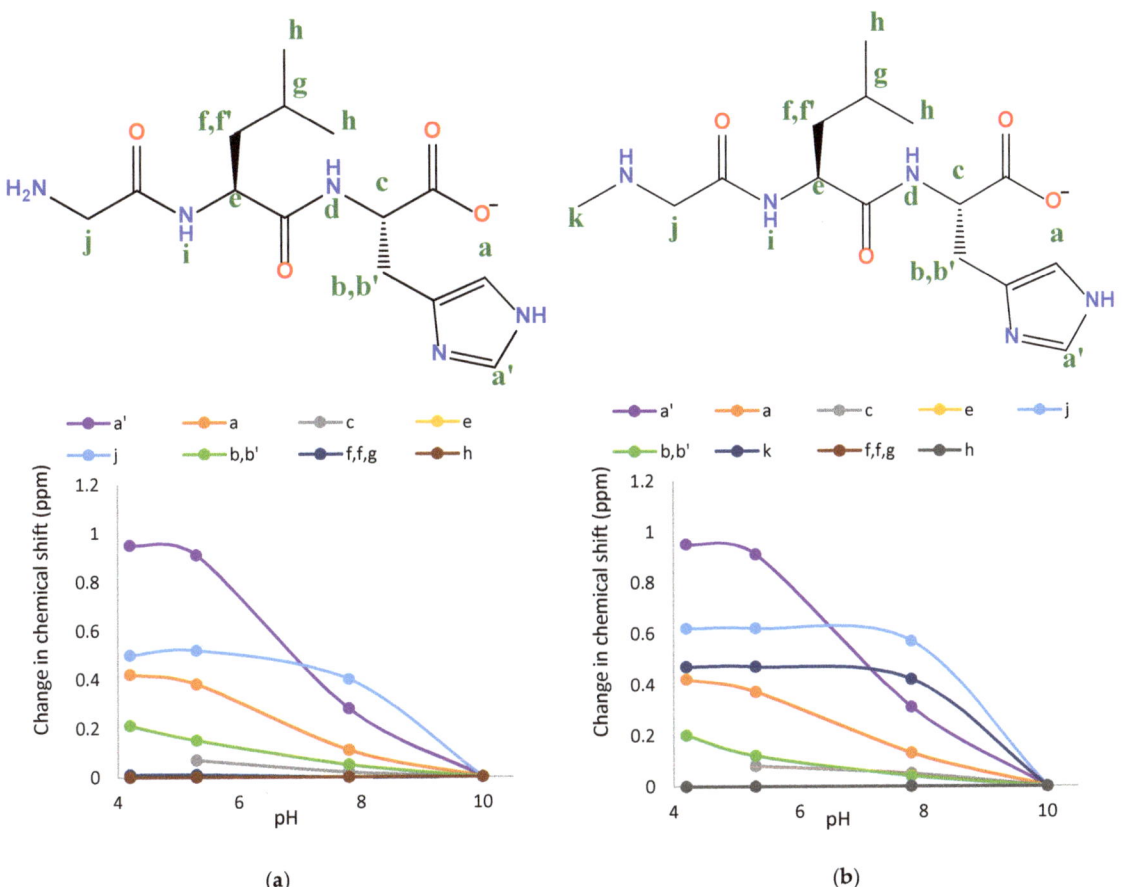

Figure 9. *Cont.*

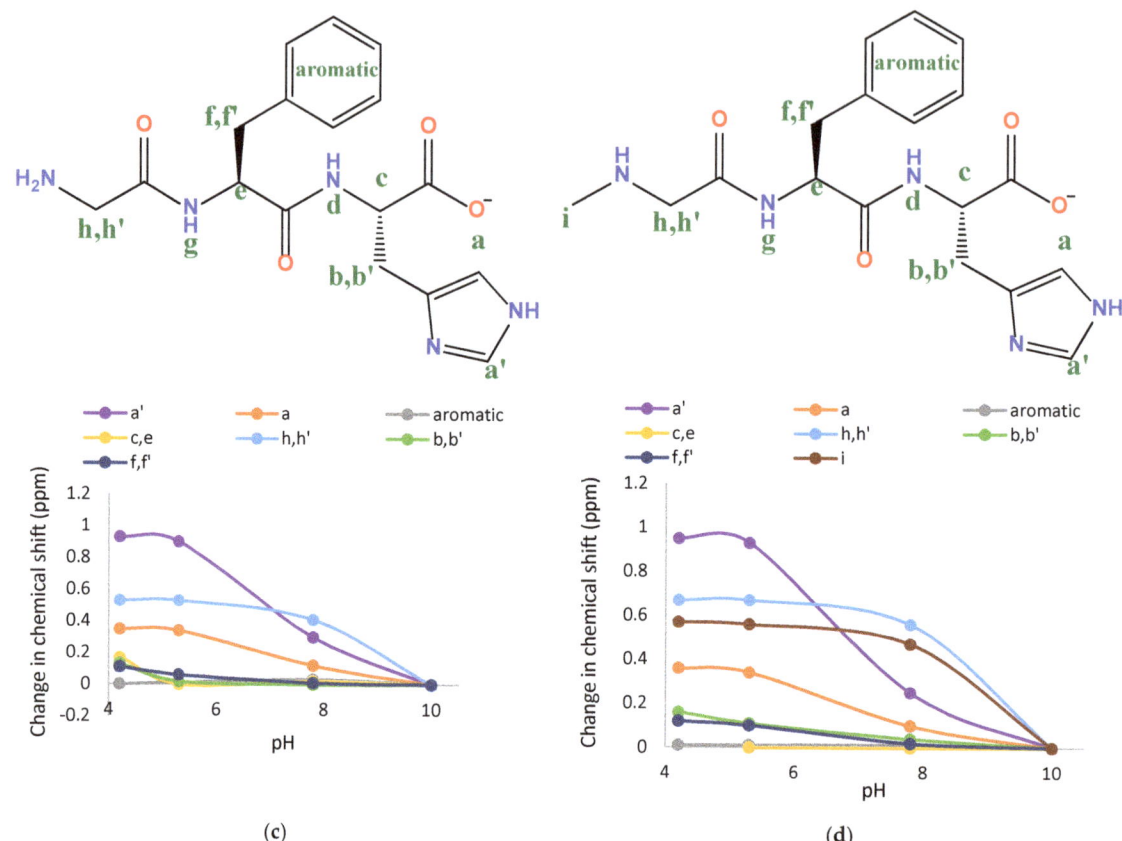

Figure 9. The ^{1}H NMR proton assignments and change in ^{1}H chemical shift as a function of pH for (**a**) GLH, (**b**) Sar-LH, (**c**) GFH and (**d**) Sar-FH.

All four copper(II) complexes form the MLH species, which, according to the species distribution diagrams, is the only species present at pH 3.5. Significant signal broadening will occur in the protons that are on carbons that are α or β to the coordination site, while other proton signals will not be affected to the same extent. The spectrum of Cu-GFH, as well as the spectrum of Cu-Sar-FH, have overlapping **b**, **b'**, **f** and **f'** peaks. By setting the pH to 4.8 and overlapping the spectra of GFH, Cu-GFH and GLH (Figure S3), as well as overlapping the spectra of Sar-FH, Cu-Sar-FH and Sar-LH (Figure S4), it was possible to distinguish between these peaks. At pH 4.8, the species present will cause peaks **b** and **b'** to broaden, while peaks **f** and **f'** will not broaden. Both GLH/Sar-LH and GFH/Sar-FH have peaks **b** and **b'** positioned at approximately 3 ppm, but only GFH/Sar-FH has peaks **f** and **f'** also positioned at approximately 3 ppm. As GFH/Sar-FH is titrated with copper(II), the comparisons show that only peaks **b** and **b'** broaden significantly.

Figure 10 shows the effect copper(II) has on the line width of the spectra of the four ligands at pH 3.5. For all ligands, the broadening of peaks **a**, **a'**, **b**, **b'** and **c** shows coordination to the imidazole-N and carboxyl-O. Peak **j** in the GLH spectra, peaks **j** and **k** in the Sar-LH spectra, peaks **e**, **h** and **h'** in the GFH spectra, and peaks **e**, **h**, **h'** and **i** in the Sar-FH spectra also broaden, which means that a second coordination to the amine-N and neighboring carbonyl-O also occurs. This "double-sided" coordination has also been suggested in the literature [32]. Note, at this pH, the amide proton signals of Sar-FH and GFH are still present, which means that the amide-N cannot be coordinated.

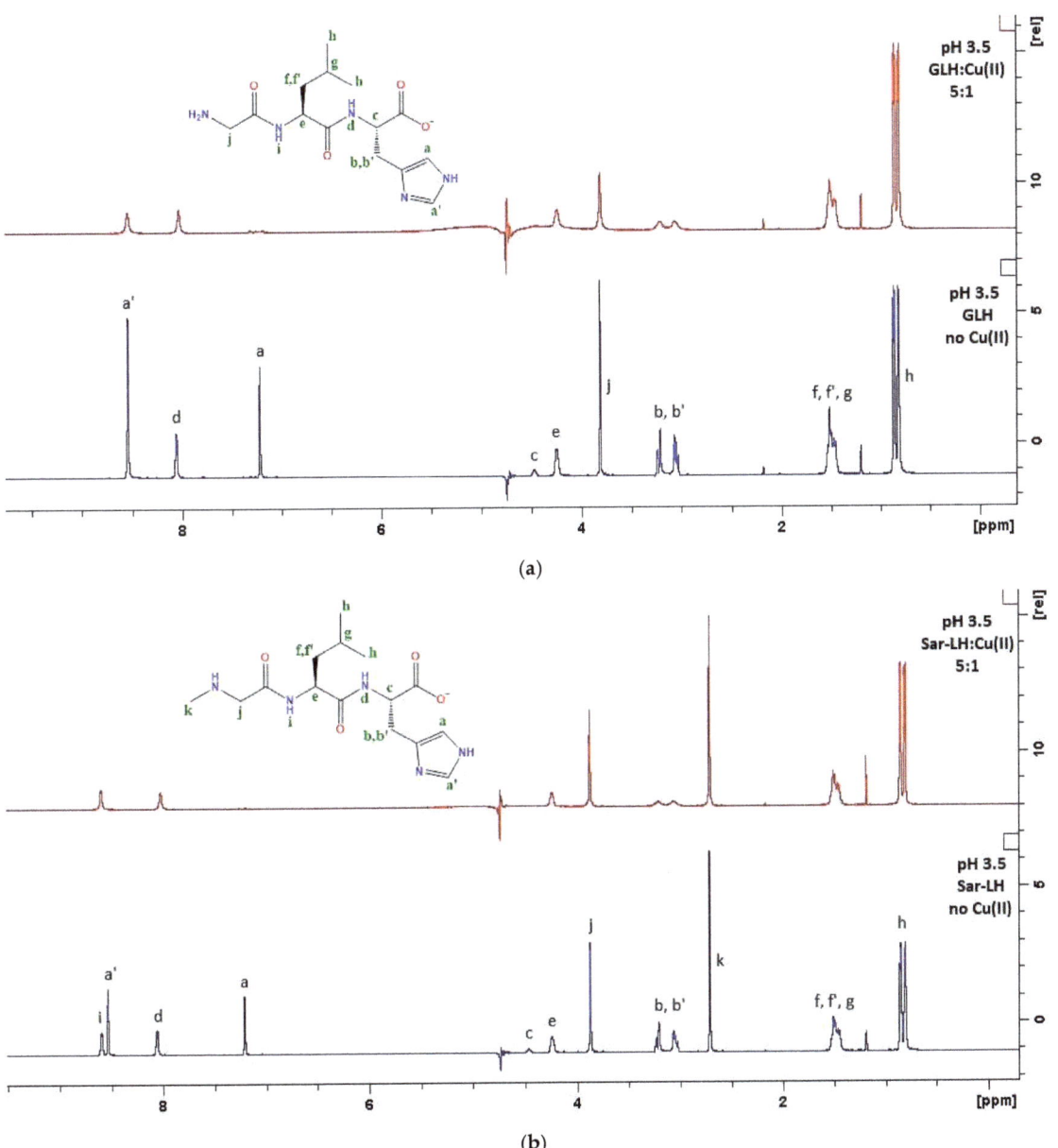

Figure 10. *Cont.*

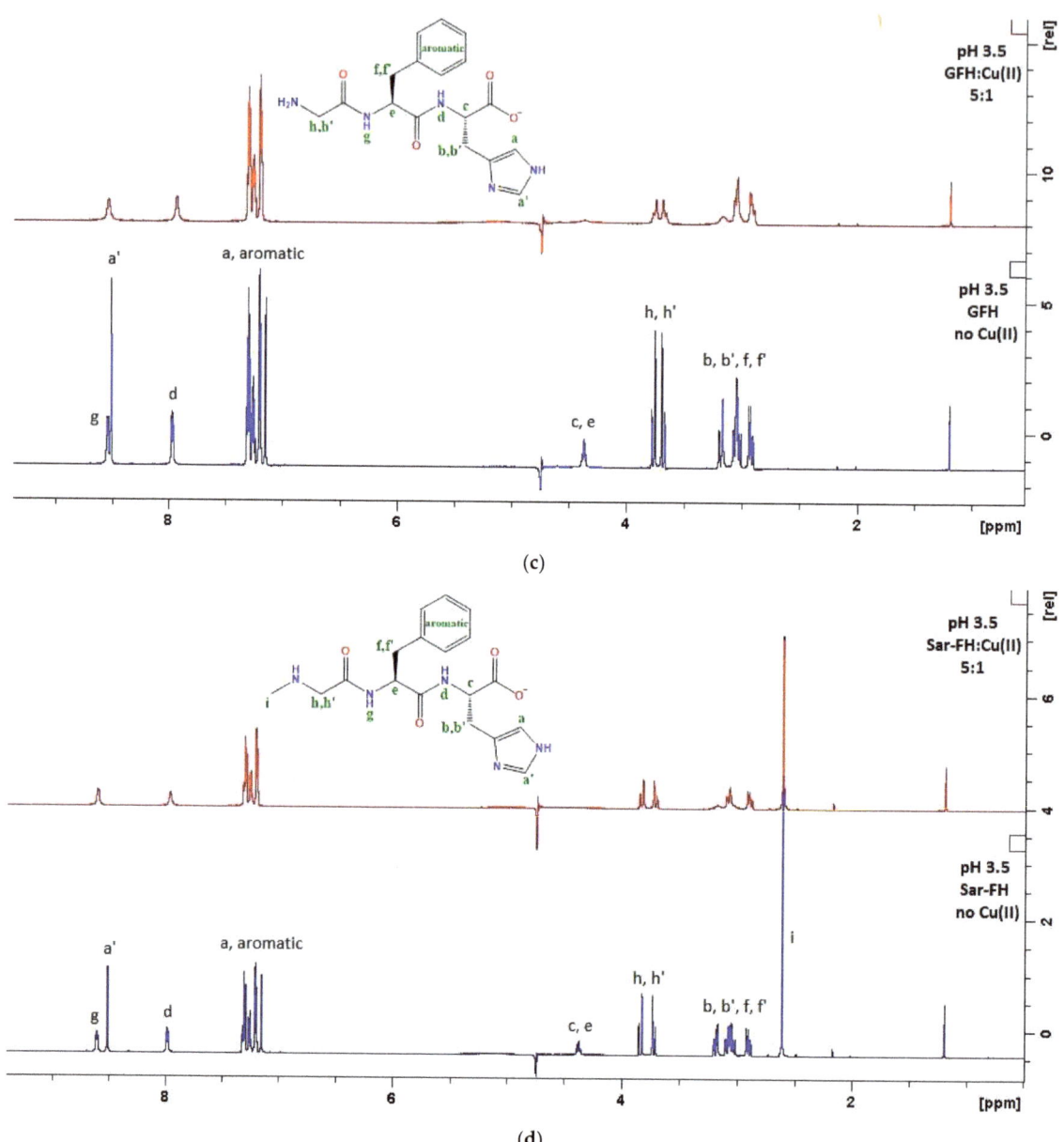

Figure 10. The ^{1}H NMR spectra (blue) of the ligand (**a**) GLH, (**b**) Sar-LH, (**c**) GFH and (**d**) Sar-FH and the ^{1}H NMR spectra (red) after (**a**) GLH, (**b**) Sar-LH, (**c**) GFH and (**d**) Sar-FH had been titrated with copper(II) to reach a 5:1 ligand:copper(II) ratio at a pH of 3.5 in 90% water and 10% D$_2$O.

Identifying the binding sites of the MLH$_{-1}$ species for Cu-GLH and the ML species for Cu-Sar-LH is not possible because the formation of these species overlaps with other species. At high pH values where the MLH$_{-2}$ species is the only present species, the peaks sharpen, which is due to the decrease in the exchange rate. Only the spectrum of the free ligand is seen, because the line widths of the complex are so broad (~457 Hz) [53] that the spectrum will disappear into the baseline. This then results in the peaks appearing to sharpen instead of broaden, since only the spectrum of the free ligand is seen [54,55].

2.5. Mass Spectrometry

Since the copper species present in this study are all labile, the soft ionization, electrospray ionization mass spectrometry (ESI-MS) technique was used [39,56,57]. While this technique generally does not cause fragmentation, radical generation can lead to fragmentation [58]. Thus, a combination of intact and fragmented complexes is expected.

Since all four ligands behaved similarly, only the Sar-LH system will be discussed. In the positive mode spectrum (Figure 11a), the base peaks 340.10 m/z, 362.08 m/z and 378.11 m/z represent the uncomplexed ligand in the form of $(LH + H)^{1+}$, $(LH + Na)^{1+}$ and $(LH + K)^{1+}$, respectively. K^+ ions were not added during the preparation of the complex, and so they most likely come from residual salts in the injector or tubing of the mass spectrometer. The 4N coordination of the MLH_{-2} species is verified by the peaks between 401.00–405.13 m/z. These peaks represent the overlap of the two different structures of the MLH_{-2} species, namely, where copper is in the Cu(II) form and where it is reduced from Cu(II)→Cu(I). The reduction of the Cu(II) occurs by collision-induced processes in the medium vacuum area of the source, which causes "inner-sphere" ligand to metal electron transfer, as well as the de-coordination of odd-electron species [59,60]. For simplicity, when copper(II) is reduced to copper(I), the symbol "M" in the complex species will be labelled as "M^I", while copper(II) will remain as "M". The base peak at 401.00 m/z represents the MLH_{-2} species in the $(MLH_{-2} + 2H)^{1+}$ form, and the base peak at 402.05 m/z represents the $(M^I LH_{-2} + 3H)^{1+}$ form. The base peak at 355.08 m/z also could represent the MLH_{-2} species, but after it has undergone fragmentation. A suggestion for fragmentation is $(MLH_{-2}$-carboxyl group$)^{1+}$. Structural assignments can be seen in Table 4. A similar scenario was seen for the other ligand complexes, and their structural assignments are given in Table S1. Note that, for Cu-GLH in Figure 11b, the base peak at 387.07 m/z could represent either the MLH_{-1} or MLH_{-2} species, or that the peak is a combination of the two species in the form of $(MLH_{-1} + H)^{1+}$ and $(MLH_{-2} + 2H)^{1+}$, respectively. Similarly, the base peak at 388.05 m/z could represent either the MLH_{-1} or MLH_{-2} species, or again a combination of both species in the form of $(M^I LH_{-1} + 2H)^{1+}$ and $(M^I LH_{-2} + 3H)^{1+}$, respectively. Additionally, note that, in Figure 11c, the base peak at 368.03 m/z for Cu-Sar-FH represents the free ligand after it has undergone fragmentation. A suggestion for how the ligand has undergone fragmentation involves the decomposition of the phenyl moiety. This has been seen in the literature where tropone first loses CO to yield a phenyl cation radical, which then leads to the decomposition of the phenyl moiety [61].

Table 4. Structural assignments of m/z base peaks that were found in the ESI-MS spectrum for Cu-Sar-LH at pH 5 (positive mode) with a 1:1 ratio and concentration of 1 mM for Sar-LH and 0.7 mM for copper(II) in aqueous solution.

Complex	m/z	Assignment
Cu-Sar-LH	340.10	$(LH + H)^{1+}$
	355.08	$(MLH_{-2}$-carboxyl group$)^{1+}$
	362.08	$(LH + Na)^{1+}$
	378.11	$(LH + K)^{1+}$
	401.00	$(MLH_{-2} + 2H)^{1+}$
	402.05	$(M^I LH_{-2} + 3H)^{1+}$

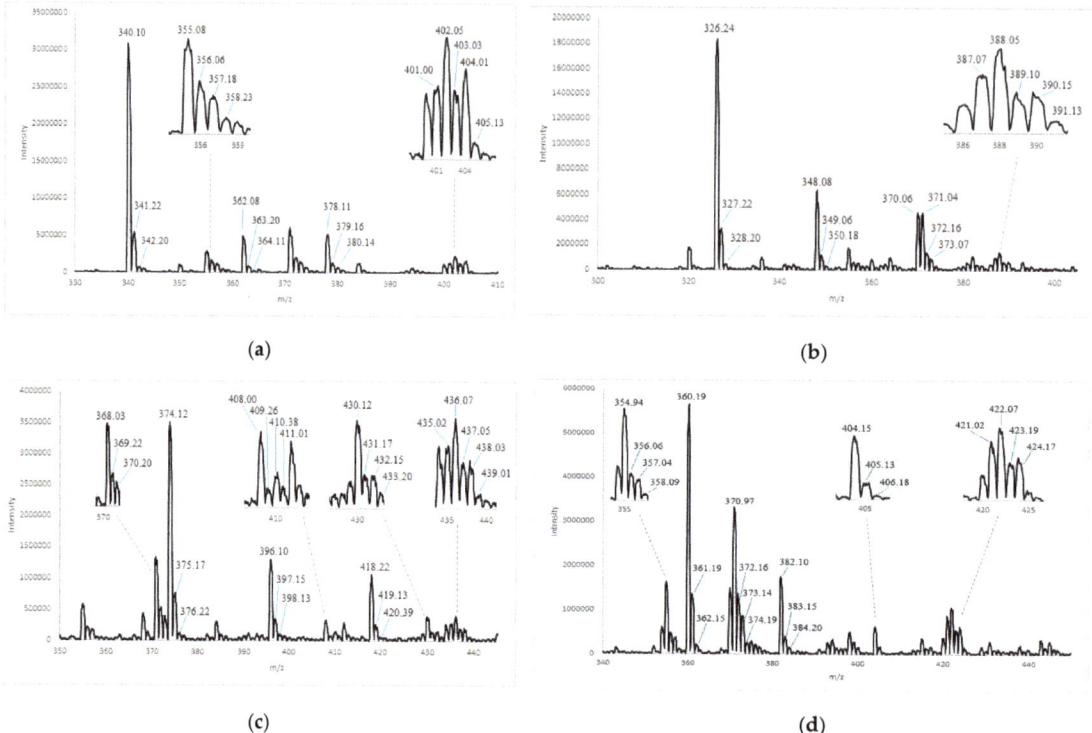

Figure 11. Section of the ESI-MS spectrum (positive mode) for the (**a**) Cu-Sar-LH, (**b**) Cu-GLH, (**c**) Cu-Sar-FH and (**d**) Cu-GFH complexes at a ratio of 1:1 and concentration of 1 mM for Sar-LH, GLH, Sar-FH and GFH, and 0.7 mM for copper(II) in aqueous solution at pH 5.

The MLH$_{-2}$ species is the dominant species for all the ligands, and therefore the identification of this species in the ESI-MS spectrum at pH 5 was relatively easy. The challenge was to try to identify the MLH species, as well as the ML species of Cu-Sar-LH and the MLH$_{-1}$ species of Cu-GLH, since they are present at much lower concentrations. The MLH and ML species could not be found. The assignment of peaks to the MLH$_{-1}$ species was debatable, and so the identification of this species was unconfirmed.

2.6. DFT Calculations

Molecular modelling using DFT calculations was used to validate the proposed structures and coordination modes for the MLH$_{-2}$ species. It was also used to indicate which coordination modes would most likely form in solution for the MLH, ML and MLH$_{-1}$ species, as well as to calculate λ_{max}. All starting structures had a copper(II) coordinate in an octahedral manner, with water molecules taking up vacant sites, and then they were optimized. If water molecules were moved out of the coordination distance to the copper atom after optimization, the resulting structures became a benchmark for additional optimization. The resultant structures for the MLH, ML, MLH$_{-1}$ and MLH$_{-2}$ species can be seen along with their geometry, the DFT calculated λ_{max} value, and in the case of MLH$_{-2}$, the experimental λ_{max} value, in Figures 12–15 respectively. All the possible coordination modes for each MLH, ML and MLH$_{-1}$ species were found to have similar ground state energies, and therefore, all coordination modes have an equal probability of forming in solution. DFT calculated the λ_{max} values of the MLH, ML and MLH$_{-1}$ species, but experimentally the absorption bands were not seen, which could be due to a combination of their low concentration prevalence, or that they are hidden by the broad absorption bands of either the MLH$_{-2}$ species (λ_{max} = 517–523 nm) and/or Cu(H$_2$O)$_6$ (λ_{max} = 800 nm) [62].

The possible coordination modes for the MLH species (Figure 12) were the two coordination modes suggested from ^{1}H NMR: coordination mode (a) has an amine-N and neighboring carbonyl-O coordination with the imidazole-N protonated; and coordination mode (b) has an imidazole-N and carboxyl-O coordination with the amine protonated. All coordination modes, except for coordination mode (b) of GLH and the coordination mode (b) of Sar-LH, became square pyramidal. The coordination mode (b) of GLH and Sar-LH became tetragonally distorted octahedral.

The possible coordination modes for the ML species of Cu-Sar-LH (Figure 13) were the two coordination modes proposed from potentiometry: coordination mode (a) has an amine-N, neighboring amide-N and carbonyl-O of leucine coordination with the imidazole-N protonated; and coordination mode (b) has an amine-N and neighboring amide-N coordination with the imidazole-N protonated. This agrees with the five lines of superhyperfine splitting that were found in EPR and indicated a CuN$_2$O$_2$ chromophore. Coordination modes (a) and (b) produced square planar and tetragonally distorted octahedral geometries, respectively.

The possible coordination modes for the MLH$_{-1}$ species of Cu-GLH (Figure 14) were the three coordination modes proposed from potentiometry: coordination mode (a) has an amine-N and two amide-Ns coordination with a protonated imidazole-N; coordination mode (b) has an amine-N, neighboring amide-N and imidazole-N coordination; and coordination mode (c) has an amine-N, two amide-Ns and the carboxyl-O coordination with a protonated imidazole-N. Coordination modes (a) and (c) resulted in a square planar geometry, whereas coordination mode (b) resulted in a distorted square pyramidal geometry with an elongated axial water bond.

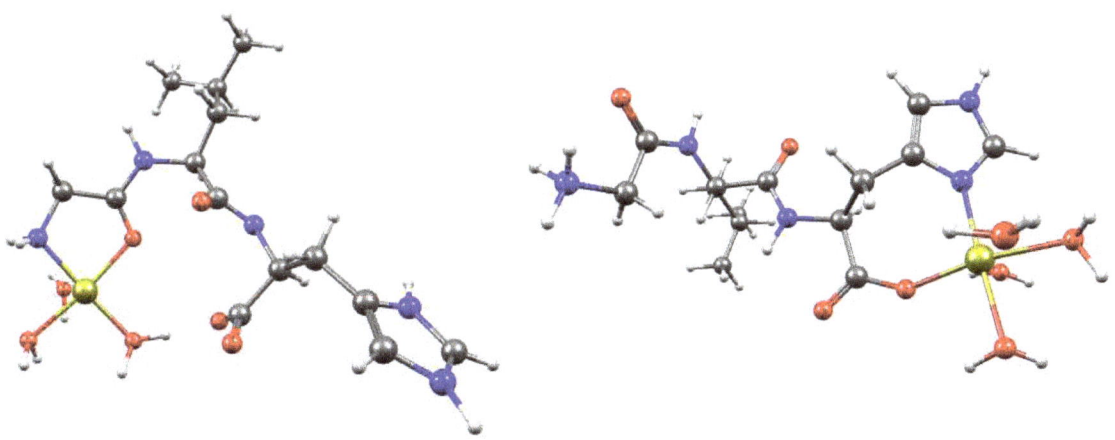

Ligand: GLH coordination mode (**a**)
Geometry: square pyramidal
λ_{max}: 710 nm

Ligand: GLH coordination mode (**b**)
Geometry: tetragonally distorted octahedral
λ_{max}: 882 nm

Figure 12. *Cont.*

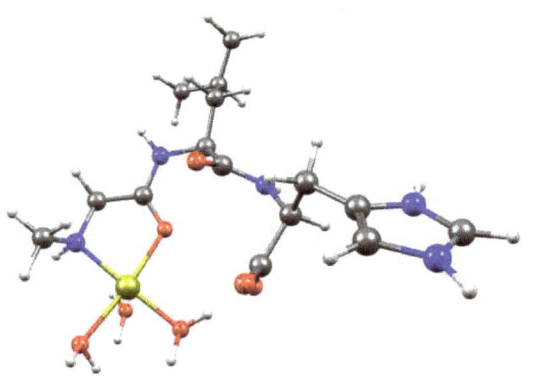

Ligand: Sar-LH coordination mode (**a**)
Geometry: square pyramidal
λ_{max}: 699 nm

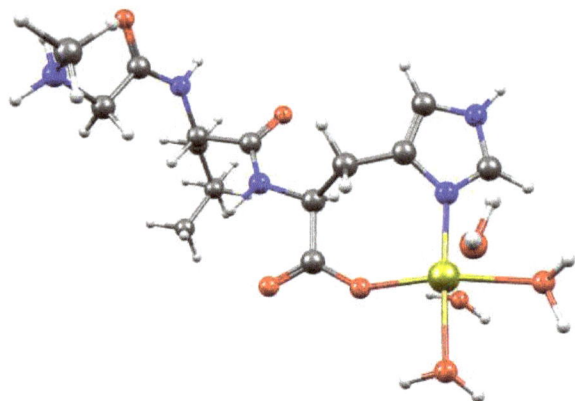

Ligand: Sar-LH coordination mode (**b**)
Geometry: tetragonally distorted octahedral
λ_{max}: 889 nm

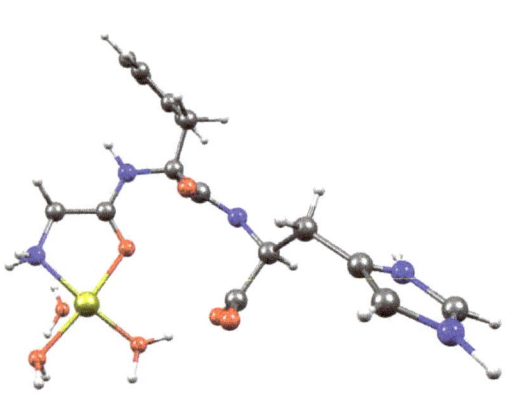

Ligand: GFH coordination mode (**a**)
Geometry: square pyramidal
λ_{max}: 837 nm

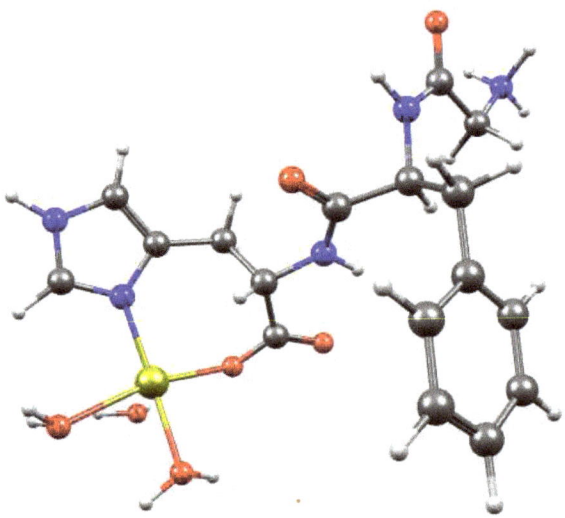

Ligand: GFH coordination mode (**b**)
Geometry: square pyramidal
λ_{max}: 875 nm

Figure 12. *Cont.*

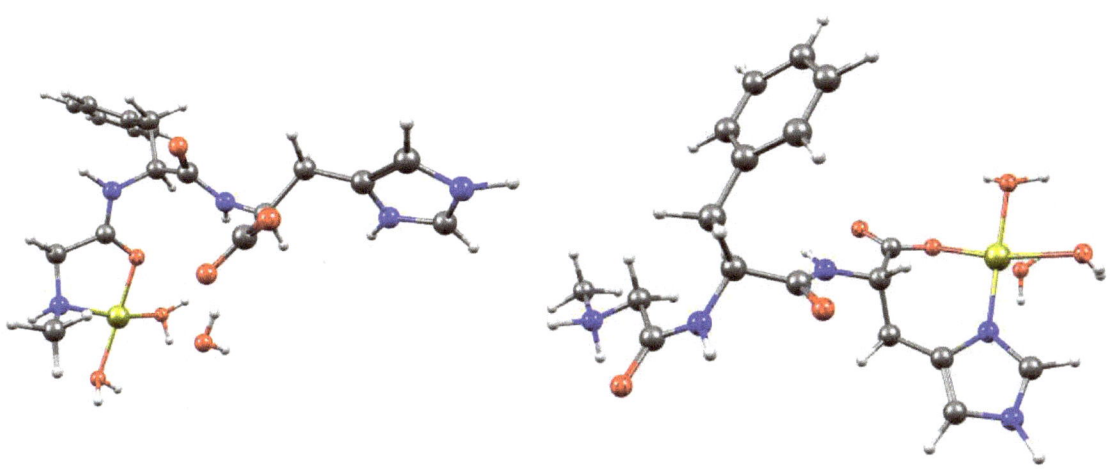

Ligand: Sar-FH coordination mode (**a**)
Geometry: square pyramidal
λ_{max}: 707 nm

(**a**)

Ligand: Sar-FH coordination mode (**b**)
Geometry: square pyramidal
λ_{max}: 868 nm

(**b**)

Figure 12. Visual representation of the proposed structures for the MLH species of (**a**) Cu-GLH, (**b**) Cu-Sar-LH, Cu-GFH and Cu-Sar-FH, as well as their geometries and calculated λ_{max} obtained at B3LYP/6-31++G** in water.

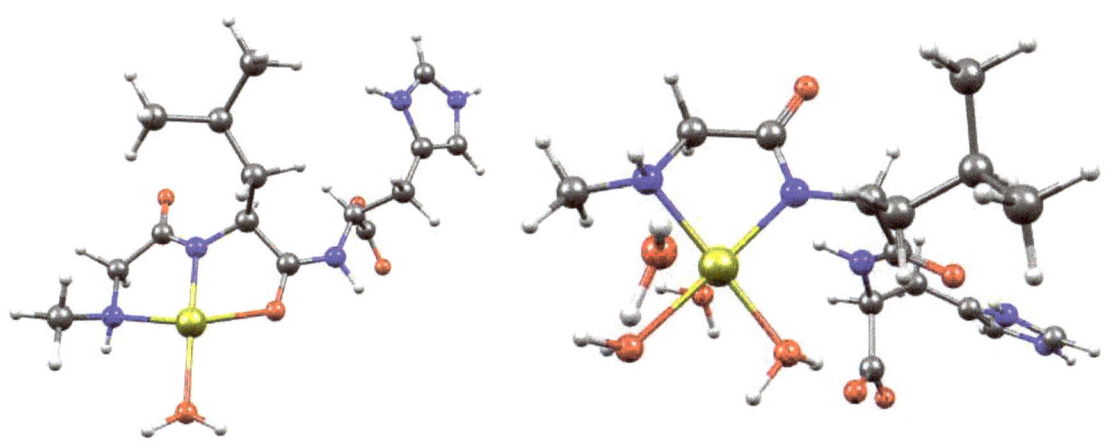

Coordination mode (**a**)
Geometry: square planar
λ_{max}: 586 nm

Coordination mode (**b**)
Geometry: tetragonally distorted
octahedral λ_{max}: 719 nm

Figure 13. Visual representation of the proposed structures for the ML species of (**a**) Cu-GLH, (**b**) Cu-Sar-LH, as well as their geometries and calculated λ_{max} obtained at B3LYP/6-31++G** in water.

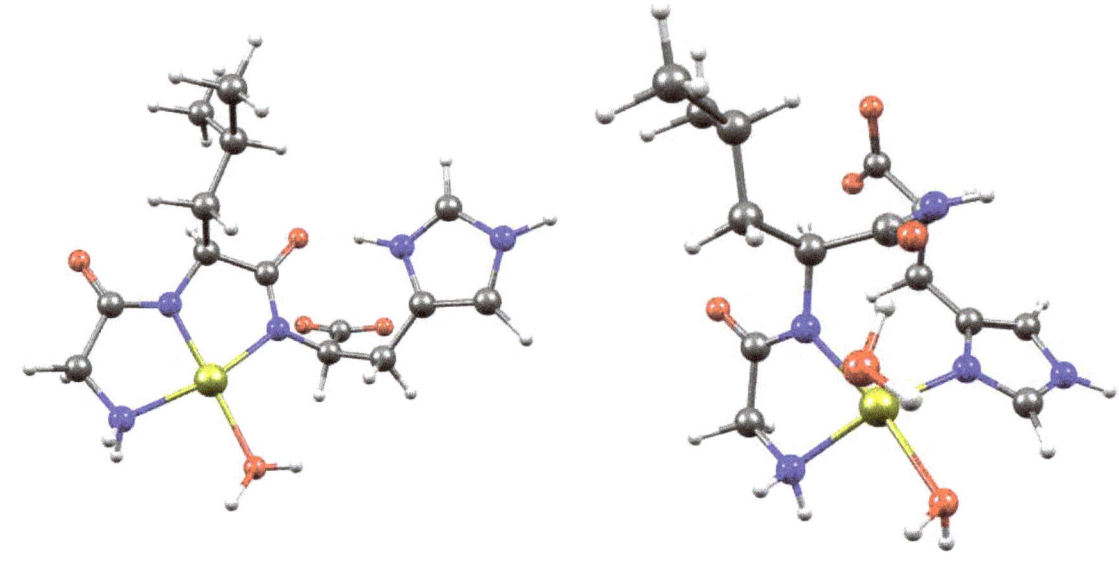

Coordination mode
Geometry: square planar
λmax: 569 nm

Coordination mode
Geometry: distorted square pyramidal
λmax: 763 nm

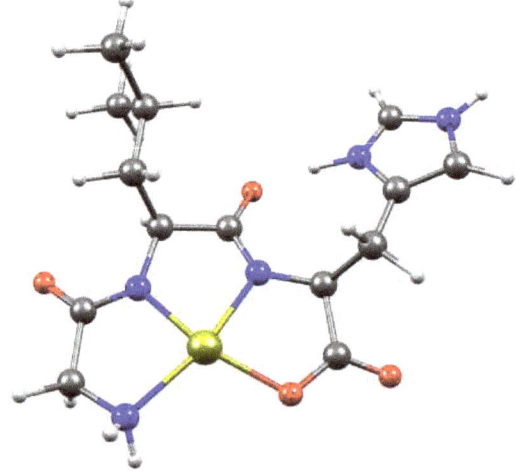

Coordination mode
Geometry: square planar
λmax: 592 nm

Figure 14. Visual representation of the proposed structures for the MLH$_{-1}$ species of Cu-GLH, as well as their geometries and calculated λ_{max} obtained at B3LYP/6-31++G** in water.

The MLH$_{-2}$ species (Figure 15) is coordinated through the amine-N, two amide-Ns and the imidazole-N and the resultant structures all formed a square planar geometry. This is the exact structure and symmetry that was proposed by EPR and UV-vis. The high extinction coefficient, the low $g_{||}$ and high $A_{||}$, as well as the high value of the superhyperfine nitrogen constant, are all in agreement with the symmetry of this molecular modelling. Out of the four species, MLH$_{-2}$ is the only species where the experimental λ_{max} values are known, and since the difference between the calculated and experimental λ_{max} is 2–18 nm, it signifies that the selected DFT levels (B3LYP/6-31++G**) and solvent effect (SMD) are valid for this system [63,64].

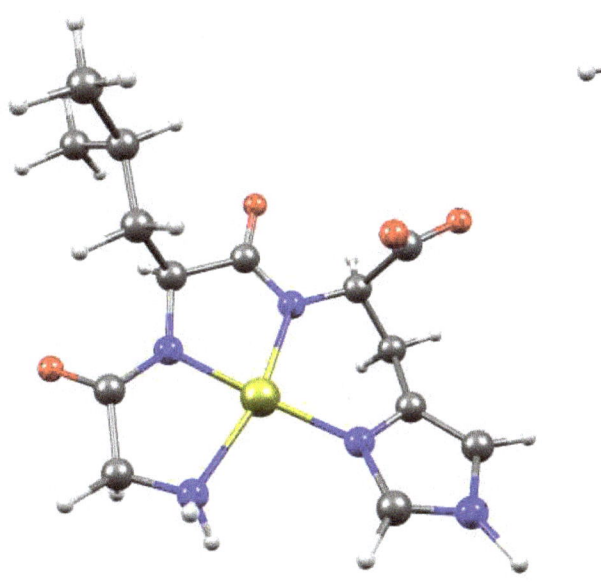

Ligand: GLH
Geometry: square planar
Calculated λ_{max}: 516 nm
Experimental λ_{max}: 518 nm

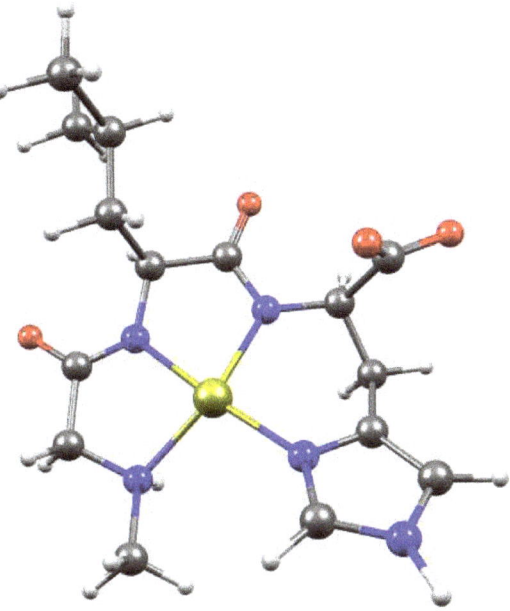

Ligand: Sar-LH
Geometry: square planar
Calculated λ_{max}: 541 nm
Experimental λ_{max}: 523 nm

Figure 15. *Cont.*

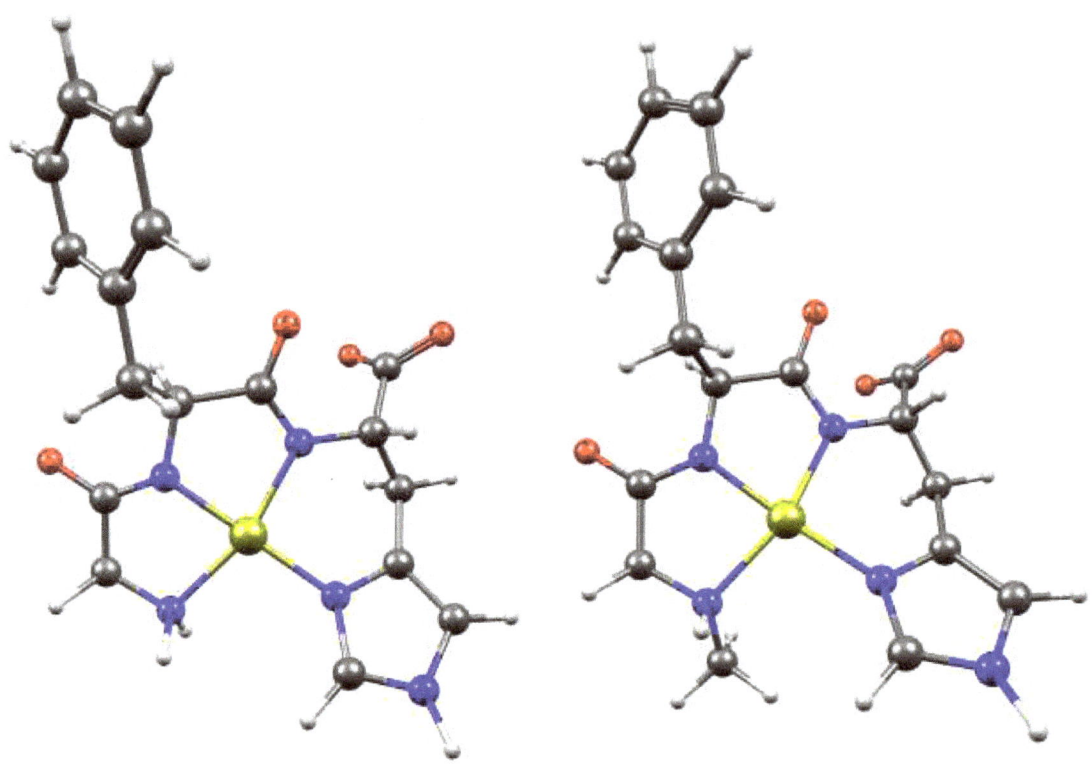

Ligand: GFH
Geometry: square planar
Calculated λ_{max}: 527 nm
Experimental λ_{max}: 517 nm

Ligand: Sar-FH
Geometry: square planar
Calculated λ_{max}: 534 nm
Experimental λ_{max}: 521 nm

Figure 15. Visual representation of the proposed structures for the MLH$_{-2}$ species of Cu-GLH, Cu-Sar-LH, Cu-GFH and Cu-Sar-FH, as well as their geometries, calculated λ_{max} at B3LYP/6-31++G** in water and experimental λ_{max}.

3. Materials and Methods

3.1. Materials

All analytical grade chemicals and reagents were commercially available and were used as received. The ligands, glycyl-L-leucyl-L-histidine (GLH), sarcosyl-L-leucyl-L-histidine (Sar-LH), glycyl-L-phenylalanyl-L-histidine (GFH) and sarcosyl-L-phenylalanyl-L-histidine (Sar-FH), were purchased from GL Biochem (Shanghai, China). Purity was checked potentiometrically and by means of HPLC-MS and found to be >98%. Boiled Milli-Q water (18.2 MΩ.cm) was used to prepare all the potentiometric titration solutions in order to remove carbon dioxide, as outlined by Vogel [65]. The background electrolyte of the titration solutions was prepared with NaCl to have an ionic strength of 0.15 M so that it matched the ionic strength found in human blood [66].

Ligand solutions (5 mM) were prepared by dissolving weighed samples of GLH, Sar-LH, GFH and Sar-FH into a standardized hydrochloric acid and background electrolyte solution. Copper(II) solutions (0.01 M) were prepared using CuCl$_2$·2H$_2$O, adding the background electrolyte to make the ionic strength 0.15 mol.dm^{-3} and standardizing the solutions with EDTA.

3.2. Potentiometric Measurements

The potentiometric titrations were performed under an inert atmosphere of purified nitrogen gas at 25 °C and at a constant ionic strength of 0.15 mol.dm^{-3} (NaCl) using a Metrohm 888 Titrando. The measurements took place in a double-walled titration vessel in a pH range of 2–11 and were kept at 25 ± 0.1 °C by a Haake thermostat bath (Thermo Fisher Scientific, Waltham, MA, USA). The amount of titrant added to the titrated solution was administered by a Metrohm 765 Dosimat automated burette (Metrohm, Herisau, Switzerland) via a capillary tip, which had a non-return valve. The amount of added titrant was controlled by the software program built into the Titrando, automatic titrator (Metrohm, Herisau, Switzerland), which also monitored the electromotive force. The titrated solution was stirred by a magnetic bar throughout the titration analysis. As an initial preparative setup, a range of Metrohm ion analysis pH buffers (pH 4, 7, and 9) was used to calibrate the slope of the electrode. Strong acid-strong base titrations (HCl/NaOH) were used to calculate the electrode potential, $E°$, and the dissociation constant of water, pK_w [67,68]. The metal ligand ratios were 1:1, and both the protonation and metal ligand titrations were titrated against NaOH over a pH range of 2–11. The ESTA suite of programs [69] was then used to analyze the data from the potentiometric titrations. Potentiometric solutions were also analyzed spectroscopically on a Shimadzu UV-1800 recording spectrophotometer (Shimadzu, Kyoto, Japan) in the range from 200–800 nm.

3.3. Continuous Wave EPR Measurements

A Bruker Elexsys E500 CW-EPR spectrometer (Bruker, Billerica, MA, USA) driven by a PC running XEpr program under Linux and equipped with a Super-X microwave bridge operating at 9.3–9.9 GHz and an SHQE cavity was used throughout this work. All the frozen solution EPR spectra of copper(II) complexes were recorded in quartz tubes at 150 K by means of an ER4131VT variable temperature apparatus. The measurements at room temperature (RT) were recorded by means of a WG-812-H flat quartz cell, and occasionally a glass capillary was inserted into a quartz tube. In the case of RT EPR spectra, the isotropic magnetic parameters were evaluated from the average distances among the peaks of the experimental spectra recorded in the 2nd derivative mode.

EPR anisotropic magnetic parameters were obtained directly from the experimental EPR spectra, calculating them from the 2nd and the 3rd line to remove second order effects [70]. Perpendicular parameters were obtained by exploiting the appearance of the extra peak due to the angular anomaly, whose field can be used in connection with the parallel parameters to calculate with a certain accuracy $g_\perp$ and $A_\perp$, as explained in the literature [71,72].

Instrumental settings of the frozen solution EPR spectra were recorded as follows: number of scans 1–5 (in the case of RT spectra, more than 10 scans were occasionally required to collect an acceptable signal to noise ratio); microwave frequency 9.46–9.48 GHz; modulation frequency 100 kHz; modulation amplitude 0.2–0.6 mT; time constant 164–327 ms; sweep time 3–6 min; microwave power 10–20 mW; and linear receiver gain 1×10^4–1×10^5. The instrumental settings of RT solution EPR spectra were substantially the same, except for the value of the microwave frequency. This was in the range of 9.70–9.80 GHz when using the flat quartz cell, and the microwave was powered up to 40 mW.

3.4. Preparation of Copper Complexes for EPR Measurements

Copper(II) complexes with these ligands were prepared by adding the appropriate amount of isotopically pure ^{63}Cu(NO$_3$)$_2$ (50 mM) to an aqueous solution containing the pertinent ligand in slight excess. The absolute copper(II) concentrations ranged from 1–4 mM. Up to 10% methanol or glycerin was added to the aqueous solution containing the copper(II) complex species in order to increase the resolution of the low temperature

(LT) frozen solution spectra. The final aqueous solution pH was adjusted by means of an Orion 9103SC combined glass microelectrode, which was connected to an Orion Star A 211 pH meter. The pH was adjusted by using concentrated NaOH or HNO_3 as required.

3.5. Nuclear Magnetic Resonance (NMR)

For the ^1H NMR spectra, 0.005 M solution for all ligands was prepared using 90% Milli-Q water and 10% D_2O. Tertiary butyl alcohol was added as an internal reference, and the pH was adjusted using a NaOH/HCl solution. The pH of each solution was recorded with an accuracy of 0.1 using a Crison micropH 2000 pH meter, which is equipped with a Ω metrohm glass electrode.

For the complexes, pH values were chosen in accordance with the species distribution diagrams. A copper(II) solution of 0.05 M was prepared using 90% Milli-Q water and 10% D_2O. The ^1H NMR spectra were recorded on a Bruker 300 MHz spectrometer (Bruker, Billerica, MA, USA) and processed using Bruker Topspin software, version 4.0.7. The residual water peak was suppressed using excitation sculpting.

For the 1D Total Correlation Spectroscopy (TOCSY) NMR spectra, 0.001 g of GLH was weighed out and added to 0.9 mL of Milli-Q water and 0.1 mL of D_2O. A phosphate buffer was also added, and the pH was adjusted to 4.5 using NaOH/HCl. The TOCSEY spectra were recorded on a Bruker 600 MHz spectrometer (Bruker, Billerica, MA, USA) and processed using Bruker Topspin software, version 4.0.7.

3.6. Mass Spectrometry

Copper(II) complexes for each of the four ligands were prepared in 10 mL of Milli-Q water (18.2 MΩ.cm), and the pH was adjusted to pH 5 with NaOH or HCl. The concentration of the ligands was 1 mM, and the concentration of copper(II) was adjusted to a slightly lower concentration of 0.7 mM to prevent precipitation. ESI-MS measurement samples were analyzed by direct infusion using the sample solvent as a carrier. Spectra were recorded on a Thermo TSQ quadrupole spectrometer (Thermo Fisher Scientific, Waltham, MA, USA) with a HESI ion source and analyzed with MS1 over a scan range of 100–600 m/z in both the positive and the negative mode. The capillary temperature was 270 °C, and nitrogen was used as a nebulizing gas. The conditions for electrospray ionization were: spray voltage 3500 V, flow rate 5 µL/min, vaporizing temperature 100 °C, auxiliary gas pressure 10 (arbitrary units) and the sheath gas pressure 5 (arbitrary units). The data were viewed using the Thermos Qual Browser (Thermo Fisher Scientific, Waltham, MA, USA).

3.7. Density Functional Theory (DFT) Calculations

All calculations were performed using facilities provided by the University of Cape Town's High-Performance Computing center (hpc.uct.ac.za) using Gaussian 09 software (Gaussian, Pittsburgh, PA, USA) [73]. The coordination modes for the MLH, ML, MLH_{-1} and MLH_{-2} species of Cu-GLH, Cu-Sar-LH, Cu-GFH and Cu-Sar-FH were built using the demo Chemcraft program [74] with the appropriate multiplicity and charge of each structure. The starting geometry of copper(II) was also realigned to an octahedral geometry. The multiplicity for all structures is a doublet since copper(II) has an unpaired electron, and the charge for the MLH, ML, MLH_{-1} and MLH_{-2} species is +2, +1, 0 and −1, respectively. Each structure was then optimized at a B3LYP/6-31++G** level using water as a solvent. The solvent effect was implemented using the solvation model density (SMD) [75]. All optimized structures were found to be a minima with no imaginary frequency and were viewed in the demo Chemcraft program [74]. Time-dependent density functional theory (TD-DFT) calculations were conducted at the same level and solvent to obtain the electronic transitions of the copper(II) complexes. At first, 20 excited states were considered for excitation calculations; however, only the first 3 excited states were involved and considered for the rest of the calculations.

4. Conclusions

As an anti-inflammatory drug, the complexation of the copper(II) complexes cannot be too strong, or the ability to release copper(II) into the bloodstream is jeopardized. The stability constants belonging to the four Cu(II)-tripeptide systems have relatively low constants and, therefore, should be able to release copper(II) in vivo.

The objective to form a stable complex between copper(II) and the ligands, GLH, Sar-LH, GFH and Sar-FH, has been achieved. At the physiological pH, all four copper(II) complexes formed the MLH_{-2} species, which is the major species in all systems. At low pH values, all four copper(II) complexes formed the MLH species, and Cu-GLH and Cu-Sar-LH also formed the MLH_{-1} and ML species, respectively. The MLH, ML and MLH_{-1} species have low concentrations and are thus minor species. The comparison between the thermodynamic stability of complexes with an N-methylated group and a non-N-methylated group was carried out. The expected increase in the stability of the complexes with the N-methylated group was not observed, and it was rationalized that the methyl group could either have steric effects or that there is a preference for the ammonium ions or charged amine groups to form hydrogen bonds with water.

In terms of structure determination, the absorption spectra in the visible region, as well as the EPR spectra recorded at room temperature and low temperature, suggested that the MLH_{-2} species formed a stable CuN_4 chromophore with a square planar geometry at physiological pH values. The broadening of the ^1H NMR signals showed that the MLH species formed two simultaneous coordination modes in solution, where the one has coordinated to the amine-N and neighboring amide-N and the second has coordinated to the imidazole-N and carboxyl-O. The detection of the coordination mode for the MLH_{-1} species was attempted by looking at the pK_a value for the transition from MLH_{-1} to MLH_{-2} (pK_a = 5.01). This value was not close enough to the literature to confirm that a second amide is deprotonated nor coordinated to the imidazole-N and, thus, three probable coordination modes were proposed. The pK_a value for the transition from the MLH to the ML species (pK_a = 4.99) is typical of a metal-assisted amide deprotonation, and thus two probable coordination modes were suggested, which consisted of a CuN_2O_2 chromophore. This chromophore was supported by the five lines of superhyperfine splitting in the EPR spectrum. The DFT calculations agreed with the experimental outputs that proposed the structure of MLH_{-2} and showed that, for all proposed coordination modes for the MLH, ML and MLH_{-1} species, each has an equal probability of forming simultaneously in solution. DFT calculations also provide an optimized structure for each coordination mode for each species and are viewed as the final structures.

Supplementary Materials: The following are available online at https://www.mdpi.com/article/10.3390/inorganics10010008/s1, Figure S1: ^1H NMR spectra of (a) GLH, (b) Sar-LH, (c) GFH and (d) Sar-FH at increasing pH values from 2–11. An arrow has been added to indicate the shifting of peaks over increasing pH values; Figure S2: The 1D selective gradient TOCSY NMR spectra (red) and ^1H NMR spectra (blue) of GLH at pH 4.5. (a) full spectrum of the irradiated amide-NH peak **d** at 8.246 ppm, (b) full spectrum of the irradiated amide-NH peak **i** at 8.511 ppm and (c) section of the spectrum of the irradiated amide-NH peak **i** at 8.511 ppm. An arrow has been added to indicate the irradiated amide-N; Figure S3: The ^1H NMR spectra (blue) of the ligand GFH and the ^1H NMR spectra (red) after GFH has been titrated with copper(II) to reach a 5:1 ligand copper(II) ratio at a pH of 4.8 in 90% water and 10% D_2O. ^1H NMR spectra (green) with arrows pointing to the significant broadening of peaks **b** and **b′**, after GFH has been titrated with copper(II); Figure S4: The 1H NMR spectra (blue) of the ligand Sar-FH and the 1H NMR spectra (red) after Sar-FH has been titrated with copper(II) to reach a 5:1 ligand copper(II) ratio at a pH of 4.8 in 90% water and 10% D2O. 1H NMR spectra (green) with arrows pointing to the significant broadening of peaks b and b′, after Sar-FH has been titrated with copper(II). Table S1: Structural assignments of m/z base peaks that were found in the ESI-MS spectrum for Cu-GLH, Cu-Sar-FH and Cu-GFH at pH 5 (positive mode) with a 1:1 ratio and concentration of 1 mM for GLH, Sar-FH and GFH, and 0.7 mM for copper(II) in aqueous solution.

Author Contributions: Conceptualization, G.M.V. and S.A.B.; formal analysis, G.E.J., G.M.V., A.N.H., R.A., R.P.B. and G.V.; investigation, G.M.V., A.N.H. and G.V.; methodology, G.E.J. and R.P.B.; project administration, G.E.J.; resources, G.E.J.; supervision, G.E.J., A.N.H. and S.A.B.; visualization, G.M.V. and R.A.; writing—original draft, G.M.V.; writing—review and editing, G.M.V., R.A. and S.A.B. All authors have read and agreed to the published version of the manuscript.

Funding: This research was funded by the National Research Foundation of South Africa (grant Nos 93450 and 85466 to G.E.J., and bursary to G.M.V.) and the University of Cape Town Research Committee. The Centre for High-Performance Computing (CHPC), South Africa, provided computational resources for this research project.

Institutional Review Board Statement: Not applicable.

Informed Consent Statement: Not applicable.

Data Availability Statement: Not applicable.

Conflicts of Interest: The authors declare no conflict of interest. The funders had no role in the design of the study; in the collection, analyses, or interpretation of data; in the writing of the manuscript; or in the decision to publish the results.

References

1. Odisitse, S.; Jackson, G.E. In vitro and in vivo studies of the dermally absorbed Cu(II) complexes of N5O2 donor ligands-Potential anti-inflammatory drugs. *Inorg. Chim. Acta* **2009**, *362*, 125–135. [CrossRef]
2. Wang, D.; Miller, S.C.; Liu, X.-M.; Anderson, B.; Wang, X.S.; Goldring, S.R. Noveldexamethasone-HPMA copolymer conjugate and its potential application in treatment of rheumatoid arthritis. *Arthritis Res. Ther.* **2007**, *9*, R2. [CrossRef] [PubMed]
3. Koopman, L.W.; Moreland, W.J. *Arthritis and Allied Conditions: A Textbook of Rheumatology*, 15th ed.; Lippincott Williams & Wilkins: Philadelphia, PA, USA, 2005.
4. Morrey, B.; Adams, R. Semiconstrained elbow replacement for distal humeral nonunion. *J. Bone Jt. Surg. Br.* **1995**, *77-B*, 67–72. [CrossRef]
5. Zvimba, J.N.; Jackson, G.E. Copper chelating anti-inflammatory agents; N1-(2-aminoethyl)-N2-(pyridin-2-ylmethyl)-ethane-1,2-diamine and N-(2-(2-aminoethylamino)ethyl)picolinamide: An in vitro and in vivo study. *J. Inorg. Biochem.* **2007**, *101*, 148–158. [CrossRef]
6. Odisitse, S.; Jackson, G.E. In vitro and in vivo studies of N,N'-bis[2 (2-pyridyl)-methyl]pyridine-2,6-dicarboxamide–copper(II) and rheumatoid arthritis. *Polyhedron* **2008**, *27*, 453–464. [CrossRef]
7. Khurana, R.; Berney, S.M. Clinical aspects of rheumatoid arthritis. *Pathophysiology* **2005**, *12*, 153–165. [CrossRef]
8. Weinblatt, M.E.; Kuritzky, L. RAPID: Rheumatoid arthritis. *J. Fam. Pract.* **2007**, *56*, S1–S7.
9. Suresh, E. Diagnosis of early rheumatoid arthritis: What the non-specialist needs to know. *JRSM* **2004**, *97*, 421–424. [CrossRef]
10. Jackson, G.E.; May, P.M.; Williams, D.R. Metal-ligand complexes involved in rheumatoid arthritis—I. *J. Inorg. Nucl. Chem.* **1978**, *40*, 1189–1194. [CrossRef]
11. Hardin, G.L.; Longenecker, J.G. *Handbook of drug therapy in rheumatic disease. Pharmacology and clinical aspects*, 1st ed.; Little, Brown and Company: London, UK, 1992.
12. Nordberg, G.F.; Fowler, B.A.; Nordberg, M. (Eds.) *Handbook on the Toxicology of Metals*, 4th ed.; Elsevier: London, UK; San Diego, CA, USA,, 2015.
13. Lahey, M.E.; Gubler, C.J.; Cartwright, G.E.; Wintrobe, M.M. Studies on copper metabolism. VI. Blood copper in normal human subjects. *J. Clin. Invest.* **1953**, *32*, 322–328. [CrossRef]
14. Weder, J.E.; Hambley, T.W.; Kennedy, B.J.; Lay, P.A.; MacLachlan, D.; Bramley, R.; Delfs, C.D.; Murray, K.S.; Moubaraki, B.; Warwick, B.; et al. Anti-Inflammatory Dinuclear Copper(II) Complexes with Indomethacin. Synthesis, Magnetism and EPR Spectroscopy. Crystal Structure of the N,N-Dimethylformamide Adduct. *Inorg. Chem.* **1999**, *38*, 1736–1744. [CrossRef] [PubMed]
15. Sorenson, J.R.J. Copper Chelates as Possible Active Forms of the Antiarthritic Agents. *J. Med. Chem.* **1976**, *19*, 135–148. [CrossRef]
16. Jackson, G.E.; May, P.M.; Williams, D.R. Metal-ligand complexes involved in rheumatoid arthritis—VI: Computer models simulating the low molecular weight complexes present in blood plasma for normal and arthritic individuals. *J. Inorg. Nucl. Chem.* **1978**, *40*, 1227–1234. [CrossRef]
17. Linder, M.C.; Hazegh-Azam, M. Copper biochemistry and molecular biology. *Am. J. Clin. Nutr.* **1996**, *63*, 797S–811S. [CrossRef]
18. Zvimba, J.N.; Jackson, G.E. Thermodynamic and spectroscopic study of the interaction of Cu(II), Ni(II), Zn(II) and Ca(II) ions with 2-amino-N-(2-oxo-2-(2-(pyridin-2-yl)ethyl amino)ethyl)acetamide, a pseudo-mimic of human serum albumin. *Polyhedron* **2007**, *26*, 2395–2404. [CrossRef]
19. Perrone, L.; Mothes, E.; Vignes, M.; Mockel, A.; Figueroa, C.; Miquel, M.-C.; Maddelein, M.-L.; Faller, P. Copper Transfer from Cu-Aβ to Human Serum Albumin Inhibits Aggregation, Radical Production and Reduces Aβ Toxicity. *ChemBioChem* **2009**, *11*, 110–118. [CrossRef]

20. Weder, J.E.; Dillon, C.T.; Hambley, T.W.; Kennedy, B.J.; Lay, P.A.; Biffin, J.R.; Regtop, H.L.; Davies, N.M. Copper complexes of non-steroidal anti-inflammatory drugs: An opportunity yet to be realized. *Coord. Chem. Rev.* **2002**, *232*, 95–126. [CrossRef]
21. Odisitse, S.; Jackson, G.E.; Govender, T.; Kruger, H.G.; Singh, A. Chemical speciation of copper(II) diaminediamide derivative of pentacycloundecane-A potential anti-inflammatory agent. *Dalt. Trans.* **2007**, 1140–1149. [CrossRef]
22. Odisitse, S. In Vivo Bio-Distribution Study of 64Cu (II)-Labelled Copper (II) Complexes of Peptides Mimics in Balb/C Mice-Development of Copper Based Anti-Inflammatory Agents. *MOJ Bioorganic Org. Chem.* **2017**, *1*, 153–157. [CrossRef]
23. Gruchlik, A.; Jurzak, M.; Chodurek, E.; Dzierzewicz, Z. Effect of Gly-Gly-His, Gly-His-Lys and their copper complexes on TNF-alpha-dependent IL-6 secretion in normal human dermal fibroblasts. *Acta Pol. Pharm.* **2012**, *69*, 1303–1306. [PubMed]
24. Hostynek, J.J.; Dreher, F.; Maibach, H.I. Human skin penetration of a copper tripeptide in vitro as a function of skin layer. *Inflamm. Res.* **2011**, *60*, 79–86. [CrossRef]
25. Elmagbari, F.M.A. Synthesis and Design of Ligand Copper Complexes as Anti-Inflammatory Drugs, Synthesis and Design of Ligand Copper Complexes as Anti-Inflammatory Drugs. Ph.D. Thesis, University of Cape Town, Cape Town, South Africa, 2015.
26. Vicatos, G.M. In Vitro Studies of Dermally Absorbed Cu(II) Tripeptide Complexes as Potential Anti-Inflammatory Drugs, In Vitro Studies of Dermally Absorbed Cu(II) Tripeptide Complexes as Potential Anti-Inflammatory Drugs. Master's Thesis, University of Cape Town, Cape Town, South Africa, 2016.
27. Hammouda, A.N. Development of Copper Peptide Complexes as Anti-Inflammatory Drugs, Development of Copper Peptide Complexes as Anti-Inflammatory Drugs. Ph.D. Thesis, University of Cape Town, Cape Town, South Africa, 2015.
28. Pickart, L.; Freedman, J.H.; Loker, W.J.; Peisach, J.; Perkins, C.M.; Stenkamp, R.E.; Weinstein, B. Growth-modulating plasma tripeptide may function by facilitating copper uptake into cells. *Nature* **1980**, *288*, 715–717. [CrossRef] [PubMed]
29. Pickart, L.; Vasquez-Soltero, J.M.; Margolina, A. The Human Tripeptide GHK-Cu in Prevention of Oxidative Stress and Degenerative Conditions of Aging: Implications for Cognitive Health. *Oxid. Med. Cell. Longev.* **2012**, *2012*, 324832. [CrossRef]
30. Hall, H.K. Potentiometric Determination of the Base Strength of Amines in Non-protolytic Solvents. *J. Phys. Chem.* **1956**, *60*, 63–70. [CrossRef]
31. Nakon, R.; Angelici, R.J. Copper(II) complexes of glycylglycine and glycylsarcosine and their methyl esters. *Inorg. Chem.* **1973**, *12*, 1269–1274. [CrossRef]
32. Várnagy, K.; Szabó, J.; Sóvágó, I.; Malandrinos, G.; Hadjiliadis, N.; Sanna, D.; Micera, G. Equilibrium and structural studies on copper(II) complexes of tetra-, penta- and hexa-peptides containing histidyl residues at the C-termini. *J. Chem. Soc. Dalt. Trans.* **2000**, 467–472. [CrossRef]
33. Turek, M.; Senar, X.L. Potentiometric and Spectroscopic Studies on Di-, Tri- and Tetraglycine with Copper (II) Ions Systems. *Food Chem. Biotechnol.* **2008**, *72*, 15–33. [CrossRef]
34. Sanna, D.; Ágoston, C.G.; Micera, G.; Sóvágó, I. The effect of the ring size of fused chelates on the thermodynamic and spectroscopic properties of peptide complexes of copper(II). *Polyhedron* **2001**, *20*, 3079–3090. [CrossRef]
35. Kozłowski, H.; Bal, W.; Dyba, M.; Kowalik-Jankowska, T. Specific structure–stability relations in metallopeptides. *Coord. Chem. Rev.* **1999**, *184*, 319–346. [CrossRef]
36. Housecroft, C.E.; Sharpe, A.G. *Inorganic Chemistry*, 3rd ed.; Pearson: London, UK, 2008.
37. Lever, A.B.P. *Inorganic Electronic Spectroscopy*, 2nd ed.; Elsevier: Amsterdam, The Netherlands, 1984.
38. Deeth, R.J.; Hearnshaw, L.J.A. Molecular modelling of Jahn–Teller distortions in Cu(II)N6 complexes: Elongations, compressions and the pathways in between. *Dalt. Trans.* **2006**, *8*, 1092–1100. [CrossRef]
39. Farkas, E.; Csapó, E.; Buglyó, P.; Damante, C.A.; Natale, G. Di Metal-binding ability of histidine-containing peptidehydroxamic acids: Imidazole versus hydroxamate coordination. *Inorg. Chim. Acta* **2009**, *362*, 753–762. [CrossRef]
40. Sigel, H.; Martin, R.B. Coordinating properties of the amide bond. Stability and structure of metal ion complexes of peptides and related ligands. *Chem. Rev.* **1982**, *82*, 385–426. [CrossRef]
41. Prenesti, E.; Daniele, P.G.; Prencipe, M.; Ostacoli, G. Spectrum–structure correlation for visible absorption spectra of copper(II) complexes in aqueous solution. *Polyhedron* **1999**, *18*, 3233–3241. [CrossRef]
42. Billo, E.J. Copper(II) chromosomes and the rule of average environment. *Inorg. Nucl. Chem. Lett.* **1974**, *10*, 613–617. [CrossRef]
43. Hathaway, B.J.; Billing, D.E. The electronic properties and stereochemistry of mono-nuclear complexes of the copper(II) ion. *Coord. Chem. Rev.* **1970**, *5*, 143–207. [CrossRef]
44. Morrison, R.T.; Boyd, R.N. Organic Chemistry. In *Organic Chemistry*; Allyn and Bacon: Boston, MA, USA, 1987; pp. 578–580.
45. Kleckner, I.R.; Foster, M.P. An introduction to NMR-based approaches for measuring protein dynamics. *Biochim. Biophys. Acta-Proteins Proteom.* **2011**, *1814*, 942–968. [CrossRef] [PubMed]
46. Liang, B.; Bushweller, J.H.; Tamm, L.K. Site-directed parallel spin-labeling and paramagnetic relaxation enhancement in structure determination of membrane proteins by solution NMR spectroscopy. *J. Am. Chem. Soc.* **2006**, *128*, 4389–4397. [CrossRef] [PubMed]
47. Ure, A.M.; Davidson, C.M. (Eds.) Chemical Speciation in the Environment. In *Chemical Speciation in the Environment*; Blackwell Science Ltd.: Oxford, UK, 2002; p. 46. ISBN 9780470988312.
48. Wells, M.A.; Jelinska, C.; Hosszu, L.L.P.; Craven, C.J.; Clarke, A.R.; Collinge, J.; Waltho, J.P.; Jackson, G.S. Multiple forms of copper (II) co-ordination occur throughout the disordered N-terminal region of the prion protein at pH 7.4. *Biochem. J.* **2006**, *400*, 501–510. [CrossRef] [PubMed]
49. Zhao, X.Z.; Jiang, T.; Wang, L.; Yang, H.; Zhang, S.; Zhou, P. Interaction of curcumin with Zn(II) and Cu(II) ions based on experiment and theoretical calculation. *J. Mol. Struct.* **2010**, *984*, 316–325. [CrossRef]

50. Hou, L.; Zagorski, M.G. NMR reveals anomalous copper(II) binding to the amyloid Aβ peptide of Alzheimer's disease. *J. Am. Chem. Soc.* **2006**, *128*, 9260–9261. [CrossRef]
51. Nuclear Magnetic Resonance: An Introduction. Available online: http://instructor.physics.lsa.umich.edu/adv-labs/NMR/Ch12_NMRTEC.pdf (accessed on 3 November 2019).
52. Marusak, R.A.; Doan, K.; Cummings, S.D. *Integrated Approach to Coordination Chemistry: An Inorganic Laboratory Guide*; John Wiley & Sons, Inc.: Hoboken, NJ, USA, 2007.
53. Elmagbari, F.M.; Hammouda, A.N.; Jackson, G.E.; Bonomo, R.P. Stability, solution structure and X-ray crystallography of a copper (II) diamide complex. *Inorg. Chim. Acta* **2019**, *498*, 119132. [CrossRef]
54. Laussac, J.P.; Haran, R.; Sarkar, B.N.m.r. and e.p.r. investigation of the interaction of copper(II) and glycyl-l-histidyl-l-lysine, a growth-modulating tripeptide from plasma. *Biochem. J.* **1983**, *209*, 533–539. [CrossRef] [PubMed]
55. Szabó, Z. Multinuclear NMR studies of the interaction of metal ions with adenine-nucleotides. *Coord. Chem. Rev.* **2008**, *252*, 2362–2380. [CrossRef]
56. Gizzi, P.; Henry, B.; Rubini, P.; Giroux, S.; Wenger, E. A multi-approach study of the interaction of the Cu(II) and Ni(II) ions with alanylglycylhistamine, a mimicking pseudopeptide of the serum albumine N-terminal residue. *J. Inorg. Biochem.* **2005**, *99*, 1182–1192. [CrossRef]
57. Ross, A.R.S.; Luettgen, S.L. Speciation of cyclo(Pro-Gly)3 and its divalent metal-ion complexes by electrospray ionization mass spectrometry. *J. Am. Soc. Mass Spectrom.* **2005**, *16*, 1536–1544. [CrossRef] [PubMed]
58. Demarque, D.P.; Crotti, A.E.M.; Vessecchi, R.; Lopes, J.L.C.; Lopes, N.P. Fragmentation reactions using electrospray ionization mass spectrometry: An important tool for the structural elucidation and characterization of synthetic and natural products. *Nat. Prod. Rep.* **2016**, *33*, 432–455. [CrossRef] [PubMed]
59. Lavanant, H.; Hecquet, E.; Hoppilliard, Y. Complexes of l-histidine with Fe^{2+}, Co^{2+}, Ni^{2+}, Cu^{2+}, Zn^{2+} studied by electrospray ionization mass spectrometry. *Int. J. Mass Spectrom.* **1999**, *185–187*, 11–23. [CrossRef]
60. Lavanant, H.; Virelizier, H.; Hoppilliard, Y. Reduction of copper(II) complexes by electron capture in an electrospray ionization source. *J. Am. Soc. Mass Spectrom.* **1998**, *9*, 1217–1221. [CrossRef]
61. Ishiwata, A.; Yamabe, S.; Minato, T.; Machiguchi, T. Norcaradiene intermediates in mass spectral fragmentations of tropone and tropothione. *J. Chem. Soc. Perkin Trans.* **2001**, *2*, 2202–2210. [CrossRef]
62. Miessler, G.L.; Tarr, D.A. *Inorganic Chemistry*, 3rd ed.; Pearson Education, Inc.: Philippines, Manila, 2004.
63. Rulíšek, L.; Havlas, Z. Theoretical Studies of Metal Ion Selectivity. 1. DFT Calculations of Interaction Energies of Amino Acid Side Chains with Selected Transition Metal Ions (Co^{2+}, Ni^{2+}, Cu^{2+}, Zn^{2+}, Cd^{2+}, and Hg^{2+}). *J. Am. Chem. Soc.* **2000**, *122*, 10428–10439. [CrossRef]
64. Robertazzi, A.; Magistrato, A.; de Hoog, P.; Carloni, P.; Reedijk, J. Density Functional Theory Studies on Copper Phenanthroline Complexes. *Inorg. Chem.* **2007**, *46*, 5873–5881. [CrossRef]
65. Vogel, A.I. *Vogel's Qualitative Inorganic Analysis*; 3rd ed.; Longman: London, UK, 1961.
66. Covington, A.K.; Robinson, R.A. References standards for the electrometric determination, with ion-selective electrodes, of potassium and calcium in blood serum. *Anal. Chim. Acta* **1975**, *78*, 219–223. [CrossRef]
67. Guilbault, G.G.; Kramer, D.N.; Goldberg, P. The application of modified Nernstian equations to the electrochemical determination of enzyme kinetics. *J. Phys. Chem.* **1963**, *67*, 1747–1749. [CrossRef]
68. Lee, Y.H.; Brosset, C. The slope of Gran's plot: A useful function in the examination of precipitation, the water-soluble part of airborne particles, and lake water. *Water. Air. Soil Pollut.* **1978**, *10*, 457–469. [CrossRef]
69. Murray, K.; May, P.M. *ESTA: Equilibrium Simulation for Titration Analysis*; University of Wales, Institute of Science and Technology (UWIST), Department of Applied Chemistry: Cardiff, UK, 1984.
70. Lund, A.; Vänngård, T. Note on the Determination of the Principal Fine and Hyperfine Coupling Constants in ESR. *J. Chem. Phys.* **1965**, *42*, 2979–2980. [CrossRef]
71. Bonomo, R.P.; Riggi, F. Study of angular anomalies in the X-band powder EPR spectra of copper (II) complexes with axial symmetry. *Lett. Al Nuovo Cim.* **1981**, *30*, 304–310. [CrossRef]
72. Bonomo, R.P.; Riggi, F. Determination of the perpendicular magnetic parameters for Cu(II) EPR spectra from angular anomalies. *Chem. Phys. Lett.* **1982**, *93*, 99–102. [CrossRef]
73. Frisch, M.J.; Trucks, G.W.; Schlegel, H.B.; Scuseria, G.E.; Robb, M.A.; Cheeseman, J.R.; Scalmani, G.; Barone, V.; Mennucci, B.; Petersson, G.A.; et al. *Gaussian 09 (Revision D.01)*; Gaussian, Inc.: Wallingford, CT, USA, 2010.
74. Chemcraft—Graphical Software for Visualization of Quantum Chemistry Computations. Available online: https://www.chemcraftprog.com (accessed on 2 February 2020).
75. Marenich, A.V.; Cramer, C.J.; Truhlar, D.G. Universal Solvation Model Based on Solute Electron Density and on a Continuum Model of the Solvent Defined by the Bulk Dielectric Constant and Atomic Surface Tensions. *J. Phys. Chem. B* **2009**, *113*, 6378–6396. [CrossRef] [PubMed]

Article

The Photochemistry of Fe$_2$(S$_2$C$_3$H$_6$)(CO)$_6$(μ-CO) and Its Oxidized Form, Two Simple [FeFe]-Hydrogenase CO-Inhibited Models. A DFT and TDDFT Investigation

Federica Arrigoni [1], Giuseppe Zampella [1], Luca De Gioia [1], Claudio Greco [2,*] and Luca Bertini [1,*]

[1] Department of Biotechnology and Biosciences, University of Milano-Bicocca, 20126 Milan, Italy; federica.arrigoni@unimib.it (F.A.); giuseppe.zampella@unimib.it (G.Z.); luca.degioia@unimib.it (L.D.G.)
[2] Department of Earth and Environmental Sciences, University of Milano-Bicocca, 20126 Milan, Italy
* Correspondence: claudio.greco@unimib.it (C.G.); luca.bertini@unimib.it (L.B.)

Abstract: Fe^{I}Fe^I Fe$_2$(S$_2$C$_3$H$_6$)(CO)$_6$(μ-CO) (**1a–CO**) and its Fe^{I}Fe^{II} cationic species (**2a$^+$–CO**) are the simplest model of the CO-inhibited [FeFe] hydrogenase active site, which is known to undergo CO photolysis within a temperature-dependent process whose products and mechanism are still a matter of debate. Using density functional theory (DFT) and time-dependent density functional theory (TDDFT) computations, the ground state and low-lying excited-state potential energy surfaces (PESs) of **1a–CO** and **2a$^+$–CO** have been explored aimed at elucidating the dynamics of the CO photolysis yielding Fe$_2$(S$_2$C$_3$H$_6$)(CO)$_6$ (**1a**) and [Fe$_2$(S$_2$C$_3$H$_6$)(CO)$_6$]$^+$ (**2a$^+$**), two simple models of the catalytic site of the enzyme. Two main results came out from these investigations. First, **a–CO** and **2a$^+$–CO** are both bound with respect to any CO dissociation with the lowest free energy barriers around 10 kcal mol^{-1}, suggesting that at least **2a$^+$–CO** may be synthesized. Second, focusing on the cationic form, we found at least two clear excited-state channels along the PESs of **2a$^+$–CO** that are unbound with respect to equatorial CO dissociation.

Keywords: metal-carbonyl complexes; [FeFe]-hydrogenases; density functional theory; time-dependent DFT; organometallic photochemistry

Citation: Arrigoni, F.; Zampella, G.; De Gioia, L.; Greco, C.; Bertini, L. The Photochemistry of Fe$_2$(S$_2$C$_3$H$_6$)(CO)$_6$(μ-CO) and Its Oxidized Form, Two Simple [FeFe]-Hydrogenase CO-Inhibited Models. A DFT and TDDFT Investigation. *Inorganics* **2021**, *9*, 16. https://doi.org/10.3390/inorganics 9020016

Academic Editor: Duncan Gregory
Received: 5 January 2021
Accepted: 4 February 2021
Published: 9 February 2021

Publisher's Note: MDPI stays neutral with regard to jurisdictional claims in published maps and institutional affiliations.

Copyright: © 2021 by the authors. Licensee MDPI, Basel, Switzerland. This article is an open access article distributed under the terms and conditions of the Creative Commons Attribution (CC BY) license (https://creativecommons.org/licenses/by/4.0/).

1. Introduction

In recent times, the study of substituted binuclear carbonyl species has gained vast popularity in the context of bioinorganic chemistry due to the fact that the hydrogenase enzymes are currently known for their specificity towards dihydrogen oxidation/evolution invariably include a binuclear carbonyl-containing moiety in their active site [1]. These enzymes, which encompass either only iron ions as metal cofactors ([FeFe]-hydrogenases), or both nickel and iron ([NiFe]-hydrogenases), have inspired the design and synthesis of a plethora of synthetic models to date [2–4], with diiron models being actually prevalent in literature. Such prevalence depends not only on the interest raised by the knowledge that [FeFe]-hydrogenases are extremely efficient [5] but also on the fact that diiron hexacarbonyls of the general formula Fe$_2$(SR)$_2$(CO)$_6$—which closely resemble the diiron portion of FeFe-hydrogenase active site, see Figure 1—had been known for seventy years before the publication of the first X-ray structure of the enzyme [6,7]. The availability of a large number of biomimetic catalysts has proved to be the main asset in the quest for a deeper understanding of hydrogenase chemistry [8]. For example, a biomimetic complex described Camara, and Rauchfuss [9] has proved highly valuable to confirm the hypothesis that H$_2$-binding and splitting in [FeFe]-hydrogenases occur on a single Fe center in the active site (the so-called "distal" iron ion, Fe$_d$ in Figure 1) [10,11]. Interestingly, the same Fe center is thought to be directly involved also in the enzyme inhibition mediated by carbon monoxide [12–14]. CO inhibition is a key topic that has bearings for the perspective [FeFe]-hydrogenases utilization for industrial purposes, as the contact of the enzyme with even traces of CO, can completely impair hydrogenase activity [15]. Still, many aspects of the chemistry of CO-inhibited [FeFe]-hydrogenases are far from being fully understood. This is true not only for the rather complex photochemistry occurring at the CO-inhibited active site (vide infra) but also with reference to the possibility of structural rearrangements occurring at the active site in concomitance with CO-binding [14,16,17]. Notwithstanding such open issues, biomimetic modeling has had a relatively limited impact so far for the elucidation of the (photo)chemical processes that can occur after CO inhibition, which is mainly due to the small number of biomimetic models of the CO-inhibited enzyme described to date. In any case, the availability of the latter [18] has stimulated positive feedback between experiments and theory, which allowed relating the hardness of ligands in biomimetic models with the stability of key stereoelectronic features in the latter [18,19]. The theory-experiment interplay proved relevant also in more recent studies in which density functional theory (DFT) calculations were used to rationalize the structural outcomes of CO-inhibition and subsequent reduction of the enzyme that gives place to over-saturated forms of the active site [20].

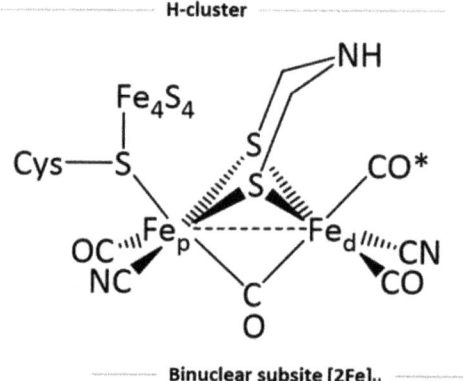

Figure 1. The active site of [FeFe]-hydrogenases, generally referred to as the H-cluster; its diiron portion "[2Fe]$_H$" directly involved in the binding of incipient substrates (H$_2$ and H$^+$) and inhibitors is highlighted. The carbonyl ligand marked with an asterisk represents an exogenous CO ligand behaving as an inhibitor. A cysteinyl sulfur bridges the diiron subsite with a [Fe$_4$S$_4$] subsite that completes the H-cluster composition. The two Fe atoms of the diiron subsite are labeled with subscripts "d" (distal) or "p" (proximal), depending on their position with respect to the [Fe$_4$S$_4$] subsite.

As far as the photochemistry of the CO inhibited form of the enzyme is concerned, it was found to be light-sensitive at cryogenic temperature. CO photolysis is a typical organometallic light-driven process [21], and this type of temperature-dependent mechanism has been already observed in Fe$_2$(CO)$_9$ photolysis, as pointed out by Chen et al. [22]. In this case, the structure of the Fe$_2$(CO)$_8$ photoproduct depends on the reaction condition: photolysis up to 35 K yields the Fe$_2$(CO)$_8$ bridged form, while at the higher temperature, the unbridged form is obtained.

This photolytic process is temperature-dependent and has been studied by EPR [23] and IR spectroscopy [22,24]. At low temperature (6–8 K), the initial axial EPR signal of the CO inhibited form is converted to that of the active form in the absence of CO. This photoproduct arises from the loss of the exogenous CO restoring the initial active form. The illumination at a higher temperature (14–30 K) yields a different photoproduct with a rhombic EPR signal. The IR spectra of this second photoproduct are characterized by the loss of the band associated with the bridging CO. According to the model proposed by Chen et al. [22], the photolyzed ligand can be the bridging one or a terminal one. In this latter case, a successive conversion of the bridging CO to terminal CO would take place. On the contrary, Rosenboom et al. [24] interpreted their IR spectra for the second photoproduct as the photolysis of two CO ligands (the bridging CO and the exogenous CO).

Among the synthetic models of the {Fe$_2$S$_2$} subcluster, Fe$_2$(S$_2$C$_3$H$_6$)(CO)$_6$ (^{1}a hereafter) can be considered the simplest one [25]. This complex has been extensively studied, and it is able to electrocatalyze proton reduction, although with a mechanism different compared with that of the enzyme [26,27]. ^{1}a has one bridging coordination position still available, which may be occupied by a CO ligand.

Fe^{I}Fe^I Fe$_2$(S$_2$C$_3$H$_6$)(CO)$_6$(μ-CO) (^{1}a–CO hereafter) and its Fe^{I}Fe^{II} cation (^{2}a$^+$–CO hereafter) have never been synthesized and represent simple biomimetic models for the CO-inhibited {Fe$_2$S$_2$} subcluster. These have been previously investigated by DFT in a study on the CO affinity of the series of Fe^{I}Fe^I and Fe^{I}Fe^{II} models of the [FeFe]-hydrogenase active site [28]. According to the 18-electron rule, these two models are oversaturated [29,30] complexes compared to the ^{1}a and ^{2}a$^+$ since at least one Fe atom counts 19 valence electrons. Despite this, their formation enthalpy and free-energy are in qualitative agreement with data obtained from the enzyme. In particular, (i) ^{1}a–CO formation results endothermic and not spontaneous, in agreement with the fact that Fe^{I}Fe^I CO inhibited form has never been observed; (ii) ^{2}a$^+$–CO formation is exothermic and spontaneous in qualitative agreement with the values obtained by Thauer et al. [31].

The aim of the present study is to outline a general mechanism for CO photolysis in oversaturated diiron systems related to the CO-inhibited [FeFe]-hydrogenase catalytic site. To do so, we investigated the ground state and excited-state potential energy surface (PES) topologies of simple model systems, namely ^{1}a–CO and ^{2}a$^+$–CO, by means of DFT and TDDFT. The main targets of this investigation are (i) to predict the stability of the complex toward CO dissociation and (ii) to shed light on the CO photolysis mechanism considering the hypothetical ^{1}a–CO and ^{2}a$^+$–CO photolysis

$$^1\text{a-CO} \rightarrow {}^1\text{a} + \text{CO}$$

$$^2\text{a}^+\text{-CO} \rightarrow {}^2\text{a}^+ + \text{CO}$$

as model photolytic processes. While the photochemistry of ^{1}a and ^{2}a$^+$ has already been investigated in detail [28,32,33], the case of the oversaturated CO forms has not been studied yet. Previous investigations have shown that the absorption spectrum of ^{1}a is characterized by an intense band at 355 nm along with a weak shoulder at 400 nm. Both features display an MLCT character that always involves the S atomic orbital, therefore indicating Fe $\rightarrow$ S as prevalent CT. For this, the dynamics of the low-energy excited states are mainly dominated by the Fe-S bond elongation/dissociation that favors the rotation of the partial Fe(CO)$_3$ group [32], while CO photolysis is mostly induced by populating higher-energy states. The question that arises on the basis of this evidence is the following: in the case of an oversaturated model such as a–CO and ^{2}a$^+$–CO, would a fully CO dissociative pathway for low-energy excited states emerge? Indeed, as rightly pointed out by Chen and coworkers [22], there is a close analogy between temperature-dependent CO photolysis processes of Fe$_2$(CO)$_9$ and of the CO inhibited [FeFe]-hydrogenase catalytic site. DFT/TDDFT investigations of the photochemistry of Fe$_2$(CO)$_9$ evidence two CO dissociation unbound pathways that evolve towards different Fe$_2$(CO)$_8$ isomers [34,35]. Although Fe$_2$(CO)$_8$ and Fe$_2$(CO)$_9$ may be considered all-CO prototypes of the {Fe$_2$S$_2$} subcluster in its active and CO-inhibited forms, the electronic and geometric structures of the former couple of species are somewhat different from those of the latter two, as we will show in the following sections, and therefore more complex systems must be considered. Moreover, recent investigations show that UVB light photo-inhibits the enzyme [36,37], whereas it is not the case for visible light, suggesting that a fully dissociative character of the lower excitation emerges only when a complex is oversaturated.

2. Results and Discussion

2.1. Ground States

2.1.1. ^{1}a–CO and ^{2}a$^+$–CO Ground State Properties

^{1}a–CO and ^{2}a$^+$–CO are binuclear complexes with a global minimum of C_s symmetry with ^{1}A′ and ^{2}A′ ground state molecular terms, respectively. As shown in Figure 2, the terminal CO (*t*-CO) ligands can be distinguished between *trans* or *cis* to the bridging propanedithiolate (μ-pdt) ligand. Each *trans* or *cis* CO group can be further distinguished between *anti* and *syn* with respect to the β carbon of CH$_2$ group of the μ-pdt. Accordingly, the two Fe atoms are classified as *syn* or *anti*.

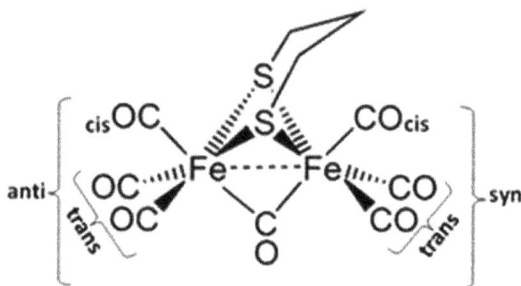

Figure 2. Schematic representation of the [Fe$_2$(S$_2$C$_3$H$_6$)(CO)$_6$(μ-CO)]^z (*z* = 0, 1) species analyzed in the present work.

The energy global minimum structures of ^{1}a–CO and ^{2}a$^+$–CO are characterized by Fe–Fe bond elongation of 0.452 Å and 0.167 Å, respectively, compared to ^{1}a and ^{2}a$^+$, recalling that this latter has a *syn* rotated C$_s$ minimum geometry [38]. The formation of ^{1}a–CO is endothermic and not spontaneous since ^{1}a is a saturated complex. On the contrary, ^{2}a$^+$–CO formation is exothermic and spontaneous [38]. Similar considerations regarding the CO-binding process under oxidative conditions have also been reported for more electron-rich diiron models [39–41].

Regarding ^{2}a$^+$–CO, which can be considered as the most promising CO-inhibited model for a successful synthesis, one could ask about the nature of the Fe–Fe bond compared to the ^{2}a$^+$ parent model. According to the NBO spin population equally distributed on both Fe atoms, the redox state is assigned as 2Fe$^{1.5}$. We investigate this issue from the point of view of the quantum theory of atoms in molecules (QTAIM) approach. The topology of ^{1}a and ^{2}a$^+$ electron density are characterized by a Fe–Fe bond critical point (BCP hereafter), which is not found for the corresponding CO oversaturated forms, an evident signal of the weakening of the Fe–Fe bond in the oversaturated moiety. We better characterize this bond by the QTAIM analysis of the electron density using delocalization indexes. The delocalization index δ(A,B) is an integral property that indicates the number of electron pairs delocalized between the two atoms (A and B) and can be considered a covalent bond order [42,43]. In Figure 3, we plotted the δ(Fe,Fe) for the four models considered here plus Fe$_2$(CO)$_9$ [44], which can be considered as the classic metal carbonyl complex where the nature of the Fe–Fe bond is easily questionable and long debated.

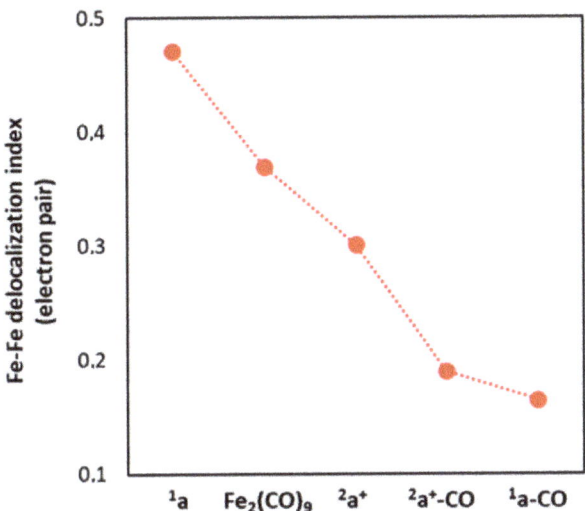

Figure 3. Values of the quantum theory of atoms in molecules (QTAIM) δ(Fe,Fe) delocalization index (in electron pair).

According to the 18-electron counting rule in ^{1}**a** and $Fe_2(CO)_9$, the Fe–Fe bond counts one electron, and this is in line with δ(Fe,Fe) values, both close to 0.5 (0.471 and 0.370, respectively). Going from ^{1}**a** to ^{2}**a**$^+$, the Fe–Fe σ bond highest occupied molecular orbital (HOMO) results singly occupied to the detriment of the Fe–Fe bond, which results weakened according to a δ(Fe,Fe) decreasing of 0.17 electron pair. When a further CO ligand is added to forms ^{1}**a** and ^{2}**a**$^+$, a decrease of δ(Fe,Fe) of 0.306 and 0.111 is observed, respectively, suggesting that the bonding density of the new Fe–C bond comes from the Fe–Fe bond. The value of the δ(Fe,Fe) ^{1}**a**–CO of 0.165 is very low and suggests a very weak or even absent Fe–Fe bond (Figure 3).

2.1.2. ^{1}**a**–CO CO Dissociation Transition States

We investigate three CO dissociation pathways: the dissociation of the *syn cis*, the *syn trans* t-CO and the dissociation of the μ-CO. For each pathway, the corresponding transition state has been characterized.

TS geometry parameters and-free energy barriers are summarized in Figure 4, Figure S1 and in Table 1. The TS structures for the *syn cis* CO dissociation (^{1}**TS1**) retain the C_s symmetry of ^{1}**a**–CO and is characterized by a Fe–C distance of the CO leaving group equal to 2.618 Å. ^{1}**TS1** further evolves first toward ^{1}**a** *syn* rotated form (which is still a TS) and finally to ^{1}**a**. The structure of the *syn trans* terminal CO ^{1}**TS2** is peculiar. Along the *syn trans* CO dissociation pathway, the μ-CO group has substituted the leaving equatorial CO group, which is partially bound to the CO ligand *cis* to the μ-pdt. ^{1}**TS2** further evolves toward a very intriguing local minimum (^{1}**a–CO(2)**). In this structure, the Fe atoms form two Fe–C bonds with a $(CO)_2$ dimer with O–C–C–O atomic disposition. The Fe–C distances are in the middle between those of a terminal and a bridged CO ligand, and the $Fe(CO)_2$ group is almost planar. Finally, along the μ-CO dissociation pathway, the leaving of the μ-CO is accompanied by a conformational rearrangement of the alkyl chain of μ-pdt (^{1}**TS3**) that brings one of the hydrogen atoms of the bidentate ligand closer to Fe. The Fe_2S_2 tetrahedrane unit becomes almost planar (S–Fe–Fe–S dihedral angle equal to 148.2°), with a Fe–Fe distance increased by 0.519 Å compared to ^{1}**a**–CO.

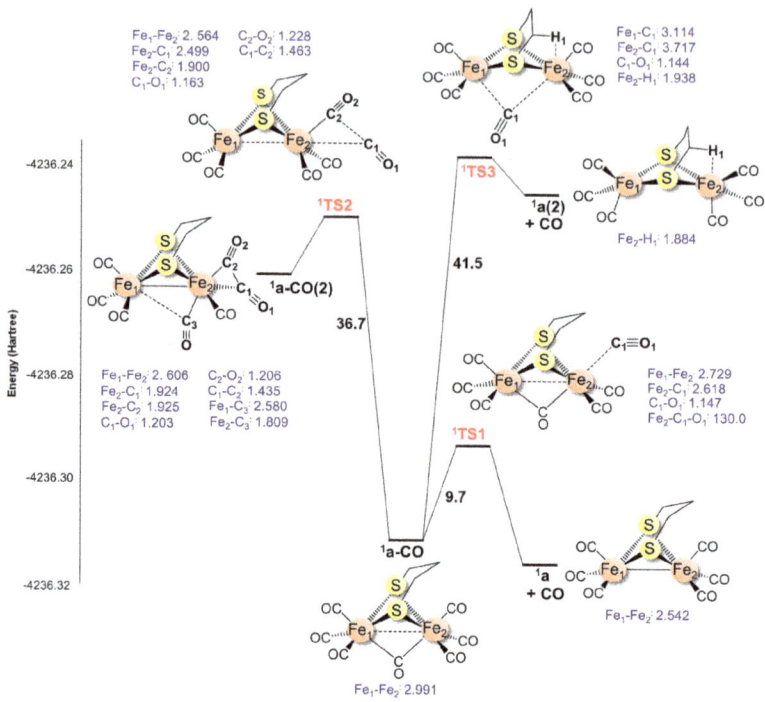

Figure 4. Total density functional theory (DFT) energy diagram and optimized main geometry parameters of ^{1}**a**–CO and *cis syn*, *trans syn* and μ-CO CO dissociation transition states and products. The energy in Hartree and free-energy barriers in kcal·mol^{-1}. Distances are in Å and angles in degrees.

Table 1. CO dissociation-free energy barriers ($\Delta G^{\ddagger}$, in kcal·mol^{-1}). Imaginary normal mode (i) frequency (in cm^{-1}).

	CO	i	$\Delta G^{\ddagger}$
	syn cis	145.5i	9.7
^{1}**a**–CO	*syn trans*	102.3i	36.7
	μ-CO	77.2i	41.5
	syn cis	76.4i	14.6
^{2}**a**$^+$–CO	*syn trans*	124.8i	53.0
	μ-CO	40.6i	33.5
	syn cis	111.4i	11.5
^{3}**a**–CO	*syn trans*	99.6i	12.5
	μ-CO	54.1i	8.2

The structure of ^{1}**TS3** is also very similar to the TS structure found by Greco et al. [27] relative to the H$_2$ evolution of the (μ-pdt)Fe$_2$(CO)$_6$H$_2$ adduct. ^{1}**TS3** further evolves to a local minimum (^{1}**a(2)**) 44.5 kcal·mol^{-1} higher in energy with respect to ^{1}**a**. This fact could be explained by considering this isomer as electronically unsaturated in the framework of the 18 electrons rule because of the breaking of the Fe–Fe bond. Finally, the structure of this isomer resembles the rhombus form of Fe$_2$S$_2$(CO)$_6$ [45], previously characterized as TS along the tetrahedrane-butterfly isomerization path. According to the free energy barrier, only *syn trans* CO dissociation is a chemically available process.

2.1.3. Transition States for $^2\text{a}^+$–CO Dissociation

The TS structures for the *syn cis* CO dissociation ($^2\text{TS1}^+$) is similar to that found on the ^1a–CO PES, but with a Fe–C distance of the leaving group 0.388 Å longer (Figure 5 and Figure S2). At variance with $^1\text{TS3}$, this TS evolves toward the *syn* rotated $^2\text{a}^+$ form, which is the global minimum of the $^2\text{a}^+$ PES. The TS for the *syn trans* CO dissociation ($^2\text{TS2}^+$) is characterized by the simple elongation of the Fe–C distance to 2.519 Å and the bending of the Fe–C–O angle to 139.3°. At the same time, the μ-CO partially loses its bridging character, moving toward the *syn* Fe atom. $^2\text{TS2}^+$ evolves as $^2\text{TS1}^+$ to $^2\text{a}^+_{syn}$. The TS for the μ-CO leaving ($^2\text{TS3}^+$) is characterized by the simple increasing of the Fe–Cμ distances and the decreasing of the Fe–Fe distance and evolving toward the $^2\text{a}^+$ all terminal ligand. Two further pathways for CO release have been characterized, although energetically impeded. The first (through $^2\text{TS4}^+$) is similar to the one found for ^1a–CO, which entails the formation of $^1\text{TS2}$ along the *syn trans* CO dissociation (Figure 4). Here, however, $^2\text{TS4}^+$ evolves towards CO release, giving $^2\text{a}^+$. The other additional pathway implies $^2\text{TS5}^+$, providing a second hypothetical route for μ-CO detachment. In analogy to $^1\text{TS3}$ (Figure 4), this is accompanied by a conformational rearrangement of the alkyl chain of μ-pdt that brings one of the hydrogen atoms of the bidentate ligand closer to Fe. The Fe–C distances of the leaving μ-CO group and the Fe–Fe distance result to be very elongated. This TS structure further evolves to a high-energy local minimum ($^2\text{a}^+(2)$).

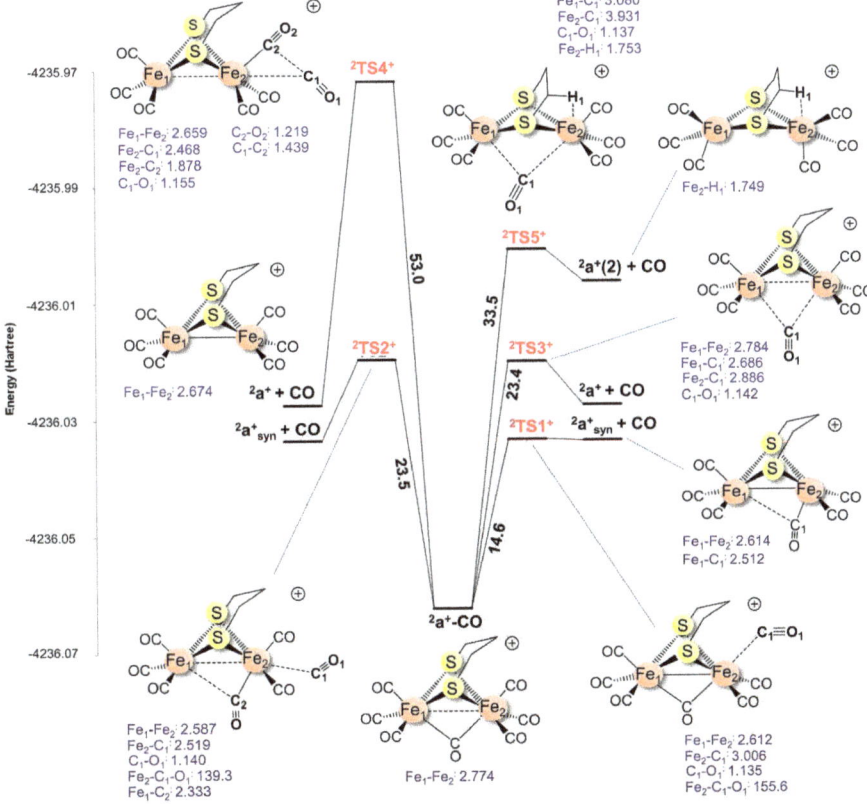

Figure 5. Total DFT energy diagram and optimized main geometry parameters of $^2\text{a}^+$–CO and *cis syn*, *trans syn* and μ-CO CO dissociation transition states and products. The energy in Hartree and free-energy barriers in kcal·mol^{-1}. Distances are in Å and angles in degrees.

2.2. Excited States

2.2.1. Electronic Transitions

The simulated spectra of ^{1}a–CO and ^{2}a$^+$–CO were computed at the optimized ground state geometry (Table 2), covering a spectral range between 350 and 700 nm (see Figure 6).

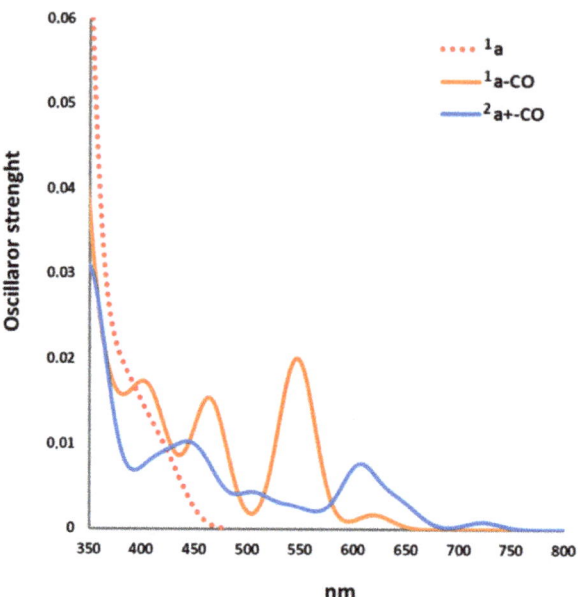

Figure 6. Time-dependent density functional theory (TDDFT) spectra of ^{1}a–CO and ^{2}a$^+$–CO. The spectrum computed for **a** (ref. [33]) at the same level of theory is also reported for comparison. Stick spectra were first generated and then convoluted (solid line) using oscillator strength weighted Gaussian distribution functions centered on the computed excitation energies (nm) with halfwidths at half-maxima of 40 nm.

^{1}a–CO singlet electronic excitations are of HOMO(67a′) → LUMO+n (n = 0–9) type. The first two lowest energy transitions involve HOMO(67a′) → LUMO(37a″) and HOMO(67a′) → LUMO+1(38a″) mono-electronic excitations, respectively. In Figure 7 are reported the HOMO, LUMO and LUMO+1 ^{1}a–CO.

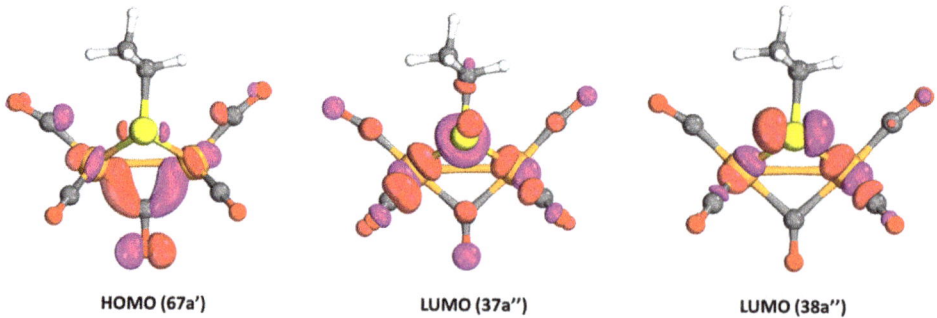

Figure 7. MO isosurfaces (0.05 a.u) for HOMO, LUMO and LUMO+1 of ^{1}a–CO.

According to the FMO shapes and atomic orbital composition (Figure 7):

(i) HOMO is a sort of three center-two electron Fe–Fe bonding MO through μ-CO; (ii) the lowest unoccupied molecular orbital (LUMO) and LUMO+1 are similar in energy and mainly made of Fe-S and *trans* Fe–C orbitals in antibonding combinations. Therefore, the first two excitations have a Fe → S and Fe → *trans* CO charge transfer (CT) character. On average, LUMO+3 to LUMO+9 are characterized by significant contributions of the *trans* CO orbital contribution, and therefore HOMO → LUMO+n (n = 3–9) are Fe,S → *trans* CO CT bands.

The ^{2}**a**$^+$–**CO** electronic spectrum is mainly dominated by the SOMO(67a′(α)) → LUMO+n (n = 0–5) and HOMO-n → 67a′(β). In detail, LUMO+4/LUMO+9 are mainly characterized by strong *trans* CO orbital contributions in Fe–C antibonding combinations. The percentage of Mulliken population of the frontier MOs involved in the excitation is reported in Table S1. None of the unoccupied MOs considered is characterized by strong μ-CO orbitals contributions, although all are antibonding or non-bonding with respect to the Fe–C$_\mu$ bonds. Finally, LUMO+6 and LUMO+7 are also characterized by significant *cis* CO contributions. These results suggest that (i) 9^2A″ and 11^2A″ states have a strong Fe,S → *trans* CO CT character and therefore could have a repulsive character with respect to the *trans* CO dissociation; (ii) none of the excited states considered presents strong indication of a μ-CO repulsive character according to FMO shape and composition; (iii) 9^2A″ state could also be repulsive with respect to *cis* CO dissociation. According to these suggestions, there may be some repulsive excited-state PES along the *trans* or *cis* Fe–C bond elongation, while all excited PESs could be characterized by an energy barrier for the μ-CO dissociation. The scan of the excited-state PESs, presented in the following, has been performed to confirm this picture.

Finally, comparing the spectrum of ^{1}**a**–**CO** with that of ^{1}**a** computed at the same level of theory, we observe that ^{1}**a** HOMO interacting with the CO LUMO in the classical back-donating scheme rises its energy (see Table S2), placing it in the middle of the ^{1}**a** HOMO-LUMO gap and lowering the wavelength of the MLCT band. From this point of view, CO acts as a chromophore. This fact also implies that ^{1}**a**–**CO** and ^{2}**a**$^+$–**CO** excited-state dynamics will be characterized by low-energy μ-CO dissociative channels.

2.2.2. ^{3}**a**–**CO** Lowest Triplet State

The investigation of the ^{3}**a**–**CO**, and in particular of the various CO dissociation TSs on such excited-state PES, was performed using ordinary DFT instead of TDDFT since the latter does not allow for the computation of frequency modes.

The lowest triplet state PES has 1^3A″ electronic term in C$_s$ symmetry, which results from the excitation 67a′ → 37a″. The same minimum structure is also found without any symmetry constraints. According to the shape of the two MOs involved, this excitation can be described as a reorganization of the electron density with a small Fe → S and Fe → *trans* CO CT. By analyzing the natural bond order (NBO) total and spin population, we verified that (i) Fe atoms electron population decreases on average by 0.1 electrons while that of the S atoms increases by 0.074 electrons; (ii) most of the NBO spin population (1.94 electrons) is localized on Fe atoms (0.96 Fe *syn*, 0.36 Fe *anti*), S atoms (0.23 each) and the μ-CO (0.16).

The optimized geometry parameters are reported in Figure 8 and Figure S3. Compared with the singlet ground state (i), the Fe–Fe distance decreases by 0.158 Å; (ii) the μ-CO slightly loses its bridging character (the difference between the shorter and the longer Fe–C distances goes from 0.007 Å (singlet) to 0.281 Å (triplet)); (iii) the *syn* Fe-S distances increase by 0.235 Å. These structural changes are in accordance with the orbital compositions and the characters of the two MOs involved in the monoelectronic excitation. The 67a′ MO has Fe–Fe antibonding and Fe–C$_\mu$ bonding character, while the 37a″ MO composition is mainly made of sulfur and iron orbitals in antibonding combinations. The latter MO also has a *syn trans* Fe–C antibonding character, which accounts for the slight *trans* Fe–C distances increase. It is interesting that no Fe–Fe BCP was found, but the δ(Fe,Fe) value equal to 0.183 is even slightly higher than in ^{1}**a**–**CO**, thus suggesting that this excitation does not involve the Fe–Fe bond.

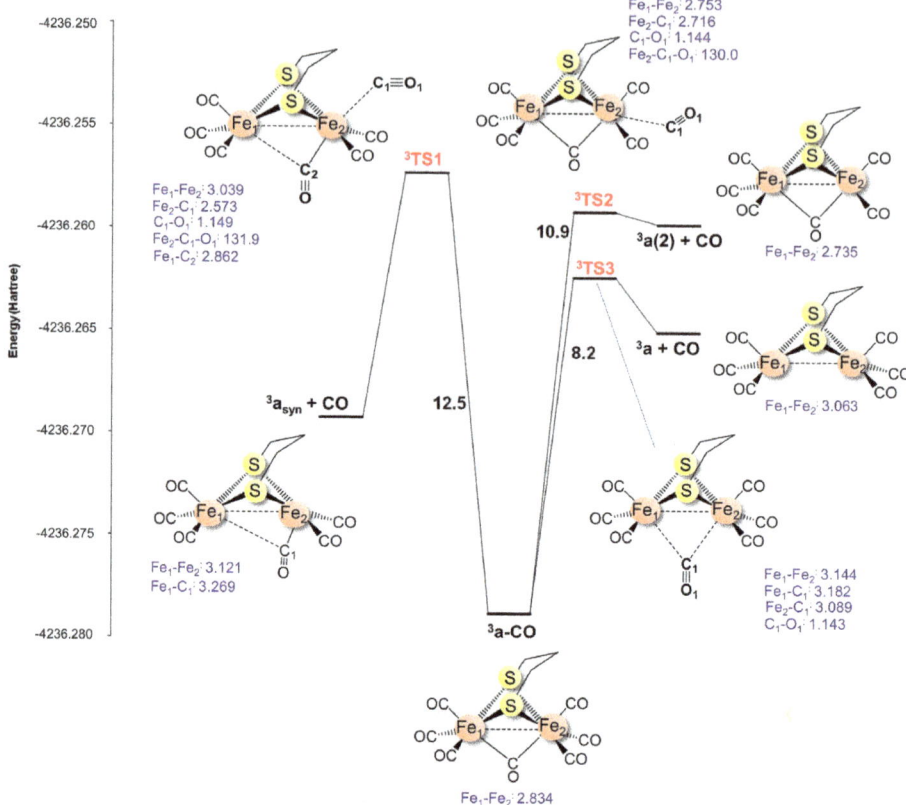

Figure 8. Total DFT energy diagram and optimized main geometry parameters of ^{3}**a**–CO and *cis syn*, *trans syn* and μ-CO CO dissociation transition states and products. The energy in Hartree and free-energy barriers in kcal·mol^{-1}. Distances are in Å and angles in degrees.

The TS structure along the *syn cis* CO dissociation (^{3}**TS1**) pathway is similar to that found on the ground state PES (Figure 8). The same also holds for the-free energy barrier (only 1.9 kcal·mol^{-1} higher in energy with respect to ground state barrier). This TS evolves to the *syn* rotated form ^{3}**a**$_{syn}$, which is also the triplet state PES global minimum. The energy barrier for the *trans* CO and the μ-CO dissociations are predicted to be lower. The TS structure of the *trans* CO dissociation is similar to that found on the cationic ^{2}**a**$^+$–CO PES. The TS along the μ-CO dissociation pathway (^{3}**TS3**) structure presents no conformational rearrangement of the alkyl chain of a bidentate group, and the corresponding-free energy barrier is dramatically lowered by 33.3 kcal·mol^{-1} with respect to ^{1}**TS3**. ^{3}**TS3** evolves to all-terminal triplet ^{3}**a**, which is a local minimum of the triplet PES 2.4 kcal·mol^{-1} higher in energy compared to the ^{3}**a**$_{syn}$ global minimum.

Table 2. Computed excited-state energies, state compositions, and oscillation strengths for **a–CO** and **^{2}a$^+$–CO**. For each transition is reported the molecular term of the corresponding excited state, the excitation energy (in nm), the oscillation strength (f), and the main mono-electronic excitations with the corresponding percentage MO composition. A representation of all relevant MOs involved in electronic transitions can be found in Figure S4.

^{1}a–CO	nm	f	1e	^{2}a$^+$–CO	nm	f	1e
1^1A″	767.8	7·10^{-8}	67a′ → 37a″	1^2A″	764.4	1·10^{-5}	67(α)a′ → 37(α)a″
2^1A″	619.1	2·10^{-3}	67a′ → 38a″	2^2A″	722.1	9·10^{-4}	67(α)a′ → 38(α)a″
1^1A′	546.4	0.02	67a′ → 68a′	1^2A′	652.4	1·10^{-3}	66(β)a′ → 67(β)a′ (65%) 65(β)a′ → 67(β)a′ (19%) 67(α)a′ → 68(α)a′ (12%)
2^1A′	464.1	1·10^{-2}	67a′ → 69a′	2^2A′	637.7	2·10^{-3}	65(β)a′ → 67(β)a′ (46%) 66(β)a′ → 67(β)a′ (33%) 67(α)a′ → 68(α)a′ (17%)
3^1A″	429.9	6·10^{-6}	67a′ → 39a″	3^2A″	608.6	1·10^{-4}	36(β)a″ → 67(β)a′
3^1A′	422.1	1·10^{-3}	67a′ → 70a′	3^2A′	604.9	7·10^{-3}	67(α)a′ → 68(α)a′ (52%) 65(β)a′ → 67(β)a′ (33%) 63(β)a′ → 67(β)a′ (11%)
4^1A″	407.5	1·10^{-3}	67a′ → 40a″	4^2A″	566.2	9·10^{-3}	35(β)a″ → 67(β)a′
5^1A″	405.6	3·10^{-4}	67a′ → 41a″ (63%) 67a′ → 40a″ (34%)	4^2A′	540.1	2·10^{-3}	64(β)a′ → 67(β)a′
4^1A′	403.4	3·10^{-3}	67a′ → 71a′ (69%) 67a′ → 72a′ (27%)	5^2A′	504.8	4·10^{-3}	63(β)a′ → 67(β)a′ (63%) 67(α)a′ → 70(α)a′ (17%) 67(α)a′ → 68(α)a′ (14%)
6^1A″	400.3	3·10^{-4}	66a′ → 37a″ (62%) 67a′ → 41a″ (33%)	5^2A″	475.6	2·10^{-7}	34(β)a″ → 67(β)a′ (81%) 67(α)a′ → 37(α)a″ (16%)
				6^2A′	463.4	4·10^{-3}	67(α)a′ → 69(α)a′
				6^2A″	458.9	2·10^{-5}	67(α)a′ → 39(α)a″ (82%) 34(β)a″ → 67(β)a′ (16%)
				7^2A′	442.5	7·10^{-3}	67(α)a′ → 70(α)a′ (78%) 63(β)a′ → 67(β)a′ (15%)
				7^2A″	433.0	5·10^{-7}	66(α)a′ → 38(α)a″ (42%) 66(β)a′ → 38(β)a″ (27%) 66(β)a′ → 37(β)a″ (14%) 65(β)a′ → 37(β)a″ (5%)
				8^2A″	430.3	3·10^{-7}	65(α)a′ → 37(α)a″ (34%) 65(β)a′ → 37(β)a″ (21%) 65(β)a′ → 38(β)a″ (13%) 66(β)a′ → 38(β)a″ (11%)
				8^2A′	412.8	2·10^{-3}	62(β)a′ → 67(β)a′ (31%) 36(α)a″ → 37(α)a″ (28%) 36(β)a″ → 37(β)a″ (18%) 36(α)a″ → 38(α)a″ (8%)
				9^2A″	414.5	1·10^{-3}	67(α)a′ → 40(α)a″
				10^2A″	410.2	2·10^{-3}	66(α)a′ → 37(α)a″ (33%) 66(β)a′ → 37(β)a″ (21%) 67(α)a′ → 41(α)″ (9%) 66(β)a′ → 38(β)a″ (9%)
				11^2A″	409.9	8·10^{-3}	67(α)a′ → 41(α)a″ (87%) 66(β)a′ → 37(β)a″ (4%)
				9^2A′	408.3	7·10^{-4}	36(α)a″ → 38(α)a″ (33%) 36(β)a″ → 38(β)a″ (31%) 36(α)a″ → 37(α)a″ (15%) 66(β)a′ → 68(β)a′ (5%)
				10^2A′	405.1	1·10^{-3}	62(β)a′ → 67(β)a′ (44%) 36(β)a″ → 37(β)a″ (27%) 36(α)a″ → 37(α)a″ (9%)

183

2.2.3. $^2a^+$–CO $1^2A''$ Excited-State PES Exploration

$1^2A''$ state is the lowest doublet excited state of $^2a^+$–CO, characterized by the 67a' (α) → 37a''(α) mono-electronic excitation. Analogously to ^3a–CO, this state has both Fe → S and Fe → CO$_{trans}$ MLCT characters. It is possible to explore the corresponding PES at DFT level by imposing the C_s symmetry and the proper MO occupancy (66a'(2)67a'(α)(0)37a''(α)(1)) in order to constrain the wavefunction to the correct electronic term. Using this approach, one can optimize the structure, providing to maintain the molecular symmetry throughout the optimizations. This last point is clearly a limitation that does not allow us to study the structure of the TS. For these reasons, the results obtained cannot be compared in the same way we did for ^1a–CO and its lowest triplet state PES but provide insights on the topology of the lowest singlet excited-state PES of $^2a^+$–CO, which is the closest model of the first $^2a^+$–CO excited state.

As for the minimum structure, starting from $^2a^+$–CO, the structure converges to a local minimum (Figure 9). The optimized geometric parameters are similar to those obtained for the ^3a–CO triplet state minimum structure since the mono-electronic excitation is identical. With respect to the $^2a^+$–CO ground state, the Fe–Fe distance decreases by 0.1 Å and *syn* Fe-S distances are increased by 0.229 Å.

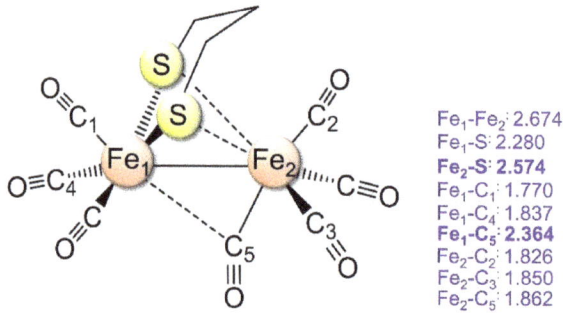

Fe$_1$-Fe$_2$: 2.674
Fe$_1$-S: 2.280
Fe$_2$-S: 2.574
Fe$_1$-C$_1$: 1.770
Fe$_1$-C$_4$: 1.837
Fe$_1$-C$_5$: 2.364
Fe$_2$-C$_2$: 1.826
Fe$_2$-C$_3$: 1.850
Fe$_2$-C$_5$: 1.862

Figure 9. DFT-optimized geometry of the $^2a^+$–CO $1^2A''$ excited state at DFT level imposing the proper MO occupancy. Distances in Å.

Moreover, we observe a clear loss of the bridging character of the μ-CO, similarly to the case of the μ-H models [46]. The analysis of the NBO atomic and spin populations enlighten the electronic nature of this excited form. Upon excitation is observed (i) a net CT of 0.4 e Fe atoms to *trans* CO ligand induced by the population of the 37a''(α); (ii) a Fe atomic charges difference of 0.32 e (compared to that of $^2a^+$–CO ground state equal 0.03 e); (iii) a spin localization on distal Fe suggesting a Fe$_d^{II}$Fe$_p^I$ electronic state. This brings to the *trans* Fe–C bonds elongation and, at the same time, determines the Fe–Fe bond shortening. As a side effect, the Fe$_d$-S is highly elongated.

2.2.4. $^2a^+$–CO Excited States PES Scanning

We considered the scanning of the first 20 excited-states PES along the *syn cis*, *trans cis* and bridged CO dissociation paths starting from the ground state geometry. In Figure 10 are reported the excitation energies along each path for the 20 excited states considered. The scanning along the *syn cis* CO dissociation path does not evidence the dissociative nature of the PESs considered. The scanning along the *syn trans* CO dissociation path evidences at least two dissociative excited-state PESs, as suggested by the complex succession of avoided MO crossing that involves the $7^2A''$ and $11^2A''$ states. Similarly, the scanning along the μ-CO dissociation path shows an avoided crossing between $1^2A''$ and the $^2A'$ ground state. The other excited PESs are bound with respect to this dissociation.

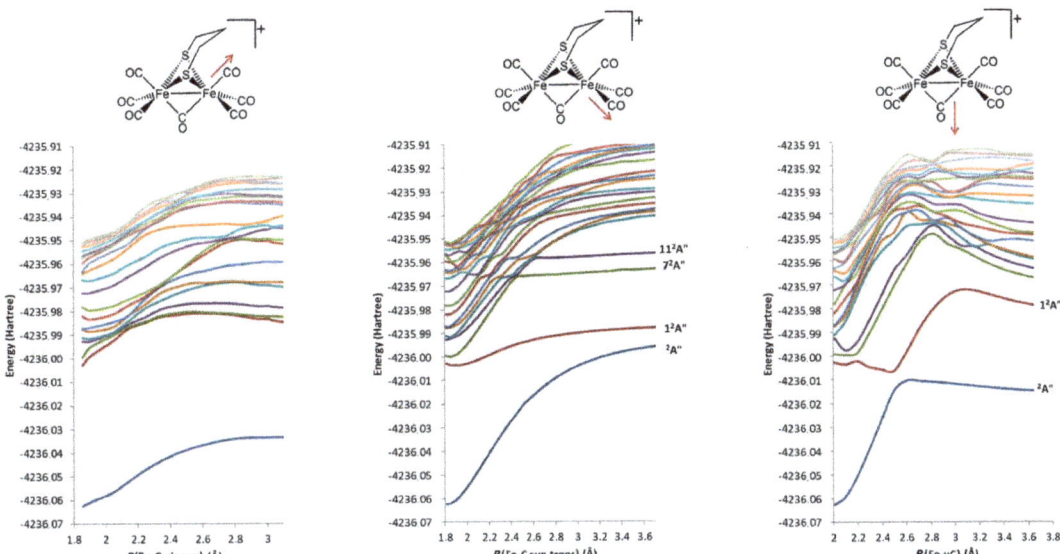

Figure 10. Potential energy surfaces of the ground state and the first 20 singlet excited states of $^2\mathbf{a}^+$–CO along the apical and equatorial Fe–CO stretching coordinates (in Å). In *syn trans* and μ Fe–CO plots are also evidenced the excitations whose dynamics lead to the CO photolysis after a series of internal conversions.

As a final step in the investigation of the first $^2\mathbf{a}^+$–CO excited state, we optimized its geometry at TDDFT level starting from the 2.2 Å *syn trans* Fe–CO elongated structure identified in the PES scans reported in Figure 10.

TDDFT-optimization energy profile is reported in Figure 11. The optimization does not converge to a stationary point, and the lowest energy structure before the conical intersection is characterized by a *syn trans* Fe–C distance equal to 1.840 Å, in line with the PES scan.

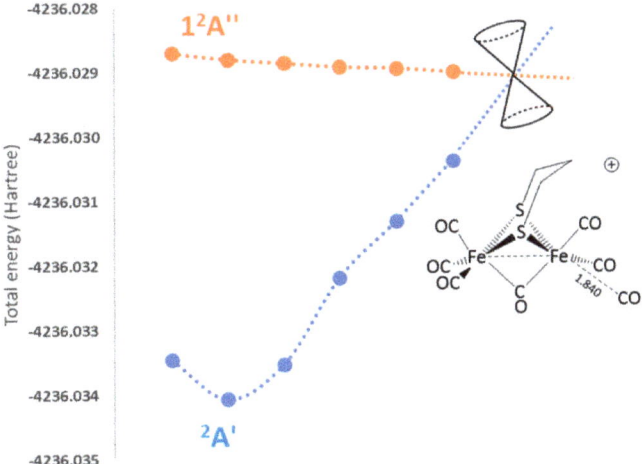

Figure 11. TDDFT-optimization energy profiles for the first excited state of $^2\mathbf{a}^+$–CO (1^2A″ in orange), starting from the 2.200 Å *syn trans* Fe–CO elongated structure, evidencing the conical intersection with the ground state (in blue) reached after 7 optimization cycles at Fe–CO 1.840 Å. Energies in Hartree.

3. Methods

All computations were carried out using the TURBOMOLE suite of programs [47,48]. The pure gradient generalized approximation (GGA) BP86 [49,50] DFT functional was chosen, in conjunction with a triple-ζ plus polarization split valence quality (TZVP) basis set [51], adopted for all atoms. The resolution-of-identity technique was also adopted [52]. The choice of such a level of theory is justified by its satisfactory performances in the modeling of hydrogenase-related systems [53–56]. The optimization of transition state structures on the ground state PESs was carried out according to a procedure based on a pseudo-Newton–Raphson method. All ground state and excited-state geometry optimizations were carried out with convergence criteria fixed 10^{-7} Hartree for the energy and 0.001 Hartree·Å^{-1} for the gradient norm vector. This computational setting provides ground state geometry parameters in concordance with experimental X-ray values and a reasonable picture for excited-state PES properties.

4. Conclusions

The ^{1}**a–CO** and ^{2}**a$^+$–CO** can be considered as simple models of the [FeFe]-hydrogenase catalytic site in its CO inhibited form. More in general, they are also quite simple models of oversaturated metallo-carbonyl complexes, and the investigation of their photochemistry can be useful to deeply understand the nature of the CO photo-dissociative channels in these moieties.

The ^{1}**a–CO** and ^{2}**a$^+$–CO** ground state PESs exploration at DFT level evidence that the two complexes are bound with respect to any CO dissociation paths, being the lowest-free energy barriers for the terminal *cis* CO dissociation as large as 9.2 and 14.6 kcal·mol^{-1}, respectively. The other two barriers are much higher and all characterized by a significant distortion of the molecular geometry. The analysis of the electron density at the QTAIM level shows that the stability of ^{1}**a–CO** and ^{2}**a$^+$–CO** is possible to the detriment of the Fe−Fe bonds: the bonding density to form the new Fe–C bond comes mainly from the Fe–Fe bond, according to the decrease of the δ(Fe,Fe) going from ^{1}**a** and ^{2}**a$^+$** to ^{1}**a–CO** and ^{2}**a$^+$–CO**. These results also suggest that at least ^{2}**a$^+$–CO** could be synthetically achieved.

The investigation of the nature and reactivity of the excited states was carried out in two ways, first at DFT level by imposing spin-triplet state MO occupancy or exploiting the symmetry features of the two models imposing their C$_s$ symmetry; then we performed a PES scan at TDDFT level with no symmetry constraints.

The lowest triplet state of ^{3}**a–CO** is characterized by the HOMO → LUMO mono-electronic excitation, and it is bound with respect to any CO dissociations. The global minimum shows a Fe–Fe bond decreasing and a *syn* Fe-S bond increasing that mirror the features of the LUMO. Since the excitation does not involve Fe–Fe bonding/antibonding MOs, the delocalization index δ(Fe,Fe) is similar to that of ^{1}**a–CO**. The-free energy barriers for *trans syn* and μ-CO dissociation result to be dramatically lowered with respect to those computed for ^{1}**a–CO**. The corresponding TS structures do not show significant geometry distortions. μ-CO dissociation presents the lowest free-energy barrier, and the TS evolves toward the ^{3}**a** all-terminal form, a local minimum on the lowest triplet state PES.

The 1^{2}A″ ^{2}**a$^+$–CO** PES was explored at the DFT level by imposing the C$_s$ symmetry and the proper MO occupancy. Geometry optimization converges to a local minimum whose structure is similar to ^{3}**a–CO**, but with a larger loss of the bridging character of the μ-CO.

The computations for the excited state at DFT level (^{3}**a–CO** 1^{2}A″ ^{2}**a$^+$–CO**) strongly suggest that upon excitation, all CO dissociation pathways are photochemically available compared to the ground state, in particular, the *trans syn* and μ-CO ones. This latter conclusion is corroborated by the scan of the ^{2}**a$^+$–CO** excited PES at the TDDFT level. We observe a clear sequence of internal conversions along the *trans syn* CO (7^{2}A″ and 11^{2}A″ states) and μ-CO dissociation pathways. Moreover, these channels are available upon irradiation in the visible range, populating a Fe,S → CO MLCT band.

A further aspect is related to the differences among ^{1}a–CO/^{1}a and ^{2}a$^+$–CO/^{2}a$^+$ excited state dynamics. Indeed, the CO inhibited models do not show any photo-isomerization processes as in the case of ^{1}a and ^{2}a$^+$. This fact suggests a sort of regioselectivity for the CO photolysis of ^{1}a–CO and ^{2}a$^+$–CO in which only *trans* and bridging CO are involved. In conclusion, we can reasonably claim that the low-energy excitations of an oversaturated metallo-carbonyl complex such as those investigated here are characterized by at least one CO dissociative channel, most likely of a terminal ligand. This result can be useful also for a better rationalization of photolysis occurring in the CO inhibited [FeFe]-hydrogenase enzyme. Indeed, it seems reasonable that CO acts as a sort of photoprotector for the H-cluster since it is able to capture visible light when bound to the [2Fe]$_H$ without affecting its functionality. Even when the light in the UV energy range is absorbed, during the excited state dynamics, the system decays through photo-dissociative channels that involve a CO ligand, preventing the irreversible enzyme photo-damage characterized by Sensi et al. [36,37].

Supplementary Materials: The following are available online at https://www.mdpi.com/2304-6740/9/2/16/s1, Figures S1–S3: optimized structures and geometrical parameters of minima and transition states involved in CO dissociation from ^{1}a–CO, ^{2}a$^+$–CO, ^{3}a–CO, respectively; Figure S4: relevant ^{1}a–CO MOs; Table S1: Mulliken population analysis of ^{2}a$^+$–CO FMOs; Table S2: MOs eigenvalues for ^{1}a–CO and ^{1}a; Cartesian coordinates of ^{1}a–CO, ^{2}a$^+$–CO, ^{3}a–CO, ^{2}a$^+$–CO 1^{2}A″.

Author Contributions: Conceptualization, L.B.; investigation and formal analysis, F.A., C.G., L.D.G., L.B.; data curation, F.A., L.B., C.G., G.Z.; writing—original draft preparation, L.B., F.A.; writing—review and editing, F.A., L.B., C.G., G.Z., L.D.G. All authors have read and agreed to the published version of the manuscript.

Funding: This research received no external funding.

Institutional Review Board Statement: No applicable.

Informed Consent Statement: No applicable.

Data Availability Statement: Data is contained within the article or supplementary material.

Acknowledgments: The CINECA consortium is gratefully acknowledged for the HPC resources provided in the context of the CINECA-UNIMIB agreement.

Conflicts of Interest: The authors declare no conflict of interest.

References

1. Lubitz, W.; Ogata, H.; Rüdiger, O.; Reijerse, E. Hydrogenases. *Chem. Rev.* **2014**, *114*, 4081–4148. [CrossRef]
2. Kaur-Ghumaan, S.; Stein, M. [NiFe] hydrogenases: How close do structural and functional mimics approach the active site? *Dalton Trans.* **2014**, *43*, 9392–9405. [CrossRef]
3. Kleinhaus, J.T.; Wittkamp, F.; Yadav, S.; Siegmund, D.; Apfel, U.-P. [FeFe]-Hydrogenases: Maturation and reactivity of enzymatic systems and overview of biomimetic models. *Chem. Soc. Rev.* **2020**. [CrossRef]
4. Tard, C.; Pickett, C.J. Structural and Functional Analogues of the Active Sites of the [Fe]-, [NiFe]-, and [FeFe]-Hydrogenases†. *Chem. Rev.* **2009**, *109*, 2245–2274. [CrossRef]
5. Frey, M. Hydrogenases: Hydrogen-activating enzymes. *Chembiochem* **2002**, *3*, 153–160. [CrossRef]
6. Reihlen, H.; Gruhl, A.; v. Hessling, G. Über den photochemischen und oxydativen Abbau von Carbonylen. *Justus Liebig's Ann. Chem.* **1929**, *472*, 268–287. [CrossRef]
7. Peters, J.W.; Lanzilotta, W.N.; Lemon, B.J.; Seefeldt, L.C. X-ray crystal structure of the Fe-only hydrogenase (CpI) from Clostridium pasteurianum to 1.8 angstrom resolution. *Science* **1998**, *282*, 1853–1858. [CrossRef]
8. Arrigoni, F.; Bertini, L.; Breglia, R.; Greco, C.; De Gioia, L.; Zampella, G. Catalytic H$_2$ evolution/oxidation in [FeFe]-hydrogenase biomimetics: Account from DFT on the interplay of related issues and proposed solutions. *New J. Chem.* **2020**, *44*, 17596–17615. [CrossRef]
9. Camara, J.M.; Rauchfuss, T.B. Combining acid-base, redox and substrate binding functionalities to give a complete model for the [FeFe]-hydrogenase. *Nat. Chem.* **2011**, *4*, 26–30. [CrossRef]
10. Greco, C. H$_2$ binding and splitting on a new-generation [FeFe]-hydrogenase model featuring a redox-active decamethylferrocenyl phosphine ligand: A theoretical investigation. *Inorg. Chem.* **2013**, *52*, 1901–1908. [CrossRef]
11. Fan, H.J.; Hall, M.B. A capable bridging ligand for Fe-only hydrogenase: Density functional calculations of a low-energy route for heterolytic cleavage and formation of dihydrogen. *J. Am. Chem. Soc.* **2001**, *123*, 3828–3829. [CrossRef] [PubMed]

12. Lacey, A.L.D.; De Lacey, A.L.; Stadler, C.; Cavazza, C.; Claude Hatchikian, E.; Fernandez, V.M. FTIR Characterization of the Active Site of the Fe-hydrogenase fromDesulfovibrio desulfuricans. *J. Am. Chem. Soc.* **2000**, *122*, 11232–11233. [CrossRef]
13. Lemon, B.J.; Peters, J.W. Binding of exogenously added carbon monoxide at the active site of the fe-only hydrogenase (CpI) from *clostridium pasteurianum*. *Biochemistry* **1999**, *38*, 12969–12973. [CrossRef]
14. Greco, C.; Bruschi, M.; Heimdal, J.; Fantucci, P.; De Gioia, L.; Ryde, U. Structural insights into the active-ready form of [FeFe]-hydrogenase and mechanistic details of its inhibition by carbon monoxide. *Inorg. Chem.* **2007**, *46*, 7256–7258. [CrossRef] [PubMed]
15. Bertsch, J.; Müller, V. Bioenergetic constraints for conversion of syngas to biofuels in acetogenic bacteria. *Biotechnol. Biofuels* **2015**, *8*, 210. [CrossRef]
16. Zilberman, S.; Stiefel, E.I.; Cohen, M.H.; Car, R. Resolving the CO/CN ligand arrangement in CO-inactivated [FeFe] hydrogenase by first principles density functional theory calculations. *Inorg. Chem.* **2006**, *45*, 5715–5717. [CrossRef]
17. Duan, J.; Mebs, S.; Laun, K.; Wittkamp, F.; Heberle, J.; Happe, T.; Hofmann, E.; Apfel, U.-P.; Winkler, M.; Senger, M.; et al. Geometry of the Catalytic Active Site in [FeFe]-Hydrogenase Is Determined by Hydrogen Bonding and Proton Transfer. *ACS Catal.* **2019**, *9*, 9140–9149. [CrossRef]
18. Razavet, M.; Borg, S.J.; George, S.J.; Best, S.P.; Fairhurst, S.A.; Pickett, C.J. Transient FTIR spectroelectrochemical and stopped-flow detection of a mixed valence Fe(i)–Fe(ii) bridging carbonyl intermediate with structural elements and spectroscopic characteristics of the di-iron sub-site of all-iron hydrogenase. *Chem. Comm.* **2002**, 700–701. [CrossRef]
19. Greco, C.; Bruschi, M.; Fantucci, P.; De Gioia, L. Relation between coordination geometry and stereoelectronic properties in DFT models of the CO-inhibited [FeFe]-hydrogenase cofactor. *J. Organomet. Chem.* **2009**, *694*, 2846–2853. [CrossRef]
20. Baffert, C.; Bertini, L.; Lautier, T.; Greco, C.; Sybirna, K.; Ezanno, P.; Etienne, E.; Soucaille, P.; Bertrand, P.; Bottin, H.; et al. CO disrupts the reduced H-cluster of FeFe hydrogenase. A combined DFT and protein film voltammetry study. *J. Am. Chem. Soc.* **2011**, *133*, 2096–2099. [CrossRef]
21. Bitterwolf, T.E. Organometallic photochemistry at the end of its first century. *J. Organomet. Chem.* **2004**, *689*, 3939–3952. [CrossRef]
22. Chen, Z.; Lemon, B.J.; Huang, S.; Swartz, D.J.; Peters, J.W.; Bagley, K.A. Infrared Studies of the CO-Inhibited Form of the Fe-Only Hydrogenase from *Clostridium pasteurianum* I: Examination of Its Light Sensitivity at Cryogenic Temperatures†. *Biochemistry* **2002**, *41*, 2036–2043. [CrossRef]
23. Albracht, S.P.J.; Roseboom, W.; Claude Hatchikian, E. The active site of the [FeFe]-hydrogenase from Desulfovibrio desulfuricans. I. Light sensitivity and magnetic hyperfine interactions as observed by electron paramagnetic resonance. *J. Biol. Inorg. Chem.* **2006**, *11*, 88–101. [CrossRef]
24. Roseboom, W.; De Lacey, A.L.; Fernandez, V.M.; Claude Hatchikian, E.; Albracht, S.P.J. The active site of the [FeFe]-hydrogenase from Desulfovibrio desulfuricans. II. Redox properties, light sensitivity and CO-ligand exchange as observed by infrared spectroscopy. *J. Biol. Inorg. Chem.* **2006**, *11*, 102–118. [CrossRef]
25. Darensbourg, M.Y.; Lyon, E.J.; Zhao, X.; Georgakaki, I.P. The organometallic active site of [Fe]hydrogenase: Models and entatic states. *Proc. Natl. Acad. Sci. USA* **2003**, *100*, 3683–3688. [CrossRef]
26. Gloaguen, F.; Rauchfuss, T.B. Small molecule mimics of hydrogenases: Hydrides and redox. *Chem. Soc. Rev.* **2009**, *38*, 100–108. [CrossRef]
27. Greco, C.; Zampella, G.; Bertini, L.; Bruschi, M.; Fantucci, P.; De Gioia, L. Insights into the mechanism of electrocatalytic hydrogen evolution mediated by $Fe_2(S_2C_3H_6)(CO)_6$: The simplest functional model of the Fe-hydrogenase active site. *Inorg. Chem.* **2007**, *46*, 108–116. [CrossRef]
28. Bertini, L.; Greco, C.; De Gioia, L.; Fantucci, P. DFT/TDDFT Exploration of the Potential Energy Surfaces of the Ground State and Excited States of $Fe_2(S_2C_3H_6)(CO)_6$: A Simple Functional Model of the [FeFe] Hydrogenase Active Site. *J. Phys. Chem. A.* **2009**, *113*, 5657–5670. [CrossRef]
29. Tyler, D.R. 19-Electron organometallic adducts. *Acc. Chem. Res.* **1991**, *24*, 325–331. [CrossRef]
30. Astruc, D. Nineteen-electron complexes and their role in organometallic mechanisms. *Chem. Rev.* **1988**, *88*, 1189–1216. [CrossRef]
31. Thauer, R.K.; Käufer, B.; Zähringer, M.; Jungermann, K. The reaction of the iron-sulfur protein hydrogenase with carbon monoxide. *Eur. J. Biochem.* **1974**, *42*, 447–452. [CrossRef]
32. Bertini, L.; Alberto, M.E.; Arrigoni, F.; Vertemara, J.; Fantucci, P.; Bruschi, M.; Zampella, G.; De Gioia, L. On the photochemistry of $Fe_2(edt)(CO)_4(PMe_3)_2$, a [FeFe]-hydrogenase model: A DFT/TDDFT investigation. *Int. J. Quantum Chem.* **2018**, *118*, e25537. [CrossRef]
33. Bertini, L.; Greco, C.; Fantucci, P.; De Gioia, L. TDDFT modeling of the CO-photolysis of $Fe_2(S_2C_3H_6)(CO)_6$, a model of the [FeFe]-hydrogenase catalytic site. *Int. J. Quantum Chem.* **2014**, *114*, 851–861. [CrossRef]
34. Fletcher, S.C.; Poliakoff, M.; Turner, J.J. Structure and reactions of octacarbonyldiiron: An IR spectroscopic study using carbon-13 monoxide, photolysis with plane-polarized light, and matrix isolation. *Inorg. Chem.* **1986**, *25*, 3597–3604. [CrossRef]
35. Bertini, L.; Bruschi, M.; De Gioia, L.; Fantucci, P. Structure and energetics of $Fe_2(CO)_8$ singlet and triplet electronic states. *J. Phys. Chem. A* **2007**, *111*, 12152–12162. [CrossRef] [PubMed]
36. Sensi, M.; Baffert, C.; Greco, C.; Caserta, G.; Gauquelin, C.; Saujet, L.; Fontecave, M.; Roy, S.; Artero, V.; Soucaille, P.; et al. Reactivity of the Excited States of the H-Cluster of FeFe Hydrogenases. *J. Am. Chem. Soc.* **2016**, *138*, 13612–13618. [CrossRef]
37. Sensi, M.; Baffert, C.; Fradale, L.; Gauquelin, C.; Soucaille, P.; Meynial-Salles, I.; Bottin, H.; de Gioia, L.; Bruschi, M.; Fourmond, V.; et al. Photoinhibition of FeFe Hydrogenase. *ACS Catal.* **2017**, *7*, 7378–7387. [CrossRef]

38. Bertini, L.; Greco, C.; Bruschi, M.; Fantucci, P.; De Gioia, L. CO Affinity and Bonding Properties of [FeFe] Hydrogenase Active Site Models. A DFT Study. *Organometallics* **2010**, *29*, 2013–2025. [CrossRef]
39. Arrigoni, F.; Mohamed Bouh, S.; De Gioia, L.; Elleouet, C.; Pétillon, F.Y.; Schollhammer, P.; Zampella, G. Influence of the Dithiolate Bridge on the Oxidative Processes of Diiron Models Related to the Active Site of [FeFe] Hydrogenases. *Chem. Eur. J.* **2017**, *23*, 4364–4372. [CrossRef]
40. Arrigoni, F.; Bouh, S.M.; Elleouet, C.; Pétillon, F.Y.; Schollhammer, P.; De Gioia, L.; Zampella, G. Electrochemical and Theoretical Investigations of the Oxidatively Induced Reactivity of the Complex [$Fe_2(CO)_4(\kappa^2$-dmpe$)(\mu$-adtBn$)$] Related to the Active Site of [FeFe] Hydrogenases. *Chem. Eur. J.* **2018**, *24*, 15036–15051. [CrossRef]
41. Chouffai, D.; Zampella, G.; Capon, J.-F.; De Gioia, L.; Le Goff, A.; Pétillon, F.Y.; Schollhammer, P.; Talarmin, J. Electrochemical and Theoretical Studies of the Impact of the Chelating Ligand on the Reactivity of [$Fe_2(CO)_4(\kappa^2$-LL$)(\mu$-pdt$)$] Complexes with Different Substrates (LL = IMe-CH_2-IMe, dppe; IMe = 1-Methylimidazol-2-ylidene). *Organometallics* **2012**, *31*, 1082–1091. [CrossRef]
42. Macchi, P. Chemical bonding in transition metal carbonyl clusters: Complementary analysis of theoretical and experimental electron densities. *Coord. Chem. Rev.* **2003**, *238–239*, 383–412. [CrossRef]
43. Gatti, C.; Lasi, D. Source function description of metal-metal bonding in d-block organometallic compounds. *Faraday Discuss.* **2007**, *135*, 55–78. [CrossRef] [PubMed]
44. Bertini, L.; Greco, C.; De Gioia, L.; Fantucci, P. Time-dependent density functional theory study of $Fe_2(CO)_9$ low-lying electronic excited states. *J. Phys. Chem. A* **2006**, *110*, 12900–12907. [CrossRef]
45. Bertini, L.; Fantucci, P.; De Gioia, L. On the Photochemistry of the Low-Lying Excited State of $Fe_2(CO)_6S_2$. A DFT and QTAIM Investigation. *Organometallics* **2011**, *30*, 487–498. [CrossRef]
46. Bertini, L.; Fantucci, P.; De Gioia, L.; Zampella, G. Excited state properties of diiron dithiolate hydrides: Implications in the unsensitized photocatalysis of H_2 evolution. *Inorg. Chem.* **2013**, *52*, 9826–9841. [CrossRef]
47. Balasubramani, S.G.; Chen, G.P.; Coriani, S.; Diedenhofen, M.; Frank, M.S.; Franzke, Y.J.; Furche, F.; Grotjahn, R.; Harding, M.E.; Hättig, C.; et al. TURBOMOLE: Modular program suite for ab initio quantum-chemical and condensed-matter simulations. *J. Chem. Phys.* **2020**, *152*, 184107. [CrossRef] [PubMed]
48. Ahlrichs, R.; Bär, M.; Häser, M.; Horn, H.; Kölmel, C. Electronic structure calculations on workstation computers: The program system turbomole. *Chem. Phys. Lett.* **1989**, *162*, 165–169. [CrossRef]
49. Becke, A.D. Density-functional exchange-energy approximation with correct asymptotic behavior. *Phys. Rev. A Gen. Phys.* **1988**, *38*, 3098–3100. [CrossRef] [PubMed]
50. Perdew, J.P. Density-functional approximation for the correlation energy of the inhomogeneous electron gas. *Phys. Rev. B Condens. Matter* **1986**, *33*, 8822–8824. [CrossRef]
51. Schäfer, A.; Huber, C.; Ahlrichs, R. Fully optimized contracted Gaussian basis sets of triple zeta valence quality for atoms Li to Kr. *J.Chem. Phys.* **1994**, *100*, 5829–5835. [CrossRef]
52. Eichkorn, K.; Weigend, F.; Treutler, O.; Ahlrichs, R. Optimized accurate auxiliary basis sets for RI-MP2 and RI-CC2 calculations for the atoms Rb to Rn. *Theor. Chem. Acc.* **1997**, *97*, 119–124. [CrossRef]
53. Goy, R.; Bertini, L.; Rudolph, T.; Lin, S.; Schulz, M.; Zampella, G.; Dietzek, B.; Schacher, F.H.; De Gioia, L.; Sakai, K.; et al. Photocatalytic Hydrogen Evolution Driven by [FeFe] Hydrogenase Models Tethered to Fluorene and Silafluorene Sensitizers. *Chem. Eur. J.* **2017**, *23*, 334–345. [CrossRef] [PubMed]
54. Arrigoni, F.; Rizza, F.; Vertemara, J.; Breglia, R.; Greco, C.; Bertini, L.; Zampella, G.; De Gioia, L. Rational Design of Fe_2 (μ-PR_2)(L)$_6$ Coordination Compounds Featuring Tailored Potential Inversion. *Chemphyschem* **2020**, *21*, 2279–2292. [CrossRef]
55. Greco, C.; Fantucci, P.; De Gioia, L.; Suarez-Bertoa, R.; Bruschi, M.; Talarmin, J.; Schollhammer, P. Electrocatalytic dihydrogen evolution mechanism of [$Fe_2(CO)_4(\kappa^2$-$Ph_2PCH_2CH_2PPh_2)(\mu$-S$(CH_2)_3$S$)$] and related models of the [FeFe]-hydrogenases active site: A DFT investigation. *Dalton Trans.* **2010**, *39*, 7320. [CrossRef] [PubMed]
56. Sicolo, S.; Bruschi, M.; Bertini, L.; Zampella, G.; Filippi, G.; Arrigoni, F.; De Gioia, L.; Greco, C. Towards biomimetic models of the reduced [FeFe]-hydrogenase that preserve the key structural features of the enzyme active site; a DFT investigation. *Int. J. Hydrog. Energy* **2014**, *39*, 18565–18573. [CrossRef]

Article

Synthesis and Characterisation of Molecular Polarised-Covalent Thorium-Rhenium and -Ruthenium Bonds

Joseph P. A. Ostrowski, Ashley J. Wooles and Stephen T. Liddle *

Department of Chemistry, The University of Manchester, Oxford Road, Manchester M13 9PL, UK; joe-ostrowski@outlook.com (J.P.A.O.); ashley.wooles@manchester.ac.uk (A.J.W.)
* Correspondence: steve.liddle@manchester.ac.uk; Tel.: +44-161-275-4612

Abstract: Separate reactions of $[Th\{N(CH_2CH_2NSiMe_2Bu^t)_2(CH_2CH_2NSi(Me)(Bu^t)(\mu-CH_2)\}]_2$ (**1**) with $[Re(\eta^5-C_5H_5)_2(H)]$ (**2**) or $[Ru(\eta^5-C_5H_5)(H)(CO)_2]$ (**3**) produced, by alkane elimination, $[Th(Tren^{DMBS})Re(\eta^5-C_5H_5)_2]$ (**ThRe**, $Tren^{DMBS} = \{N(CH_2CH_2NSiMe_2Bu^t)_3\}^{3-}$), and $[Th(Tren^{DMBS})Ru(\eta^5-C_5H_5)(CO)_2]$ (**ThRu**), which were isolated in crystalline yields of 71% and 62%, respectively. Complex **ThRe** is the first example of a molecular Th-Re bond to be structurally characterised, and **ThRu** is only the second example of a structurally authenticated Th-Ru bond. By comparison to isostructural U-analogues, quantum chemical calculations, which are validated by IR and Raman spectroscopic data, suggest that the Th-Re and Th-Ru bonds reported here are more ionic than the corresponding U-Re and U-Ru bonds.

Keywords: thorium; rhenium; ruthenium; metal–metal bonds; uranium; density functional theory

Citation: Ostrowski, J.P.A.; Wooles, A.J.; Liddle, S.T. Synthesis and Characterisation of Molecular Polarised-Covalent Thorium-Rhenium and -Ruthenium Bonds. Inorganics 2021, 9, 30. https://doi.org/10.3390/inorganics9050030

Academic Editor: Duncan H. Gregory

Received: 13 April 2021
Accepted: 20 April 2021
Published: 21 April 2021

Publisher's Note: MDPI stays neutral with regard to jurisdictional claims in published maps and institutional affiliations.

Copyright: © 2021 by the authors. Licensee MDPI, Basel, Switzerland. This article is an open access article distributed under the terms and conditions of the Creative Commons Attribution (CC BY) license (https:// creativecommons.org/licenses/by/ 4.0/).

1. Introduction

The field of metal–metal bonds is mature and vast with many applications [1–5]. Whilst homopolymetallic d transition metal systems dominate this still burgeoning field [1–3], heteropolymetallic derivatives have been gaining prominence [3–5]. Where actinide elements are concerned, in the absence of isolable homopolymetallic species under normal conditions [6], a growing number of heteropolymetallic derivatives have been isolated and characterised [7–11], including Th and U complexes exhibiting bonds to d transition metals such as Mo [12], Fe [13–19], Ru [13,14,19,20], Re [21–23], Co [24–26], Rh [27–29], Ni [30–35], Pd [31], Pt [31,36], Cu [37,38], and Ag [39], and p main group metals and metalloids such as Al [40,41], Ga [41,42], Si [43–46], Ge [47], Sn [48,49], Sb [50], and Bi [50–52].

For some time now we have had an interest in the study of f-block-metal bonds, but most of our efforts have focussed on the synthesis and characterisation of U-derivatives. Of particular pertinence to this paper are the U-Re and U-Ru complexes $[U(Tren^{DMBS})Re(\eta^5-C_5H_5)_2]$ (**URe**) [22] and $[U(Tren^{DMBS})Ru(\eta^5-C_5H_5)(CO)_2]$ (**URu**) [20], respectively ($Tren^{DMBS} = \{N(CH_2CH_2NSiMe_2Bu^t)_3\}^{3-}$). During our studies, we noted that there are no examples of structurally authenticated molecular Th-Re complexes to date, and that there is only one example of a structurally characterised Th-Ru complex, namely, $[Th(\eta^5-C_5Me_5)_2(I)Ru(\eta^5-C_5H_5)(CO)_2]$ (**I**) reported by Marks in 1985 [19]. Since **URe** and **URu** had proven to be accessible by salt and alkane elimination strategies [20,22], we sought to prepare and characterise the corresponding Th-derivatives so that comparisons between Th and U could be made.

Here, we report the synthesis and characterisation of two new Th-complexes that exhibit bonds to Re and Ru and compare them to the previously reported U-congeners. We find that these Th-complexes exhibit more ionic Th-Re/-Ru bonds than the U-analogues; by validating quantum chemical calculations with spectroscopic vibrational data, we have identified the strength of these Th-Re/-Ru bonds indirectly using Ru-carbonyls as reporters, and we have probed the Th-Re and Th-Ru bonds directly using Raman spectroscopy.

2. Results and Discussion

2.1. Synthesis and Isolation of the Th-Re and Th-Ru Complexes **ThRe** and **ThRu**

Treatment of the colourless thorium-cyclometallate complex [Th{N(CH$_2$CH$_2$NSiMe$_2$Bu^t)$_2$(CH$_2$CH$_2$NSi(Me)(Bu^t)(μ-CH$_2$)]$_2$ (**1**) [53] with two equivalents of the pale yellow rhenium complex [Re(η^5-C$_5$H$_5$)$_2$(H)] (**2**) [54] in toluene, Scheme 1, afforded a dark yellow solution. Work-up afforded yellow crystals of the Th-Re complex [Th(Tren^{DMBS})Re(η^5-C$_5$H$_5$)$_2$] (**ThRe**) in 71% isolated yield. Thus, as was found for analogous uranium chemistry,[22] alkane elimination is an effective method for constructing thorium-metal bonds.

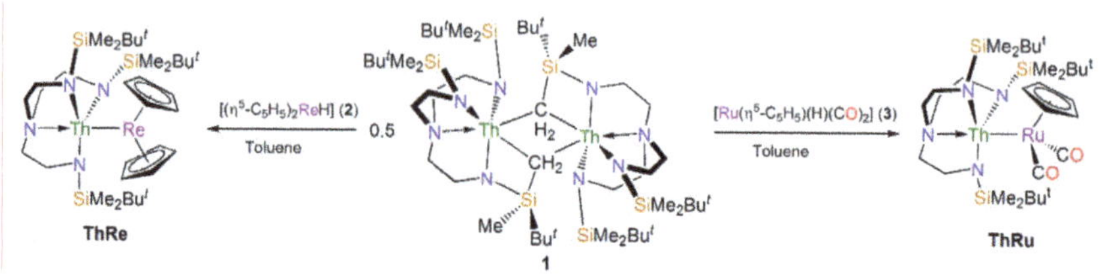

Scheme 1. Synthesis of **ThRe** and **ThRu** from **1**, **2**, and **3**, respectively.

Encouraged by the successful synthesis of **ThRe** by alkane elimination, we also treated **1** with two equivalents of the yellow ruthenium complex [Ru(η^5-C$_5$H$_5$)(H)(CO)$_2$] (**3**) [55] in benzene, Scheme 1. Subsequent removal of the reaction mother liquor from the resulting pale yellow precipitate and recrystallisation afforded the Th-Ru complex [Th(Tren^{DMBS})Ru(η^5-C$_5$H$_5$)(CO)$_2$] (**ThRu**), isolated as light brown crystals of the benzene solvate in 62% yield.

2.2. Solid-State Structures of the Th-Re and Th-Ru Complexes **ThRe** and **ThRu**

The solid-state structures of **ThRe** and **ThRu** were determined by X-ray Diffraction, Figure 1. The structure of **ThRe**, Figure 1a, confirms the anticipated trigonal bipyramidal thorium ion and Th-Re bond, which is the first example of a molecular Th-Re bond. The Th-Re distance is found to be 3.1117(2) Å. By definition there are no other Th-Re bonds for comparison, but this distance is marginally (~0.05 Å) longer than the sum of the single covalent bond radii and Th and Re (3.06 Å) [56], and is ~0.06 Å longer than the U-Re distance of 3.0479(6) Å in **URe** [22], which is in-line with the different single bond covalent radii of Th (1.75 Å) and U (1.70 Å). The N$_{amine}$-Th-Re angle is 156.37(9)°, and whilst this is approximately *trans* in nature, the deviation perhaps reflects that the DMBS substituents allow the Tren^{DMBS} unit to flex from C$_3$ symmetry towards a C$_s$ symmetry to accommodate the rhenocene fragment. All other distances and angles are unremarkable.

The structure of **ThRu**, Figure 1b, confirms the presence of a Th-Ru bond and thus the absence of any isocarbonyl linkages. The Th-Ru distance is found to be 3.1227(3) Å, which compares to a value of 3.00 Å for the sum of the single bond covalent radii of Th and Ru [56]. There is only one other example of a molecular Th-Ru bond, which is **I** reported in 1985, that exhibits a Th-Ru distance of 3.0277(6) Å [19]. Thus, the Th-Ru distance in **ThRu** can be considered to be long, likely reflecting charge polarisation into the Ru-CO back-bonding orbitals rendering the Ru a poorer donor site than the Re in rhenocene; support for this comes from the spectroscopic data (see below) and the fact that the U-Ru distance of 3.0739(2) Å in **URu** is ~0.05 Å shorter, in-line with the respective metal single bond covalent radii [56]. The Ru-C$_{CO}$ and C-O distances in **ThRu** average 1.844(4) and 1.165(4) Å, respectively. The latter is slightly longer than the C-O distance in free CO (1.128 Å), and the Ru-C$_{CO}$ and C-O distances are similar to the analogous distances in **URu** [20]. The N$_{amine}$-Th-Ru angle is 156.60(6)°, which is rather similar to the N$_{amine}$-Th-Re angle in **ThRe**.

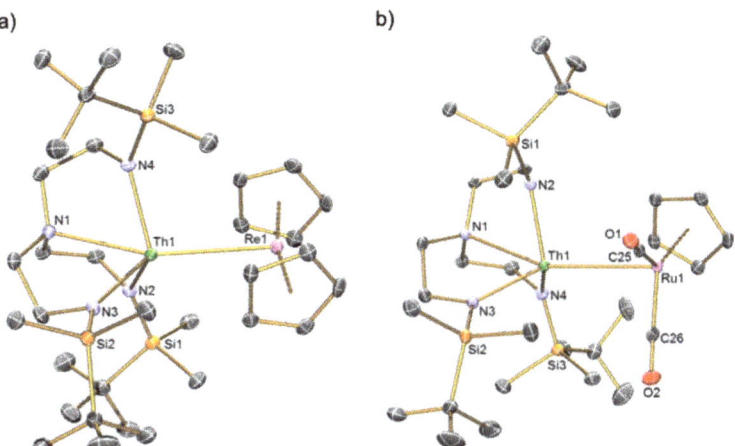

Figure 1. Solid-state structures of (**a**) **ThRe** and (**b**) **ThRu** at 150 K with displacement ellipsoids set to 30%. Hydrogen atoms, disorder components, and benzene lattice solvent for **ThRu** are omitted for clarity.

*2.3. Spectroscopic and Analytical Characterisation of the Th-Re and Th-Ru Complexes **ThRe** and **ThRu***

The ^{1}H and ^{13}C{^{1}H} NMR spectra of **ThRe** and **ThRu** exhibit resonances in the ranges 0–4 and 0–5 ppm, consistent with the diamagnetic, closed-shell formulations of Th(IV) and the Re and Ru ions. The NMR spectra of **ThRe** and **ThRu** are largely as anticipated, though the CO C-resonances for **ThRu** and Cp C-resonances could not be located despite extensive attempts to locate them. The NMR and IR data for **ThRe** and **ThRu** do not provide any evidence for hydrides, and their isostructural natures compared to the corresponding U-complexes support the view that there are not any hydrides in **ThRe** and **ThRu**.

The IR spectrum of **ThRu** exhibits strong carbonyl absorptions at 1943 and 1869 cm^{-1}, which can be compared to those of **URu** (1944, 1872 cm^{-1}) [20], **I** (1968, 1900 cm^{-1}),[19] and [Th(η^5-C$_5$H$_5$)$_3$Ru(η^5-C$_5$H$_5$)(CO)$_2$] (1915, 1847 cm^{-1}) [13]. It can thus be noted that there is more Ru-CO back-bonding in **ThRu** compared to **I** [19] but less than [Th(η^5-C$_5$H$_5$)$_3$Ru(η^5-C$_5$H$_5$)(CO)$_2$] [13], which would be expected to render the Ru-centre in **ThRu** a weaker donor; in that scenario, a longer Th-Ru bond would be predicted in **ThRu** compared to **I**, which is the only other Th-Ru complex to be structurally characterised, and that is indeed the experimental observation. These data suggest that the [Th(Tren^{DMBS})]$^+$ fragment is slightly more electron-deficient than [Th(η^5-C$_5$H$_5$)$_3$]$^+$, since an An-fragment that polarises charge from the Ru-component most will deplete the extent of Ru-CO back-bonding the most resulting in a higher CO stretching frequency. Conversely, the exchange for a Cp ligand for iodide, even with two cyclopentadienyl ligands replaced by the more strongly donating pentamethylcyclopentadienyl ligand, renders [Th(η^5-C$_5$Me$_5$)$_2$(I)]$^+$ the most electron deficient fragment. The magnitude of the shift of the asymmetric CO stretching frequency to high frequency compared to [Ru(η^5-C$_5$H$_5$)(CO)$_2$]$^-$ (1749 cm^{-1}) reveals a shift of 120 cm^{-1} [20], which is lower than the shift observed for group 4-group 8 bonds (140–150 cm^{-1}) [4] but higher than for analogous lanthanide-group 8 bonds (110 cm^{-1}) [4,8,10], suggesting that the Th-Ru linkage in **ThRu** is more covalent than lanthanide analogues but still largely dominated by ionic character.

We also recorded Raman spectra of **ThRe** and **ThRu**, and of most pertinence are inelastic scattering bands at 103/136 and 112 cm^{-1}, respectively. On the basis of analytical frequencies calculations on **ThRu** and **ThRu** (see below), these are assigned as corresponding to Th-Re and Th-Ru stretches, respectively. For comparison, the W-W stretch for [W$_2$(CO)$_{10}$][K(18-crown−6)(THF)$_2$] was recently reported to be found at 97 cm^{-1} by Raman spectroscopy [57].

*2.4. Quantum Chemical Computational Analysis of the Th-Re and Th-Ru Complexes **ThRe** and **ThRu***

In order to probe the Th-Re/-Ru linkages in **ThRe** and **ThRu**, we performed scalar relativistic DFT calculations on the full structures of these two complexes. The optimised gas-phase geometries (see Supplementary Materials for final coordinates and energies) compare well to the experimental solid-state structures, for example, returning Th-Re and Th-Ru distances of 3.1293 and 3.1139 Å, which are in excellent agreement with the experimental values of 3.1117(2) and 3.1227(3) Å. We therefore conclude that the computed models of **ThRe** and **ThRu** provide qualitative and representative models of the electronic structures of these complexes.

The computed MDC-q charges for the two Th, Re, and Ru ions in **ThRe** and **ThRu** are 1.87, 1.41, 0.38, and 0.71, respectively. These values can be compared to the two U, Re, and Ru values of 1.88, 1.50, 0.38, and 0.67 [20,22], respectively, for the analogous U-complexes, showing negligible variations. The calculated Nalewajski–Mrozek Th-Re and Th-Ru bond orders for **ThRe** and **ThRu** are 0.72 and 0.52. Interestingly, in contrast to the largely invariant computed charges, the calculated bond orders for **ThRe** and **ThRu** are lower than for **URe** and **URu**, which have computed bond orders of 1.13 and 0.76, respectively [20,22], suggesting more ionic metal–metal bonds for the thorium complexes compared to the uranium ones. For comparison, the Th-N$_{amide}$ and Th-N$_{amine}$ bond orders are calculated to average 0.95 and 0.28. The Ru-C$_{CO}$ and C-O bond orders for **ThRu** average 1.24 and 2.32, respectively, reflecting the back-bonding of electron density from Ru to C with concomitant reduction of the CO bond order from 3 in free CO.

The HOMO-1 and HOMO-2 for complex **ThRe** represent the principal possible Th-Re bonding interactions on symmetry grounds, Figure 2a,b. HOMO-2 constitutes the Th-Re σ-bond, being derived from the 2a$_1$ orbital from bent metallocene molecular orbital considerations. HOMO-2 contains only 5.9% Th-character, compared to 10.8% U-character in the isostructural U-derivative [22]. HOMO-1 is the corresponding b$_2$ orbital, but unlike the U-analogue which shows a π-bond, in **ThRe** this MO remains largely localised on the rhenocene fragment (<2% Th character, *cf* 9.9% U character in the U-analogue). Together, these observations nicely account for the reduced Th-Re bond order compared to the larger U-Re bond order in the isostructural U-congener [22]. The HOMO of **ThRe** is the 1a$_1$ orbital, and it is essentially non-bonding with respect to the Th-Re bonding interaction; the energy ordering of the 1a$_1$, b$_2$, and 2a$_1$ orbitals in **ThRe** can thus be related to the degree of stabilisation afforded to those orbitals by how strongly they donate to the Th ion in **ThRe**.

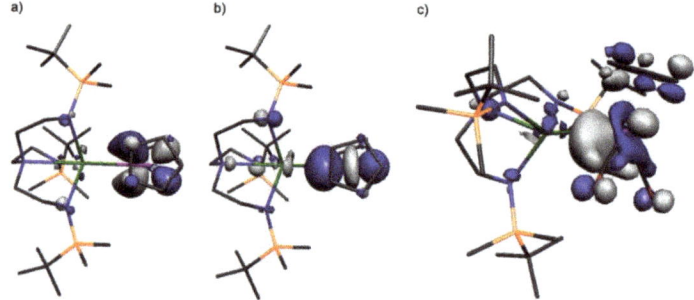

Figure 2. (a) HOMO-1 (252, −4.054 eV) of **ThRe**. (b) HOMO-2 (251, −4.161 eV) of **ThRe**. (c) HOMO (234, −4.409 eV) of **ThRu**. Hydrogen atoms are omitted for clarity.

The HOMO of **ThRu** represents the Th-Ru interaction, Figure 2c. As found analogously for **URu** [20], this MO is [Ru(η^5-C$_5$H$_5$)(CO)$_2$]-centred (60%), with ~29% 4d character, and 9.6 and 5.8% 5p and 5s character, respectively, with the remaining 15.6% involving the CO π^* orbital coefficients. The rest of this MO is delocalised over the Tren^{DMBS} ligand or Th (0.7%), revealing a largely ionic Th-Ru interaction and one where the CO stretching frequency is confirmed as a reporter of the Th-Ru bonding interaction, as discussed above.

Lastly, an analytical frequencies calculation predicts CO stretching frequencies of 1924 and 1864 cm^{-1} for **ThRu**, which are in excellent agreement with the experimentally observed values of 1943 and 1869 cm^{-1}. This experimental confirmation of the accuracy of the DFT calculations then permits us to address the Th-Re and Th-Ru bond vibrations in **ThRe** and **ThRu**. The Th-Ru stretch for **ThRu** is computed to occur at 126 cm^{-1}, which is in good agreement with the corresponding experimental Raman value of 112 cm^{-1}. The Th-Re stretch of **ThRe** appears to be split by coupling to other vibrational modes, principally characterised by bands computed at 107 and 138 cm^{-1}, which compares to experimentally observed bands at 103 and 136 cm^{-1}.

3. Materials and Methods

3.1. General Materials and Methods

All manipulations were carried out using Schlenk techniques, or an MBraun UniLab glovebox, under an atmosphere of dry nitrogen. Solvents were dried by passage through activated alumina towers and degassed before use. Deuterated solvents were dried over NaK$_2$, distilled, and stored over NaK$_2$. Crystals were examined using either a Rigaku FR-X diffractometer, equipped with a HyPix 6000HE photon counting pixel array detector with mirror-monochromated Cu Kα (λ = 1.5418 Å) radiation or a Rigaku Xcalibur2 diffractometer, equipped with an Atlas CCD area detector and a sealed tube source with graphite-monochromated Mo Kα radiation (λ = 0.71073 Å). Intensities were integrated from a sphere of data recorded on narrow (1.0°) frames by ω rotation. Cell parameters were refined from the observed positions of all strong reflections in each data set. Gaussian grid face-indexed absorption corrections with a beam profile correction were applied. The structures were solved by dual methods, and all non-hydrogen atoms were refined by full-matrix least-squares on all unique F^2 values with anisotropic displacement parameters with exceptions noted in the respective cif files. Except where noted, hydrogen atoms were refined with constrained geometries and riding thermal parameters. CrysAlisPro [58] was used for control and integration, SHELXT [59] was used for structure solution, and SHELXL [60] and Olex2 [61] were employed for structure refinement. ORTEP-3 [62] and POV-Ray [63] were employed for molecular graphics. FTIR spectra were recorded on a Bruker Alpha spectrometer with Platinum-ATR module. A Horiba XploRA Plus Raman microscope with a 638 nm laser (power: $\leq$150 mW) was used to obtain all Raman spectra. The power of the laser was adjusted for each sample using a filter to prevent sample decomposition. Elemental microanalyses were carried out by Mr Martin Jennings at the Micro Analytical Laboratory, Department of Chemistry, The University of Manchester. Complexes **1–3** were prepared as described previously [53–55].

3.2. Quantum Chemical Calculations

Geometry optimisations for **ThRe** and **ThRu** were performed using coordinates derived from their respective crystal structures as the starting points. No constraints were imposed on the structures during the geometry optimisations. The calculations were performed using the Amsterdam Density Functional (ADF) suite version 2017 with standard convergence criteria [64,65]. The DFT geometry optimisations employed Slater type orbital (STO) triple-ζ-plus polarisation all electron basis sets (from the Dirac and ZORA/TZP database of the ADF suite). Scalar relativistic approaches (spin-orbit neglected) were used within the ZORA Hamiltonian [66–68] for the inclusion of relativistic effects, and the local density approximation (LDA) with the correlation potential due to Vosko et al. was used in all of the calculations [69]. Generalised gradient approximation corrections were performed using the functionals of Becke and Perdew [70,71]. MOLEKEL [72] was used to prepare the three-dimensional plots of the electron density. Frequencies were computed using the analytical frequencies routine in ADF.

3.3. Preparation of [Th(Tren^{DMBS})ReCp$_2$] (ThRe)

Toluene (20 mL) was added to a precooled (−78 °C) mixture of **1** (0.359 g, 0.25 mmol) and **2** (0.154 g, 0.5 mmol). The resultant pale-yellow suspension was allowed to warm to RT and then stirred for 16 h, giving a dark yellow solution. Volatiles were removed in vacuo, and the resulting brown solid was extracted into toluene, concentrated to 5 mL, and cooled to −30 °C to afford **ThRe** as pale brown crystals. Yield: 0.260 g, 71%. Anal. Calc'd for C$_{34}$H$_{67}$N$_4$ReSi$_3$Th: C 39.48; H 6.53; N 5.42%. Found: C 39.48; H 6.67; N 5.40%. ^{1}H NMR (C$_6$D$_6$) δ: 4.42 (10H, s, Cp-*H*), 3.21 (6H, t, C*H*$_2$), 2.45 (6H, t, C*H*$_2$), 1.09 (27H, s, C*H*$_3$), 0.38 (18H, s, C*H*$_3$) ppm. ^{13}C{^{1}H} NMR (C$_6$D$_6$) δ: 67.93 (CH$_2$), 46.65 (CH$_2$), 28.90 (C-CH$_3$), 21.96 (C-CH$_3$), −1.78 (Si-CH$_3$) ppm. FTIR ν/cm^{-1} (ATR): 2950 (w), 2850 (m), 1462 (m), 1245 (m), 1058 (m), 929 (s), 805 (s). Raman ν/cm^{-1} (Neat, ≤15 mW): 3103 (m, br), 2952 (m), 2853 (m), 1084 (s), 567 (m), 338 (s), 136 (w), 103 (w), and 63 (s).

3.4. Preparation of [Th(Tren^{DMBS})RuCp-(CO)$_2$] (ThRu)

Benzene (10 mL) was added to a mixture of **1** (0.210 g, 0.15 mmol) and **3** (0.067 g, 0.30 mmol) in a glovebox. The resultant light brown solution quickly gave a pale-yellow precipitate, which was separated from the solution, washed with pentane (2 × 5 mL), and allowed to dry. The solid was extracted into minimal THF, which afforded **ThRu** as pale brown crystals; these crystals were found to be the benzene solvate **ThRu**.C$_6$H$_6$, but drying under vacuum results in their desolvation, as reflected by the other characterisation data. Yield: 0.152 g, 62 %. Anal. Calc'd for C$_{31}$H$_{62}$N$_4$O$_2$RuSi$_3$Th: C 39.60; H 6.65; N 5.96%. Found: C 39.45; H 6.70; N 5.36%. ^{1}H NMR (C$_6$D$_6$) δ: 5.08 (5H, s, Cp-*H*), 3.20 (6H, t, J_{HH} = 4.96 Hz, C*H*$_2$), 2.38 (6H, t, J_{HH} = 4.96 Hz, C*H*$_2$), 1.04 (27H, s, C-C*H*$_3$), 0.51 (18H, s, Si-C*H*$_3$) ppm. ^{13}C{^{1}H} NMR (C$_6$D$_6$) δ: 86.52 (CH), 64.39 (CH$_2$), 47.46 (CH$_2$), 28.51 (C-CH$_3$), 21.41 (C-CH$_3$), −3.25 (Si-CH$_3$) ppm. FTIR ν/cm^{-1} (ATR): 2924 (w), 2848 (w), 1943 (s), 1869 (s), 1462 (w), 1248 (m), 1072 (m), 936 (m). Raman ν/cm^{-1} (Neat, ≤15 mW): 2957 (m), 2853 (s), 1935 (s), 1869 (s), 1200 (s), 1102 (s), 567 (s), 518 (s), 303 (s), 112 (m), and 73 (s).

4. Conclusions

To conclude, we have prepared two new triamidoamine complexes of Th that possess Th-Re and Th-Ru bonds using an alkane elimination strategy. The Th-Re bond in **ThRe** is the first structurally authenticated example of a molecular Th-Re bond, and the Th-Ru bond in **ThRu** is only the second example of a structurally characterised Th-Ru bond, with the other example reported 36 years ago. Structural, spectroscopic, and computational characterisation of these two complexes, along with comparisons to previously reported isostructural U-Re and U-Ru complexes, has permitted an analysis of these metal–metal bonds. In-line with the current understanding, these Th-metal bonds are like-for-like more ionic than the U-congeners. By being able to experimentally validate the calculations using spectroscopic vibrational data, as well as gauging the strength of these metal–metal bonds indirectly using the carbonyls as reporters, we have been able to probe with certainty the Th-Re and Th-Ru bonds directly using Raman spectroscopy.

Supplementary Materials: The following are available online at https://www.mdpi.com/article/10.3390/inorganics9050030/s1, Table S1: Final Coordinates and Energy from a Single Point Energy Calculation on Geometry Optimised ThRe, Table S2: Final Coordinates and Energy from a Single Point Energy Calculation on Geometry Optimised ThRu. CIFs and checkCIF output files for **ThRe** and **ThRu**.

Author Contributions: Conceptualisation, S.T.L.; methodology, S.T.L. and J.P.A.O.; formal analysis, J.P.A.O. and A.J.W.; investigation, J.P.A.O.; data curation, A.J.W.; writing—original draft preparation, S.T.L.; writing—review and editing, S.T.L., J.P.A.O., and A.J.W.; supervision, S.T.L.; project administration, S.T.L.; funding acquisition, S.T.L. All authors have read and agreed to the published version of the manuscript.

Funding: We thank the Royal Society (UF071260, UF110005, and RG080285), EPSRC (EP/G051763/1, EP/M027015/1, and EP/P001386/1), ERC (StG23921 and CoG612724), and The University of Manchester for funding and support, and the Alexander von Humboldt Foundation for the award of a Friedrich Wilhelm Bessel Research Award to S.T.L.

Institutional Review Board Statement: Not applicable.

Informed Consent Statement: Not applicable.

Data Availability Statement: The crystallographic data for **ThRe** and **ThRu** have been deposited with the Cambridge Crystallographic Data Centre, CCDC numbers 2076926 and 2076927, and all other data are available from the authors on request.

Conflicts of Interest: The authors declare no conflict of interest. The funders had no role in the design of the study; in the collection, analyses, or interpretation of data; in the writing of the manuscript; or in the decision to publish the results.

References

1. Cotton, F.A.; Murillo, C.A.; Walton, R.A. (Eds.) *Multiple Bonds between Metal Atoms*, 3rd ed.; Springer: New York, NY, USA, 2005.
2. Parkin, G. (Ed.) *Metal-Metal Bonding*; Springer: New York, NY, USA, 2010.
3. Liddle, S.T. (Ed.) *Molecular Metal-Metal Bonds: Compounds, Synthesis, Properties*; Wiley VCH: Weinheim, Germany, 2015.
4. Wheatley, N.; Kalck, P. Structure and Reactivity of Early-Late Heterobimetallic Complexes. *Chem. Rev.* **1999**, *99*, 3379–3420. [CrossRef] [PubMed]
5. Chipman, J.A.; Berry, J.F. Paramagnetic Metal-Metal Bonded Heterometallic Complexes. *Chem. Rev.* **2020**, *120*, 2409–2447. [CrossRef] [PubMed]
6. Zhang, X.; Wang, Y.; Morales-Martínez, R.; Zhong, J.; de Graaf, C.; Rodríguez-Fortea, A.; Poblet, J.M.; Echegoyen, L.; Feng, L.; Chen, N. $U_2@I_h(7)$-C_{80}: Crystallographic Characterization of a Long-Sought Dimetallic Actinide Endohedral Fullerene. *J. Am. Chem. Soc.* **2018**, *140*, 3907–3915. [CrossRef] [PubMed]
7. Liddle, S.T.; Mills, D.P. Metal-Metal Bonds in f-Element Chemistry. *Dalton Trans.* **2009**, 5592–5605. [CrossRef]
8. Oelkers, B.; Butovskii, M.V.; Kempe, R. f-Element-Metal Bonding and the Use of the Bond Polarity to Build Molecular Intermetalloids. *Chem. Eur. J.* **2012**, *18*, 13566–13579. [CrossRef]
9. Patel, D.; Liddle, S.T. f-Element-Metal Bond Chemistry. *Rev. Inorg. Chem.* **2012**, *32*, 1–22. [CrossRef]
10. Butovskii, M.V.; Kempe, R. Rare earth-metal bonding in molecular compounds: Recent advances, challenges, and perspectives. *New J. Chem.* **2015**, *39*, 7544–7558. [CrossRef]
11. Réant, B.L.L.; Liddle, S.T.; Mills, D.P. f-Element silicon and heavy tetrel chemistry. *Chem. Sci.* **2020**, *11*, 10871–10886. [CrossRef]
12. Ayres, A.J.; Zegke, M.; Ostrowski, J.P.A.; Tuna, F.; McInnes, E.J.L.; Wooles, A.J.; Liddle, S.T. Actinide-transition metal bonding in heterobimetallic uranium− and thorium−molybdenum paddlewheel complexes. *Chem. Commun.* **2018**, *54*, 13515–13518. [CrossRef]
13. Sternal, R.S.; Marks, T.J. Actinide-to-transition metal bonds. Synthesis, characterization, and properties of metal-metal bonded systems having the tris(cyclopentadienyl)actinide fragment. *Organometallics* **1987**, *6*, 2621–2623. [CrossRef]
14. Nolan, S.P.; Porchia, M.; Marks, T.J. Organo-f-element thermochemistry. Actinide-Group 14 element and actinide-transition-element bond disruption enthalpies and stoichiometric/catalytic chemical implications thereof in heterobimetallic tris(cyclopentadienyl)uranium(IV) compounds. *Organometallics* **1991**, *10*, 1450–1457. [CrossRef]
15. Bucaille, A.; Le Borgne, T.; Ephritikhine, M.; Daran, J.C. Synthesis and X-ray Crystal Structure of a Urana[1]ferrocenophane, the First Tris(1,1′-ferrocenylene) Metal Compound. *Organometallics* **2000**, *19*, 4912–4914. [CrossRef]
16. Monreal, M.J.; Carver, C.T.; Diaconescu, P.L. Redox processes in a uranium bis(1,1′-diamidoferrocene) complex. *Inorg. Chem.* **2007**, *46*, 7226–7228. [CrossRef]
17. Monreal, M.J.; Diaconescu, P.L. A weak interaction between iron and uranium in uranium alkyl complexes supported by ferrocene diamide ligands. *Organometallics* **2008**, *27*, 1702–1706. [CrossRef]
18. Fortier, S.; Aguilar-Calderon, J.R.; Vlaisavljevich, B.; Metta-Magana, A.J.; Goos, A.G.; Botez, C.E. An N-Tethered Uranium(III) Arene Complex and the Synthesis of an Unsupported U-Fe Bond. *Organometallics* **2017**, *36*, 4591–4599. [CrossRef]
19. Sternal, R.S.; Brock, C.P.; Marks, T.J. Metal-metal bonds involving actinides. Synthesis and characterization of a complex having an unsupported actinide to transition metal bond. *J. Am. Chem. Soc.* **1985**, *107*, 8270–8272. [CrossRef]
20. Gardner, B.M.; Patel, D.; Cornish, A.D.; McMaster, J.; Lewis, W.; Blake, A.J.; Liddle, S.T. The Nature of Unsupported Uranium-Ruthenium Bonds: A Combined Experimental and Theoretical Study. *Chem. Eur. J.* **2011**, *17*, 11266–11273. [CrossRef]

21. Gardner, B.M.; McMaster, J.; Lewis, W.; Liddle, S.T. Synthesis and structure of {N(CH$_2$CH$_2$NSiMe$_3$)$_3$}URe(η^5-C$_5$H$_5$)$_2$: A heterobimetallic complex with an unsupported uranium-rhenium bond. *Chem. Commun.* **2009**, 2851–2853. [CrossRef]
22. Gardner, B.M.; McMaster, J.; Moro, F.; Lewis, W.; Blake, A.J.; Liddle, S.T. An Unsupported Uranium-Rhenium Complex Prepared by Alkane Elimination. *Chem. Eur. J.* **2011**, *17*, 6909–6912. [CrossRef]
23. Patel, D.; King, D.M.; Gardner, B.M.; McMaster, J.; Lewis, W.; Blake, A.J.; Liddle, S.T. Structural and theoretical insights into the perturbation of uranium-rhenium bonds by dative Lewis base ancillary ligands. *Chem. Commun.* **2011**, *47*, 295–297. [CrossRef]
24. Patel, D.; Moro, F.; McMaster, J.; Lewis, W.; Blake, A.J.; Liddle, S.T. A Formal High Oxidation State Inverse-Sandwich Diuranium Complex: A New Route to f-Block-Metal Bonds. *Angew. Chem. Int. Ed.* **2011**, *50*, 10388–10392. [CrossRef]
25. Napoline, J.W.; Kraft, S.J.; Matson, E.M.; Fanwick, P.E.; Bart, S.C.; Thomas, C.M. Tris(phosphinoamide)-Supported Uranium-Cobalt Heterobimetallic Complexes Featuring Co → U Dative Interactions. *Inorg. Chem.* **2013**, *52*, 12170–12177. [CrossRef] [PubMed]
26. Ward, A.L.; Lukens, W.W.; Lu, C.C.; Arnold, J. Photochemical Route to Actinide-Transition Metal Bonds: Synthesis, Characterization and Reactivity of a Series of Thorium and Uranium Heterobimetallic Complexes. *J. Am. Chem. Soc.* **2014**, *136*, 3647–3654. [CrossRef]
27. Lu, E.; Wooles, A.J.; Gregson, M.; Cobb, P.J.; Liddle, S.T. A Very Short Uranium(IV)-Rhodium(I) Bond with Net Double-Dative Bonding Character. *Angew. Chem. Int. Ed.* **2018**, *57*, 6587–6591. [CrossRef]
28. Hlina, J.A.; Wells, J.A.L.; Pankhurst, J.R.; Love, J.B.; Arnold, P.L. Uranium rhodium bonding in herometallic complexes. *Dalton Trans.* **2017**, *46*, 5540–5545. [CrossRef]
29. Geng, G.; Zhang, M.; Wang, P.; Wang, S.; Maron, L.; Zhu, C. Identification of a uranium-rhodium triple bond in a heterometallic cluster. *Proc. Natl. Acad. Sci. USA* **2019**, *116*, 17654–17658.
30. Ritchey, J.M.; Zozulin, A.J.; Wrobleski, D.A.; Ryan, R.R.; Wasserman, H.J.; Moody, D.C.; Paine, R.T. An organothorium-nickel phosphido complex with a short thorium-nickel distance. The structure of Th(η^5-C$_5$Me$_5$)$_2$(μ-PPh$_2$)$_2$Ni(C)(CO)$_2$. *J. Am. Chem. Soc.* **1985**, *107*, 501–503. [CrossRef]
31. Hlina, J.A.; Pankhurst, J.R.; Kaltsoyannis, N.; Arnold, P.L. Metal−Metal Bonding in Uranium−Group 10 Complexes. *J. Am. Chem. Soc.* **2016**, *138*, 3333–3345. [CrossRef] [PubMed]
32. Camp, C.; Toniolo, D.; Andrez, J.; Pecaut, J.; Mazzanti, M. A versatile route to homo- and hetero-bimetallic 5f-5f and 3d-5f complexes supported by a redox active ligand framework. *Dalton Trans.* **2017**, *46*, 11145–11148. [CrossRef]
33. Kozimor, S.A.; Bartlett, B.M.; Rinehart, J.D.; Long, J.R. Magnetic exchange coupling in chloride-bridged 5f-3d heterometallic complexes generated via insertion into a Uranium(IV) dimethylpyrazolate dimer. *J. Am. Chem. Soc.* **2007**, *129*, 10672–10674. [CrossRef]
34. Feng, G.; Zhang, M.; Shao, S.; Wang, X.; Wang, S.; Maron, L.; Zhu, C. Transition-metal-bridged bimetallic clusters with multiple uranium-metal bonds. *Nat. Chem.* **2019**, *11*, 248–253. [CrossRef] [PubMed]
35. Feng, G.; McCabe, K.N.; Wang, S.; Maron, L.; Zhu, C. Construction of heterometallic clusters with multiple uranium-metal bonds by using dianionic nitrogen-phosphorus ligands. *Chem. Sci.* **2020**, *11*, 7585–7592. [CrossRef]
36. Hay, P.J.; Ryan, R.R.; Salazar, K.V.; Wrobleski, D.A.; Sattelberger, A.P. Synthesis and x-ray structure of (C$_5$Me$_5$)$_2$Th(μ-PPh$_2$)$_2$Pt(PMe$_3$): A complex with a thorium-platinum bond. *J. Am. Chem. Soc.* **1986**, *108*, 313–315. [CrossRef]
37. Leverd, P.C.; Lance, M.; Nierlich, M.; Vigner, J.; Ephritikhine, M. Synthesis and crystal structure of homoleptic uranium hexathiolates: [NEt$_2$H$_2$]$_2$[U(SPh)$_6$] and [(Ph$_3$P)Cu(μ-SPh)$_3$-U(μ-SPh)$_3$Cu(PPh$_3$)]. *J. Chem. Soc. Dalton Trans.* **1994**, 3563–3567. [CrossRef]
38. Yang, P.; Zhou, E.; Hou, G.; Zi, G.; Ding, W.; Walter, M.D. Experimental and Computational Studies on the Formation of Thorium-Copper Heterobimetallics. *Chem. Eur. J.* **2016**, *22*, 13845–13849. [CrossRef]
39. Fortier, S.; Walensky, J.R.; Wu, G.; Hayton, T.W. High-Valent Uranium Alkyls: Evidence for the Formation of U^{VI}(CH$_2$SiMe$_3$)$_6$. *J. Am. Chem. Soc.* **2011**, *133*, 11732–11743. [CrossRef]
40. Minasian, S.G.; Krinsky, J.L.; Williams, V.A.; Arnold, J. A heterobimetallic complex with an unsupported Uranium(III)-Aluminum(I) bond: (CpSiMe$_3$)$_3$U-AlCp* (Cp* = C$_5$Me$_5$). *J. Am. Chem. Soc.* **2008**, *130*, 10086–10087. [CrossRef]
41. Minasian, S.G.; Krinsky, J.L.; Rinehart, J.D.; Copping, R.; Tyliszczak, T.; Janousch, M.; Shuh, D.K.; Arnold, J. A Comparison of 4f vs 5f Metal-Metal Bonds in (CpSiMe$_3$)$_3$M-ECp* (M = Nd, U; E = Al, Ga; Cp* = C$_5$Me$_5$): Synthesis, Thermodynamics, Magnetism, and Electronic Structure. *J. Am. Chem. Soc.* **2009**, *131*, 13767–13783. [CrossRef]
42. Liddle, S.T.; McMaster, J.; Mills, D.P.; Blake, A.J.; Jones, C.; Woodul, W.D. σ and π Donation in an Unsupported Uranium-Gallium Bond. *Angew. Chem. Int. Ed.* **2009**, *48*, 1077–1080. [CrossRef]
43. Diaconescu, P.L.; Odum, A.L.; Agapie, T.; Cummins, C.C. Uranium-Group 14 Element Single Bonds: Isolation and Characterization of a Uranium(IV) Silyl Species. *Organometallics* **2001**, *20*, 4993–4995. [CrossRef]
44. Porchia, M.; Brianese, N.; Casellato, U.; Ossola, F.; Rossetto, G.; Zanella, P. Tri(η-cyclopentadienyl)uranium(IV) silyl and siloxide compounds: Crystal structure of [U(η^5-C$_5$H$_5$)$_3$(OSiPh$_3$)]: Insertion of Isocyanide into a uranium-silicon bond. *J. Chem. Soc. Dalton Trans.* **1989**, 677–681. [CrossRef]
45. Brackbill, I.J.; Douair, I.; Lussier, D.J.; Boreen, M.A.; Maron, L.; Arnold, J. Synthesis and Structure of Uranium-Silylene Complexes. *Chem. Eur. J.* **2020**, *26*, 2360–2364. [CrossRef]
46. Réant, B.L.L.; Berryman, V.E.J.; Seed, J.A.; Basford, A.R.; Formanuik, A.; Wooles, A.J.; Kaltsoyannis, N.; Liddle, S.T.; Mills, D.P. Polarised Covalent Thorium(IV)- and Uranium(IV)-Silicon Bonds. *Chem. Commun.* **2020**, *56*, 12620–12623. [CrossRef]

47. Porchia, M.; Ossola, F.; Rossetto, G.; Zanella, P.; Brianese, N. Synthesis of triscyclopentadienyl(triphenylgermyl)uranium and facile isonitrile insertion into the uranium-germanium bond. *J. Chem. Soc. Chem. Commun.* **1987**, 550–551. [CrossRef]
48. Porchia, M.; Casellato, U.; Ossola, F.; Rossetto, G.; Zanella, P.; Graziani, R. Synthesis and crystal structure of triscyclopentadienyl-(triphenyltin)uranium: The first example of a uranium-tin bond. *J. Chem. Soc. Chem. Commun.* **1986**, 1034–1035. [CrossRef]
49. Winston, M.S.; Batista, E.R.; Yang, P.; Tondreau, A.M.; Boncella, J.M. Extending Stannyl Anion Chemistry to the Actinides: Synthesis and Characterization of a Uranium−Tin Bond. *Inorg. Chem.* **2016**, *55*, 5534–5539. [CrossRef]
50. Rookes, T.M.; Wildman, E.P.; Balazs, G.; Gardner, B.M.; Wooles, A.J.; Gregson, M.; Tuna, F.; Scheer, M.; Liddle, S.T. Actinide-Pnictide (An-Pn) Bonds Spanning Non-Metal, Metalloid, and Metal Combinations (An = U, Th; Pn = P, As, Sb, Bi). *Angew. Chem. Int. Ed.* **2018**, *57*, 1332–1336. [CrossRef]
51. Lichtenberger, N.; Wilson, R.J.; Eulenstein, A.R.; Massa, W.; Clérac, R.; Weigend, F.; Dehnen, S. Main Group Metal-Actinide Magnetic Coupling and Structural Response Upon U^{4+} Inclusion into Bi, Tl/Bi, or Pb/Bi Cages. *J. Am. Chem. Soc.* **2016**, *138*, 9033–9036. [CrossRef]
52. Eulenstein, A.R.; Franzke, Y.J.; Lichtenberger, N.; Wilson, R.J.; Deubner, H.L.; Kraus, F.; Clérac, R.; Weigand, F.; Dehnen, S. Substantial π-aromaticity in the anionic heavy-metal cluster $[Th@Bi_{12}]^{4-}$. *Nat. Chem.* **2021**, *13*, 149–155. [CrossRef]
53. Gardner, B.M.; Lewis, W.; Blake, A.J.; Liddle, S.T. Thorium Triamidoamine Complexes: Synthesis of an Unusual Dinuclear Tuck-In-Tuck-Over Thorium Metallacycle Featuring the Longest Known Thorium-σ-Alkyl Bond. *Organometallics* **2015**, *34*, 2386–2394. [CrossRef]
54. Green, M.L.H.; Pratt, L.; Wilkinson, G. Biscyclopentadienylrhenium hydride. *J. Chem. Soc.* **1958**, 3916–3922. [CrossRef]
55. Karunananda, M.K.; Mankad, N.P. Heterobimetallic H_2 Addition and Alkene/Alkane Elimination Reactions Related to the Mechanism of E-Selective Alkyne Semihydrogenation. *Organometallics* **2017**, *36*, 220–227. [CrossRef]
56. Pyykkö, P. Additive Covalent Radii for Single-, Double-, and Triple-Bonded Molecules and Tetrahedrally Bonded Crystals: A Summary. *J. Phys. Chem. A* **2015**, *119*, 2326–2337. [CrossRef] [PubMed]
57. Ostrowski, J.P.A.; Atkinson, B.E.; Doyle, L.R.; Wooles, A.J.; Kaltsoyannis, N.; Liddle, S.T. The Ditungsten Decacarbonyl Dianion. *Dalton Trans.* **2020**, *49*, 9330–9355. [CrossRef]
58. Oxford Diffraction/Agilent Technologies UK Ltd. *CrysAlisPRO Version 39.46*; Oxford Diffraction/Agilent Technologies UK Ltd.: Yarnton, UK, 2013.
59. Sheldrick, G.M. *SHELXT*—Integrated space-group and crystal structure determination. *Acta Cryst. Sect. A* **2015**, *71*, 3–8. [CrossRef]
60. Sheldrick, G.M. Crystal structure refinement with *SHELXL*. *Acta Cryst. Sect. C* **2015**, *71*, 3–8. [CrossRef]
61. Dolomanov, O.V.; Bourhis, L.J.; Gildea, R.J.; Howard, J.A.K.; Puschmann, H. *OLEX2*: A complete structure solution, refinement and analysis program. *J. Appl. Cryst.* **2009**, *42*, 339–341. [CrossRef]
62. Farugia, L.J. *WinGX* and *ORTEP* for Windows: An update. *J. Appl. Cryst.* **2012**, *45*, 849–854. [CrossRef]
63. Persistence of Vision Pty Ltd. *Persistence of Vision (TM) Raytracer*; Persistence of Vision Pty. Ltd.: Williamstown, Melbourne, VIC, Australia, 2020.
64. Fonseca Guerra, C.; Snijders, J.G.; Te Velde, G.; Baerends, E.J. Towards an order-N DFT Method. *Theor. Chem. Acc.* **1998**, *99*, 391–403. [CrossRef]
65. Te Velde, G.; Bickelhaupt, F.M.; Van Gisbergen, S.J.; Fonseca Guerra, A.C.; Baerends, E.J.; Snijders, J.G.; Ziegler, T. Chemistry with ADF. *J. Comput. Chem.* **2001**, *22*, 931–967. [CrossRef]
66. Van Lenthe, E.; Baerends, E.J.; Snijders, J.G. Relativistic regular two-component Hamiltonians. *J. Chem. Phys.* **1993**, *99*, 4597–4610. [CrossRef]
67. Van Lenthe, E.; Baerends, E.J.; Snijders, J.G. Relativistic total energy using regular approximations. *J. Chem. Phys.* **1994**, *101*, 9783–9792. [CrossRef]
68. Van Lenthe, E.; Ehlers, A.E.; Baerends, E.J. Geometry optimization in the Zero Order Regular Approximation for relativistic effects. *J. Chem. Phys.* **1999**, *110*, 8943–8953. [CrossRef]
69. Vosko, S.H.; Wilk, L.; Nusair, M. Accurate spin-dependent electron liquid correlation energies for local spin density calculations: A critical analysis. *Can. J. Phys.* **1980**, *58*, 1200–1211. [CrossRef]
70. Becke, A.D. Density-functional exchange-energy approximation with correct asymptotic behaviour. *Phys. Rev. A* **1988**, *38*, 3098. [CrossRef]
71. Perdew, J.P. Density-functional approximation for the correlation energy of the inhomogeneous electron gas. *Phys. Rev. B* **1986**, *33*, 8822. [CrossRef]
72. Portmann, S.; Luthi, H.P. MOLEKEL: An interactive molecular graphics tool. *Chimia* **2000**, *54*, 766–770.

Article

Functionalized Tris(anilido)triazacyclononanes as Hexadentate Ligands for the Encapsulation of U(III), U(IV) and La(III) Cations

Alasdair Formanuik [1], Fabrizio Ortu [1,2], Iñigo J. Vitorica-Yrezabal [1], Floriana Tuna [1], Eric J. L. McInnes [1], Louise S. Natrajan [1,*] and David P. Mills [1,*]

1 Department of Chemistry, The University of Manchester, Oxford Road, Manchester M13 9PL, UK; aformanuik@outlook.com (A.F.); fabrizio.ortu@leicester.ac.uk (F.O.); inigo.vitorica@manchester.ac.uk (I.J.V.-Y.); floriana.tuna@manchester.ac.uk (F.T.); eric.mcinnes@manchester.ac.uk (E.J.L.M.)
2 School of Chemistry, University of Leicester, University Road, Leicester LE1 7RH, UK
* Correspondence: louise.natrajan@manchester.ac.uk (L.S.N.); david.mills@manchester.ac.uk (D.P.M.); Tel.: +44-161-275-1426 (L.S.N.); +44-161-275-4606 (D.P.M.)

Abstract: Tripodal multidentate ligands have become increasingly popular in f-element chemistry for stabilizing unusual bonding motifs and supporting small molecule activation processes. The steric and electronic effects of ligand donor atom substituents have proved crucial in both of these applications. In this study we functionalized the previously reported tris-anilide ligand {tacn(SiMe$_2$NPh)$_3$} (tacn = 1,3,7-triazacyclononane) to incorporate substituted aromatic rings, with the aim of modifying f-element complex solubility and ligand steric effects. We report the synthesis of two proligands, {tacn(SiMe$_2$NHAr)$_3$} (Ar = C$_6$H$_3$Me$_2$-3,5 or C$_6$H$_4$Me-4), and their respective group 1 transfer agents—{tacn(SiMe$_2$NKAr)$_3$}, M(III) complexes [M{tacn(SiMe$_2$NAr)$_3$}] for M = La and U, and U(IV) complexes [M{tacn(SiMe$_2$NAr)$_3$}(Cl)]. These compounds were characterized by multinuclear NMR and FTIR spectroscopy and elemental analysis. The paramagnetic uranium complexes were also characterized by solid state magnetic measurements and UV/Vis/NIR spectroscopy. U(III) complexes were additionally studied by EPR spectroscopy. The solid state structures of all f-block complexes were authenticated by single-crystal X-ray diffraction (XRD), together with a minor byproduct [U{tacn(SiMe$_2$NC$_6$H$_4$Me-4)$_3$}(I)]. Comparisons of the characterization data of our f-element complexes with similar literature examples containing the {tacn(SiMe$_2$NPh)$_3$} ligand set showed minor changes in physicochemical properties resulting from the different aromatic ring substitution patterns we investigated.

Keywords: f-element; lanthanide; actinide; multidentate ligand; macrocycle

1. Introduction

Coordinatively unsaturated metal complexes with well-defined reactive sites are ideal candidates for systematic reactivity studies, and as with the rest of the periodic table, such systems can be accessed for f-elements through the use of judiciously selected ligands [1]. Given the predominantly electrostatic bonding regimes and large ionic radii of f-elements, sterically demanding ligands with hard donor atoms provide the kinetic and electronic stabilization required to give robust complexes [2]. Multidentate and macrocyclic ligands form a privileged subset over their monodentate analogues in f-element coordination chemistry, as they can impart additional thermodynamic stability [1,2]. Tripodal examples, which exhibit approximate C_3 symmetry in solution, have proved particularly effective in controlling the coordination sphere of uranium by encapsulating the metal ion in a single well-defined "steric pocket" along the C_3 axis. Such motifs have supported unusual uranium oxidation states and bonding regimes, and coordinatively unsaturated U(III) centers that support rich small molecule activation chemistry [3–7].

Over the last 25 years, the tripodal ligands that have proved most popular for generating landmark uranium complexes include the anchored tris-aryloxides {N(CH$_2$OAr)$_3$} (Ar = substituted aryl), {tacn(CH$_2$OAr)$_3$} (tacn = 1,3,7-triazacyclononane) and {Mes(CH$_2$OAr)$_3$} (Mes = C$_6$H$_2$Me$_3$-2,4,6); and the tris-amides {N(CH$_2$CH$_2$NSiR$_3$)$_3$} (SiR$_3$ = SiMe$_3$, Si^{t}BuMe$_2$, Si^{i}Pr$_3$) [3–7]. Several related macrocyclic ligand systems have more recently been applied in f-element chemistry [8–10]. f-Element complexes of tripodal ligands that impose approximate C$_3$ symmetry have been applied in the electrocatalytic reduction of water [11–13], and it has been shown that the steric and electronic effects of substituents on donor atoms of these ligands are crucial for dictating the physicochemical properties of resultant complexes. For example, the reaction of CO$_2$ with [U{tacn(CH$_2$Ar^{tBu,tBu}O)$_3$}] (Ar^{tBu,tBu}O = 3,5-di-*tert*-butyl-2-oxybenzyl) gives the dinuclear U(IV) bridging oxo complex [{U[tacn(CH$_2$Ar^{tBu,tBu}O)$_3$]}$_2$(μ-O)] [14], whereas the more sterically demanding [U{tacn (CH$_2$Ar^{Ad,Ad}O)$_3$}] (Ar^{Ad,Ad}O = 3-adamantyl-5-*tert*-butyl-2-oxybenzyl) reacts with CO$_2$ to yield the remarkable terminal CO$_2$ U(IV) complex [U{tacn(Ar^{Ad,Ad}O)$_3$}(CO$_2$)] [15]. Those examples showcase how the outcomes of small molecule activation reactions for tripodal U(III) complexes are dependent upon the size of the axial reactivity pocket [3–7].

By contrast, there have only been a handful of reports to date of the synthesis and physicochemical properties of uranium complexes of the related tris-anilido ligand {tacn(SiMe$_2$NPh)$_3$} [16–20]. However, it has already been shown that the SiMe$_2$ linkers engender considerable flexibility compared to the more rigid substituted aryl groups in [U{tacn(CH$_2$ArO)$_3$}] [3–7]. For example, the nearly trigonal prismatic arrangement of N-donor atoms in the U(III) complex [U{tacn(SiMe$_2$NPh)$_3$}] [16] is lost upon oxidation by elemental sulfur to give the distorted bicapped trigonal bipyramidal dinuclear U(IV) complex [{U[tacn(SiMe$_2$NPh)$_3$]}$_2$(μ-S)] [17], whilst approximate C$_3$ symmetry is typically retained when the U(III) complexes [U{tacn(CH$_2$ArO)$_3$}] are converted to U(IV/V/VI) products [3–7].

Herein we report the synthesis of two novel tris-anilido ligands with substituted aryl rings, {tacn(SiMe$_2$Nar)$_3$} (Ar = C$_6$H$_3$Me$_2$-3,5; C$_6$H$_4$Me-4). We utilized group 1 transfer agents of these ligands to synthesize U(III), La(III) and U(IV) complexes; and we characterized these complexes by a variety of techniques in order to compare these data with respective literature examples of lanthanum and uranium complexes of the unsubstituted ligand {tacn(SiMe$_2$NPh)$_3$}.

2. Results

2.1. Synthesis

The ligand precursors {tacn(SiMe$_2$NHAr)$_3$} (Ar = C$_6$H$_4$Me-4, **1-H$_3$**; C$_6$H$_3$Me$_2$-3,5, **2-H$_3$**) were prepared by modifications of the reported synthesis of {tacn(SiMe$_2$NHPh)$_3$} [21] (Scheme 1). The substituted anilines NH$_2$C$_6$H$_4$Me-4 and NH$_2$C$_6$H$_3$Me$_2$-3,5 were separately reacted with excess NaH in THF to provide the sodium anilides NaNHC$_6$H$_4$Me-4 and NaNHC$_6$H$_3$Me$_2$-3,5. These salts were filtered from the remaining NaH and were separately added dropwise to toluene solutions of {tacn(SiMe$_2$Cl)$_3$} [21] to give **1-H$_3$** and **2-H$_3$**, respectively, as pale yellow oils in very good yields after removal of volatiles under reduced pressure. The proligands were both reacted with excess KH in THF without further purification to afford the group 1 transfer agents {tacn(SiMe$_2$NKC$_6$H$_4$Me-4)$_3$}·2THF (**1-K$_3$**) and {tacn(SiMe$_2$NKC$_6$H$_3$Me$_2$-3,5)$_3$}·THF (**2-K$_3$**), respectively, in fair–good yields as off-white powders (Scheme 1). Despite repeated attempts, crystals of **1-K$_3$** and **2-K$_3$** suitable for analysis by single-crystal X-ray diffraction could not be obtained, but elemental analysis values obtained from powders dried in vacuo for 1 h indicated that bound THF molecules remained, as is the case for the Li and Na salts of {tacn(SiMe$_2$NPh)$_3$} [21].

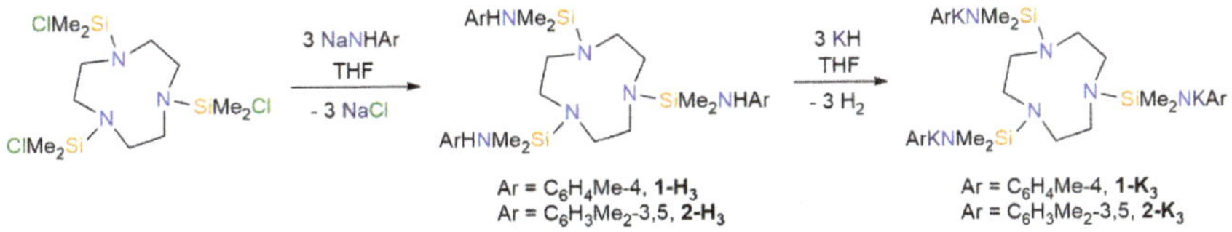

Scheme 1. Synthesis of proligands **1-H₃** and **2-H₃**, and the group 1 transfer agents **1-K₃** and **2-K₃**.

With **1-K₃** and **2-K₃** in hand, we adapted the syntheses of [La{tacn(SiMe₂NPh)₃}(THF)] and [U{tacn(SiMe₂NPh)₃}] [16] to perform separate salt metathesis reactions with [La(I)₃(THF)₄] [22] and [U(I)₃(THF)₄] [23] in THF to give [M{tacn(SiMe₂NC₆H₄Me-4)₃}] (M = La, **3-La**, U, **3-U**) and [M{tacn(SiMe₂NC₆H₃Me₂-3,5)₃}] (M = La, **4-La**, U, **4-U**) in fair yields following work-up and recrystallization (Scheme 2). We first performed the lanthanum reactions, as these have no radiological hazard and La(III) is diamagnetic, providing NMR spectra that are often more easily interpreted than 5f³ U(III) [3–7]. Additionally, La(III) is a reasonable surrogate for U(III) due to their comparable ionic radii [6-coordinate, La(III) = 1.032 Å, U(III) = 1.025 Å] [24]. It is noteworthy that on one occasion during the synthesis of **3-U**, we observed several crystals of the U(IV) complex [U{tacn(SiMe₂NC₆H₄Me-4)₃}(I)] (**5**), which we attribute to the presence of traces of UI₄, but the low yield precluded solution phase characterization (see below). Instead of optimizing conditions for the synthesis of **5**, we decided to target analogous U(IV) chloride complexes, as straightforward salt metathesis routes to [U{tacn(SiMe₂NPh)₃}(Cl)] have been reported previously [10]. By adapting these procedures, the separate reactions of UCl₄ [25,26] with **1-K₃** and **2-K₃** gave the U(IV) complexes [U{tacn(SiMe₂NC₆H₄Me-4)₃}(Cl)] (**6**) and [U{tacn(SiMe₂NC₆H₃Me₂-3,5)₃}(Cl)] (**7**) in poor yields, following work-up and recrystallization from THF or DME, respectively (Scheme 2). We found that **6** and **7** could also be synthesized by the respective oxidation of **3-U** or **4-U** with excess ᵗBuCl. This oxidative route gave improved yields of **6** and **7** over salt metathesis from UCl₄. We anticipate that the U(III) complexes **3-U** and **4-U** could give rich reactivity with small molecules and unsaturated organic compounds, as previously seen for similar tripodal U(III) complexes [3–7]; however, we were unable to isolate any uranium-containing products from the separate 1:1 reactions of **4-U** with either benzophenone or 4,4′-bipyridine in toluene. Since [U{tacn(SiMe₂NPh)₃}] has been shown to react with elemental sulfur [17], and the 2e⁻ oxidation reactions of other tripodal U(III) complexes have been shown to give U(V) products [3–7], we attempted the synthesis of terminal U(V) oxo complexes by the separate reactions of **3-U** with pyridine N-oxide in toluene and **4-U** with 4-methyl morpholine N-oxide in THF; no products could be identified from either of these reaction mixtures.

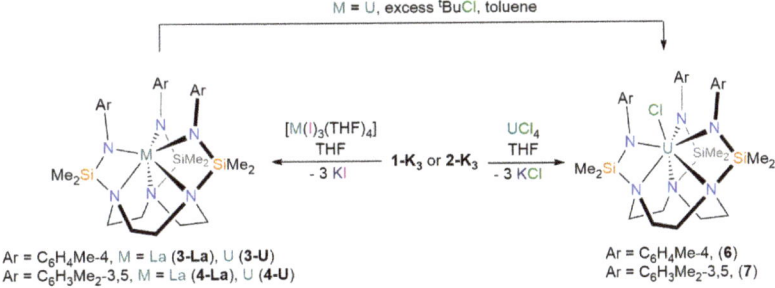

Ar = C₆H₄Me-4, M = La (**3-La**), U (**3-U**)
Ar = C₆H₃Me₂-3,5, M = La (**4-La**), U (**4-U**)

Ar = C₆H₄Me-4, (**6**)
Ar = C₆H₃Me₂-3,5, (**7**)

Scheme 2. Synthesis of complexes **3-M**, **4-M**, **6** and **7** (M = La, U).

2.2. NMR and IR Spectroscopy

NMR spectra are presented in the ESI Figures S13–S42. The ^{1}H and ^{13}C{^{1}H} NMR spectra of **1-H$_3$** in C$_6$D$_6$ indicate that approximate C_3 symmetry was present on the NMR timescale at 298 K. One signal was observed for each proton environment in the ^{1}H NMR spectrum apart from the *para*-Me resonances, where two signals were observed in a 3:6 ratio (δ_H: 2.11 and 2.20 ppm). The three NH protons were obscured by the broad multiplets of the tacn protons (δ_H: 2.98–3.19 ppm). Similar features were observed for **2-H$_3$**, though with the *meta*-Me resonances in a 12:6 ratio (δ_H: 2.23 and 2.24 ppm) in the ^{1}H NMR spectrum. In both **1-H$_3$** and **2-H$_3$** there are also more Ar-*H* resonances than would be expected from ideal C_3 symmetry, and this deviation is tentatively attributed to conformational rigidity of the tacn ring in solution rendering the CH$_2$ protons diastereotopic. The low solubility of **1-K$_3$** and **2-K$_3$** in benzene and toluene necessitated the employment of deuterated THF for NMR spectroscopy, precluding accurate integration of the proton THF resonances to validate elemental analysis data. The addition of more polar solvents to **1-K$_3$** and **2-K$_3$**, such as pyridine, lead to decomposition. The doubling of the number of resonances in the ^{13}C and ^{29}Si NMR spectra of **1-K$_3$** (δ_{Si}: −18.48 and −15.33 ppm) and **2-K$_3$** (δ_{Si}: −17.88 and −14.74 ppm) compared to the parent anilines **1-H$_3$** (δ_{Si}: −8.02 ppm) and **2-H$_3$** (δ_{Si}: −7.97 ppm) indicates that either coordinated THF reduced these systems to approximate C_s symmetry or that two different approximately C_3 symmetric complexes formed in solution. The former interpretation is the most plausible, as coordinating solvents should disfavor oligomerization.

The most noteworthy difference between the ^{1}H NMR spectra of **3-M** and **4-M** and their parent proligands is that the tacn methylene group signals split into two multiplets with AA'BB' patterns and a relative intensity of 1:1 upon complexation. This pattern of signals is due to a significant deviation from ideal C_3 symmetry and fluxional behavior, and was previously seen for [M{tacn(SiMe$_2$NPh)$_3$}] (M = Y, Eu, Yb, U) and [La{tacn(SiMe$_2$NPh)$_3$}(THF)] by Marques and co-workers [16]. Following variable temperature ^{1}H NMR spectroscopy experiments on those literature complexes, the authors postulated that the two N$_3$ vertices of an approximate trigonal prism twist with respect to each other with a concomitant inversion of the tacn chelate rings; thus, we infer that similar processes are in operation for **3-M** and **4-M**. As expected, the ^{1}H NMR spectra of **3-U** and **4-U** are paramagnetically shifted, with signals from −30 to +10 ppm that show significant line broadening [27]. The ^{1}H NMR spectra of **6** and **7** displayed similar features to those of **3-U** and **4-U**, but the signals were broadened to a greater extent, with some FWHM values being as high as 1200 Hz for **6**, and the aromatic ring protons of **7** were not observed. The bulk features of these spectra are similar to those previously seen for [U{tacn(SiMe$_2$NPh)$_3$}(Cl)] [16]. Only one resonance was seen for the methylene groups in the ^{13}C{^{1}H} NMR spectra of **3-La** and **4-La**, but no signals could be observed in the ^{13}C{^{1}H} NMR spectra of **3-U**, **4-U**, **6** and **7** due to paramagnetism. The ^{29}Si{^{1}H} NMR spectra of **3-La** (δ_{Si}: −5.10 ppm) and **4-La** (δ_{Si}: −5.18 ppm) are deshielded compared to **1-H$_3$** and **2-H$_3$**; whilst no signals were seen for the U(IV) complexes **6** and **7**, the U(III) complexes exhibited remarkably shielded resonances (δ_{Si}: −263.80 ppm, **3-U**; −270.90 ppm, **4-U**) that to the best of our knowledge are the most negative values reported for any U(III) complex to date [28].

The FTIR spectra of all complexes containing {tacn(SiMe$_2$NAr)$_3$} scaffolds exhibit strong absorptions at $\tilde{\nu}$ ~ 1600 cm^{-1} due to the substituted anilide groups, as has been seen by other authors for f-block complexes containing {tacn(SiMe$_2$Nar)$_3$} [16–20]. In addition, **1-H$_3$** ($\tilde{\nu}$ = 3406 cm^{-1}) and **2-H$_3$** ($\tilde{\nu}$ = 3404 and 3377 cm^{-1}) exhibit N–H stretching absorptions, as does {tacn(SiMe$_2$HNHPh)$_3$} ($\tilde{\nu}$ = 3381 cm^{-1}) [16].

2.3. NIR/Vis/UV Spectroscopy

The electronic absorption spectra of **3-U**, **4-U**, **6** and **7** were recorded as 0.5 mM toluene solutions (Figure 1). The similar spectra of the pale green U(IV) complexes **6** and **7** are dominated by intense charge transfer bands in the UV region that tail into the visible region, and are otherwise essentially featureless. The charge transfer bands are less prominent at higher energy for red **3-U** and **4-U** solutions, as they both exhibit a series of strong absorptions with shoulders in the visible region ($\tilde{\nu}_{max}$, cm^{-1} (ε, M^{-1} cm^{-1}): ~17,000 (~628, **3-U**; ~692, **4-U**); ~20,000 (~1,444, **3-U**; ~1,758, **4-U**); ~23,000 (~1,308, **3-U**; ~1,612, **4-U**)) that are mainly assigned as $5f^26d^1 \leftarrow 5f^3$ transitions, consistent with U(III) centers [6]. In the NIR region, **3-U** and **4-U** exhibit nearly identical series of weak absorptions (ε < 250 M^{-1} cm^{-1}) that are assigned as $5f \leftarrow 5f$ transitions with intensities that are typical of U(III) complexes [29].

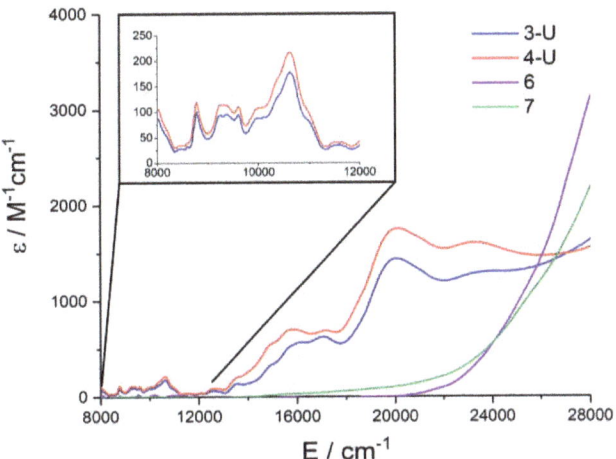

Figure 1. Electronic spectra of **3-U**, **4-U**, **6** and **7** in toluene (0.5 mM) at 8000–28,000 cm^{-1}.

2.4. Structural Characterization

The molecular structures of **3-M·0.5C$_7$H$_8$**, **4-M**, **5**, **6·THF** and **7** were determined by single-crystal X-ray diffraction (selected bond lengths and angles are listed in Table 1, and supporting crystallographic information is compiled in the ESI Tables S1 and S2). The structures of **3-U·0.5C$_7$H$_8$**, **4-U**, **6·THF** and **7** are shown in Figures 2 and 3. As **3-La·0.5C$_7$H$_8$** and **4-La** are structurally analogous with **3-U·0.5C$_7$H$_8$** and **4-U**, respectively, these are discussed together with U(III) congeners for brevity; and their structures are depicted in the ESI together with **5**, which shows similar features to **6·THF** (Figures S1–S3).

There are two independent complex molecules in the unit cells of **3-M·0.5C$_7$H$_8$**. This was seen previously for [U{tacn(SiMe$_2$NPh)$_3$}]·0.5C$_7$H$_8$ and was attributed to the presence of two isomers [16]. As the metrical parameters of the isomers are similar, we describe one of the molecules of **3-M** here. In each case, the metal centers exhibit approximate trigonal prismatic geometries defined by approximately parallel and near-eclipsed amine (N$_1$-N$_3$) and anilide (N$_4$-N$_6$) planes, and lie much closer to the anilide [M···N$_4$-N$_6$ (Å): 0.468(2) (**3-La**); 0.513(4) (**3-U**)] than the amine [M···N$_1$-N$_3$ (Å): 2.064(2) (**3-La**); 2.030(4) (**3-U**)] planes. Despite differences in 6-coordinate ionic radii (La(III) = 1.032 Å, U(III) = 1.025 Å) [18], significant disorder in the dataset for **3-U** led to poor resolution of M–N distances, to the extent that these were within statistical error of the corresponding distances in **3-La**. The bulk features of **4-M** are similar to those of their **3-M** analogs; for example, in **4-U** the position of the uranium center within the N$_6$ prism [U···N$_4$-N$_6$: 0.509(8) Å; U···N$_1$-N$_3$: 2.002(8) Å; mean U–N$_{amine}$: 2.63(2) Å; U–N$_{anilide}$: 2.36(2) Å] is almost identical to those seen for [U{tacn(SiMe$_2$NPh)$_3$}] [16] and **3-U** (see Figure S4 for space-filling representations of [U{tacn(SiMe$_2$NPh)$_3$}], **3-U** and **4-U** viewed along the approximate C_3 axes above the N-aryl rings).

Table 1. Selected bond lengths (Å) and angles (°) for 3-M·0.5C$_7$H$_8$, 4-M, 5, 6·THF and 7 (M = La, U; X = Cl, I). Symmetry operations to generate equivalent atoms: i 1−x, +y, +z; ii = +x, $\frac{1}{2}$−y, +z.

Complex	M–N$_{amine}$	M–N$_{anilide}$	N–M–N$_{amine}$	N–M–N$_{anilide}$	M–X
3-La·0.5C$_7$H$_8$	2.705(3) 2.705(3) 2.654(3)	2.395(3) 2.414(3) 2.436(3)	66.21(8) 68.00(9) 67.54(7)	120.06(9) 113.46(8) 115.46(10)	-
3-U·0.5C$_7$H$_8$	2.637(9) 2.659(8) 2.658(6)	2.381(8) 2.364(7) 2.396(10)	68.2(2) 67.4(3) 67.5(3)	119.0(3) 116.6(3) 110.8(3)	-
4-La	2.674(2) 2.676(3) 2.707(3)	2.421(3) 2.426(2) 2.400(3)	67.80(7) 67.09(8) 68.02(8)	115.74(9) 116.64(9) 115.99(8)	-
4-U	2.616(12) 2.646(10) 2.63(2)	2.348(10) 2.31(2) 2.418(14)	68.2(4) 68.3(4) 68.6(5)	117.9(6) 115.6(4) 112.9(5)	-
5	2.74(4) 2.67(2) 2.59(4)	2.36(2) 2.32(2) 2.32(2)^i	66.0(1) 64.8.(8) 67.7(9)	96.6(4) 154.5(7) 96.6(4)^i	3.143(2)
6·THF	2.659(8) 2.683(7) 2.635(7)	2.285(7) 2.295(8) 2.327(8)	66.6(3) 67.1(2) 66.0(2)	95.4(3) 160.3(3) 94.1(3)	2.706(3)
7	2.69(2) 2.667(7) 2.62(2)	2.305(5) 2.286(5) 2.305(5)^{ii}	66.8(5) 67.8(5) 65.4(6)	95.77(10) 158.2(3)(10) 95.77(10)^{ii}	2.687(2)

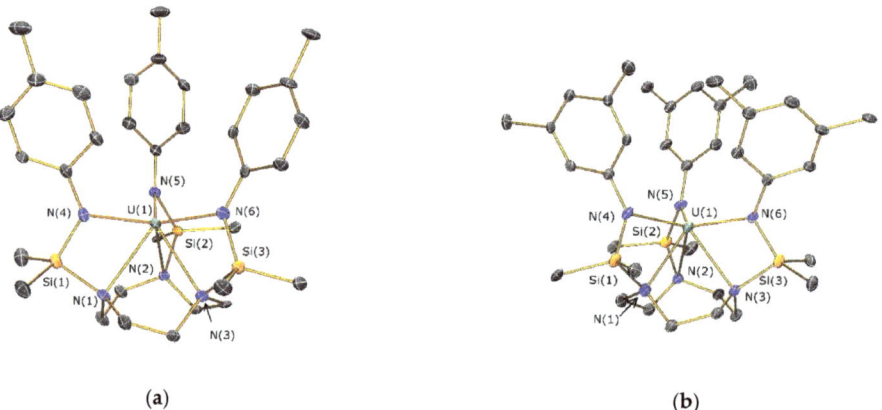

Figure 2. Molecular structures of (**a**) [U{tacn(SiMe$_2$NC$_6$H$_4$Me-4)$_3$}] (**3-U·0.5C$_7$H$_8$**) and (**b**) [U{tacn(SiMe$_2$NC$_6$H$_3$Me$_2$-3,5)$_3$}] (**4-U**) with selective atom labelling. Displacement ellipsoids set at 30% probability level; hydrogen atoms and lattice toluene were omitted for clarity.

Although **5–7** all crystallize in different space groups and **6·THF** additionally incorporates a molecule of THF in the unit cell, the metrical parameters of **5–7** are similar to those of [U{tacn(SiMe$_2$NPh)$_3$}(Cl)]. Whilst [U{tacn(SiMe$_2$NPh)$_3$}(I)] was also synthesized previously by oxidation of [U{tacn(SiMe$_2$NPh)$_3$}] with iodine, its solid state structure was not reported [16]. As with [U{tacn(SiMe$_2$NPh)$_3$}(Cl)] [16], the positions of the donor atoms in **5–7** can be described as bicapped trigonal bipyramids with one anilide and two amine donors forming the equatorial plane; the halide and remaining amine at the axial positions [N(1)–U(1)–X(1): 169.7(5)° (**5**); 171.7(2)° (**6**); 171.7(2)° (**7**)]; and the other two anilides capping two faces near the halide apex. The uranium centers are situated above the N$_3$-tacn plane [U···N$_1$-N$_3$ (Å): 1.04(2) (**5**); 1.021(5) (**6**); 1.011(6) (**7**)] such that for **6** and **7** the apical U–N distances [2.683(7) Å (**6**); 2.667(9) Å (**7**)] are similar to the U–Cl distances [2.706(3) Å (**6**); 2.687(2) Å (**7**)].

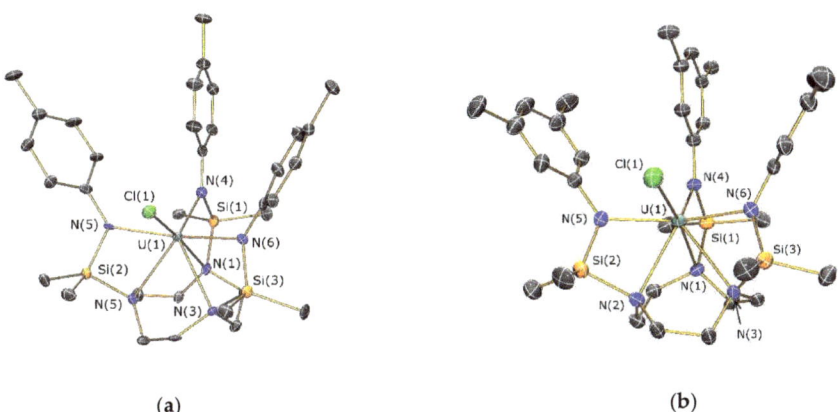

Figure 3. Molecular structures of (**a**) [U{tacn(SiMe$_2$NC$_6$H$_4$Me-4)$_3$}(Cl)] (**6·THF**) and (**b**) [U{tacn(SiMe$_2$NC$_6$H$_3$Me$_2$-3,5)$_3$}(Cl)] (**7**) with selective atom labelling. Displacement ellipsoids set at 30% probability level; hydrogen atoms omitted for clarity.

2.5. Magnetism and EPR Spectroscopy

Solution magnetic susceptibility measurements of U(III) and U(IV) complexes, measured in C$_6$D$_6$ at 298 K using the Evans method [30], gave similar χT values: **3-U** (1.05 cm^3 K mol^{-1}), **4-U** (0.95 cm^3 K mol^{-1}), **6** (0.98 cm^3 K mol^{-1}) and **7** (1.13 cm^3 K mol^{-1}). Whilst these molar susceptibilities are all lower than the free ion values [U(III) 5f^3 ^{4}I$_{9/2}$ = 1.70 cm^3 K mol^{-1}; U(IV) 5f^2 ^{3}H$_4$ = 1.60 cm^3 K mol^{-1}], this was expected, as not all crystal field levels are thermally occupied at this temperature [31]. Indeed, the values obtained are around the middle of the previously reported ranges for monometallic U(III) (0.38–1.81 cm^3 K mol^{-1}) and U(IV) (0.48–1.34 cm^3 K mol^{-1}) complexes [32]. Variable temperature solid state magnetic susceptibility measurements for powdered samples of **3-U**, **4-U**, **6** and **7** suspended in eicosane were recorded using a SQUID from 300–2 K (ESI Figures S5–S9). At 300 K, the magnetic susceptibility temperature products [χT (cm^3 K mol^{-1}): 0.84 (**3-U**); 0.70 (**4-U**); 0.60 (**6**); 0.80 (**7**)] were consistently lower than solution values by 0.2–0.4 cm^3 K mol^{-1}, but such discrepancies should be expected from changes in phase, together with variable errors associated with sample masses and diamagnetic corrections. At 298 K, the solid state χT value for **3-U** is similar to that previously reported for [U{tacn(SiMe$_2$NPh)$_3$}] (0.94 cm^3 K mol^{-1}) [19]. U(IV) complexes **6** and **7** showed χT temperature dependence that correlates with metal oxidation state, tending towards zero at low temperatures due to a non-magnetic singlet ground state [χT (cm^3 K mol^{-1}) at 2 K: 0.07 (**6**); 0.10 (**7**)]. The U(III) complexes, **3-U** and **4-U**, reached values that are higher than those of **6** and **7** [χT (cm^3 K mol^{-1}) at 2 K: 0.20 (**3-U**); 0.15 (**4-U**)], but did not reach their low temperature limit, which suggests occupation of low lying excited states [33–35]. This is consistent with the magnetization data (see below).

The U(III) formulations of **3-U** and **4-U** were confirmed by X- and Q-band EPR spectroscopy on powdered samples at 5 K (X-band spectrum of **4-U** shown in Figure 4a; see ESI Figures S10–S12 for other spectra). The spectra of **4-U** are characteristic of an effective spin $\frac{1}{2}$, consistent with a Kramers doublet, with g_{eff} = 2.80, 2.37 and 1.17. Complex **3-U** shows greater rhombicity in the g_{eff} values, but they are complicated by what appears to be an impurity, and we were unable to model these spectra. As U(III) complexes can exhibit single-molecule magnet (SMM) behavior in small applied magnetic fields [33–35], we investigated the dynamic magnetic properties of **3-U** and **4-U**. Both complexes exhibited hysteresis at 1.8 K, but this was more pronounced for **3-U** (Figure 4b) than for **4-U** (see ESI Figure S6d); in both cases, magnetic saturation was not achieved at 7 T, the physical limit of the magnetometer. Ac susceptibility measurements were performed on **3-U** and **4-U** under an applied dc field of 600 G at 1.8 K. Whilst **3-U** did not exhibit in-phase or out-of-phase components, **4-U** displayed clear frequency dependent behavior (Figure 4c,d).

An Arrhenius plot of the data gave an effective energy barrier to magnetic reversal, U_{eff}, of 16.2 ($\pm$0.8) K (see ESI Figure S9).

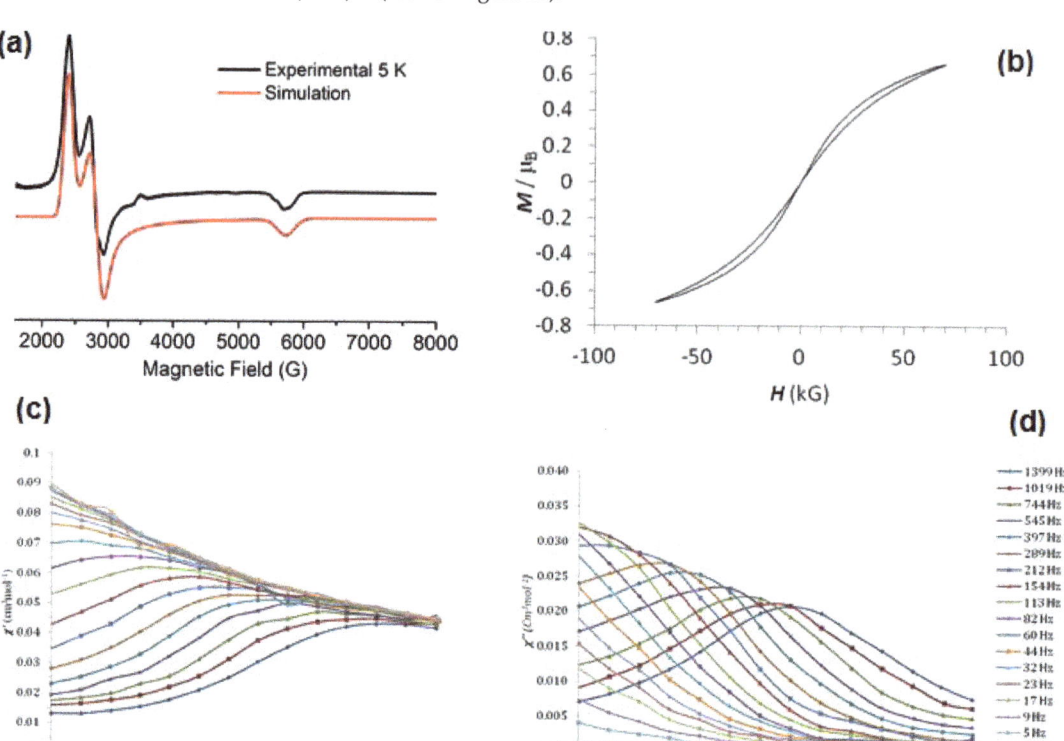

Figure 4. (a) (black) Powder X-band EPR spectrum of **4-U** at 5 K, and (red) its simulation with g_{eff} = 2.80, 2.27 and 1.17. (b) Magnetization (M) hysteresis of **3-U** at 1.8 K, sweep rate 13 G s^{-1}. (c) In phase (χ') and (d) out of phase (χ'') components of the ac susceptibility of **4-U** in an applied field of 600 G and an oscillating field of 1.55 G at 1.8 K.

3. Discussion

Comparisons of the multinuclear NMR spectra of **3-M** and **4-M** with each other and analogous M(III) {tacn(SiMe$_2$NPh)$_3$} complexes [16] indicate that minor variations in anilide substitution patterns do not considerably affect dynamic solution behavior. The solubility and solution stability of complexes also did not appear to vary significantly upon ligand substitution. In the solid state, the single-crystal XRD data show that most of the metrical parameters of **3-U** (e.g., mean U–N$_{\text{amine}}$: 2.651(13) Å; U–N$_{\text{anilide}}$: 2.38(2) Å) are nearly identical to those reported previously for [U{tacn(SiMe$_2$NPh)$_3$}] [U···N$_1$–N$_3$: 0.52 Å; U···N$_4$–N$_6$: 2.02 Å; mean U–N$_{\text{amine}}$: 2.66(3) Å; U–N$_{\text{anilide}}$: 2.35(3) Å] [10]. However, the ranges of N$_{\text{anilide}}$–U–N$_{\text{anilide}}$ [**3-U**: 110.8(3)–119.0(3)°; **4-U**: 112.9(5)–117.9(6)°] and U–N$_{\text{anilide}}$–C$_{\text{ipso}}$ [**3-U**: 115.8(6)–127.0(7)°; **4-U**: 117.7(10)–125.9(9)°] angles are greater for **3-U** than for **4-U**, consistent with nearer-axial EPR spectra of **4-U**, and showing that substitution of the N-aryl groups can influence both the shape and size of the apical channel in the solid state. Structural differences are even more pronounced for the La(III) homologs **3-La** and **4-La**, which do not contain bound THF molecules, in contrast with the 7-coordinate La(III) unsubstituted analog [La{tacn(SiMe$_2$NPh)$_3$}(THF)] [16]. This change in coordination number leads to more significant changes in metrical parameters; for example, the mean La–N$_{\text{amine}}$ [2.685(5) Å] and La–N$_{\text{anilide}}$ [2.415(5) Å] distances for **3-La** differ markedly from [La{tacn(SiMe$_2$NPh)$_3$}(THF)] [mean La–N$_{\text{amine}}$: 2.751(8) Å; La–N$_{\text{anilide}}$: 2.453(8) Å] [16].

The main differences between the solid state structures of the U(IV) and U(III) complexes reported herein are that the approximate C_3 symmetry is broken upon oxidation of the metal center, with the seventh coordination site being occupied by a halide ion. Despite the increase in metal oxidation state, the respective mean U–N$_{anilide}$ and U–N$_{amine}$ distances of **5–7** were similar to those seen for **3-U** and **4-U**, which we attribute to the considerable rearrangement of metal coordination spheres. This facile reorganization is a distinguishing feature of the flexible SiMe$_2$ linkers in {tacn(SiMe$_2$NAr)$_3$} frameworks [16–20] compared to the more rigid tethered aryloxides in {tacn(CH$_2$ArO)$_3$)}, where oxidation of U(III) starting materials to U(IV/V/VI) products has been shown to proceed with retention of approximate C_3 symmetry [3–7]. As {tacn(SiMe$_2$NAr)$_3$} scaffolds readily reorganize to accommodate an additional donor atom, we posit that THF may bind to **3-M** and **4-M** in solution, but that this molecule is readily displaced upon exposure to vacuum during workup and recrystallization. A detailed reactivity study would be required to determine if the fluxionality of {tacn(SiMe$_2$NAr)$_3$} frameworks in solution reduces the steric effect of anilide substituents compared to analogous R-group variation in more rigid tripodal ligands.

A comparison of the characterization data for the U(III) complexes **3-U**, **4-U** and [U{tacn(SiMe$_2$NPh)$_3$}] [16,19] indicates that minor changes in the anilide substituents can lead to subtle changes in the physicochemical properties of complexes. The UV/Vis/NIR electronic absorption spectrum of **4-U** shows more intense absorption intensities than that of **3-U**. To the best of our knowledge, the corresponding data for [U{tacn(SiMe$_2$NPh)$_3$}] have not been published. Similarly, as we were only able to model the EPR spectra of **4-U** (g_{eff} = 2.80, 2.37 and 1.17) and we could not find literature EPR data for [U{tacn(SiMe$_2$NPh)$_3$}], we cannot make a detailed comparison of structurally similar complexes, though we note that g_{eff} = 3.54(5), 2.042(4) and 1.66(5) were determined for [U{tacn(SiMe$_2$NPh)$_3$}(OPPh$_3$)], where the phosphine oxide occupies the apical position [20]. We previously reported the EPR spectra of the planar U(III) tris-amide [U{N(SiBuMe$_2$)$_2$}$_3$] [36]. They gave g_{eff} (=3.55, 2.97, 0.55) that approach those expected for the $|m_J|$ = 1/2 doublet of a $^4I_{9/2}$ term (3.65, 3.65, 0.73), which is stabilized by the in-plane crystal field. For **4-U**, the uranium center is close to the anilide N$_3$ plane, which likewise would stabilize the $|m_J|$ = 1/2 state, but the out-of-plane crystal field arising from the three amine donors leads to significant mixing. Assuming three-fold symmetry, the m_J = ±1/2 state can mix with both the ±5/2 and ±7/2 states (assuming a $^4I_{9/2}$ term) leading to very different g-values. A similar result was concluded for the parent [U{tacn(SiMe$_2$NPh)$_3$}], where crystal field calculations gave a very mixed ground state comprising almost equal fractions of $|m_J|$ = 5/2 and 1/2 (EPR spectra were not reported) [19]. Finally, we note that the SMM behavior of U(III) {tacn(SiMe$_2$NAr)$_3$} complexes appears to be quite sensitive to anilide substituent variation. No effective barrier to magnetic reversal was seen for [U{tacn(SiMe$_2$NPh)$_3$}] [19] and **3-U**, whereas U_{eff} = 16.2 (±0.8) K for **4-U** is similar to values reported previously for U(III) complexes [33–35]; e.g., [U{tacn(SiMe$_2$NPh)$_3$}(OPPh$_3$)], U_{eff} = 21.9(7) K [20].

4. Materials and Methods

General Information

Caution—uranium-238 ($t_{1/2}$ = 4.47 × 10^9 years) is a weak α-emitter; therefore, all manipulations should be performed in suitable laboratories that have been designated for radiochemical use, and α-counting equipment should be available. All manipulations were carried out using standard Schlenk and glove box techniques under an atmosphere of dry argon. THF, toluene, *n*-hexane and DME were dried by refluxing over potassium. All solvents were stored over K mirrors (with the exception of THF and DME, which were stored over activated 4 Å molecular sieves), and were degassed prior to use. Deuterated solvents were dried over K, degassed by three freeze-pump-thaw cycles and stored under Ar. tacn(SiMe$_2$Cl)$_3$ [21], [La(I)$_3$(THF)$_4$] [22], [U(I)$_3$(THF)$_4$] [23] and UCl$_4$ [25,26] were prepared according to literature procedures, and all other reagents were purchased. *p*-Toluidine was dried for 4 h under vacuum before use, whilst 3,5-dimethylaniline was refluxed over CaH$_2$ and distilled.

KH and NaH were obtained as suspensions in mineral oil and were washed three times with *n*-hexane before use.

^{1}H, ^{13}C{^{1}H} and ^{29}Si{^{1}H} NMR spectra were recorded on a Bruker Avance III 400 MHz spectrometer operating at 400.1, 100.6 and 79.5 MHz, respectively; chemical shifts are quoted in ppm and are relative to TMS. FTIR spectra were recorded as Nujol mulls in KBr discs on a Perkin Elmer Spectrum RX1 spectrometer (Perkin Elmer, Waltham, MA, USA). Elemental microanalyses were carried out by Stephen Boyer at the Microanalysis Service, London Metropolitan University, or by Martin Jennings and Anne Davies at The University of Manchester. Low carbon values were consistently obtained in elemental analyses of **2-K$_3$**, **3-La** and **4-La**. We attribute this observation to silicon carbide formation, as <5% protic impurities were observed by ^{1}H NMR spectroscopy, and we note that low C values were intermittently obtained in microanalysis experiments for f-element complexes of the related ligand {tacn(SiMe$_2$NPh)$_3$} [16]. UV-Vis-NIR spectroscopy was performed on samples in Youngs tap appended 10 mm pathlength quartz cuvettes on an Agilent Technologies Cary Series UV-Vis-NIR Spectrophotometer (Agilent Technologies, Santa Clara, CA, USA) from 175–3300 nm. X- and Q-band EPR spectroscopy was performed on powdered samples in quartz tubes at 5 K sealed under vacuum. Magnetic measurements were made using a Quantum Design MPMS-XL7 SQUID magnetometer on ground crystalline samples suspended in eicosane in borosilicate tubes sealed under vacuum.

Crystals of **3-M·0.5C$_7$H$_8$**, **4-M** (M = La and U) and **5–8** were examined on a Rigaku Oxford Diffraction SuperNova CCD area detector diffractometer (Rigaku, Tokyo, Japan) using mirror-monochromated Mo Kα radiation (λ = 0.71073 Å). Intensities were integrated from data recorded on 1° frames by ω rotation. Cell parameters were refined from the observed positions of all strong reflections in each data set. A Gaussian grid face-indexed absorption correction with a beam profile correction was applied [37]. The structures were solved variously by direct and heavy atom methods using *SHELXS* [38] or *SHELXT* [39] and were refined by full-matrix least-squares on all unique F^2 values [38], with anisotropic displacement parameters for all non-hydrogen atoms, and with constrained riding hydrogen geometries; U_{iso}(H) was set at 1.2 (1.5 for methyl groups) times U_{eq} of the parent atom. The largest features in final difference syntheses were close to those of heavy atoms and were of no chemical significance. *CrysAlisPRO* [37] was used for control and integration, *SHELX* [38] and was employed through *OLEX2* [40] for structure solution and refinement and *ORTEP-3* [41] and *POVRAY* [42] were used for molecular graphics.

Synthesis of {tacn(SiMe$_2$NHC$_6$H$_4$Me-4)$_3$} (**1-H$_3$**): *p*-toluidine (5.36 g, 50.0 mmol) was added to a slurry of NaH (2.16 g, 90.0 mmol) in THF (30 mL) and was heated to 50 °C for 18 h. The resultant brown suspension was allowed to cool to room temperature and filtered. The dark brown supernatant was added dropwise to a solution of tacn(SiMe$_2$Cl)$_3$ (6.30 g, 15.5 mmol) in toluene (50 mL) at −78 °C, allowed to slowly warm to room temperature and stirred for 18 h. Volatiles were removed under reduced pressure and the resultant yellow oil was extracted with hexane (3 × 50 mL). The solvent was removed under reduced pressure to give crude **1-H$_3$** as a viscous yellow oil (7.56 g, 79%), which was used without further purification. Anal. Calcd. for C$_{33}$H$_{54}$N$_6$Si$_3$: C, 64.02; H, 8.79; N, 13.58. Found: C, 64.22; H, 8.94; N, 13.44. ^{1}H NMR (C$_6$D$_6$, 298 K): δ = 0.21 (18H, s, Si(CH$_3$)$_2$), 2.11 (3H, s, Ar-CH$_3$), 2.20 (6H, s, Ar-CH$_3$), 2.98–3.19 (15H, br m, NCH$_2$ and NH), 6.61 (d, 4H, J_{HH} = 8.0 Hz, Ar-H), 7.01 (m, 6H, J_{HH} = 8.0 Hz, Ar-H), 7.13 (2H, br m, J_{HH} = 8.0 Hz, Ar-H). ^{13}C{^{1}H} NMR (C$_6$D$_6$, 298 K): δ = −0.69 (Si(CH$_3$)$_2$), 20.99 (Ar-CH$_3$), 50.96 (NCH$_2$), 117.25 (*p*-Ar-CH), 127.25 (Ar-CH), 130.45 (Ar-CH), 145.33 (*ipso*-Ar-C). ^{29}Si{^{1}H} NMR (C$_6$D$_6$, 298 K): δ = −8.02 (*Si*Me$_2$). FTIR (Nujol, $\tilde{\nu}$/cm^{-1}): 3406 (m), 1614 (s), 1512 (w), 1285 (s), 1167 (m), 1148 (m), 1109 (m), 995 (s), 968 (s), 895 (s), 692 (m), 638 (m).

Synthesis of {tacn(SiMe$_2$NKC$_6$H$_4$Me-4)$_3$}·2THF (**1-K$_3$**): **1-H$_3$** (7.56 g, 12.2 mmol) was dissolved in THF (20 mL) and added dropwise to a suspension of KH (2.79 g, 70 mmol) in THF (10 mL) at −78 °C. The reaction mixture was allowed to warm to room temperature, heated to 50 °C overnight, allowed to cool and filtered. Volatiles were removed under reduced pressure and the resultant solid was washed with hexane (20 mL) and dried in vacuo to yield **1-K₃** as an off-white powder (6.10 g, 62%). Anal. Calcd. for $C_{41}H_{67}K_3N_6O_2Si_3$: C, 56.11; H, 7.70; N, 9.52. Found: C, 55.95; H, 7.80; N, 9.47. ^{1}H NMR (C_4D_8O, 298 K): δ = 0.01 (s, 12H, Si(C*H*₃)₂), 0.11 (s, 6H, Si(C*H*₃)₂), 1.78 (m, 4H, OCH₂C*H*₂), 2.05 (s, 9H, Ar-C*H*₃), 2.72–2.91 (m, 12H, NC*H*₂), 3.62 (m, 4H, OC*H*₂CH₂), 6.23 (d, 4H, J_{HH} = 8.0 Hz, Ar-*H*), 6.56 (d, 6H, J_{HH} = 8.0 Hz, Ar-*H*), 6.63 (d, 2H, J_{HH} = 8.0 Hz, Ar-*H*). ^{13}C{^{1}H} NMR (C_4D_8O, 298 K): δ = 0.22 (Si(*C*H₃)₂), 0.39 (Si(*C*H₃)₂), 20.74 (Ar-*C*H₃), 20.83 (Ar-*C*H₃), 26.55 (O*C*H₂CH₂), 49.19 (N*C*H₂), 49.32 (N*C*H₂), 68.39 (OCH₂*C*H₂), 117.46 (*p*-Ar-*C*), 118.82 (*p*-Ar-*C*), 122.23 (Ar-*C*H), 122.82 (Ar-*C*H), 130.66 (Ar-*C*H), 130.90 (Ar-*C*H), 160.09 (*ipso*-Ar-*C*), 160.94 (*ipso*-Ar-*C*). ^{29}Si DEPT NMR (C_4D_8O, 298 K): δ = −18.48, −15.33 (*Si*Me₂). FTIR (Nujol, $\tilde{\nu}$/cm^{-1}): 1595 (s), 1172 (m), 1057 (m), 995 (s), 955 (m), 893 (m), 870 (m), 756 (m), 683 (w), 635 (w).

Synthesis of {tacn(SiMe₂NC₆H₃Me-3,5)₃} (**2-H₃**): 3,5-dimethylaniline (7.23 g, 59.7 mmol) was added to a slurry of NaH (3.46 g, 144.0 mmol) in THF (30 mL) and heated to 50 °C for 18 h. The resultant brown suspension was allowed to cool to room temperature and filtered. The dark brown supernatant was added dropwise to a solution of tacn(SiMe₂Cl)₃ (7.51 g, 18.5 mmol) in toluene (50 mL) at −78 °C, allowed to slowly warm to room temperature and stirred for 18 h. Volatiles were removed under reduced pressure and the resultant yellow oil was extracted with hexane (3 × 50 mL). The solvent was removed under reduced pressure to give crude **2-H₃** as a viscous yellow oil, which was used without further purification (10.40 g, 85%). Anal. Calcd. for $C_{36}H_{60}N_6Si_3$: C, 65.40; H, 9.15; N, 12.71. Found: C, 64.94; H, 9.14; N, 12.38. ^{1}H NMR (C_6D_6, 298 K): δ = 0.23 (s, 12H, s, Si(C*H*₃)₂), 0.27 (s, 6H, s, Si(C*H*₃)₂), 2.23 (s, 12H, Ar-C*H*₃), 2.24 (6H, s, Ar-C*H*₃), 3.02–3.25 (15H, br m, NC*H*₂ and N*H*), 6.32 (s, 4H, Ar-*H*), 6.35 (s, 2H, Ar-*H*), 6.45 (s, 3H, Ar-*H*). ^{13}C{^{1}H} NMR (C_6D_6, 298 K): δ = 0.68 (Si(*C*H₃)₂), 22.08 (Ar-*C*H₃), 51.03 (N*C*H₂), 115.35 (Ar-*C*H), 120.62 (Ar-*C*H), 138.93 (*m*-Ar-*C*), 147.70 (*ipso*-Ar-*C*). ^{29}Si{^{1}H} NMR (C_6D_6, 298 K): δ = −7.97 (*Si*Me₂). FTIR (Nujol, $\tilde{\nu}$/cm^{-1}): 3404 (m), 3377 (m), 1599 (s), 1406 (m), 1325 (m), 11776 (s), 1120 (s), 1109 (m), 1057 (s), 966 (m), 821 (br, s), 691 (s).

Synthesis of {tacn(SiMe₂NKC₆H₃Me₂-3,5)₃}·THF (**2-K₃**): **2-H₃** (10.40 g, 15.7 mmol) was dissolved in THF (20 mL) and added dropwise to a suspension of KH (2.16 g, 54.0 mmol) in THF (10 mL) at −78 °C. The reaction mixture was allowed to warm to room temperature and heated to 50 °C overnight, allowed to cool and filtered. Volatiles were removed under reduced pressure, and the resultant solid was washed with hexane (20 mL) and dried in vacuo to yield **2-K₃** as an off-white powder (12.50 g, 86%). Anal. Calcd. for $C_{40}H_{65}K_3N_6OSi_3$: C, 58.69; H, 7.73; N, 9.92. Found: C, 54.96; H, 7.72; N, 9.95. ^{1}H NMR (C_4D_8O, 298 K): δ = 0.04 (s, 12H, s, Si(C*H*₃)₂), 0.13 (s, 6H, s, Si(C*H*₃)₂), 1.78 (m, OCH₂C*H*₂), 2.00 (s, 12H, Ar-C*H*₃), 2.06 (6H, s, Ar-C*H*₃), 2.72-2.93 (12H, br m, NC*H*₂), 3.62 (m, OC*H*₂CH₂), 5.70 (s, 2H, Ar-*H*), 5.78 (s, 1H, *p*-Ar-*H*), 6.00 (s, 4H, Ar-*H*), 6.29 (s, 2H, Ar-*H*). ^{13}C{^{1}H} NMR (C_4D_8O, 298 K): δ = 0.35 (Si(*C*H₃)₂), 0.43 (Si(*C*H₃)₂), 22.31 (Ar-*C*H₃), 22.36 (Ar-*C*H₃), 26.55 (O*C*H₂CH₂), 49.16 (N*C*H₂), 49.39 (N*C*H₂), 68.39 (OCH₂*C*H₂), 112.52 (Ar-*C*H), 113.67 (Ar-*C*H), 120.32 (Ar-*C*H), 120.91 (Ar-*C*H), 138.34 (*m*-Ar-*C*), 138.74 (*m*-Ar-*C*), 162.82 (*ipso*-Ar-*C*), 163.25 (*ipso*-Ar-*C*). ^{29}Si DEPT NMR (C_4D_8O, 298 K): δ = −17.88, −14.74 (*Si*Me₂). FTIR (Nujol, $\tilde{\nu}$/cm^{-1}): 1557 (s), 1352 (m), 1240 (m), 1194 (m), 1173 (m), 1117 (m), 1057 (m), 978 (w), 916 (m), 893 (m), 843 (m), 760 (m), 654 (m).

Synthesis of [La{tacn(SiMe₂NC₆H₄Me-4)₃}] (**3-La**): THF (20 mL) was added to a precooled (−78 °C) mixture of [La(I)₃(THF)₄] (0.40 g, 0.5 mmol) and **1-K₃** (0.44 g, 0.5 mmol). The pale yellow reaction mixture was allowed to warm to room temperature and stirred for 16 h. Volatiles were removed under reduced pressure and the residue was extracted with toluene (2 × 10 mL), reduced in volume to ca. 2 mL and stored at −30 °C overnight to give colorless crystals of **3-La·0.5C₇H₈** (0.13 g, 32%). Anal. Calcd. for $C_{33}H_{51}LaN_6Si_3 \cdot C_7H_8$: C, 56.71; H, 7.02; N, 9.92. Found: C, 51.64; H, 6.93; N, 9.54. ^{1}H NMR (C_6D_6, 298 K): δ = 0.27 (s, 18H, Si(C*H*₃)₂), 2.21 (s, 9H, Ar-C*H*₃), 2.23-2.34 (m, 6H, NC*H*₂), 2.81-2.89 (m, 6H, NC*H*₂), 6.64 (d, 6H, J_{HH} = 8.0 Hz, Ar-*H*), 6.98 (d, 6H, J_{HH} = 8.0 Hz, Ar-*H*). ^{13}C{^{1}H} NMR (C_6D_6, 298 K): δ = 0.80 (Si(*C*H₃)₂), 21.13 (Ar-*C*H₃), 48.65 (N*C*H₂), 121.04 (Ar-*C*H), 126.14 (*p*-Ar-*C*), 131.33 (Ar-*C*H), 151.10 (*ipso*-Ar-*C*). ^{29}Si{^{1}H} NMR (C_6D_6, 298 K): δ = −5.10 (s, *Si*Me₂). FTIR

(Nujol, $\tilde{v}/\text{cm}^{-1}$): 1606 (s), 1503 (s), 1297 (m), 1178 (w), 1041 (m), 1017 (m), 958 (m), 894 (m), 816 (s), 771 (s).

Synthesis of [U{tacn(SiMe$_2$NC$_6$H$_4$Me-4)$_3$}] (**3-U**) *and [U{tacn(SiMe$_2$NC$_6$H$_4$Me-4)$_3$}(I)]* (**5**): THF (20 mL) was added to a pre-cooled (−78 °C) mixture of [U(I)$_3$(THF)$_4$] (0.45 g, 0.5 mmol) and **1-K$_3$** (0.44 g, 0.5 mmol). The dark red reaction mixture was allowed to warm to room temperature and stirred for 16 h. The volatiles were removed in vacuo and the residue was extracted with toluene (2 × 10 mL) and filtered. The solution was reduced in volume to ca. 2 mL and stored at −30 °C overnight to give dark red crystals of **3-U·0.5C$_7$H$_8$** (0.25 g, 56%). On one occasion, several crystals of **5** formed. The solid state structure of **5** was determined, but no other characterization data could be obtained due to the low yield. Data for **3-U**: Anal. Calcd. for C$_{33}$H$_{51}$N$_6$Si$_3$U·0.5C$_7$H$_8$: C, 48.70; H, 6.16; N, 9.34. Found: C, 48.53; H, 6.09; N, 9.47. μ_{eff} = 2.97 μ_B (Evans method). ^{1}H NMR (C$_6$D$_6$, 298 K): δ = −30.29 (br, $v_{\frac{1}{2}}$ ~ 150 Hz, 6H, NCH_2), −7.50 (br, $v_{\frac{1}{2}}$ ~ 150 Hz, 6H, NCH_2), −0.65 (br, $v_{\frac{1}{2}}$ ~ 20 Hz, 6H, Ar-CH), −0.40 (br, $v_{\frac{1}{2}}$ ~ 30 Hz, 9H, Ar-CH_3), 2.67 (br, $v_{\frac{1}{2}}$ ~ 30 Hz, 6H, Ar-CH), 7.74 (br, $v_{\frac{1}{2}}$ ~ 50 Hz, 18H, Si(CH_3)$_2$). ^{13}C{^{1}H} NMR (C$_6$D$_6$, 298 K): not observed. ^{29}Si{^{1}H} NMR (C$_6$D$_6$, 298 K): δ = −263.80 (*Si*Me$_2$). FTIR (Nujol, $\tilde{v}/\text{cm}^{-1}$): 1605 (s), 1506 (m), 1290 (s), 1260 (m), 1177 (m), 1034 (m), 951 (s), 912 (w), 895 (m), 862 (m), 816 (s), 770 (s).

Synthesis of [La{tacn(SiMe$_2$NC$_6$H$_3$Me$_2$-3,5)$_3$}] (**4-La**): THF (20 mL) was added to a pre-cooled (−78 °C) mixture of [La(I)$_3$(THF)$_4$] (0.40 g, 0.5 mmol) and **2-K$_3$** (0.46 g, 0.5 mmol). The pale yellow reaction mixture was allowed to warm to room temperature and stirred for 16 h. Volatiles were removed under reduced pressure and the residue was extracted with toluene (2 × 10 mL), reduced in volume to ca. 1 mL and stored at −30 °C overnight to give colorless crystals of **4-La** (0.27 g, 68%). Anal. Calcd. for C$_{36}$H$_{57}$LaN$_6$Si$_3$: C, 54.25; H, 7.21; N, 10.54. Found: C, 53.06; H, 7.23; N, 10.03. ^{1}H NMR (C$_6$D$_6$, 298 K): δ = 0.30 (s, 18H, s, Si(CH_3)$_2$), 2.13 (s, 18H, Ar-CH_3), 2.23-2.33 (br m, 6H, NCH_2), 2.79-2.87 (br m, 6H, NCH_2), 6.39 (s, 6H, *o*-Ar-H), 6.42 (s, 3H, *p*-Ar-H). ^{13}C{^{1}H} NMR (C$_6$D$_6$, 298 K): δ = 0.82 (Si(CH_3)$_2$), 21.84 (Ar-CH_3), 48.66 (NCH_2), 118.08 (Ar-CH), 119.95 (Ar-CH), 140.01 (*m*-Ar-*C*), 153.33 (*ipso*-Ar-*C*). ^{29}Si{^{1}H} NMR (C$_6$D$_6$, 298 K): δ = −5.18 (s, *Si*Me$_2$). FTIR (Nujol, $\tilde{v}/\text{cm}^{-1}$): 1585 (s), 1350 (m), 1246 (m), 1196 (s), 1070 (m), 1041 (m), 966 (m), 988 (m), 896 (s), 866 (s), 769 (s).

Synthesis of [U{tacn(SiMe$_2$NC$_6$H$_3$Me-3,5)$_3$}] (**4-U**): THF (20 mL) was added to a pre-cooled (−78 °C) mixture of [U(I)$_3$(THF)$_4$] (0.45 g, 0.5 mmol) and **2-K$_3$** (0.46 g, 0.5 mmol). The dark red reaction mixture was allowed to warm to room temperature and stirred for 16 h. Volatiles were removed under reduced pressure and the residue was extracted with toluene (2 × 10 mL) and filtered. The solution was reduced in volume to ca. 2 mL and stored at −30 °C overnight to give dark red crystals of **4-U** (0.29 g, 65%). Anal. Calcd. for C$_{36}$H$_{57}$N$_6$Si$_3$U: C, 48.25; H, 6.64; N, 9.38. Found: C, 47.74; H, 5.96; N, 9.53. μ_{eff} = 2.61 μ_B (Evans method). ^{1}H NMR (C$_6$D$_6$, 298 K): δ = −30.84 (br, $v_{\frac{1}{2}}$ ~ 60 Hz, 6H, NCH_2), −9.53 (br, $v_{\frac{1}{2}}$ ~ 60 Hz, 6H, NCH_2), −3.64 (br, $v_{\frac{1}{2}}$ ~ 20 Hz, 18H, Ar-CH_3), −0.75 (br, $v_{\frac{1}{2}}$ ~ 40 Hz, 6H, *o*-Ar-H), 0.51 (s, 3H, *p*-Ar-H), 8.09 (br, $v_{\frac{1}{2}}$ ~ 40 Hz, 18H, Si(CH_3)$_2$). ^{13}C{^{1}H} NMR (C$_6$D$_6$, 298 K): not observed. ^{29}Si{^{1}H} NMR (C$_6$D$_6$, 298 K): δ = −270.90 (*Si*Me$_2$). FTIR (Nujol, $\tilde{v}/\text{cm}^{-1}$): 1584 (s), 1344 (s), 1289 (w), 1069 (m), 1038 (m), 989 (w), 966 (m), 912 (m), 897 (s), 866 (s), 772 (s), 673 (m).

Synthesis of [U{tacn(SiMe$_2$NC$_6$H$_3$Me-4)$_3$}(Cl)] (**6**): *Method 1*: THF (20 mL) was added to a pre-cooled (−78 °C) mixture of UCl$_4$ (0.38 g, 1 mmol) and **1-K$_3$** (0.88 g, 1 mmol). The dark green reaction mixture was allowed to warm to room temperature and stirred for 16 h. Volatiles were removed *in vacuo* and the residue was extracted with THF (2 × 30 mL) and filtered. The solution was reduced in volume to ca. 2 mL and stored at −30 °C overnight to give green crystals of **6·THF** (0.16 g, 16%). *Method 2*: An excess of ^{t}BuCl (0.023 g, 0.25 mmol) was added to a pre-cooled (−10 °C) solution of **3-U** (0.050 g, 0.056 mmol) in toluene. Upon warming to room temperature, a color change to green was observed. Volatiles were removed under reduced pressure to yield **6** in quantitative crude yield as a green solid. Anal. Calcd. for C$_{33}$H$_{51}$ClN$_6$Si$_3$U: C, 44.56; H, 5.78; N, 9.45. Found: C, 44.56; H, 5.73; N, 9.33. μ_{eff} = 2.76 μ_B (Evans method). ^{1}H NMR (C$_4$D$_8$O, 298 K): δ = −46.31 (br, $v_{\frac{1}{2}}$ ~ 70 Hz,

6H, Ar-CH), −15.17 (br, $\nu_{\frac{1}{2}}$ ~ 1,200 Hz, 6H, NCH_2), 5.86 (br, $\nu_{\frac{1}{2}}$ ~ 100 Hz, 9H, Ar-CH_3), 12.34 (br, $\nu_{\frac{1}{2}}$ ~ 150 Hz, 6H, Ar-CH), 16.08 (br, $\nu_{\frac{1}{2}}$ ~ 600 Hz, 18H, Si(CH_3)$_2$), 23.82 (br, $\nu_{\frac{1}{2}}$ ~ 1,000 Hz, 6H, NCH_2). ^{13}C{^{1}H} and ^{29}Si{^{1}H} NMR (C$_4$D$_8$O, 298 K): not observed. FTIR (Nujol, $\tilde{\nu}$/cm^{-1}): 1605 (s), 1501 (m), 1247 (m), 1176 (w), 1109 (w), 1085 (m), 1045 (s), 1013 (s), 945 (m), 921 (m), 905 (m), 831 (s), 776 (m), 730 (s), 708 (m).

*Synthesis of [U{tacn(SiMe$_2$NC$_6$H$_3$Me$_2$-3,5)$_3$}(Cl)] (**7**): Method 1:* THF (20 mL) was added to a pre-cooled (−78 °C) mixture of UCl$_4$ (0.38 g, 0.5 mmol) and **2-K$_3$** (0.92 g, 1 mmol). The dark green reaction mixture was allowed to warm to room temperature and stirred for 16 h. Volatiles were removed in vacuo and the residue was extracted with DME (2 × 20 mL) and filtered. The solution was reduced in volume to ca. 2 mL and stored at −30 °C overnight to give green crystals of **7** (0.09 g, 10%). *Method 2:* An excess of ^{t}BuCl (0.079 g, 0.85 mmol) was added to a pre-cooled (−10 °C) solution of **4-U** (0.150 g, 0.16 mmol) in toluene. Upon warming to room temperature, a color change to green was observed. Volatiles were removed under reduced pressure and the green solid was dissolved in DME (1 mL) and stored −30 °C for 8 h to give green needles of **7** (0.063 g, 42%). Anal. Calcd. for C$_{36}$ClH$_{57}$N$_6$Si$_3$U: C, 46.41; H, 6.17; N, 9.02. Found: C, 46.34; H, 6.30; N, 8.93. μ_{eff} = 3.20 μ_B (Evans method). ^{1}H NMR (C$_6$D$_6$, 298 K): δ = −50.42 (br, $\nu_{\frac{1}{2}}$ ~ 200 Hz, 2H, Ar-CH), 1.15 (br, $\nu_{\frac{1}{2}}$ ~ 300 Hz, 18H, Si(CH_3)$_2$), 12.67 (br, $\nu_{\frac{1}{2}}$ ~ 1,000 Hz, 12H, NCH_2), most Ar-CH and Ar-CH_3 not observed. ^{13}C{^{1}H} and ^{29}Si{^{1}H} NMR (C$_6$D$_6$, 298 K): not observed. FTIR (Nujol, $\tilde{\nu}$/cm^{-1}): 1587 (s), 1317 (s), 1184 (s), 1042 (m), 962 (w), 920 (m), 903 (m), 889 (m), 679 (m).

Supplementary Materials: The following are available online at https://www.mdpi.com/article/10.3390/inorganics9120086/s1. Figures S1–S4: structures of **3-La·0.5C$_7$H$_8$**, **4-La** and **5**, together with a space-filling diagrams of **3-U**, **4-U** and [U{tacn(SiMe$_2$NPh)$_3$}]. Tables S1 and S2: supplementary crystallographic data for all complexes. Figures S5–S9: solid state magnetic data for **3-U** and **4-U**. Figures S10–S12: EPR spectra for **3-U** and **4-U**. Figures S13–S42: NMR spectra for all complexes. CIF and checkCIF output files for the solid state structures of **3-La·0.5C$_7$H$_8$**, **3-U·0.5C$_7$H$_8$**, **4-La**, **4-U**, **5**, **6·THF** and **7**.

Author Contributions: Conceptualization, A.F., L.S.N. and D.P.M.; methodology, A.F. and D.P.M.; validation, A.F. and D.P.M.; formal analysis, all authors; investigation, A.F., F.O., I.J.V.-Y. and F.T.; data curation, A.F. and D.P.M.; writing—original draft preparation, A.F. and D.P.M.; writing—review and editing, all authors; visualization, A.F., F.T. and D.P.M.; supervision, L.S.N. and D.P.M.; project administration, A.F., D.P.M. and L.S.N.; funding acquisition, A.F., E.J.L.M., D.P.M. and L.S.N. All authors have read and agreed to the published version of the manuscript.

Funding: This research was funded by the University of Manchester and the UK Engineering and Physical Sciences Research Council (EPSRC), grant numbers EP/G037140/1 (Nuclear FiRST DTC), EP/L014416/1 and EP/K039547/1; the EPSRC UK National Electron Paramagnetic Resonance Service provided access to the EPR Facility and the SQUID magnetometer (grant number EP/S033181/1).

Data Availability Statement: The data presented in this study are openly available in FigShare at doi:10.6084/m9.figshare.16680670 and the Cambridge Structural Database, deposition numbers CCDC 1833228-1833234.

Acknowledgments: We would like to thank Ian Fallis and Thomas Tatchell from Cardiff University for helpful advice on the synthesis of tacn.

Conflicts of Interest: The authors declare no conflict of interest.

References

1. Mills, D.P.; Liddle, S.T. Ligand Design in Modern Lanthanide Chemistry. In *Ligand Design in Metal Chemistry: Reactivity and Catalysis*; Lundgren, R., Stradiotto, M., Eds.; John Wiley & Sons, Ltd.: Chichester, UK, 2016; pp. 330–363. [CrossRef]
2. Cotton, S. *Lanthanide and Actinide Chemistry*; John Wiley & Sons, Ltd.: Chichester, UK, 2006. [CrossRef]
3. Ephritikhine, M. The vitality of uranium molecular chemistry at the dawn of the XXIst century. *Dalton Trans.* **2006**, 2501–2516. [CrossRef] [PubMed]
4. Bart, S.C.; Meyer, K. Highlights in Uranium Coordination Chemistry. *Struct. Bond.* **2008**, *127*, 119–176. [CrossRef]
5. Meyer, K.; Bart, S.C. Tripodal carbene and aryloxide ligands for small-molecule activation at electron-rich uranium and transition metal centers. *Adv. Inorg. Chem.* **2008**, *60*, 1–30. [CrossRef]

6. Liddle, S.T. The Renaissance of Non-Aqueous Uranium Chemistry. *Angew. Chem. Int. Ed.* **2015**, *54*, 8604–8641. [CrossRef]
7. Gardner, B.M.; Liddle, S.T. Uranium Triamidoamine Chemistry. *Chem. Commun.* **2015**, *51*, 10589–10607. [CrossRef]
8. Maria, L.; Santos, I.C.; Sousa, V.R.; Marçalo, J. Uranium(III) Redox Chemistry Assisted by a Hemilabile Bis(phenolate) Cyclam Ligand: Uranium–Nitrogen Multiple Bond Formation Comprising a trans-{RN=U(VI)=NR}$^{2+}$ Complex. *Inorg. Chem.* **2015**, *54*, 9115–9126. [CrossRef]
9. Maria, L.; Bandeira, N.A.G.; Santos, I.C.; Marçalo, J.; Gibson, J.K. CO_2 conversion to phenyl isocyanates by uranium(vi) bis(imido) complexes. *Chem. Commun.* **2020**, *56*, 431–434. [CrossRef]
10. Xin, T.; Wang, X.; Yang, K.; Liang, J.; Huang, W. Rare Earth Metal Complexes Supported by a Tripodal Tris(amido) Ligand System Featuring an Arene Anchor. *Inorg. Chem.* **2021**, *60*, 15321–15329. [CrossRef]
11. Halter, D.P.; Heinemann, F.W.; Bachman, J.; Meyer, K. Uranium-mediated electrocatalytic dihydrogen production from water. *Nature* **2016**, *530*, 317–321. [CrossRef]
12. Halter, D.P.; Heinemann, F.W.; Maron, L.; Meyer, K. The role of uranium–arene bonding in H_2O reduction catalysis. *Nat. Chem.* **2018**, *10*, 259–267. [CrossRef] [PubMed]
13. Halter, D.P.; Palumbo, C.T.; Ziller, J.W.; Gembicky, M.; Rheingold, A.L.; Evans, W.J.; Meyer, K. Electrocatalytic H_2O Reduction with f-Elements: Mechanistic Insight and Overpotential Tuning in a Series of Lanthanide Complexes. *J. Am. Chem. Soc.* **2018**, *140*, 2587–2594. [CrossRef] [PubMed]
14. Castro-Rodríguez, I.; Meyer, K. Carbon Dioxide Reduction and Carbon Monoxide Activation Employing a Reactive Uranium(III) Complex. *J. Am. Chem. Soc.* **2005**, *127*, 11242–11243. [CrossRef] [PubMed]
15. Castro-Rodríguez, I.; Nakai, H.; Zacharov, L.N.; Rheingold, A.L.; Meyer, K. A Linear, O-Coordinated η^1-CO_2 Bound to Uranium. *Science* **2004**, *305*, 1757–1760. [CrossRef]
16. Monteiro, B.; Roitershtein, D.; Ferreira, H.; Ascenso, J.R.; Martins, A.M.; Domingos, A.; Marques, N. Triamidotriazacyclononane Complexes of Group 3 Metals. Synthesis and Crystal Structures. *Inorg. Chem.* **2003**, *42*, 4223–4231. [CrossRef] [PubMed]
17. Antunes, M.A.; Dias, M.; Monteiro, B.; Domingos, A.; Santos, I.C.; Marques, N. Synthesis and reactivity of uranium(iv) amide complexes supported by a triamidotriazacyclononane ligand. *Dalton Trans.* **2006**, 3368–3374. [CrossRef]
18. Camp, C.; Antunes, M.A.; García, G.; Ciofini, I.; Santos, I.C.; Pécaut, J.; Almeida, M.; Marçalo, J.; Mazzanti, M. Two-electron versus one-electron reduction of chalcogens by uranium(iii): Synthesis of a terminal U(v) persulfide complex. *Chem. Sci.* **2014**, *5*, 841–850. [CrossRef]
19. Antunes, M.A.; Coutinho, J.T.; Santos, I.C.; Marçalo, J.; Almeida, M.; Baldoví, J.J.; Pereira, L.C.J.; Gaita-Ariño, A.; Coronado, E. A Mononuclear Uranium(IV) Single-Molecule Magnet with an Azobenzene Radical Ligand. *Chem. Eur. J.* **2015**, *21*, 17817–17826. [CrossRef] [PubMed]
20. Coutinho, J.T.; Perfetti, M.; Baldoví, J.J.; Antunes, M.A.; Hallmen, P.P.; Bamberger, H.; Crassee, I.; Orlita, M.; Almeida, M.; van Slageren, J.; et al. Spectroscopic Determination of the Electronic Structure of a Uranium Single-Ion Magnet. *Chem. Eur. J.* **2019**, *25*, 1758–1766. [CrossRef]
21. Dias, A.R.; Martins, A.M.; Ascenso, J.R.; Ferreria, H.; Duarte, M.T.; Henriques, R.T. Li, Ti(III), and Ti(IV) Trisamidotriazacyclononane Complexes. Syntheses, Reactivity, and Structures. *Inorg. Chem.* **2003**, *42*, 2675–2682. [CrossRef]
22. Izod, K.; Liddle, S.T.; Clegg, W. A Convenient Route to Lanthanide Triiodide THF Solvates. Crystal Structures of $LnI_3(THF)_4$ [Ln = Pr] and $LnI_3(THF)_{3.5}$ [Ln = Nd, Gd, Y]. *Inorg. Chem.* **2004**, *43*, 214–218. [CrossRef]
23. Avens, L.R.; Bott, S.G.; Clark, D.L.; Sattelberger, A.P.; Watkin, J.G.; Zwick, B.D. A Convenient Entry into Trivalent Actinide Chemistry: Synthesis and Characterization of $AnI_3(THF)_4$ and $An[N(SiMe_3)_2]_3$ (An = U, Np, Pu). *Inorg. Chem.* **1994**, *33*, 2248–2256. [CrossRef]
24. Shannon, R.D. Revised effective ionic radii and systematic studies of interatomic distances in halides and chalcogenides. *Acta Cryst. Sect. A* **1976**, *32*, 751–767. [CrossRef]
25. Patel, D.; Wooles, A.J.; Hashem, E.; Omorodion, H.; Baker, R.J.; Liddle, S.T. Comments on reactions of oxide derivatives of uranium with hexachloropropene to give UCl_4. *New J. Chem.* **2015**, *39*, 7559–7562. [CrossRef]
26. Kiplinger, J.L.; Morris, D.E.; Scott, B.L.; Burns, C.J. Convenient Synthesis, Structure, and Reactivity of $(C_5Me_5)U(CH_2C_6H_5)_3$: A Simple Strategy for the Preparation of Monopentamethylcyclopentadienyl Uranium(IV) Complexes. *Organometallics* **2002**, *21*, 5978–5982. [CrossRef]
27. Hashem, E.; Swinburne, A.N.; Schulzke, C.; Evans, R.C.; Platts, J.A.; Kerridge, A.; Natrajan, L.S.; Baker, R.J. Emission spectroscopy of uranium(IV) compounds: A combined synthetic, spectroscopic and computational study. *RSC Adv.* **2013**, *3*, 4350–4361. [CrossRef]
28. Windorff, C.J.; Evans, W.J. ^{29}Si NMR Spectra of Silicon-Containing Uranium Complexes. *Organometallics* **2014**, *33*, 3786–3791. [CrossRef]
29. Carnall, W.T. A systematic analysis of the spectra of trivalent actinide chlorides in D_{3h} site symmetry. *J. Chem. Phys.* **1992**, *96*, 8713. [CrossRef]
30. Sur, S.K. Measurement of magnetic susceptibility and magnetic moment of paramagnetic molecules in solution by high-field fourier transform NMR spectroscopy. *J. Magn. Reson.* **1989**, *82*, 169–173. [CrossRef]
31. Jones, E.R., Jr.; Hendricks, M.E.; Stone, J.A.; Karraker, D.G. Magnetic properties of the trichlorides, tribromides, and triiodides of U(III), Np(III), and Pu(III). *J. Chem. Phys.* **1974**, *60*, 2088. [CrossRef]
32. Kindra, D.R.; Evans, W.J. Magnetic Susceptibility of Uranium Complexes. *Chem. Rev.* **2014**, *114*, 8865–8882. [CrossRef]
33. Liddle, S.T.; van Slageren, J. Actinide Single Molecule Magnets. In *Lanthanides and Actinides in Molecular Magnetism*; Layfield, R.A., Murugesu, M., Eds.; Wiley-VCH: Weinheim, Germany, 2015; pp. 315–340. [CrossRef]

34. Liddle, S.T.; van Slageren, J. Improving f-element single-molecule magnets. *Chem. Soc. Rev.* **2015**, *44*, 6655–6670. [CrossRef]
35. Moro, F.; Mills, D.P.; Liddle, S.T.; van Slageren, J. The Inherent Single-Molecule Magnet Character of Trivalent Uranium. *Angew. Chem. Int. Ed.* **2013**, *52*, 3430–3433. [CrossRef]
36. Goodwin, C.A.P.; Tuna, F.; McInnes, E.J.L.; Liddle, S.T.; McMaster, J.; Vitorica-Yrezabal, I.J.; Mills, D.P. [U^{III}{N(SiMe$_2$*t*Bu)$_2$}$_3$]: A Structurally Authenticated Trigonal Planar Actinide Complex. *Chem. Eur. J.* **2014**, *20*, 14579–14582. [CrossRef] [PubMed]
37. *CrysAlisPRO*; Agilent Technologies UK Ltd.: Yarnton, UK, 2010.
38. Sheldrick, G.M. A short history of *SHELX*. *Acta Cryst.* **2008**, *A64*, 112–122. [CrossRef] [PubMed]
39. Sheldrick, G.M. Crystal structure refinement with *SHELXL*. *Acta Cryst.* **2015**, *C71*, 3–8. [CrossRef]
40. Dolomanov, O.V.; Bourhis, L.J.; Gildea, R.J.; Howard, J.A.K.; Puschmann, H. *OLEX2*: A complete structure solution, refinement and analysis program. *J. Appl. Cryst.* **2009**, *42*, 339–341. [CrossRef]
41. Farrugia, L.J. *WinGX* and *ORTEP* for Windows: An update. *J. Appl. Cryst.* **2012**, *45*, 849–854. [CrossRef]
42. *POV-Ray*; Persistence of Vision Raytracer Pty., Ltd.: Williamstown, Australia, 2004.

Review

Solvothermal Synthesis Routes to Substituted Cerium Dioxide Materials

James W. Annis [1], Janet M. Fisher [2], David Thompsett [2] and Richard I. Walton [1,*]

1. Department of Chemistry, University of Warwick, Coventry CV4 7AL, UK; J.Annis@warwick.ac.uk
2. Johnson Matthey Technology Centre, Sonning Common, Reading RG4 9NH, UK; janet.fisher@matthey.com (J.M.F.); thompd@matthey.com (D.T.)
* Correspondence: r.i.walton@warwick.ac.uk

Abstract: We review the solution-based synthesis routes to cerium oxide materials where one or more elements are included in place of a proportion of the cerium, i.e., substitution of cerium is performed. The focus is on the solvothermal method, where reagents are heated above the boiling point of the solvent to induce crystallisation directly from the solution. This yields unusual compositions with crystal morphology often on the nanoscale. Chemical elements from all parts of the periodic table are considered, from transition metals to main group elements and the rare earths, including isovalent and aliovalent cations, and surveyed using the literature published in the past ten years. We illustrate the versatility of this synthesis method to allow the formation of functional materials with applications in contemporary applications such as heterogeneous catalysis, electrodes for solid oxide fuel cells, photocatalysis, luminescence and biomedicine. We pick out emerging trends towards control of crystal habit by use of non-aqueous solvents and solution additives and identify challenges still remaining, including in detailed structural characterisation, the understanding of crystallisation mechanisms and the scale-up of synthesis.

Keywords: ceria; hydrothermal; catalysis; nanomaterials; crystallisation

1. Introduction

Cerium dioxide, ceria, is well known for significant applications in heterogeneous catalysis, where it is frequently used as a redox-active support [1,2]. This chemistry is exploited in numerous practical applications, exemplified by its use as the oxygen storage component in three-way catalytic convertors, as a support for precious metals for destruction of CO, NO_x and hydrocarbon emissions in automotive exhausts [3–5]. There are many emerging applications of ceria in catalysis and these include use in steam reforming [6], soot oxidation [7], water–gas shift [8], oxidation of volatile organics [9] and thermochemical water splitting [10]. All of these rely on the ease of reversible release of oxygen from the solid that has the cubic fluorite structure type, which is well known for its oxide ion migration properties, accompanied by the reversible reduction of Ce^{4+} to Ce^{3+}.

Another important developing application of ceria is as an oxide-conducting electrolyte in solid oxide fuel cells [11]. Ceria-based solid electrolytes show promising conductivity in the temperature range 500–700 °C, and here, oxide ion vacancies are vital to optimise the hopping of ions to carry charge. Beyond heterogeneous catalysis and ionic conductors, biomedical applications of ceria are under development, such as in antioxidant defense [12–14]. The material also finds use in UV shielding [15–17], as an abrasive agent [18] and in humidity sensors [19].

For most real samples of ceria, the ideal formula CeO_2 is not found: in reality, there are typically some levels of oxide defects and hence the presence of Ce^{3+} to give the formula $CeO_{2-\delta}$ [20]. These intrinsic defects are inherently beneficial for the oxide ion migration that gives rise to many of the materials' useful properties, since bulk oxygen mobility is enhanced, but in order to further enhance defects, partial substitution of Ce by a second cation, M, to give ternary materials $Ce_{1-x}M_xO_{2-\delta}$ is a common strategy. (Note that in this article, we refer to this replacement as substitution to indicate that a proportion of cerium is swapped for a second cation, rather than "doping", which implies the addition of a second species.) If M is aliovalent (a trivalent cation or divalent cation), then a greater proportion of oxide ion vacancies are needed to balance the charge, Figure 1. However, even replacement of Ce by an isovalent cation (such as Zr^{4+} or Hf^{4+}) engenders useful properties: the local distortion of the structure due to the size mismatch of the ions introduces high lattice strain [5]. This makes these solid solutions highly reducible with a significant proportion of the bulk cerium being reduced at moderate temperatures, with redox properties maintained over many cycles of reduction–oxidation. Hence, ceria-zirconia has become one of the most widely studied materials for redox catalysis, especially as a support for precious metals [21].

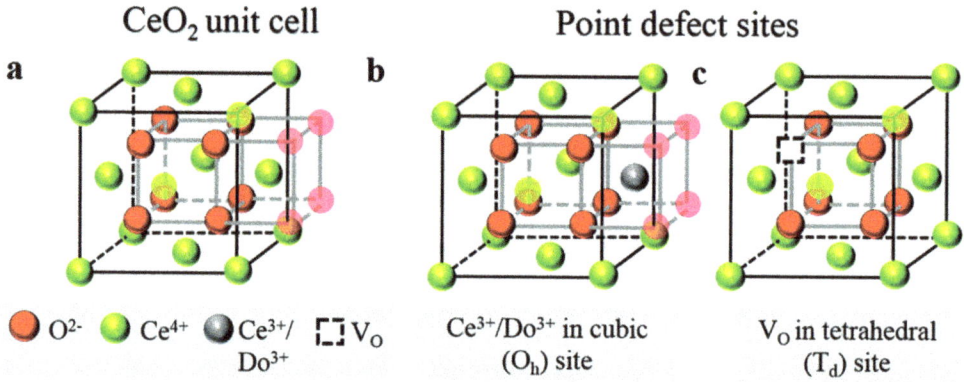

Figure 1. Schematic of crystal structure and defects for ceria and substituted variant [20]. (**a**) The fluorite ceria unit cell, containing 8 oxygen atoms (red) forming a simple cube embedded inside a face-centred cubic (FCC) array of Ce cations (green). Every second oxygen cube contains a Ce cation, as illustrated by the 4 semi-transparent oxygen atoms on the right, belonging to the next unit cell. Overall, the fluorite unit cell has 8 oxygen atoms and 4 cerium atoms. (**b**,**c**) Illustration of the two main point defects which may or may not be preferentially nearest neighbours, depending on the dopant type and prevalence of oxygen deficiency. Cation defects, Ce^{3+} (in oxygen-deficient ceria) and Do^{3+} (in substituted ceria), sit in a Ce^{4+} cubic site of Oh symmetry with 8-fold oxygen coordination. Every two cation defects in the structure are balanced by one oxygen vacancy defect, V_O, which sits in a tetrahedral site of Td symmetry, with 4-fold cation coordination. Reproduced from Schmitt, R.; Nenning, A.; Kraynis, O.; Korobko, R.; Frenkel, A.I.; Lubomirsky, I.; Haile, S.M.; Rupp, J.L.M. A review of defect structure and chemistry in ceria and its solid solutions. *Chem. Soc. Rev.* **2020**, *49*, 554–592, published by The Royal Society of Chemistry [20].

Coupled with substitutional chemistry is the effect of texture and crystal morphology on the properties of cerium oxide materials. For catalysis, the high surface area offered by small crystallites is desirable, but the shape and crystal habit can also affect properties, and for ceria, this includes the redox associated with oxygen loss and gain. Thus, the activity of ceria in catalysis may depend crucially on crystal habit, particularly when nanoscale crystallites are considered [22–26], and when the material is used as a support for precious metals, when specific crystal faces may offer favourable interactions. Similar relationships have been found for ceria-zirconia solid solutions [27,28].

In general, there is a need to develop reproducible methods for control of the synthesis of nanostructured ceria materials [29], and this article will review the recent literature that deals with the solvothermal synthesis of substituted cerium dioxide materials. We will focus on the literature from the past decade, since 2010, and for the previous literature, we direct the reader to a previous review article that considered the earlier literature on solvothermal crystallisation of various oxides of cerium [30].

2. Solvothermal Synthesis of Oxides

The term hydrothermal refers to the use of water as a reaction medium when heated at or above its boiling point in a closed vessel, where autogenous pressure is introduced [31,32]. The method has been explored since the 1840s as a means of preparation of analogues of naturally found minerals, is well known for the synthesis of zeolites and as such has been intensely developed since the 1940s [33]. Preparative hydrothermal synthesis was also developed for the growth of large crystals of dense materials, such as quartz [34]. The term solvothermal was later introduced by Demazeau to encompass the use of any solvent heated above its boiling point [35]. In terms of the synthesis of oxides, solvothermal synthesis methods (in fact, commonly hydrothermal) have been applied for a variety of materials from binary phases to more complex multinary compositions. Here, the emphasis is often on the use of rather mild conditions (typically around 200 °C, where 10 s of bar pressure may be developed) for the formation of fine powders of a material [36–40]. Earlier review articles described the changes in water properties that occur when it is heated in sealed vessels, and these may include changes in viscosity and dielectric constant such that otherwise insoluble reagents are solubilised and dissolution–recrystallisation may take place [31,32]. Often, a mineraliser is added to the crystallisation, whose role is to aid the dissolution of reagents, and commonly alkali-metal hydroxides or fluorides are used.

The advantage of solvothermal synthesis over other methods is that crystallisation of the desired phase occurs directly from the solution, offering the potential for control of crystal habit directly at the point of synthesis, without any high temperature annealing necessary to induce crystallisation and that might destroy any intricate crystal form. We will thus focus one such one-step synthesis, without annealing or firing to bring about crystallisation. Commonly, however, calcination up to 500 °C is employed to remove any excess solvent before study, particularly when catalytic properties are to be investigated, but this is a considerably lower temperature method than employed in other methods, such as precipitation firing, combustion or thermal decomposition of precursor salts. Interestingly, early work on ceria synthesis in the 1980s used extreme hydrothermal conditions (up to 700 °C and 100 MPa) with the aim of growth of large crystals (100 μm in dimension) [41], but the focus since then has almost entirely been on the formation of fine powders, particularly with nanostructure. Under these milder conditions, synthesis is conveniently carried out in the laboratory using Teflon™-lined steel reactors.

3. Survey of Substituted Cerium Oxides

3.1. Ceria-Zirconia

As noted above, the replacement of Ce^{4+} by isovalent Zr^{4+} is one the most widely studied substitutions used to tune the properties of ceria. In fact, this gives a complete substitutional series, $Ce_{1-x}Zr_xO_2$, from cubic cerium-rich compositions with $x < 0.1$, to tetragonal zirconium-rich phases with $x > 0.1$, and monoclinic for the most zirconia-rich materials, $x > 0.88$ [2]. Note that these are equilibrium crystal structures for materials prepared at high temperature, and the degree of tetragonality may be small for $x < 0.5$ and can also be difficult to detect when a diffraction profile is broadened due to the small crystal domain size, giving rise to pseudo-cubic materials. There is also the possibility of various metastable phases that depend on the composition and preparative method, and coupled with the possibility of atomic-scale inhomogeneity and nanoscale phase segregation, this makes determination of the precise structure of ceria-zirconia solid solutions challenging [42]. Even in the pseudo-cubic materials, the smaller ionic radius of zirconium compared to cerium induces local distortion of the structure, which is believed to be responsible for the enhanced oxygen storage capacity and more favourable reducibility, and this is also accompanied by enhanced thermal stability [5]. The precise redox and other catalytic properties depend on composition and which crystal phase is present, and here, the low temperatures of solvothermal synthesis offer the possibility of control over polymorphism, or the formation of mixed-phase samples with desirable properties. For example, Yang et al. used a solvothermal synthesis method from a mixed solvent of ethanol, N,N-dimethylformamide and ethylene glycol in the presence of lauric acid to form a ceria-zirconia with a Ce/Zr molar ratio of ~7/3; even in this ceria-rich composition, a mixed cubic (41%) and tetragonal (59%) sample was formed with distinct cubic–tetragonal interfaces on the nanoscale, Figure 2 [43]. Typically, to engineer such interfaces, high-temperature treatment above 1000 °C is needed to induce phase separation, and the advantage of the solvothermal approach is the high surface areas of the materials, which Yang et al. showed offered good activity for soot oxidation, ascribed to the presence of cubic–tetragonal interfaces, which significantly enhance the formation of oxygen vacancies from release of surface lattice oxygen.

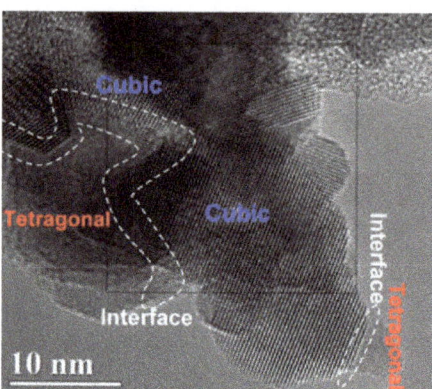

Figure 2. Cubic–tetragonal interfaces in ceria-zirconia prepared by a mixed-solvent solvothermal synthesis [43]. Reprinted from Yang, Z.; Hu, W.; Zhang, N.; Li, Y.; Liao, Y. Facile synthesis of ceria–zirconia solid solutions with cubic–tetragonal interfaces and their enhanced catalytic performance in diesel soot oxidation. *J. Catal.* **2019**, *377*, 98–109, with permission from Elsevier.

In general, solvothermal synthesis allows surface properties of ceria materials to be modified, compared to materials prepared by conventional methods. Luciani et al. prepared $Ce_{0.75}Zr_{0.25}O_2$ by two hydrothermal routes, one with 1-propanol as a co-solvent and the other with urea as an additive, and X-ray photoelectron and electron paramagnetic resonance spectroscopies showed an increase in both surface and bulk Ce^{3+} content, corresponding to a higher oxygen mobility [44]. Wang et al. also used urea-assisted hydrothermal synthesis to prepare defect-rich ceria-zirconia materials, with the highest levels of defects seen for the zirconia-rich phases, and used these materials as catalysts for the oxidative dehydrogenation of ethylbenzene with CO_2 [45].

The benefits of solvothermal synthesis in controlling crystal morphology have been explored for ceria-zirconia materials, and Table 1 summarises some of the key examples. These illustrate how the crystal shape on the nanoscale can be adjusted from isotropic to anisotropic plates and rods by choice of reagents and solvent. Figure 3 shows an example of how morphology can be selected from facetted polyhedral to rod-shaped and plate-like with the crystal dimension on the nanoscale by choice of reagents, solution additives and temperature (and/or time of reaction) [46].

An orientated attachment mechanism was proposed by Liu et al., who formed hollow, truncated octahedral agglomerates of primary particles of cerium-rich compositions [47]. However, in general, the crystallisation mechanism and the origin of the shape control that is possible by synthesis have not been determined. Clearly, this is an area where further work is needed. Some observations have been made on the relationship between crystal morphology and the resulting properties. For example, CeO_2 nanorods were shown to have higher activity towards CO oxidation than CeO_2 nanoparticles, primarily because of the preferential exposure of the reactive {110} planes, but ZrO_2 doping promoted the reducibility of the nanoparticles more significantly than that of the nanorods [48].

Table 1. Summary of solvothermal conditions used to prepare various crystal morphologies of $Ce_xZr_{1-x}O_2$.

Crystal Morphology	Composition $Ce_xZr_{1-x}O_2$	Solvent Temperature Time [a]	Reagents	Reference
Facetted polyhedral 3–5 nm diameter	$0 \leq x \leq 1$	6 M NaOH(aq) 150 °C 48 h	$Ce(NO_3)_3 \cdot 6H_2O$ $ZrO(NO_3)_2 \cdot xH_2O$	[49]
Hollow, truncated octahedral ~80 nm edge	$x = 0.01, 0.1, 0.2$	water 100 °C 18 h	$CeCl_3 \cdot 7H_2O$ $ZrOCl_2 \cdot 8H_2O$ Poly(vinylpyrrolidone) H_2O_2 formic acid tert-butylamine	[47]
Rods diameter ~8 nm length ~40 nm	$0 \leq x \leq 0.2$	10 M NaOH(aq) 100 °C 20 h	$Ce(NO_3)_3 \cdot 6H_2O$ $Zr(NO_3)_4 \cdot 5H_2O$	[48]
Facetted <10 nm diameter	$0 \leq x \leq 0.2$	water 95 °C 8 h	$(NH_4)_2Ce(NO_3)_6$ $Zr(NO_3)_4 \cdot 5H_2O$ urea	[48]
Polyhedral 10–15 nm diameter	$x = 0.2$	water 180 °C 12 h	$Ce(NO_3)_3 \cdot 6H_2O$ $ZrO(NO_3)_2 \cdot 6H_2O$ poly(vinylpyrrolidone) hydrazine	[46]
Rods diameter ~10 nm length ~20–70 nm	$x = 0.2$	6 M NaOH(aq) 100 °C 24 h	$Ce(NO_3)_3 \cdot 6H_2O$ $Zr(NO_3)_4 \cdot 5H_2O$	[46]
Plates 6–12 nm diameter	$x = 0.2$	NH_3(aq) 100 °C 24 h	$Ce(NO_3)_3 \cdot 6H_2O$ $ZrO(NO_3)_2 \cdot 6H_2O$ CTAB [b]	[46]

[a]: A post-synthesis calcination ~600 °C was commonly applied to yield the studied product. [b]: CTAB = cetyltrimethylammonium bromide.

The high surface areas offered by small particles of ceria-zirconia prepared by solvothermal routes are suited to act as supports for precious metal nanoparticles in heterogeneous catalysis, and, indeed, some of the examples reported in Table 1 were used for this purpose, with the precious metals added in a subsequent step [46]. Taking a different approach, Das et al. used a one-step approach to prepare such catalysts, by inclusion of ruthenium chloride in the hydrothermal reaction of cerium and zirconium chlorides, with hydrazine present as a reducing agent and CTAB as a structure-influencing agent [50]. After a mild calcination, the material consisted of $Ce_{0.5}Zr_{0.5}O_2$ nanorods with a fine dispersal of Ru nanoparticles; these proved effective catalysts for the partial oxidation of methane. This illustrates how solution chemistry offers the possibility to form multicomponent systems with some control of their form.

The use of ceria as a redox catalyst in solar reactors for the upgrading of carbon dioxide and/or water to carbon monoxide and/or hydrogen in a thermochemical splitting cycle has received some attention, and the design of active materials based on ceria-zirconia has proved possible using hydrothermal synthesis. Luciani et al. showed how $Ce_{0.75}Zr_{0.25}O_2$, when prepared in 1-propanol as a solvent at 120 °C with urea as an additive, yields materials with an increase in both surface and bulk Ce^{3+}, leading to a higher oxygen mobility from and towards the catalyst surface, and, notably, even after several cycles of high-temperature treatment, these defects were maintained, leading to high activity for solar thermochemical splitting of water and carbon dioxide [44]. It should be noted that in such high-temperature applications, one possible disadvantage of materials prepared by a low-temperature solvothermal route is that any metastable composition or structure may irreversibly phase separate upon heating; this is an area in which future research is needed.

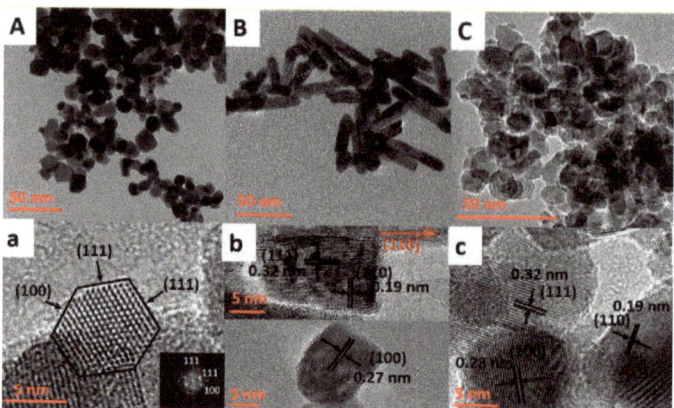

Figure 3. TEM (**top**) and HRTEM (**bottom**) images of $Ce_xZr_{1-x}O_2$ prepared using hydrothermal synthesis: (**A,a**) polyhedral, (**B,b**) rod-shaped and (**C,c**) plates [46]. Reprinted from Yang, Z.; Hu, W.; Zhang, N.; Li, Y.; Liao, Y. Facile synthesis of ceria–zirconia solid solutions with cubic–tetragonal interfaces and their enhanced catalytic performance in diesel soot oxidation. *J. Catal.* **2019**, *377*, 98–109, with permission from Elsevier.

Solvothermal synthesis methods lend themselves to scaling for industrial manufacture, especially when continuous flow reactors can be designed to make use of superheated solvents. Rapid-mixing devices may also be used to induce crystallisation in confined geometries, allowing control over the crystal form. Ceria-zirconia solid solutions have been prepared in this way, offering the prospect of production of catalyst materials for real application. Kim et al. used supercritical water (250 bar, 550 °C) as a reaction medium and produced fine powders of $Ce_{1-x}Zr_xO_2$ (x = 0.35, 0.5, 0.8) that were much less agglomerated than those formed by co-precipitation [51] and could be used to support Rh for catalytic reduction of NO by CO [52]. Auxéméry et al. produced nanostructured $Ce_xZr_{1-x}O_2$ ($0 \leq x \leq 0.75$) in flow reactors using an ethanol/water mixture (25 MPa, 400 °C). This yielded nanocrystallites that proved to be effective supports for Ni in dry methane reforming catalysis [53].

3.2. Transition Metal-Substituted Cerias

Through the addition of divalent (D^{2+}) and trivalent (T^{3+}) transition metal ions, compositions of the form $Ce_{1-x}D_xO_{2-x}$ or $Ce_{1-x}T_xO_{2-x/2}$, respectively, can be expected. The oxygen site vacancies that this produces can modify lattice oxide mobility, making them attractive materials for use as electrolytes in solid oxide fuel cells, and may also be of benefit in oxygen storage materials. Some transition metals are of particular interest if they can form redox pairs along with Ce^{4+}/Ce^{3+}. This is because these materials are more capable of undergoing reduction and oxidation reactions, due to the possibility of cooperative electron exchange between the two metals. This is of relevance for applications in heterogeneous catalysis. An example of this effect can be found when doping CeO_2 with CuO. The Cu^{2+}/Cu^{+} redox cycle has been shown to couple with the Ce^{4+}/Ce^{3+} pair, helping to promote the redox features [54]. These materials have also been shown to have an increased number of oxygen vacancies associated with the Cu^{2+} ions [55].

Wang et al. demonstrated that Cu-doped ceria can be synthesised hydrothermally using Ce and Cu nitrates in NaOH, heated to 180 °C for 24 h [56]. The resulting material was shown to be single phase and indexed to the $Fm\overline{3}m$ cubic space group associated with ceria. TEM showed that the synthesised particles had sizes of between 4 and 5 nm. Other groups have shown that morphology control is possible: samples with 5 and 10 mol% Cu were made using chlorides and H_2O_2 heated at 180 °C, and 5 h of heating produced spheres with diameters of 60–80 nm, Figure 4, [57], while addition of PVP, ammonia and urea to the mixture with heating for 6 h gave spheres of ~100 nm in diameter [58].

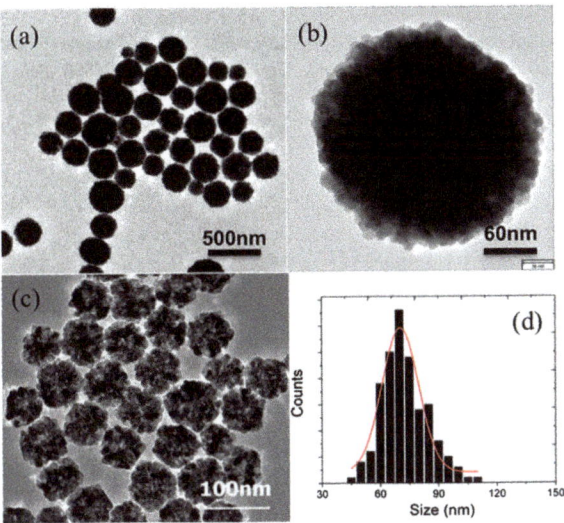

Figure 4. (**a,b**) TEM images of the CeO$_2$ nanospheres synthesised without the presence of Cu^{2+}; (**c**) TEM image of the Cu$_{0.10}$Ce$_{0.90}$O$_2$ nanospheres; (**d**) is the corresponding size distribution of the Cu$_{0.10}$Ce$_{0.90}$O$_2$ nanospheres [57]. Reproduced with permission from: Yang, F.; Wei, J.; Liu, W.; Guo, J.; Yang, Y. Copper doped ceria nanospheres: surface defects promoted catalytic activity and a versatile approach. *J. Mater. Chem. A* **2014**, *2*, 5662–5667—Published by The Royal Society of Chemistry.

Rood et al. were able to synthesise Cu-containing ceria nanotubes that were single phase up to 7 mol% Cu [59]. Their method involved using Ce(NO$_3$)$_3$ and CuSO$_4$ in NaOH solution heated to 100 °C for 10 h. In the same work, Cr-containing ceria nanotubes, up to a concentration of 5 mol%, were reported where a similar method was used, with Cr(NO$_3$)$_3$ as a precursor. These materials were tested as catalysts for both CO oxidation and NO reduction, with the Cu samples outperforming pure CeO$_2$ in terms of CO oxidation, but no overall improvement for NO reduction, with the Cr samples showing the opposite effect, an improvement in NO reduction, with no noticeable improvement in CO oxidation. XPS data suggested that Cu inclusion led to an increase in surface Ce^{3+}, while Cr increased the percentage of Ce^{4+} on the surface. A Cu-Cr-co-doped sample showed an improvement in conversion of CO and NO over pure ceria, and only a dopant level of 1% Cu and 1% Cr was required to reach a similar CO oxidation to a sample of 7% Cu doping, suggesting a synergistic effect in the co-doped samples. Separately, Zhang et al. showed that the catalytic performance of hydrothermally prepared Cu-CeO$_2$ remains constant over six cycles of CO oxidation, suggesting that the material has the potential to be a stable, recyclable oxidation catalyst [60].

Kurajica et al. produced ceria samples doped with ~10 mol% Mn, Cu or Zn from sulfate precursors and 8 M aqueous NaOH heated at 120 °C for 16 h. Elemental analysis using electron energy loss spectroscopy (EELS) and energy-dispersive X-ray spectroscopy (EDS) analysis showed that Mn substitution had occurred to the expected concentration, but the Cu and Zn concentration was lower than expected [61]. These materials appeared single phase by powder X-ray diffraction (XRD) and had average crystallite sizes between 4.2 and 6.3 nm, with Mn doped > pure ceria > Cu doped > Zn doped, as confirmed using scanning transmission electron microscopy (STEM). Prior to testing the materials' catalytic activity, the samples were fired to 500 °C to test for the presence of excess metal oxides on the surface of the materials. XRD of the materials after firing showed no extra peaks, suggesting that the as-made samples were unlikely to have amorphous metal oxides present. A trend that was noted, with regard to crystallite size, was that the samples with larger as-made crystallites showed only minor growth relative to pure ceria, whereas the smaller as-made crystallites showed greater growth. This resulted in the Cu and Zn samples having a lower specific surface area (SSA) than pure CeO_2, while Mn had a greater SSA. To test the catalytic activity of these samples, they were used to oxidise toluene. All samples reached 50% conversion at lower temperatures than pure CeO_2.

Syed Khadar et al. demonstrated a one-step hydrothermal synthesis of 2–8 mol% Co-doped ceria [62]. Their method used $Ce(NO_3)_3$ and $CoCl_2$ in water in the presence of Na_3PO_4, heated to 180 °C for 15 h. The role of the phosphate additive was not proposed, but these nanocrystalline samples consisted of cuboid particles, which were used in their as-prepared form for their antibacterial properties. Zhang et al. hydrothermally synthesised Mn-doped CeO_2 using Ce and Mn nitrates, in a mixture of ethanol, formic acid, ammonia and peroxide heated at 150 °C for 16 h [63]. The resulting nanospheres showed single-phase XRD patterns and had an average diameter of 130 nm, as seen from electron microscopy. EDS analysis shows that there is an even distribution of all elements across the particles, Figure 5. The catalytic performance of these materials was tested to measure CO oxidation. The sample with 7 mol% Mn was cycled six times, and after an initial increase in activity at low temperatures from first to second cycle, the activity remained stable at low temperatures. The initial change in activity was associated with the removal of impurities from the surface of the nanospheres.

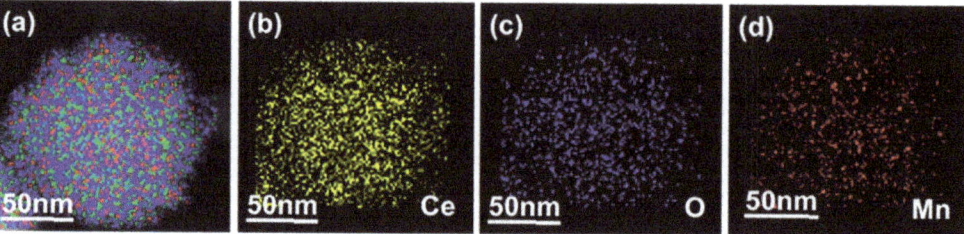

Figure 5. Energy-dispersive spectroscopy maps of a 7% Mn-substituted ceria nanosphere [63]. (**a**) shows an overlay of signals from all elements and (**b**–**d**) show the signals from individual elements. Reproduced with permission of John Wiley and Sons.

Zhang et al. produced ceria nanocubes doped with 0.5–7 mol% Mn, using nitrates and NaOH and heating to 180 °C for 24 h [64]. They observed that the size of the cubes decreased with increasing Mn concentration, but no additional peaks were observed in the XRD patterns. Temperature-programmed reduction was used to demonstrate that these materials were more reducible than pure ceria nanocubes, and the materials were used for the selective oxidation of styrene *tert*-butyl hydroperoxide.

The work of Jampaiah et al. on manganese-, cobalt- and nickel-containing ceria nanorods (10 mol% transition metal added) demonstrated that Co doping increased the concentration of surface Ce^{3+} and that these materials showed greater oxidation of CO than CeO_2 produced using the same method [65]. The reason for the formation of extremely anisotropic crystal forms, Figure 6, is not apparent, since no solution additives were employed: the synthesis method simply used nitrate salts in sodium hydroxide solution heated at 100 °C. Song et al. used a similar synthesis approach to Fe- and Co-containing ceria samples and also found rod-like crystal morphologies; in this case, beneficial activity for the catalysis for the total oxidation of propane was found [66]. Xing et al. also reported rod-shape particles upon inclusion of Mn into CeO_2 by a direct hydrothermal crystallisation [67].

Figure 6. Nanorods of ceria materials prepared by Jampaiah: TEM images of (**a**) CeO_2, (**b**) $Ce_{0.9}Mn_{0.1}O_{2-\delta}$ (Mn–CeO_2), (**c**) $Ce_{0.9}Co_{0.1}O_{2-\delta}$ (Co–CeO_2) and (**d**) $Ce_{0.9}Ni_{0.1}O_{2-\delta}$ (Ni–CeO_2) [65]. Reproduced from Jampaiah, D.; Venkataswamy, P.; Coyle, V.E.; Reddy, B.M.; Bhargava, S.K. Low-temperature CO oxidation over manganese, cobalt, and nickel doped CeO_2 nanorods. *RSC Adv.* **2016**, *6*, 80541–80548—Published by The Royal Society of Chemistry.

Liu et al. demonstrated the synthesis of 2 mol% Co-doped nanospheres from $(NH_4)_2Ce(NO_3)_6$ and $CoCl_2$ in acetic acid and ethylene glycol heated at 180 °C for 8 h [68]. When tested as a catalyst for CO oxidation, the Co-doped sample outperformed pure CeO_2. The same method was used to prepare Ni-containing samples, and these outperformed the Co samples at CO oxidation at lower temperatures, but as the temperature increased, the Co samples reached maximum conversion faster. The work of Du et al. focussed on Ni- and Co-doped ceria nanosheets, synthesised hydrothermally using nitrates, acrylamide, glucose and ammonia, heated to 180 °C for 96 h and then fired at 400 °C for 4 h, testing the materials' ability to reduce NO [69]. It was found that the Ni-doped samples performed better, likely due to them having more oxygen vacancies than other comparable samples. Venkataswamy et al. produced CeO_2 doped at 5, 10 and 15 mol% with Cr [70]. This was achieved by mixing Ce and Cr nitrates in NaOH and heating to 100 °C for 24 h. They then fired the sample to 350 °C for 2 h. They found that the addition of Cr caused an increase in the catalytic oxidation of CO at lower temperatures up to 10 mol% The 15 mol% Cr-doped CeO_2 was shown to be less efficient than the 5 mol% sample. This is likely due to Cr being less effective as an oxidation catalyst. At lower amounts, the lattice distortion created by the dopant enhances the catalytic effect of Ce, but as the concentration increases, the loss of Ce from the sample eventually outweighs the benefit from the distortion.

Fe-substituted CeO$_2$ was synthesised by Phokha et al. by mixing nitrates in PVP and heating at 200 °C for 12 h [71]. They produced samples up to 7 mol% Fe that appeared single phase by XRD and Raman spectroscopy, but higher concentrations of Fe gave Fe$_2$O$_3$ as an impurity. Wang et al. were successful in producing Fe-doped ceria, up to a concentration of 20 mol% [72]. Their method involved using nitrates in ethylene glycol, heated to 180 °C for 28 h. They followed this up by heating the sample to 450 °C for 4 h. The resulting materials all matched the XRD pattern associated with CeO$_2$, rather than Fe$_2$O$_3$. Zheng et al. also produced Fe-doped ceria, forming the materials in a "nanoflower" morphology, Figure 7 [73]. This required mixing nitrates in urea, citric acid and ammonia, heating to 155 °C for 26 h and then firing to 500 °C for a further 3 h. The materials synthesised had an iron content of between 4 and 12 mol% and were applied for the selective oxidation of H$_2$S.

Figure 7. SEM images of Fe-containing ceria materials produced from solutions containing urea and citric acid. Fe%: (**A**) = 0, (**B**) = 4, (**C**) = 8, (**D**) = 12 [73]. Reprinted with permission from Zheng, X.; Li, Y.; Zheng, Y.; Shen, L.; Xiao, Y.; Cao, Y.; Zhang, Y.; Au, C.; Jiang, L. Highly Efficient Porous Fe$_x$Ce$_{1-x}$O$_{2-\delta}$ with Three-Dimensional Hierarchical Nanoflower Morphology for H$_2$S-Selective Oxidation. *ACS Catal.* **2020**, *10*, 3968–3983. Copyright (2020) American Chemical Society.

The use of additives in solution has been shown to result in unusual crystal forms. In the case of Ni-containing ceria, Liu et al. used a mixed water–ethylene glycol solvent with the addition of acetic acid to form particles with spherical-like nanocluster morphology that possessed hollow centres [74]. The porosity of the particles was dependent on the volume of Ni(NO$_3$)$_2$ solution used in synthesis, and materials with ~6.5 mol% Ni showed high catalytic performance in carbon monoxide oxidation. Mesoporous cerium–iron binary oxides prepared by a hydrothermal technique using CTAB as a template were tested as catalysts in methanol decomposition and total oxidation of ethyl acetate: low levels of Fe were dispersed in the fluorite structure, but higher levels gave clustering of Fe and ultimately phase separation above 50% Fe [75].

As well as properties in catalysis, transition metal-substituted cerias have been prepared by hydrothermal synthesis for other applications. Barbosa et al. synthesised Ce$_{1-x}$M$_x$O$_2$ (M = Mn, Cr, Co and Fe) from aqueous sodium hydroxide solution and studied the magnetic behaviour [76]. Radović et al. synthesised ceria materials with up to 5 mol% iron by crystallisation from aqueous ammonia solution, and optical properties were studied by spectroscopic ellipsometry [77]. Fe-substituted CeO$_2$ prepared from nitrate salts in aqueous ammonia was found to have a high level of oxide vacancies useful for photocatalytic ozonation [78]. Recently, Fe-substituted ceria has been examined for use in rechargeable lithium batteries: Ce$_{0.9}$Fe$_{0.1}$O$_2$ was prepared by a simple hydrothermal reaction of nitrates salts in aqueous sodium hydroxide, and its behaviour towards lithium insertion and removal was tested [79]. It was proposed that reversible reduction to metallic iron occurred on cycling, and an anode was fabricated to produce a high-powered lithium-ion battery with cycling stability and high power density; Figure 8 shows a proposal for the structural chemistry at play in this material, where the off-centring of Fe in the ideal CeO_2 structure provides space for Li insertion.

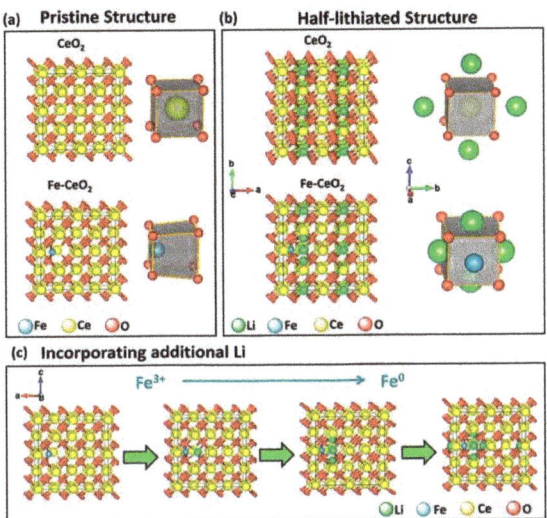

Figure 8. (a) Crystal structure of CeO_2 and Fe-CeO_2 and their local environments. (b) Prototype structure for a half-lithiated, ordered, Fe-free CeO_2, and Fe-containing structure. (c) Prototype structures for incorporating additional Li within the structure of Fe-containing CeO_2. Reproduced from Ma et al. [79].

Dual substitution of more than one transition metal into ceria is also of interest to tune properties. Zhang et al. prepared $Ce_{1-x}(Fe_{0.5}Zn_{0.5})_xO_{2-\delta}$ with x up to 0.018 at 200 °C and found a contraction of the lattice parameter and reduction in the crystallite size with increasing transition metal content, and XPS proved the presence of Fe^{3+} [80]. Dosa et al. prepared $Ce_{0.95}Mn_{0.025}Cu_{0.025}O_{2-\delta}$ alongside samples that contained only Cu or Mn using similar methods and found that the Mn-containing sample was more effective for soot oxidation and the Cu-containing sample showed the highest activity for CO oxidation: in this case, the co-substitution offered no benefit over CeO_2 itself [81].

A surfactant-assisted hydrothermal method was used by Zhu et al. to produce materials $Mn_xCu_yCe_{1-x-y}O_{2-\delta}$ [82]. Cetyltrimethylammonium bromide (CTAB) was added to aqueous solutions of metal salts and the pH was adjusted to 11 with NaOH before heating to 200 °C. Calcination at 550 °C yielded samples with high surface areas, >74 m^{2}g^1, and with evidence of mesoporosity, as seen by gas adsorption measurements. The materials showed enhanced catalytic activity for CO oxidation.

It is also worth drawing attention to the possibility of forming composite materials by hydrothermal chemistry. For example, Ji et al. produced some striking hierarchically structured particles of CuO-CeO_2 composites, Figure 9 [83]. While powder XRD showed signals from both crystalline phases, implying limited inclusion of Cu into the CeO_2 structure, electron microscopy showed an intimate association of the two phases to give composite particles of agglomerated CuO needles with surface CeO_2. Although strictly not homogeneous single phases, this work illustrates the possibilities in the preparation of novel materials with unique morphologies by choice of reagents and conditions in hydrothermal reactions.

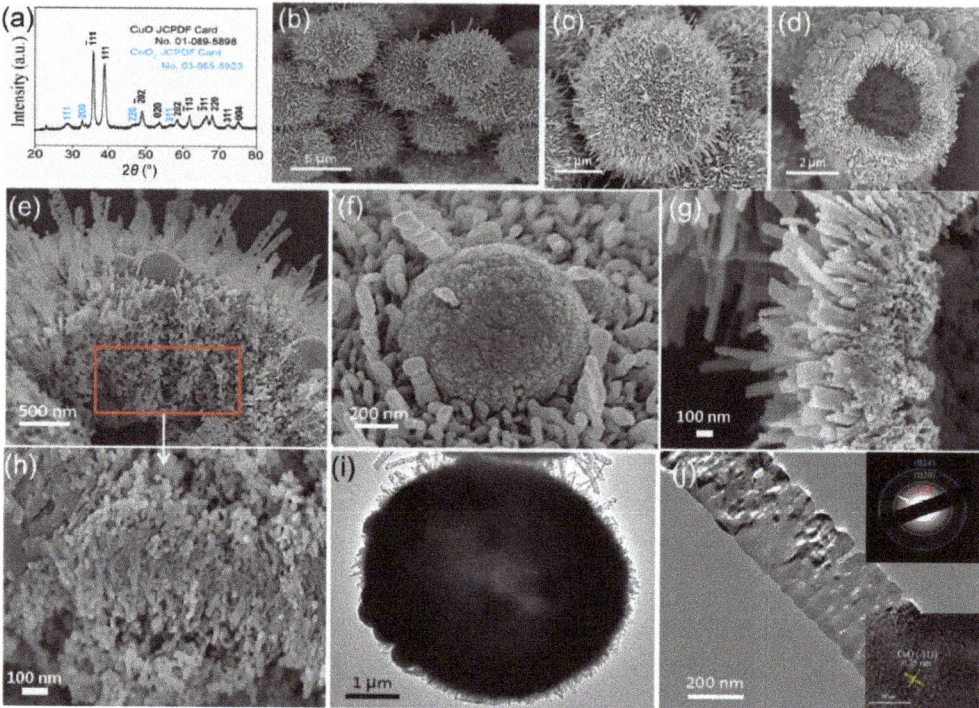

Figure 9. (**a**) Powder X-ray diffraction pattern, (**b–h**) FESEM images, (**i,j**) TEM images of CeO$_2$-CuO hollow microspheres. The inset of panel (**j**) (**top right**) shows the corresponding SAED pattern, with the HRTEM image of CuO nanosheets shown at the bottom right [83]. Reprinted with permission from Springer Nature Customer Service Centre GmbH: Springer. Ji, Y.; Jin, Z.; Li, J.; Zhang, Y.; Liu, H.; Shi, L.; Zhong, Z.; Su, F. Rambutan-like hierarchically heterostructured CeO$_2$-CuO hollow microspheres: Facile hydrothermal synthesis and applications. *Nano Res.* **2017**, *10*, 381–396.

As described thus far, much of the work on transition metal substitution has focussed on the first-row series, and few examples of inclusion of the second-row and third-row elements in ceria by solvothermal routes have also been proven. The case of precious metals is important to consider since ceria is often used as a support for precious metal particles. Hiley et al. showed the inclusion of Pd^{2+} into the cubic fluorite structure of CeO$_2$ was possible for up to 20 mol% substitution from CeCl$_3$·7H$_2$O and PdCl$_2$ in aqueous NaOH/H$_2$O$_2$ [84]. As well as inducing charge-balancing oxide vacancies, the square planar preference of Pd^{2+} induces a structural distortion, as evidenced from EXAFS spectroscopy, Figure 10a. This material is unstable on heating in air above 700 °C, and the Pd is extruded to give metallic particles on CeO$_2$, which is, in itself, of interest for heterogeneous catalysis. The case of niobium substitution is an interesting situation as its preferred oxidation of +5 requires a different charge-balancing mechanism than substituents with lower charge. Hydrothermal synthesis from NbCl$_5$ and CeCl$_3$·7H$_2$O in aqueous NaOH solutions produced substituted ceria materials in which charge compensation was achieved by co-inclusion of Na$^+$ to give materials of composition (Ce$_{1-x}$Nb$_x$)$_{1-y}$Na$_y$O$_{2-\delta}$ (where $x \leq 0.30$ and $y \geq \sim x/3$) [85]. Here, neutron pair distribution function analysis showed how the lower coordination number preference of Nb^{5+} leads to a local distortion of the structure, Figure 10b, which can be used to explain the enhanced oxygen storage properties of the material.

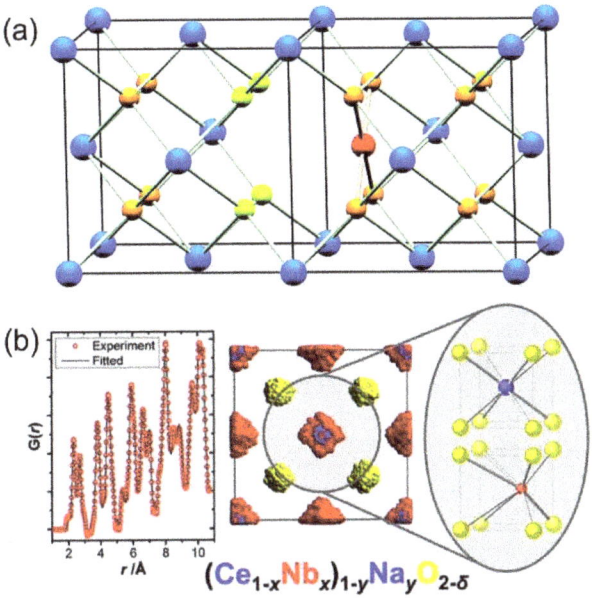

Figure 10. Local environment of (**a**) Pd^{2+} and (**b**) Nb^{5+} in substituted ceria materials prepared by hydrothermal synthesis. In (**a**), cerium is represented by the blue spheres, Pd by the red and the oxide anions by orange or yellow, with the oxide anions being adjacent to a vacant Ce site caused by the square planar preference of Pd^{2+} [84]. In (**b**), the left panel shows the fitted pair distribution function used to derive the structural model shown on the right, where the colour scheme is indicated by the chemical formula [85]. (**a**) Reproduced from Hiley, C.I.; Fisher, J.M.; Thompsett, D.; Kashtiban, R.J.; Sloan, J.; Walton, R.I. Incorporation of square-planar Pd^{2+} in fluorite CeO_2: hydrothermal preparation, local structure, redox properties and stability. *J. Mater. Chem. A* **2015**, *3*, 13072–13079—Published by The Royal Society of Chemistry. (**b**) Reprinted with permission from Hiley, C.I.; Playford, H.Y.; Fisher, J.M.; Felix, N.C.; Thompsett, D.; Kashtiban, R.J.; Walton, R.I. Pair Distribution Function Analysis of Structural Disorder by Nb^{5+} Inclusion in Ceria: Evidence for Enhanced Oxygen Storage Capacity from Under-Coordinated Oxide. *J. Am. Chem. Soc.* **2018**, *140*, 1588–1591. Copyright (2018) American Chemical Society.

3.3. Main Group Substituents

The main group elements include those from the s and p blocks of the periodic table and cover a wide range of possible substituents for cerium dioxide, ranging from the hard cations of Groups 1 and 2 to the more covalently bound atoms from the p block. Hydrothermal synthesis provides a convenient way to explore this substitutional chemistry since phases that might otherwise phase separate during high-temperature synthesis may crystallise to yield unique compositions with unusual structural features and, potentially, properties.

From the alkali earths in Group 2, Ma et al. reported magnesium-containing materials $Ce_{1-x}Mg_xO_2$ (x = 0.00, 0.02, 0.04, 0.06, 0.08) in the form of nanowires, prepared in a water–ethylenediamine mixture with NaOH as a mineraliser and $CeCl_3 \cdot 7H_2O$ and $MgCl_2 \cdot 2H_2O$ heated at 180 °C for 14 h [86]. Calcium- and strontium-containing CeO_2 with nominal compositions $Ce_{0.90}Ca_{0.10}O_{1.90}$, $Ce_{0.90}Ca_{0.05}Sr_{0.05}O_{1.90}$ and $Ce_{0.90}Sr_{0.10}O_{1.90}$ was prepared from Ce(III) nitrate and salts of Ca^{2+} and Sr^{2+} in aqueous NaOH at pH = 14, heated at 220 °C for up to 8 h [87]. Only a low level of incorporation of Sr into the ceria structure was found, explained by the larger size of Sr^{2+} compared to Ce^{4+}. The small crystal size (12–16 nm) of the powders allowed effective sintering into dense ceramic pellets at 1350 °C for 5 h, and for the Sr-containing material, a secondary phase of $SrCeO_3$ was then seen, consistent with the fact that Sr is not effectively incorporated into ceria. The Ca-containing sample showed favourable ionic conductivity after sintering.

From Group 12, zinc-containing ceria with up to 20% of the cerium replaced by the substituent was prepared by a continuous hydrothermal synthesis approach using supercritical water as a reaction medium (450 °C and 24.1 MPa) and KOH as a mineraliser. The samples were formed as nanocrystalline powders with crystallite sizes less than 5 nm [88]. A 10% Zn-containing CeO_2 was found to be an effective catalyst for the CO_2 hydrogenation reaction, ascribed to the highest level of Ce^{3+} defects [89]. Although not an elemental substitution, the formation of ZnO-CeO_2 heterojunctions by hydrothermal synthesis has also been studied, where the purposeful growth of the two phases in intimate contact leads to superior visible-light photocatalytic performance [90].

Of the Group 13 elements, only indium has been introduced in ceria via hydrothermal conditions. Younis et al. published a synthesis of indium-containing CeO_2 that appears single phase up to 15 mol% In [91]. This involved mixing aqueous solutions of Ce and In nitrates and adding toluene, oleic acid and *tert*-butylamine. The reaction mixture was heated to 200 °C for 36 h and the solid product calcined at 180 °C for 1 h. The lattice parameters of the synthesised samples decreased with increasing In^{3+} concentration, attributed to the smaller ionic radius of In. The 5 and 10 mol% In samples outperformed pure CeO_2 in the photocatalytic degradation of organic dyes.

From Group 14, the inclusion of Sn and Pb into CeO_2 has been examined by solvothermal routes. Abbas et al. produced phase-pure samples on Sn-containing ceria up to 7% Sn, designed for use as a cytotoxic nanomedicine using $Ce(NO_3)_3$ and $SnCl_2$ in polyethylene glycol and NaOH heated to 95 °C for 12 h and then fired to 300 °C for 2 h [92]. The addition of Sn changed the morphology of the material from spherical nanoparticles to a heterogeneous nanowires–nanosheets structure. A contraction of the unit cell was reported, although the oxidation state of Sn was not measured. Shajahan et al. successfully synthesised single-phase Pb-containing CeO_2 using nitrates in NaOH, heated to 140 °C for 14 h followed by heat treatment at 350 °C for 3 h [93]. The samples appeared single phase under XRD in samples up to 12% Pb, and it was found that up to 6% Pb, the band gap of the material was lower, making it a better photocatalyst than pure ceria, while higher Pb content increased the materials' bandgaps. The oxidation state of Pb was proved to be +2 using X-ray photoelectron spectroscopy.

From the elements in Group 15, there are several examples of how Bi^{3+} has been studied as a way of modifying CeO_2, and hydrothermal synthesis allows a convenient route to the introduction of high levels of substitution with up to 60% of the cerium replaced. The hydrothermal oxidation of Ce^{3+} in basic aqueous solution by $NaBiO_3$ at 240 °C provides a redox synthetic method, where the product phase contains Ce^{4+} and Bi^{3+}, to form nanocrystalline powders (Figure 11) that show high reducibility, although the extrusion of the Bi metal will limit the practical application of the materials [94]. Bi-substituted CeO_2 has frequently been investigated for use in photocatalysis, although it has also been looked at for uses such as antibacterial activity: Frerichs et al. produced single, cubic-phase samples up to 20 mol% Bi by reacting Bi and Ce nitrates in NaOH and heating to 80 °C for 25 h [95]. Samples made using a higher level of Bi were a two-phase mix of the cubic CeO_2 structure and the monoclinic Bi_2O_3 material. Houlberg et al. synthesised materials using $Bi(NO_3)_3$ and $(NH_4)_2Ce(NO_3)_6$ in NaOH, heated to 200 °C for 20 h, and produced samples that were single cubic phase up to 60% Bi [96]. This study employed in situ X-ray diffraction to track the formation of the product and proved that mixed metal oxide phases were formed at the earliest stages of reaction even before heating under hydrothermal conditions. Other methods that have been used to create phase-pure samples up to ~10% Bi have been reported. Shanavas et al. used nitrates in NaOH, heated to 140 °C for 14 h and then firing at 350 °C for 3 h to remove surface impurities [97]. Veedu et al. used nitrates in ammonia and HCl, heated to 100 °C and then firing to 500 °C [98].

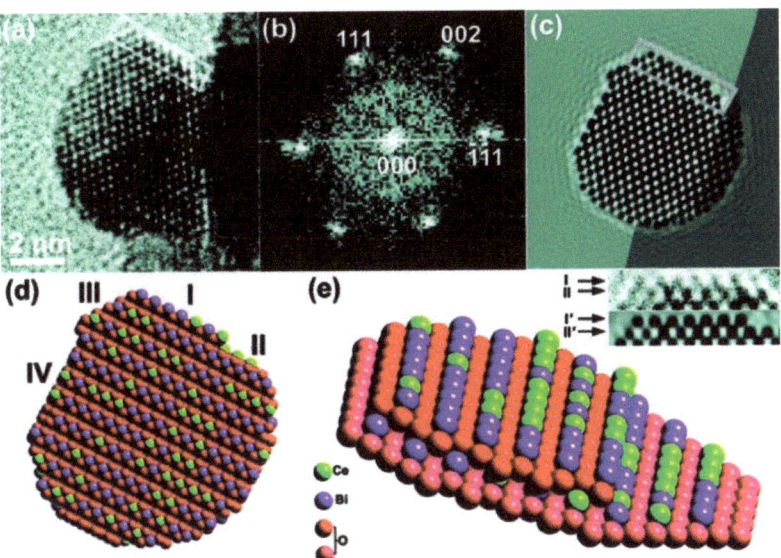

Figure 11. (a) High-resolution TEM image of a faceted crystallite of $Ce_{0.5}Bi_{0.5}O_{1.75}$ with one facet highlighted. (b) Indexed FFT obtained from the lattice image in Part (a). (c) A multislice image simulation produced for a similar fragment as in (a) with the same facet highlighted. (d) Faceted structure model used to produce the simulation in (c). I, II, III and IV indicate (002), (002)′, (111) and (220) facets, respectively, assigned relative to (b). (e) Schematic surface structure model showing the (002) step and (002) terrace. Insets are details from the corresponding lattice image (i.e., I and II) and simulation (i.e., I′ and II′). Surface oxygens are omitted for clarity [94]. Reprinted with permission from Sardar, K.; Playford, H.Y.; Darton, R.J.; Barney, E.R.; Hannon, A.C.; Tompsett, D.; Fisher, J.; Kashtiban, R.J.; Sloan, J.; Ramos, S.; Cibin, G.; Walton, R.I. Nanocrystalline Cerium−Bismuth Oxides: Synthesis, Structural Characterization, and Redox Properties. *Chem. Mater.* **2010**, *22*, 6191–6201. Copyright (2010) American Chemical Society.

Unlike bismuth, which is found in the +3 oxidation state when included in ceria, the earlier member of Group 15, antimony, has been shown to be present in the +5 oxidation state: this is the case whether $SbCl_3$ or $SbCl_5$ is used under hydrothermal conditions in the presence of NaOH and H_2O_2 [99]. Here, like the case of niobium mentioned above, the co-substitution of sodium is found to balance the charge of the pentavalent substituent, resulting in chemical formulae $(Ce_{1-x}Sb_x)_{1-y}Na_yO_{2-\delta}$ (where $x < 0.4$ and $y \geq x/3$). As with bismuth, the antimony-containing samples are unstable in highly reducing conditions and phase separate to yield elemental antimony.

Finally, we consider a more unusual case, where nitrogen can partially replace oxygen in CeO_2 to give an oxynitride analogue of ceria. Xu and Li produced samples from $Ce(NO_3)_3$ and $(NH_4)_2S_2O_8$ in water heated to 200 °C for 24 h [100]. XRD recorded before and after a calcination at 500 °C for 2 h shows that the sample adopts the cubic ceria structure, with no additional peaks visible. Large excesses of nitrogen, from the ammonium persulfate reagent, were needed to achieve substitution, but the highest Ce/N ratio achieved in the product was 1:0.052, which is close to the solid solubility limit. The lattice parameters of all doped samples were larger than the expected value for pure Ce, due to the lattice distortion caused by the N^{3-} being larger than O^{2-}. This was found to enhance the oxygen storage capacity of these materials.

3.4. Rare Earth-Substituted Ceria

The inclusion of rare earths (the other lanthanides and yttrium) into ceria has attracted much attention by various synthetic approaches, and these include numerous solvothermal routes. All of the available rare earths have been studied in this respect and a complete survey of materials prepared solvothermally in the past decade are included as Supporting Information (Table S1). Typically, up to 50% of the Ce can be replaced by a trivalent substituent, before impurity phases, such as $Ln(OH)_3$ or $LnO(OH)$, are seen, particularly for the rare earths with smaller ionic radii [101–103]. Most synthesis is performed under hydrothermal conditions from aqueous solutions of metal salts: typically nitrates of the trivalent cations, but occasionally chlorides or acetates, and in the case of cerium, Ce(IV) salt $(NH_4)_2Ce(NO_3)_6$ has also been used. In the simplest cases, the pH is increased by addition of NaOH, KOH or ammonia, and temperatures spanning 80–250 °C have been reported to produce a solid product. In many cases, the solid formed is crystalline, but often a subsequent heat treatment up to 500 °C is applied, simply to remove any surface-bound species. Under these conditions, the typical morphology of CeO_2 itself is of cube-shaped crystallites, Figure 12 [104], which may appear spherical when the size is 10 nm or less. The size of the crystals may depend on pH, temperature and time of reaction.

Upon introduction of small amounts of a second lanthanide, the cube-shaped crystal is usually maintained, but it has been reported that at higher substituent levels, and under otherwise identical reaction conditions, the crystal morphology becomes increasingly rod-shaped, with a distribution of both crystal forms at intermediate compositions. This is exemplified by the case of praseodymium, where the same phenomenon has been reported independently by several different research groups: [105–108] at 25% Pr substitution and above, the rod-like crystal morphology is detected, and an example is shown in Figure 13. Praseodymium may also be found in the +4 oxidation state, and this has been inferred from the redox properties of some of these samples [106,108] and detected by X-ray spectroscopy in others [105,107]. This, however, does not explain the formation of the rod-like crystallites. Interestingly, Mendiuk et al. found for Ln = Pr, Sm, Gd, Tb, at higher Ln substitutions, the formation of $Ln(OH)_3$ rod-like crystals [109]. It may well be possible that the formation of these anisotropic crystallites has a seeding effect in the subsequent growth of the oxide, or that when a second step of firing is applied, even at moderate temperature, the dehydration of the hydroxide occurs with retention of crystal morphology. Consistent with this idea, Hong et al. separately showed the anisotropic crystallites of phase-pure $(Ce,Gd)(OH)_3$ could be dehydrated to the mixed oxide on heating at just 400 °C with crystal morphology maintained [110]. Yoshida and Fujihara found that for Sm substitution, hydrothermal treatment at 100 °C prior to heat treatment gave rod-shaped crystals, even for very low Sm content, which may imply the initial formation of a hydroxide phase at mild conditions [111].

Regarding crystallisation mechanisms, Mendiuk et al. proposed that for the cube-shaped crystallites, their assembly occurred by an oriented attachment mechanism, owing to the observation of a bimodal crystallite size distribution [102,109]. A careful study of Sm-CeO_2 nanocubes using spatially resolved electron energy loss spectroscopy on the TEM showed the that presence of the Ln^{3+} substituent did not affect the Ce^{3+} concentration at the surface of the crystallites [112].

The addition of solution additives has been explored as a way to modify the crystal morphology of lanthanide-substituted materials. In one of the simplest cases, Ke et al. added NaCl to the crystallisation of a set of lanthanide-substituted cerias (Ln = La-Lu with 10 mol% substituent) from chloride precursors in 12 M NaOH and found nanowires ~5 nm in diameter with an aspect ratio of ~500, Figure 14 [113]. In the processing of these samples, heating to only 300 °C was performed, which does imply that the ionic strength of the solution may play a role in the crystallisation process.

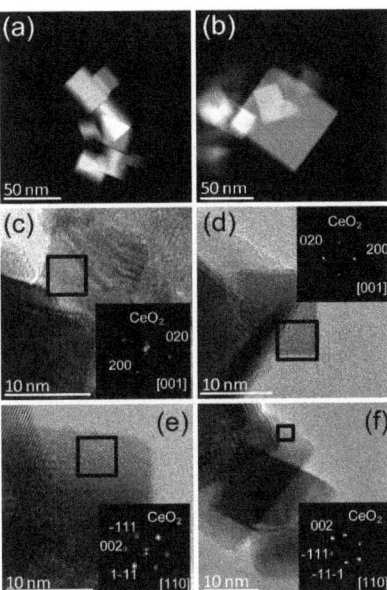

Figure 12. HAADF-STEM images of (**a**) CeO$_2$ nanocrystals and (**b**) 10% La-CeO$_2$ nanocrystals; HREM images of (**c,e**) CeO$_2$ nanocrystals and (**d**), (**f**) 10%La-CeO$_2$ nanocrystal samples in the [001] and [110] zone axis [104]. Reprinted with permission from Fernandez-Garcia, S.; Jiang, L.; Tinoco, M.; Hungria, A.B.; Han, J.; Blanco, G.; Calvino, J.J.; Chen, X. Enhanced Hydroxyl Radical Scavenging Activity by Doping Lanthanum in Ceria Nanocubes. *J. Phys. Chem. C* **2016**, *120*, 1891–1901. Copyright (2016) American Chemical Society.

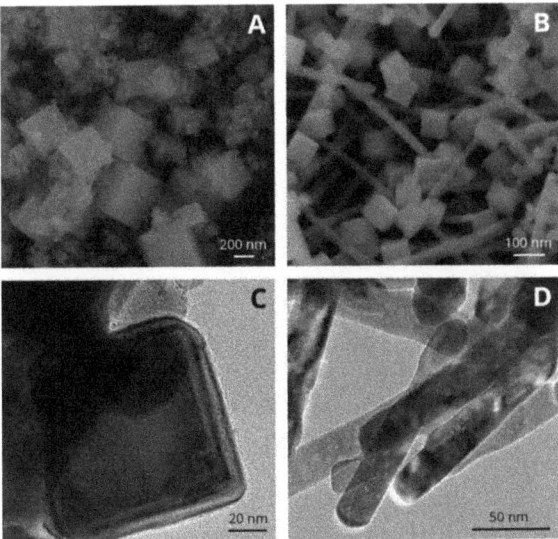

Figure 13. FE-SEM and TEM images of CeO$_2$ (**A,C**) and 50% Pr-CeO$_2$ (**B,D**) [107]. Reprinted from Andana, T.; Piumetti, M.; Bensaid, S.; Veyre, L.; Thieuleux, C.; Russo, N.; Fino, D.; Quadrelli, E.A.; Pirone, R. Nanostructured equimolar ceria-praseodymia for NO*x*-assisted soot oxidation: Insight into Pr dominance over Pt nanoparticles and metal–support interaction. *Appl. Catal. B Environ.* **2018**, *226*, 147–161, with permission from Elsevier.

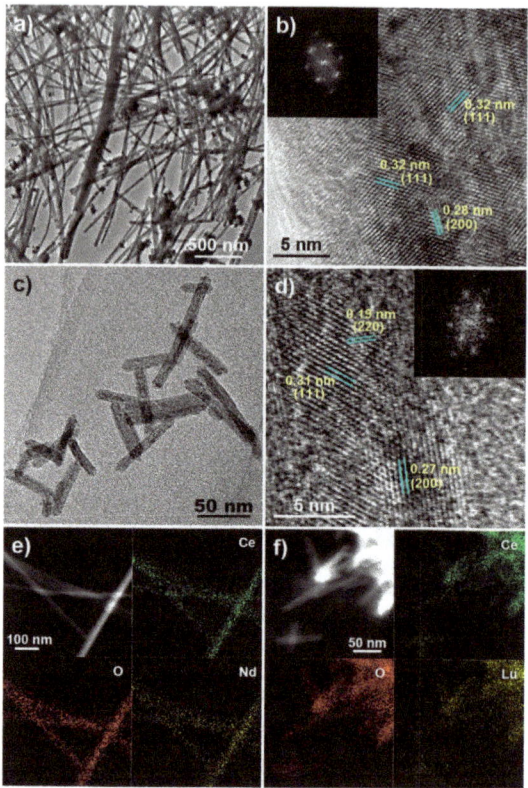

Figure 14. TEM (**a,c**), HRTEM (**b,d**) and HAADF-STEM EDS elemental mapping (**e,f**) images of CeO$_2$:Nd (**a,b,e**) and CeO$_2$:Lu nanocrystals (**c,d,f**). Insets in Panels (**b**) and (**d**) are the fast Fourier transition analyses, indicating that (110) surfaces are exposed for both samples. In Panels **e** and **f**, the up-left images are the HAADF-STEM images. The upper right and bottom left and right ones in Panels (**e,f**) are the EDS elemental mapping images representing Ce, O and the dopant (Nd for panel (**e**) and Lu for panel (**f**)) [113]. Reprinted with permission from Ke, J.; Xiao, J.-W.; Zhu, W.; Liu, H.; Si, R.; Zhang, Y.-W.; Yan, C.-H. Dopant-Induced Modification of Active Site Structure and Surface Bonding Mode for High-Performance Nanocatalysts: CO Oxidation on Capping-free (110)-oriented CeO$_2$:Ln (Ln = La–Lu) Nanowires. *J. Am. Chem. Soc.* **2013**, *135*, 15191–15200. Copyright (2013) American Chemical Society.

An additive that has been well established to affect crystal morphology in hydrothermal formation of cerium oxides is Na$_3$PO$_4$·12H$_2$O. Xu et al. found for yttrium substitution that octahedral crystals were produced but with surface nanorods with increasing addition of sodium phosphate [114]. The same octahedral crystal morphology was found by Małecka et al., who studied the formation of ytterbium-substituted CeO$_2$ [103,115], Roh et al., who examined europium substitution [116], Bo et al. for Er-CeO$_2$ [117] and Yang et al., who prepared erbium-substituted materials, some of which were co-substituted with lithium, Figure 15, where morphology is maintained even after a high-temperature annealing of 800 °C [118]. Małecka also observed that rod-shaped crystals were found at high concentrations of Na$_3$PO$_4$·12H$_2$O [115]. The octahedral morphology, bounded by {111} faces, is known to be a stable form for large crystals of CeO$_2$ itself [119], and while earlier work showed the effectiveness of Na$_3$PO$_4$ as a mineraliser for growth of octahedral nanocrystals of CeO$_2$, the mechanism of this precise morphological control has not been established, beyond discussion of how it might control pH and affect the dissolution–recrystallisation processes leading to crystallisation [120].

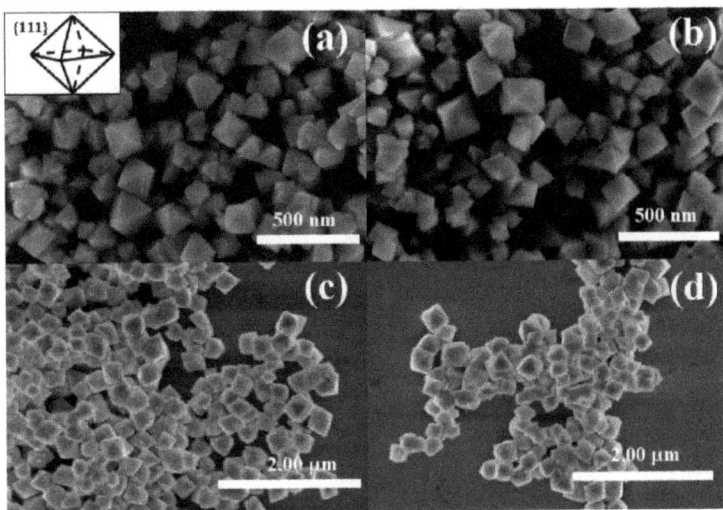

Figure 15. Octahedra of Er-CeO$_2$ prepared using Na$_3$PO$_4$·12H$_2$O as an additive in hydrothermal crystallisation (**a**), CeO$_2$: 3 mol% Er^{3+}, 10 mol% Li$^+$ (**b**) and the SEM images of the above nanocrystals with post-calcination temperatures of 800 °C (**c**,**d**) [118]. Reprinted from Yang, Y.; Cong, Y.; Dong, D.P.; Xiao, Y.; Shang, J.Y.; Tong, Y.; Zhang, H.M.; He, M.; Zhang, J.H. Structural and excitation dependent emission properties of octahedral CeO$_2$:Er^{3+} nanocrystal. *J. Lumin.* **2019**, *213*, 427–432, with permission from Elsevier.

Urea has been investigated as an additive for when crystallising ceria with substituents Pr, Gd [121] and Tb [122,123], and this may be observed to have the general effect of forming small, ~10 nm, crystallites that are often agglomerated. Muñoz et al. used microwave heating (see below) and produced Gd-CeO$_2$ materials that consisted of spherical assemblies of the primary particles, Figure 16. The combination of urea and sodium citrate was studied by Xu et al. for La, Nd, Sm and Eu substitution, who found "broom-like" agglomerates of needle-like crystals [124]. Urea is known to decompose under hydrothermal conditions into ammonium and cyanate, and this has been speculated to control the nucleation by pH control, in the case of ceria itself leading to smaller crystallites at higher urea concentrations [125].

Other additives that have been investigated for control of crystal morphology of substituted cerias include hexamethylenetetramine [126–128] and tetramethyl ammonium hydroxide [129–131]. The role of these additives has not been determined, and often simply the outcome of their influence on crystal morphology is described. Sato et al. speculated that tetramethylammonium ions cap the growing crystal faces of Gd-CeO$_2$ to avoid agglomeration and ripening, thus forming cuboidal crystals of only 4 nm in dimension [131].

Mixtures of additives have also been investigated, such as polyvinylpyrrolidone, HCl, KClO$_3$ and *N,N*-dimethylformamide [132], octadecylamine, ethanol and ethylenediamine [133] or 6-amino hexanoic acid and *tert*-butylamine [134], and non-aqueous solvents have been studied, such as ethylene glycol [135–138], ethanol [139] and triethylene glycol [140]. In some cases, the choice of combination of additive and solvent has led to some striking crystal morphologies; for example, the combination of epigallocatechin-3-gallate in ethanol as a solvent allowed the formation of hierarchical nanostructures of Eu-containing CeO$_2$ with flower-like shapes of intergrown plates, microns in thickness, Figure 17 [141].

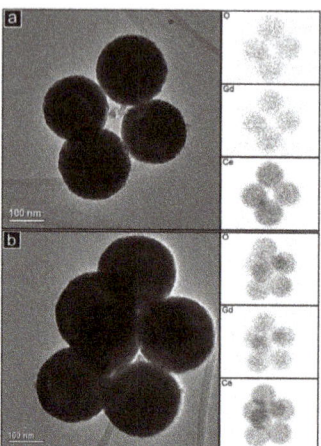

Figure 16. TEM images of groups of spherical particles of gadolinium-substituted CeO$_2$ prepared in the presence of urea with (**a**) 10% Gd and (**b**) 20% Gd, with corresponding energy-dispersive spectroscopy maps of the elements shown on the right [121]. Reproduced from Muñoz, F.F.; Acuña, L.M.; Albornoz, C.A.; Leyva, A.G.; Baker, R.T.; Fuentes, R.O. Redox properties of nanostructured lanthanide-doped ceria spheres prepared by microwave assisted hydrothermal homogeneous co-precipitation. *Nanoscale* **2015**, *7*, 271–281—Published by The Royal Society of Chemistry.

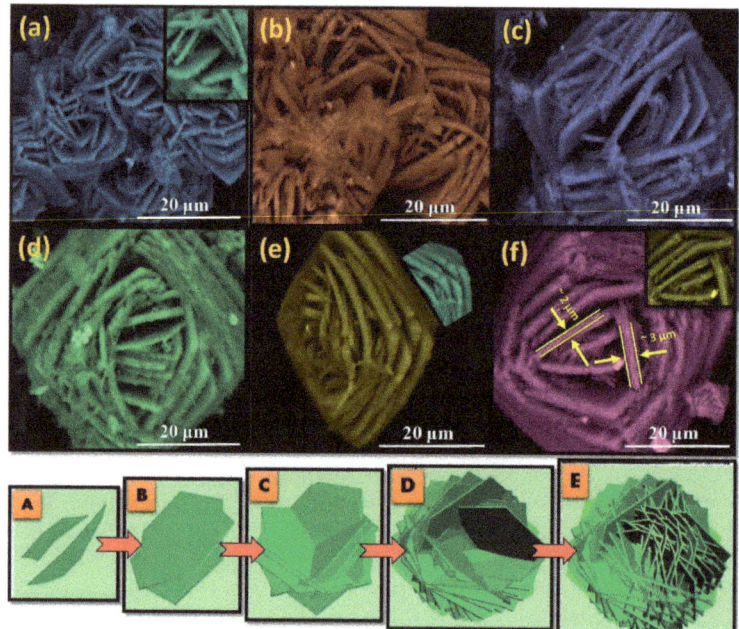

Figure 17. EM images of CeO$_2$:Eu^{3+} (3 mol%) NPs prepared with different reaction times in ethanol (**a**) 2 h, (**b**) 4 h, (**c**) 6 h, (**d**) 8 h, (**e**) 10 h and (**f**) 12 h with 30% w/v of EGCG at 180 °C, and (**A–E**) depict the proposed growth mechanism of the hierarchical morphology. EGCG is epigallocatechin-3-gallate [141]. Reprinted from Deepthi, N.H.; Darshan, G.P.; Basavaraj, R.B.; Prasad, B.D.; Nagabhushana, H. Large-scale controlled bio-inspired fabrication of 3D CeO$_2$:Eu^{3+} hierarchical structures for evaluation of highly sensitive visualization of latent fingerprints. *Sens. Actuators B Chem.* **2018**, *255*, 3127–3147, with permission from Elsevier.

As well as conventional heating, microwaves have been used to facilitate solvothermal crystallisations of rare earth-substituted cerias. The advantage here lies in the short reaction times, from a few hours down to just a few minutes [102,103,115,121–123,128,142–146].

Continuous flow synthesis has also been used for some rare earth-substituted ceria materials. Slostowski et al. produced La- and Pr-substituted ceria in a custom-built, continuous flow reactor in supercritical water, up to 400 °C, and formed crystallites less than 10 nm in size, observing that higher temperature allowed a higher level of element substitution [147]. Xu et al. used supercritical water at 396 °C and rapid mixing, ~29 s, with solutions of metal salts and KOH to yield Gd-CeO_2 [148]. Their apparatus, Figure 18, had a further two stages of heating points before allowing the product to be tapped as a slurry. Depending on pH, the crystallites produced were between 6 and 40 nm in size with a polyhedral or octahedral shape.

Applications of rare earth-substituted ceria materials span the types of fields expected for these materials, from oxide ion conductors in solid oxide fuel cells, where the enhanced sinterability of fine powders into ceramics is beneficial [136,137,143,149,150], to catalysts for CO and soot oxidation, where the high surface and exposure of specific crystal faces may offer some benefits [103,105,107,108,113,115,132,133,135,150–153]. Photocatalytic properties are of increasing focus for degradation of pollutants or organic transformations, and some of the rare earth-substituted materials have been screened for these applications [101,124,141,154].

Photoluminescence is a property that has commonly been used to characterise rare earth-substituted ceria materials [111,114,117,144,155–160], while Eu-substituted materials are of particular interest for their luminescent properties. Eu-CeO_2 materials prepared by solvothermal routes have provided some useful materials: this includes in applications such as solar cells, Figure 19 [116], and detection of fingerprints [141]. The use of solvothermal conditions to prepare nanocrystals of ceria materials has allowed specimens to be prepared for study of their toxicity: Dunnick et al. used gadolinium-substituted samples and examined their properties in vivo in rats [129,130].

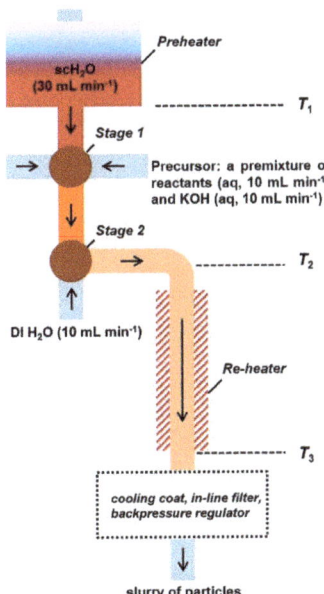

Figure 18. A schematic representation of the two-stage continuous hydrothermal flow synthesis reactor used by Xu et al. for the preparation of Gd-CeO_2 using pressure controlled at 26 MPa. The flow rate of each H_2O/solution stream is given in parentheses; arrows indicate the flow direction [148]. Reproduced with permission from John Wiley and sons.

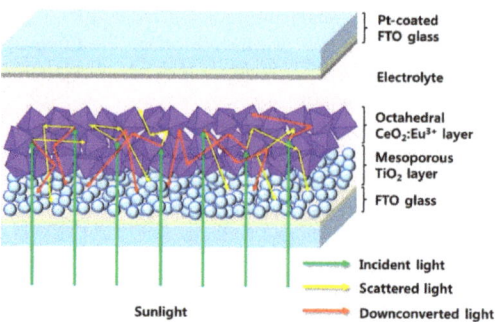

Figure 19. Schematic of a bilayer dye-sensitised solar cell device fabricated from hydrothermally synthesised octahedral CeO$_2$:Eu^{3+} nanocrystals for light scattering and down conversion [116]. Reprinted with permission from Roh, J.; Hwang, S.H.; Jang, J. Dual-Functional CeO$_2$:Eu^{3+} Nanocrystals for Performance-Enhanced Dye-Sensitized Solar Cells. *ACS Appl. Mater. Interfaces* **2014**, *6*, 19825–19832. Copyright (2014) American Chemical Society.

In terms of practical applications, the formation of composites is of relevance, and the solvothermal routes have allowed some useful possibility of accessing novel materials with practical applications. Pr-doped CeO$_2$ nanorods were introduced into a Sr-doped LaMnO$_3$–Y$_2$O$_3$-stabilized ZrO$_2$ composite by an in situ hydrothermal growth process: this gave an enhanced performance of the material for use as an electrode in solid oxide fuel cell applications [161]. Jiang et al. produced gadolinium-substituted ceria crystallites on a graphene support, by including the support material in the hydrothermal reaction [162]. This was tested for application in non-aqueous lithium–oxygen batteries, and a favourable catalytic effect on Li$_2$O$_2$ formation and decomposition was found. In a similar manner, Er-CeO$_2$ was loaded onto graphene oxide directly via in situ hydrothermal crystallisation, and then in a second hydrothermal step, CuO was co-loaded; these composite materials showed enhanced photocatalytic reduction of CO$_2$ to methanol compared to the individual components [163]. Xu et al. used polystyrene microspheres as templates to produce hollow microspheres constructed from agglomerated crystallites of Y-substituted CeO$_2$: these showed enhanced activity for the photocatalytic decomposition of acetaldehyde [158].

3.5. Multi-Element Substitutions

Given the wide range of chemical elements that can easily be added to ceria using the very mild conditions offered by solvothermal approaches, it is not surprising to find that multi-element inclusion has been increasingly explored as a way of fine-tuning properties such as oxygen storage or photocatalysis and perhaps also introducing the possibility of multifunctionality in the solid materials. This includes combining elements of similar chemistry that might provide some cooperative effect on resulting properties, such as more than one of the rare earths [164–168]. Mixing elements from different parts of the periodic table is also increasingly being explored, and one important example is the inclusion of other rare earths in ceria-zirconia solid solutions that may further tune redox properties for heterogeneous catalysis. For example, by addition of both Zr and Pr to ceria, via a hydrothermal route, it was shown that praseodymium contributes positively towards the soot oxidation reaction, but it has an adverse effect on CO oxidation over the same catalysts, as compared to pure ceria [169]. The inclusion of both Nd and La to ceria-zirconia, to give a phase with 40% cerium, 50% zirconium, 5% lanthanum and 5% neodymium, via a mild hydrothermal crystallisation, gave materials with enhanced low-temperature reducibility and thermal stability [170]. Materials in the CeO$_2$-Y$_2$O$_3$-ZrO$_2$ system have been prepared under solvothermal conditions. Wang et al. found that the activation energy of crystal growth is significantly dependent on the CeO$_2$ concentration [171], and continuous flow synthesis (410 °C, 27 MPa) has been used to form ultrafine stabilised tetragonal zirconia particles [172], while a wide range of nanoscale crystal forms, with intricate morphologies, were possible by inclusion of surfactants, Figure 20 [173]. In the even more complex system ceria-zirconia-yttria-lanthana, high-surface area nanostructured materials have been formed with the aid of lauric acid and dodecylamine [174].

Europium has been added to ceria-zirconia in low concentrations (<10%) using aqueous ammonia as a reaction medium and potassium oleate as an additive to produce nanocrystalline materials with photoemission and photoluminescence properties [175,176]. Hydrothermal synthesis using poly block copolymer surfactants provides high-surface La-containing ceria zirconia [177].

Mixed rare earth ceria materials with other substituents have been prepared via hydrothermal methods, such as Sr,Gd-ceria [178] and Bi,Gd-ceria [179,180], and sintered into ceramics for solid oxide fuel cell applications.

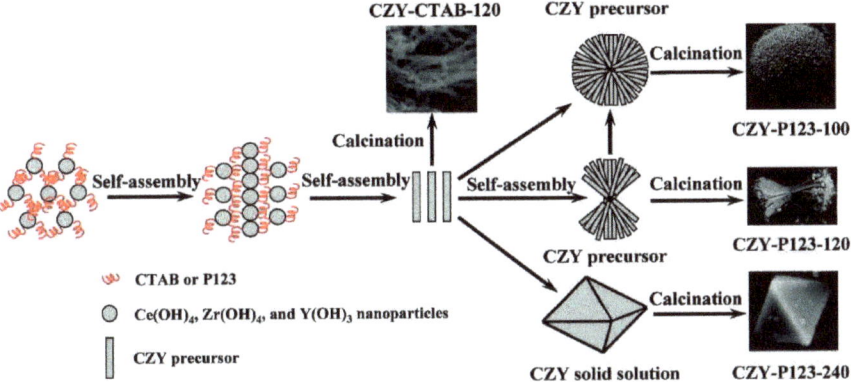

Figure 20. A schematic illustration of formation mechanisms of the ceria-zirconia-yttria precursors and ceria-zirconia-yttria solid solutions under surfactant-assisted (CTAB or P123) hydrothermal and calcination conditions [173]. Reprinted with permission from Zhang, Y.; Zhang, L.; Deng, J.; Dai, H.; He, H. Controlled Synthesis, Characterization, and Morphology-Dependent Reducibility of Ceria−Zirconia−Yttria Solid Solutions with Nanorod-like, Microspherical, Microbowknot-like, and Micro-octahedral Shapes. *Inorg. Chem.* **2009**, *48*, 2181–2192. Copyright (2009) American Chemical Society.

4. Summary, Challenges and Future Directions

Our survey of the past decade of research on solvothermal synthesis methods for producing cerium dioxide materials has shown how elements from every part of the periodic table can be introduced into the cubic fluorite lattice of CeO_2. Our earlier review 10 years ago found relatively few examples of substitutional chemistry, with the main emphasis being on ceria-zirconia solid solutions and rare-earth substituents, in all cases using water as the majority solvent [30]. This work has now been broadened to encompass an extensive set of element substitutions, including mixtures of elements, and using a variety of solvents, solvent mixtures and solution additives. As with other oxide materials, the advantage of the solvothermal method lies in the accessing of novel substitutional chemistry and in the control of crystal form, both of which are difficult when higher temperatures are employed, such as in classical solid-state chemistry, or combustion methods, or even in simple precipitation routes that require a firing step to induce crystallisation [181]. The solvothermal synthesis method typically yields materials in a crystalline state, even if a mild heat treatment (up to 500 °C) may be used to remove any surface-bound hydroxide, solvent or additive, particularly when the materials are to be used for applications in heterogeneous catalysis. Much of the research reported is towards practical applications and as well as classical fields of research, such as in redox catalysis, photocatalysis and oxide ion conductivity, novel directions are being explored and, indeed, enabled by the availability of nanocrystalline forms of ceria. This includes applications as components in solar devices, as electrodes in rechargeable batteries and as catalysts for thermochemical water splitting and has allowed properties such as luminescent sensing, magnetism and toxicity to be studied.

The nanoscale crystal morphologies are important where high surface areas are needed but also provide highly sinterable powders for forming ceramics, such as electrodes for solid oxide fuel cells. In this respect, it is important to consider the level of control over crystal morphology under solvothermal reaction conditions and whether this has truly been demonstrated with any degree of predictability. As this review has shown, a wide variety of solution additives and solvents have now been employed, and, certainly, in some cases, enough examples are provided, independently by different research groups, to identify certain conditions that yield specific crystal morphologies. For example, $Na_3PO_4 \cdot 12H_2O$ has been used to form octahedral-shaped crystals for CeO_2 itself, as well as for various substituted versions. However, the mechanism of its action has not been uncovered, beyond speculation of control of the solution pH. The formation of nanorod morphologies is often related to the formation of lanthanide trihydroxides, implying either a seeding effect of $Ln(OH)_3$ which forms early in the synthesis, or thermal decomposition yielding oxides with morphological control. In many cases, the role of the solution additives (along with the choice of solvent) is unknown and many authors reported the outcome of the use of the additive without any consideration of its role.

From our review, three trends are emerging in tuning the properties of materials for applications which make use of the full advantages of solvothermal chemistry, and these are ripe for future development.

(1) Multi-element substitution is now increasingly being used to adjust properties, and this has been extended from mixtures of rare-earth substituents to combinations of, for example, rare earths in ceria-zirconia to mixtures of transition metal ions. This includes compositions that might be expected to phase separate in alternative synthesis approaches, particularly those that use an annealing to induce crystallisation. Here, computer simulation, which has long been used to understand the effect of M^{3+} substituents in ceria [182], may provide important guiding principles in narrowing down likely candidate combinations of substituting elements.

(2) The formation of composite materials is being tuned to introduce dual functionality or enhanced properties beyond the individual components: this includes the purposeful growth of mixed-phase materials where one oxide is grown in intimate proximity to a second. This may give interesting properties from the interfaces present between the two phases but may also yield morphologies not possible by other routes, and the example of CeO_2-CuO illustrates this.

(3) The use of in situ growth on templates or substrates allows fabrication of materials in a form suited for application. For example, impregnation of a porous support material with a reagent solution, followed by hydrothermal crystallisation, provides a convenient method for the formation of a device, and an example of ceria nanorods grown within a ceramic matrix for an electrode in solid oxide fuel cells illustrates this concept. The same idea could be used to produce coating or films of an immobilised material. This is of importance in emerging applications of ceria materials, such as in electrocatalysis, where the role of a support is crucial to determine properties [183].

In consideration of the directed synthesis of active materials for applications, there are some outstanding issues that are apparent from the literature we have reviewed:

(1) A detailed atomic-scale characterisation of materials is needed to understand fully structure–property relationships to then justify and predict choices of elemental substitution. The vast majority of work reviewed above has relied on the use of powder X-ray diffraction patterns and electron microscopy to prove sample identity and purity, in some cases supplemented by Raman spectroscopy as an additional fingerprint. Very few works have performed an in-depth analysis of the local atomic structure in the substituted materials, which is likely to be highly distorted from the ideal fluorite structure, when the coordination preferences of cations are taken into consideration: for example, the Jahn–Teller distortion expected for Cu^{2+}, or the asymmetric environment expected for Bi^{3+}

in an oxide. In multi-element substitutions, the issue of local atomic clustering, i.e., the homogeneity of element mixing, must be considered. Even for the case where an isovalent substitution is made, such as in ceria-zirconia solid solutions, there are complex structure–composition phase diagrams, and nanoscale phase segregation is possible. Here, methods such as EXAFS spectroscopy and pair distribution function analysis must be employed to determine these structural subtleties that are hidden in conventional diffraction measurements, especially when powder diffraction profiles are broadened by a small crystallite domain size.

(2) Crystallisation mechanisms, including the mode of operation of crystal habit modifiers mentioned above, must be understood in order to plan synthesis effectively and to develop tailored synthesis to target materials with a desired structure and properties. The use of high-energy X-rays has proved highly beneficial in this respect to track the evolution of a crystalline material from within a realistic reaction vessel as a function of temperature and time [96]. This yields quantitative kinetic information and also the possibility of observation of any intermediate phases, and an example of such a study of Bi-CeO_2 is provided above. If pair distribution functions can be measured in situ, then this provides local structure information, without the requirement for a crystalline structure, and such an approach has already been applied for CeO_2 itself where nucleation was observed from previously unknown cerium dimer complexes [184]. The more widespread use of such methods will lead to the mechanistic detail needed for predictive synthesis, and this will include a proper understanding of the role of additives in modifying crystal habit. Computer simulation may also play a part in such efforts.

(3) Scale-up of synthesis for applications is crucial if materials' properties are to be put into real use. As was highlighted above several times, the use of continuous flow reactors has been the focus of development for a number of years, where a flow of reagents is rapidly mixed with a heated solvent, often under supercritical conditions, to give rapid nucleation and an output stream of a slurry of a product. This has already been put to use on a commercial scale, and cerium oxide products are available commercially [185]. Much more development needs to be conducted on developing scaled synthesis when the use of additives and crystal habit modifiers is used, and with the much more complex chemical compositions that are now being investigated; here, mixing and inclusion of additives must be carefully considered.

(4) The choice of precursors in the preparation of any functional material is of crucial importance since contaminant ions can modify the properties of the solid. In the case of ceria, alkali metal cations can act as a poison when used in selective catalytic oxidation of NO_x or NH_3 [186,187]. On the other hand, the presence of sodium can have a beneficial effect in other applications; for example, it can act as a promoter when ceria is used for ethane oxidation [188], and its inclusion in ceria-zirconia can lead to enhanced low-temperature reproducibility [189]. Similarly, chloride ions can influence the surface properties of ceria, in some cases enhancing interaction with supported species [190], while in others having a detrimental effect by the formation of CeOCl that results in loss of surface area [191]. It is also the case that the choice of precipitant can affect the sintering behaviour of ceria powders [192]. In an industrial setting, extensive washing of powders must be avoided, and therefore to achieve reproducibility in the properties of ceria materials prepared from solution, future work must consider the role of surface contaminants that arise from the choice of chemical reagents; this is true whether the contaminants have a detrimental or a beneficial effect.

(5) The toxicity of nanocrystals to humans and other organisms is a growing concern, and ceria has attracted some attention in this respect, especially given its biological activity arising from its antioxidant properties. With a growing range of specimens available from solvothermal chemistry, each with their own distinctive chemistry from the substituent cations, clearly consideration of biotoxicity must be the focus of greater attention. As well as studies of the action of nanoparticles in biological systems, their transport into the human body is another important field of research, whether by inhalation or through the

skin. This has recently been examined for CeO_2 itself [193], but there is much further work to be conducted in this respect.

Finally, we note that in the case of CeO_2 itself, a recent article by Xu et al. extensively reviewed the various solution methods for synthesis, including solvothermal methods, and attempted to classify the various influences on crystal growth with the aim of having some predictability in the control of the crystal form, Figure 21 [194]. While the addition of a substituting cation, and indeed habit-modifying additives, introduces further complexity, it is highly likely that the same principles put forward will apply to the wider family of cerium oxide materials. This accumulation of experimental evidence leads to the possibility of controlled synthesis of materials with highly tuned properties for applications.

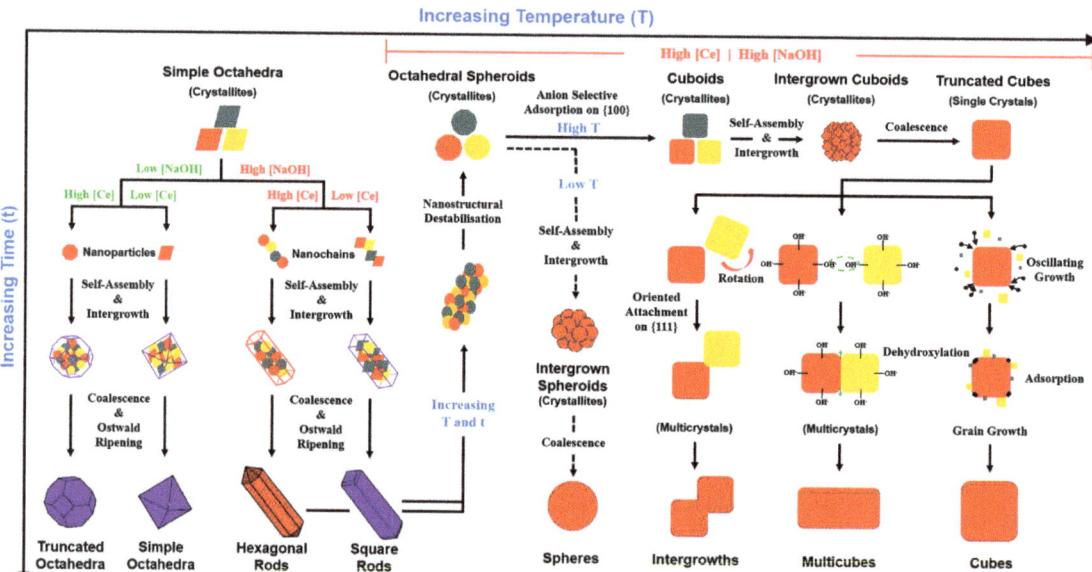

Figure 21. Schematic of solution growth mechanisms of CeO_{2-x} nanomorphologies (red/yellow = investigated by Xu et al. [194]; purple = from the literature; [Ce]: low < 0.50 M ≤ high; [NaOH]: low < 5.00 M ≤ high; T: low < 150 °C ≤ high; t: short ≤ 2 h < long) [194]. Reproduced from Xu, Y.; Mofarah, S.S.; Mehmood, R.; Cazorla, C.; Koshy, P.; Sorrell, C.C. Design strategies for ceria nanomaterials: untangling key mechanistic concepts. *Mater. Horiz.* **2021**, *8*, 102–123—Published by The Royal Society of Chemistry.

5. Conclusions

Solvothermal synthesis routes to substituted cerium oxides provide versatile conditions for the preparation and, ultimately, manufacture of functional materials for use in established and emerging applications. The accumulated literature points towards a move to control of the chemical composition and crystal form, with particular emphasis on the nanostructure. We have emphasised developing trends in targeting complex compositions with multi-element replacements, and in the preparation of composite phases, including in situ growth onto substrates to fabricate devices for application. There are also a number of challenges in developing future targeted synthesis of novel materials, which include more in-depth structural characterisation of existing materials to provide proper structure–property relationships, and in situ studies of crystallisation to provide the formation mechanism with a firmer basis. Finally, scale-up of synthesis requires further consideration, bearing in mind the complex mixtures of reagents being used to tailor the crystal form.

Supplementary Materials: The following are available online at https://www.mdpi.com/article/10.3390/inorganics9060040/s1. Table S1: Summary of rare-earth substituted cerium dioxide materials prepared by solvothermal synthesis. Please see Supplementary File.

Author Contributions: Conceptualization, J.M.F., D.T., R.I.W.; writing—original draft preparation, J.W.A., R.I.W.; writing—reviewing and editing, J.W.A., R.I.W.; funding acquisition, R.I.W. All authors have read and agreed to the published version of the manuscript.

Funding: This work was supported by the Royal Society via PhD position for J.W.A. and an Industry Fellowship in collaboration with Johnson Matthey for R.I.W.

Institutional Review Board Statement: Not applicable.

Informed Consent Statement: Not applicable.

Data Availability Statement: Data sharing not applicable.

Acknowledgments: The authors acknowledge the financial support of the Royal Society.

Conflicts of Interest: The authors declare no conflict of interest.

References

1. Trovarelli, A. Catalytic properties of ceria and CeO_2-containing materials. *Catal. Rev. Sci. Eng.* **1996**, *38*, 439–520. [CrossRef]
2. Montini, T.; Melchionna, M.; Monai, M.; Fornasiero, P. Fundamentals and Catalytic Applications of CeO_2-Based Materials. *Chem. Rev.* **2016**, *116*, 5987–6041. [CrossRef] [PubMed]
3. Kim, G. Ceria-promoted three-way catalysts for auto exhaust emission control. *Ind. Eng. Chem. Prod. Res. Dev.* **1982**, *21*, 267–274. [CrossRef]
4. Diwell, A.F.; Rajaram, R.R.; Shaw, H.A.; Truex, T.J. The Role of Ceria in Three-Way Catalysts. In *Studies in Surface Science and Catalysis*; Crucq, A., Ed.; Elsevier: Amsterdam, The Netherlands, 1991; Volume 71, pp. 139–152.
5. Li, P.; Chen, X.; Li, Y.; Schwank, J.W. A review on oxygen storage capacity of CeO_2-based materials: Influence factors, measurement techniques, and applications in reactions related to catalytic automotive emissions control. *Catal. Today* **2019**, *327*, 90–115. [CrossRef]
6. Soykal, I.I.; Sohn, H.; Singh, D.; Miller, J.T.; Ozkan, U.S. Reduction Characteristics of Ceria under Ethanol Steam Reforming Conditions: Effect of the Particle Size. *ACS Catal.* **2014**, *4*, 585–592. [CrossRef]
7. Liu, S.; Wu, X.; Weng, D.; Ran, R. Ceria-based catalysts for soot oxidation: A review. *J. Rare Earths* **2015**, *33*, 567–590. [CrossRef]
8. Gorte, R.J.; Zhao, S. Studies of the water-gas-shift reaction with ceria-supported precious metals. *Catal. Today* **2005**, *104*, 18–24. [CrossRef]
9. Wang, Q.; Yeung, K.L.; Bañares, M.A. Ceria and its related materials for VOC catalytic combustion: A review. *Catal. Today* **2020**, *356*, 141–154. [CrossRef]
10. Le Gal, A.; Abanades, S.; Bion, N.; Le Mercier, T.; Harlé, V. Reactivity of Doped Ceria-Based Mixed Oxides for Solar Thermochemical Hydrogen Generation via Two-Step Water-Splitting Cycles. *Energy Fuels* **2013**, *27*, 6068–6078. [CrossRef]
11. Jaiswal, N.; Tanwar, K.; Suman, R.; Kumar, D.; Upadhyay, S.; Parkash, O. A brief review on ceria based solid electrolytes for solid oxide fuel cells. *J. Alloys Compd.* **2019**, *781*, 984–1005. [CrossRef]
12. Shcherbakov, A.B.; Ivanov, V.K.; Zholobak, N.M.; Ivanova, O.S.; Krysanov, E.Y.; Baranchikov, A.E.; Spivak, N.Y.; Tretyakov, Y.D. Nanocrystalline ceria based materials—Perspectives for biomedical application. *Biophysics* **2011**, *56*, 987–1004. [CrossRef]
13. Xu, C.; Qu, X.G. Cerium oxide nanoparticle: A remarkably versatile rare earth nanomaterial for biological applications. *NPG Asia Mater.* **2014**, *6*, e90. [CrossRef]
14. Casals, E.; Zeng, M.; Parra-Robert, M.; Fernández-Varo, G.; Morales-Ruiz, M.; Jiménez, W.; Puntes, V.; Casals, G. Cerium Oxide Nanoparticles: Advances in Biodistribution, Toxicity, and Preclinical Exploration. *Small* **2020**, *16*, 1907322. [CrossRef] [PubMed]
15. Yabe, S.; Yamashita, M.; Momose, S.; Tahira, K.; Yoshida, S.; Li, R.; Yin, S.; Sato, T. Synthesis and UV-shielding properties of metal oxide doped ceria via soft solution chemical processes. *Int. J. Inorg. Mater.* **2001**, *3*, 1003–1008. [CrossRef]
16. Yabe, S.; Sato, T. Cerium oxide for sunscreen cosmetics. *J. Solid State Chem.* **2003**, *171*, 7–11. [CrossRef]
17. Caputo, F.; De Nicola, M.; Sienkiewicz, A.; Giovanetti, A.; Bejarano, I.; Licoccia, S.; Traversa, E.; Ghibelli, L. Cerium oxide nanoparticles, combining antioxidant and UV shielding properties, prevent UV-induced cell damage and mutagenesis. *Nanoscale* **2015**, *7*, 15643–15656. [CrossRef]
18. Feng, X.; Sayle, D.C.; Wang, Z.L.; Paras, M.S.; Santora, B.; Sutorik, A.C.; Sayle, T.X.T.; Yang, Y.; Ding, Y.; Wang, X.; et al. Converting ceria polyhedral nanoparticles into single-crystal nanospheres. *Science* **2006**, *312*, 1504–1508. [CrossRef] [PubMed]
19. Fu, X.Q.; Wang, C.; Yu, H.C.; Wang, Y.G.; Wang, T.H. Fast humidity sensors based on CeO_2 nanowires. *Nanotechnology* **2007**, *18*, 145503. [CrossRef]
20. Schmitt, R.; Nenning, A.; Kraynis, O.; Korobko, R.; Frenkel, A.I.; Lubomirsky, I.; Haile, S.M.; Rupp, J.L.M. A review of defect structure and chemistry in ceria and its solid solutions. *Chem. Soc. Rev.* **2020**, *49*, 554–592. [CrossRef]
21. Di Monte, R.; Kašpar, J. Heterogeneous environmental catalysis—A gentle art: CeO_2–ZrO_2 mixed oxides as a case history. *Catal. Today* **2005**, *100*, 27–35. [CrossRef]

22. Gao, Y.X.; Wang, W.D.; Chang, S.J.; Huang, W.X. Morphology Effect of CeO_2 Support in the Preparation, Metal-Support Interaction, and Catalytic Performance of Pt/CeO_2 Catalysts. *Chemcatchem* **2013**, *5*, 3610–3620. [CrossRef]
23. Qiao, Z.A.; Wu, Z.L.; Dai, S. Shape-Controlled Ceria-based Nanostructures for Catalysis Applications. *ChemSusChem* **2013**, *6*, 1821–1833. [CrossRef] [PubMed]
24. Wang, S.; Zhao, L.; Wang, W.; Zhao, Y.; Zhang, G.; Ma, X.; Gong, J. Morphology control of ceria nanocrystals for catalytic conversion of CO_2 with methanol. *Nanoscale* **2013**, *5*, 5582–5588. [CrossRef] [PubMed]
25. Aneggi, E.; Wiater, D.; de Leitenburg, C.; Llorca, J.; Trovarelli, A. Shape-Dependent Activity of Ceria in Soot Combustion. *ACS Catal.* **2014**, *4*, 172–181. [CrossRef]
26. Trovarelli, A.; Llorca, J. Ceria Catalysts at Nanoscale: How Do Crystal Shapes Shape Catalysis? *ACS Catal.* **2017**, *7*, 4716–4735. [CrossRef]
27. Kašpar, J.; Di Monte, R.; Fornasiero, P.; Graziani, M.; Bradshaw, H.; Norman, C. Dependency of the Oxygen Storage Capacity in Zirconia–Ceria Solid Solutions upon Textural Properties. *Top. Catal.* **2001**, *16*, 83–87. [CrossRef]
28. Monte, R.D.; Kašpar, J. Nanostructured CeO_2–ZrO_2 mixed oxides. *J. Mater. Chem.* **2005**, *15*, 633–648. [CrossRef]
29. Sun, C.; Li, H.; Chen, L. Nanostructured ceria-based materials: Synthesis, properties, and applications. *Energy Environ. Sci.* **2012**, *5*, 8475–8505. [CrossRef]
30. Walton, R.I. Solvothermal synthesis of cerium oxides. *Prog. Cryst. Growth Charact. Mater.* **2011**, *57*, 93–108. [CrossRef]
31. Rabenau, A. The Role of Hydrothermal Synthesis in Preparative Chemistry. *Angew. Chem. Int. Ed.* **1985**, *24*, 1026–1040. [CrossRef]
32. Yoshimura, M.; Byrappa, K. Hydrothermal processing of materials: Past, present and future. *J. Mater. Sci.* **2008**, *43*, 2085–2103. [CrossRef]
33. Cundy, C.S.; Cox, P.A. The hydrothermal synthesis of zeolites: History and development from the earliest days to the present time. *Chem. Rev.* **2003**, *103*, 663–701. [CrossRef] [PubMed]
34. Laudise, R.A. Hydrothermal Synthesis of Crystals. *Chem. Eng. News* **1987**, *65*, 30–43. [CrossRef]
35. Demazeau, G. Solvothermal reactions: An original route for the synthesis of novel materials. *J. Mater. Sci.* **2008**, *43*, 2104–2114. [CrossRef]
36. Sōmiya, S.; Roy, R. Hydrothermal synthesis of fine oxide powders. *Bull. Mater. Sci.* **2000**, *23*, 453–460. [CrossRef]
37. Riman, R.E.; Suchanek, W.L.; Lencka, M.M. Hydrothermal crystallization of ceramics. *Ann. Chim. Sci. Mater.* **2002**, *27*, 15–36. [CrossRef]
38. Walton, R.I. Subcritical solvothermal synthesis of condensed inorganic materials. *Chem. Soc. Rev.* **2002**, *31*, 230–238. [CrossRef]
39. Komarneni, S. Nanophase materials by hydrothermal, microwave-hydrothermal and microwave-solvothermal methods. *Curr. Sci.* **2003**, *85*, 1730–1734.
40. Kumada, N. Preparation and crystal structure of new inorganic compounds by hydrothermal reaction. *J. Ceram. Soc. Jpn.* **2013**, *121*, 135–141. [CrossRef]
41. Tani, E.; Yoshimura, M.; Sōmiya, S. Crystallization and crystal growth of CeO_2 under hydrothermal conditions. *J. Mater. Sci. Lett.* **1982**, *1*, 461–462. [CrossRef]
42. Montini, T.; Speghini, A.; Rogatis, L.D.; Lorenzut, B.; Bettinelli, M.; Graziani, M.; Fornasiero, P. Identification of the Structural Phases of $Ce_xZr_{1-x}O_2$ by Eu(III) Luminescence Studies. *J. Am. Chem. Soc.* **2009**, *131*, 13155–13160. [CrossRef]
43. Yang, Z.; Hu, W.; Zhang, N.; Li, Y.; Liao, Y. Facile synthesis of ceria–zirconia solid solutions with cubic–tetragonal interfaces and their enhanced catalytic performance in diesel soot oxidation. *J. Catal.* **2019**, *377*, 98–109. [CrossRef]
44. Luciani, G.; Landi, G.; Imparato, C.; Vitiello, G.; Deorsola, F.A.; Di Benedetto, A.; Aronne, A. Improvement of splitting performance of $Ce_{0.75}Zr_{0.25}O_2$ material: Tuning bulk and surface properties by hydrothermal synthesis. *Int. J. Hydrogen Energy* **2019**, *44*, 17565–17577. [CrossRef]
45. Wang, H.; Yang, G.-Q.; Song, Y.-H.; Liu, Z.-T.; Liu, Z.-W. Defect-rich $Ce_{1-x}Zr_xO_2$ solid solutions for oxidative dehydrogenation of ethylbenzene with CO_2. *Catal. Today* **2019**, *324*, 39–48. [CrossRef]
46. Wan, J.; Lin, J.; Guo, X.; Wang, T.; Zhou, R. Morphology effect on the structure-activity relationship of Rh/CeO_2-ZrO_2 catalysts. *Chem. Eng. J.* **2019**, *368*, 719–729. [CrossRef]
47. Liu, X.; Liu, W.; Zhang, X.; Han, L.; Zhang, C.; Yang, Y. Zr-doped CeO_2 Hollow slightly-truncated nano-octahedrons: One-pot synthesis, characterization and their application in catalysis of CO oxidation. *Cryst. Res. Technol.* **2014**, *49*, 383–392. [CrossRef]
48. Chen, A.; Zhou, Y.; Ta, N.; Li, Y.; Shen, W. Redox properties and catalytic performance of ceria–zirconia nanorods. *Catal. Sci. Technol.* **2015**, *5*, 4184–4192. [CrossRef]
49. Wang, R.; Mutinda, S.I.; Fang, M. One-pot hydrothermal synthesis and high temperature thermal stability of $Ce_xZr_{1-x}O_2$ nanocrystals. *RSC Adv.* **2013**, *3*, 19508–19514. [CrossRef]
50. Das, S.; Gupta, R.; Kumar, A.; Shah, M.; Sengupta, M.; Bhandari, S.; Bordoloi, A. Facile Synthesis of Ruthenium Decorated $Zr_{0.5}Ce_{0.5}O_2$ Nanorods for Catalytic Partial Oxidation of Methane. *ACS Appl. Nano Mater.* **2018**, *1*, 2953–2961. [CrossRef]
51. Kim, J.-R.; Myeong, W.-J.; Ihm, S.-K. Characteristics in oxygen storage capacity of ceria–zirconia mixed oxides prepared by continuous hydrothermal synthesis in supercritical water. *Appl. Catal. B Environ.* **2007**, *71*, 57–63. [CrossRef]
52. Kim, J.-R.; Lee, K.-Y.; Suh, M.-J.; Ihm, S.-K. Ceria–zirconia mixed oxide prepared by continuous hydrothermal synthesis in supercritical water as catalyst support. *Catal. Today* **2012**, *185*, 25–34. [CrossRef]

53. Auxéméry, A.; Frias, B.B.; Smal, E.; Dziadek, K.; Philippot, G.; Legutko, P.; Simonov, M.; Thomas, S.; Adamski, A.; Sadykov, V.; et al. Continuous supercritical solvothermal preparation of nanostructured ceria-zirconia as supports for dry methane reforming catalysts. *J. Supercrit. Fluids* **2020**, *162*, 104855. [CrossRef]
54. Tang, X.; Zhang, B.; Li, Y.; Xu, Y.; Xin, Q.; Shen, W. CuO/CeO$_2$ catalysts: Redox features and catalytic behaviors. *Appl. Catal. A Gen.* **2005**, *288*, 116–125. [CrossRef]
55. Bernardi, M.I.B.; Mesquita, A.; Béron, F.; Pirota, K.R.; De Zevallos, A.O.; Doriguetto, A.C.; De Carvalho, H.B. The role of oxygen vacancies and their location in the magnetic properties of Ce$_{1-x}$Cu$_x$O$_{2-\delta}$ nanorods. *Phys. Chem. Chem. Phys.* **2015**, *17*, 3072–3080. [CrossRef]
56. Wang, J.; Liu, Q.; Liu, Q. Ceria- and Cu-doped ceria nanocrystals synthesized by the hydrothermal methods. *J. Am. Ceram. Soc.* **2008**, *91*, 2706–2708. [CrossRef]
57. Yang, F.; Wei, J.; Liu, W.; Guo, J.; Yang, Y. Copper doped ceria nanospheres: Surface defects promoted catalytic activity and a versatile approach. *J. Mater. Chem. A* **2014**, *2*, 5662–5667. [CrossRef]
58. Yang, H.; Pan, Y.; Xu, Y.; Yang, Y.; Sun, G. Enhanced catalytic performance of (CuO)$_x$/Ce$_{0.9}$Cu$_{0.1}$O$_2$ nanospheres: Combined contribution of the synergistic effect and surface defects. *ChemPlusChem* **2015**, *80*, 886–894. [CrossRef]
59. Rood, S.C.; Pastor-Algaba, O.; Tosca-Princep, A.; Pinho, B.; Isaacs, M.; Torrente-Murciano, L.; Eslava, S. Synergistic Effect of Simultaneous Doping of Ceria Nanorods with Cu and Cr on CO Oxidation and NO Reduction. *Chem. Eur. J.* **2021**, *27*, 2165–2174. [CrossRef]
60. Zhang, J.; Guo, J.; Liu, W.; Wang, S.; Xie, A.; Liu, X.; Wang, J.; Yang, Y. Facile preparation of Mn^{n+}-Doped (M = Cu, Co, Ni, Mn) hierarchically mesoporous CeO$_2$ nanoparticles with enhanced catalytic activity for CO oxidation. *Eur. J. Inorg. Chem.* **2015**, *2015*, 969–976. [CrossRef]
61. Kurajica, S.; Mužina, K.; Dražić, G.; Matijašić, G.; Duplančić, M.; Mandić, V.; Župančić, M.; Munda, I.K. A comparative study of hydrothermally derived Mn, Fe, Co, Ni, Cu and Zn doped ceria nanocatalysts. *Mater. Chem. Phys.* **2020**, *244*, 122689. [CrossRef]
62. Syed Khadar, Y.A.; Balamurugan, A.; Devarajan, V.P.; Subramanian, R.; Dinesh Kumar, S. Synthesis, characterization and antibacterial activity of cobalt doped cerium oxide (CeO$_2$:Co) nanoparticles by using hydrothermal method. *J. Mater. Res.* **2019**, *8*, 267–274. [CrossRef]
63. Zhang, X.; Wei, J.; Yang, H.; Liu, X.; Liu, W.; Zhang, C.; Yang, Y. One-Pot Synthesis of Mn-Doped CeO$_2$ Nanospheres for CO Oxidation. *Eur. J. Inorg. Chem.* **2013**, *2013*, 4443–4449. [CrossRef]
64. Zhang, Y.; Yang, F.; Gao, R.; Dai, W.L. Manganese-doped CeO$_2$ nanocubes as highly efficient catalysts for styrene epoxidation with TBHP. *App.Surf. Sci.* **2019**, *471*, 767–775.
65. Jampaiah, D.; Venkataswamy, P.; Coyle, V.E.; Reddy, B.M.; Bhargava, S.K. Low-temperature CO oxidation over manganese, cobalt, and nickel doped CeO$_2$ nanorods. *RSC Adv.* **2016**, *6*, 80541–80548. [CrossRef]
66. Song, Q.; Ran, R.; Ding, J.; Wu, X.; Si, Z.; Weng, D. The controlled preparation and performance of Fe, Co-modified porous ceria nanorods for the total oxidation of propane. *Mol. Catal.* **2020**, *480*, 110663. [CrossRef]
67. Xing, X.; Cai, Y.; Chen, N.; Li, Y.; Deng, D.; Wang, Y. Synthesis of mixed Mn–Ce–O$_x$ one dimensional nanostructures and their catalytic activity for CO oxidation. *Ceram. Int.* **2015**, *41*, 4675–4682. [CrossRef]
68. Liu, X.; Han, L.; Liu, W.; Yang, Y. Synthesis of Co/Ni unitary- or binary-doped CeO$_2$ mesoporous nanospheres and their catalytic performance for CO oxidation. *Eur. J. Inorg. Chem.* **2014**, *2014*, 5370–5377.
69. Du, X.; Dai, Q.; Wei, Q.; Huang, Y. Nanosheets-assembled Ni (Co) doped CeO$_2$ microspheres toward NO + CO reaction. *Appl. Catal. A Gen.* **2020**, *602*, 117728. [CrossRef]
70. Venkataswamy, P.; Damma, D.; Jampaiah, D.; Mukherjee, D.; Vithal, M.; Reddy, B.M. Cr-Doped CeO$_2$ Nanorods for CO Oxidation: Insights into Promotional Effect of Cr on Structure and Catalytic Performance. *Catal. Lett.* **2020**, *150*, 948–962.
71. Phokha, S.; Pinitsoontorn, S.; Maensiri, S. Structure and Magnetic Properties of Monodisperse Fe3+-doped CeO2 Nanospheres. *Nano-Micro Lett.* **2013**, *3*, 223–233.
72. Wang, W.; Zhu, Q.; Qin, F.; Dai, Q.; Wang, X. Fe doped CeO$_2$ nanosheets as Fenton-like heterogeneous catalysts for degradation of salicylic acid. *Chem. Eng. J.* **2018**, *333*, 226–239. [CrossRef]
73. Zheng, X.; Li, Y.; Zheng, Y.; Shen, L.; Xiao, Y.; Cao, Y.; Zhang, Y.; Au, C.; Jiang, L. Highly Efficient Porous Fe$_x$Ce$_{1-x}$O$_{2-\delta}$ with Three-Dimensional Hierarchical Nanoflower Morphology for H$_2$S-Selective Oxidation. *ACS Catal.* **2020**, *10*, 3968–3983. [CrossRef]
74. Liu, W.; Wang, W.; Tang, K.; Guo, J.; Ren, Y.; Wang, S.; Feng, L.; Yang, Y. The promoting influence of nickel species in the controllable synthesis and catalytic properties of nickel–ceria catalysts. *Catal. Sci. Tech.* **2016**, *6*, 2427–2434. [CrossRef]
75. Tsoncheva, T.; Rosmini, C.; Dimitrov, M.; Issa, G.; Henych, J.; Němečková, Z.; Kovacheva, D.; Velinov, N.; Atanasova, G.; Spassova, I. Formation of Catalytic Active Sites in Hydrothermally Obtained Binary Ceria–Iron Oxides: Composition and Preparation Effects. *ACS Appl. Mater. Inter.* **2021**, *13*, 1838–1852. [CrossRef] [PubMed]
76. Barbosa, C.C.S.; Peixoto, E.B.; Jesus, A.C.B.; Jesus, J.R.; Fabian, F.A.; Costa, I.M.; Almeida, J.M.A.; Duque, J.G.S.; Meneses, C.T. Effect of doping in Ce$_{1-x}$TM$_x$O$_2$ (TM = Mn, Cr, Co and Fe) nanoparticles obtained by hydrothermal method. *Mater. Chem. Phys.* **2019**, *225*, 187–191. [CrossRef]
77. Radović, M.; Dohčević-Mitrović, Z.; Golubović, A.; Fruth, V.; Preda, S.; Šćepanović, M.; Popović, Z.V. Influence of Fe^{3+}-doping on optical properties of CeO$_{2-y}$ nanopowders. *Ceram. Int.* **2013**, *39*, 4929–4936. [CrossRef]
78. Chen, X.; Zhan, S.; Chen, D.; He, C.; Tian, S.; Xiong, Y. Grey Fe-CeO$_{2-\sigma}$ for boosting photocatalytic ozonation of refractory pollutants: Roles of surface and bulk oxygen vacancies. *Appl. Catal. B Environ.* **2021**, *286*, 119928. [CrossRef]

79. Ma, Y.; Ma, Y.; Giuli, G.; Euchner, H.; Groß, A.; Lepore, G.O.; d'Acapito, F.; Geiger, D.; Biskupek, J.; Kaiser, U.; et al. Introducing Highly Redox-Active Atomic Centers into Insertion-Type Electrodes for Lithium-Ion Batteries. *Adv. Ener. Mater.* **2020**, *10*, 2000783. [CrossRef]
80. Zhang, G.; Li, Y.; Zhao, X.; Xu, J.; Zhang, Y. The variation of microstructures, spectral characteristics and catalysis effects of Fe^{3+} and Zn^{2+} co-doped CeO_2 solid solutions. *J. Rare Earths* **2020**, *38*, 241–249. [CrossRef]
81. Dosa, M.; Piumetti, M.; Bensaid, S.; Andana, T.; Novara, C.; Giorgis, F.; Fino, D.; Russo, N. Novel Mn–Cu-Containing CeO_2 Nanopolyhedra for the Oxidation of CO and Diesel Soot: Effect of Dopants on the Nanostructure and Catalytic Activity. *Catal. Lett.* **2018**, *148*, 298–311. [CrossRef]
82. Zhu, H.; Chen, Y.; Gao, Y.; Liu, W.; Wang, Z.; Cui, C.; Liu, W.; Wang, L. Catalytic oxidation of CO on mesoporous codoped ceria catalysts: Insights into correlation of physicochemical property and catalytic activity. *J. Rare Earths* **2019**, *37*, 961–969. [CrossRef]
83. Ji, Y.; Jin, Z.; Li, J.; Zhang, Y.; Liu, H.; Shi, L.; Zhong, Z.; Su, F. Rambutan-like hierarchically heterostructured CeO_2-CuO hollow microspheres: Facile hydrothermal synthesis and applications. *Nano Res.* **2017**, *10*, 381–396. [CrossRef]
84. Hiley, C.I.; Fisher, J.M.; Thompsett, D.; Kashtiban, R.J.; Sloan, J.; Walton, R.I. Incorporation of square-planar Pd^{2+} in fluorite CeO_2: Hydrothermal preparation, local structure, redox properties and stability. *J. Mater. Chem. A* **2015**, *3*, 13072–13079. [CrossRef]
85. Hiley, C.I.; Playford, H.Y.; Fisher, J.M.; Felix, N.C.; Thompsett, D.; Kashtiban, R.J.; Walton, R.I. Pair Distribution Function Analysis of Structural Disorder by Nb^{5+} Inclusion in Ceria: Evidence for Enhanced Oxygen Storage Capacity from Under-Coordinated Oxide. *J. Am. Chem. Soc.* **2018**, *140*, 1588–1591. [CrossRef]
86. Ma, X.; Lu, P.; Wu, P. Structural, optical and magnetic properties of CeO_2 nanowires with nonmagnetic Mg^{2+} doping. *J. Alloys Compd.* **2018**, *734*, 22–28. [CrossRef]
87. Siqueira Júnior, J.M.; Brum Malta, L.F.; Garrido, F.M.S.; Ogasawara, T.; Medeiros, M.E. Raman and Rietveld structural characterization of sintered alkaline earth doped ceria. *Mater. Chem. Phys.* **2012**, *135*, 957–964. [CrossRef]
88. Tighe, C.J.; Cabrera, R.Q.; Gruar, R.I.; Darr, J.A. Scale Up Production of Nanoparticles: Continuous Supercritical Water Synthesis of Ce-Zn Oxides. *Ind. Eng. Chem. Res.* **2013**, *52*, 5522–5528. [CrossRef]
89. Rajkumar, T.; Sápi, A.; Ábel, M.; Kiss, J.; Szenti, I.; Baán, K.; Gómez-Pérez, J.F.; Kukovecz, Á.; Kónya, Z. Surface Engineering of CeO_2 Catalysts: Differences Between Solid Solution Based and Interfacially Designed $Ce_{1-x}M_xO_2$ and MO/CeO_2 (M = Zn, Mn) in CO_2 Hydrogenation Reaction. *Catal. Lett.* **2021**. [CrossRef]
90. Das, A.; Patra, M.; Bhagavathiachari, M.; Nair, R.G. Defect-induced visible-light-driven photocatalytic and photoelectrochemical performance of ZnO–CeO_2 nanoheterojunctions. *J. Alloys Compd.* **2021**, *858*, 157730. [CrossRef]
91. Younis, A.; Chu, D.; Kaneti, Y.V.; Li, S. Tuning the surface oxygen concentration of {111} surrounded ceria nanocrystals for enhanced photocatalytic activities. *Nanoscale* **2016**, *8*, 378–387. [CrossRef]
92. Abbas, F.; Iqbal, J.; Maqbool, Q.; Jan, T.; Ullah, M.O.; Nawaz, B. ROS mediated malignancy cure performance of morphological, optical, and electrically tuned Sn doped CeO_2 nanostructures. *AIP Adv.* **2017**, *7*, 095205. [CrossRef]
93. Shajahan, S.; Arumugam, P.; Rajendran, R.; Munusamy, A.P. Optimization and detailed stability study on Pb doped ceria nanocubes for enhanced photodegradation of several anionic and cationic organic pollutants. *Arab. J. Chem.* **2020**, *13*, 1309–1322. [CrossRef]
94. Sardar, K.; Playford, H.Y.; Darton, R.J.; Barney, E.R.; Hannon, A.C.; Tompsett, D.; Fisher, J.; Kashtiban, R.J.; Sloan, J.; Ramos, S.; et al. Nanocrystalline Cerium−Bismuth Oxides: Synthesis, Structural Characterization, and Redox Properties. *Chem. Mater.* **2010**, *22*, 6191–6201.
95. Frerichs, H.; Pütz, E.; Reich, T.; Gazanis, A.; Panthöfer, M.; Hartmann, J.; Jegel, O.; Tremel, W. Nanocomposite antimicrobials prevent bacterial growth through the enzyme-like activity of Bi-doped cerium dioxide ($Ce_{1-x}Bi_xO_{2-\delta}$). *Nanoscale* **2020**, *12*, 21344–21358. [CrossRef]
96. Houlberg, K.; Bøjesen, E.D.; Mamakhel, A.; Wang, X.; Su, R.; Besenbacher, F.; Iversen, B.B. Hydrothermal Synthesis and in Situ Powder X-ray Diffraction Study of Bismuth-Substituted Ceria Nanoparticles. *Cryst. Growth Des.* **2015**, *15*, 3628–3636.
97. Shanavas, S.; Priyadharsan, A.; Dharmaboopathi, K. Ultrasonically and Photonically Simulatable Bi-Ceria Nanocubes for Enhanced Catalytic Degradation of Aqueous Dyes: A Detailed Study on Optimization, Mechanism and Stability. *ChemistrySelect* **2018**, *3*, 12841–12853. [CrossRef]
98. Veedu, S.N.; Jose, S.; Narendranath, S.B.; Prathapachandra Kurup, M.R.; Periyat, P. Visible light-driven photocatalytic degradation of methylene blue dye over bismuth-doped cerium oxide mesoporous nanoparticles. *Environ. Sci. Pollut. Res.* **2021**, *28*, 4147–4155. [CrossRef]
99. Hiley, C.I.; Fisher, J.M.; Kashtiban, R.J.; Cibin, G.; Thompsett, D.; Walton, R.I. Incorporation of Sb^{5+} into CeO_2: Local structural distortion of the fluorite structure from a pentavalent substituent. *Dalton Trans.* **2018**, *47*, 9693–9700.
100. Xu, Y.; Li, R. Wet-chemical synthesis and characterization of nitrogen-doped CeO_2 powders for oxygen storage capacity. *Appl. Surf. Sci.* **2018**, *455*, 997–1004. [CrossRef]
101. Liyanage, A.D.; Perera, S.D.; Tan, K.; Chabal, Y.; Balkus, K.J. Synthesis, Characterization, and Photocatalytic Activity of Y-Doped CeO_2 Nanorods. *ACS Catal.* **2014**, *4*, 577–584.
102. Mendiuk, O.; Kepinski, L. Synthesis of $Ce_{1-x}Er_xO_{2-y}$ nanoparticles by the hydrothermal method: Effect of microwave radiation on morphology and phase composition. *Ceram. Int.* **2014**, *40*, 14833–14843. [CrossRef]
103. Małecka, M.A. Characterization and thermal stability of Yb-doped ceria prepared by methods enabling control of the crystal morphology. *CrystEngComm* **2017**, *19*, 6199–6207. [CrossRef]

104. Fernandez-Garcia, S.; Jiang, L.; Tinoco, M.; Hungria, A.B.; Han, J.; Blanco, G.; Calvino, J.J.; Chen, X. Enhanced Hydroxyl Radical Scavenging Activity by Doping Lanthanum in Ceria Nanocubes. *J. Phys. Chem. C* **2016**, *120*, 1891–1901. [CrossRef]
105. Andana, T.; Piumetti, M.; Bensaid, S.; Russo, N.; Fino, D.; Pirone, R. Nanostructured ceria-praseodymia catalysts for diesel soot combustion. *Appl. Catal. B Environ.* **2016**, *197*, 125–137. [CrossRef]
106. Jiang, L.; Fernandez-Garcia, S.; Tinoco, M.; Yan, Z.; Xue, Q.; Blanco, G.; Calvino, J.J.; Hungria, A.B.; Chen, X. Improved Oxidase Mimetic Activity by Praseodymium Incorporation into Ceria Nanocubes. *ACS Appl. Mater. Inter.* **2017**, *9*, 18595–18608. [CrossRef] [PubMed]
107. Andana, T.; Piumetti, M.; Bensaid, S.; Veyre, L.; Thieuleux, C.; Russo, N.; Fino, D.; Quadrelli, E.A.; Pirone, R. Nanostructured equimolar ceria-praseodymia for NOx-assisted soot oxidation: Insight into Pr dominance over Pt nanoparticles and metal–support interaction. *Appl. Catal. B Environ.* **2018**, *226*, 147–161. [CrossRef]
108. Sartoretti, E.; Martini, F.; Piumetti, M.; Bensaid, S.; Russo, N.; Fino, D. Nanostructured Equimolar Ceria-Praseodymia for Total Oxidations in Low-O_2 Conditions. *Catalysts* **2020**, *10*, 165. [CrossRef]
109. Mendiuk, O.; Nawrocki, M.; Kepinski, L. The synthesis of $Ce_{1-x}Ln_xO_{2-y}$ (Ln = Pr, Sm, Gd, Tb) nanocubes by hydrothermal methods. *Ceram. Int.* **2016**, *42*, 1998–2012. [CrossRef]
110. Hong, S.; Lee, D.; Yang, H.; Kim, Y.-B. Direct hydrothermal growth of GDC nanorods for low temperature solid oxide fuel cells. *Appl. Surf. Sci.* **2018**, *444*, 430–435. [CrossRef]
111. Yoshida, Y.; Fujihara, S. Shape-Controlled Synthesis and Luminescent Properties of $CeO_2:Sm^{3+}$ Nanophosphors. *Eur. J. Inorg. Chem.* **2011**, *2011*, 1577–1583. [CrossRef]
112. Bhatta, U.M.; Reid, D.; Sakthivel, T.; Sayle, T.X.T.; Sayle, D.; Molinari, M.; Parker, S.C.; Ross, I.M.; Seal, S.; Möbus, G. Morphology and Surface Analysis of Pure and Doped Cuboidal Ceria Nanoparticles. *J. Phys. Chem. C* **2013**, *117*, 24561–24569. [CrossRef]
113. Ke, J.; Xiao, J.-W.; Zhu, W.; Liu, H.; Si, R.; Zhang, Y.-W.; Yan, C.-H. Dopant-Induced Modification of Active Site Structure and Surface Bonding Mode for High-Performance Nanocatalysts: CO Oxidation on Capping-free (110)-oriented CeO_2:Ln (Ln = La–Lu) Nanowires. *J. Am. Chem. Soc.* **2013**, *135*, 15191–15200. [CrossRef] [PubMed]
114. Xu, B.; Zhang, Q.; Yuan, S.; Zhang, M.; Ohno, T. Morphology control and photocatalytic characterization of yttrium-doped hedgehog-like CeO_2. *Appl. Catal. B Environ.* **2015**, *164*, 120–127. [CrossRef]
115. Małecka, M.A.; Kraszkiewicz, P.; Bezkrovnyi, O. Catalysis by shapely nanocrystals of the $Ce_{1-x}Yb_xO_{2-x/2}$ mixed oxides—Synthesis and phase stability. *Mater. Char.* **2019**, *155*, 109796. [CrossRef]
116. Roh, J.; Hwang, S.H.; Jang, J. Dual-Functional $CeO_2:Eu^{3+}$ Nanocrystals for Performance-Enhanced Dye-Sensitized Solar Cells. *ACS Appl. Mater. Interfaces* **2014**, *6*, 19825–19832. [CrossRef]
117. Bo, Q.; Wang, J. Structural and optical properties of erbium doped ceria nanoparticles synthesized by hydrothermal method. *Spectrosc. Lett.* **2021**, *54*, 165–170. [CrossRef]
118. Yang, Y.; Cong, Y.; Dong, D.P.; Xiao, Y.; Shang, J.Y.; Tong, Y.; Zhang, H.M.; He, M.; Zhang, J.H. Structural and excitation dependent emission properties of octahedral $CeO_2:Er^{3+}$ nanocrystal. *J. Lumin.* **2019**, *213*, 427–432. [CrossRef]
119. Wang, Z.L.; Feng, X. Polyhedral Shapes of CeO_2 Nanoparticles. *J. Phys. Chem. B* **2003**, *107*, 13563–13566. [CrossRef]
120. Yan, L.; Yu, R.; Chen, J.; Xing, X. Template-Free Hydrothermal Synthesis of CeO_2 Nano-octahedrons and Nanorods: Investigation of the Morphology Evolution. *Cryst. Growth Des.* **2008**, *8*, 1474–1477. [CrossRef]
121. Muñoz, F.F.; Acuña, L.M.; Albornoz, C.A.; Leyva, A.G.; Baker, R.T.; Fuentes, R.O. Redox properties of nanostructured lanthanide-doped ceria spheres prepared by microwave assisted hydrothermal homogeneous co-precipitation. *Nanoscale* **2015**, *7*, 271–281. [CrossRef]
122. Llusar, M.; Vitásková, L.; Šulcová, P.; Tena, M.A.; Badenes, J.A.; Monrós, G. Red ceramic pigments of terbium-doped ceria prepared through classical and non-conventional coprecipitation routes. *J. Eur. Ceram. Soc.* **2010**, *30*, 37–52. [CrossRef]
123. Acuña, L.M.; Muñoz, F.F.; Albornoz, C.A.; Leyva, A.G.; Baker, R.T.; Fuentes, R.O. Nanostructured terbium-doped ceria spheres: Effect of dopants on their physical and chemical properties under reducing and oxidizing conditions. *J. Mater. Chem. A* **2015**, *3*, 16120–16131. [CrossRef]
124. Xu, B.; Yang, H.; Zhang, Q.; Yuan, S.; Xie, A.; Zhang, M.; Ohno, T. Design and Synthesis of Sm, Y, La and Nd-doped CeO_2 with a broom-like hierarchical structure: A photocatalyst with enhanced oxidation performance. *ChemCatChem* **2020**, *12*, 2638–2646. [CrossRef]
125. Hirano, M.; Kato, E. Hydrothermal Synthesis of Nanocrystalline Cerium(IV) Oxide Powders. *J. Am. Ceram. Soc.* **1999**, *82*, 786–788. [CrossRef]
126. Basu, J.; Divakar, R.; Winterstein, J.P.; Carter, C.B. Low-temperature and ambient-pressure synthesis and shape evolution of nanocrystalline pure, La-doped and Gd-doped CeO_2. *Appl. Surf. Sci.* **2010**, *256*, 3772–3777. [CrossRef]
127. Polezhaeva, O.S.; Ivanov, V.K.; Dolgopolova, E.A.; Baranchikov, A.E.; Shcherbakov, A.B.; Tret'yakov, Y.D. Synthesis of Nanocrystalline Solid Solutions $Ce_{1-x}R_xO_2$-(R = Nd, Eu) by the Homogeneous Hydrolysis Method. *Dokl. Chem.* **2010**, *433*, 183–185. [CrossRef]
128. Dolgopolova, E.A.; Ivanova, O.S.; Ivanov, V.K.; Sharikov, F.Y.; Baranchikov, A.E.; Shcherbakov, A.B.; Trietyakov, Y.D. Microwave-hydrothermal synthesis of gadolinium-doped nanocrystalline ceria in the presence of hexamethylenetetramine. *Russ. J. Inorg. Chem.* **2012**, *57*, 1303–1307. [CrossRef]
129. Dunnick, K.M.; Pillai, R.; Pisane, K.L.; Stefaniak, A.B.; Sabolsky, E.M.; Leonard, S.S. The Effect of Cerium Oxide Nanoparticle Valence State on Reactive Oxygen Species and Toxicity. *Biol. Trace Elem. Res.* **2015**, *166*, 96–107. [CrossRef]

130. Dunnick, K.M.; Morris, A.M.; Badding, M.A.; Barger, M.; Stefaniak, A.B.; Sabolsky, E.M.; Leonard, S.S. Evaluation of the effect of valence state on cerium oxide nanoparticle toxicity following intratracheal instillation in rats. *Nanotoxicology* **2016**, *10*, 992–1000. [CrossRef]
131. Sato, K.; Arai, M.; Valmalette, J.-C.; Abe, H. Surface Capping-Assisted Hydrothermal Growth of Gadolinium-Doped CeO_2 Nanocrystals Dispersible in Aqueous Solutions. *Langmuir* **2014**, *30*, 12049–12056. [CrossRef] [PubMed]
132. Deng, T.; Zhang, C.; Xiao, Y.; Xie, A.; Pang, Y.U.E.; Yang, Y. One-step synthesis of samarium-doped ceria and its CO catalysis. *Bull. Mater. Sci.* **2015**, *38*, 1149–1154. [CrossRef]
133. Anantharaman, A.P.; Gadiyar, H.J.; Surendran, M.; Rao, A.S.; Dasari, H.P.; Dasari, H.; Babu, G.U.B. Effect of synthesis method on structural properties and soot oxidation activity of gadolinium-doped ceria. *Chem. Pap.* **2018**, *72*, 3179–3188. [CrossRef]
134. Yamamoto, K.; Hashishin, T.; Matsuda, M.; Qiu, N.; Tan, Z.; Ohara, S. High-performance Ni nanocomposite anode fabricated from Gd-doped ceria nanocubes for low-temperature solid-oxide fuel cells. *Nano Energy* **2014**, *6*, 103–108. [CrossRef]
135. Kempaiah, D.M.; Yin, S.; Sato, T. A facile and quick solvothermal synthesis of 3D microflower CeO_2 and $Gd:CeO_2$ under subcritical and supercritical conditions for catalytic applications. *CrystEngComm* **2011**, *13*, 741–746. [CrossRef]
136. Öksüzömer, M.A.F.; Dönmez, G.; Sariboğa, V.; Altınçekiç, T.G. Microstructure and ionic conductivity properties of gadolinia doped ceria ($Gd_xCe_{1-x}O_{2-x/2}$) electrolytes for intermediate temperature SOFCs prepared by the polyol method. *Ceram. Int.* **2013**, *39*, 7305–7315. [CrossRef]
137. Dönmez, G.; Sarıboğa, V.; Gürkaynak Altınçekiç, T.; Öksüzömer, M.A.F. Polyol Synthesis and Investigation of $Ce_{1-x}RE_xO_{2-x/2}$ (RE = Sm, Gd, Nd, La, $0 \leq x \leq 0.25$) Electrolytes for IT-SOFCs. *J. Am. Ceram. Soc.* **2015**, *98*, 501–509. [CrossRef]
138. Yang, Y.; Yin, Z.; Zhao, Z.; Yu, J.; Li, J.; Ren, Z.; Yu, G. Morphologies and magnetic properties of La-doped CeO_2 nanoparticles by the solvothermal method in a low magnetic field. *Mater. Chem. Phys.* **2020**, *240*, 122148. [CrossRef]
139. Thorat, A.V.; Ghoshal, T.; Carolan, P.; Holmes, J.D.; Morris, M.A. Defect Chemistry and Vacancy Concentration of Luminescent Europium Doped Ceria Nanoparticles by the Solvothermal Method. *J. Phys. Chem. C* **2014**, *118*, 10700–10710. [CrossRef]
140. Karaca, T.; Altınçekiç, T.G.; Faruk Öksüzömer, M. Synthesis of nanocrystalline samarium-doped CeO_2 (SDC) powders as a solid electrolyte by using a simple solvothermal route. *Ceram. Int.* **2010**, *36*, 1101–1107. [CrossRef]
141. Deepthi, N.H.; Darshan, G.P.; Basavaraj, R.B.; Prasad, B.D.; Nagabhushana, H. Large-scale controlled bio-inspired fabrication of 3D $CeO_2:Eu^{3+}$ hierarchical structures for evaluation of highly sensitive visualization of latent fingerprints. *Sens. Actuators B Chem.* **2018**, *255*, 3127–3147. [CrossRef]
142. Godinho, M.; Gonçalves, R.d.F.; Leite, E.R.; Raubach, C.W.; Carreño, N.L.V.; Probst, L.F.D.; Longo, E.; Fajardo, H.V. Gadolinium-doped cerium oxide nanorods: Novel active catalysts for ethanol reforming. *J. Mater. Sci.* **2010**, *45*, 593–598. [CrossRef]
143. Prado-Gonjal, J.; Schmidt, R.; Espíndola-Canuto, J.; Ramos-Alvarez, P.; Morán, E. Increased ionic conductivity in microwave hydrothermally synthesized rare-earth doped ceria $Ce_{1-x}RE_xO_{2-(x/2)}$. *J. Power Source* **2012**, *209*, 163–171. [CrossRef]
144. Deus, R.C.; Cortés, J.A.; Ramirez, M.A.; Ponce, M.A.; Andres, J.; Rocha, L.S.R.; Longo, E.; Simões, A.Z. Photoluminescence properties of cerium oxide nanoparticles as a function of lanthanum content. *Mater. Res. Bull.* **2015**, *70*, 416–423. [CrossRef]
145. Silva, A.G.M.; Rodrigues, T.S.; Dias, A.; Fajardo, H.V.; Gonçalves, R.F.; Godinho, M.; Robles-Dutenhefner, P.A. $Ce_{1-x}Sm_xO_{1.9-\delta}$ nanoparticles obtained by microwave-assisted hydrothermal processing: An efficient application for catalytic oxidation of α-bisabolol. *Catal. Sci. Tech.* **2014**, *4*, 814–821. [CrossRef]
146. Bezkrovnyi, O.; Małecka, M.A.; Lisiecki, R.; Ostroushko, V.; Thomas, A.G.; Gorantla, S.; Kepinski, L. The effect of Eu doping on the growth, structure and red-ox activity of ceria nanocubes. *CrystEngComm* **2018**, *20*, 1698–1704. [CrossRef]
147. Slostowski, C.; Marre, S.; Bassat, J.-M.; Aymonier, C. Synthesis of cerium oxide-based nanostructures in near- and supercritical fluids. *J. Supercrit. Fluids* **2013**, *84*, 89–97. [CrossRef]
148. Xu, Y.; Farandos, N.; Rosa, M.; Zielke, P.; Esposito, V.; Vang Hendriksen, P.; Jensen, S.H.; Li, T.; Kelsall, G.; Kiebach, R. Continuous hydrothermal flow synthesis of Gd-doped CeO_2 (GDC) nanoparticles for inkjet printing of SOFC electrolytes. *Int. J. Appl. Ceram. Technol.* **2018**, *15*, 315–327. [CrossRef]
149. Karl Chinnu, M.; Vijai Anand, K.; Mohan Kumar, R.; Alagesan, T.; Jayavel, R. Synthesis and enhanced electrochemical properties of $Sm:CeO_2$ nanostructure by hydrothermal route. *Mater. Lett.* **2013**, *113*, 170–173. [CrossRef]
150. Dell'Agli, G.; Spiridigliozzi, L.; Marocco, A.; Accardo, G.; Ferone, C.; Cioffi, R. Effect of the Mineralizer Solution in the Hydrothermal Synthesis of Gadolinium-Doped (10% mol Gd) Ceria Nanopowders. *J. Appl. Biomater. Funct. Mater.* **2016**, *14*, 189–196. [CrossRef]
151. Wang, Z.; Wang, Q.; Liao, Y.; Shen, G.; Gong, X.; Han, N.; Liu, H.; Chen, Y. Comparative Study of CeO_2 and Doped CeO_2 with Tailored Oxygen Vacancies for CO Oxidation. *ChemPhysChem* **2011**, *12*, 2763–2770. [CrossRef]
152. Kim, G.; Lee, N.; Kim, K.-B.; Kim, B.-K.; Chang, H.; Song, S.-J.; Park, J.-Y. Various synthesis methods of aliovalent-doped ceria and their electrical properties for intermediate temperature solid oxide electrolytes. *Int. J. Hydrogen Energy* **2013**, *38*, 1571–1587. [CrossRef]
153. Sohn, Y. Lanthanide (III) (La, Pr, Nd, Sm, Eu, Gd, Tb, Dy, Ho, Er, Tm, and Yb) Ions Loaded in CeO_2 Support; Fundamental Natures, Hydrogen Reduction, and CO Oxidation Activities. *Appl. Sci. Converg. Technol.* **2019**, *28*, 35–40. [CrossRef]
154. Gnanam, S.; Gajendiran, J.; Ramana Ramya, J.; Ramachandran, K.; Gokul Raj, S. Glycine-assisted hydrothermal synthesis of pure and europium doped CeO_2 nanoparticles and their structural, optical, photoluminescence, photocatalytic and antibacterial properties. *Chem. Phys. Lett.* **2021**, *763*, 138217. [CrossRef]

155. Palard, M.; Balencie, J.; Maguer, A.; Hochepied, J.-F. Effect of hydrothermal ripening on the photoluminescence properties of pure and doped cerium oxide nanoparticles. *Mater. Chem. Phys.* **2010**, *120*, 79–88. [CrossRef]
156. Cabral, A.C.; Cavalcante, L.S.; Deus, R.C.; Longo, E.; Simões, A.Z.; Moura, F. Photoluminescence properties of praseodymium doped cerium oxide nanocrystals. *Ceram. Int.* **2014**, *40*, 4445–4453. [CrossRef]
157. Wang, P.; Kobiro, K. Synthetic versatility of nanoparticles: A new, rapid, one-pot, single-step synthetic approach to spherical mesoporous (metal) oxide nanoparticles using supercritical alcohols. *Pure Appl. Chem.* **2014**, *86*, 785–800. [CrossRef]
158. Xu, B.; Zhang, Q.; Yuan, S.; Liu, S.; Zhang, M.; Ohno, T. Synthesis and photocatalytic performance of yttrium-doped CeO_2 with a hollow sphere structure. *Catal. Today* **2017**, *281*, 135–143. [CrossRef]
159. Maria Magdalane, C.; Kaviyarasu, K.; Raja, A.; Arularasu, M.V.; Mola, G.T.; Isaev, A.B.; Al-Dhabi, N.A.; Arasu, M.V.; Jeyaraj, B.; Kennedy, J.; et al. Photocatalytic decomposition effect of erbium doped cerium oxide nanostructures driven by visible light irradiation: Investigation of cytotoxicity, antibacterial growth inhibition using catalyst. *J. Photochem. Photobiol. B* **2018**, *185*, 275–282. [CrossRef]
160. De Oliveira, R.C.; Cabral, L.; Cabral, A.C.; Almeida, P.B.; Tibaldi, N.; Sambrano, J.R.; Simões, A.Z.; Macchi, C.E.; Moura, F.; Marques, G.E.; et al. Charge transfer in Pr-Doped cerium oxide: Experimental and theoretical investigations. *Mater. Chem. Phys.* **2020**, *249*, 122967. [CrossRef]
161. Ren, Y.; Ma, J.; Ai, D.; Zan, Q.; Lin, X.; Deng, C. Fabrication and performance of Pr-doped CeO_2 nanorods-impregnated Sr-doped $LaMnO_3$–Y_2O_3-stabilized ZrO_2 composite cathodes for intermediate temperature solid oxide fuel cells. *J. Mater. Chem.* **2012**, *22*, 25042–25049. [CrossRef]
162. Jiang, Y.; Zou, L.; Cheng, J.; Huang, Y.; Wang, Z.; Chi, B.; Pu, J.; Li, J. Gadolinium doped ceria on graphene cathode with enhanced cycle stability for non-aqueous lithium-oxygen batteries. *J. Power Source* **2018**, *400*, 1–8. [CrossRef]
163. Shi, S.-J.; Zhou, S.-S.; Liu, S.-Q.; Chen, Z.-G. Photocatalytic activity of erbium-doped CeO_2 enhanced by reduced graphene Oxide/CuO cocatalyst for the reduction of CO_2 to methanol. *Environ. Prog. Sustain. Energy* **2018**, *37*, 655–662. [CrossRef]
164. Wang, Z.; Gu, F.; Wang, Z.; Han, D. Solvothermal synthesis of CeO_2:Er/Yb nanorods and upconversion luminescence characterization. *Mater. Res. Bull.* **2014**, *53*, 141–144. [CrossRef]
165. Aouissi, M.L.; Ghrib, A. Effect of Synthesis Route on Structural, Morphological Properties of $Ce_{0.8}Y_{0.2-x}La_xO_2$ (x = 0, 0.1, 0.2). *Bull. Chem. Soc. Jpn.* **2015**, *88*, 1159–1163. [CrossRef]
166. Zhao, R.; Huan, L.; Gu, P.; Guo, R.; Chen, M.; Diao, G. Yb,Er-doped CeO_2 nanotubes as an assistant layer for photoconversion-enhanced dye-sensitized solar cells. *J. Power Source* **2016**, *331*, 527–534. [CrossRef]
167. Han, D.; Yang, Y.; Gu, F.; Wang, Z. Tuning the morphology and upconversion fluorescence of CeO_2: Er/Yb nano-octahedra. *J. Alloys Compd.* **2016**, *656*, 524–529. [CrossRef]
168. Shirbhate, S.C.; Singh, K.; Acharya, S.A.; Yadav, A.K. Review on local structural properties of ceria-based electrolytes for IT-SOFC. *Ionics* **2017**, *23*, 1049–1057. [CrossRef]
169. Piumetti, M.; Andana, T.; Bensaid, S.; Fino, D.; Russo, N.; Pirone, R. Ceria-based nanomaterials as catalysts for CO oxidation and soot combustion: Effect of Zr-Pr doping and structural properties on the catalytic activity. *AIChE J.* **2017**, *63*, 216–225. [CrossRef]
170. Zhou, Y.; Deng, J.; Lan, L.; Wang, J.; Yuan, S.; Gong, M.; Chen, Y. Remarkably promoted low-temperature reducibility and thermal stability of CeO_2–ZrO_2–La_2O_3–Nd_2O_3 by a urea-assisted low-temperature (90 °C) hydrothermal procedure. *J. Mater. Sci.* **2017**, *52*, 5894–5907. [CrossRef]
171. Wang, C.; Wang, Y.; Huang, W.; Zou, B.; Khan, Z.S.; Zhao, Y.; Yang, J.; Cao, X. Influence of CeO_2 addition on crystal growth behavior of CeO_2–Y_2O_3–ZrO_2 solid solution. *Ceram. Int.* **2012**, *38*, 2087–2094. [CrossRef]
172. Khajavi, P.; Xu, Y.; Frandsen, H.L.; Chevalier, J.; Gremillard, L.; Kiebach, R.; Hendriksen, P.V. Tetragonal phase stability maps of ceria-yttria co-doped zirconia: From powders to sintered ceramics. *Ceram. Int.* **2020**, *46*, 9396–9405. [CrossRef]
173. Zhang, Y.; Zhang, L.; Deng, J.; Dai, H.; He, H. Controlled Synthesis, Characterization, and Morphology-Dependent Reducibility of Ceria−Zirconia−Yttria Solid Solutions with Nanorod-like, Microspherical, Microbowknot-like, and Micro-octahedral Shapes. *Inorg. Chem.* **2009**, *48*, 2181–2192. [PubMed]
174. Zhou, Y.; Xiong, L.; Deng, J.; Wang, J.; Yuan, S.; Lan, L.; Chen, Y. Facile synthesis of high surface area nanostructured ceria-zirconia-yttria-lanthana solid solutions with the assistance of lauric acid and dodecylamine. *Mater. Res. Bull.* **2018**, *99*, 281–291. [CrossRef]
175. Ozawa, M.; Yoshimura, Y.; Kobayashi, K. Structure and photoluminescence properties of $Ce_{0.5}Zr_{0.5}O_2$:Eu^{3+} nanoparticles synthesized by hydrothermal method. *Jpn. J. Appl. Phys.* **2017**, *56*, 01ae07. [CrossRef]
176. Ozawa, M.; Matsumoto, M.; Hattori, M. Photoemission properties of Eu-doped $Zr_{1-x}Ce_xO_2$ (x=0-0.2) nanoparticles prepared by hydrothermal method. *Jpn. J. Appl. Phys.* **2018**, *57*, 01ae04. [CrossRef]
177. Chen, H.; Shi, J.; Hua, Z.; Zhao, J.; Gao, J.; Lei, L.; Yan, D. Synthesis and Characteristics of La Doped Ceria–Zirconia Composite with Uniform Nano-Crystallite Dispersion. *Sci. Adv. Mater.* **2010**, *2*, 43–50.
178. Dudek, M. Ceramic Electrolytes in the CeO_2-Gd_2O_3-SrO System—Preparation, Properties and Application for Solid Oxide Fuel Cells. *Int. J. Electrochem. Sci.* **2012**, *7*, 2874–2889.
179. Dikmen, S. Effect of co-doping with Sm^{3+}, Bi^{3+}, La^{3+}, and Nd^{3+} on the electrochemical properties of hydrothermally prepared gadolinium-doped ceria ceramics. *J. Alloys Compd.* **2010**, *491*, 106–112. [CrossRef]

180. Dikmen, S.; Aslanbay, H.; Dikmen, E.; Şahin, O. Hydrothermal preparation and electrochemical properties of Gd^{3+} and Bi^{3+}, Sm^{3+}, La^{3+}, and Nd^{3+} codoped ceria-based electrolytes for intermediate temperature-solid oxide fuel cell. *J. Power Source* **2010**, *195*, 2488–2495. [CrossRef]
181. Hiley, C.I.; Walton, R.I. Controlling the crystallisation of oxide materials by solvothermal chemistry: Tuning composition, substitution and morphology of functional solids. *CrystEngComm* **2016**, *18*, 7656–7670. [CrossRef]
182. Minervini, L.; Zacate, M.O.; Grimes, R.W. Defect cluster formation in M_2O_3-doped CeO_2. *Solid State Ion.* **1999**, *116*, 339–349. [CrossRef]
183. Wang, J.D.; Xiao, X.; Liu, Y.; Pan, K.M.; Pang, H.; Wei, S.Z. The application of CeO_2-based materials in electrocatalysis. *J. Mater. Chem. A* **2019**, *7*, 17675–17702. [CrossRef]
184. Tyrsted, C.; Jensen, K.M.Ø.; Bøjesen, E.D.; Lock, N.; Christensen, M.; Billinge, S.J.L.; Iversen, B.B. Understanding the Formation and Evolution of Ceria Nanoparticles Under Hydrothermal Conditions. *Angew. Chem. Int. Ed.* **2012**, *51*, 9030–9033. [CrossRef]
185. Promethean Particles Products: Metal Oxides. Available online: https://prometheanparticles.co.uk/metal-oxides/ (accessed on 29 March 2021).
186. Wang, H.; Chen, X.; Gao, S.; Wu, Z.; Liu, Y.; Weng, X. Deactivation mechanism of Ce/TiO_2 selective catalytic reduction catalysts by the loading of sodium and calcium salts. *Catal. Sci. Tech.* **2013**, *3*, 715–722. [CrossRef]
187. Li, J.; Peng, Y.; Chang, H.; Li, X.; Crittenden, J.C.; Hao, J. Chemical poison and regeneration of SCR catalysts for NOx removal from stationary sources. *Front. Environ. Sci. Eng.* **2016**, *10*, 413–427. [CrossRef]
188. Kennedy, E.M.; Cant, N.W. Oxidative dehydrogenation of ethane and the coupling of methane over sodium containing cerium oxides. *Appl. Catal. A Gen.* **1992**, *87*, 171–183. [CrossRef]
189. Wright, C.S.; Walton, R.I.; Thompsett, D.; Fisher, J.; Ashbrook, S.E. One-step hydrothermal synthesis of nanocrystalline ceria-zirconia mixed oxides: The beneficial effect of sodium inclusion on redox properties. *Adv. Mater.* **2007**, *19*, 4500–4504. [CrossRef]
190. Soria, J.; Martinez-Arias, A.; Coronado, J.M.; Conesa, J.C. Chloride-induced modifications of the properties of rhodia/ceria catalysts. *Top. Catal.* **2000**, *11*, 205–212. [CrossRef]
191. Kępiński, L.; Okal, J. Occurrence and Mechanism of Formation of CeOCl in Pd/CeO_2 Catalysts. *J. Catal.* **2000**, *192*, 48–53. [CrossRef]
192. Spiridigliozzi, L.; Dell'Agli, G.; Biesuz, M.; Sglavo, V.M.; Pansini, M. Effect of the Precipitating Agent on the Synthesis and Sintering Behavior of 20 mol% Sm-Doped Ceria. *Adv. Mater. Sci. Eng.* **2016**, *2016*, 6096123. [CrossRef]
193. Mauro, M.; Crosera, M.; Monai, M.; Montini, T.; Fornasiero, P.; Bovenzi, M.; Adami, G.; Turco, G.; Filon, F.L. Cerium Oxide Nanoparticles Absorption through Intact and Damaged Human Skin. *Molecules* **2019**, *24*, 3759. [CrossRef] [PubMed]
194. Xu, Y.; Mofarah, S.S.; Mehmood, R.; Cazorla, C.; Koshy, P.; Sorrell, C.C. Design strategies for ceria nanomaterials: Untangling key mechanistic concepts. *Mater. Horiz.* **2021**, *8*, 102–123. [CrossRef]

Review

Pyridinesilver Tetraoxometallate Complexes: Overview of the Synthesis, Structure, and Properties of Pyridine Complexed AgXO₄ (X = Cl, Mn, Re) Compounds

Fernanda Paiva Franguelli [1,2], Kende Attila Béres [2,3] and Laszló Kótai [2,*]

[1] Department of Inorganic and Analytical Chemistry, Budapest University of Technology and Economics, Muegyetem Rakpart 3, 1111 Budapest, Hungary; fernandapaivafranguelli@edu.bme.hu
[2] Institute of Materials and Environmental Chemistry, Research Centre for Natural Sciences, ELKH, 1117 Budapest, Hungary; beres.kende.attila@ttk.hu
[3] Institute of Chemistry, ELTE Eötvös Loránd University, Pázmány Péter s. 1/A, 1117 Budapest, Hungary
* Correspondence: kotai.laszlo@ttk.hu

Abstract: We reviewed the synthesis, structure, and properties of pyridine complexes of AgXO₄ (X = Cl, Mn, and Re) compounds with various compositions ([AgPy₂] XO₄, [AgPy₂XO₄]·0.5Py, [AgPy₄] XO₄, and 4 [AgPy₂XO₄] [AgPy₄] XO₄). We also clarified the controversial information about the existence and composition of pyridine complexes of silver permanganate, used widely as mild and selective oxidants in organic chemistry. We discussed in detail the available structural and spectroscopic (IR, Raman, and UV) data and thermal behavior, including the existence and consequence of quasi-intramolecular reactions between the reducing ligand and anions containing oxygen.

Keywords: silver complexes; permanganates; perchlorates; perrhenate; pyridine; structure; spectroscopy; thermal behavior

1. Introduction

Silver permanganate, perchlorate, and perrhenate (AgXO₄, X = Cl, Mn, and Re) are well soluble in pyridine, and various complexes with studied and unknown chemical nature were isolated even in the 19th and the beginning of the 20th century [1–3]. The composition and purity of the complexes strongly depend on the conditions of synthesis, and even small changes in reaction conditions lead to the formation of other (or a mixture of) compounds [4]. In principle, silver can formally form two, three, or four-coordinated pyridine complexes. However, single-crystal studies [5,6] showed that the complexes with 2.4 or 2.5 pyridine/silver atom ratios always contain 2- and 4-coordinated silver atoms.

Considering the wide use of silver permanganate complexes in organic chemistry, as mild and selective oxidants, and that literature sources contain controversial or incorrect information about these compounds, we discuss the chemistry of pyridine complexes of AgXO₄ (X = Cl, Mn, and Re) in detail. We also comprehensively summarized their synthesis, composition, structure, and redox properties, including the occurring quasi-intramolecular redox reactions between the reducing pyridine ligand and oxidizing XO₄⁻ anions and the existence of hydrogen bonds between the acidic C–H and polarized X–O bonds.

The compounds were prepared and their labels are given in Table 1.

Table 1. Labels and Ag: Py ratios in pyridine complexes of AgXO$_4$ (X = Mn (a), Cl (b), and Re (c)) complexes.

Compound	Label	Py:Ag Ratio
AgMnO$_4$	1a	-
AgClO$_4$	1b	-
AgReO$_4$	1c	-
AgPy$_2$MnO$_4$	2a	2
[AgPy$_2$]ClO$_4$	2b	2
[AgPy$_2$]ReO$_4$	2c	2
[AgPy$_2$MnO$_4$]·0.5Py	3a	2.5
[AgPy$_2$ClO$_4$]·0.5Py	3b	2.5
AgPy$_2$ReO$_4$·0.5Py	3c	2.5
4[AgPy$_2$MnO$_4$]·[AgPy$_4$]MnO$_4$	4a	2.4
4[AgPy$_2$ClO$_4$]·[AgPy$_4$]ClO$_4$	4b	2.4
4AgPy$_2$ReO$_4$·AgPy$_4$ReO$_4$	4c	2.4
[AgPy$_4$]MnO$_4$	5a	4
[AgPy$_4$]ClO$_4$	5b	4
[AgPy$_4$]ReO$_4$	5c	4

2. Synthesis and Composition of AgPy$_n$XO$_4$ (X = Mn, Cl, and Re) Complexes

2.1. Pyridine Complexes of Silver Permanganate

The first pyridine complexes of silver permanganate were described by Klobb in 1886 [1], when a small amount of pyridine (20% aq.) was dissolved in an equivalent solution of silver sulfate and potassium permanganate. Based on chemical analysis, two complexes were isolated with formulas AgPy$_2$MnO$_4$ and AgPy$_{2.5}$MnO$_4$ with decomposition points of 65 and 103 °C, respectively. Almost one hundred years later, Firouzabadi et al. (1982) [4] reported an easy preparation of a purple crystalline material in high yield (90%), which was believed to be AgPy$_2$MnO$_4$. They prepared this compound by adding AgNO$_3$ dissolved in ten times more water and three equivalent of pyridine to a KMnO$_4$ solution. The crystalline precipitate was recrystallized from acetone–benzene and its decomposition point was found to be 104–105 °C. Lee (2001) [7] reproduced the method described previously [4] to check the oxidation property of this material in several organic reactions.

Kotai et al. [8] reproduced the methods described by Klobb [1] and Firouzabadi [4] and found that AgPy$_2$MnO$_4$ (compound **2a**) and 4[AgPy$_2$MnO$_4$] [AgPy$_4$]MnO$_4$ (compound **4a**) (it is equal to the formula AgPy$_{2.4}$MnO$_4$) could be isolated as the main products in Klobb's methods, but a mixture containing **2a**, **3a** (**2a** 0.5Py), and **4a** formed in Firouzabadi's experiments. Recrystallization of the reaction product from acetone–benzene depending on the acetone–benzene solvent concentration and recrystallization conditions, led to pure **2a**, **3a**, or **4a**, or mixtures of these with different **2a:3a:4a** ratios.

To study the interconversion of compounds **2a–5a**, and further characterize these complexes, Sajó et al. (2018) [9] created new methods for the isolation of pure compounds (Figure 1). They isolated bis (pyridine)silver permanganate (**2a**) in a pure form by reacting [AgPy$_2$]NO$_3$ with a concentrated sodium permanganate solution and crystallizing it with a high temperature gradient. To prepare pure 4[AgPy$_2$MnO$_4$]·[AgPy$_4$]MnO$_4$ (**4a**), they added 10 volumes of water to a pyridine solution containing AgMnO$_4$ (Py:Ag = 50) [9]. First, silver (I) permanganate was obtained from a saturated solution of AgNO$_3$ and 40% aq. NaMnO$_4$ in an Ag:MnO$_4$ ratio of 1:1. The freshly prepared compound **1a** (wet) bookmark0 was dissolved in pure pyridine, resulting in a saturated purple solution (the surface of the dried (old form) AgMnO$_4$ crystals (decomposed by sunlight with the formation of catalytically active Ag-Mn oxides) can catalyze the oxidation of heterocycles containing nitrogen. Sometimes it resulted in the inflammation of the pyridine on dissolution of "old" dry AgMnO$_4$ in dry pyridine). Sajó et al. immediately diluted it with water to reach a pyridine content of 10%, allowing the formation of the pure double salt (**4a**). Using one volume of benzene as a salting-out solvent, however, they prepared compound [AgPy$_4$] MnO$_4$ (**5a**) in a pure state. Another way to prepare compound **5a** is to cool down a concentrated solution of AgMnO$_4$ (**1a**) in pyridine to −18 °C overnight. The authors [8] reported a recrystallization route to synthesize pure [AgPy$_2$MnO$_4$]·0.5Py (**3a**). They recrystallized the mixture of AgPy$_2$MnO$_4$ and 4[AgPy$_2$MnO$_4$][AgPy$_4$] MnO$_4$ in various ratios from a 1:1 (v/v) acetone–benzene mixture. This led to the formation of a large quantity of platelet-like purple monoclinic single crystals, and some amount of amorphous brown decomposition product.

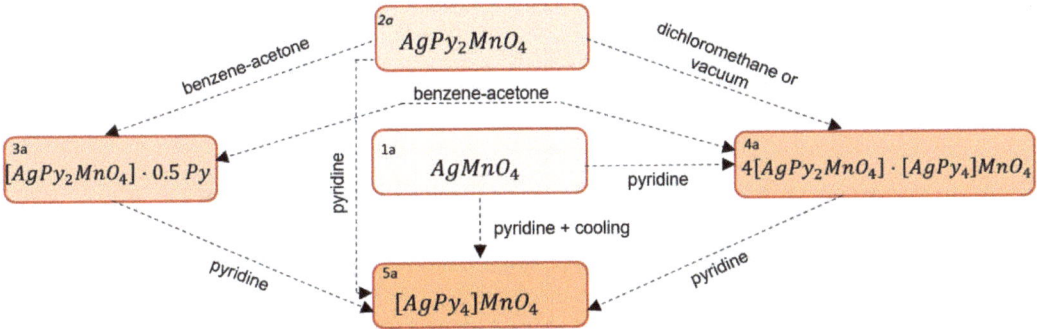

Figure 1. Interconversion scheme of permanganate complexes [9].

Compound **5a**, containing the highest amount of pyridine (Py/Ag = 4), decomposes into the double salt-type compound (**4a**) (Py/Ag = 2.4), which contains less pyridine, or into its mixture with **2a** (Py/Ag = 2) by controlled evaporation of the solvent. The removal of pyridine from compound **4a** in a vacuum, however, did not give the expected compound **2a**. Surprisingly, compound **2a** decomposed into compound **4a** in vacuum or by recrystallization from dichloromethane. It means that compound **2a** decomposes in a vacuum or on dissolution with liberation of free pyridine, which recombines with the silver and permanganate ions into **4a**.

2.2. Pyridine Complexes of Silver Perchlorate

Macy (1925) [3] constructed the silver perchlorate-pyridine-water ternary phase diagram at 25 °C. Four compounds, including three pyridine complexes, namely [AgPy$_2$]ClO$_4$ (**2b**) AgPy$_{2.25}$ClO$_4$ (later it was proved to be compound **4b** with the formula AgPy$_{2.4}$ClO$_4$), [AgPy$_4$]ClO$_4$ (**5b**), and AgClO$_4$·H$_2$O (**1b**.H$_2$O) were described in the system as shown in Figure 2. The author claimed that these materials could be isolated under different temperatures from a saturated solution of silver perchlorate in pyridine. The important features regarding the solubility curve are presented in Figure 2.

There are four invariant equilibriums (1–4) and an unsaturated solution (L). The amount of water in the system can directly affect the solubility of AgClO$_4$ in pyridine. Line d–e specifies the solubility of AgClO$_4$, whereas line b–c represents the composition of the solid phases. Macy found that [AgPy$_2$]ClO$_4$ (**2b**) was the most stable solid phase with a melting point between 144 and 147 °C, whereas AgPy$_{2.25}$ClO$_4$ and [AgPy$_4$]ClO$_4$ (**5b**) are unstable solid phases, which melt at 68 °C and 95.6 °C, respectively.

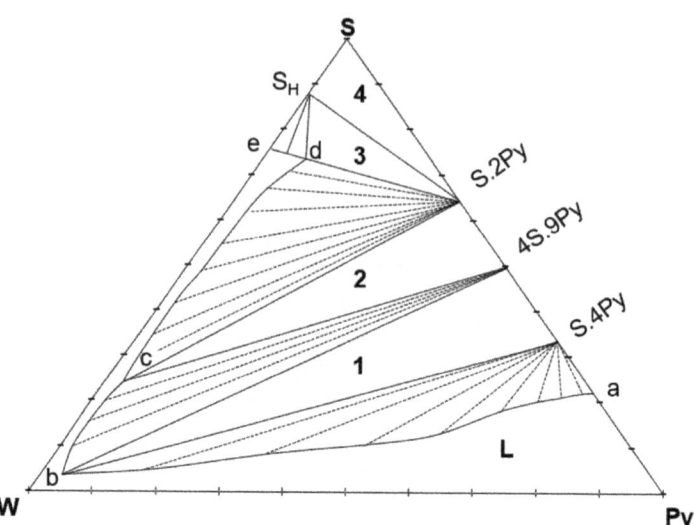

Figure 2. Ternary phase diagram of the AgClO$_4$–pyridine–H$_2$O system [3] (W—water, Py—pyridine, S—salt (AgClO$_4$)). Areas of invariant equilibria are marked with letters a–e.

The complex [AgPy$_2$]ClO$_4$ (**2b**) was prepared by Kauffman and Pinnell (1960) [10] in the reaction of AgNO$_3$ with NaClO$_4$ in an aqueous pyridine solution in an Ag$^+$:ClO$_4^-$:Py = 1:1.3:6 (molar) and pyridine/H$_2$O = 5/9 (v/v) ratio. Cooling the solution at 10 °C produced a white precipitate, which was recrystallized from a chloroform–pyridine mixture (5:1 (v/v)). Dyason et al. (1985) [11] determined the composition of this product and found it to be compound **4b** instead of **2b**. However, vacuum treatment of compound **4b** produced pure compound **2b**. Chen et al. (2007) [12] emphasized that different stoichiometry of silver (I) pyridine adducts could be obtained depending on preparation conditions and also reinforced the importance of a vacuum system to form compound **2b** during the drying process [10,11]. Hollo et al. (2019) [13] reported that a mixture containing compound **4** and compound **5** could be easily prepared from an aqueous solution of Ag$_2$SO$_4$, pyridine, and NaClO$_4$ (20% aq. pyridine and 0.2 M of aq. NaClO$_4$) at lower temperatures (5–8 °C). If the container is left open, [AgPy$_4$] ClO$_4$ (compound **5b**) completely transforms into [AgPy$_{2.4}$] ClO$_4$ (compound **4b**) at lower temperatures, while at room temperature or vacuum, compound **2b** was the final product.

Nilson and Oskarsson (1982) [5] reported the preparation of pure [AgPy$_4$] ClO$_4$ by dissolving silver (I) perchlorate in pyridine solution at −18 °C. It slowly loses pyridine at room temperature and transforms into **2b**. Dyason et al. and Bowmaker et al. also proved that compound **4b** is also converted to compound **2b** [11,12].

Sajó et al. (2018) [9] reproduced the method described by Kauffman and Pinnell [10]. They found that the raw product consists of a mixture of compounds **2a** and **4a**. A new compound was also identified as [AgPy$_2$ClO$_4$]·0.5Py (compound **3b**), which formed when a mixture of compounds **2b** and **4b** were triturated (1:1 mixture) with an acetone:benzene solution (1:1, v/v). They mentioned that the free pyridine to form **3b** from **2b** or **4b** came from the decomposition of compound **4b**. Compound **3b** is stable only in the presence of the mother liquor, but decomposes to [AgPy$_2$]ClO$_4$ when it is kept in air.

The interconversions of compound **1b**–**5b** can be seen in Figure 3.

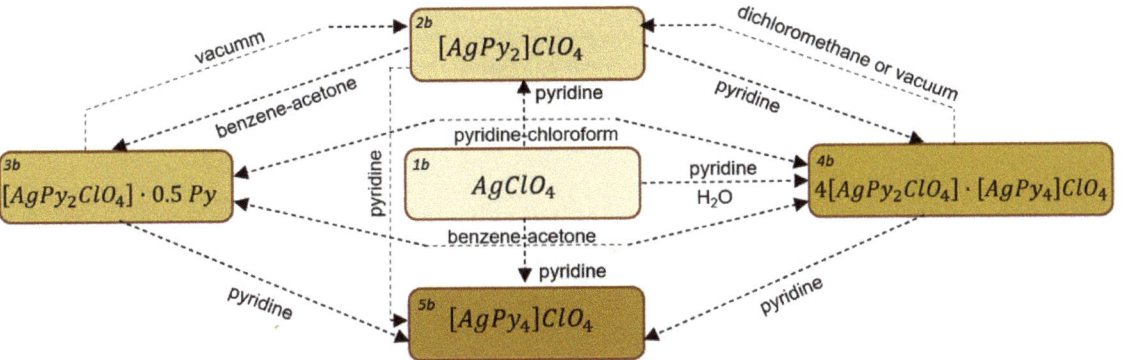

Figure 3. Interconversion scheme of perchlorate complexes [9].

2.3. Pyridine Complexes of Silver Perrhenate

This group of complexes has not been extensively studied so far. Wilke-Dörfurt and Gunzert (1933) [14] performed the first attempt to prepare it. [AgPy$_4$]ReO$_4$ (**5c**) formed when they combined silver nitrate, pyridine, and perrhenic acid in a ratio of 1:10:1 at the low temperature. However, since that time, no detailed information regarding its structure and properties has been available in the literature.

Recently, Sajó et al. (2018) [8] synthesized the previously unknown [AgPy$_2$]ReO$_4$ (**2c**) in a pure form, and studied its properties. This complex was formed from a mixture of Ag$_2$SO$_4$, 0.2 M aq. NaReO$_4$, and 10% aq. pyridine. A white precipitate was obtained after the reaction mixture was cooled to 0 °C. Another way to produce the same compound was to dissolve silver (I) perrhenate (**1c**) in pyridine, and completely remove the solvent by vacuum treatment at room temperature. Efforts to isolate complexes (**3c**) and (**4c**) have been unsuccessful until now. The existence of two AgPy$_n$ReO$_4$ compounds with 3 and 5 pyridines was mentioned without any characterization by Woolf [15].

3. Thermal Analysis of Pyridine Complexes of AgXO$_4$ Compounds (X = Mn, Cl, and Re)

3.1. Pyridine Complexes of AgMnO$_4$

Pyridine complexes of silver permanganate decompose in exothermic reactions on slow heating. Fast heating causes explosion-like decomposition with flames. Decomposition points of compounds **2a** and **4a** are 65 and 103 °C, respectively [1,2].

Kotai et al. performed thermogravimetric and mass spectrometry analysis on compound **3a** obtained by the recrystallization of a mixture of compound **2a** and **4a** from acetone–benzene [8], under an Ar atmosphere. A small amount of benzene was evolved at the beginning of thermal decomposition, with a maximum rate at 73 °C. They attributed the presence of benzene to the presence of a pyridine solvate substituent in the lattice. They also reported the thermal behaviors of compounds **2a** and [AgPy$_{2.25}$]MnO$_4$ (it was later proved to be a pyridine-deficient compound **4a** [6]) under an inert atmosphere. Both complexes started to decompose at 50 °C with a maximum mass loss achieved at 78. They observed unusual phenomena, such as the lack of oxygen evolution and the appearance of free pyridine (m/z = 79) and pyridine oxidation products, such as CO$_2$ (m/z = 44), CO (m/z = 28), H$_2$O (m/z = 18), and NO (m/z = 30). They attributed the presence of the pyridine ring oxidation products to the partial oxidation of the pyridine ligands by the permanganate ion (the experiments were done in an inert atmosphere). This quasi-intramolecular redox reaction took place in the solid phase between the permanganate anion and the coordinated pyridine, even below the temperature of the pyridine loss of the AgPy$_2^+$ cation (T_{dec} > 147 °C) of other compounds containing an [AgPy$_2$]-cation with thermally stable and non-reactive anions [10,14]. The mechanism proposed is shown in Figure 4. The decomposition center is the hydrogen bond between the α-CH of the pyridine ring and an oxygen of permanganate ion. Rearrangement via transition states TS$_1$ and TS$_2$ due to thermal activation results in a 2-hydroxypyridine derivative (TS$_2$). They assumed tautomerization and a Dewar benzene-type ring (TS$_4$) and the transformation of Dewar-C$_5$H$_5$NO to HNCO and "C$_4$H$_4$". The decomposition of these intermediates result in CO, CO$_2$, NO, and H$_2$O as reaction products [16,17].

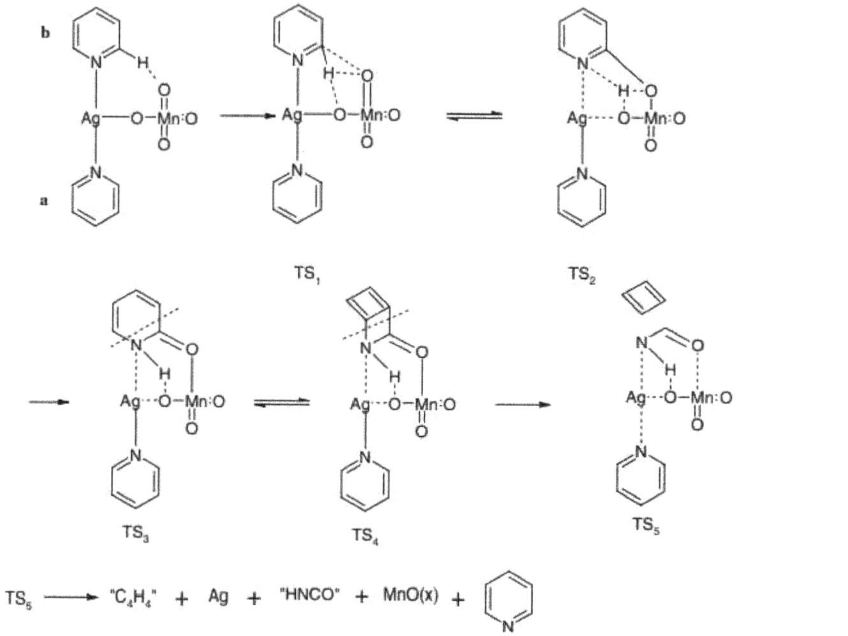

Figure 4. Proposed mechanism for the thermal decomposition of AgPy$_2$MnO$_4$ [16]. a -non-hydrogen bound pyridine ring: b-hydrogen bound pyridine ring.

The thermal analysis of **4a** was studied in detail by Kovács et al. [6] under an inert (He) and an oxidizing atmosphere. They observed a strongly exothermic reaction during the decomposition process even under an inert atmosphere. It indicates that the O$_2$ present in the gas phase does not play a key role in starting the decomposition reaction. Thermal decomposition occurred mainly in one step at 85 °C. Metallic silver and manganese oxides formed as solid decomposition products. An intermediate that formed at 300 °C contained metallic silver, MnO, Mn$_3$O$_4$, and Mn$_2$O$_3$. Based on the decomposition temperatures of MnO$_2$ → Mn$_2$O$_3$ → Mn$_3$O$_4$ → MnO [18] reactions, which are 542, 918, and 1027 °C, respectively, they concluded that these oxide phases could form only in a redox reaction and not in the thermal decomposition of the intermediate manganese oxide phases.

3.2. Pyridine Complexes of AgClO$_4$

The thermal properties of **4b** were investigated by Holló et al. (2019) [13]. The authors studied the decomposition of the compound under an argon and an air atmosphere. They observed that in the inert atmosphere, the compound started to decompose above 45 °C. The first decomposition intermediate was proved to be stable and was isolated at 85 °C as [AgPy$_2$]ClO$_4$ (**2b**). Compound **2b** was found to be stable up to 105 °C. Decomposition continued with the formation of pyridine and pyridine oxidation products (NO, H$_2$O, and CO$_2$). The authors also investigated in detail the decomposition products prepared at 350 °C and 500 °C. The first decomposition product consist of mainly AgClO$_4$ and AgCl. The final residue was AgCl. The m/z = 44 (CO$_2$) sign in the third decomposition peak around 320 °C was attributed to the oxidation of carbon residues that is formed from pyridine degradation by AgClO$_4$ [19]. The unreacted AgClO$_4$ decomposed into AgCl and O$_2$, which suggests the presence of a complex decomposition process with simultaneous pyridine ligand loss with AgClO$_4$ formation, and a redox reaction between pyridine and AgClO$_4$ in the solid phase.

The thermal decomposition of **4b** was also investigated under an oxidizing atmosphere [13]. The presence of O$_2$ in the gas phase accelerated the decomposition reaction and provided the required activation energy (~60 kJ mol^{-1}) [19] for that decomposition. Another essential feature reported is the low mass of the residue obtained at 400 °C, which was less than the total amount of Ag in the initial compound. This suggests that AgCl partly evaporated in the decomposition process [13]. The proposed reaction route is shown in Figure 5.

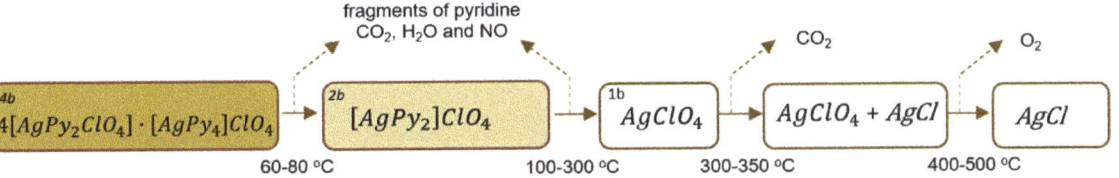

Figure 5. Proposed mechanism for the thermal decomposition of **4b** in argon [13].

3.3. Pyridine Complexes of AgReO$_4$

No available literature data about thermal decomposition of the pyridine complexes of silver perrhenate. Our preliminary studies on the thermal decomposition of **2c** showed the lack of the redox reactions between the pyridine ligand and perrhenate ion.

4. Crystallographic Structure of Pyridine Complexes of AgXO$_4$ Compounds (X = Mn, Cl, and Re)

The crystallographic parameters of pyridine complexes of AgXO$_4$ are summarized in Table 2.

Holló et al. [13] and Dyason et al. [11] crystallized compound **4b** in a tetragonal system containing one [AgPy$_2$ClO$_4$] and one-quarter of [AgPy$_4$]ClO$_4$ units in the asymmetric unit. The permanganate analog of this compound (compound **4a**) [6] is isostructural with compound **4b**. The Cα–H⋯O permanganate distance (3.121 Å) is longer (consequently bond strength is weaker) than the Cα–H⋯O perchlorate distance in the perchlorate (compound **4b**, 2.645 Å), as presented in Table 2.

Compound **3a** ([AgPy$_2$MnO$_4$]·0.5Py) crystallizes as platelets [9] belonging to the monoclinic crystal system. The shortest argentophilic interaction between the neighboring fragments in the unit cell (Ag⋯Ag = 3.421 Å) is stronger than what was found in [4AgPy$_2$MnO$_4$]·[AgPy$_4$]MnO$_4$ (Ag⋯Ag = 4.822 Å) [11,13].

The compound [AgPy$_2$]ClO$_4$ (**2b**) crystallizes in an orthorhombic crystal system [20]. The argentophilic interaction in this compound (Ag⋯Ag = 2.9997) provides a closer contact between [AgPy$_2$]$^+$ ions than in compounds **3a**, **4a**, or **4b**. The two coordinated pyridine ligands seem to be coplanar and the Ag-N bond distances were reported as 2.126 (4) and 2.133 (4) Å. The linear N-Ag-N bond angles are 173.83 (17°).

Table 2. Crystallographic parameters of Ag-Py-XO$_4$ (X = Mn, Cl, and Re) compounds.

Empirical Formula	Label	Space Group	Unit Cell Dimensions, (Å or °)	Z	D (g·cm^{-3})	T (K)	V (Å)^3	Reference
Agpy$_2$MnO$_4$	2a	Cc	a = 22.875 b = 12.266 c = 20.225 β = 62.361	16	1.970	298	5191.2	[9]
[Agpy$_2$]ClO$_4$	2b	Pnn2/Pnnm	a = 20.138 b = 12.694 c = 10.125	8	1.876	298	2588.3	[9]
[Agpy$_2$]ClO$_4$	2b	Pbcn	a = 19.958 (2) b = 10.0034 (13) c = 12.3082 (16)	8	1.976	150	2457.3 (5)	[20]
[Agpy$_2$]ReO$_4$	2c	-	a = 7.140 b = 8.616 c = 10.827 α = 102.20 β = 96.25 γ = 105.58	2	2.655	298	645.85	[9]
4[Agpy$_2$MnO$_4$]·[Agpy$_4$]MnO$_4$	4a	$I\bar{4}$	a = 22.01 c = 7.6075	10 (1)	1.877	298	3685.4	[9]
4[Agpy$_2$MnO$_4$]·[Agpy$_4$]MnO$_4$	4a	$I\bar{4}$	a = 21.982 (3) b = 21.982 (3) c = 7.5974 (15)	2	1.885	293	3671.13	[6]
4[Agpy$_2$ClO$_4$]·[Agpy$_4$]ClO$_4$	4b	$I\bar{4}$	a = 21.95 (1) c = 7.6843 (3)	-	1.78	295	3702 (2)	[11]
4[Agpy$_2$ClO$_4$]·[Agpy$_4$]ClO$_4$	4b	$I\bar{4}$	-	-	-	-	-	[13]
[Agpy$_2$MnO$_4$]·0.5Py	3a	C$_2$/c	a = 19.410 (1) b = 7.788 (1) c = 21.177 (1) α = 90.00 β = 104.20 (1) γ = 90.00	4	1.817	293	3103.4 (5)	[9]
[Ag(py)$_4$]MnO$_4$	5a	P$_2$1	a = 15.24 b = 13.89 c = 5.31 β = 84.13°	2	-	298	1117	[9]
[Ag(py)$_4$]ClO$_4$	5b	$I\bar{4}$, I$_4$/m	a = 12.874 (1) c = 6.748 (4)	2	1.55	260	1118.4	[5]

Nilsson and Oskarsson (1982) [4] described the crystal system of [AgPy$_4$]ClO$_4$ (compound **5b**) as tetragonal. The structure of this compound is shown in Figure 6, where pyridine molecules are in almost parallel planar dispositions. The structure and packing of compounds **3a**, 3b, and **4a**, **4b** can be seen in Figures 7 and 8, respectively.

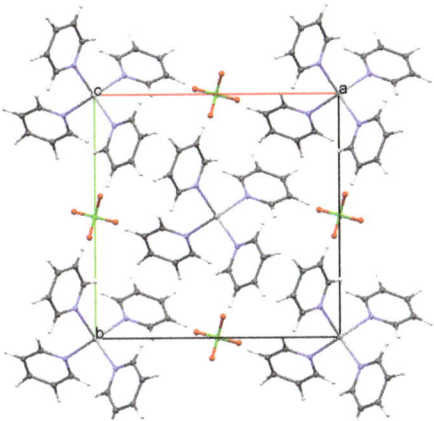

Figure 6. Structure and packing in compound **5b** [9].

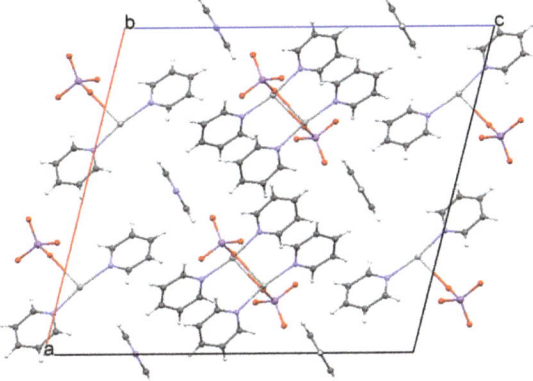

Figure 7. Structure and packing of compounds **3a** and **3b** [9].

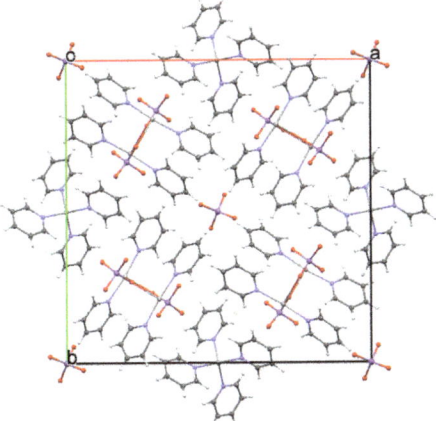

Figure 8. Structure and packing of compounds **4a** and **4b** [9].

The [Agpy$_2$]$^+$ units in compounds **2b, 3a, 3b, 4a,** and **4b** show weak Ag . . . O coordination with anions, and π–π stacking between the neighboring pyridine rings, whereas the [Agpy$_4$]$^+$ units in **3a, 3b,** and **4b** do not show these kinds of interactions.

The lattice parameters of the known AgXO$_4$ complexes with pyridine were studied by Sajó et al. [9]. The single crystal study results are summarized in Table 3.

Table 3. The main crystallographic parameters of [Ag(py)$_2$]MnO$_4$, [Ag(py)$_2$]ClO$_4$, [Ag(py)$_2$]ReO$_4$, and [Ag(py)$_{2.4}$]MnO$_4$.

Compounds/Pace Group.	Label	Ag-Ag (Å)	Ag-N (Å)	α-CH...O-X/F-X (Å)	Reference
4[Agpy$_2$]ClO$_4$·[Agpy$_4$]ClO$_4$, I-4	4b	4.843	2.740 2.15	2.645 2.645	[13]
4[Agpy$_2$]ClO$_4$·[Agpy$_4$]ClO$_4$, I-4	4b	4.843	2.16 2.30	2.753 2.781	[11]
4[Agpy$_2$]MnO$_4$·[Agpy$_4$]MnO$_4$, I-4	4a	4.822	2.601	3.121	[6]
[Agpy$_2$MnO$_4$]·0.5Py C$_2$/c	3a	3.421	2.166 2.174	2.602 2.770	[9]
[Agpy$_2$]ClO$_4$, Pbcn	2b	2.999	2.126 2.133	2.672 2.581 2.700 2.566	[20]
[Agpy$_4$]ClO$_4$, I-4	2b	6.748	2.322	2.712 3.237	[5]

5. Spectroscopic Properties of Pyridine Complexes of AgXO$_4$ Compounds (X = Mn, Cl, and Re)

5.1. Infrared and Raman Spectra

Sajó et al. assigned and evaluated, in detail, the IR and Raman band frequencies of the known pyridine complexes of AgXO$_4$ (X = Mn, Cl, Re) compounds [9]. Kovács et al. assigned Ag–N vibrations, arising from the di- and tetracoordinated [AgPy$_n$]$^+$ cations in compound **4a** (2019) [6]. The far-IR spectra revealed the symmetric and the antisymmetric ν (AgN) modes of the [AgPy$_2$]$^+$ ion at 246 and 166 cm^{-1}, respectively. Possibly, the deviation along with the ideal (linear) N–Ag–N angle of [Agpy$_2$]$^+$ ion was sufficient to activate the ν$_s$ (AgN) mode in the IR spectrum [9,12]. Similar spectral characteristics were found in the case of compound **4b** [12,13]. The far-IR and Raman spectrum of compound **4b** showed three Ag-N modes, which correspond to the asymmetric and symmetric Ag-N modes of [AgPy$_2$]+ and ν (AgN) of [AgPy$_4$]$^+$ [12,13]. The coordinated perchlorate ion (C$_1$ site) and non-coordinated (S$_4$ site) perchlorate ions were distinguished spectroscopically. The symmetric deformations modes of perchlorates ν$_2$ (E) appeared as a singlet band (459 cm^{-1}) in IR spectra and as a doublet in Raman analysis (460 and 417 cm^{-1}) in which the higher wave number was correlated to coordinate perchlorate ions (C$_1$ site) and the lower to perchlorates in the S$_4$ site [12]. The asymmetric Cl–O stretching bands ν$_3$ (F$_2$) resulted in two weak and two very intensive bands in the IR spectra [13].

5.2. UV-Vis Spectra

There are not many studies reporting on the spectral characteristics of pyridine complexes of AgXO$_4$ (X = Cl, Mn, Re) compounds [17]. Holló et al. [13] reported the diffuse reflection UV-Vis of compound **4b** and confirmed that [AgPy$_2$]$^+$ and [AgPy$_4$]$^+$ cations are present in the lattice of this compound. However, they assigned a band system containing pyridine n–π* and Ag$^+$-Py electronic charge transfer from metal to ligand (MLCT) [9,21,22]. The bands with band maxima at 218.9 and 295.2 nm are the CT band, and the maximum at 251.5 nm is a pyridine ring (1A$_{1g}$–1B$_{2u}$ (π–π*)) transition.

Kovács et al. [6] reported the diffuse reflection UV-Vis spectroscopic data for compound **4b** and found the presence of the same band system as mentioned above (pyridine n–π* and Ag$^+$-Py MLCT), and also the permanganate transitions t$_1$-4t$_2$, 3t$_2$-2e. The authors reported three different groups of band maxima, namely (i) 219.9 nm assigned to combined bands of Ag-Py (CT) and MnO$_4^-$ ((^{1}A$_1$-^{1}T$_2$(t$_1$-4t$_2$)); (ii) 258.4 nm regarding the components of pyridine ((^{1}A$_1$-^{1}B$_2$(n–π*)) and MnO$_4^-$ (^{1}A$_1$-^{1}T$_2$(3t$_2$-2e)) transitions; (iii) and 521.9 and 710.1 nm assigned to the components of MnO$_4^-$, (^{1}A$_1$-^{1}T$_2$(t$_1$-2e)) and ((^{1}A$_1$-^{1}T$_1$(t$_1$-2e)), respectively.

6. Organic Oxidation Reactions with Pyridine-Silver Permanganate Complexes

A large number of studies report the use of AgPy$_2$MnO$_4$ as an oxidative medium in organic reactions. However, in almost all publications, Firouzabadi's procedure [4] was used to synthesize compound **2a**. Thus, the reaction products were probably mixtures of compounds **2a**, **3a**, and **4a**. The first approach was made by Firouzabadi et al. [4,23], and later, Lee [7] used compound **2** in various oxidation reactions. The conversion of polycyclic aromatic hydrocarbons (e.g., anthracene, phenanthrene) [4,23] in dichloromethane resulted in diketones with high yield. The oxidation of diphenylacetylene resulted in 95% benzyl, and in the case of phenylacetylene (nonterminal alkyne), coupling reactions took place with the formation of 1,4-diphenyl-1,3-butadiyne. The conversion of benzocyclobutadiene resulted in 2,3-dibenzoylnaphthalene accompanied by carbon–carbon bond cleavage.

Compound **2a** (or its mixtures with **3a** and **4a**) converted primary and secondary alcohols into aldehydes and ketones, respectively, e.g., piperonol into piperonal (81%) and benzyl alcohol into benzaldehyde (100%). Coupling reactions were performed with 4-chloroaniline and 4-nitroaniline [4,23]. Besedin et al. (2000) [24] studied nucleophilic substitution of 6,8-dimethylpyrimido[4,5-c]pyridazine-5,7(6H,8H)-dione by reacting it with α,ω-diamines, as a result of which, polycyclic heterocycles formed. Gulevskaya et al. [25] reported the oxidative substitution of the pyridazine ring at C-H sites with a secondary amine in the presence of compound **2a** (or its mixture with **3a** and **4a**). The complete reaction was performed at 20 °C for 168 h and the final product was 6,8-dimethyl-3-ethylpyrrolo[1,2,2,3]pyridazino[6,5-d]pyrimidine7,9(6H,8H)-dione with a yield of 42%.

Compound **2a** (or its mixture with **3a** and **4a**) was used by Kesenheimer and Growth (2006) [26] in the synthesis of a natural antibiotic (-)-8-O-methyltetrangomycin, in a multi-step process. The oxidation of tetrahydrobenz[a]anthracene into tetrahydrobenz[a]anthra quinone in this reaction had a yield of 65%.

Banerji et al. (2012) [27] reported the oxidation of sulfides into sulfoxides in aqueous acetic acid solutions and achieved a yield of 80% over the conversion of dimethyl sulfide into dimethyl sulfoxide.

Kovács et al. [6] studied the oxidation abilities of 2a, 3a, and 4a in the conversion of benzyl alcohol to benzaldehyde and benzoic acid in various solvents. A higher yield of benzaldehyde was achieved in chloroform as a solvent medium.

7. Conclusions

We comprehensively reviewed the synthesis, structure, and properties of four series of pyridine complexes of AgXO$_4$ with 2, 2.4, 2.5, and 4 pyridines (X = Cl, Mn) and 2 or 4 pyridines (X = Re) per AgXO$_4$ unit. The controversial pieces of information about the existence and composition of pyridine complexes of silver permanganate, used widely as mild and selective oxidants in organic chemistry, was evaluated and clarified in detail. We discussed the available structural and spectroscopic (IR, Raman, and UV) data and thermal behavior, including the existence and consequence of quasi-intramolecular reactions between the reducing ligand and anions containing oxygen. The present review clearly shows that two members of the pyridine complexes of AgXO$_4$ (X = Cl, Mn, and Re, Py/Ag = 2, 2.4, 2.5, and 4) compounds have not been prepared and characterized yet (X = Re and Py/Ag = 2.4 and 2.5). For the members of AgXO$_4$-pyridine complexes with unknown crystal structures and thermal properties, structure elucidation and thermal studies are planned.

Clarification of the phase relations and the chemical nature of AgMnO4 pyridine complexes provide new and prosperous perspectives for studying AgMnO4-pyridine complexes in organic reactions, in which only a mixture of these compounds was used earlier.

Author Contributions: Conceptualization, L.K. and F.P.F.; resources, K.A.B.; writing—original draft preparation, F.P.F.; writing—review and editing, K.A.B. and L.K.; visualization, F.P.F. and K.A.B.; supervision, L.K.; project administration, F.P.F. and L.K.; funding acquisition, L.K. and K.A.B. All authors have read and agreed to the published version of the manuscript.

Funding: This research was funded by the European Union and the State of Hungary, co-financed by the European Regional Development Fund, grant number VEKOP-2.3.2-16-2017-00013 and New National Excellence program of the Ministry for Innovation and Technology from the source of the National Research, Development, and Innovation Fund, grant number ÚNKP-21-3.

Institutional Review Board Statement: Not applicable.

Informed Consent Statement: Not applicable.

Data Availability Statement: Not applicable.

Conflicts of Interest: The authors declare no conflict of interest.

References

1. Klobb, T. Combinaisons de la pyridine avec les permanganates. *C. R. Chim.* **1886**, *118*, 1271–1273.
2. Klobb, T. Compounds of pyridine and permanganates. *Bull. Soc. Chim. Paris* **1894**, *11*, 604–609.
3. Macy, R. The ternary system: Silver perchlorate, pyridine and water. *J. Am. Chem. Soc.* **1925**, *47*, 1031–1036. [CrossRef]
4. Firouzabadi, H.; Vessal, B.; Naderi, M. Bispyridinesilver permanganate [Ag(C_5H_5N)$_2$]MnO$_4$: An efficient oxidizing reagent for organic substrates. *Tetrahedron Lett.* **1982**, *23*, 1847–1850. [CrossRef]
5. Nilson, K.; Oskarsson, A. The crystal structure of tetrapyridine copper (I) perchlorate and tetrapyridine silver (I) perchlorate at 260 K. *Acta Chem. Scand. Ser. A Phys. Inorg. Chem.* **1982**, *37*, 605–610. [CrossRef]
6. Kovács, G.B.; May, N.V.; Bombicz, P.A.; Klébert, S.; Németh, P.; Menyhárd, A.; Novodárszki, G.; Petrusevski, V.; Franguelli, F.P.; Magyari, J.; et al. An unknown component of a selective and mild oxidant: Structure and oxidative ability of a double salt-type complex having κ^1O-coordinated permanganate anions and three- and four-fold coordinated silver cations. *RSC Adv.* **2019**, *9*, 28387–28398. [CrossRef]
7. Lee, D.G. Bis(pyridine) silver(I) Permanganate. In *Encyclopedia of Reagents for Organic Synthesis*; John Wiley and Sons Inc.: New York, NY, USA, 2001; Print ISBN: 9780471936237 | Online ISBN: 9780470842898. [CrossRef]
8. Kótai, L.; Sajó, I.; Fodor, J.; Szabó, P.; Jakab, E.; Argay, G.; Holly, S.; Gács, I.; Banerji, K.K. Reasons for and Consequences of the Mysterious Behaviour of Newly Prepared Hemipyridine Solvate of Bis(pyridine)silver(I) Permanganate, Agpy$_2$MnO$_4$*0.5py. *Trans. Met. Chem.* **2005**, *30*, 939–943. [CrossRef]
9. Sajó, I.E.; Kovács, G.B.; Pasinszki, T.; Bombicz, P.A.; May, Z.; Szilágyi, I.M.; Jánosity, A.; Banerji, K.K.; Kant, R.; Kótai, L. The chemical identity of "[Ag(py)$_2$]MnO$_4$" organic solvent soluble oxidizing agent and new synthetic routes for the preparation of [Ag(py)$_n$]XO$_4$ (X = Mn, Cl, and Re, n = 2–4) complexes. *J. Coord. Chem.* **2018**, *71*, 2884–2904. [CrossRef]
10. Kauffman, G.B.; Pinnell, R.P.; Stone, R.D. Dipyridinesilver(I) Perchlorate. In *Inorganic Synthesis*, 6th ed.; McGraw-Hill: New York, NY, USA, 1925.
11. Dyason, J.; Healy, P.; Engelhardt, L.; White, A. Lewis-Base Adducts of Group 1B Metal(I) Compounds. XXII. Crystal Structure of 'Bis(pyridine)silver(I) Perchlorate. *Aust. J. Chem.* **1985**, *38*, 1325–1328. [CrossRef]
12. Bowmaker, G.A.; Effendy, M.S.; Skelton, B.W.; White, A.H. Syntheses, structures and vibrational spectroscopy of some 1:1 and 1:2 adducts of silver(I) oxyanion salts with 2,2'-bis(pyridine) chelates. *Inorg. Chim. Acta* **2005**, *358*, 4371–4388. [CrossRef]
13. Holló, B.B.; Petrusevski, V.M.; Kovács, G.B.; Franguelli, F.P.; Farkas, A.; Menyhárd, A.; Lendvay, G.; Sajó, I.E.; Nagy-Bereczki, L.; Pawar, R.P.; et al. Thermal and spectroscopic studies on a double-salt-type pyridine–silver perchlorate complex having κ^1-O coordinated perchlorate ions. *J. Therm. Anal. Calorim.* **2019**, *138*, 1193–1205. [CrossRef]
14. Wilke-Dörfurt, E.; Gunzert, T. Über Neue Salze der Perrheniumsäure. *Z. Anorg. Allg. Chem.* **1933**, *215*, 369–387. [CrossRef]
15. Woolf, A.A. A comparison of silver perrhenate with silver perchlorate. *J. Less Common Met.* **1978**, *61*, 151–160. [CrossRef]
16. Kótai, L.; Fodor, J.; Jakab, E.; Sajó, I.E.; Sazbó, P.; Lónyi, F.; Valyon, J.; Gács, I.; Argay, G.; Banerji, K.K. A Thermally Induced Low-temperature Intramolecular Redox Reaction of bis(pyridine)silver(I) Permanganate and its Hemipyridine Solvate. *Trans. Met. Chem.* **2006**, *31*, 30–34. [CrossRef]
17. Buck, R.P.; Singhadeja, S.; Rogers, L.B. Ultraviolet Absorption Spectra of Some Inorganic Ions in Aqueous Solutions. *Anal. Chem.* **1954**, *26*, 1240–1242. [CrossRef]
18. Larbi, T.; Doll, K.; Manoubi, T. Density functional theory study of ferromagnetically and ferrimagnetically ordered spinel oxide Mn$_3$O$_4$. A quantum mechanical simulation of their IR and Raman spectra. *J. Alloys Compd.* **2016**, *688*, 692–698. [CrossRef]

19. Solymosi, F. The Thermal Stability and Some Physical Properties of Silver Chlorite, Chlorate and Perchlorate. *Z. Phys. Chem.* **1968**, *57*, 1–18. [CrossRef]
20. Chen, T.Y.; Zeng, J.Y.; Lee, H.M. Argentophilic interaction and anionic control of supramolecular structures in simple silver pyridine complexes. *Inorg. Chem. Acta* **2007**, *360*, 21–30. [CrossRef]
21. Bando, Y.; Nagakura, S. The electronic structure and spectrum of the silver(I)perchlorate-pyridine complex. *Theor. Chem. Acta* **1968**, *9*, 210–221. [CrossRef]
22. Boopalachandran, P.; Laane, J. Ultraviolet absorption spectra of pyridine-d0 and -d5 and their ring-bending potential energy function in the S1 (n,π*) state. *Chem. Phys. Lett.* **2008**, *462*, 178–182. [CrossRef]
23. Firouzabadi, H.; Sardarian, A.R. Facile oxidation od polycyclic arenes and acetylenic hydrocarbons with Bis(pyridine)silver permanganate and Bis(2,2'-bipyridil)copper(II)permanganate under mild and neutral conditions. *Synth* **1986**, 946–948. [CrossRef]
24. Besedin, D.V.; Gulevskaya, A.V.; Pozharskii, A.F. Reaction of 6,8-dimethylpyrimido[4,5- c]pyridazine-5,7(6 H,8 H)-dione with α,ω-diamines as the first example of tandem nucleophilic substitution in neutral azines. *Mendeleev Commun.* **2000**, *10*, 150–151. [CrossRef]
25. Gulevskaya, A.V.; Besedin, D.V.; Pozharskii, A.F.; Starikova, Z.A. 6,8-Dimethylpyrimido[4, 5-c]pyridazine-5,7(6H,8H)-dione: A novel method of pyrrole-ring annulation to an azine nucleus based on a tandem S_N^H–S_N^H process. *Tetrahedron Lett.* **2001**, *42*, 5981–5983. [CrossRef]
26. Kesenheimer, C.; Groth, U. Total synthesis of (-)-8-O-methyltetrangomycin (MM47755). *Org. Lett.* **2006**, *8*, 2507–2510. [CrossRef]
27. Banerji, J.; Kótai, L.; Sharma, P.K.; Banerji, K.K. Kinetics and mechanisms of the oxidation of substituted benzaldehydes with bis(pyridine) silver permanganate. *Eur. Chem. Bull.* **2012**, *1*, 135–140.

Review

Lanthanide-Based Single-Molecule Magnets Derived from Schiff Base Ligands of Salicylaldehyde Derivatives

Mamo Gebrezgiabher [1,2], Yosef Bayeh [1,2], Tesfay Gebretsadik [1,2], Gebrehiwot Gebreslassie [1,2], Fikre Elemo [1,2], Madhu Thomas [1,2,*] and Wolfgang Linert [3,*]

1. Department of Industrial Chemistry, College of Applied Sciences, Addis Ababa Science and Technology University, Addis Ababa P.O. Box 16417, Ethiopia; mamo.gebrezgiabher@aastu.edu.et (M.G.); yosef.bayeh@aastu.edu.et (Y.B.); tesfay.gebretsadik@aastu.edu.et (T.G.); gebrehiwot.gebreslassie@aastu.edu.et (G.G.); fikre.elemo@aastu.edu.et (F.E.)
2. Center of Excellence in Nanotechnology, Addis Ababa Science and Technology University, Addis Ababa P.O. Box 16417, Ethiopia
3. Institute of Applied Synthetic Chemistry, Vienna University of Technology, Getreidemarkt 9/163-AC, 1060 Vienna, Austria
* Correspondence: madhu.thomas@aastu.edu.et (M.T.); wolfgang.linert@tuwien.ac.at (W.L.); Tel.: +251-96689-7628 (M.T.); +43-1-58801-163613 (W.L.)

Received: 20 October 2020; Accepted: 24 November 2020; Published: 4 December 2020

Abstract: The breakthrough in Ln(III)-based SMMs with Schiff base ligands have been occurred for the last decade on account of their magnetic behavior, anisotropy and relaxation pathways. Herein, we review the synthetic strategy, from a structural point of view and magnetic properties of mono, di, tri and polynuclear Ln(III)-based single-molecule magnets mainly with Schiff bases of Salicylaldehyde origin. Special attention has been given to some important breakthroughs that are changing the perspective of this field with a special emphasis on slow magnetic relaxation. An overview of 50 Ln(III)-Schiff base complexes with SMM behavior, covering the period 2008–2020, which have been critical in understanding the magnetic interactions between the Ln(III)-centers, are presented and discussed in detail.

Keywords: lanthanides; Schiff base; synthetic strategies; single-molecule magnets

1. Introduction

1.1. General Introduction to Single-Molecule Magnets (SMMs)

SMMs are subunits of metal-organic compounds that show superparamagnetic behavior below a certain blocking temperature (T_B), purely of molecular origin [1]. Since the discovery of the manganese coordination cluster, $[Mn_{12}O_{12}(O_2CMe)_{16}(H_2O)_4]$ [2], behaving as a single-domain magnet, numerous discoveries have been devoted to the "hot" area of molecular magnetism [3–5]. Because of the large magnetic moment and single-ion anisotropy, lanthanides(III) (Ln(III)) have entered into this area. Among them, some of the mononuclear complexes of Ln(III) have drawn maximum attention compared to the polynuclear ones, owing to their small size and bistable nature and so could be ideal candidates for high-density storage and quantum computing [6–9].

After the report of the first SMM, $[Mn_{12}O_{12}(O_2CMe)_{16}(H_2O)_4]$, a tremendous amount of SMMs were investigated with 3d transition metals on account of their strong coupling while less interest was with Ln(III) systems due to their week exchange interaction when they are in their most stable trivalent oxidation state. Furthermore, much work has been done on a single-ion magnet (SIM)/SMMs with 4d and 5d transition metal, owing to their magnetic anisotropy, which received much attention in the area of high T_B molecular magnets. However, due to the synthetic difficulties, the SMMs/SIM based on 4d and 5d are still very limited compared to 3d analog [10,11].

However, in recent years SMMs with f-block elements are commonly based on the interaction between the electron density of the 4f-ions and the crystal field environment in which they are placed. For example, dysprosium (III)(Dy(III)) has a spin-orbit coupled ground state, $^6H_{15/2}$, that is more informative in understanding how 4f-ions are useful in constructing SMM than simply representing the electronic structure by the number of valence electron ($4f^9$) (Figure 1) [12,13].

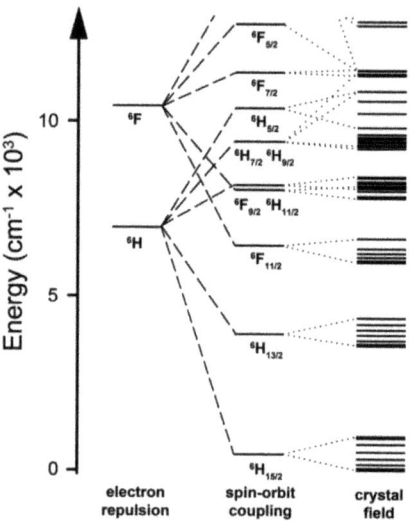

Figure 1. Low energy electronic structure of Dy(III) ions, with the crystal field-splitting modeled on the Dy[Me$_3$Si)$_2$N]$_3$ complexes. Reproduced with permission from [13]; Published by Royal Society of Chemistry, 2011.

1.2. The Anisotropy of Lanthanide Ions-Oblate/Prolate Model

In Ln(III) systems, the magnetic anisotropy typically arises from the extensive splitting of M_J ground state caused by the ligand field [14] and is quantified by a g-factor, which characterizes the shape of the ions and the amplitude of anisotropy. In most of the SMM's, there is an anisotropic axis in a "hard plane" (Figure 2d). When the magnetic axis of the metal ion is in line with the anisotropic axis, and the value of gz will be maximized [15]. This anisotropy is also can be called an Ising type anisotropy, which is an ideal condition for an SMM. The higher the "pure" excited state, the bigger the effective energy barrier will be obtained [16]. In some cases, the easy axis can be a plane, where it is possible to find magnetization, which can be called "easy plane", where $g_z < g_x \approx g_y$.

In other words, there is a way to maximize the single-ion anisotropy by modifying the coordination environment of the Ln(III)-ion. Rinehart and Long suggested simple rules in order to optimize 4f-SMMs, simply by exploiting their single-ion anisotropy [13]. For increasing the anisotropy of the oblate ions which are equatorially expanded (Figure 2a), we should place it in a ligand field for which the ligand electron density is located above and below the xy plane ("sandwich" type ligand geometry), which means donor atoms should be on the axial position (Figure 2b). It is not a coincidence that most of the geometries of mononuclear Tb(III) and/or Dy(III) SMMs with the highest energy barriers are square-antiprismatic (SARP-8) [15]. However, for prolate ions, which are axially expanded (Figure 2a), equatorial coordination geometry is preferred (Figure 2c), and here, the donor atoms should be on the equatorial plane.

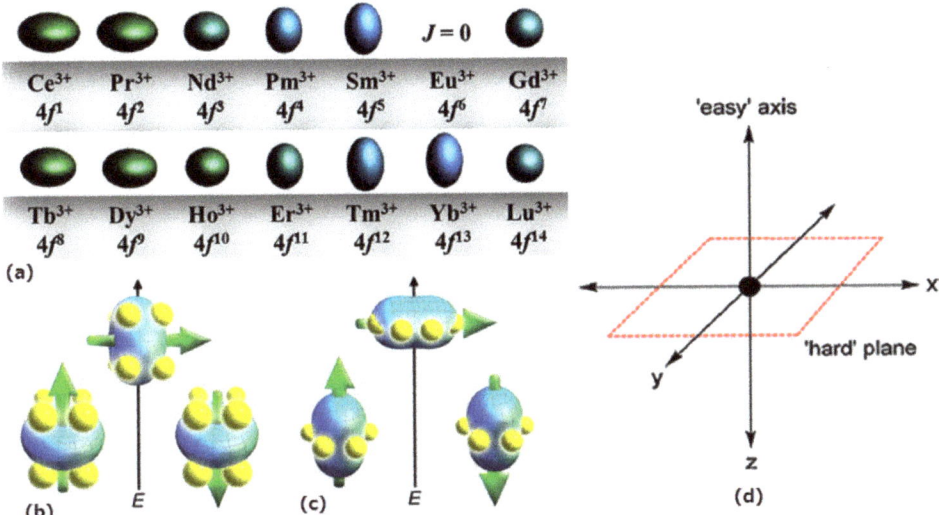

Figure 2. (a) Shapes of 4f electron densities in Ln(III). Europium is not shown as it has a J = 0 ground state; low and high energy configurations of the f-orbital electron coupled to the orbital moment (J): density with respect to the crystal field environment for an oblate (left) (b) and a prolate (right) (c) electron density of Ln(III) ions. The green arrow represents the orientation of the spin angular momentum (L), (d) the representation of the easy axis and hard plane in three-dimensional space. Reproduced with permission from [13,15]; Published by Royal Society of Chemistry, 2011 and Elsevier, 2014.

For Dy(III) single-ion magnet (SIM) with ground state J = 15/2 under extremely axial crystal field, the energy landscape of the magnetic microstates could resemble the time-reversal symmetric double-well potential as depicted in Figure 3a. It is important mentioning the difference between the height of the double-well potential and the effective energy barrier (U_{eff}) for magnetization reversal. Provided all, but Orbach mechanisms of relaxation (vide infra, Figure 3a) are prohibited, the molecules in principle can revert their magnetization moment (i.e., jumping from one potential well to other in the double-well potential) via climbing through all the possible M_J states as shown by dashed green arrows in Figure 3a. The energy required for such a magnetization reversal equals the height of the double-well potential (U = U_{eff}). However, in practice, spin-lattice relaxations (direct/Raman) and quantum tunneling of magnetization accompany the Orbach process. The former processes are more prone in the excited states. Therefore, it is not needed for the system to climb all the possible M_J states for magnetization reversal. In most of the Ln(III) based SIMs/SMMs, magnetization reversal takes place through first (e.g., M_J = +15/2 → M_J = +13/2 → M_J = 13/2 → M_J = 15/2; Figure 3a) or second (e.g., M_J = +15/2 → M_J = +13/2 → M_J = +11/2 → M_J = 11/2 → M_J = 13/2 → M_J = 15/2; Figure 3a) excited state [13,17–20]. Therefore, the effective energy (U_{eff}) required for the magnetization reversal is the energy between the ground state and first excited state where M_J = +15/2 and +13/2, respectively or the ground state (M_J = +15/2) and the second excited state (M_J = +11/2), respectively in the above cases.

It is worthwhile to note that the eigenstates are not necessarily organized following the decreasing/increasing order of the M_J values. As shown in Figure 3b, the ground, the first- and second excited states for Tb analog are associated with M_J = ±6, ±5 and 0, respectively. On the other hand, those states for Dy analog correspond to M_J = ±13/2, ±11/2 and ±9/2, respectively. However, the best performing SMMs are indeed those for which magnetization relaxes via the third [21], fourth [22] and even fifth [23] excited states since this provides a larger anisotropy barrier.

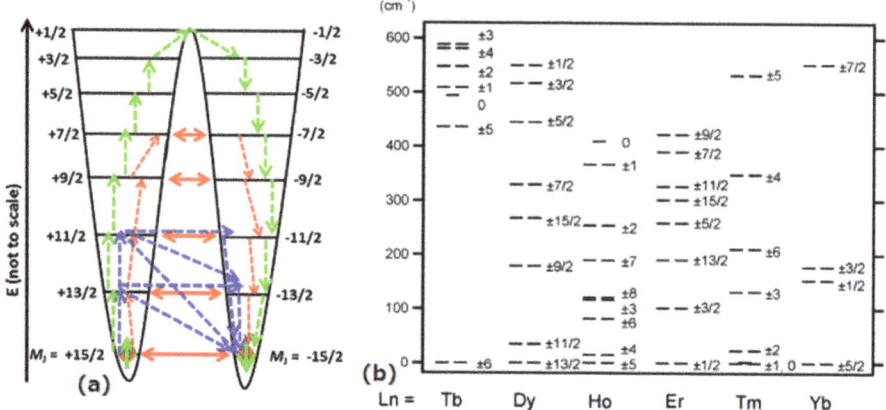

Figure 3. (**a**) The most common mechanisms involved in the magnetization relaxation of magnetically bi-stable systems. Color codes: green = thermally activated (Orbach) process; red = quantum tunneling of magnetization (QTM) or thermally assisted (TA) QTM; blue = phonon-triggered direct (Raman) spin-lattice relaxation. (**b**) calculated angular dependence (approximate) of every lanthanide magnetic state. Reproduced with permission from [5,24,25]; Published by Elsevier, 2018, American Chemical Society, 2003 and 2004.

1.3. Phthalocyanine Double-Decker SIM and Dy(III) Triangular SIM/SMM

In 2003, Ishikawa et al. [26] (Figure 4b) reported the first example of SIM comprising Ln(III)-ions ([Pc_2Ln]$^-$·TBA$^+$; Pc = phthalocyanine (Figure 4a); Ln = Dy, Tb; TBA$^+$ = $N(C_4H_9)_4{}^+$). The arrangement of the Pc ligands around the Ln(III) induces a sandwich structure in which the Ln(III) ions are in a D_4d symmetrical environment. Their energy barriers for spin reversal are reported to be 28 and 230 cm^{-1} for the Dy(III) and Tb(III) compounds, respectively. Solution ^{1}HNMR studies of the electronic spin dynamics of [Pc_2Tb]$^-$ molecules produced a much larger barrier of U_{eff} = 584 cm^{-1}, and solid samples diluted with the diamagnetic salt (Bu_4N)Br also indicated considerably high barriers of 641 cm^{-1} [26]. The radical analog [$TbPc_2$] has shown a strong frequency dependence of out-of-phase signal, with a peak maximum observed at 50 K. The resulting Arrhenius analysis revealed a remarkable U_{eff} = 410 cm^{-1}, which is considerably higher than any analogous value extracted for [Pc_2Ln]$^-$ from ac susceptibility data [27].

Furthermore, in 2006, Powell et al. reported the discovery of the exotic Dy(III) triangular clusters having the formulae [$Dy_3(\mu_3\text{-OH})_2L_3Cl_2(H_2O)_4$][$Dy_3(\mu_3\text{-OH})_2L_3Cl(H_2O)_5$]$Cl_5\cdot19H_2O$ (**1**) and [$Dy_3(\mu_3\text{-OH})_2L_3Cl(H_2O)_5$]$Cl_3\cdot4H_2O\cdot2MeOH\cdot0.7MeCN$ (**2**) in Figure 4c using o-vanillin as a ligand (Figure 4d). Ever since, pure Dy(III) based SMMs have drawn the attention of many investigators, as evident by the plethora of contributions with Dy(III) to this interesting field of molecular magnetism. In the Dy$_3$ triangle, despite an almost nonmagnetic ground state, all the characteristics of SMM have been observed with an effective energy barrier of 61.7 K, possibly derived from the thermally populated excited state. Antiferromagnetic linking of two Dy$_3$ triangles to form Dy$_6$ gave an increase in the temperature at which the magnetization is observed from 8 to 25 K, suggesting a promising strategy to increase the blocking temperature of lanthanide-based SMMs [28]. The above pioneering investigations shed light on the further development of lanthanide-based mononuclear and cluster-based SIMs and SMMs.

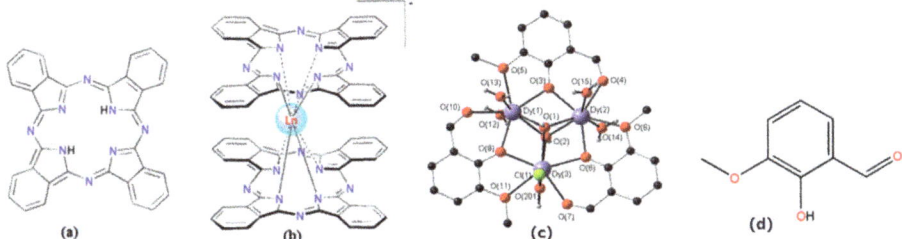

Figure 4. (a) The structure of the phthalocyanine molecule, and (b) structure of [Pc$_2$Ln]–TBA$^+$ (Pc = dianion of phthalocyanine, where TBA$^+$ = $N(C_4H_9)_4^+$ and Ln = Tb, Dy, Ho, Er, Tm and Yb. (c) structure of the triangular units in **1** and **2**, (d) the structure of *o*-vanillin. Reproduced with permission from [26,29]; Published by American Chemical Society, 2003 and Viley, 2006.

1.4. The Current Progresses in Ln(III)-Based SMMs

After the published work of Ishikawa [26], a number of Ln(III)-SMMs have been reported. Two main features are continuously optimized, namely, the energy barrier value (U_{eff}) and T_B. Before 2016, the largest record of energy barrier was 938 K [30], and the highest T_B was 14 K [31]. In 2016, a striking development of the value of the energy barrier was pushed forward. Here the complex [Dy(bbpen)Br] (H$_2$bbpen = *N,N'*-bis(2-hydroxybenzyl)-*N,N'*-bis(2 methylpyridyl)ethylenediamine) synthesized by Liu et al. [32], breaks the record of the energy barrier, surpassing 1000 K. The Dy(III) ions is in a pseudo-D$_5$h symmetry, surrounded by four neutral atoms and one bromide ion in the equatorial plane, and two oxygens occupying the capping positions. The spin flips from the ground state to the third excited doublet, which increases the effective energy barrier dramatically to 1025 K, accompanied by a magnetic hysteresis of up to 14 K.

In 2018 Guo et al. also reported a dysprosium metallocene cation [(Cp^{i}Pr$_5$)Dy(Cp*)]$^+$ (Cp^{i}Pr$_5$, penta-iso-propylcyclopentadienyl; Cp*, pentamethylcyclopentadienyl), which displays magnetic hysteresis above liquid-nitrogen temperatures with an effective energy barrier to a reversal of the magnetization of U_{eff} = 1541 cm^{-1} and having magnetic blocking temperature of T_B = 80 K. In the same year, McClain and coworkers have synthesized and reported dysprosium metallocenium SMMS, [Dy(Cp^{i}Pr$_4$R)$_2$] [B(C$_6$F$_5$)$_4$], where R = H (**3**), Me (**4**), Et (**5**), ^{i}Pr(**6**) [33]. A slight variation of the cyclopentadienyl ring substituents resulted in large changes of the molecular structures and in an increase of 45 K in the operating temperature for **3–6** and also led to the manifestation of the highest 100 s blocking temperatures reported for an SMM till now. Complex **3** has the highest operating temperature along the series, having a 100 s T_B at 62 K. It is interesting to note that **4** displays an energy barrier of 1468 cm^{-1} with hysteresis at 72 K [33].

By comparing the success of different SMMs, the bigger the magnitude of the anisotropic barrier, the more prominent the SMM properties at higher temperatures. A breakthrough in SMM based technology can only be achieved when these two major problems get addressed. One of the imposing challenges in this field is to design and synthesis well-organized SMMs, that operating at ambient temperature for practical uses [6,34].

2. Schiff Base Ligands

Schiff bases are prepared from amines and aldehyde precursors by condensation reaction [35] (Scheme 1). The ligand designs can be tailored by incorporating new functionalities to either the aldehyde or the amine precursor component. This makes them ideal candidates for the development of a library of ligands for metal aggregate synthesis. The number of amino and keto precursors that can be selected for condensation reactions, leading to azomethine compounds, are numerous and practically limitless [35]. The careful selection of both aldehydic and amino precursors enables us to perfectly switch the donating ability of the resulting ligands, the nature of the donor atoms and the number of the chelating motif. Moreover, one or both precursors can be decorated with bulky groups that can further influence the stereochemistry of the metal ions.

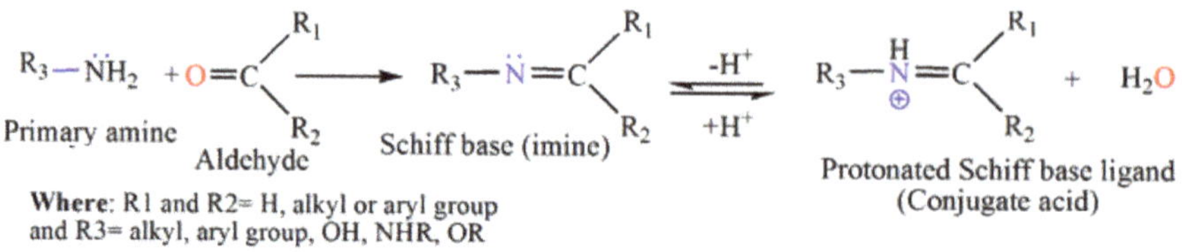

Scheme 1. The condensation mechanism of carbonyl compounds with primary amines. Reproduced from [35]; Published by Viley, 1864.

3. Designing Schiff Bases in Ln(III)-Based SMM Systems

Schiff base ligands have been widely used for the purpose of synthesizing Ln(III)-based SMMs [36–39]. The growing interest in Ln(III)-SMMs leads to a great demand for ligand architecture since the coordination environment is the main factor in the properties of metal aggregates. Usually, Schiff base ligands are derived from aldehydes and primary amines, reactions commonly taking place in alcohol through condensation [40]. The straightforward synthesis could be one of the reasons that Schiff base ligands are so popular, and also the basic imine nitrogen, which exhibits pi-acceptor property, shows affinity to Ln(III) ions, making it the preferred choice.

Additionally, Schiff base ligands can be easily modified by controlling the amines and aldehydes of different sizes, flexibilities and basicities [41,42], which provides different "pockets" for Ln(III) ions to occupy the resulting dinuclear, trinuclear, tetranuclear and even more polynuclear structures [43–46]. Additionally, the flexibility of the diamines or the linkers potentially introduces chirality to the complex by increasing the chances of forming helicate or mesocate structures [47]. Schiff base ligands derived from o-vanillin have proven to be particularly suitable for the synthesis of Ln-SMMs [37–39,45,46,48]. In his recent review, Andruh has discussed several relevant structural types of heterobinuclear 3d–3d' and 3d–4f complexes obtained from o-vanillin-based Schiff ligands, which show interesting magnetic, luminescence, catalytic, cytotoxic, and ferroelectric properties [49].

The ligands can be designed to provide a coordination environment favored by a 3d metal such as Mn(III) or Cu(II) (relatively N rich) as well as a site (relatively O rich) more favored by a "hard" metal ion such as Fe(III) or any Ln(III). For example, combining o-vanillin and tris (hydroxymethyl) aminomethane to give the ligand (H_4L, Chart 1) provides a system that captures two Ln(III) and transition metal ions [50]. While this ligand had been previously employed to prepare homometallic clusters [51], it seemed reasonable to expect that the oxygen donor rich tripodaltris(hydroxymethyl) aminoethane derived group would capture the oxophilic Ln(III), while the transition metal in +2 oxidation state like ions would be coordinated by both the imine nitrogen and oxygen donors [51]. Schiff base ligands can also be designed with the aim of getting spin-crossover (SCO) materials [52].

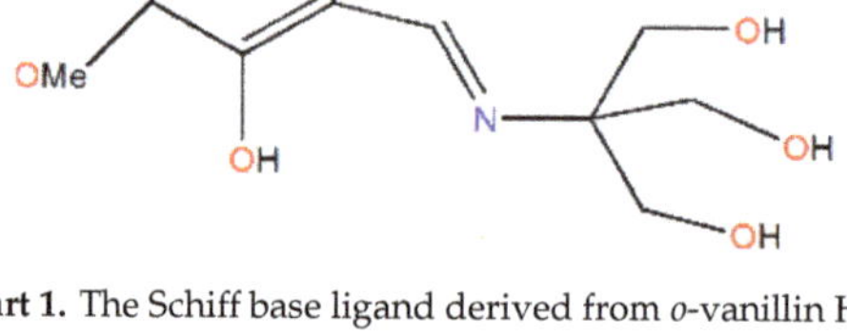

Chart 1. The Schiff base ligand derived from o-vanillin H_4L.

Some important theoretical and experimental studies on Ln(III)-based Schiff base complexes, namely Dy-(trenovan), have been done by Lucaccini et al. The studies reached the conclusion that the crucial parameter in determining the slow relaxation of the magnetization is the number of unpaired electrons (only Kramers ions showing in-field slow relaxation) than the shape of the charge distribution for different Ln(III) ions when the complex exhibits trigonal symmetry and the Ln(III) ion is heptacoordinated [53].

With respect to the scope of the review, we have attempted an overview on the synthesis, structural and magnetic properties of Ln(III)-based complexes with the following list of Schiff base ligands (HL1–H$_2$L20), mainly of salicylaldehyde origin (Chart 2).

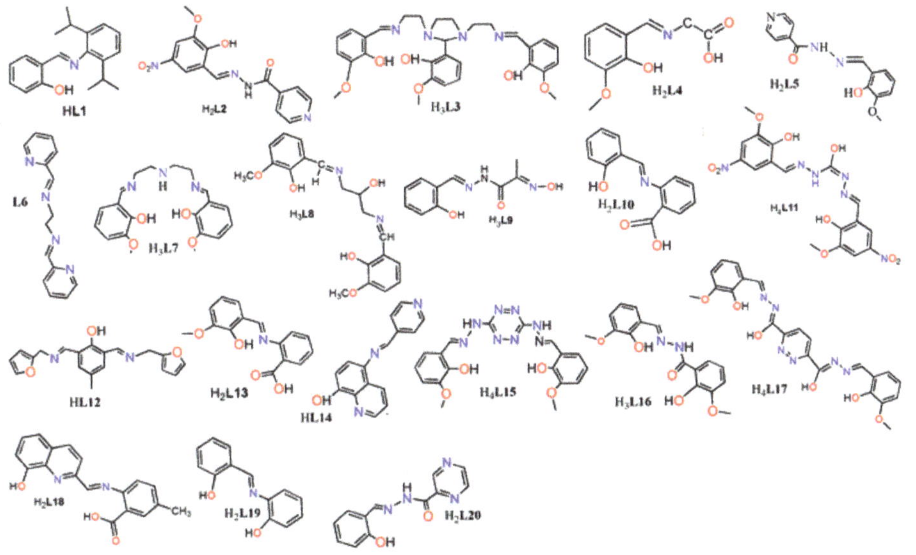

Chart 2. List of selected Schiff base ligands mainly of salicylaldehyde origin in the present review.

4. Ln(III)-Based Schiff Base SMMs of Different Nuclearities

In the following sections, we are discussing the lanthanide complexes of Schiff base ligands listed in Chart 2 of different nulcearity varying from mononuclear to hexanuclear, with a special reference to their synthetic strategy, structure and magnetic behavior.

4.1. Mononuclear Schiff Base Ln(III)-Complexes of SMMs

Gou and Tang have reported six coordinate Ln(III) complexes with the general formula [Ln(**L1**)$_3$] (Ln = Dy (**7**) and Er (**8**), having isostructure with the metal ion in distorted trigonal-prismatic coordination geometry, Where H**L1** = 2-(((2,6-diisopropylphenyl)imino)methyl)-phenol) (Figure 5a). Complexes **7** and **8** are isostructural (Figure 5b) with an RT; the χ_MT values (Figure 5c) of **7** and **8** are 13.77 and 11.38 cm^3 K mol^{-1}, respectively. For **7**, χ_MT gradually decreases on decreasing T till 2 K, reaching the value of 11.60 cm^3 K mol^{-1} for **7** [54]. The τ vs. T^{-1} plots the presence of more than one relaxation pathway with a crossover from a linear increase of thermally activated to a temperature-independent regime of QTM (Figure 5d) [54].

Both in-phase (χ') and out-of-phase (χ'') ac susceptibilities for complex **7** show frequency and temperature dependence (Figure 6a). However, no maximum peaks of the temperature dependence of the out-of-phase (χ'') signal are observed in the range of 1–1488 Hz, which may be caused by the quantum tunneling of the magnetization (QTM), as indicated by strong temperature-independent peaks below 9 K (Figure 6a) [54]. For **8**, there are no out-of-phase (χ'') signals observed above 1.9 K at

997 Hz, attributed to the quick quantum tunneling of the magnetization at zero dc field. The rather different magnetic properties of **7** and **8** are correlated with the axial ligand field of trigonal-prismatic coordination geometry, as Dy(III) is oblate and Er(III) is prolate [13].

Both **7** and **8** show frequency and temperature-dependent and temperature-dependent χ' and χ signals (Figure 6a) at low-temperature [54], proving its field-induced SMM behavior. The Cole–Cole plots (Figure 6b) of both **7** and **8** are asymmetrical semicircular in shape and can be well fitted by the generalized Debye model, with a series of α parameters below 0.11 from 1.9 to 13 K and 0.13 from 1.9 to 3.7 K, respectively, which shows a narrow distribution of the relaxation time for both complexes [54]. Complex **7** and shows the features of SMMs under zero dc field with an effective energy barrier of 31.4 K while **8** shows SMM characteristic under an external field of 400 Oe with an effective energy barrier of 23.96 K.

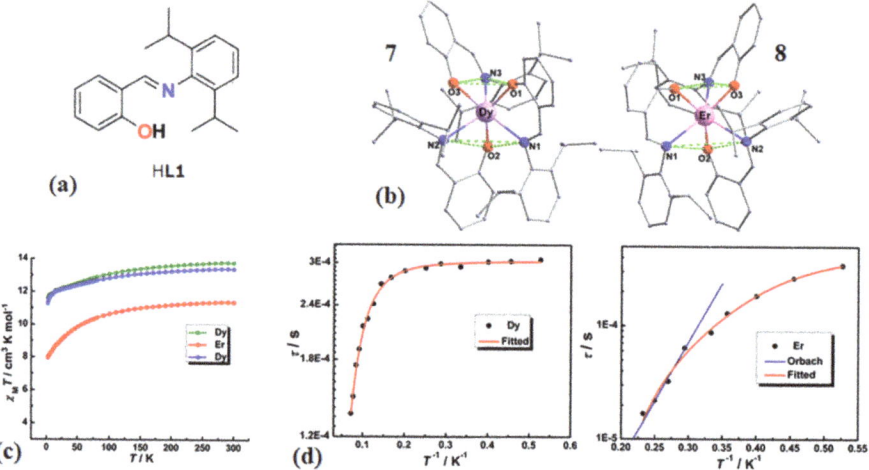

Figure 5. (a) The structure of HL1, (b) the molecular structures of complexes **7** (left) and **8** (right), (c) plots of the $\chi_M T$ vs. T for **7** and **8** in an applied field of 1 kOe, (d) plots of τ vs. T^{-1} for **7** (left) and **8** (right), respectively. Reproduced from [54]; Published by MDPI, 2018.

Cheng and coworkers synthesized and characterized two mononuclear complexes with the general formula, [Ln(**L3**)(NO$_3$)(DMF)$_2$]·DMF (**9**·DMF) and [Ln(**L2**)$_2$(H$_2$O)$_2$]·NO$_3$·EtOH (**10**·NO$_3$·EtOH) [55], where Ln = Dy, with two Schiff bases with an electron-withdrawing group NO$_2$ where (H$_2$**L2** = 2-(hydroxyl-3-methoxy-5 nitrophenyl)methylene(isonictino)hydrazine (Figure 7d).

Complex **9** (Figure 7a) is a nine coordinated species having a spherical tricapped trigonal prism geometry with one H$_2$**L2**, two nitrates and two DMF molecules in the coordination sphere. The Schiff base ligand H$_2$**L2** coordinates through both oxygen atoms (O1 and O2) and one N atom (N3) to the central metal ion. Upon comparing with the earlier report, [Dy(hmb)(NO$_3$)$_2$(DMF)$_2$] (Hhmb: (N'-(2-hydroxy-3-methoxybenzylidene)benzohydrazide) [56], it is interesting to note that the electron-withdrawing nature of the nitro group para to the phenoxide group in H$_2$**L2** has a pronounced effect on the coordination nature of the ligand. Complex **10** is 8 coordinated species with a spherical triangular dodecahedron geometry with two ligands as well as two water molecules. The ligands are coordinated in keto form and NO$_3^-$ are also present in the crystal lattice as a counter anion (Figure 7b) [55].

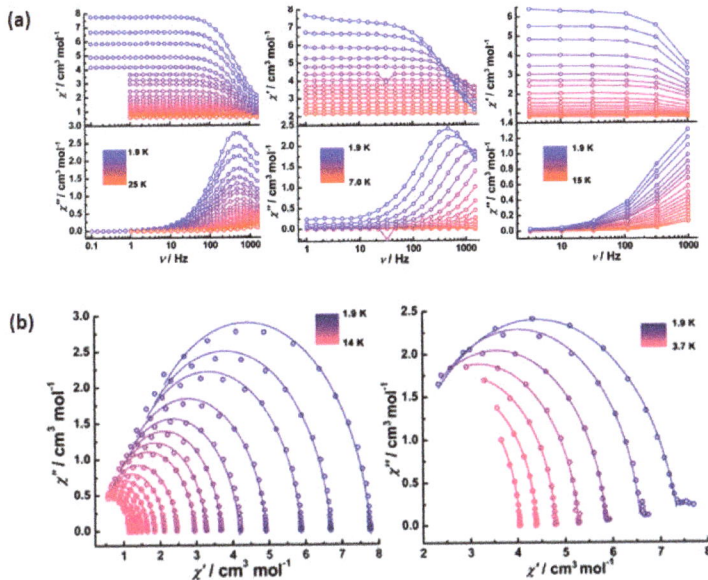

Figure 6. (a) Frequency-dependent in-phase χ′ (top) and out-of-phase χ″ (bottom) AC susceptibilities for **7** (**left**) under 0 Oe DC field and **8** (**right**) under 400 Oe DC field, (b) Cole–Cole plots for **7** (left) and **8** (right) at the indicated temperature. Reproduced from [54]; Published by MDPI, 2018.

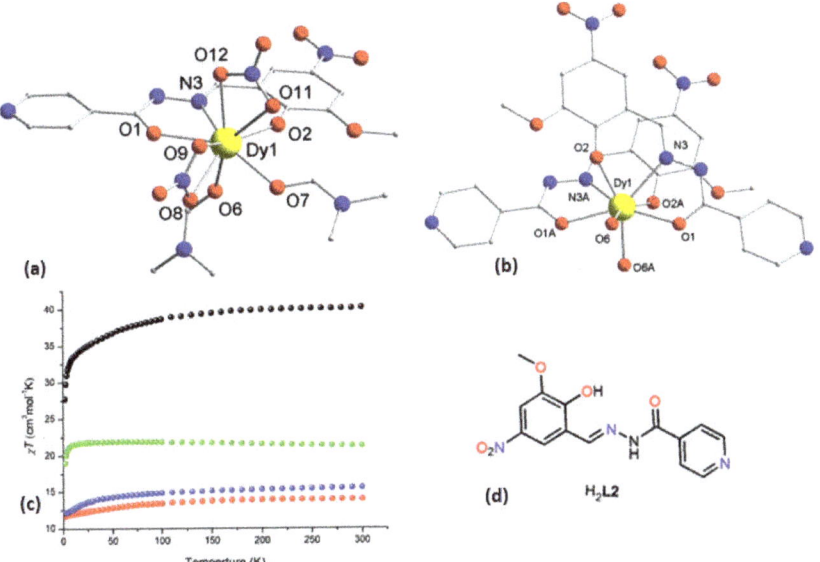

Figure 7. (a) Molecular structures of complex **9**, (b) molecular structure of complex **10**, (c) χT product of complexes **9** (red) and **10** (blue) at 1000 Oe, (d) the structures of H_2L2. Reproduced with permission from [55]; Published by Royal Society of Chemistry, 2019.

The RT value of χT for complexes **9** and **10** are 13.96 and 14.69, cm^3 K mol^{-1}, respectively (Figure 7c), and a gradual decrease with temperature for both complexes may be due to inherent magnetic anisotropy from the Dy(III) ion, the stark level depopulation and/or occurring of SMM behavior [55]. The AC measurements show out of phase signal χ'' below 11 K for **9** indicate low-temperature SMM behavior (Figure 8a,b). Compound **9** exhibits an anisotropic energy barrier of U_{eff} = 34 K [55].

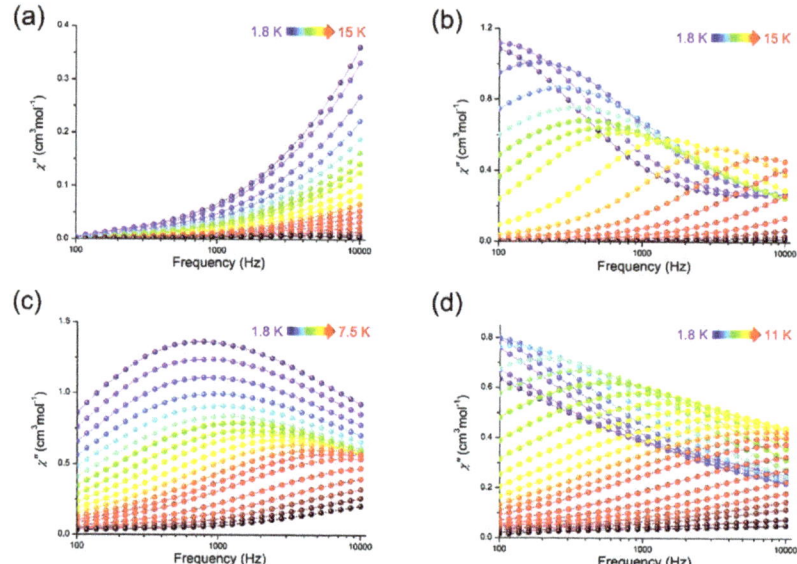

Figure 8. AC measurement showing out of phase signal (χ'') under zero DC field for complex **9** (**a,b** (400 Oe)) and **10** (**c,d** (800 Oe)). Reproduced with permission from [55]; Published by Royal Society of Chemistry, 2019.

For **10**, the dynamics of magnetization shows a frequency χ'' in the absence of DC field (Figure 8c). The effective energy barrier and relaxation time are U_{eff} = 19 K and τ_0 = 3.8 × 10^{-7} s, respectively, which is larger than that of complex **9**. The AC measurements of various DC fields show that the broad peaks under 200 and 400 Oe, is an indication of multiple exchange interactions with tunneling electrons [55]. For further investigation of QTM, the AC measurements were done under the magnetic field of 800 Oe. However, the relaxation got slower with a higher energy barrier of U_{eff} = 41 k [55].

4.2. Dinuclear and Trinuclear Schiff Base Ln(III) SMM Complexes

The investigations on multinuclear Ln(II) systems were very crucial with respect to the advancement of T_B, and these types of compounds generated showed significant progress to quench QT effects. Hence, designing ligands with appropriate symmetry, and incorporating Ln(III) centers, may enhance exchange coupling and quench the QT effects. Controlling intermolecular interactions using bulky counter anions/ligands and then utilizing enriched lanthanides to avoid hyperfine couplings is a hot research area in molecular magnetism in general and Ln(III) based Schiff base SMMs in particular. Here we are attempting to show how these different multinuclear Schiff base Ln(III) compounds derived from salicylaldehyde derivatives will have an effect on quenching QT and thereby generating SMMs with improved T_B.

Tang and coworkers have prepared isomorphous dinuclear Ln(III)-complexes [57] having the formula of [Ln$_2$(HL$_3$)$_2$]·nCH$_3$CN, where Ln = Gd (**11**), Tb (**12**) with n = 0 and Ln = Dy(**13**) with n = 4 and H$_3$L3 = N1,N2,N3,N4-tri(3-methoxysalicylidene)triethylenetetraamine) (Figure 9d). H$_3$L3 behaves as a heptadentate ligand coordinating through three phenoxide oxygens and four azomethine nitrogen. It is interesting to note that the three methoxy oxygen of H$_3$L3 is not coordinated in the complex [57].

The Ln(III)-coordination sphere of **11**–**13** are slightly longitudinally compressed with comparable parameters of skew angles (φ) 56.03, 56.12 and 55.64°, respectively. As shown in Figure 9a, the ligand binds to the central metal ion through the N$_4$O$_4$ coordination environment generating a square-antiprismatic geometry (SAP). The α angles shown in Figure 9c corresponds to the magic angle, 54.74° for **13** using H$_3$L3 [57]. Here the obtuse and acute angles are in accordance with the compression and elongation along the tetragonal axis [12]. However, the φ values, 42.9, 36.9 and 42.9° of **11**–**13** show that in **12**, the coordination sphere deviates from the ideal square-antiprismatic than **11** and **13** (Figure 9b). The shortest intermolecular distance between Dy(III)-ions (from different dinuclear units) is 9.142 Å, and this shows that there are no significant intermolecular interactions [57].

For **11**, the $\chi_M T$ remains constant till 50 K, then shows a sharp decrease to a minimum value of 10.64 cm^3 K mol^{-1} at 2 K, suggesting dominant intramolecular antiferromagnetic interaction between Gd(III) ions (Figure 10b). The RT DC magnetic susceptibilities of **11**–**13** in a magnetic field of 1000 Oe are 14.98, 21.82 and 26.99 cm^3 K mol^{-1}, respectively (Figure 10a). The variation in these values from the expected theoretical values is due to the weak magnetic exchange interaction between the metal centers through the bridging phenoxy group. The magnetization (M) data for **11**–**13** in 0–70 kOe field below 5 K shows the occurrence of weak antiferromagnetic coupling for **11** (Figure 10b). The AC measurement of **13** exhibits a frequency-dependent out-of-phase signal under 800 Oe DC filed, showing slow relaxation of magnetization having an energy barrier of 18.9 K (Figure 10c) [57]. For **13**, above 3.5 k, the magnetic relaxation follows a thermally activated Orbach mechanism having an energy gap of 18.9 k (Figure 10d).

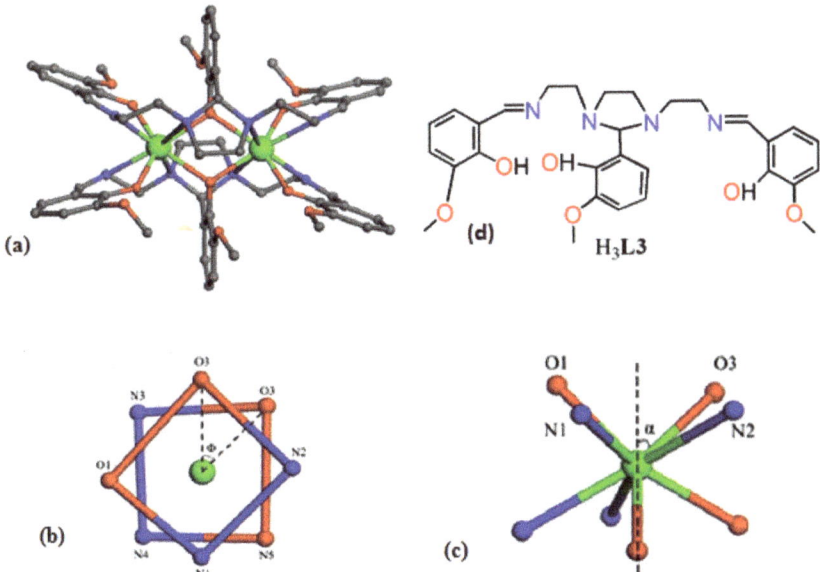

Figure 9. (a) Molecular structure of **13**, (b) SAP environment with skew angle φ between the diagonals of the two squares of **11**, (c) angle between the S$_8$ axis and Dy-L vector (α) of **13**, (d) the structure of H$_3$L3. Reproduced with permission from [57]; Published by Royal Society of Chemistry, 2013.

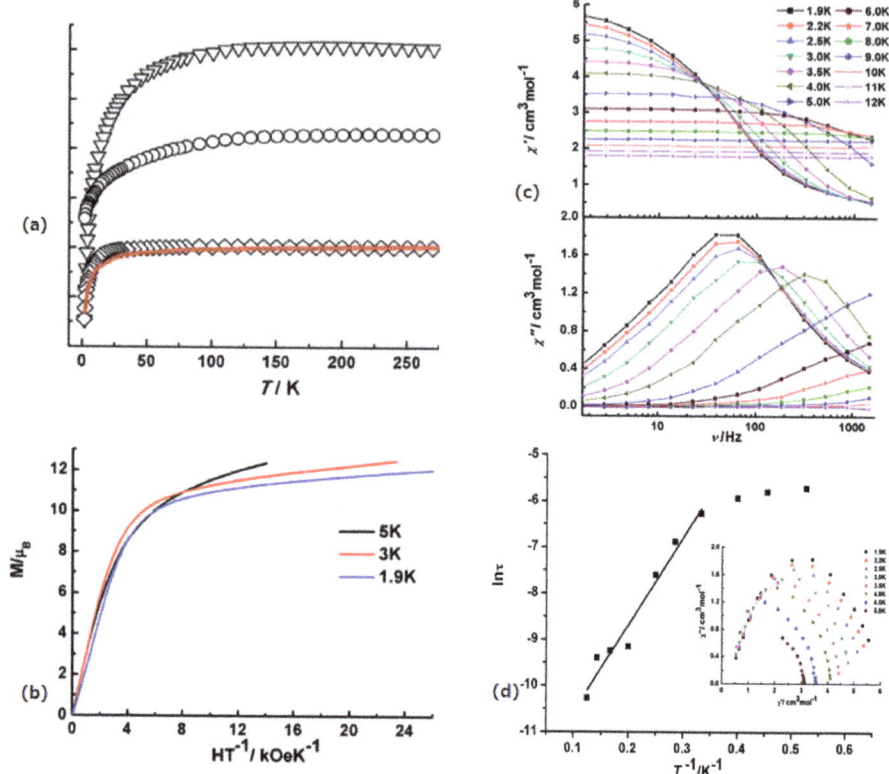

Figure 10. (a) $\chi_M T$ vs. T for **11–13** (b) M vs. H/T plots for **13** at indicated temperatures, (c) AC susceptibility measurement of **13** (d) ln τ vs. T^{-1} plot for **13**. Reproduced with permission from [57]; Published by Royal Society of Chemistry, 2013.

In this class of compounds, another dinuclear mixed ligand complex, [Dy$_2$(**L4**)$_2$(DBM)$_2$ (DMF)$_2$]·3CH$_3$OH(**14**), (Figure 11b), has been synthesized by Zhang et al. (Where H$_2$**L4** = 2-((2-hydroxy-3-methoxybenzylidene)amino)acetic acid, HDBM = dibenzoylmethane)) (Chart 3). The structure of **14** consists of two crystallographically independent units of neutral centrosymmetric dinuclear complex, with two Dy(III), two dianionic (**L4**$^{2-}$) (Figure 11a), two bidentate monoanionic DBM$^-$ ligands and two terminal DMF by generating a NO$_7$ coordination environment with a square-antiprismatic coordination geometry [58].

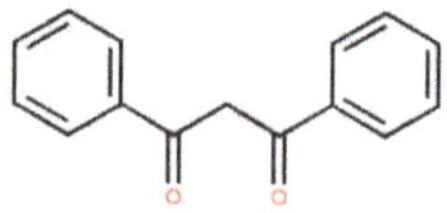

HDBM

Chart 3. The structures of HDBM.

The M vs. H plots for **14** below 5 K (Figure 11c) gives a relatively fast increase of the magnetization at low fields followed by a slow linear increase at high fields reaching the values of 10.7 N$_\beta$ up to 70 kOe at 2 K, which is supported by the observation that the M vs. HT^{-1} curves at different fields (Figure 11c, bottom), are not superimposed on a single master-curve [59]. The complex **14** shows a frequency-dependent χ′ and χ″, exhibiting slow relaxation. The maximum value of χ″ for 1488 Hz is observed approximately at 3 K [59]. From the χ″ vs. χ′, (Figure 11d), an effective energy barrier of 11 K is obtained for **14** (Figure 11e) [60].

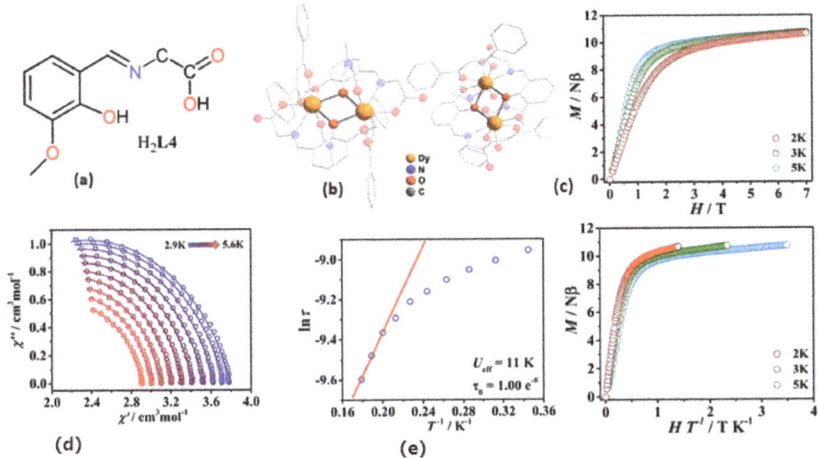

Figure 11. (a) The structure of H$_2$L4, (b) molecular structure of complex **14**, (c) field dependence of the magnetization (M) at 2, 3 and 5 K for complex **14** plotted as M vs. H (left) and M vs. HT^{-1} (right), (d) the Cole–Cole plot for complex **14** (e) plot of lnτ vs. 1/T plot for **14**. Reproduced with permission from [58]; Published by American Chemical Society, 2003.

Murugesu et al. have discussed the use of the pro-ligand (2-hydroxy-3-methoxyphenyl) methylene(isonicotino)hydrazine(H$_2$L5) (Figure 12c), in reactions with Dy(III), because of its ability to act as a rigid chelate [38]. By reacting penta-aqua Dy-nitrate, Dy(NO$_3$)$_3$·5H$_2$O, with H$_2$L5 in methanol (with the presence of triethylamine) in a 3:1 mixture of acetonitrile and methanol (in the presence of pyridine), pale orange crystals of [Dy$_2$(HL5)$_2$(NO$_3$)$_2$(MeOH)$_2$](**15**) and [Dy$_2$(HL5)$_2$(NO$_3$)$_2$(MeOH)$_2$]$_\infty$(MeCN)·(2MeCN)(**16**) were obtained (Figure 12a,b). Phenoxide-bridged Dy-dimers resulted from this reaction, where the positioning of the pyridyl groups helps the formation of an extended network that can control the arrangement of the SMM units in a three-dimensional way. However, with regard to complex **16**, the pyridyl N atoms further coordinate to the Dy atoms of the adjacent complexes forming a two- dimensional network of the dimetallic Dy complexes. Dy(III) is eight coordinated in **15** and **16**, and a square-antiprismatic geometry may be assigned in both cases [38].

The magnetic properties of **15** and **16** are somewhat similar as they possess the same coordination environment (Figure 12d). At RT, the $\chi_M T$ values of 30.4 and 30.0 cm^3 K mol^{-1} for **15** and **16**, respectively and they are in accordance with the expected value of 28.34 cm^3 K mol^{-1} for two Dy(III) ions [61,62]. For both complexes, the $\chi_M T$ product remains roughly constant before reaching a minimum value of 29.3 cm^3 K mol^{-1} at 23 K. The $\chi_M T$ then sharply increases to a maximum value of 38.4 cm^3 K mol^{-1} for **15** and 36.6 cm^3 K mol^{-1} for **16** at 1.8 K, which confirms the presence of intramolecular ferromagnetic interactions between metal centers [38].

Repeated magnetic measurements on **16** showed that this compound possessed one of the largest energy barriers, at 71 K, reported for an Ln(III)-complex at that time with Schiff base ligands. The magnetization curve below 10 K exhibits a rapid increase at low field, which is expected for ferromagnetically coupled spins. Magnetization increases linearly up to 11.9 μB (**15**) and 11.6 μB (**16**) at 1.8 K and 7 T without clear saturation. From the M vs. H/T data inset (Figure 12d), we can reach a conclusion that there is a significant magnetic anisotropy and/or low-lying excited states in these compounds. It is interesting to note that the M vs. H data does not exhibit a hysteresis effect above 1.8 K, but below 12 K (at about 1500 Hz), the indication of out-of-phase AC signal reveals a slow relaxation of the magnetization (Figure 12e) [38].

From the frequency dependence measurement, the relaxation time (t) is derived and is plotted as a function of 1/T (in the range 1.8–10 K) (Figure 12e). The dynamics of **15** and **16** below 2 K are temperature-independent as expected in a pure quantum regime with a τ value of 0.3×10^{-2} s and 1.2×10^{-2} s for **15** and **16**, respectively and above 2 K, the relaxation becomes thermally activated. Above 8 K, remarkably big energy barriers are observed at 56 K and 71 K, and the pre-exponential factors of the Arrhenius laws (τ_0) are 3×10^{-7} s and 7×10^{-8} s for **15** and **16**, respectively.

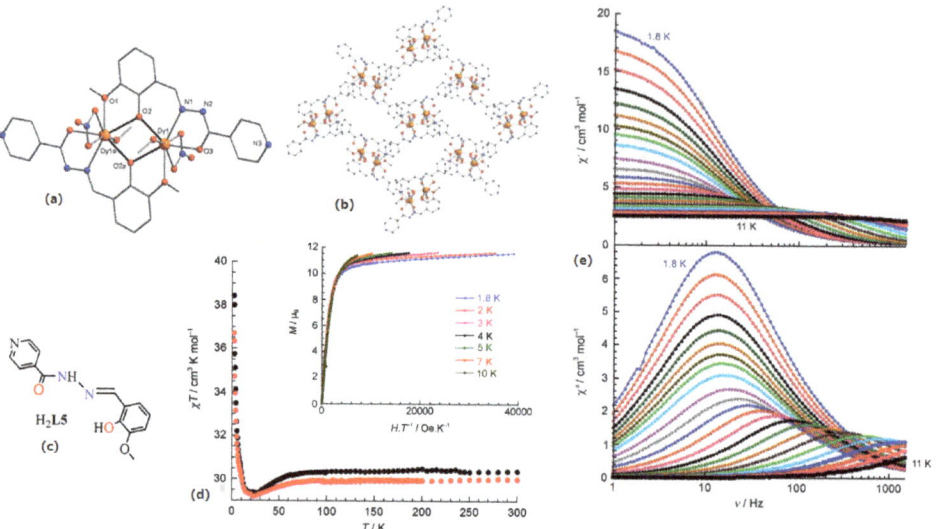

Figure 12. (a) the molecular structure of the centrosymmetric complex **15**, (b) 2D network of complex **16** as viewed along the c axis, (c) the structure of H$_2$L5, (d) temperature dependence of the χT product at 1000 Oe for **15** (black) and **16** (red) (with χ = M/H normalized per mol): Inset M vs. H/T plot at low-temperatures for **16**, (e) frequency dependence of the in-phase (χ′) and out-of-phase (χ″) AC susceptibility from 1.8 to 3.2 K at an interval of 0.2 K and from 3.5 to 11 K at an interval of 0.5 K under zero DC field for **16**. Reproduced with permission from [38]; Published by Viley, 2008.

In 2016 a new family of five new isostructural Ln(III)-complexes with the general formula [Ln$_2$(DBM)$_6$(**L6**)] (where Ln = Sm, Eu, Gd, Dy, Yb; **L6** = N,N′- bis(pyridin-2-ylmethylene)ethane-1,2-diamine (Figure 13a), DBM = dibenzoylmethane have been reported by Sun et al. (Chart 3). Based on the single-crystal analysis, all the complexes are isomorphic and for discussion, let us take the example of Dy(**17**) and Yb(**18**) complexes. (Figure 13b,c) which have one ligand (**L6**), two Yb(III) ions and six DBM molecules, generating a distorted square antiprismatic geometry [63].

Compound **17** and **18** have the RT $\chi_M T$ values of 28.21 and 5.53 cm^3 K mol^{-1}, respectively (Figure 13d). For **17**, there is the first drop of $\chi_M T$ vs. T occurring due to the thermal depopulation of the magnetic energy levels split by the crystal field. Below 60 K, a second drop occurs may be because of intramolecular antiferromagnetic coupling. The $\chi_M T$ for **18** declined along with the temperature and went down to 1.85 cm^3 K mol^{-1} at 2 K. For **17** and **18**, the magnetization curves show a saturation value of 11.1 μβ (top) and 3.5 μβ (bottom), respectively at 2 K and 70 kOe, which is lower than the expected saturation value. This may be due to the anisotropy and important crystal field effects at the Dy(III)-ion that eliminate the 16-fold degeneracy of the $^6H_{15/2}$ ground state [63]. The energy barriers for **17** and **18** are 46.8 K and 23.0 K, respectively (Figure 13e) [63].

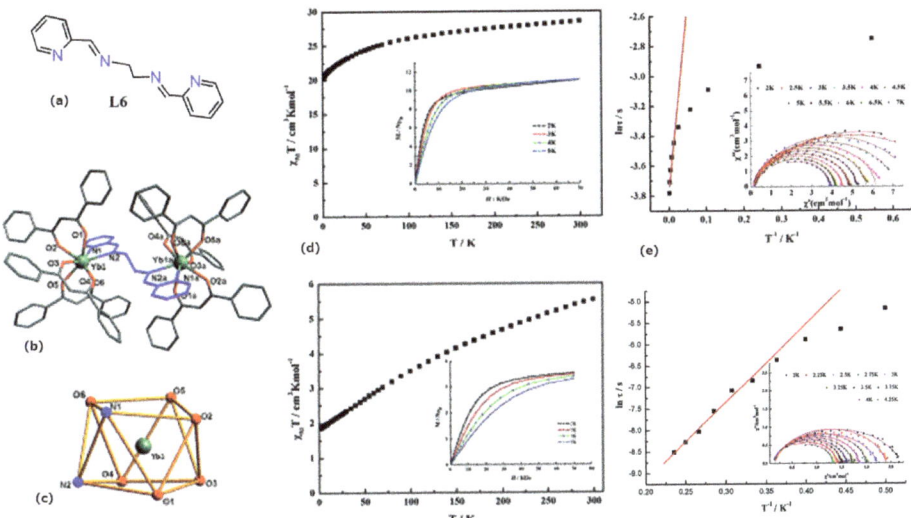

Figure 13. (a) The structure of **L6**, (b) the molecular structure of **18**, (c) coordination polyhedron for **18**, (d) $\chi_M T$ vs. T at 1000 Oe field for **17** (top) and **18** (bottom): the field dependence of magnetization for **17** (top) and **18** (bottom) at 2.0–5.0 K (insets), (e) magnetization relaxation time, $\ln\tau$ vs. T^{-1} for **17** (top) and **18** (bottom) under a DC field of 2000 Oe. Reproduced with permission from [63]; Published by Royal Society of Chemistry, 2016.

In 2011 a family of five dinuclear Ln(III)-complexes having the general formula [Ln(III)$_2$(**L7**)$_2$(NO$_3$)$_2$], where (H$_3$**L7** = N1,N3-bis(3-methoxy salicylidene) diethylene triamine) and Ln(III) = Eu(III) (**19**), Gd(III) (**20**), Tb(III) (**21**), Dy(III) (**22**), and Ho(III) (**23**) were reported by Murugesu et al. having an intermediate coordination geometry between square antiprism (D$_4$d) and dodecahedron (D$_2$d) for all complexes [37]. The Ligand H$_3$**L7** (Figure 14a), with a large inner compartment having N$_3$O$_2$ coordination sites, is particularly appropriate to accommodate a sizable 4f-ion. While in a previous study of 2007, Dou et al. structurally characterized two mononuclear complexes with the same ligand H$_3$**L7** with large La(III) and Nd(III) ions where the outer donor O$_4$ set of H$_3$**L7** was involved in the coordination [64], but in Murugesu's report, the synthetic methods they promptly employed the functionality to promote coordination in both compartments of H$_3$**L7**. For a matter of discussion, the structure of the Dy-analog, compound **22**, was described as in-depth as a representative of the other families (Figure 14b).

For all compounds **19**–**23** (Figure 14c) at RT, the χT values are in good agreement with the expected theoretical values for two non-interacting Ln(III)-ions. For the europium analog **19**, the nonmagnetic ground state (^{7}F$_0$) is observed at low-temperatures as indicated by the χT value of 0.04 cm^3 K mol^{-1} at 1.8 K [2]. While in the gadolinium analog **20**, the decrease of the χT when lowering the temperature reveals the presence of antiferromagnetic interaction between the Gd(III) ions. For complexes, **21**–**23**, the χT vs. T curves for complexes **21**–**23** reach a value of 10.2, 5.2, and 7.1 cm^{3}K mol^{-1} at 1.8 K, respectively (Figure 14c). The frequency dependence of the maximum of **22** related only with a single relaxation process and which appears clearly on a tridimensional plot of the variation of χ'' vs. the temperature and the frequency of the oscillating field between 1 and 1500 Hz (Figure 14e), confirming the slow magnetic relaxation. Here, the existence of a single relaxation process agrees with the presence of a unique crystallographic Dy(III) ion in the dinuclear structure. The relaxation process in the tridimensional plot gives two regimes of relaxation, as indicated in Figure 14e. The χ'' vs. χ' in the temperature range 2–12 K additionally confirms the single relaxation process (Figure 14d). The effective energy barrier obtained from fitting, for **22** (Figure 14f), is U_{eff} = 76 K [37].

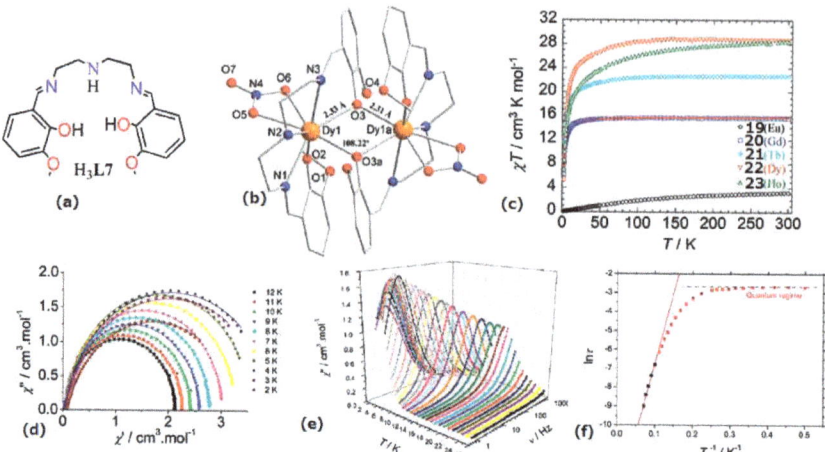

Figure 14. (a) The structure of H$_3$L7, (b) the molecular structure of **22**, (c) temperature dependence of the χT product at 1000 Oe for complexes **21–23**, (d) Cole–Cole plot for **22** obtained using the AC susceptibility data, (e) out-of-phase susceptibility χ'' vs. frequency v (logarithmic scale) in the temperature range 2–25 K for **22**, (f) relaxation time of the magnetization lnτ vs. T^{-1} (Arrhenius plot using AC data) for **22**. Reproduced with permission from [37]; Published by American Chemical Society, 2011.

Zhang and coworkers have prepared two new mixed ligand dinuclear Ln(III)-complexes, [Ln$_2$(HL8)(DBM)$_4$]·2CH$_3$OH (Ln = Gd (**24**) and Dy (**25**) [65], H$_3$L8 = N,N′-bis(3-methoxysalicylidene)-1,3-diamino-2-propanol (Figure 15a), and Hdbm = 1,3-diphenyl-1,3-propanedione) (Chart 3). As a representative, complex **24** is discussed, and its crystal structure is depicted in Figure 15b. The asymmetric unit consists of two Gd(III)- ions, one HL8^{2-}, four dbm$^-$ ligands and two free methanol molecules. The two Gd(III) ions in the Gd$_2$ dimeric unit are bridged by three μ$_2$-O atoms from two HL8^{2-} (O2 and O4) and one dbm$^-$ (O8), respectively (Figure 15c) [65]. The Gd1 located in an N$_2$O$_6$ pocket (N1, N2, O2, O4, O6, O7, O8 and O9), is eight coordinated with a distorted dodecahedron coordination geometry, and Gd2 has a nine coordination environment with O9 set (O1, O2, O4, O5, O8, O10, O11, O12 and O13), exhibiting a three-capped trigonal prism [65].

Variable temperature DC magnetic susceptibility for **24** and **25** was done under an applied magnetic field of 1 kOe and in between 2–300 K (Figure 15d). The RT, $\chi_M T$ values for **24** and **25** are 15.70(3) and 28.28(4) cm^3 Kmol^{-1}, respectively. For **24**, on decreasing the temperature, the $\chi_M T$ values almost keep constant up to 25 K, then decrease to a minimum of 6.78 cm^3 K mol^{-1} at 2 K, proving the existence of weak antiferromagnetic exchange between the gadolinium ions. While in **25**, the $\chi_M T$ values drop gradually over the temperature range from 300 to about 50 K, then drop abruptly to the minimum value 8.63(0) cm^3 K mol^{-1} at 2 K, which may be due to either the depopulation of excited Stark sublevels and/or a weak antiferromagnetic interaction of Dy(III) ions [66]. The magnetization data of **24** is collected in the temperature range from 2.0 to 10.0 K under the external magnetic field of 0–80 kOe. The M vs. H plots in Figure 15e show a continuous increase with the increasing of the magnetic field and reach the saturation value of 14.05(1) Nβ at 80 kOe and 2.0 K, which is in good agreement with the expected value of 14.0 Nβ for two isolated Gd(III) (g = 2, $^8S_{7/2}$) ions [65]. Further to investigate the dynamics of the magnetization, the AC-susceptibility measurements for **25** were performed as a function of temperature and frequency under zero DC field with an oscillation of 3.0 Oe. On increasing the frequency (111–2311 Hz), the frequency dependence below 20 K cannot be clearly observed from the (χ') vs. T plots (Figure 15f (top)). The frequency dependence of out-of-phase AC signal below 12 K suggests slow magnetization relaxation, indicating the presence of QTM (Figure 15f (bottom)).

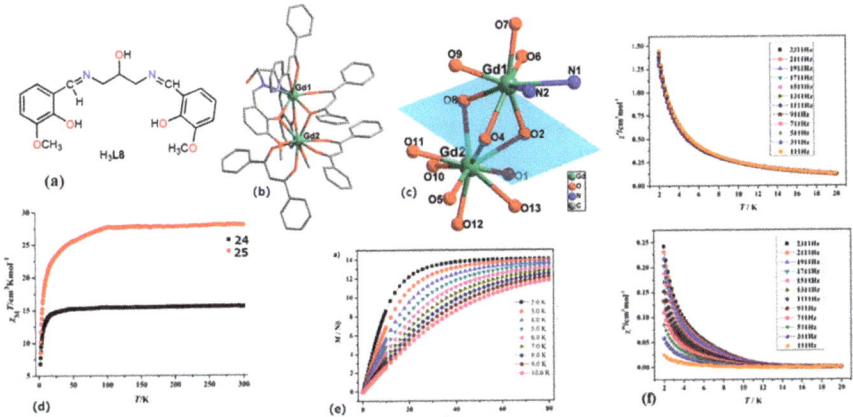

Figure 15. (a) The structure of H$_3$L8, (b) the molecular structure for **24** (c) the [Gd$_2$] cluster bridged by three μ$_2$-O atoms, (d) the plots of $\chi_M T$ vs. T for **24** and **25** under an applied field of 1 kOe between 2 and 300 K, (e) M vs. H curves for **24**, where T = 2.0–10.0 K and H = 0–80 kOe (f) χ' (top) vs. χ'' AC-susceptibility for **25** under zero DC field with an oscillation of 3.0 Oe (bottom). Reproduced with permission from [65]; Published by Elsevier, 2019.

Most recently, Li and coworkers have prepared and characterized three dinuclear mixed ligand Dy(III)-compounds, [Dy$_2$(dbm)$_2$(HL9)$_2$(H$_2$O)$_2$]·CH$_3$CN (**26**), [Dy$_2$(dbm)$_3$(HL9)(H$_2$L9) (CH$_3$OH)]·CH$_3$OH (**27**) and [Dy$_2$(dbm)$_2$(HL9)$_2$(C$_2$H$_5$OH)$_2$] (**28**) [67] (H$_3$L9 = N'-(2-hydroxybenzylidene)-2-(hydroxyimino)-propanohydrazide (Figure 16d), and HDBM = 1,3-diphenyl-1,3-propanedione) (Chart 3). The compound **26** is centro-symmetric, with two Dy(III), two doubly deprotonated HL9^{2-}, two singly deprotonated co-ligands dbm$^-$, two water molecule and one free acetonitrile molecule (Figure 16a) [67]. Each independent Dy(III) ion is eight-coordinated with an N$_2$O$_6$ coordination environment displaying a distorted bicapped trigonal-prismatic geometry. Compound **27** (Figure 16b) has a structure different from that of **26** and **28**. The asymmetric unit is composed of two Dy(III) ions, with singly deprotonated Schiff base ligand H$_2$L9$^-$, doubly deprotonated Schiff base ligand HL9^{2-}, three singly deprotonated DBM$^-$, one coordinated MeOH molecule and one free methanol molecule [67].

The coordination sphere of N$_2$O$_6$ for Dy1 ion is constructed by two oxygen atoms O5 and O6 from dbm$^-$, two phenoxide oxygen atoms from O7 and O12, two carbonyl oxygen atoms O8 and O11 also two imidogen nitrogen atoms N1 and N4 from H$_2$L9$^-$ and HL9^{2-}. Dy2 has a different coordination environment compared with Dy1. It is located in O$_7$N set completed by four oxygen atoms O1, O2, O3 and O4 of two dbm$^-$, one oximido nitrogen atom N6, one carbonyl oxygen atom O11 of HL9^{2-}, phenoxide oxygen atom O7 of H$_2$L9$^-$ ligand as well as the oxygen atom O13 of methanol molecule. The coordination geometries of Dy1 and Dy2 show a distorted bicapped trigonal-prismatic geometry [67].

The variable temperature DC magnetic susceptibility for **26–28** was collected between 2–300 K under 1000 Oe (Figure 16c) [67] and their $\chi_M T$ values at RT are 28.15 (**26**), 28.08 (**27**), 28.11 (**28**) cm^3 K mol^{-1} [67], which are in accordance with the expected values of two isolated Dy(II) ions. On lowering the temperature, the $\chi_M T$ values for **26–28** are almost unchanged from 300 to 150 K, then drop abruptly to the minimum values of 15.08 (**26**), 14.25 (**27**), and 14.50 (**28**) cm^3 K mol^{-1} at 2.0 K, respectively. The decreasing trend points to weak antiferromagnetic coupling between the neighboring Dy(III)-ions in **26–28** [67]. The plots of χ_M^{-1} vs. T for **26–28** obey the Curie-Weiss law, with a Weiss constants θ = −3.96 ± 0.59 (**26**), −5.09 ± 0.92 (**27**) and −8.26 ± 1.12 (**28**) [67]. The negative values of the parameter θ further suggesting the presence of the antiferromagnetic interaction between the Dy(III) in **26–28**, respectively [68].

The lnτ vs. T^{-1} plot for **28** is shown in Figure 16e, with spin-reversal energy barrier ($\Delta E/k_B$) of (45.6 ± 3.24) K having pre-exponential factor τ_0 = (5.14 ± 1.61) × 10^{-8}. However, the values of $\Delta E/k_B$ for **26** and **27** cannot be extracted by the Arrhenius law due to the lack of χ'' maxima. It is assumed that only one relaxation exists in **26** and **27**, thus their susceptibilities could be fitted to the Debye function ln(χ''/χ') = ln($\omega\tau_0$) + E_a/k_BT (Figure 16g,h) [38], resulting in the energy barrier $\Delta E/k_B$ of (2.29 ± 1.17) K for **26** and 1.31 ± 1.05 K for **27**, and the pre-exponential factor τ_0 of (0.31 ± 0.21) × 10^{-5} s for **26** and (0.20 ± 0.09) × 10^{-5} s for **27** [67]. The Cole–Cole plots of **28** having a semicircular shape (Figure 16f) were fitted by the generalized Debye function between 2.0–16.0 K [69], and the obtained parameter α fall in the range of 0.06–0.27. The small α values suggest a narrow distribution of magnetic relaxation time under a 5 kOe DC field.

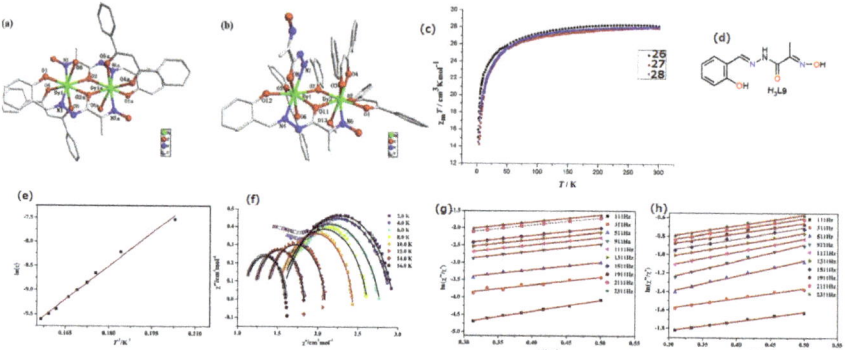

Figure 16. Molecular structures for **26** (**a**), and **27** (**b**), (**c**) temperature dependence of $\chi_M T$ for compounds **26–28** under 1000 Oe, (**d**) the structure of H$_3$**L9**, (**e**) the plots of lnτ vs. T^{-1} for **28**: the solid red line corresponds to the fitting of the experimental data to the Arrhenius equation, (**f**) the Cole–Cole plots of **28** measured from 2.0 to 16.0 K in H$_{dc}$ = 5 kOe, plots of the natural logarithm of χ''/χ' vs. T^{-1} for **26** (**g**) and **27** (**h**). Reproduced with permission from [67]; Published by Elsevier, 2020.

Very recently, Ln(III) based dinuclear SMMs with the Schiff base ligand, H$_2$**L10** (Figure 17a), were reported by Anastasiadis et al. The complexes are of the formulae [Ln$_2$(NO$_3$)$_4$(H**L10**)$_2$(MeOH)$_4$]·MeOH, Ln = Gd (**29**·MeOH) and [Ln$_2$(NO$_3$)$_2$(H**L10**)$_4$]·6MeCN, Ln = Dy (**30**·6MeCN) [70] having interesting magnetic optical, and catalytic properties [70]. Complex **29**·MeOH contains two crystallographically independent, centrosymmetric molecules (Figure 17b). The two Gd(III) atoms are bridged by the syn, syn-carboxylate groups of two η^1:$\eta^1\eta^1\mu_2$ of H**L10**$^-$ ligands. Two bidentate chelating nitrato groups, two-terminal MeOH molecules, and one terminal phenolate O atom complete 9-coordination at each metal ion. The H atom of the phenol –OH group of H**L10**$^-$ has "emigrated" to the imine N atom, and the latter is thus protonated without coordination.

The two Dy(III) atoms in the representative centrosymmetric molecule **30** (Figure 17c) are bridged by the syn, syn-carboxylate groups of the four η^1:$\eta^1\eta^1\mu^2$ H**L10**$^-$ ligands. In addition to this, there is a bidentate chelating nitrato group and two terminal phenolate O atoms, which complete 8-coordination, resulting in a square-antiprismatic geometry for the metal ion [70]. The DC molar magnetic susceptibility (χ_M) data on **29** and **30** were collected at 0.03 T in the temperature range 300–2.0 K and plotted as $\chi_M T$ vs. T in Figure 17d, left; The Ln(III) center in **29**, which is bridged by two carboxylate groups, is not interacting. A weak intramolecular ferromagnetic interaction is observed in **30**, where the two Dy(III) centers are bridged by four carboxylate groups. The AC magnetic susceptibility studies performed on a polycrystalline sample of **30** measured in the 2.0–12 K range in a 4.0 G AC field oscillating at 1490–10 Hz. The χ_M'' signals are observed at low-temperatures under a DC field of 1500 G only for **30** (Figure 17d, right), providing evidence of field-induced slow magnetic relaxation for **30**. Analysis of the experimental data afforded an effective barrier for the magnetization reversal (U_{eff}) of 8.0 cm^{-1} and a τ_0 value of 3.5 × 10^{-6} s [70].

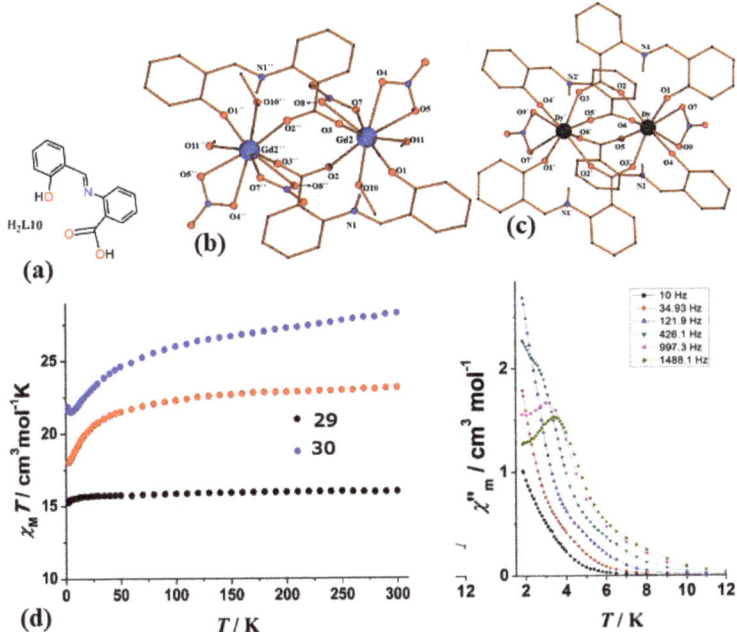

Figure 17. (a) The structure of H$_2$L10, (b) structure of the Gd2/2″—containing dinuclear molecule, 29·MeOH. (c) structure of the dinuclear molecule, 30·6MeCN. (d) (Left) $\chi_M T$ vs. T plots for complexes **29** (black) and **30** (blue). (Right) out-of-phase (χ_M'') vs. T AC susceptibility signals for **30** in a 4.0 G AC field oscillating between 1488 and 10 Hz, under a DC field of 1500 G. Reproduced with permission from [70]; Published by American Chemical Society, 2019.

Cheng and coworkers synthesized and characterized two trinuclear complexes with the general formulae Dy$_3$(HL11)$_3$(DMF)$_6$] (**31**), and [Gd$_3$(HL11)$_3$(DMF)$_6$] (**32**) [55], with Schiff base ligands containing an electron-withdrawing group (NO$_2$), H$_4$L11(Where H$_4$L11 = 1,5-bis(2-hydroxy-3-methoxy-5-nitrobenzylidene)carbonohydrazide(Figure 18f). Complexes **31** and **32** are isostructural. A perspective view of the molecular structure of **31** is represented in Figure 18a. Dy(III) in **31** was bonded to two H$_4$L11 and two DMF terminal solvents, leading to an eight coordinated triangular dodecahedron [55]. The coordination modes of the three ligands between any two Dy(III) ions are similar, where the two Dy(III) ions are held together by μ-κ^3:κ^3-HL11^{3-}.

In contrast to the report of similar ligands [71], the trans-trans conformations were switched to cis-trans conformation in **31**, leading to the N–N pathway (Figure 18b) (N2–N3, N2a–N3a and N2b–N3b between Dy1–Dy1a, Dy1a–Dy1b and Dy1b–Dy1, respectively) [55]. On account of this, the three intramolecular Dy–Dy distances are 5.862 Å (Figure 18c), which are longer than other literature triangular complexes with Schiff base ligands [72] with Dy–Dy–Dy angles close to 60.0°, giving a nearly perfect equilateral triangle. In addition to this, N3, N3a and N3b are also settled on the triangular plane, where the three O, N, N coordination environments of the 3 ligands are all above the plane, as shown in Figure 18d [55].

The RT χT of **31** and **32** are 40.23 and 21.89 cm^3 K mol^{-1}, respectively (Figure 18e). The χT product shows weak antiferromagnetic interactions between the Dy(III) ions. The gradual decrease with temperature for **31** till 100 K, then a rapid decrease to a value of 27.73 cm^3 K mol^{-1} at 1.8 K can be due to weak antiferromagnetic interactions between the Dy(III) ions. In Gd(III) analog **32**, the χT value reach of a minimum of 18.99 cm^3 K mol^{-1} at 1.8 K shows the presence of weak antiferromagnetic coupling between the metal centers [55].

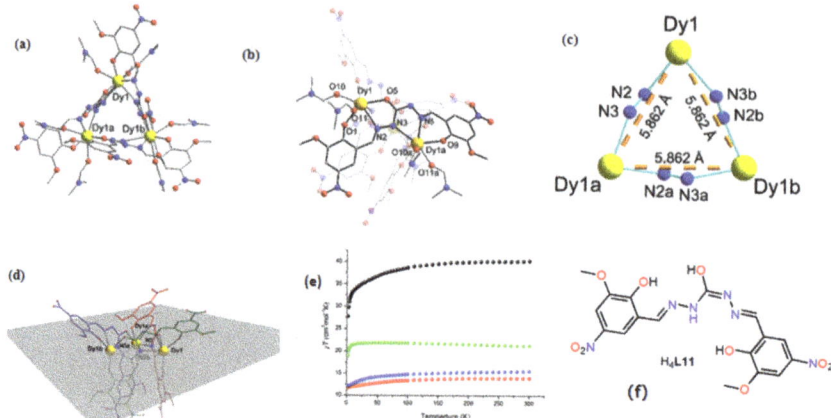

Figure 18. (a) Molecular structure of **31**, (b) the structure of [Dy₃] (c) the core **31** with Dy–Dy distances, (d) side view of **31**, (e) temperature dependence of the χT product of complexes **31** (black) and **32** (green) at 1000 Oe, (f) the structures of H$_4$**L11**. Reproduced with permission from [55]; Published by Royal Society of Chemistry, 2019.

The dynamics of magnetization for **31** between 30 and 1.8 K show frequency-dependent χ'' above 6 K, indicating SMM properties (Figure 19a) [55]. The ac susceptibility at 10 K in an applied field of 5000 Oe was done to investigate the feasibility of lowering the relaxation probability via the quantum pathway. At 1.8 K, the relaxation dynamics of **31** was affected by a static field with a shift of the peaks towards the left and having the tail of the peak in the 100–10,000 Hz range. Above 7 K out-of-phase signal was observed having a shift of the peak maxima with an energy barrier of 81 K (Figure 19a). The χ'' vs. χ' plot of **31** 6–13 K and 5–15 K also supports the relaxation process (Figure 19b) [55]. The longer Dy(III)–Dy(III) distances and Dy$_3$ plane of **31** are shown in Figure 19c,d [55].

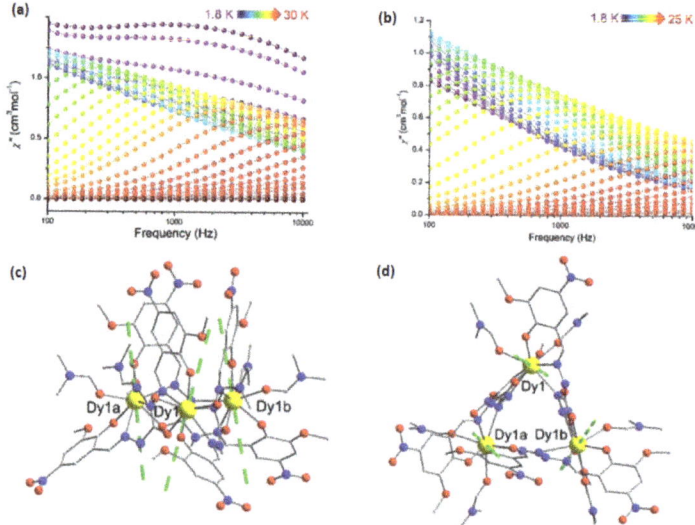

Figure 19. (a) The χ'' under zero, (b) the χ'' under 2600 Oe: DC field for complex **31** exhibiting anisotropy axes, (green dashed line) for all Dy(III) ions (side view: left (**c**), top view: right (**d**)). Reproduced with permission from [55]; Published by Royal Society of Chemistry, 2019.

4.3. Tetranuclear Schiff Base Ln(III)-Complexes of SMMs

In the construction of tetranuclear complexes, Tang and coworkers have successfully isolated two discrete linear Ln$_4$ isostructural complexes [Dy$_4$(**L12**)$_2$(C$_6$H$_5$COO)$_{12}$(MeOH)$_4$] (**33**) and [Gd$_4$(**L12**)$_2$(C$_6$H$_5$COO)$_{12}$(MeOH)$_4$] (**34**), by introducing steric, potentially pentadentate Schiff base ligand HL12. The ligand HL12 (Figure 20a) may provide effective hindrances to prevent the formation of extended structures and result in tetranuclear complexes. The structural description of Dy analog is taken as a phototype [73].

Complex **34** is nearly linear with Dy1–Dy2 and Dy1–Dy1A distances of 4.241 and 4.055 Å, respectively. Two different coordinating modes were observed for the benzoate ligands in its syn-syn η^1:η^1-μ_2 and non-bridging chelating forms. Two central Dy atoms of Dy1 and Dy1A are coordinated by one non-bridging chelating benzoate ligand, while two peripheral Dy2 and Dy2A are coordinated by two benzoates. Every two neighboring Dy ions (Dy2 and Dy1; Dy1 and Dy1A) are bridged by syn-syn η^1:η^1-μ_2-benzoate in pairs (by μ_2-O8–C41–O9 and μ_2-O10–C34–O11; μ_2-O14–C48–O17A and μ_2-O14A–C48A–O17) by generating a linear array. There are two methanol molecules in the coordination sphere generating eight-coordinate centers with a distorted bicapped trigonal-prismatic geometry [73].

Magnetization dynamics also reveal mainly one thermally-activated discrete linear Ln$_4$ complexes, instead of 1D Ln(III) chains (Figure 20b). The temperature-dependent magnetic susceptibility measurements (Figure 20d) were performed on the polycrystalline samples in the range of 300–2 K under 1000 Oe of the external field. For compound **34**, $\chi_M T$ of 32.58 cm^3 K mol^{-1} at 300 K is consistent with the spin-only value based on Gd(III) [73] ions ($^8S_{7/2}$, S = 7/2; L = 0, g = 2) and almost remains constant till 14 K. Upon cooling, the $\chi_M T$ value shows small decrease below 10 K, reaching a minimum value of 30.63 cm^3 K mol^{-1} at 2 K, showing the existence of antiferromagnetic coupling between Gd(III) ions (Figure 20c) for **34**. The $\chi_M T$ product of **33** at RT is 57.5 cm^3 K mol^{-1}, and it decreases gradually over the operating temperature range to reach a minimum of 47.8 cm^3 K mol^{-1} at 2 K (Figure 20c). This behavior can be related to the thermal depopulation of the Dy(III) excited states because the Dy–Dy exchange interactions are insignificant by comparison [73]. The χ'' vs. χ' plots of **33** (Figure 20e) show multiple relaxations associated with distinct anisotropic centers indicating the Dy(III) complex exhibits SMM characteristics.

The relaxation times of **33** at different temperatures were obtained from frequency-dependent out-of-phase AC susceptibility measurements and plotted as a function of lnτ vs. 1/T between 1.9 and 5.0 K (Figure 20f). Above 4 K, the relaxation follows a thermally activated mechanism and can be determined by the Arrhenius law (lnτ = lnτ_0 + Δ/k$_B$T) with an energy barrier of 17.2 K and a pre-exponential factor of τ_0 = 6.7 × 10^{-6} s, which is consistent with the expected τ_0 of 10^{-6}–10^{-11} for an SMM [74]. At lower temperatures, a gradual crossover to a temperature-independent regime is observed, while below 2.5 K, a dominant temperature-independent quantum regime is observed with a characteristic time of 0.005 s. This behavior is predictable for an SMM when the quantum tunneling of the magnetization becomes dominant [38].

Other tetranuclear and octanuclear Dy clusters having the formulae [Dy$_4$(HL13)$_4$(C$_6$H$_4$NH$_2$COO)$_2$(μ_3-OH)$_4$(OH)$_2$(H$_2$O)$_4$]·4CH$_3$CN$_3$·12H$_2$O] (**35**) and [Dy$_8$(HL13)$_{10}$(C$_6$H$_4$NH$_2$COO)$_2$(μ_3-OH)$_8$(OH)$_2$(NO$_3$)$_2$(H$_2$O)$_4$] (**36**), where H$_2$L13 = 2-{[(2-hydroxy-3-methoxyphenyl) methylidene]amino}benzoic acid (Figure 21d) have also been reported by Tang et al. [75].

The single-crystal X-ray diffraction studies reveal that compound **35** exhibits a tetranuclear cubane-like structure (Figure 21a) and that **36** is a dimerized version of the tetranuclear analog resulting in octanuclear, bis-cubane complex (Figure 21b). The [Dy$_4$(μ_3-OH)$_4$] cubane cores of **35** and **36** are structurally related. For compound **36**, the two cubane units are doubly bridged by μ-O,O'-carboxylato ligands (oxygen atoms O26, O26a, O24, and O24a; orange bonds). The blue bonds symbolize the μ_3-hydroxido ligands [75].

As shown in Figure 21c, the $\chi_M T$ value at 300 K is 53.6 cm^3 K mol^{-1} for **35** and 50.2 cm^3 K mol^{-1} for **36**. $\chi_M T$ slowly decreases until 50 K and then additionally decreases, reaching a minimum of

23.1 cm^3 K mol^{-1} for **35** and 23.8 cm^3 K mol^{-1} for **36** at 2 K, indicating a progressive depopulation of excited Stark sublevels [75]. The non-superimposition of the M vs. H/T data on a single master curve (Figure 21e,f) suggests the presence of a significant magnetic anisotropy and/or low-lying excited states. The magnetization eventually reaches the value of 24.4 µ$_\beta$ for **35** and 22.2 µ$_\beta$ for **36** at 2 K and 70 kOe without clear saturation. This value is much smaller than the expected saturation value of 40 µ$_\beta$ for four non-interacting Dy(III) ions, which may be due to the crystal-field effect of Dy(III), which cancels the 16-fold degeneracy of the ^{6}H$_{15/2}$ ground state.

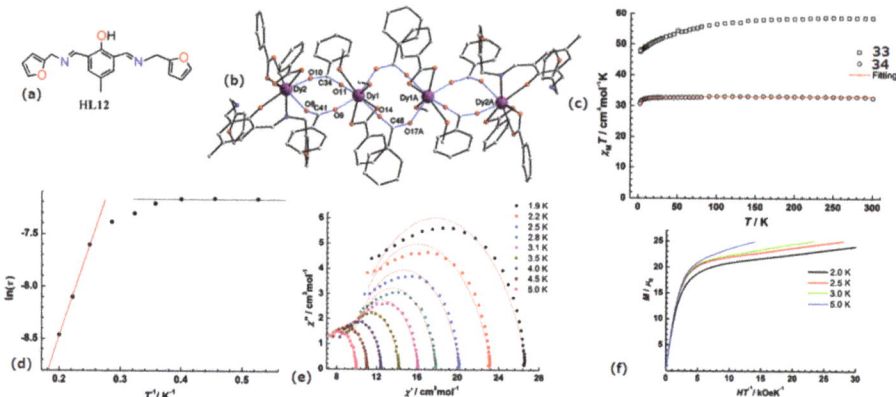

Figure 20. (a) The structure of HL12, (b) tetranuclear unit of **33** with the central core highlighted by blue bonds, (c) $\chi_M T$ vs. T products at 1 kOe for **33** (open square) and **34** (open cycle) and fitting of data for **34** (red line), (d) magnetization relaxation time, lnτ vs. T^{-1} plot for **34** in zero DC field, (e) χ'' vs. χ' plots measured below 9 K and zero-DC field (right side) for **33**, (f) the plot of the reduced magnetization vs. H/T for **33** in the field range 0–7 T and temperature range 2.0–5.0 K. Reproduced with permission from [73]; Published by Royal Society of Chemistry, 2012.

The AC susceptibility measurements were performed for **35** and **36** under a zero-DC field to investigate the dynamics of the magnetization. Checking the bond distances and angles of the respective [Dy$_4$(µ$_3$-OH)$_4$] cores reveal small but apparently important disparities in the M–O–M angles. These angle differences (induced by the presence of additional µ-OH– bridges in **35**) clearly affect the orbital overlaps between the metal centers and the µ$_3$-hydroxido ligands, as well as the local tensor of anisotropy on each Dy(III) site and their relative orientations, generating dissimilar dynamic magnetic behavior [75].

More recently in 2018, Hong-Ling and coworkers have reported five tetranuclear Ln(III)-complexes represented by the following chemical formulae [76], [Ln$_4$(dbm)$_4$(**L14**)$_6$(µ$_3$-OH)$_2$]·5CH$_3$CN (**37**), [Ln$_4$(dbm)$_4$(**L14**)$_6$(µ$_3$-OH)$_2$]·2CH$_3$CH$_2$OH·4CH$_3$CN (**38**), [Ln$_4$(dbm)$_4$(**L14**)$_6$(µ$_3$-OH)$_2$]·4CH$_3$CN·2H$_2$O (**39**), [ln$_4$(dbm)$_4$(**L14**)$_6$(µ$_3$ OH)$_2$]·CH$_3$CH$_2$OH·4CH$_3$CN·2H$_2$O (**40**) and [Ln$_4$(dbm)$_4$(**L9**)$_6$(µ$_3$-OH)$_2$] (**41**), where Ln = Gd(**37**), Tb(**38**), Dy(**39**), Ho(**40**), Er(**41**) and HL14 = 5-(4-pyridinecarboxaldehyde)amino-8 hydroxylquinoline (Figure 22d).

In **39**, there are four Dy(III) centers with **L14**$^-$ and four DBM$^-$ ligands with crystallographic inversion symmetry. The complexes **37–41** adopt a typical butterfly topology and have a distorted square-antiprismatic geometry (Figure 22b).

The $\chi_M T$ values of **37–41** at RT are 33.58, 47.20, 56.52, 57.16 and 44.37 cm^3 K mol^{-1}, respectively (Figure 22c). For **37**, the $\chi_M T$ value remains nearly constant between 300 and 60 K and decreases sharply, reaching a minimum of 17.36 cm^3 K mol^{-1} at 2 K showing weak intramolecular antiferromagnetic interaction between the Gd(III) centers. For **39**, the $\chi_M T$ increase as temperature decreases to 70 K to a value of 58.96 cm^3 K mol^{-1} is an indication of weak ferromagnetic interaction between metal ions. On further cooling of **39** to 2 K, the $\chi_M T$ reaches a value of 35.91 cm^3 K mol^{-1}. For **38**, **40**, and

41, the $\chi_M T$ product slowly decreases from 300 to 60 K, and then reaches minima of 39.61, 30.05 and 18.90 cm^3 K mol^{-1}, respectively, at 2 K [76].

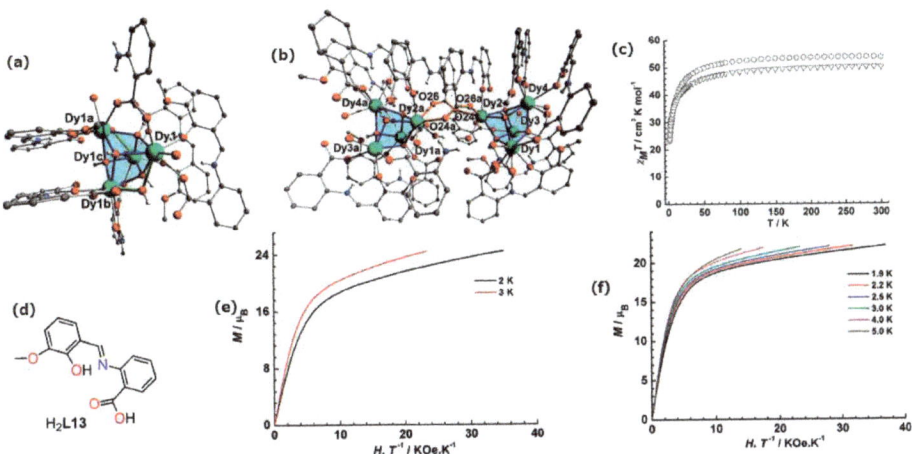

Figure 21. (a) Molecular structure of **35**, illustrating the Dy$_4$ cubane core, (b) molecular structure of octanuclear **36**, (c) temperature dependence of the $\chi_M T$ products at one kOe for **35** (open cycle) and **36** (open triangle) (d) structure of the ligand H$_2$**L13**, (e) M vs. H/T plots for **35** at 2 and 3 K, (f) M vs. H/T plots for **36** at different temperatures below 5 K, by normalizing the data per Dy$_4$ unit. Reproduced with permission from [75]; Published by American Chemical Society, 2010.

The dynamics of the magnetization of **39** were investigated in both temperature and frequency-dependent AC fields under a zero DC field. Here the relaxation parameters from lnτ vs. 1/T plots came out with an energy barrier of 89.38 K (Figure 22f) [77]. From (Figure 22e), complex **39** shows the presence of strong frequency-dependent in-phase (χ') and out-of-phase (χ'') signals, with slow magnetic relaxation at a lower temperature, proving its SMM nature [76].

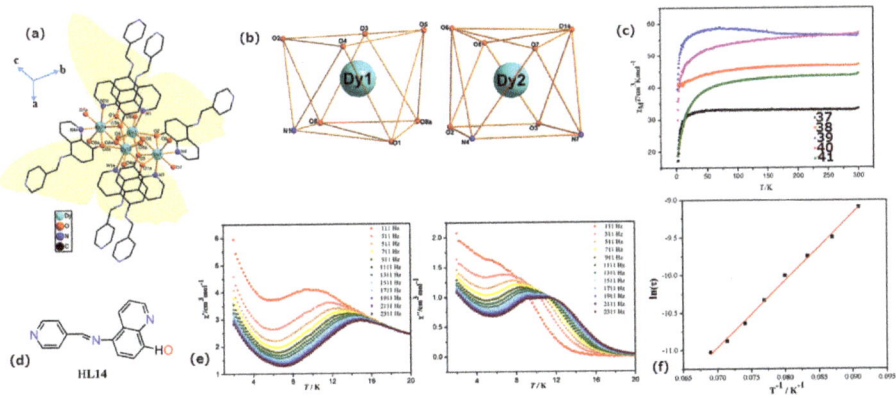

Figure 22. (a) Molecular structure of **39**, (b) geometrical representation of Dy1 and Dy2 ions in **39**, (c) $\chi_M T$ vs. T plots for complexes **37–41** between 2 and 300 K, (d) the structure of H**L14**, (e) temperature dependence of the χ' (left) and χ'' (right) components of the AC magnetic susceptibility for **39** (f) Plot of lnτ vs. T^{-1} fitted to the Arrhenius law for complex **39**. Reproduced with permission from [76]; Published by Royal Society of Chemistry, 2018.

Two analogous tetranuclear Ln(III)-complexes with Dy(III) (**42**), Gd(III)(**43**) with tetrazine-centered hydrazone Schiff base ligand having the general formula, [Ln$_4$(H$_2$**L15**)$_4$(MeOH)$_8$](NO$_3$)$_4$·a MeOH·bH$_2$O, were reported by Murugesu et al. [78], where H$_4$**L15** = (3,6-bis(vanillidenehydrazinyl)-1,2,4,5-tetrazine) (Figure 23c) and **a** = 8.07 and **b** = 0.65 for **42** and **a** = 8.19 and **b** = 0.91 for **43**. The ligand H$_4$**L15** has a tetrazine ring at the center and two identical hydrazone moieties [78]. The ligand has four coordination pockets comprising of N1, N4a and O1 tridentate coordination sites, and a bidentate pocket consists of O1 and O2 coordination sites. A distorted spherical capped square antiprism geometry is assigned for Dy1 and a spherical tricapped trigonal prism for Dy2 in **42** (Figure 23a).

The χT vs. T product exhibits the presence of non-negligible ferromagnetic coupling between spin carriers (Figure 23b). On lowering from RT, the χT values of **42** and **43** remain unchanged till 12 K and abruptly increased to a value of 69.86 cm^3 K mol^{-1} for **42** and 33.73 cm^3 K mol^{-1} for **43** at 1.9 K, showing an indication of intramolecular ferromagnetic exchange [78]. The AC susceptibility of **42** under zero applied DC field (Figure 23d) shows both the in-phase (χ') and out-of-phase (χ'') signals, and the shifting of peak maxima shows slow relaxation of the magnetization with an energy barrier of 158 K. The micro-SQUID measurement of **42** at below 0.5 K with a sweep rate of 14 T s^{-1} shows hysteretic behavior (Figure 23e). The width of the magnetic hysteresis loop of complex **42** indicates a strong dependence on temperature and moderate dependence on sweep rate [78].

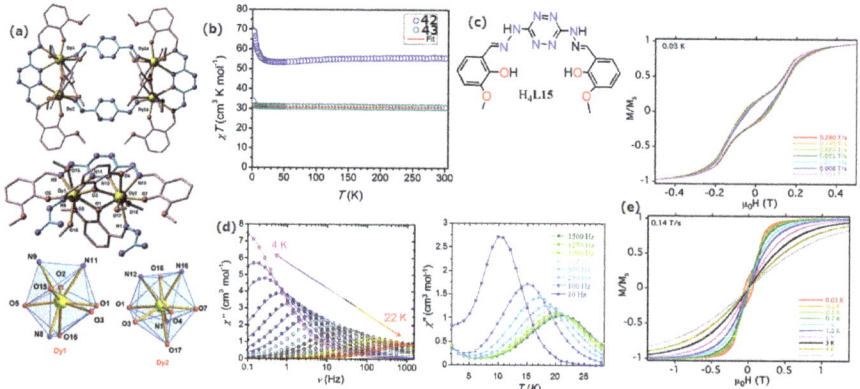

Figure 23. (**a**) Molecular structures (top), dinuclear complex (middle) and coordination polyhedra of **42** (bottom), (**b**) Temperature dependence of χT for **42** and **43**, (**c**) the structure of H$_4$**L15**, (**d**) χ'' vs. ν(Hz)(left) and χ'' vs. T (right) for **43**, (**e**) magnetic hysteresis measurements of a single crystal of **43** on a micro-SQUID **43** (top) and temperatures (bottom). Reproduced with permission from [78]; Published by Royal Society of Chemistry, 2017.

A tetranuclear Dy-cluster having the structural formula, [Dy$_4$(H**L16**)$_4$(MeOH)$_6$]·2MeOH (**44**) was reported by Tang et al., whose molecular structure is depicted in Figure 24a, where H$_3$**L16** = 2-hydroxy-3-methoxybenzioc acid [2-hydroxy-3-methoxyphenyl) methylene] hydrazide (Figure 24d) [44]. The central Dy(III) ions of the Dy$_4$ core are connected by two μ-O units. Strong inter-and intramolecular hydrogen bonding interactions give a two-dimensional supramolecular array with a zigzag arrangement of the molecules. The Dy$_4$ compound derived from a rigid hydrazone ligand shows a nearly linear Dy$_4$ core, one being in a distorted bicapped trigonal-prismatic geometry and the other in a nearly perfect mono-capped square-antiprismatic environment [44].

Direct-current magnetic susceptibility studies of **44** were carried out in an applied magnetic field of 1 kOe in the temperature range 300–2 K Figure 24e. The $\chi_M T$ value of 54.9 cm^3 K mol^{-1} at 300 K observed is slightly lower than the value of 56.7 cm^3 K mol^{-1}. The value of $\chi_M T$ gradually decreases until ~30 K, where it drops abruptly to a minimum of 31.3 cm^3 K mol^{-1} at 2 K, indicating a progressive depopulation of excited Stark sublevels. Magnetization data are shown in the inset of (Figure 24e,f).

The non-super position of the $\chi_M T$ vs. H/T data on a single master curve suggests the existence of significant magnetic anisotropy or/and low-lying excited states [44].

It is interesting to note that more than one peak is seen in the temperature-dependent ac magnetic susceptibility curves (Figure 24b), showing an unusual multiple relaxation mechanism operating in **44**. Additionally, in terms of the χ'' vs. frequency plots characterized by two clear maxima, two effective energy barriers identified are 19.7 and 173 K corresponding to fast and slow relaxation phases, respectively [44], which was confirmed by the Cole–Cole plots, that clearly indicate the evolution from fast relaxation to slow relaxation phases with the changing of temperature observed at 7 and 8 K (Figure 24b). The two different relaxation processes might be associated with distinct anisotropic centers, that is, two Dy(III) ions with different geometries [44]. The (χ'') vs. frequency plot of **44** at 7 K (Figure 24c) exhibiting two peaks centered at 1.2 and 1200 Hz, respectively, could be because of the spin noncollinearity of two types of Dy(III) ions in the weakly coupled molecular system. It is interesting to note that the peaks in the frequency-dependent AC susceptibility are quite distorted, exhibiting a unique double-ridge structure (Figure 24c) as opposed to the shoulder structure in the Dy$_3$ system [29].

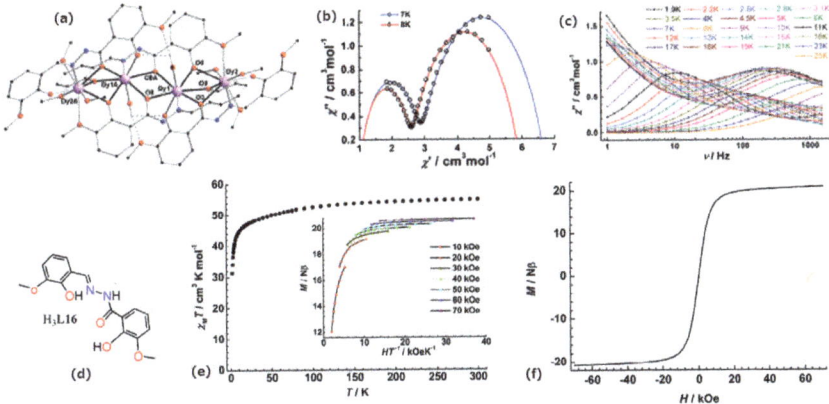

Figure 24. (a) Molecular structure of **44**, (b) dynamical susceptibility (ω) at 7 and 8 K for **44**, (c) out-of-phase AC-susceptibility $''$ vs. frequency ν on a logarithmic scale for **44** over the temperature range 1.9–25 K, (d) the structure of H$_3$**L16**, (e) plot of $\chi_M T$ vs. T for **44**. The inset is a plot of the reduced magnetization (M/Nβ) vs. H/T for **44** in the field range 1–7 T and temperature range 1.9–4 K, (f) $\chi_M T$ vs. H data of **44** at 1.9 K. Reproduced with permission from [44]; Published by American Chemical Society, 2010.

In another communication, Tang and coworkers have reported a tetranuclear Dy cluster having the formula [Dy$_4$(**L10**)$_4$(H**L10**)$_2$(C$_6$H$_4$NH$_2$COO)$_2$(CH$_3$OH)$_4$]·5CH$_3$OH (**45**) where H$_2$**L10** = N-(2-carboxyphnyl) salicylaldimine (Figure 25d), whose molecular structure is in Figure 25a [79]. Here H$_2$**L10** has three different binding modes (Chart 4). Two full deprotonated tetradentate **L10**$^{2-}$ wrap Dy atoms in $\eta^1{:}\eta^1{:}\eta^2{:}\eta^2{:}\eta^2{:}\mu_3$- fashion and two peripheral fully deprotonated **L10**$^{2-}$ ligands bind in a $\eta^1{:}\eta^1{:}\eta^2{:}\eta^1{:}\mu_2$- way. Furthermore, two peripheral zwitterionic ligands coordinate in $\eta^1{:}\eta^1{:}\eta^1{:}\mu_2$ way generating a linear metal array. Lastly, two peripheral zwitterionic-tridentate ligands bind in a $\eta^1{:}\eta^1{:}\eta^1{:}\mu_2$- condition, and a linear metal array is generated. The peripheral zwitterionic ionic H**L10**$^-$ ligands block the N-coordination to the metal ions.

A molecule of methanol is also coordinated to Dy1, generating an 8-coordinated species with a square-antiprismatic geometry, while for Dy2, a molecule of methanol and an anthranilato ligand is also coordinated for an 8-coordination around Dy2 in between a bi-capped trigonal prism and a square antiprism.

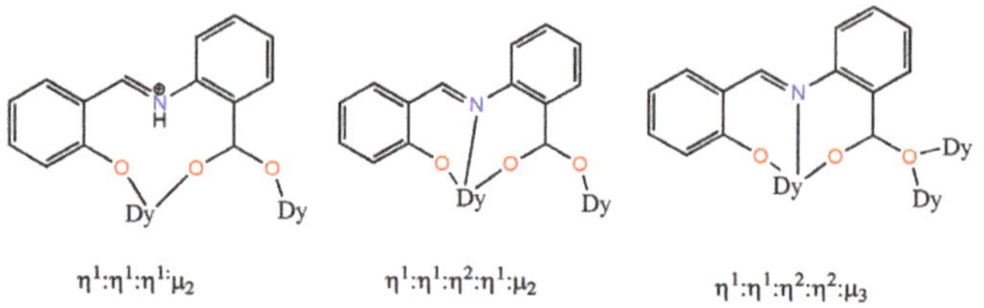

Chart 4. Coordination modes of **HL10**$^-$ (η^1:η^1:η^1:μ_2) and **L10**$^{2-}$ (η^1:η^1:η^2:η^1:μ_2 and η^1:η^1:η^2:η^2:μ_3) in compound **45**. Reproduced with permission from [79]; Published by Royal Society of Chemistry, 2010.

The $\chi_M T$ vs. T curve of **45** shows a value of 56.3 cm^3 K mol^{-1} at 300 K, and it decreases gradually to reach a minimum of 31.7 cm^3 K mol^{-1} at 2 K (Figure 25b) [80]. The magnetization data collected in the 0–70 kOe shows significant magnetic anisotropy, which rapidly increases and reaches 27.0 μB at 1.9 K and 70 kOe without clear saturation (Figure 25b), which is lower than the expected saturation value for four non-interacting Dy(III) ions [81]. The relaxation time was extracted from the frequency-dependent data between 1.9 and 9 K (Figure 25c). Below 3 K, a temperature-independent relaxation regime is observed with a characteristic time of 0.00068 s. Such behavior is expected for an SMM when the quantum tunneling of the magnetization becomes dominant [38,82].

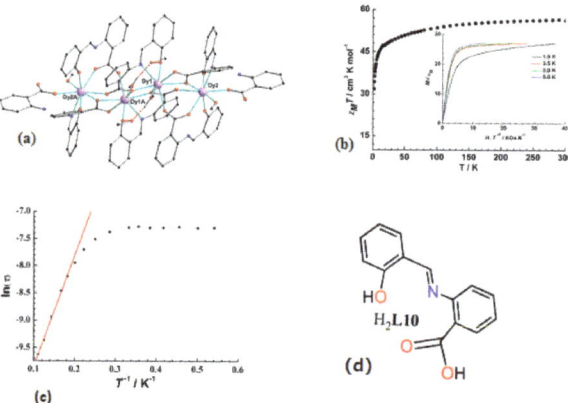

Figure 25. (a) The molecular structure of compound **45**, (b) $\chi_M T$ vs. T plot at 1 kOe; inset: M vs. H/T plots at different temperatures below 5 K for **45**, (c) relaxation time, lnt vs. T^{-1} plot for **45** under zero-DC field: the solid line is fitted with the Arrhenius law, (d) the structure of H$_2$**L10**. Reproduced with permission from [79]; Published by Royal Society of Chemistry, 2010.

In 2019 Tang and coworkers prepared and characterized a Dy-based linear helicate cluster, [Dy$_4$(H**L17**)$_2$**L17** (DMF)$_8$]·2ClO$_4$·CH$_2$Cl$_2$·4DMF·(CH$_3$CH$_2$)·2H$_2$O (**46**), with pyridazine-based Schiff base, ([3,6-bis(2-hydroxy-3-methoxybenzylidene)hydrazinecarbonyl]-pyridazine), H$_4$**L17**, (Figure 26a), was synthesized by the condensation of pyridazine-3,6-dicarbohydrazide and o-vanillin [83,84] and it exhibits flexible coordination modes (Chart 5), owing to structural tautomerism. It is interesting to note that the coordination centers always reside on the same side of H$_4$**L17**, which favors the formation of polynuclear clusters [85]. With respect to the scope of this review, the pure Ln(III) compound, [Dy$_4$(H**L17**)$_2$**L17**(DMF)$_8$]·2ClO$_4$·CH$_2$Cl$_2$·4DMF·(CH$_3$CH$_2$)$_2$O·H$_2$O (**46**) is discussed.

The compound **46** is a linear array of Dy$_4$ core (Figure 26b). In **46**, among the three ligands, one is completely deprotonated and connects with four Dy(III) ions with the binding mode indicated in Chart 5 (left) by utilizing NO-bidentate and ONO-tridentate coordination nature. The other two are tri-deprotonated and coordinate to four Dy(III) ions, as shown in Chart 5 (right) using three kinds of coordination pockets [86]. For compound **46**, the smallest intermolecular Dy–Dy distance is 9.495, indicating relatively weak intermolecular magnetic interactions [86].

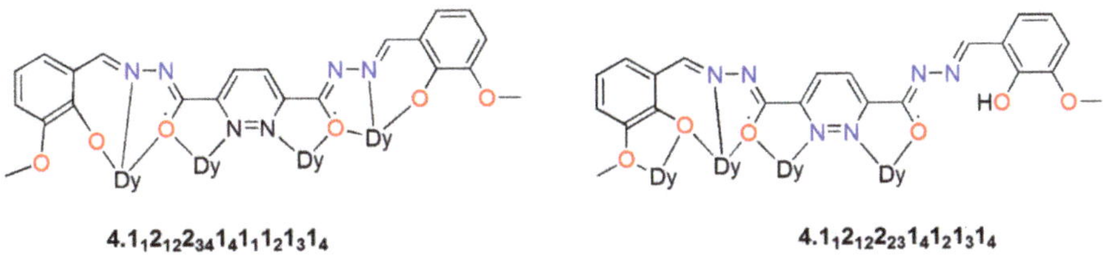

$4.1_12_{12}2_{34}1_41_11_21_31_4$ $4.1_12_{12}2_{23}1_41_21_31_4$

Chart 5. Coordination nature of the H$_4$L17 with different deprotonated conditions in **46** (Harris notation). Reproduced with permission from [86]; Published by Royal Society of Chemistry, 2020.

The DC susceptibility measurement was carried out for compound **46** from 2 to 300 K in an applied field of 1 kOe (Figure 26c). The $\chi_M T$ products at 300 K, for compound **46** is 57.31 cm^3 K mol^{-1}. When the temperature is lowered, the $\chi_M T$ product of **46** progressively decreases and reaches a value of 48.37 cm^3 K mol^{-1} at 12 K, and then sharply increases to a value of 63.36 cm^3 K mol^{-1} at 2 K, suggesting dominant ferromagnetic interactions [39,87,88].

The field-dependent magnetization measurements were done for compound **46** between 0–70 kOe at 1.9, 3 and 5 K, respectively (Figure 26e). For compound **46**, the magnetizations quickly rise to 10 kOe and reach the value of 20.96 μB, at 70 kOe at 1.9 K, which is much lower than the expected saturation value of 40 μB for four independent Dy(III) ions [86], which may be due to the significant crystal-field effect [29,89]. The non-superposition of magnetization plots over a single master curve proves the presence of considerable magnetic anisotropy and/or low-lying excited states [31,90]. In order to investigate the dynamics of magnetization of **46**, the AC measurements were conducted under a zero DC field (Figure 26d). The χ'' signals of **46** show frequency dependence below 20 K, indicating the slow relaxation of magnetization. On cooling, a remarkable increase without well-defined peaks indicates a fast QTM effect at low-temperatures, as observed in most of the early reports [89–94]. The U_{eff} of **46** is reported as ~4 k [86].

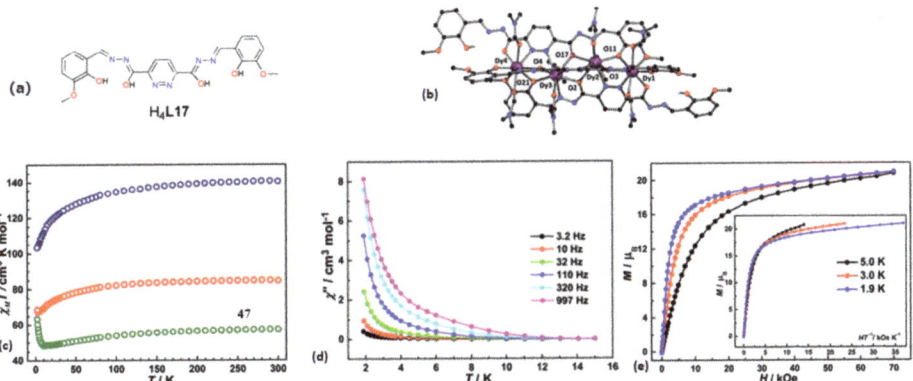

Figure 26. (a) The structure of H$_4$L17 (b) crystal structure of **46**, (c) $\chi_M T$ vs. T of **46** (blue) under a 1 kOe DC field, (d) temperature dependences of out-of-phase (χ'') signals for **46** under zero DC field, (e) field dependences of magnetization between 0 and 70 kOe and at temperatures of 1.9 and 3.0 K. Insets: plots of the reduced magnetization M vs. HT^{-1} for **46**. Reproduced with permission from [86]; Published by Royal Society of Chemistry, 2020.

In the search for tetranuclear aggregates, another Dy(III)$_4$ aggregate has been reported by Hou et al. having the formula [Dy$_4$(TTA)$_4$(**L18**)$_4$(H$_2$O)$_2$]·4CH$_3$OH (**47**) [95], with the multidentate 8 hydroxyquinoline-based Schiff base ligand (H$_2$**L18**) (Figure 27a), where H$_2$**L18** is 2-[(4-methoxyphenyl)imino] methyl]-8-hydroxy-quinolone and TTA is 2-thenoyltrifluoroacetone (Chart 6). The ligand H$_2$**L18**, having flexible coordination modes of N and O atoms, acts as a good candidate to construct Ln(III)-based high nuclearity clusters. On reacting Dy(TTA)$_3$·2H$_2$O, with H$_2$**L18** compound **47** was obtained.

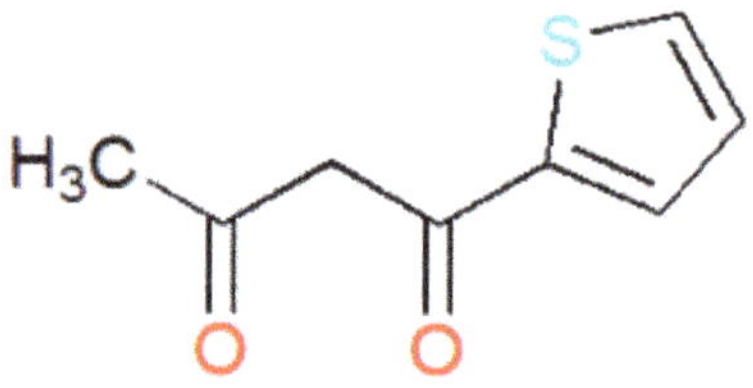

Chart 6. The structures of TTA.

The coordination modes of H$_2$**L18** and TTA in cluster **47** are shown in Figure 27c. As shown in Figure 27b, the molecular structure of **47** mainly consists of four Dy(III) ions, four TTA$^-$, four **L18**$^{2-}$ and two coordinated waters. Six oxygen atoms (O1, O2, O4, O7, O8 and O11) and two nitrogen atoms (N1and N2) are coordinated to the central Dy(III) ion with the N$_2$O$_6$ coordination environment. The four Dy(III) ions are bridged by two carboxyl oxygen atoms and four μ_2-O atoms from four **L18**$^{2-}$, resulting in a Dy$_4$ parallelogram core. The coordination polyhedrons for both 8-coordinate Dy1 and Dy2 central ions are described as a distorted square-antiprismatic geometry with a quasi-D$_4$d symmetry, which was calculated using Shape 2.0 software [96].

As shown in Figure 27d, for **47**, at RT $\chi_M T$ value is 56.64 cm^3 K mol^{-1} [95], which is in conformity with the expected value of four non-interacting Dy(III) ions. On decreasing the temperature, the $\chi_M T$ values decrease slowly between 300–50 K, and then rapidly falls to a minimum of 38.15 cm^3 K mol^{-1} at 2.0 K [95]. This behavior generally can be attributed to the weak antiferromagnetic exchange between the adjacent Dy(III) ions in the system and/or the thermal depopulation of the Dy(III) Stark sub-levels [92]. In the 0–80 kOe magnetic field range and at T = 2.0 K, the M vs. H curve for **47** was investigated [95]. M value increases quickly at low field and then increases slowly without complete saturation till H = 80 kOe. The M value of **47** is 23.05 Nβ at 80 kOe, which is much lower than the theoretical saturated value of 40 Nβ for four free Dy(III) ions. Furthermore, like shown in Figure 27e, the M vs. HT^{-1} curves at 2.0–8.0 K show non-superimposed magnetization curves for cluster **47**, which also suggests the existence of significant anisotropy and/or low-lying excited states of Dy(III) ions [97].

In order to understand the magnetic relaxation dynamics of **47**, AC susceptibility measurement was done at zero DC magnetic field in the temperature range 2.0–15.0 K and frequency 111–3111 Hz. As shown in Figure 27f [95], there is no obvious frequency dependence below 15.0 K in the in-of-phase (χ') component susceptibility for **47**, however, the out-of-phase susceptibility (χ'') clearly displays frequency-dependent signals below 10 K, but no well-defined peaks are seen till the temperature drops to 2.0 K, which may be due to quantum tunneling of the magnetization(QTM) [98].

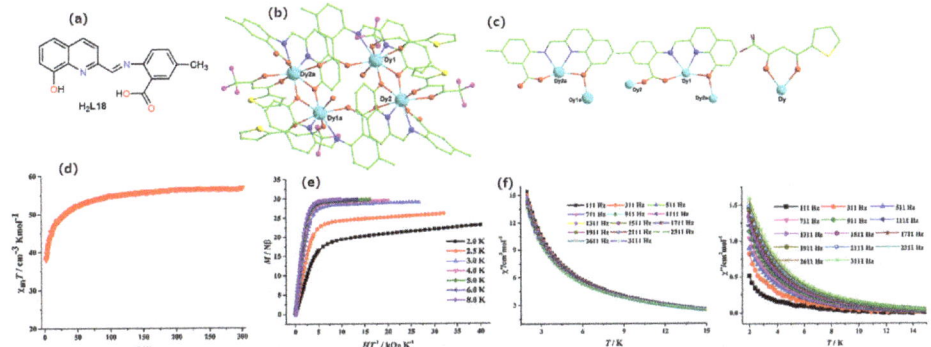

Figure 27. (a) The structure of H$_2$L18, (b) molecular structure for 47, (c) coordination modes of H$_2$L18 and TTA in 47, (d) $\chi_M T$ vs. T for 47 under an applied DC field of 1000 Oe, (e) M vs. HT^{-1} for 47 in the field range 0–80 kOe, (f) χ' (left) and χ'' (right) components of the AC magnetic susceptibility measurement for 47. Reproduced with permission from [95]; Published by Elsevier, 2020.

4.4. Hexanuclear Schiff Base Ln(III)-Complexes of SMMs

Ghosh and coworkers have reported a hexanuclear Dy(III) cluster [Dy$_6$(**L19**)$_7$(H**L19**)(MeOH)$_2$(H$_2$O)(OH)$_2$(OAc)] (**48**) [99], where (H$_2$**L19**) are o-phenolsalicylimine (comprising of two similar coordination pockets for Dy encapsulation) (Figure 28a). This unique Dy(III)$_6$ cluster, formed by the exclusive combination of two vertex-sharing and one edge-sharing high-anisotropy Dy$_3$ triangles (Figure 28b), gives rise to an unprecedentedly asymmetric Dy(III)$_6$ homometallic core. The crystal packing reveals that the molecules of **48** are in contact through π–π interactions, generating an infinite supramolecular array, where there is a strong π–π interaction (d = 3.268 Å) between two ligands of the two closely situated molecules, generating intermolecular π–π interactions. The hexanuclear Dy(III) complex is represented in Figure 28b. Each Dy(III) in the hexanuclear aggregate is 8-coordinated, and a square-antiprismatic geometry may be assigned around the metal ion [99].

The dynamics of the magnetization measurements operating in a 3.0 Oe AC field oscillating at frequencies of 3–1200 Hz and with a zero DC field for Dy$_6$ is shown in Figure 28d, as the plots of χ' vs. T and χ'' vs. T. The DC magnetic susceptibility studies of a polycrystalline sample (Figure 28c) gives a room-temperature $\chi_M T$ value equal to 82.39 cm^3 K mol^{-1}. The $\chi_M T$ values decrease gradually with decreasing the temperature. The M vs. H data at different temperatures show a swift rise in the magnetization at low fields, reaching values of 32.09 μ$_B$ at 1.9 K and 7 T for Dy$_6$ (Figure 28f). The non-superimposed curves validate the existence of anisotropy and/or low-lying excited states [99]. Linear fitting of the experimental ln(χ''/χ') data to the equation ln(χ''/χ') = ln(ωτ) + U_{eff}/kT generating the parameters $U_{eff} \approx 3.0$ K and $\tau_0 \approx 8.3 \times 10^{-6}$ s can be seen in Figure 28e. The frequency-dependent out-of-phase signals signify the onset of slow magnetization relaxation. The nonexistence of frequency-dependent peaks in the out-of-phase susceptibility signals for this Dy$_6$ system is most probably attributed to the fast quantum tunneling of the magnetization [99].

Continuing the search for hexanuclear aggregates, a distinctive hexanuclear Dy(III) compound having the formula [Dy$_6$(μ$_3$-OH)$_3$(μ$_3$-CO$_3$)(μ-OMe)(**L20**)$_6$(MeOH)$_4$-(H$_2$O)$_2$]·3MeOH·2H$_2$O (**49**), Figure 29a, was reported by Tang et al. with the polydentate Schiff base ligand(H$_2$**L20**) (Figure 29d). The hexanuclear core of complex **49** contains six Dy(III) ions, which can be considered as the amalgamation of three capped triangular Dy$_3$ units [100]. The structure of **49** consists of two crystallographically unique but structurally same, Dy$_6$ units in the unit cell, as shown in Figure 29a. A total of six polydentate Schiff-base ligands surround the Dy$_6$ cluster core and exhibit three different binding modes in its di-deprotonated forms. Four methanol molecules, two water molecules, and one CO$_3^{2-}$ anion occupy the remaining coordination sites of Dy(III) ions. Importantly, the CO$_3^{2-}$ anion coordinated to the three Dy(III) ions in a η^2:η^2-μ$_3$ bidentate fashion. Each metal center in **49** is

8-coordinated, and a square-antiprismatic geometry may be assigned around the Dy(III) ions in the aggregate [100].

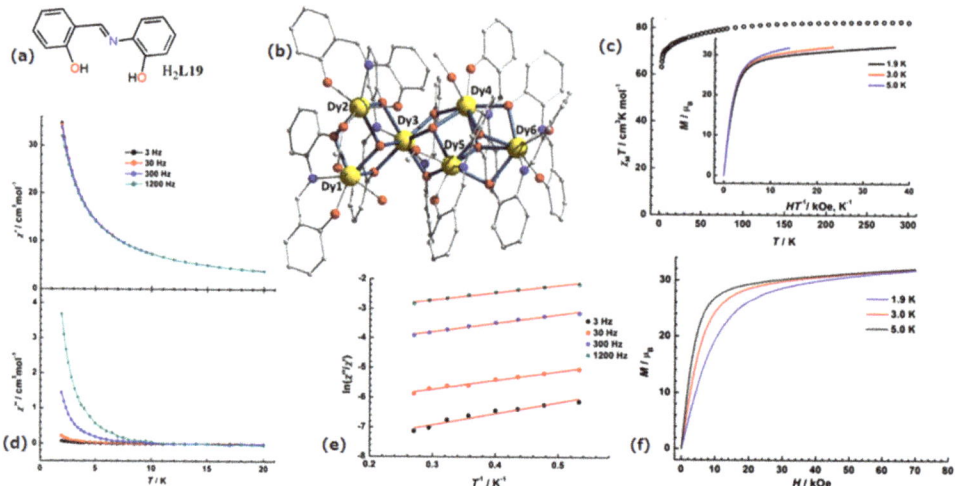

Figure 28. (a) The structure of H2L19, (b) hexanuclear unit of **48** with the central core highlighted by blue bonds, (c) $\chi_M T$ vs. T for **48** at 1 kOe (with χ = M/H normalized per mol). Inset: M vs. H/T plot at various temperatures between 1.9 and 5 K for **48**, (d) temperature-dependent AC susceptibility data for compound **48**, obtained under zero DC field at the indicated frequency, (e) natural logarithm of the ratio of χ'' over χ' vs. 1/T of the data for **48**, (f) field dependence of the magnetization at various temperatures between 1.9 and 5 K. Reproduced with permission from [99]; Published by Elsevier, 2013.

The DC magnetic susceptibility studies for complex **49** was performed in a magnetic field of 1000 Oe in the temperature range 300–2 K (Figure 29b). The room temperature $\chi_M T$ value of **49** is 84.8 cm^3 K mol^{-1}, which corresponds to the anticipated value of 85.02 cm^3 K mol^{-1} for six uncoupled Dy(III)-ions. The $\chi_M T$ values decrease up to 50 K with an additional drop at 2 K and reach a minimum of 68.6 cm^3 K mol^{-1} probably due to the progressive depopulation of excited stark sub-levels and the additional drop at 2 K may be due to the competition between the ligand field effect and the ferromagnetic interaction between the Dy (III) ions. The M vs. H/T (Figure 29b), inset data at different temperatures disclose a prompt surge of the magnetization at low magnetic fields, which finally reaches a value of 30.9 μ_B at 1.9 K and 7 T without the saturation value of 60 μB (six no-interacting Dy(III) ions). This could be due to the anisotropy and the crystal field effects of Dy(III) ions.

The non-superposition of the M vs. H/T data on a single master curve refers to the existence of noteworthy magnetic anisotropy and/or low-lying excited states in compound **49** [100]. The dynamics of the magnetization by AC susceptibility measurements at zero static fields and a 3.0 Oe AC field oscillating from 1 to 1500 Hz are shown in Figure 29c. At temperatures below ~30 K, a frequency-dependent out-of-phase (χ'') AC signal reveals the onset of slow relaxation of the magnetization. The relaxation time was calculated from the frequency-dependent data between 1.9 and 17 K, with the Arrhenius plot (Figure 29c). It is important to note that the two relaxation processes, which are clearly observed, may be due to the single-ion interaction of individual Dy(III) ions and the weak coupling at high and low-temperatures, respectively [100]. It is interesting to note the energy gap (Δ) between the two relaxation regimes are 5.6 and 37.9 K with pre-exponential factors (τ_0) of 4.2×10^{-5} and 3.8×10^{-6} s for the low- and high-temperature domain, respectively [100].

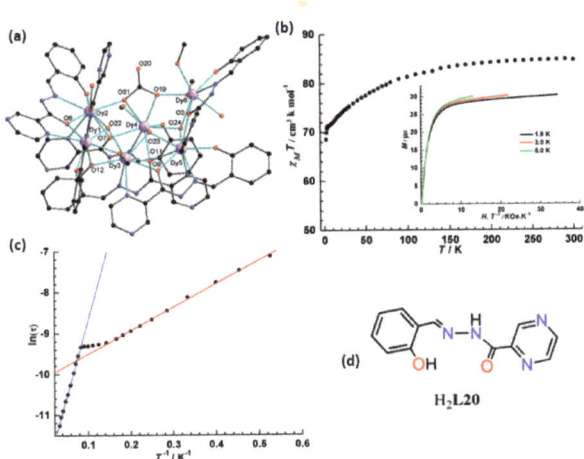

Figure 29. (a) Molecular structure of complex **49**, (b) $\chi_M T$ vs. T plot at 1 kOe. Inset: M vs. H/T plots at different temperatures below 5 K for complex **49**, (c) magnetization relaxation time, $\ln\tau$ vs. T^{-1} plot under zero DC field, (d) the structure of H_2L20. Reproduced with permission from [100]; Published by American Chemical Society, 2011.

Another Dy-based linear helicate Dy_6 cluster with H_4L17 have been prepared and characterized by Tang et al. The compound is of the formula $[Dy_6(L17)_3(PhCOO)_6(CH_3OH)_6]\cdot 11CH_3OH\cdot H_2O$ (**50**), where the ligand, H_4L17, is ([3,6-bis(2-hydroxy-3-methoxybenzylidene)hydrazinecarbonyl]-pyridazine) (Figure 30a).

In compound **50**, three H_4L17 wrap around six-Dy(III), generating a linear hexanuclear triple helical structure (Figure 30b) [86]. The asymmetric unit of **50** consists of three fully deprotonated H_4L17, six-coordinated benzoate and methanol molecules also having solvent molecules in the lattice. H_4L17 coordinated with six Dy(III), as shown in Chart 7 [86].

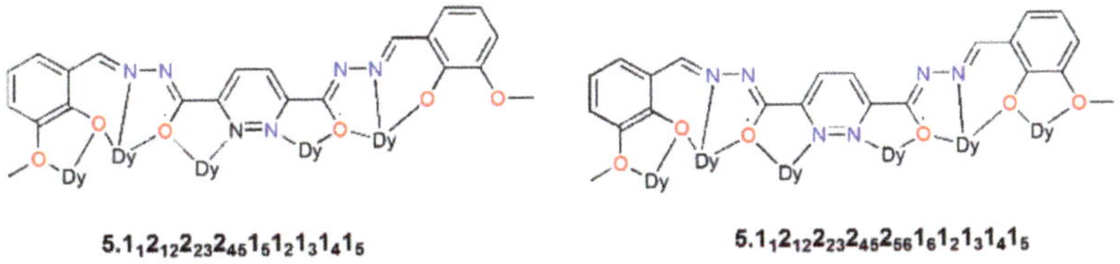

$5.1_12_12_22_23_24_61_51_21_31_41_5$ $5.1_12_12_22_23_24_52_56_1_61_21_31_41_5$

Chart 7. Coordination nature of the H_4L17 with various deprotonated forms in compound **50** (Harris notation). Reproduced with permission from [86]; Published by Royal Society of Chemistry, 2020.

The DC susceptibility was done on **50**, between 2 and 300 K in an applied field of 1 kOe (Figure 30c). The $\chi_M T$ products at 300 K, for **50** is 84.77 cm^3 K mol^{-1}, which slightly decreases on decreasing temperature to a value of 78.44 cm^3 K mol^{-1} at 50 K, and then decreases rapidly to a value of 68.61 cm^3 K mol^{-1} at 2 K. The lowering of $\chi_M T$ products can be attributed to the thermal depopulation of excited Stark sub-levels, with the possibility of weak antiferromagnetic interactions between the Dy(III) ions at low-temperatures [86].

The AC measurements were done on **50** under a zero DC field (Figure 30d). It has been observed that the out-of-phase (χ'') signals for **50** at 11 K indicate the slow magnetic relaxation behavior. χ'' component shows a significant increase in cooling without well-defined peaks, probably induced by the fast QTM effect at low-temperatures. The field-dependent magnetization measurements were done in the range of 0–70 kOe and at 1.9, 3 and 5 K, respectively, for **50** (Figure 30e). It has been observed that up to 10 KOe, the magnetization rises quickly and reaches a maximum value of 33.14 µB with 70 kOe at 1.9 K, which is less than the expected saturation value of six non-interacting DY(III) ions is 60 µB [86]. Further, the non-superposition of the M vs. H/T plots over a single master curve shows the presence of magnetic anisotropy and **50** exhibits an effective energy barrier around 2 K [86].

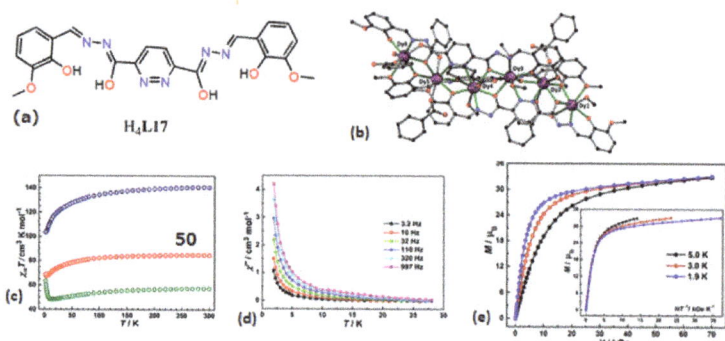

Figure 30. (**a**) The structure of H$_4$L17 (**b**) crystal structure of **50**, (**c**) $\chi_M T$ vs. T curve for **50** (red), (**d**) temperature dependence of out-of-phase (χ'') signals for **50** under zero DC field, (**e**) field dependences of magnetization between 0 and 70 kOe and at temperatures of 1.9 and 3.0 K. Insets: plots of the reduced magnetization M vs. HT^{-1} for **50**. Reproduced with permission from [86]; Published by Royal Society of Chemistry, 2020.

The above discussed Ln(III)-based Schiff base complexes are categorized in Table 1 according to their nuclearities, coordination environment and polyhedra, as well as their indicative SMMs characteristics like energy barrier (U_{eff}).

Table 1. The coordination environment and energy barriers of Ln(III)-based Schiff base single-molecule magnets (SSMMs) of different nuclearities.

Complex Formula and Number	Nuclearity	Coordination Environments	Coordination Polyhedra	Energy Barrier (U_{eff}/K)	Ref.
[Ln(**L1**)$_3$] (**7**) (Ln = Dy)	Mononuclear	N_3O_3	Distorted trigonal-prismatic	31.4 K	[54]
[Ln(**L1**)$_3$] (**8**) (Ln = Er)	Mononuclear	N_3O_3	Distorted trigonal-prismatic	24 K	[54]
[Dy(**L2**)(NO$_3$)(DMF)$_2$]·DMF (**9**·DMF)	Mononuclear	N_2O_7	Spherical triangular dodecahedron	34 K	[55]
[Dy(**L2**)$_2$(H$_2$O)$_2$]·NO$_3$·EtOH (**10**·NO$_3$·EtOH)	Mononuclear	N_2O_6	Spherical triangular dodecahedron	19 K	[55]
[Ln$_2$(H**L3**)$_2$]·nCH$_3$CN (Ln = Gd) (**11**)	Dinuclear	N_4O_4	Square-antiprismatic	N/A	[57]
[Ln$_2$(H**L3**)$_2$]·nCH$_3$CN (Ln = Tb) (**12**)	Dinuclear	N_4O_4	Square-anti-prismatic	N/A	[57]
[Ln$_2$(H**L3**)$_2$]·nCH$_3$CN(Ln = Dy) (**13**)	Dinuclear	N_4O_4	Square-antiprismatic	19 K	[57]
[Dy$_2$(**L4**)$_2$(DBM)$_2$(DMF)$_2$]·3CH$_3$OH(**14**)	Dinuclear	NO_7	Square-antiprismatic	11 K	[58]
[Dy$_2$(H**L5**)$_2$(NO$_3$)$_2$(MeOH)$_2$](**15**)	Dinuclear	NO_7	Square-antiprismatic	56 K	[38]
[Dy$_2$(H**L5**)$_2$(NO$_3$)$_2$(MeOH)$_2$]$_\infty$(MeCN)·(2MeCN)(**16**)	Dinuclear	NO_7	Square-antiprismatic	71 K	[38]
Ln$_2$(DBM)$_6$(**L6**)] (Ln = Dy(**17**) and Ln = Yb(**18**)	Dinuclear	N_2O_6	Distorted square antiprism	47 (Dy)	[63]
[Ln(III)$_2$(**L7**)$_2$(NO$_3$)$_2$] (Ln(III) = Dy(III)) (**22**)	Dinuclear	N_3O_5	Square antiprism or/and dodecahedron	76 K	[37]
[Ln$_2$(H**L8**)(dbm)$_4$]·2CH$_3$OH (Ln = Dy) (**25**),	Dinuclear	N_2O_6	Distorted dodecahedron	0.73 K	[65]
[Dy$_2$(dbm)$_2$(H**L9**)$_2$(C$_2$H$_5$OH)$_2$] (**28**)	Dinuclear	N_2O_6	Distorted bicapped trigonal-prismatic	46 ± 3.2 K	[67]
[Ln$_2$(NO$_3$)$_4$(H**L10**)$_2$(MeOH)$_4$]·MeOH, Ln = Gd (**29**·MeOH) and [Ln$_2$(NO$_3$)$_2$(H**L10**)$_4$]·6MeCN, Ln = Dy (**30**·6MeCN)	Dinuclear	N_2O_7	Square-antiprismatic	N/A	[70]
[Dy$_3$(H**L11**)$_3$(DMF)$_6$] (**31**)	Trinuclear	N_2O_4	Triangular dodecahedron	80 K	[55]

Table 1. *Cont.*

Complex Formula and Number	Nuclearity	Coordination Environments	Coordination Polyhedra	Energy Barrier (U_{eff}/K)	Ref.
[Dy$_4$(**L12**)$_2$(C$_6$H$_5$COO)$_{12}$(MeOH)$_4$](**33**) and [Gd$_4$(**L12**)$_2$·(C$_6$H$_5$COO)$_{12}$(MeOH)$_4$](**34**)	Tetranuclear	NO$_7$	Distorted bicapped trigonal-prismatic	17 K	[73]
[H**L13**)$_4$(C$_6$H$_4$NH$_2$COO)$_2$(μ_3-OH)$_4$(μ-OH)$_2$ (H$_2$O)$_4$]·4CH$_3$CN$_3$·12H$_2$O(**35**)	Tetranuclear	O$_8$	Tetranuclear cubane for 35 and octanuclear, bis-cubane for 36	N/A	[75]
[Dy$_4$(dbm)$_4$(**L14**)$_6$(μ_3-OH)$_2$]·4CH$_3$CN·2H$_2$O (**39**)	Tetranuclear	N$_2$O$_6$	Square-antiprismatic	89 K	[76]
[Ln$_4$(H$_2$**L15**)$_4$(MeOH)$_8$](NO$_3$)$_4$·aMeOH·bH$_2$O, (Ln = Dy(III)) (**42**)	Tetranuclear	N$_3$O$_6$	Distorted spherical capped square antiprism	158 K	[78]
[Dy$_4$(H**L16**)$_4$(MeOH)$_6$]·2MeOH (**44**)	Tetranuclear	NO$_7$	Distorted bicapped trigonal-prismatic	173 K	[44]
[Dy$_4$(**L10**)$_4$(H**L10**)$_2$(C$_6$H$_4$NH$_2$COO)$_2$ (CH$_3$OH)$_4$]·5CH$_3$OH (**45**)	Tetranuclear	NO$_7$	Bi-capped trigonal prism or/and a square antiprism	20 K	[79]
[Dy$_4$(H**L17**)$_2$**L17**(DMF)$_8$]·2ClO$_4$·CH$_2$Cl$_2$ ·4DMF·(CH$_3$CH$_2$)·2H$_2$O(**46**)	Tetranuclear	O$_7$N$_2$/O$_6$N$_3$	Distorted spherical tricapped trigonal prism/spherical capped square antiprism	~4 K	[86]
[Dy$_4$(TTA)$_4$(**L18**)$_4$(H$_2$O)$_2$]·4CH$_3$OH (**47**)	Tetranuclear	N$_2$O$_6$	Distorted square-antiprismatic	1.5 K	[95]
[Dy$_6$(**L19**)$_7$(H**L19**)(MeOH)$_2$(H$_2$O)(OH)$_2$(OAc)] (**48**)	Hexanuclear	N$_2$O$_6$	square-antiprismatic	3.0 K	[99]
[Dy$_6$(μ_3-OH)$_3$(μ_3-CO$_3$)(μ-OMe)(**L20**)$_6$(MeOH)$_4$- (H$_2$O)$_2$]·3MeOH·2H$_2$O (**49**)	Hexanuclear	NO$_7$	square-antiprismatic	5.6 and 38 K	[100]
[Dy$_6$(**L17**)$_3$(PhCOO)$_6$(CH$_3$OH)$_6$]·11CH$_3$OH·H$_2$O (**50**)	Hexanuclear	N$_5$O$_3$/O$_7$N$_2$/O$_6$N$_3$	Triple-stranded helical structure	2 K	[86]

N/A = Not applicable.

5. Conclusions

To summarize, we have done a brief up to date review of important, notable work on pure Ln(III) based SMMs, mainly with Schiff bases of salicylaldehyde. There is a remarkable interest in Ln(III)-based SMMs in the quest to synthesize SMMs with higher effective energy barriers and blocking temperatures, whereby the synthetic strategies can play an important role. However, custom tuning of the SMM properties still remains a big challenge. Emphasis has been given for Dy(III) SMMs in the present discussions as it shows superiority in magnetism, resulting from high anisotropy of the Dy(III)-ions so as to reach the limit of the effective energy barrier.

The great potential of Schiff base ligands was used to achieve this goal by incorporating new functionalities in both amines as well as aldehydic precursors in the ligand synthesis. It was inferred from the structures discussed that by varying precursors of Schiff base condensation, we could create coordination pockets or compartments that can be used for a particular lanthanide ion to occupy, facilitating suitable magnetic exchange interactions in the clusters so generated with SMM behavior. In spite of the numerous complexes synthesized to date, the Schiff base chemistry is far from being exhausted. Among the various Schiff bases and the lanthanide (III) SMMs discussed, the Dy_4 cluster reported by Tang et al. from 2-hydroxy-3-methoxybenzioc acid (2-hydroxy-3-methoxyphenyl) methylene) hydrazide H_3**L16** exhibits the highest energy barrier (U_{eff} = 173 K). While not having exhaustively reviewed the results of the past twelve years from various research groups, what we have discussed will give an insight into the very promising field of SMMs based on Ln(III)-ions with Schiff base ligands, mainly from salicylaldehyde derivatives.

Author Contributions: The major work for this review, writing the original draft, was done by the first author (M.G.), which is part of his Ph.D. work. The second to fourth authors (Y.B.; T.G.; G.G. and F.E.) participated in execution and drawing. The last two authors (M.T. and W.L.) were responsible for supervising and editing the article. All authors have read and agreed to the published version of the manuscript.

Funding: This research received no external funding.

Acknowledgments: We are thankful to Addis Ababa Science and Technology University, Ethiopia, for a Ph.D. studentship to one of us (M.G.) We are thankful to Jinkui Tang, State Key Laboratory of Rare Earth Resource Utilization, Changchun Institute of Applied Chemistry, China, for all valuable suggestions and help during the preparation of this manuscript. We are also highly thankful to Sindhu Thomas for the careful reading of the manuscript and the necessary modifications and suggestions.

Conflicts of Interest: The authors declare no conflict of interest regarding the publication of this paper.

References

1. Ritter, S.K. Single-molecule magnets evolve. *Chem. Eng. News* **2004**, *82*, 29–32. [CrossRef]
2. Kahn, O. *Molecular Magnetism*; VCH Pubilishers: New York, NY, USA, 1993; ISBN 1560815663.
3. Caneschi, A.; Gatteschi, D.; Sessoli, R.; Barra, A.L.; Brunel, L.C.; Guillot, M. Alternating Current Susceptibility, High Field Magnetization, and Millimeter Band EPR Evidence for a Ground S = 10 State in [$Mn_{12}O_{12}(CH_3COO)_{16}(H_2O)_4$]·$2CH_3COOH·4H_2O$. *J. Am. Chem. Soc.* **1991**, *113*, 5873–5874. [CrossRef]
4. Bagai, R.; Christou, G. The *Drosophila* of single-molecule magnetism: [$Mn_{12}O_{12}(O_2CR)_{16}(H_2O)_4$]. *Chem. Soc. Rev.* **2009**, *38*, 1011–1026. [CrossRef] [PubMed]
5. Bar, A.-K.; Kalita, P.; Singh, M.-K.; Rajaraman, G.; Chandrasekhar, V. Low-coordinate mononuclear lanthanide complexes as molecular nanomagnets. *Coord. Chem. Rev.* **2018**, *367*, 163–216. [CrossRef]
6. Woodruff, D.N.; Winpenny, R.E.P.; Layfield, R.A. Lanthanide Single-Molecule Magnets. *Chem. Rev.* **2013**, *113*, 5110–5148. [CrossRef]
7. Liu, J.L.; Chen, Y.C.; Tong, M.L. Symmetry strategies for high performance lanthanide-based single-molecule magnets. *Chem. Soc. Rev.* **2018**, *47*, 2431–2453. [CrossRef]
8. Guo, Y.-N.; Xu, G.-F.; Guo, Y.; Tang, J. Relaxation dynamics of dysprosium(III) single molecule magnets. *Dalton Trans.* **2011**, *40*, 9953–9963. [CrossRef]
9. Liddle, S.T.; Slageren, J. van Improving f-element single molecule magnets. *Chem. Soc. Rev.* **2015**, 6655–6669. [CrossRef]

10. Pedersen, K.S.; Schau-Magnussen, M.; Bendix, J.; Weihe, H.; Palii, A.V.; Klokishner, S.I.; Ostrovsky, S.; Reu, O.S.; Mutka, H.; Tregenna-Piggott, P.L.W. Enhancing the blocking temperature in single-molecule magnets by incorporating 3d–5d exchange interactions. *Chem. Eur. J.* **2010**, *16*, 13458–13464. [CrossRef]
11. Pedersen, K.S.; Dreiser, J.; Nehrkorn, J.; Gysler, M.; Schau-Magnussen, M.; Schnegg, A.; Holldack, K.; Bittl, R.; Piligkos, S.; Weihe, H.; et al. A linear single-molecule magnet based on [Ru^{III}(CN)$_6$]$^{3-}$. *Chem. Commun.* **2011**, *47*, 6918–6920. [CrossRef]
12. Sorace, L.; Benelli, C.; Gatteschi, D. Lanthanides in molecular magnetism: Old tools in a new field. *Chem. Soc. Rev.* **2011**, *40*, 3092–3104. [CrossRef] [PubMed]
13. Rinehart, J.D.; Long, J.R. Exploiting single-ion anisotropy in the design of f-element single-molecule magnets. *Chem. Sci.* **2011**, *2*, 2078–2085. [CrossRef]
14. Baldoví, J.J.; Coronado, E.; Gaita-ariço, A.; Gamer, C.; Gimønez-marquøs, M.; Mínguez, G. A SIM-MOF: Three-Dimensional Organisation of Single-Ion Magnets with Anion-Exchange Capabilities. *Chem. Eur. J.* **2014**, *20*, 10695–10702. [CrossRef] [PubMed]
15. Feltham, H.L.C.; Brooker, S. Review of purely 4f and mixed-metal nd–4f single-molecule magnets containing only one lanthanide ion. *Coord. Chem. Rev.* **2014**, *276*, 1–33. [CrossRef]
16. Boulon, M.; Cucinotta, G.; Luzon, J.; Innocenti, C.D.; Perfetti, M.; Bernot, K.; Calvez, G.; Caneschi, A.; Sessoli, R. Magnetic Anisotropy and Spin-Parity Effect Along the Series of Lanthanide Complexes with DOTA. *Angew. Chem. Int. Ed.* **2013**, *52*, 350–354. [CrossRef] [PubMed]
17. Liu, K.; Zhang, X.; Meng, X.; Shi, W.; Cheng, P.; Powell, A.K. Constraining the coordination geometries of lanthanide centers and magnetic building blocks in frameworks: A new strategy for molecular nanomagnets. *Chem. Soc. Rev.* **2016**, *45*, 2423–2439. [CrossRef]
18. Ungur, L.; Chibotaru, L.F. Strategies toward High-Temperature Lanthanide-Based Single-Molecule Magnets. *Inorg. Chem.* **2016**, *55*, 10043–10056. [CrossRef]
19. Katoh, K.; Komeda, T.; Yamashita, M. The Frontier of Molecular Spintronics Based on Multiple-Decker Phthalocyaninato TbIII Single-Molecule Magnets. *Chem. Rec.* **2016**, *16*, 987–1016. [CrossRef]
20. Zhang, P.; Zhang, L.; Tang, J. Lanthanide single molecule magnets: Progress and perspective. *Dalton Trans.* **2015**, *44*, 3923–3929. [CrossRef]
21. Gregson, M.; Chilton, N.F.; Ariciu, A.M.; Tuna, F.; Crowe, I.F.; Lewis, W.; Blake, A.J.; Collison, D.; McInnes, E.J.L.; Winpenny, R.E.P.; et al. A monometallic lanthanide bis(methanediide) single molecule magnet with a large energy barrier and complex spin relaxation behaviour. *Chem. Sci.* **2016**, *7*, 155–165. [CrossRef]
22. Singh, M.K.; Rajaraman, G. Acquiring a record barrier height for magnetization reversal in lanthanide encapsulated fullerene molecules using DFT and: Ab initio calculations. *Chem. Commun.* **2016**, *52*, 14047–14050. [CrossRef] [PubMed]
23. Guo, F.; Day, B.M.; Chen, Y.; Tong, M.; Mansikkam, A.; Layfield, R.A. A Dysprosium Metallocene Single-Molecule Magnet Functioning at the Axial Limit. *Angew. Chem. Int. Ed.* **2017**, *56*, 11445–11449. [CrossRef] [PubMed]
24. Ishikawa, N.; Sugita, M.; Okubo, T.; Tanaka, N.; Iino, T.; Kaizu, Y. Determination of ligand-field parameters and f-electronic structures of double-decker bis(phthalocyaninato)lanthanide complexes. *Inorg. Chem.* **2003**, *42*, 2440–2446. [CrossRef] [PubMed]
25. Ishikawa, N.; Sugita, M.; Ishikawa, T.; Koshihara, S.Y.; Kaizu, Y. Mononuclear lanthanide complexes with a long magnetization relaxation time at high temperatures: A new category of magnets at the single-molecular level. *J. Phys. Chem. B* **2004**, *108*, 11265–11271. [CrossRef]
26. Ishikawa, N.; Sugita, M.; Ishikawa, T.; Koshihara, S.; Kaizu, Y. Lanthanide Double-Decker Complexes Functioning as Magnets at the Single-Molecular Level. *J. Am. Chem. Soc.* **2003**, *125*, 8694–8695. [CrossRef]
27. Ishikawa, N.; Sugita, M.; Tanaka, N.; Ishikawa, T.; Koshihara, S.; Kaizu, Y. Upward Temperature Shift of the Intrinsic Phase Lag of the Magnetization of Bis(phthalocyaninato) terbium by Ligand Oxidation Creating an S = 1/2 Spin. *Inorg. Chem.* **2004**, *43*, 5498–5500. [CrossRef]
28. Hewitt, I.J.; Tang, J.; Madhu, N.T.; Anson, C.E.; Lan, Y.; Luzon, J.; Etienne, M.; Sessoli, R.; Powell, A.K. Coupling Dy3 triangles enhances their slow magnetic relaxation. *Angew. Chem. Int. Ed.* **2010**, *49*, 6352–6356. [CrossRef]

29. Tang, J.; Hewitt, I.; Madhu, N.T.; Chastanet, G.; Wernsdorfer, W.; Anson, C.E.; Benelli, C.; Sessoli, R.; Powell, A.K. Dysprosium Triangles Showing Single-Molecule Magnet Behavior of Thermally Excited Spin States. *Angew. Chem. Int. Ed.* **2006**, *45*, 1729–1733. [CrossRef]
30. Ungur, L.; Chibotaru, L.F. Magnetic anisotropy in the excited states of low symmetry lanthanide complexes. *Phys. Chem. Chem. Phys.* **2011**, *13*, 20086–20090. [CrossRef]
31. Rinehart, J.D.; Fang, M.; Evans, W.J.; Long, J.R.; Rinehart, D.; Fang, M.; Evans, W.J.; Long, R. A N23-Radical-Bridged Terbium Complex Exhibiting Magnetic Hysteresis at 14 K. *J. Am. Chem. Soc.* **2011**, *133*, 14236–14239. [CrossRef]
32. Liu, J.; Chen, Y.; Liu, J.; Vieru, V.; Ungur, L.; Jia, J.; Tong, M. A Stable Pentagonal Bipyramidal Dy(III) Single-Ion Magnet with a Record Magnetization Reversal Barrier over 1000 K. *J. Am. Chem. Soc.* **2016**, *138*, 5441–5450. [CrossRef]
33. McClain, K.R.; Gould, C.A.; Chakarawet, K.; Teat, S.J.; Groshens, T.J.; Long, J.R.; Harvey, B.G. High-temperature magnetic blocking and magneto-structural correlations in a series of dysprosium(III) metallocenium single-molecule magnets. *Chem. Sci.* **2018**, *9*, 8492–8503. [CrossRef] [PubMed]
34. Carolina, R.G.; Ballesteros, B.; de la Torre, G.; Clemente-Juan, J.M.; Coronado, E.; Torres, T. Influence of Peripheral Substitution on the Magnetic Behavior of Single-Ion Magnets Based on Homo- and Heteroleptic TbIII Bis(phthalocyaninate). *Chem. Eur. J.* **2013**, *19*, 1457–1465. [CrossRef]
35. Schiff, H. Mittheilungen aus dem Universitätslaboratorium in Pisa: Eine neue Reihe organischer Basen. *Justus Liebigs Annalen der Chemie* **1864**, *131*, 118–119. [CrossRef]
36. Lin, P.-H.; Sun, W.-B.; Yu, M.-F.; Li, G.-M.; Yan, P.-F.; Murugesu, M. An unsymmetrical coordination environment leading to two slow relaxation modes in a Dy2 single-molecule magnet. *Chem. Commun.* **2011**, *47*, 10993–10995. [CrossRef] [PubMed]
37. Long, J.; Habib, F.; Lin, P.H.; Korobkov, I.; Enright, G.; Ungur, L.; Wernsdorfer, W.; Chibotaru, L.F.; Murugesu, M. Single-molecule magnet behavior for an antiferromagnetically superexchange-coupled dinuclear dysprosium(III) complex. *J. Am. Chem. Soc.* **2011**, *133*, 5319–5328. [CrossRef] [PubMed]
38. Lin, P.H.; Burchell, T.J.; Clérac, R.; Murugesu, M. Dinuclear dysprosium(III) single-molecule magnets with a large anisotropic barrier. *Angew. Chem. Int. Ed.* **2008**, *47*, 8848–8851. [CrossRef] [PubMed]
39. Guo, Y.; Xu, G.; Wernsdorfer, W.; Ungur, L.; Guo, Y.; Tang, J.; Zhang, H.-J.; Chibotaru, L.F.; Powell, A. Strong Axiality and Ising Exchange Interaction Suppress Zero-Field Tunneling of Magnetization of an Asymmetric Dy2 Single-Molecule Magnet. *J. Am. Chem. Soc.* **2011**, *133*, 11948–11951. [CrossRef] [PubMed]
40. Dhar, D.N.; Taploo, C.L. Schiff-bases and their applications. *J. Sci. Ind. Res.* **1982**, *41*, 501–506.
41. Pearson, R.G. Hard and Soft Acids and Bases, HSAB, Part I Fundamental principles. *J. Chem. Educ.* **1968**, *45*, 581–587. [CrossRef]
42. Pearson, R. Hard and Soft Acids and Bases. *J. Am. Chem. Soc.* **1963**, *85*, 3533–3539. [CrossRef]
43. Anwar, M.U.; Thompson, L.K.; Dawe, L.N.; Habib, F.; Murugesu, M. Predictable self-assembled [2x2] Ln(III)4 square grids (Ln = Dy,Tb)—SMM behaviour in a new lanthanide cluster motif. *Chem. Commun.* **2012**, *48*, 4576–4578. [CrossRef] [PubMed]
44. Guo, Y.; Xu, G.; Gamez, P.; Zhao, L.; Lin, S.; Deng, R.; Tang, J.; Zang, H.-J. Two-Step Relaxation in a Linear Tetranuclear Dysprosium(III) Aggregate Showing Single-Molecule Magnet Behavior. *J. Am. Chem. Soc.* **2010**, *132*, 8538–8539. [CrossRef]
45. Tian, H.; Zhao, L.; Guo, Y.; Guo, Y.; Liu, Z. Quadruple-CO_3^{2-} bridged octanuclear dysprosium(III) compound showing single-molecule magnet behaviour. *Chem. Commun.* **2012**, *6346*, 708–710. [CrossRef] [PubMed]
46. Guo, Y.; Chen, X.; Xue, S.; Tang, J. Molecular Assembly and Magnetic Dynamics of Two Novel Dy6 and Dy8 Aggregates. *Inorg. Chem.* **2012**, *51*, 4035–4042. [CrossRef]
47. Habib, F.; Long, J.; Lin, P.-H.; Korobkov, I.; Ungur, L.; Wernsdorfer, W.; Chibotaru, L.F.; Murugesu, M. Supramolecular architectures for controlling slow magnetic relaxation in field-induced single-molecule magnets. *Chem. Sci.* **2012**, *3*, 2158–2164. [CrossRef]
48. Wang, H.; Liu, C.; Liu, T.; Zeng, S.; Cao, W.; Duan, C.; Dou, J.; Jiang, J. Mixed (phthalocyaninato)(Schiff-base) di-dysprosium sandwich complexes. Effect of magnetic coupling on the SMM behavior. *Dalton Trans.* **2013**, *42*, 15355–15360. [CrossRef]
49. Andruh, M. The exceptionally rich coordination chemistry generated by Schiff-base ligands derived from o-vanillin. *Dalton Trans.* **2015**, *44*, 16633–16653. [CrossRef]

50. Wu, G.; Hewitt, I.J.; Mameri, S.; Lan, Y.; Clérac, R.; Anson, C.E.; Qiu, S.; Powell, A.K. Bifunctional Ligand Approach for Constructing 3d–4f Heterometallic Clusters. *Inorg. Chem.* **2007**, *46*, 7229–7231. [CrossRef]
51. Dey, M.; Rao, C.P.; Saarenketo, P.K.; Rissanen, K. Mono-, di- and tri-nuclear Ni(II) complexes of N–, O–donor ligands: Structural diversity and reactivity. *Inorg. Chem. Commun.* **2002**, *5*, 924–928. [CrossRef]
52. Senthil Kumar, K.; Bayeh, Y.; Gebretsadik, T.; Elemo, F.; Gebrezgiabher, M.; Thomas, M.; Ruben, M. Spin-crossover in iron(ii)-Schiff base complexes. *Dalton Trans.* **2019**, *48*, 15321–15337. [CrossRef] [PubMed]
53. Lucaccini, E.; Baldoví, J.J.; Chelazzi, L.; Barra, A.L.; Grepioni, F.; Costes, J.P.; Sorace, L. Electronic Structure and Magnetic Anisotropy in Lanthanoid Single-Ion Magnets with C3 Symmetry: The Ln(trenovan) Series. *Inorg. Chem.* **2017**, *56*, 4728–4738. [CrossRef] [PubMed]
54. Guo, M.; Tang, J. Six-Coordinate Ln(III) Complexes with Various Magnetic Properties. *Inorganics* **2018**, *6*, 16. [CrossRef]
55. Cheng, L.W.; Zhang, C.L.; Wei, J.Y.; Lin, P.H. Mononuclear and trinuclear DyIII SMMs with Schiff-base ligands modified by nitro-groups: First triangular complex with a N–N pathway. *Dalton Trans.* **2019**, *48*, 17331–17339. [CrossRef] [PubMed]
56. Lin, P.H.; Korobkov, I.; Burchell, T.J.; Murugesu, M. Connecting single-ion magnets through ligand dimerisation. *Dalton Trans.* **2012**, *41*, 13649–13655. [CrossRef] [PubMed]
57. Zhao, L.; Wu, J.; Ke, H.; Tang, J. Three dinuclear lanthanide (III) compounds of a relaxation behaviour of the DyIII derivative. *CrystEngComm* **2013**, *15*, 5301–5306. [CrossRef]
58. Zhang, K.; Wang, Y. A dinuclear dysprosium single-molecule magnet constructed by o-vanillin Schiff base ligand: Magnetic properties and solution behaviour. *Inorg. Chem. Commun.* **2017**, *76*, 103–107. [CrossRef]
59. Habib, F.; Brunet, G.; Vieru, V.; Korobkov, I.; Chibotaru, L.F.; Murugesu, M. Significant Enhancement of Energy Barriers in Dinuclear Dysprosium Single-Molecule Magnets Through Electron-Withdrawing Effects. *J. Am. Chem. Soc.* **2013**, *135*, 13242–13245. [CrossRef]
60. Zou, L.F.; Zhao, L.; Guo, Y.N.; Yu, G.M.; Guo, Y.; Tang, J.; Li, Y.H. A dodecanuclear heterometallic dysprosium-cobalt wheel exhibiting single-molecule magnet behaviour. *Chem. Commun.* **2011**, *47*, 8659–8661. [CrossRef]
61. Benelli, C.; Gatteschi, D. Magnetism of Lanthanides in Molecular Materials with Transition-Metal Ions and Organic Radicals. *Chem. Rev.* **2002**, *102*, 2369–2387. [CrossRef]
62. Rogers, M.T. *Magnetochemistry*; Springer: Berlin/Heidelberg, Germany, 1957; Volume 79, ISBN 9783642707353.
63. Sun, O.; Chen, P.; Li, H.; Gao, T.; Sun, W.; Li, G.; Yan, P. A series of dinuclear lanthanide(III) complexes constructed from Schiff base and β-diketonate ligands: Synthesis, structure, luminescence and SMM behavior. *CrystEngComm* **2016**, *18*, 4627–4635. [CrossRef]
64. Dou, W.; Yao, J.; Liu, W.; Wang, Y.; Zheng, J.-R.; Wang, D.-Q. Crystal structure and luminescent properties of new rare earth complexes with a flexible Salen-type ligand. *Inorg. Chem. Commun.* **2007**, *10*, 105–108. [CrossRef]
65. Zhang, Y.X.; Cheng, X.Y.; Tang, Y.T.; Zhang, Y.H.; Wang, S.C.; Wei, H.Y.; Wu, Z.L. Two dinuclear lanthanide(III) clusters (Gd2 and Dy2) constructed by bis-(o-vanillin) schiff base ligand exhibiting fascinating magnetic behaviors. *Polyhedron* **2019**, *166*, 23–27. [CrossRef]
66. Sharples, J.W.; Zheng, Y.Z.; Tuna, F.; McInnes, E.J.L.; Collison, D. Lanthanide discs chill well and relax slowly. *Chem. Commun.* **2011**, *47*, 7650–7652. [CrossRef] [PubMed]
67. Li, X.H.; Li, M.; Gao, N.; Wei, H.Y.; Hou, Y.L.; Wu, Z.L. Solvent-driven structures and slow magnetic relaxation behaviors of dinuclear dysprosium clusters. *Inorg. Chim. Acta* **2020**, *500*, 119242. [CrossRef]
68. Liu, S.J.; Yao, S.L.; Cao, C.; Zheng, T.F.; Liu, C.; Wang, Z.X.; Zhao, Q.; Liao, J.S.; Chen, J.L.; Wen, H.R. Two di- and trinuclear Gd(III) clusters derived from monocarboxylates exhibiting significant magnetic entropy changes. *Polyhedron* **2017**, *121*, 180–184. [CrossRef]
69. Cole, K.S.; Cole, R.H. Dispersion and absorption in dielectrics I. Alternating current characteristics. *J. Chem. Phys.* **1941**, *9*, 341–351. [CrossRef]
70. Anastasiadis, N.C.; Granadeiro, C.M.; Mayans, J.; Raptopoulou, C.P.; Bekiari, V.; Cunha-Silva, L.; Psycharis, V.; Escuer, A.; Balula, S.S.; Konidaris, K.F.; et al. Multifunctionality in two families of dinuclear lanthanide(III) complexes with a tridentate schiff-base ligand. *Inorg. Chem.* **2019**, *58*, 9581–9585. [CrossRef]
71. Holmberg, R.J.; Kuo, C.J.; Gabidullin, B.; Wang, C.W.; Clérac, R.; Murugesu, M.; Lin, P.H. A propeller-shaped µ4-carbonate hexanuclear dysprosium complex with a high energetic barrier to magnetisation relaxation. *Dalton Trans.* **2016**, *45*, 16769–16773. [CrossRef]

72. Zhang, L.; Zhang, P.; Zhao, L.; Wu, J.; Guo, M.; Tang, J. Anions influence the relaxation dynamics of mono-3-OH-capped triangular dysprosium aggregates. *Inorg. Chem.* **2015**, *54*, 5571–5578. [CrossRef]
73. Lin, S.Y.; Zhao, L.; Ke, H.; Guo, Y.N.; Tang, J.; Guo, Y.; Dou, J. Steric hindrances create a discrete linear Dy4complex exhibiting SMM behaviour. *Dalton Trans.* **2012**, *41*, 3248–3252. [CrossRef]
74. Chen, G.-J.; Gao, C.-Y.; Tian, J.-L.; Tang, J.; Gu, W.; Liu, X.; Yan, S.-P.; Liao, D.-Z.; Cheng, P. Coordination-perturbed single-molecule magnet behaviour of mononuclear dysprosium complexes. *Dalton Trans.* **2011**, *40*, 5579–5583. [CrossRef]
75. Ke, H.; Gamez, P.; Zhao, L.; Xu, G.F.; Xue, S.; Tang, J. Magnetic properties of dysprosium cubanes dictated by the M–O–M angles of the [Dy4(μ3-OH)$_4$] core. *Inorg. Chem.* **2010**, *49*, 7549–7557. [CrossRef] [PubMed]
76. Gao, H.; Huang, S.; Zhou, X.; Liu, Z.; Cui, J. Magnetic properties and structure of tetranuclear lanthanide complexes based on 8-hydroxyquinoline Schiff base derivative and β-diketone coligand. *Dalton Trans.* **2018**, *47*, 3503–3511. [CrossRef] [PubMed]
77. Gavrilenko, K.S.; Cador, O.; Bernot, K.; Rosa, P.; Sessoli, R.; Golhen, S.; Pavlishchuk, V.V.; Ouahab, L. Delicate Crystal Structure Changes Govern the Magnetic Properties of 1D Coordination Polymers Based on 3d Metal Carboxylates. *Chem. Eur. J.* **2008**, *14*, 2034–2043. [CrossRef] [PubMed]
78. Lacelle, T.; Brunet, G.; Pialat, A.; Holmberg, R.J.; Lan, Y.; Gabidullin, B.; Korobkov, I.; Wernsdorfer, W.; Murugesu, M. Single-molecule magnet behaviour in a tetranuclear DyIII complex formed from a novel tetrazine-centered hydrazone Schiff base ligand. *Dalton Trans.* **2017**, *46*, 2471–2478. [CrossRef]
79. Ke, H.; Xu, G.; Guo, Y.; Gamez, P.; Beavers, C.M.; Teat, S.J.; Tang, J. A linear tetranuclear dysprosium(III) compound showing single-molecule magnet behaviour. *Chem. Commun.* **2010**, *46*, 6057–6059. [CrossRef]
80. Kahn, M.L.; Ballou, R.; Porcher, P.; Kahn, O.; Sutter, J.P. Analytical determination of the {Ln–aminoxyl radical} exchange interaction taking into account both the ligand-field effect and the spin - Orbit coupling of the lanthanide ion (Ln = DyIII and HoIII). *Chem. Eur. J.* **2002**, *8*, 525–531. [CrossRef]
81. Osa, S.; Kido, T.; Matsumoto, N.; Re, N.; Pochaba, A.; Mrozinsk, J. A Tetranuclear 3d–4f Single Molecule Magnet: [Cu^{II}LTb^{III}(hfac)$_2$]$_2$. *J. Am. Chem. Soc.* **2004**, 420–421. [CrossRef]
82. Dekker, C.; Arts, A.F.M.; Wijn, H.W.d.; Duyneveldt, A.J.v.; Mydosh, J.A. Activated dynamics in a two-dimensional Ising spin glass: Rb$_2$Cu$_{1-x}$Co$_x$F$_4$. *Phys. Rev. B Condens. Matter* **1989**, *40*, 11243–11251. [CrossRef]
83. Spencer, G.H.; Cross, P.C.; Wiberg, K.B. 5-tetrazine. II. Infrared spectra. *J. Chem. Phys.* **1961**, *35*, 1939–1945. [CrossRef]
84. Sueur, S.; Lagrenee, M.; Abraham, F.; Bremard, C. New syntheses of polyaza derivatives. Crystal structure of pyridazine-3,6-dicarboxylic acid. *J. Hetrocycl. Cham.* **1987**, *24*, 1285–1289. [CrossRef]
85. Lu, J.; Li, X.L.; Zhu, Z.; Liu, S.; Yang, Q.; Tang, J. Linear hexanuclear helical dysprosium single-molecule magnets: The effect of axial substitution on magnetic interactions and relaxation dynamics. *Dalton Trans.* **2019**, *48*, 14062–14068. [CrossRef]
86. Lu, J.; Li, X.-L.; Jin, C.; Yu, Y.; Tang, J. Dysprosium-based linear helicate clusters: Syntheses, structures, and magnetism. *New J. Chem.* **2020**, 994–1000. [CrossRef]
87. Zhang, L.; Zhang, Y.Q.; Zhang, P.; Zhao, L.; Guo, M.; Tang, J. Single-Molecule Magnet Behavior Enhanced by Synergic Effect of Single-Ion Anisotropy and Magnetic Interactions. *Inorg. Chem.* **2017**, *56*, 7882–7889. [CrossRef]
88. Lu, J.; Zhang, Y.Q.; Li, X.L.; Guo, M.; Wu, J.; Zhao, L.; Tang, J. Influence of Magnetic Interactions and Single-Ion Anisotropy on Magnetic Relaxation within a Family of Tetranuclear Dysprosium Complexes. *Inorg. Chem.* **2019**, *58*, 5715–5724. [CrossRef]
89. Wu, J.; Zhao, L.; Zhang, L.; Li, X.L.; Guo, M.; Powell, A.K.; Tang, J. Macroscopic Hexagonal Tubes of 3d–4 f Metallocycles. *Angew. Chem. Int. Ed.* **2016**, *55*, 15574–15578. [CrossRef]
90. Mukherjee, S.; Lu, J.; Velmurugan, G.; Singh, S.; Rajaraman, G.; Tang, J.; Ghosh, S.K. Influence of Tuned Linker Functionality on Modulation of Magnetic Properties and Relaxation Dynamics in a Family of Six Isotypic Ln2 (Ln = Dy and Gd) Complexes. *Inorg. Chem.* **2016**, *55*, 11283–11298. [CrossRef] [PubMed]
91. Lin, S.Y.; Xu, G.F.; Zhao, L.; Guo, Y.N.; Guo, Y.; Tang, J. Observation of slow magnetic relaxation in triple-stranded lanthanide helicates. *Dalton Trans.* **2011**, *40*, 8213–8217. [CrossRef] [PubMed]
92. Wang, W.M.; Kang, X.M.; Shen, H.Y.; Wu, Z.L.; Gao, H.L.; Cui, J.Z. Modulating single-molecule magnet behavior towards multiple magnetic relaxation processes through structural variation in Dy4 clusters. *Inorg. Chem. Front.* **2018**, *5*, 1876–1885. [CrossRef]

93. Wu, J.; Cador, O.; Li, X.L.; Zhao, L.; Le Guennic, B.; Tang, J. Axial Ligand Field in D4d Coordination Symmetry: Magnetic Relaxation of Dy SMMs Perturbed by Counteranions. *Inorg. Chem.* **2017**, *56*, 11211–11219. [CrossRef] [PubMed]
94. Wu, J.; Li, X.L.; Zhao, L.; Guo, M.; Tang, J. Enhancement of Magnetocaloric Effect through Fixation of Carbon Dioxide: Molecular Assembly from Ln4 to Ln4 Cluster Pairs. *Inorg. Chem.* **2017**, *56*, 4104–4111. [CrossRef] [PubMed]
95. Hou, Y.L.; Liu, L.; Hou, Z.T.; Sheng, C.Y.; Wang, D.T.; Ji, J.; Shi, Y.; Wang, W.M. A novel Dy4III cluster constructed by a multidentate 8-hydroxyquinoline Schiff base ligand: Structure and slow magnetic relaxation behavior. *Inorg. Chem. Commun.* **2020**, *112*, 107691. [CrossRef]
96. Qin, L.; Yu, Y.Z.; Liao, P.Q.; Xue, W.; Zheng, Z.; Chen, X.M.; Zheng, Y.Z. A "Molecular Water Pipe": A Giant Tubular Cluster {Dy72} Exhibits Fast Proton Transport and Slow Magnetic Relaxation. *Adv. Mater.* **2016**, *28*, 10772–10779. [CrossRef]
97. Wang, W.M.; Qiao, W.Z.; Zhang, H.X.; Wang, S.Y.; Nie, Y.Y.; Chen, H.M.; Liu, Z.; Gao, H.L.; Cui, J.Z.; Zhao, B. Structures and magnetic properties of several phenoxo-O bridged dinuclear lanthanide complexes: Dy derivatives displaying substituent dependent magnetic relaxation behavior. *Dalton Trans.* **2016**, *45*, 8182–8191. [CrossRef]
98. Lu, J.; Montigaud, V.; Cador, O.; Wu, J.; Zhao, L.; Li, X.L.; Guo, M.; Le Guennic, B.; Tang, J. Lanthanide(III) Hexanuclear Circular Helicates: Slow Magnetic Relaxation, Toroidal Arrangement of Magnetic Moments, and Magnetocaloric Effects. *Inorg. Chem.* **2019**, *58*, 11903–11911. [CrossRef]
99. Mukherjee, S.; Chaudhari, A.K.; Xue, S.; Tang, J.; Ghosh, S.K. An asymmetrically connected hexanuclear DyIII6 cluster exhibiting slow magnetic relaxation. *Inorg. Chim. Acta* **2013**, *35*, 144–148. [CrossRef]
100. Tian, H.; Guo, Y.; Zhao, L.; Tang, J.; Liu, Z. Hexanuclear Dysprosium(III) Compound Incorporating Vertex- and Edge-Sharing Dy 3 Triangles Exhibiting Single-Molecule-Magnet Behavior. *Inorg. Chem.* **2011**, *50*, 8688–8690. [CrossRef] [PubMed]

Publisher's Note: MDPI stays neutral with regard to jurisdictional claims in published maps and institutional affiliations.

© 2020 by the authors. Licensee MDPI, Basel, Switzerland. This article is an open access article distributed under the terms and conditions of the Creative Commons Attribution (CC BY) license (http://creativecommons.org/licenses/by/4.0/).

Review

Metal Organic Frameworks as Heterogeneous Catalysts in Olefin Epoxidation and Carbon Dioxide Cycloaddition

Alessia Tombesi [1] and Claudio Pettinari [2,*]

[1] Chemistry Section, School of Science and Technology, University of Camerino, Via S. Agostino 1, 62032 Camerino, Italy; alessia.tombesi@unicam.it
[2] Chemistry Section, School of Pharmacy, University of Camerino, Via S. Agostino 1, 62032 Camerino, Italy
* Correspondence: claudio.pettinari@unicam.it

Abstract: Metal–organic frameworks (MOFs) are a family of porous crystalline materials that serve in some cases as versatile platforms for catalysis. In this review, we overview the recent developments about the use of these species as heterogeneous catalysts in olefin epoxidation and carbon dioxide cycloaddition. We report the most important results obtained in this field relating them to the presence of specific organic linkers, metal nodes or clusters and mixed-metal species. Recent advances obtained with MOF nanocomposites were also described. Finally we compare the results and summarize the major insights in specific Tables, outlining the major challenges for this emerging field. This work could promote new research aimed at producing coordination polymers and MOFs able to catalyse a broader range of CO_2 consuming reactions.

Keywords: heterogenous catalyst; metal–organic framework (MOF); olefin epoxidation; carbon dioxide cycloaddition

1. Introduction

International Union of Pure and Applied Chemistry (IUPAC) defines MOFs as a coordination network with an open framework containing potential voids [1]. This emerging class of porous coordination polymers are formed by metal ion or cluster nodes and functional organic ligands, all connected through coordination bonds to form 1D, 2D o 3D networks (Figure 1) [2–6]. MOFs can be easily obtained by several different synthetic methods, such as electrochemical [7], solvothermal [8] and mechanochemical [9], slow diffusion [10], and more recently also by microwave-assisted heating [11].

The crystal structures of MOFs can be customized depending on the metal and ligand choice as also on the solvents and reaction conditions employed. [12] Due to the high surface areas [13] and ultrahigh porosity they are attractive for CH_4, CO_2, and H_2 sorption and storage. Most MOFs have higher volumetric H_2 and CH_4 storage capacities concerning traditional porous materials.

In recent years nanoscale MOFs have been also investigated for their potential applications in biomedicine, for example for drug delivery [14] and biological imaging [15], mainly for the possibility to use biocompatible building blocks. MOFs were employed as electrode materials for supercapacitors using Co-based coordination polymers [16], for magnetic and electronic devices [17], for water harvesting where H_2O is extracted from the air by solar energy [18], and finally also for non-linear optics [19].

The use of MOFs as a catalyst has been widely explored and several applications have been developed, for example in the production of fine chemicals [20], or the definition of possible new green protocols replacing non-eco-friendly catalysts [21]. Differences in activity and selectivity toward specific organic reactions are significantly dependent on the MOFs structure [22]. The main MOFs advantage, when we consider their use in catalysis, is in the possibility to design and predict the structural properties based on of linker features, coordination number and geometry of the metal.

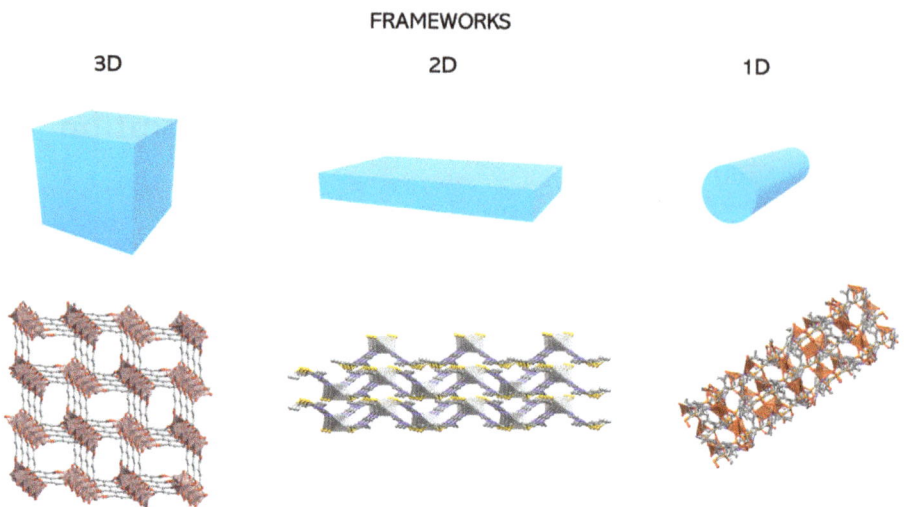

Figure 1. Schematic representation of MOFs frameworks with different dimensionalities (3D, 2D, 1D).

The presence of coordinatively unsaturated metal sites, the variety of basic linkers available, the stability to solvents and to reaction conditions, the possibility to host guest molecules within the pores makes MOFs perspective materials for heterogenous catalysis. They have also a lot of advantages concerning other inorganic systems as zeolites and aluminophosphates, i.e., they can be modified using organic synthesis, being possible to decorate their pores with catalytic sites. MOFs can be tailored by a simple change in the initial synthetic conditions or by using post-synthetic reactions. These modifications make MOFs excellent candidates for designing functional materials to allow the attachment of different catalysts [23].

While the characterization of deposited species upon conventional catalyst supports, such as metal oxides, tends to be challenging due to the non-uniform surface and pore structures of the support, the crystalline nature of MOFs enables visualization of the catalytically active species within the framework, which leads to a detailed characterization of active catalytic sites and provides insight into structure−activity relationships.

In this review we want to focus on the most recent progress in two reactions MOFs-catalyzed, i.e., the olefin epoxidation and the cycloaddition of CO_2 to epoxides to yield organic carbonates as a final product, by performing a rigorous analysis of the best MOFs in terms of conversion and selectivity. Specifically, we examined MOF-based catalytic materials producing epoxide and cyclic carbonates with percentages of conversion and selectivity exceeding 70, in the 2015 to 2021 period. Moreover, few relevant papers on the heterogeneous MOFs catalysts published before 2015, for a useful comparison have been considered.

Epoxides are important species and intermediates in the production of pharmaceuticals, agrochemicals, and relevant industrial chemicals. In the global market, the production of propylene oxide achieves 8 million tons per year with an expected annual increase of 5% [24]. Due to the industrial relevance of catalytic oxidation of olefins to fine chemicals, numerous studies have been devoted to the development of efficient homogeneous [24] and heterogeneous catalysts [24]. However, high selectivity and enantioselectivity in epoxidation reactions remain a challenge. While recovery and product separation are the main drawbacks for homogenous catalysts, MOFs used as heterogeneous catalysts in the oxidation of olefins have attracted significant attention (Figure 2) [25].

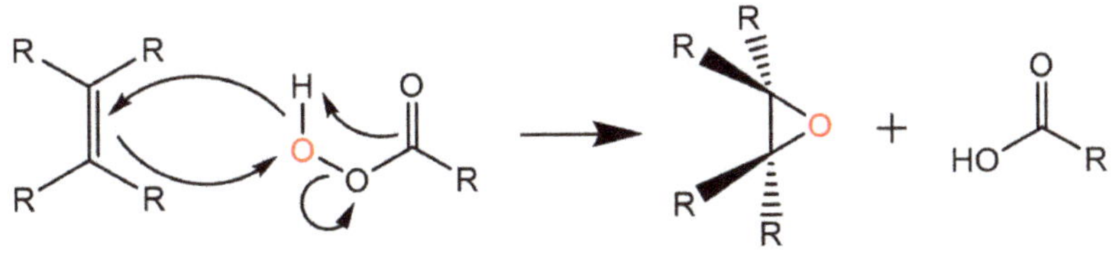

Figure 2. General mechanism of epoxidation of alkenes with peroxycarboxylic acid as co-catalyst.

CO_2 is the primary greenhouse gas in the atmosphere, and it is the cause of environmental and energy-related problems in the world. Nowadays, the development of new methods is fundamental to capture and convert CO_2 into useful chemical products to improve the environment and promote sustainable development. Several studies have been carried out on MOF's efficiency to capture CO_2. The linkers that connect the MOFs metal nodes are the major sites for CO_2 binding. The linkers that connect the MOFs metal nodes are the major sites for CO_2 binding, and they can be chemically modified with functional groups to increase their interaction with CO_2. Moreover, unsaturated metals ions can be introduced in the MOFs structure. A significantly benefit generated from the possibility to have adequate quantities of CO_2 in concentrated form within a MOF is the possible use of CO_2 as a chemical reagent (Figure 3).

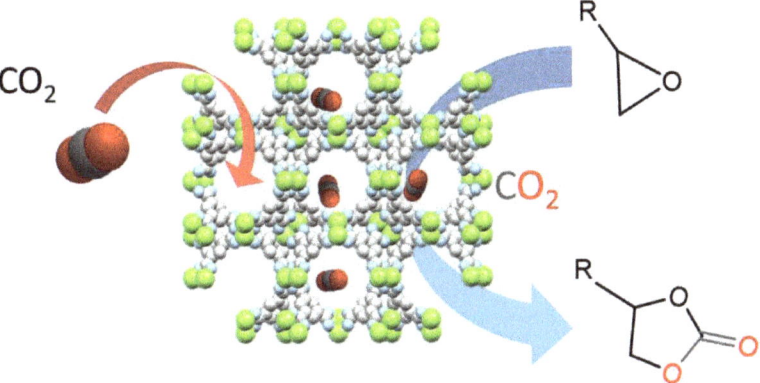

Figure 3. Representation of cycloaddition reaction of CO_2, captured by MOFs, to epoxides.

A significant number of MOFs has been recently reported to catalyse the CO_2 cycloaddition reaction to epoxides to give cyclic organic carbonates (OCs) and several papers describe the potential and effectiveness of MOFs in this important process, so it is necessary to identify better strategies to build new advanced materials as MOFs or MOF-based species to grow selectivity, capacity, and conversion of this catalytic reaction.

2. Olefin Epoxidation

C=C bond epoxidation is an attractive reaction for industrial process to obtain raw materials for epoxy resin, polymers, and pharmaceutical intermediates. Although homogenous catalysts in the epoxidation of alkene have been largely studied in the past few decades, the separation from the reaction mixture and its subsequent reusability remain open challenges [26,27].

Very recently, taking advantage of the tunability of MOFs, several transition metal-based epoxidation catalysts have been developed using MOFs synthesis in combination with post-synthetic modification. Several literature reports require utilization of expensive transition metals, but in the last period also metals like Cu, a classic non-noble transition metal, abundant, inexpensive, and non-toxic, become appealing catalyst sources.

2.1. Metal Nodes/Clusters as Catalytically Active Sites

MOFs can be applied as ideal platforms for heterogeneous catalysis towards olefin epoxidation thanks to several structural features with intrinsic catalytic activity such as the coordinatively unsaturated sites on MOFs nodes, defects, and catalytically active organic linkers.

Three pillared-layered Co_6-MOFs were utilized as heterogeneous catalysts for the selective oxidation of styrene using air, and benzyl alcohol with oxygen. The hexaprismatic $[Co_6(\mu_3-OH)_6]$ cluster with different variable valences activate the oxygen molecule for aerobic epoxidation of alkenes [28]. In Co6-MOF-3, large pores facilitated the mass transfer giving the fastest reaction rate with high conversion and good selectivity for oxidation of both styrene and benzyl alcohol [29].

The 2D-cobalt (II)-based coordination polymer, $\{(Co(L2)H_2O))_2 \cdot H_2O)\}_n$, have been obtained by hydrothermal synthesis using the histidine derivative 4-((1-carboxy-2-(1H-imidazol-4-yl)ethylamino)methyl)benzoic acid (H_2L2) as ligand. It has been investigated as heterogeneous catalysts on the allylic oxidation of cyclohexene (Appendix A).

The presence of a Co(II) open site on the surface maximizes the catalytic productivity, giving 82.56% of conversion and 71% of ter-butyl-2-cyclohexenyl-1-peroxide. Moreover, a Co(II)-based catalyst exhibits similar activity over five cycles without metal leaching [30] (Table 1).

The static and rotary hydrothermally synthetic method could affect significantly both the process of crystallization and heterogeneous catalytic activity of MOFs in the epoxidation reaction. For example, Co-MOF-150-2, hydrothermally synthesized by rotary crystallization at 150 rpm for 2 h, has reached 95.7% yield of 2,3-epoxypinane from α-pinene in aerobic conditions. The high catalytic activity of Co-MOF-150-2 is due to the better exposure of the metal active in the high crystalline structure, where the lamellar layer was more homogenous. The thinner Co-MOF-150-2 was also investigated in the epoxidation of the other olefins. Additionally, the catalytic activity was relevant for cyclic olefins like cyclooctene (78.5% of conversion after 5h of reaction) and for linear olefins (after 12 h 87.2% of 1-decene was transformed into the epoxide) [31].

A Co-MOF has been prepared under surfactant-thermal condition: NTUZ30 has been obtained with two different secondary building units (SBU), i.e., the unusual trinuclear $[Co_3(\mu_3-OH)(COO)_7]$ and $[Co(COO)_4]$. The cobalt sites onto the surface can convert *trans*-stilbene into the corresponding epoxide with excellent selectivity and high conversion [32].

A new 8-connected cobalt network $(NH_4)_2[Co_3(Ina)(BDC)_3(HCOO)]$ has been generated from dicarboxylate (BDC = 1,4-benzenedicarboxylate) and pillar isonicotinate (Ina = isonicotinate) ligands with unusual Co_3 paddle-wheel cluster. The high number of unsaturated Co active sites available and the regular crystalline structure gave a great catalytic performance for cyclooctene epoxidation with a satisfactory TOF (turnover frequency) of 1370 [33].

Cu-containing MOFs look like promising catalysts for selective oxidation reactions. Generally, the selective oxidation of alkene goes through a radical reaction pathway in which molecular oxygen, or an oxidizing agent is present. Upon coordination of the oxidizing agent to Cu(II) (Figure 4), peroxyl radicals are formed which then react with olefin to form the oxidized products [34,35].

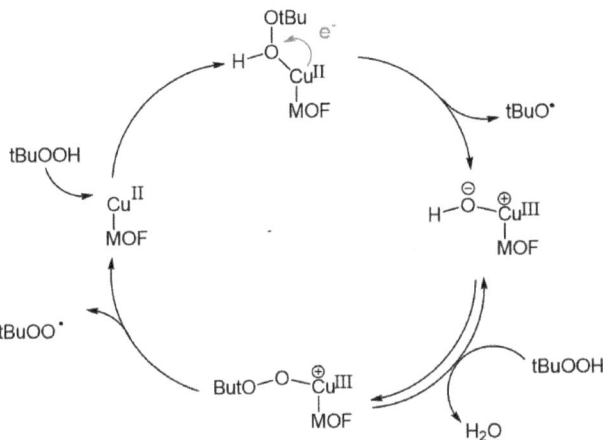

Figure 4. Schematic representation of the formation of *tert*-butylperoxyl (tBuOO·) and *tert*-butoxyl radicals (tBuO) catalyzed by the Cu(II) sites of MOF.

By using $Cu_4O(OH)_2(Me_2trz-pba)_4$ (Me_2trz-pba = 4-(3,5-dimethyl-4-H-1,2,4-triazol-4-yl)benzoate) and Cu(Me-4py-trz-ia) (Me-4py-trz-ia = 5-(3methyl-5-(pyridine-4-yl)4H-1,2,4-triazol-4-yl)isophthalate) a significantly higher catalytic activity in the epoxidation of cyclooctane with respect to $Cu_3(BTC)_2$ (HKUST-1) has been found, due to the different coordination environment at the catalytically active Cu sites [36]. The catalytic performances of Cu-MOF nanosheets for cyclooctene and 1-hexene epoxidation were nearly twice higher than that of bulk $Cu_3(BDC)_2$ crystals. This behaviour is attributed to the better exposure of a greater number of active sites on the surface of the Cu(BDC) nanosheets, which become more available during the reaction. The synthetic procedure must regulates the nanosheet thickness by controlling the dissolution rate of Cu^{2+} from $Cu(OH)_2$ precursor and tuning the solvent composition. Moreover, the epoxide yield, after 5 cycles with CuMOF nanosheets, remains 96% [37] (Table 1).

In $Cu_4O(OH)_2(Me_2trz-pba)_4$, the $Cu_4(\mu_4$-O$)(\mu_2$-OH$)$ tetrahedral node possesses two Cu^{2+} ions bridged by hydroxyl group, which take part in the activation of oxidating agent TBHP, promoting a quicker formation of *tert*-butoxyl and *tert*-butylperoxyl radicals, whereas in Cu(Me-4py-trz-ia) the asymmetric unit contains two crystallographically independent Cu^{2+} ions. One of them possesses two unsaturated sites that could cause a change in Lewis acidity and generate different redox properties [36].

Oxidation of nonterminal olefin, such as cis-stilbene and cyclooctene, occurs with 94% and 98% conversion, when the activated {[Cu(L3-H)(DMA)]·DMA·2H$_2$O}$_\infty$ MOF (H$_3$L3 = tris(4'-carboxybiphenyl)amine; DMA = N,N-dimethylacetamide) was used. The activation was carried out at 200° C for 8 h under vacuum to remove DMA coordinated molecules and produce unsaturated Cu sites that act as Lewis acid [38].

Nbo-type Cu-MOFs, synthesized from the meta-substituted ligand 2,2',6,6'-tetramethoxy-4,4'-biphenyldicarboxylic acid (H$_2$L4) and copper nitrate [Cu$_3$(L4)$_3$(H$_2$O)$_2$(DMF)]$_n$, possess a high density of catalytic sites in optimal position within the channels, in which oxidation of nonterminal olefins (e.g., norbornene, *trans*-β-methylstyrene, *cis*-β-methylstyrene, and *trans*-stilbene) occurs with 99% conversion and 99% selectivity [39]. Moreover, the less reactive aliphatic alkenes such as 1-octene and *trans*-4-octene showed moderate conversions with good selectivity.

Epoxidation of cyclohexene achieves the 100% of conversion in presence of H$_2$O$_2$ after 8h when 2D metal carboxylate framework {2(Him)·[Cu(pdc)$_2$]}$_n$ has been involved as a heterogeneous catalyst. {2(Him)·[Cu(pdc)$_2$]}$_n$ (H$_2$pdc = pyridine-2,5-dicarboxylic acid, Him = imidazole) was obtained through structural inter-conversions starting from {[Mg(H$_2$O)$_6$][Cu(pdc)$_2$]·2H$_2$O}$_n$ increasing the imidazole concentration by hydrothermal treatment. The structure of {2(Him)·[Cu(pdc)$_2$]}$_n$ derives from the connection of {[Cu(pdc)$_2$]}$_n$$^-$ ribbon-like 1D chains by intermolecular H-bonding between hydrogen in the imidazolium ion and the free carboxylate oxygens of pdc^{2-}, this 2D supramolecular structure being crucial to ensure the reaction heterogeneity. Likewise, 1-hexene showed almost complete conversion but increasing the chain length of alkene, the double bond becomes sterically hindered limiting the approach to the active site, and the catalytic activity decreases [40] (Table 1).

High stable zirconium-based MOFs are largely used as active and recyclable catalysts for a variety of catalytic transformations. The catalytic activity of UiO-66 and other Zr-MOFs can be greatly attributed to the presence of random defects in their crystalline structure [41–43]. These accessible Lewis acid centers, sometimes in conjunction with Lewis basic sites (e.g., amine groups) in functionalized linker, lead to a significant increase in the catalytic activity [44,45].

Recently, the reaction mechanism underlying both thioether oxidation in nonprotic solvents and epoxidation of electron-deficient C=C bonds in α,β-unsaturated ketones, catalysed by UiO-66 and UiO-67 has been exhaustively investigated [46]. This study suggests the formation of hydroperoxo zirconium species as an oxidant. As already known, the oxidation of less-reactive α,β-unsaturated carbonyl compounds was accompanied by oxidation of MeCN solvent and H$_2$O$_2$ under basic conditions [47], but this nucleophilic peroxo species derived from H$_2$O$_2$ and Zr-MOF can contribute to the epoxidation of the electron-deficient C=C bonds because the reaction readily proceeds even in ethyl acetate.

Table 1. MOFs with metal Nodes/clusters active in olefin epoxidation.

MOF	Substrate	Reaction Data T (°C) P (atm) Time (h)			Oxidant/Cocatalyst/ Solvent [a]	Conversion %	Epoxide Selectivity%	Ref.
Co6-MOF-3	Styrene	100	1	14	Air/-/DMF	99	90	[29]
Co-MOF-150-2	α-Pinene	90	1	5	Air/CHP/-	99.5	96.2	[31]
	Cyclooctene	90	1	5	Air/CHP/-	78.5	-	[31]
	1-Decene	90	1	5	Air/CHP/-	87.2	-	[31]
NTUZ30	trans-Stilbene	100	7	1	O$_2$/-/-	98.2	95.6	[32]
(NH$_4$)$_2$[Co$_3$(Ina)(BDC)$_3$(HCOO)]	Cyclooctene	35	1.5	1	IBA/-/CH$_3$CN	98	92	[33]
Cu$_3$(BTC)$_2$	Cyclooctene	75	1	24	Air/TBHP/Toluene	20	-	[36]
	1-Hexene	25	1		Air/TBHP/Toluene	30.5	-	[36]
Cu-MOF nanosheets	Cyclooctene	25	1	12	O$_2$/-/CH$_3$CN	100	-	[37]
	1-Hexene	25	1		O$_2$/-/CH$_3$CN	67.2	-	[37]
Cu$_4$O(OH)$_2$(Me$_2$trz-pba)$_4$	Cyclooctene	75	1	24	Air/TBHP/Toluene	90	80	[36]
Cu(Me-4py-trz-ia)	Cyclooctene	75	1	24	Air/TBHP/Toluene	38	60	[36]
{[Cu(L3-H)(DMA)]·DMA·2H$_2$O}$_∞$	cis-Stilbene	60	1	24	t-BuOOH/-/CH$_3$CN	94.4	-	[38]
	Cyclooctene	60	1	24	t-BuOOH/-/CH$_3$CN	98	-	[38]
{2(Him)·[Cu(pdc)$_2$]}$_n$	Cyclohexene	60	1	8	Air/H$_2$O$_2$/EtOH	100	-	[38]
	Cyclooctene	60	1	8	Air/H$_2$O$_2$/EtOH	100	-	[38]
[Cu$_3$(L4)$_3$(H$_2$O)$_2$(DMF)]$_n$	Styrene	40	1	6	O$_2$/TMA/CH$_3$CN	90	88	[39]
	Cyclooctene	40	1	6	O$_2$/TMA/CH$_3$CN	99	99	[39]
{(Co(L2)H$_2$O))$_2$·H$_2$O)}$_n$	Cyclohexene	60	1	6	t-BuOOH/-/-	82.56	71.93	[30]
UiO-66	2-Cyclohexen-1-one	70	1	1	H$_2$O$_2$/-/CH$_3$CN	20	60	[46]
	2-Cyclohexen-1-one	70	1	2	H$_2$O$_2$/-/EtOAc	18	45	[46]
	Chalcone	70	1	0.5	H$_2$O$_2$/-/EtOAc	30	50	[46]
UiO-67	2-Cyclohexen-1-one	70	1	1	H$_2$O$_2$/-/CH$_3$CN	20	55	[46]
	2-Cyclohexen-1-one	70	1	2	H$_2$O$_2$/-/EtOAC	20	55	[46]
	Chalcone	70	1	0.5	H$_2$O$_2$/-/CH$_3$CN	40	40	[46]

[a] tBuOOH = tert-butyl hydroperoxide; CHP = cumene hydroperoxide; TBHP = tert-butylhydroperoxide; IBA = isobutyraldehyde; TMA = trimethylacetaldehyde.

2.2. Mixed-Metal Species

Many efforts have been made to improve the catalytic performance of MOFs, and one possible way is the construction of bimetallic clusters by functionalization of metal nodes/clusters with active transition metals to afford MOF-based catalysts with high performance.

A hydrothermal reaction has been used to synthesise Cu_x-Co_y-MOF, where $Co(NO_3)_2 \cdot 6H_2O$ and $Cu(NO_3)_2 \cdot 3H_2O$ inorganic metal salts have been one-pot added to a ligand solution in different molar ratios. In addition to the high catalytic activity by doping Cu-MOF with Co, a better selectivity to produce styrene oxide is achieved. At the optimal reaction conditions, the conversion and the selectivity of styrene to styrene oxide increased to 97.81% and 83.04%, respectively, by using $Cu_{0.25}$-$Co_{0.75}$-MOF, the catalyst of this series with higher content of Co^{2+} [48]. Another study showed how the conversion of styrene-to-styrene oxide increased rapidly when Mn ions were introduced into a Cu-MOF with the two ligands 2,5-dihydroxyterephthalic acid (H_4DHTA) and 2-picolinic acid (PCA). $Mn_{0.1}Cu_{0.9}$-MOF exhibits interesting catalytic activity for the epoxidations of various aromatic and cyclic olefins and a weak activity on decomposition of H_2O_2. Styrene can be oxidized by H_2O_2, through peroxybicarbonate-assisted catalysis, the styrene oxide yield achieving 85% in the presence of $Mn_{0.1}Cu_{0.9}$-MOF at 0 °C for 6 h [49].

To increase conversion and selectivity in the solvent-free aerobic oxidation of olefins, MOF catalysts based on *3d* metal copper (II), cobalt (II) and H_2ODA (oxydiacetic acid) containing lanthanum (III) as *4f* ions {[$La_2Cu_3(\mu$-$H_2O)(ODA)_6(H_2O)_3$]$\cdot 3H_2O$}$_n$ (LaCuODA) and {[$La_2Co_3(ODA)_6(H_2O)_6$]$\cdot 12H_2O$}$_n$ (LaCoODA) were employed. Catalytic studies pointed out the difference in aerobic oxidation of cyclohexene performances due to different physicochemical properties, surface area and redox properties of the metals (Table 2). [50] LaCoODA, based on Co(II), showed better conversion and selectivity for 2-cyclohexen-1-one. This is due to the structural differences between the square planar LaCuODA and the octahedral LaCoODA, in the latter case the water molecules could easily leave the channels to foster interaction between the active sites and the oxidant/catalyst. Moreover, the acid properties of the copper(II) ions are less effective than the redox properties of cobalt(II) ones, as far as the catalytic performances [34,51].

In NU-1000 single-ion-based iron(III) species have been incorporated using solution-phase post-synthetic metalation with two different iron(III) precursors. The resulting NU-1000-Fe-NO_3 and NU-1000-Fe-Cl frameworks show two crystallographically independent Fe sites (Fe1 resides in the c-pore and Fe2 in the hexagonal mesopore), coordinated to the bridging and terminal oxygens of the Zr_6 node, with Fe−O distances in NU-1000-Fe-Cl being much longer than those of NU-1000-Fe-NO_3 (Figure 5) [52]. Epoxidation of cyclohexene in vapour H_2O_2 with NU-1000-Fe-NO_3 as catalysts initially yields cyclohexene epoxide derived from heterolytic activation of H_2O_2, which in turn hydrolyzes rapidly to *trans*-cyclohexanediol. Otherwise, NU-1000-Fe-Cl yields a mixture of products and by-products, derived from the radical oxidation products due to homolytic activation of H_2O_2 [53,54]. This behaviour is probably due to the difference in the metal−node distance between the frameworks, the active site rearranging differently.

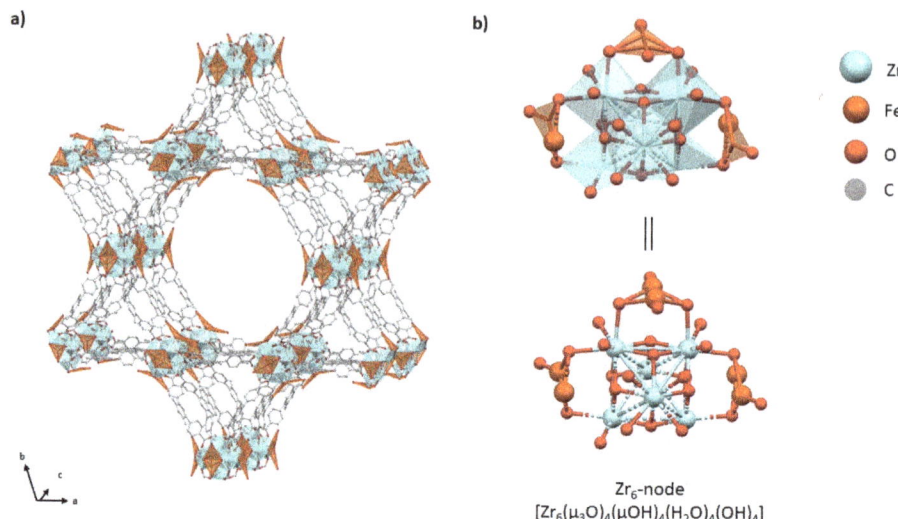

Figure 5. (a) Crystal structures of NU-1000-Fe-NO$_3$ (b) Structures of the inorganic Zr$_6$-nodes.

One-step template-free synthesis of ultrathin (~5 nm) mixed-valence {V16} clusters-based MOF nanosheets [Ni(4,4'-bpy)$_2$]$_2$ [V$_7^{IV}$V$_9^{V}$O$_{38}$Cl]·(4,4'-bpy)·6H$_2$O (NENU-MV-1) has been also reported. A large number of vanadium catalytically active sites in the NENU-MV-1 nanosheet allowed excellent cyclohexene oxidation under air exhibiting a conversion of 95%. Moreover, the nanometer scale of the catalyst increased the catalytic activity 2.7 times compared to the bulk crystal (0.25 mm) for olefin epoxidation. Excellent catalytic performances have been shown for different olefin substrates [55].

Table 2. Mixed Metal MOFs in Olefin Epoxidation.

MOF	Substrate	Reaction Data T (°C)	P (atm)	Time (h)	Oxidant/Cocatalyst/Solvent [a]	Conversion %	Epoxide Selectivity%	Ref.
Mn$_{0.1}$Cu$_{0.9}$-MOF	Styrene	0	1	6	H$_2$O$_2$/-/DMF	90.2	94.3	[49]
Cu$_{0.25}$-Co$_{0.75}$-MOF	Styrene	80	1	8	Air/TBHP/t-BuOH/H$_2$O$_2$	97.81	83.4	[48]
Zn1Co1-ZIF	Styrene	100	1	24	TBHP/-/DMF	99	71.31	[56]
LaCoODA	Cyclohexene	75	1	24	O$_2$ flow/-/-	85	75	[50]
LaCuODA	Cyclohexene	75	1	24	O$_2$ flow/-/-	67	55	[50]
NENU-MV-1	Cyclohexene	35	1	4	Air/IBA/CH$_3$CN	95	86	[55]
NU-1000-Fe-Cl	Cyclohexene	120	0.03	3	H$_2$O$_2$/-/-	-	70	[52]
NU-1000-Fe-Cl	Cyclohexene	120	0.03	3	H$_2$O$_2$/-/-	-	70	[52]

[a] tBuOH = *tert*-butyl alcohol; TBHP = *tert*-butylhydroperoxide; IBA = isobutyraldehyde.

2.3. Organic Linkers with Functional Catalytically Active Sites

Functional groups such as amino, pyridyl, amide, sulfonic acid, etc. present in organic linkers serve as active sites for catalysis and strongly influence the intrinsic catalytic activity of the MOFs through inductive effects. In addition, organic linkers can be catalytically active when organic functional groups and/or functional molecular catalysts (e.g., metalloporphyrins, salen and related ligands, chiral molecules, Schiff-base complexes, etc.) are introduced by post-synthetic ways. Alternatively, the same functional molecular catalysts can also be used as building units to fabricate MOFs.

Molybdenum complexes have been widely applied as homogeneous catalysts for the epoxidation of alkenes by H$_2$O$_2$ and organic hydroperoxide, a complete conversion and selectivity being reported. To overcome the recoverability and reusability issues correlated to the use of homogeneous molybdenum catalysts, molybdenylacetylacetonate has been supported on TMU-16-NH$_2$ [Zn$_2$(NH$_2$-BDC)$_2$(4-bpdh)]·3DMF, an amine-functionalized two-fold interpenetrated MOF via dative and combined covalent and dative post-synthetic modification [57].

A high porous NU-1000 MOF has been post-modified with the chiral L-tartaric acid, by SALI (solvent-assisted ligand incorporation) to build a chiral Zr-based MOF [C-NU-1000] [58]. Moreover, another active catalytic site, molybdenyl acetylacetonate, MoO$_2$(acac)$_2$, was incorporated on chiral NU-1000 to explore catalytic performance in the asymmetric epoxidation of olefins (Figure 6a) [59]. When olefins approach by pro-S- or R-face to the catalytic active center, they interact with the OH group of the tartrate through H-bond which induces chirality generating two chiral intermediates. The [C-NU-1000-Mo] catalyst, used in the epoxidation of styrene and 1-decene, can discriminate the S configuration in epoxides (Figure 6b).

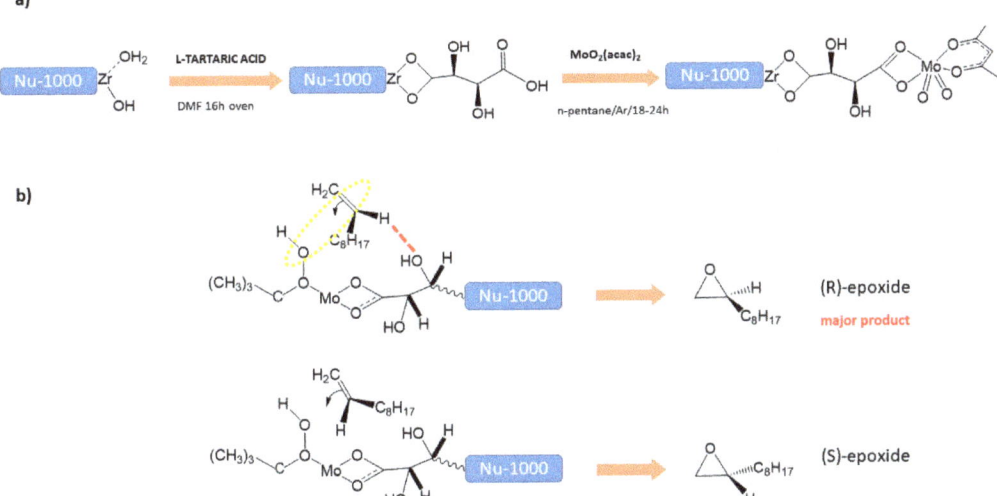

Figure 6. (a) The preparation of [C-NU-1000-Mo] catalyst with solvent-assisted ligand incorporation (SALI); (b) simplification of the chirality induction mechanism of [C-NU-1000-Mo] when olefins approach by pro-S- or R-face to the catalytic active center.

The free amine group available on UiO-66-NH$_2$ has been post-synthetically modified with salicylaldehyde (SA) or thiophene-2-carbaldehyde (TC) to graft a Schiff base in which MoO$_2$(acac)$_2$ could be immobilized. Efficient olefin epoxidation catalysed by UiO-66-NH$_2$-SA-Mo and UiO-66-NH$_2$-TC-Mo has been described and no Mo active site leaching was detected [60]. In the same way, MoO(O$_2$)$_2$·2DMF was immobilized onto UiO-66(NH$_2$) functionalized with salicylaldehyde (Sal) (UiO-66-sal-MoD), pyridine-2-aldehyde (PI) (UiO-66-PI-MoD) and 2-pyridine chloride (PC) (UiO-66-PC-MoD). All of them allowed a high dispersion of Mo catalyst, the large pores of MOFs guarantee adequate contact between the substrate and the catalytic active center, thus improving the efficiency of cyclic olefins epoxidation [61].

Molybdenum(VI) oxide was deposited on the eight-connected Zr$_6$(µ$_3$-O)$_4$(µ$_3$-OH)$_4$(H$_2$O)$_4$(OH)$_4$ nodes connected by 1,3,5,8-(p-benzoate) pyrene linkers (TBAPy^{4-}) of the mesoporous NU-1000, via condensation phase through solvothermal deposition in MOF (SIM) [62]. The stable Mo-SIM system exhibits a high conversion for cyclohexene epoxidation without leaching of molybdenum catalyst compared to Mo supported on bulk zirconia (Mo-ZrO2), in which significant leaching of the catalytic species has been observed [60].

Molybdenum tricarbonyl complexes are known to be effective catalysts for the epoxidation of olefins. They form an oxomolybdenum (VI) species in the presence of *tert*-butyl hydroperoxide (TBHP) as an oxidant which acts as highly active catalytic sites for the epoxidation of olefin. M(CO)$_6$ was deposited on UiO-66 and UiO-67 by chemical vapor deposition (CVD) treatment, UiO-66-Mo(CO)$_3$, and UiO-67-Mo(CO)$_3$ heterogeneous catalysts being fabricated. Herein, the larger tetrahedral and octahedral cavities of UiO-67 enable more accessibility of cyclooctene to catalytically active sites showing higher catalytic activity for the cyclooctene conversion than UiO-66-M(CO)$_3$ [33].

Several attempts have been made to immobilize oxovanadium(IV) complexes on different solid materials and create heterogeneous catalytic systems for the epoxidation of allylic alcohols [63–66]. A catalyst has been designed by immobilizing oxovanadium(IV) species on UiO-66(NH$_2$) via post-synthetic modification and by using two different pathways. At first, the amino-functionalized UiO-66(NH$_2$) was modified with salicylaldehyde to produce salicylideneimine modified UiO-66 (UiO-66-SI), subsequently [VO(acac)$_2$] was reacted with UiO-66-SI to give UiO-66-SI/VO(acac). In another pathway, UiO-66(NH$_2$) directly reacted with [VO(acac)$_2$] to produce UiO-66-N/VO(acac)$_2$ [67]. Excellent catalytic activity in the regioselective epoxidation of geraniol was obtained when UiO-66-SI/VO(acac) and UiO-66-N/VO(acac)$_2$ systems were employed by using reaction times of 60 and 120 min, respectively. The proper pore size of and the high dispersion of the catalytic sites on UiO-66-SI/VO(acac) and UiO-66-N/VO(acac)$_2$ guarantee good access of the substrate to the active sites.

Metallosalen-based crystalline porous materials have been realized for heterogeneous catalytic applications towards cyclopropanation, alkene epoxidation and hydrolytic kinetic resolution of epoxides with interesting enantioselectivities [5]. UiO-68-Me has been modified via post-synthetic exchange (PSE) with single- and mixed-M(salen) linker (M = Cu, Fe, Cr, V, Mn) to fabricate UiO-66(NH$_2$) attractive species for heterogeneous asymmetric catalysis, useful to overcome the problem of metal leaching. It was found that the single-M(salen) chiral MOFs (R)-UiO-68-Mn and (R)-UiO-68-Fe catalyse the epoxidation of alkenes to epoxides with up to a 98% ee of epoxide and 97% ee, respectively. The different catalytic metal centers in the mixed-(M)salen species UiO-68-Mn-Cr gave consecutive reactions starting from the epoxidation of alkene followed by ring–opening reaction of epoxide to produce the desired amino alcohol in 80−85% yields with 80−99.5% ee. Catalytic activity and enantioselectivity of all chiral UiO-68 catalysts remain unchanged for 10 cycles [68].

Encapsulation of Cu- or Ni-salen species in NH$_2$-MIL-101(Cr) through one-pot method gave a series of effective heterogeneous catalysts in the styrene oxidation under mild conditions. Specifically, the styrene conversion obtained using TBHP was 98.78%. The concentrated electronic density around Cu(II) in the Cu salen@NH$_2$-MIL-101(Cr) catalyst promoted the formation of tBuOOCu(III)-salen enhancing the selectivity to epoxide [69].

A recent synthetic strategy resides in the incorporation of different functionality in one single framework to generate a multivariate MOF (MTV-MOFs). On this basis, a chiral MOF based on multiple metallosalen bridging ligands has been synthesised. Firstly, M(salen)-derived dicarboxylate ligands H$_2$L5^M [M = Cu, VO, CrCl, MnCl, Fe(OAc), and Co(OAc)] were synthesized by reactions of N,N'-bis(3-*tert*-butyl-5-(carboxyl)salicylide (H$_4$L5) and the corresponding metal salts in MeOH at room temperature. Secondly, the crystals of binary or ternary MTV-MOFs (CuV, CuMn, CuCr, CuFe, and CuCo) were obtained by heating a 1:1 or 1:1:1 mixture of H$_2$L^M with Zn(NO$_3$)$_2$·6H$_2$O at 80 °C in DMF, [Zn$_4$O(L5$^{M,M'}$)$_3$] species is obtained. Both [Zn$_4$O(L5^{Cu,Mn})$_3$] and [Zn$_4$O(L5^{Cu,Fe})$_3$] showed efficient catalytic performances for asymmetric epoxidation of alkenes, affording up to 93% and 90% ee of the epoxides, respectively. Moreover, in the ternary heterogeneous catalyst [Zn$_4$O(L5^{Cu,Mn,Co})$_3$], the combination of Mn^{3+} and Co^{3+} promotes the epoxidation of alkene followed by enantioselective hydrolysis of epoxide to afford diols [70].

The achiral Zr-MOF, [PCN-224(Mn(Cl)], based on tetratopic ligand [manganese (chloride) tetrakis(4-carboxyphenyl)porphyrin [Mn(Cl)-TCPP], has been post-synthetically modified with tartrate anion, as a chiral auxiliary. The final chiral PCN-224-Mn(tart) contains two active metal sites (Zr and Mn) as Lewis acid centers and the chiral tartrate counterion, as Brønsted acid sites (OH functional group) has been investigated as chiral nucleophile catalyst towards both asymmetric epoxidation and CO_2 fixation. Asymmetric epoxidation of several aromatic and aliphatic olefins like styrene, *trans*-stilbene, 4-methylstyrene, α-methylstyrene, 1-phenyl-1-cyclohexene, 1-decene, and 1-octene has been tested by using PCN-224-Mn(tart) with aldehyde as co-catalyst, CH_3CN, and O_2. The all-reaction conversions were completed with an optimum range of epoxide selectivity 83–100% and high ee (84−100%). Several factors allow high enantioselectivity in the formation of the epoxide: the framework porosity, the active Mn center, a preferred face of the olefin (pro-S or -R face) close to produce the more stable configuration, and the noncovalent interactions between H atom of the olefinic double bond of the preferred face and chiral centers (Table 3) [71].

Table 3. MOFs with functionalized organic linker in olefin epoxidation.

MOF	Substrate	Reaction Data T (°C)	P (atm)	Time (h)	Oxidant/Cocatalyst/ Solvent [a]	Conversion %	Epoxide Selectivity%	Ref.
UiO-66-SI/VO(acac)	Geraniol	40	1	1	TBHP/-/CH_2Cl_2	100	100	[67]
UiO-66-N/VO(acac)$_2$	Geraniol	40	1	2	TBHP/-/CH_2Cl_2	100	100	[67]
UiO-66-sal-MoD	*cis*-Cyclooctene	80	24	1	TBHP/-/CH_3CN	99	99	[61]
PCN-224-Mn(tart)	Styrene	60	1	4	O_2/IBA/CH_3CN	100	89	[71]
	trans-Stilbene	60	1	4	O_2/IBA/CH_3CN	80	100	[71]
	1-Phenyl-1-cyclohexene	60	1	4	O_2/IBA/CH_3CN	75	100	[71]
	1-Octene	60	1	4	O_2/IBA/CH_3CN	70	100	[71]
UiO-67-Mo(CO)$_3$	Cyclooctene	55	3	1	TBHP/-/toluene	100	99	[33]
UiO-66-Mo(CO)$_3$	Cyclooctene	55	3	1	TBHP/-/toluene	92	99	[33]
[C-NU-1000-Mo]	Styrene	120	0.03	5	H_2O_2/-/$CH_2CH_2Cl_2$	100	86	[58]
	1-Octene	120	0.03	8	H_2O_2/-/$CH_2CH_2Cl_2$	72	100	[58]
UiO-66-NH$_2$-SA-Mo	Cyclooctene	83	0.75	1	TBHP/-/$CH_2CH_2Cl_2$	97	100	[60]
	Cyclohexene	83	1.5	1	TBHP/-/$CH_2CH_2Cl_2$	93	100	[60]
	Styrene	83	5	1	TBHP/-/$CH_2CH_2Cl_2$	87	92	[60]
	1-Octene	83	8	1	TBHP/-/$CH_2CH_2Cl_2$	78	100	[60]
	1-Decene	83	10	1	TBHP/-/$CH_2CH_2Cl_2$	79	100	[60]
UiO-66-NH$_2$-TC-Mo	Cyclooctene	83	1	1	TBHP/-/$CH_2CH_2Cl_2$	94	100	[60]
	Cyclohexene	83	2	1	TBHP/-/$CH_2CH_2Cl_2$	90	100	[60]
	Styrene	83	5	1	TBHP/-/$CH_2CH_2Cl_2$	86	90	[60]
	1-Octene	83	8	1	TBHP/-/$CH_2CH_2Cl_2$	75	100	[60]
	1-Decene	83	10.5	1	TBHP/-/$CH_2CH_2Cl_2$	75	100	[60]
[Zn$_4$O(L5^{Cu,Fe})$_3$]	2,2-Dimethyl-2*H*-chromene	−20	1	36	MesPhIO/-/$CHCl_3$	94	87 ee	[70]
[Zn$_4$O(L5^{Cu,Mn,Co})$_3$]	3-Chloropropene	0	1	10	sPhIO/-/$CHCl_3$	92	-	[70]
	Styrene	0	1	24	sPhIO/-/CH_2Cl_2	63	-	[70]
UiO-66-PC-MoD	*cis*-Cyclooctene	80	24	1	TBHP/-/CH_3CN	90.7	99	[61]
Mo-SIM	Cyclohexene	60	1	7	TBHP/-/toluene	93	99	[59]
(R)-UiO-68-Mn	2,2-Dimethyl-2*H*-chromene	0	1	10	sPhIO/-/CH_2Cl_2	91	88 ee	[68]
UiO-66-PI-MoD	*cis*-Cyclooctene	80	24	1	TBHP/-/CH_3CN	77.5	99	[61]
(R)-UiO-68-Fe	2,2-Dimethyl-2*H*-chromene	−20	1	36	MesPhIO/-/$CHCl_3$	84	86 ee	[68]
Cusalen@NH$_2$-MIL-101(Cr)	Styrene	80	1	6	TBHP/-/CH_3CN	98.78	89.58	[69]
[Zn$_4$O(L5^{Cu,Mn})$_3$]	2,2-Dimethyl-2*H*-chromene	−20	1	36	MesPhIO/-/CH_2Cl_2	86	86 ee	[70]
PCN-224-Mn(tart)	Styrene	60 °C	1	4	O_2/IBA/CH_3CN	100	89	[71]

Table 3. Cont.

MOF	Substrate	Reaction Data T (°C) P (atm) Time (h)			Oxidant/Cocatalyst/ Solvent [a]	Conversion %	Epoxide Selectivity%	Ref.
	trans-Stilbene	60 °C	1	4	O_2/IBA/CH_3CN	80	100	[71]
	1-Phenyl-1-cyclohexene	60 °C	1	4	O_2/IBA/CH_3CN	75	100	[71]
	1-Octene	60 °C	1	4	O_2/IBA/CH_3CN	70	100	[71]
TMU-16-NH_2	Cyclohexene	60	1	40	TBHP/-/$CHCl_3$	66	74	[57]
	Styrene	60	1	51	TBHP/-/$CHCl_3$	88	98	[57]
	Cyclooctene	60	1	24	TBHP/-/$CHCl_3$	83	83	[57]

[a] sPhIO = 2-(tertbutylsulfonyl)iodosylbenzene; IBA = isobutyraldehyde; CHP = cumene hydroperoxide; TBHP = *tert*-butylhydroperoxide.

3. Epoxidation with MOF-Based Composites

One possible way to improve the chemical and mechanical stability of MOFs as potentially heterogeneous catalysts is their immobilization onto/into supports. In this contest, solid polymer, graphene, and inorganic particles [72] or inorganic polymers [73] are largely employed as supports.

To overcome the poor hydrostability of [Cu_3-BTC_2] [74], a porous dendrimer-like porous silica nanoparticles (DPSNs) has been utilized as a carrier to support Cu-BTC Nps. The nanocomposites DPSNs@Cu-BTC were prepared by growing Cu_2O NPs in the center-radial porous channels of DPSNs. After that, Cu_2O NPs were dissolved in the presence of acid, oxidant and 1,3,5-benzenetricarboxylic acid (H_3BTC) [75]. The obtained Cu-BTC NPs have shown limited growth and a uniform distribution without agglomeration. The small size of Cu-BTC NPs (40 ± 25 nm) is useful in the aerobic epoxidation of various cyclic olefins achieving high catalytic activity without by-products. Good yield and selectivity were detected with inert terminal linear alkenes. Otherwise, epoxidation of styrene only achieved 65% of conversion due to the kinetic instability of styrene oxide (Table 4) [76].

The amphiphilic MIL-101-GH, a porous hierarchical material, has been explored as catalyst for the biphasic epoxidation reaction of 1-octene with H_2O_2. MIL-101-GH hydrogel was obtained by dispersing MIL-101 nanoparticles homogeneously in aqueous graphene oxide (GO) solutions. The TS-1 catalyst, commercially used in this biphasic reaction, was then introduced in MIL-101-GH. The resulting system, MIL-101-GH-TS-1, overcame the lower activity toward olefin epoxidation of TS-1, and the amphiphilic MIL-101-GH increased the contact areas of TS-1 with both H_2O_2 and 1-octene. The catalytic performance of MIL-101-GH-TS-1 has been much higher than that of single TS-1 and the 1,2-epoxyoctane was obtained without other by-products [77].

Polyoxometalate-based (POMs) heterogeneous catalysts are attractive species in the catalytic epoxidation of olefin. They have got great catalytic activity, selectivity, and easy separation but their leaching mainly due to the strong complexing capability of solvent and H_2O_2 oxidants, represents the major obstacle in the possible applications [78,79]. To overcome the stability issue of POMs, the polyoxomolybdic cobalt (CoPMA) and polyoxomolybdic acid (PMA) species were incorporated into UiO-bpy, a Zr-based MOFs, through self-assembly process under solvothermal condition [80]. CoPMA@UiO-bpy showed the highest catalytic activity for cyclooctene oxidation with H_2O_2 and also for the oxidation of styrene and 1-octene with O_2 as oxidant and *tert*-butyl hydroperoxide (t-BuOOH) as initiator. This is due to the uniform distribution and better immobilization of POM clusters within the size-matched cages of Zr-MOFs owing to the presence of bipyridine groups in the UiO-bpy framework. It is noteworthy that CoPMA@UiO-bpy shows excellent recyclability and stability against the leaching of active POM species.

Composite material has been obtained by encapsulating H_5-$PMo_{10}V_2O_{40}$ polyoxometalates (POMs) and 1-octyl-3-methylimidazolium bromide, ionic liquids (ILs), in the mesoporous cages and large surface area of MIL-100 (Fe). The synergic effect of ILs, Lewis and Brønsted acid sites in both $PMo_{10}V_2$ species and MOF created a PMo10V2-ILs@MIL-100(Fe) hybrid with significant catalytic properties in cycloolefins epoxidation. Indeed, the PMo10V2 was activated by the imidazolium cations originated from ILs and the incorporation on MIL-100(Fe) prevented the leaching of POMs [81]. This composite is easily regenerated for 12 cycles without loss catalytic performance [82].

MIL-100(Fe) combined with the polyoxometalate $(C_{16}H_{36}N)_6K_2[\gamma\text{-SiW}_{10}O_{36}]$ has been reported to catalyse epoxidation of 3Z,6Z,9Z-octadecatriene to the corresponding 6,7-epoxide with high site selectivity (82.35%). The conversion catalysed by POM/MIL-100(Fe) exhibits a greater performance when the MOF contains unsaturated Lewis acid iron ions [83]. The main product of this epoxidation is a sex pheromone of *E. obliqua* Prout and can be potentially used in pest insect control with environmental friendliness.

Two POMs-based MOFs, $[Cu_6(bip)_{12}(PMo^{VI}_{12}O_{40})_2(PMo^VMo^{VI}_{11}O_{40}O_2)]\cdot 8H_2O$ and $[Co_3^{II}Co_2^{III}(H_2bib)_2(Hbib)_2(PW_9O_{34})_2(H_2O)_6]\cdot 6H_2O$ (H_2bip = 1,3-bis(imidazolyl)propane; bib = 1,4-bis(imidazol)butane)), have been fabricated using a flexible N-containing bidentate ligands via hydrothermal condition. They have been employed in the catalytic processes for selective alkene epoxidation and recycled four times without loss of quality (Figure 7) [84].

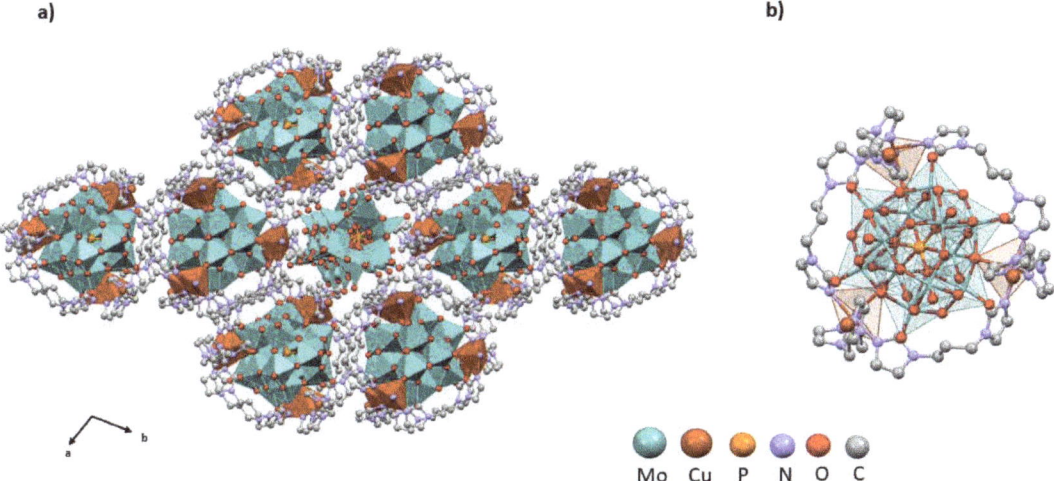

Figure 7. (a) The 2D structure of $[Cu_6(bip)_{12}(PMo^{VI}_{12}O_{40})_2(PMo^VMo^{VI}_{11}O_{40}O_2)]\cdot 8H_2O$; (b) the coordination environment of the Cu(II) cations. Hydrogens and hydroxyls are omitted for clarity. Light-blue polyhedral correspond to the (PMo_{12}) polyanion.

Metal nanoparticles can grow without agglomeration in a porous matrix to produce a stable and active heterogeneous catalyst. Pd NPs have been loaded on the pre-synthesized UiO-66-NH$_2$ using a simple solution impregnation method and NaBH$_4$ reduction. The amino groups in the linkers allow a strong interaction with Pd (II) ions which is essential to yielding well-dispersed Pd/UiO-66-NH$_2$ catalyst. The experiments suggest that the best catalytic activity for styrene epoxidation has been found under Pd NPs loadings of 3.69 wt% [85].

A dually functionalized catalytic system for the tandem H_2O_2-generation/alkene-oxidation reaction has been realized. A microcrystal of UiO-66-NH$_2$ has been used as a platform to encapsulate Au and Pd metal NPs and later Pd/Au@UiO-66-NH$_2$ surfaces have been post-synthetically modified with a (sal)Mo^{VI} (sal = salicylaldimine) molecular epoxidation catalyst. The porosity of Pd@UiO-66-sal(Mo) allows H_2 and O_2 gases to come into contact with the encapsulated NPs to generate H_2O_2. The synergic effect of the generated H_2O_2 and (sal)Mo^{VI} in a MOF enhanced epoxide productivity reducing alkene hydrogenation side reaction. This study showed that (sal)Mo moieties in Pd@UiO-66-NH$_2$

epoxidize cis-cyclooctene substrate faster, leading to the more effective usage of the H_2O_2 oxidant [86].

Systems composed of a magnetic uniform Fe_3O_4(PAA) microspheres core and of a copper-doped MOF shell demonstrated an easily catalyst recovery approach improving turnover number and turnover frequency. In addition, these magnetic core–shell heterogeneous catalysts improve both stability of the metal active site and dispersity of catalyst materials reducing the metal leaching. Two interesting magnetic core-shell copper-doped catalysts, Fe_3O_4@P4VP@ZIF-8 and Fe_3O_4/$Cu_3(BTC)_2$ have been prepared by combining the solvothermal method with layer-by-layer assembly. Initially, monodispersed PAA-modified Fe_3O_4 particles were synthesized by solvothermal methods [87]. In the case of Fe_3O_4/$Cu_3(BTC)_2$, Fe_3O_4 particles were alternately immersed in solutions containing $Cu(CH_3COO)_2 \cdot H_2O$ and H_3BTC such that $Cu_3(BTC)_2$ nanocrystals grow layer-by-layer on the surface of PAA- modified Fe_3O_4 particles. This nanosized porous structure increases the contact between the Cu(II) active sites present in the $Cu_3(BTC)_2$ shell and the catalytic substrates [88]. In Fe_3O_4@P4VP@ZIF-8 catalyst, on the other hand, the Fe_3O_4(PAA) core has been coated with P4VP middle layer to adsorb a large number of Zn^{2+} for the growth of the ZIF-8 shell thickness on the surface of the core–shell Fe_3O_4(PAA)@P4VP. Then, the Zn^{2+} ions were partially substituted by Cu^{2+} ions in the ZIF-8 shell framework. The ions exchange allowed a well-dispersed copper active site in the resulting copper-doped ZIF-8 structure, avoiding their leaching [89].

Aerobic epoxidation of cyclic olefins (e.g., cyclohexene, norbornene) using both magnetic core–shell copper-doped Fe_3O_4@P4VP@ZIF-8 and Fe_3O_4/$Cu_3(BTC)_2$ as heterogeneous catalyst achieved high conversion and selectivity (99%) in the formation of the epoxide under mild reaction conditions. Epoxidation of styrene by using Fe_3O_4@P4VP@ZIF-8 as a catalyst has brought only 54% selectivity of the desired epoxide owing to the kinetic instability of styrene oxide and its oxidation into benzaldehyde [90].

A series of Zr-based core-shell MOF composites with mesoporous cores and microporous shells have been synthesized by solvothermal under kinetic control. PCN-222(Fe) crystals have been synthesized and used as seed crystals to grow the Zr-BPDC(UiO-67) crystals. Meso- and micro-porosity inside of PCN-222(Fe)@Zr-BPDC(UiO-67) drives the catalytic performances for olefin epoxidation reaction [91]. Indeed, the core MOF with Fe-porphyrin moieties represents the catalytic center, while the shell controls the selectivity of the substrate through tuneable pore size. This size-selective catalyst showed almost complete conversions for small olefins.

Table 4. MOF-based composites for epoxidation reaction.

MOF	Substrate	Reaction Data T (°C)	P (atm)	Time (h)	Oxidant/Cocatalyst/ Solvent [a]	Conversion %	Epoxide Selectivity%	Ref.
DPSNs@Cu-BTC	Cyclooctene	40	1	4	O_2/TMA/CH_3CN	99	99	[75]
	Styrene	40	1	6	O_2/TMA/CH_3CN	62	65	[75]
Fe_3O_4@P4VP@ZIF-8	Cyclohexene Cyclooctene Norbornene	60	1	12	O_2/TMA/CH_3CN	99	99	[90]
Fe_3O_4/$Cu_3(BTC)_2$	Cyclohexene Cyclooctene Norbornene	40	1	6–8	O_2/IBA/CH_3CN	99	99	[88]
	Styrene	40	1	6–8	O_2/IBA/CH_3CN	99	84	[88]
PCN-222(Fe)@Zr-BPDC(UiO-67)	1-Hexene	r.t	1	12	PhIO/-/CH_3CN	99	-	[91]
	Cyclopentene	r.t	1	12	PhIO/-/CH_3CN	99	-	[91]
	Cyclohexene	r.t	1	12	PhIO/-/CH_3CN	99	-	[91]
CoPMA@UiO-bpy	Cyclooctene	70	1	6	H_2O_2/-/CH_3CN	91	99	[80]
	Styrene	80	1	6	O_2/t-BuOOH/-	80	56	[80]
PMo10V2-ILs@MIL-100(Fe)	Cyclohexene	60	1	4	H_2O_2/-/CH_3CN	92	93	[91]

Table 4. Cont.

MOF	Substrate	Reaction Data T (°C)	P (atm)	Time (h)	Oxidant/Cocatalyst/ Solvent [a]	Conversion %	Epoxide Selectivity%	Ref.
[Cu$_6$(bip)$_{12}$(PMoVI$_{12}$O$_{40}$)$_2$ (PMoVMoVI$_{11}$O$_{40}$O$_2$)]·8H$_2$O	Cyclooctene	20	1	4	H$_2$O$_2$/tBuOH/CH$_3$CN	>99	74.1	[84]
	1−Hexene	20	1	4	H$_2$O$_2$/tBuOH/CH$_3$CN	>99	91.9	[84]
	1−Octene	20	1	4	H$_2$O$_2$/tBuOH/CH$_3$CN	>99	71.5	[84]
Pd/UiO-66-NH$_2$	Styrene	80	1	12	N$_2$/TBHP/CH$_3$CN	90.8	96.5	[85]
[Co$_3^{II}$Co$_2^{III}$(H$_2$bib)$_2$(Hbib)$_2$ (PW$_9$O$_{34}$)$_2$(H$_2$O)$_6$]·6H$_2$O	Cyclohexene	20	1	4	H$_2$O$_2$/tBuOH/CH$_3$CN	72.9	95.3	[84]
	1−Hexene	20	1	4	H$_2$O$_2$/tBuOH/CH$_3$CN	>99	85.9	[84]
	1−Octene	20	1	4	H$_2$O$_2$/tBuOH/CH$_3$CN	95.5	70.1	[84]
POM/MIL-100(Fe)	3Z,6Z,9Z-Octadecatriene	40	1	24	H$_2$O$_2$/-/CH$_3$CN	30	82	[83]
MIL-101-GH-TS-1	Octane	40	1	12	H$_2$O$_2$ (30%)/-/-	15	-	[77]
Pd@UiO-66-sal(Mo)	cis-Cyclooctene	r.t	1	6	H$_2$O$_2$/CH$_3$OH/H$_2$O	-	-	[86]

[a] tBuOH = *tert*-butyl alcohol; TMA = trimethylacetaldehyde; IBA = isobutyraldehyde.

4. CO$_2$ Epoxide Cycloaddition to Cyclic Carbonates

Cycloaddition reaction of CO$_2$ with epoxides represents one of the most economically efficient approaches in the production of cyclic organic carbonates with relevant applications ranging from raw materials in the pharmaceuticals industry, polar aprotic solvents, electrolytes in lithium batteries, lubricants, precursors for polycarbonate materials, and other fine chemicals.

CO$_2$, being a C1 feedstock has, in fact, a high potential from the chemical point of view [92]. CO$_2$ can be employed in the highly atom-economical acid-catalysed epoxides cycloaddition to give cyclic organic carbonates, relevant species for industrial applications [93]. The cyclic carbonates (OCs) have been also used as intermediates for engineered polymers, as a lubricant (in 1987 Agip Petroli added dialkylcarbonates as lubricant in a formulation of semisynthetic gasoline engine oil components), and more recently found application in varnish production, green solvents or electrolytes in lithium-ion batteries.

The CO$_2$ cycloaddition mechanism involves an acid catalyst (Lewis or Brønsted acid) that coordinates to the epoxide substrate activating it toward nucleophilic attack by the co-catalyst (e.g., typically a tetraalkylammonium halide). The resulting halo-alkoxide intermediate reacts with carbon dioxide to generate the cyclic carbonate and subsequently regeneration of both catalyst and co-catalyst [93].

The CO$_2$ fixation reaction catalysed by homogeneous or heterogeneous catalysts has been extensively investigated, however some drawbacks remain. Differently from the homogeneous, heterogeneous catalysts (e.g., ionic liquid-supported solids [94–96], polymers [96,97], and porous organic frameworks [98]) have the advantages of easy separation and regeneration of the catalyst, but they often required rough conditions (high temperature, pressure, and time) due to a lack of accessible surface area for accelerating interactions of CO$_2$ and reagents with active sites. Therefore, the high surface area, tunability, and CO$_2$ sorption capacity of MOFs can be beneficial for improving the efficiency of the CO$_2$ cycloaddition reaction. Lewis acid metal centers and Brønsted acid groups in MOFs can promote the activation of the epoxide ring, while the functional groups in the ligands can act as Lewis/basic sites improving not only the CO$_2$ affinity inside the pore but also can fulfil the role of co-catalyst (Figure 8) [99].

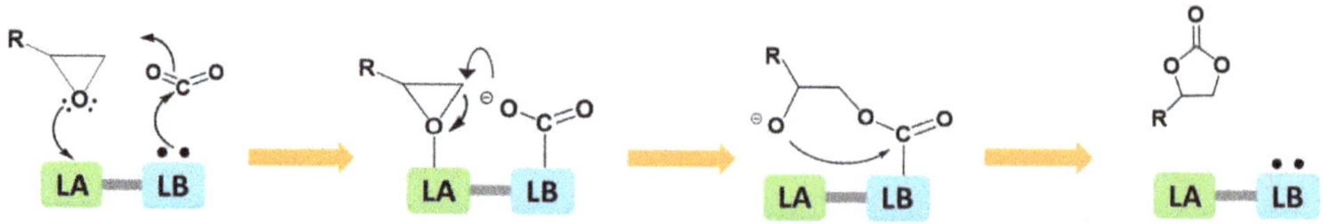

Figure 8. Proposed mechanism for the cycloaddition of CO_2 to epoxide with Lewis acid and base catalysts.

The Hf-cluster-based NU-1000 (Hf-NU-1000) demonstrated excellent catalytic activity, greater than the Zr-cluster-based NU-1000 under the same mild reaction conditions [100]. Indeed, the presence of high density stronger acidic Brønsted sites, due to stronger M−O bonds, gave a complete and quantitative conversion of styrene oxide and propylene oxide to form cyclic carbonates. Moreover, high yields have been detected for the cycloaddition reaction of CO_2 with industrially important epoxide divinylbenzene dioxide (DVBDO) [101].

Large pores in the MOFs, easily functionalized by polar groups, can promote CO_2 fixation in a short reaction time under ambient CO_2 pressure and moderate temperature without the use of solvent. Within the mesoporous M-MOF-184 series (M = Co, Ni, Mg, Zn), Zn-MOF-184 achieved efficient catalysis performances to convert CO_2 to cyclic carbonates under ambient conditions for several epoxy substrate, due to the presence of high concentration of accessibly acidic metals, basic 2-oxidobenzoate anion sites and to the high polarity induced by C≡C bonds and π systems from the phenyl rings in the linkers. Low conversion has been detected for larger epoxides due to limit diffusion into the MOF pores of reactants toward the active sites [102]. The hydrothermally synthesized flexible Zn-based $\{[Zn_2(TBIB)_2(HTCPB)_2]\cdot 9DMF\cdot 19H_2O\}_n$, has been synthesized employing two types of large linkers 1,3,5-tri(1H-benzo[d]imidazol-1-yl)benzene (TBIB) and 1,3,5-tris(4'-carboxyphenyl-)benzene (H_3TCPB). A porous structure with 1D channels was generated via noncovalent supramolecular interactions between the layers. The presence of free protonated carboxylic acid groups(−COOH), carbonyl groups (−C=O), and the presence of Lewis basic sites from the rich N-containing TBIB on the surface pores enhance the selectivity toward CO_2. Moreover, the COOH group helps in catalysing the CO_2 cycloaddition reaction efficiently through noncovalent interaction with the epoxide substrate, followed by ring-opening upon nucleophilic attack of co-catalyst [103].

Excellent conversions of epichlorohydrin and 2-vinyloxirane have been obtained using as heterogeneous catalyst $[Zn_4OL4_3]_n$ based on the meta-substituted 2,2',6,6'-tetramethoxy-4,4'-biphenyldicarboxyate ligand [39].

Zeolitic imidazolate frameworks are known for their high CO_2 solubility and capture ability [104], especially the chloro-functionalized ZIF-95 [105]. The CO_2 cycloaddition to propylene oxide by using ZIF-95 and a quaternary ammonium salt as cocatalyst procured over 99% selectivity to the desired propylene carbonate product under moderate conditions [106]. Also the imidazolate-containing species Im-UiO66(Zr)MOF reacts with methyl iodine to produce (I^-)MeIm-UiO-66 that demonstrate efficiency in the CO_2 cycloaddition reaction toward a broad range of substrates, in this case without the addition of co-catalyst [107].

Conversely, imidazolium-based IL units were grafted and immobilized into UiO-67 via direct ligand functionalization that, considering the post synthetic approach, is a quantitative method. The obtained species show a high density of IL sites. UiO-67-IL converts epichlorohydrin substrate in 95% yield under co-catalyst and solvent-free conditions. The yield increases to 99% in a shorter time when TBAB was employed (TBAB = tetrabutylammonium bromide) [108] (Table 5).

UiO-66-NH_2 pores were modified with ILs such as methylimidazolium bromide and methylbenzimidazolium bromide by coupling reactions, to generate ILA@U6N and ILB@U6N MOFs. The Lewis acid sites (for activation of the epoxide) and the IL functional sites (for epoxide ring-opening) efficiently catalyse the epichlorohydrin conversion under mild conditions [109].

A linear ionic polymer was inserted inside the MIL-101(Cr) via in situ polymerization to form polyILs@MIL-101(Cr) stable heterogeneous composites. This polyILs@MIL-101 is able to catalyse the CO_2 cycloaddition reaction with various epoxides with good to excellent conversions, including terminal epoxides with both electron-withdrawing and electron-donating substituents without the need of co-catalyst [110].

A new multimodal catalytic system has been designed via two steps post-synthetic modification of the metal nodes in the NU-1000 framework. A tandem functionalization was performed starting from the incorporation of *ortho-*, *meta-*, and *para-*pyridinecarboxylic acids into the framework of NU-1000(M), then the pyridine moieties were alkylated with various haloalkanes (CH_3I, C_4H_9I, C_4H_9Br, and $C_6H_4F_9I$) to introduce co-catalyst moieties near to the inorganic node [111]. Among catalysts, NU-1000(Zr) functionalized with 4-PyCOOH and CH_3I, i.e., SALI-4-Py-I-(Zr), showed the highest styrene carbonate yield without co-catalyst, the epoxy ring being activated upon coordination to Zr^{4+} center (Lewis acid site) and the halogen anion opening the epoxy ring by nucleophilic attack on the less sterically hindered carbon atom [111].

Two 3D metal-cyclam-based zirconium MOFs $[Zr_6(\mu_3\text{-}OH)_8(OH)_8(M\text{-}L)_4]$ (where M = Cu(II) or Ni(II), L1 = 6,13-dicarboxy-1,4,8,11-tetraazacyclotetradecane) were prepared, namely VPI-100 (Cu) and VPI-100 (Ni) (VPI = Virginia Polytechnic Institute), respectively. A two-step solvothermal synthesis has been necessary to build the MOFs. Initially, a zirconium-oxo cluster was assembled, then cyclam was added. The presence of accessible Cu^{2+}/Ni^{2+} metal active sites in the metallocyclams and of the coordinatively unsaturated Zr^{4+} sites in the equatorial plane of the Zr_6 cluster in VPI-100 improved their catalytic activity toward CO_2 cycloaddition to various organic epoxides [112].

Another strategy developed to increase the catalytic performances is based on the incorporation of an amine group in MOFs. Essentially, the amino group has the dual advantage of acting as an electron donor (Lewis base) toward CO_2 and increasing the local concentration of CO_2 near catalytic centres through a high CO_2 adsorption [113,114].

The amine-functionalized NH_2-MIL-101(Al) has been synthesized using a solvothermal or microwave method and its catalytic activity in the solvent-free cycloaddition of CO_2 to styrene oxide achieved nearly total conversion and selectivity in 96% yield, with a TOF of 23.5 h^{-1} [115]. The coordinatively unsaturated aluminium centers present in the SBUs (Lewis acidic sites) bind the epoxides and activate them toward ring-opening, this step is immediately followed by the attack of the bulky bromide ions of TBAB. The pendant amino groups polarize the CO_2 molecules, through the nucleophilic attack at the carbon atom, and facilitate CO_2 insertion and cycloaddition (Figure 9). During the catalytic reaction, the micro and mesoporous of the framework facilitate the diffusion of substrates and reactants to enhance their interactions [116].

Recently, the acid-base pair UiO-66-NH_2 has been used to synthesize bio-based five-membered cyclic carbonate from vegetable oil methyl ester by CO_2 fixation. At first, 95% of double bonds in the O-acetyl methyl ricinoleate starting material were converted to epoxide through an enzyme-catalyzed process. Then, the cycloaddition of epoxy fatty acid methyl esters was performed in the presence of UiO-66-NH_2 as catalyst and TBAB as co-catalyst for CO_2 fixation. At 120 °C under 3 MPa CO_2 pressure for 12 h, the reaction conversion reached 94.4% [117] (Table 5).

A series of diamino-tagged zinc bipyrazolate MOFs have been investigated as heterogeneous catalyst in the reaction of CO_2 with the epoxides epichlorohydrin and epibromohydrin to give the corresponding cyclic carbonates at 393 K and pCO_2 5 bar under relatively mild conditions (solvent and co-catalyst-free) [118]. The presence of amino group in the MOFs pores increased the CO_2 storage capacity as well as the catalytic performances. The epoxide has been activated through halogen-amine interaction which was observed in structure of the [epibromohydrin@Zn(3,3'-$(NH_2)_2$BPZ)] adduct. The isomeric Lewis basic site (NH_2) in Zn(3,5 NH_2-Bpz) (64% yield) improves more than twice the catalytic transformation of epichlorohydrin compared to its mono(amino) parent Zn(BPZNH_2) (32% yield) [118].

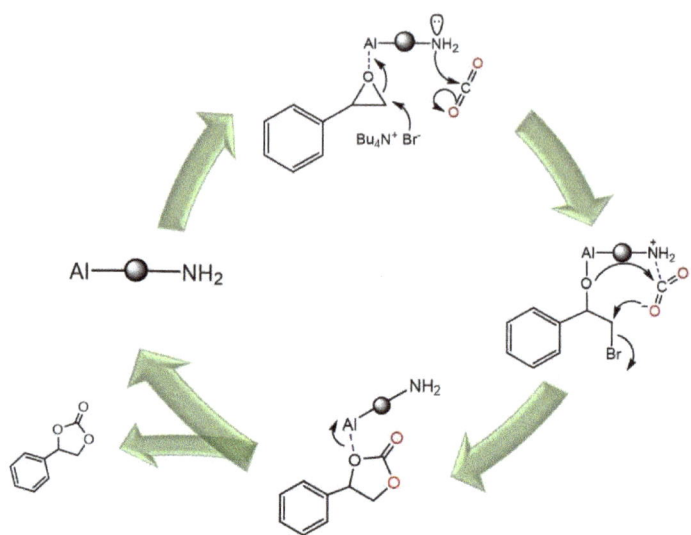

Figure 9. Proposed mechanism for the cycloaddition of styrene oxide and CO_2 using tetrabutylammonium bromide.

Post-synthetic metalation of organic linkers is employed strategically to tailor the MOFs' properties. In Hf-Bipy-UiO-67, the 2,2-bipyridine-5,5-dicarboxylate ligand was grafted with $Mn(OAc)_2$ and the resulting Hf-Bipy-UiO-67($Mn(OAc)_2$) showed that synergy of the binary Lewis acid function significantly enhances the CO_2 uptake capacity and catalytic performance of the cycloaddition reaction under mild conditions [119].

Vanadium chlorides have been used to produce the post-metalated Zr-based MOF-VCl3 and MOF-VCl4, with biphenyl-4,4′-dicarboxylic acid, and 2,2′-bipyridine- 5,5′-dicarboxylic acid, respectively, which provide Lewis basic sites. Their high catalytic activity in the CO_2 cycloaddition to various organic epoxides was attributed to the accessible Cu^{2+}/Ni^{2+} metal active sites in the metallocyclams and the presence of coordinatively unsaturated Zr^{4+} sites in the equatorial plane of the Zr_6 cluster in VPI-100 MOFs [112].

UiO-type MOFs become susceptible to water and alkaline solution when the length of the carboxylic linker increase. A series of UiO-type MOF named ZSF, incorporating chiral metallosalen as linker has been produced. The chemically stable ZSF-1 MOF, synthesized by dissolving a mixture of $ZrCl_4$, Cy-salen-Ni, and modulators (trifluoroacetic acid), showed excellent catalytic performance for the conversion of CO_2 with epoxides into cyclic carbonates. The tetrahedral cages of ZSF-1 decorated with salen-Ni moieties entrap efficiently CO_2 and activate the substrate. ZSF-1 catalyses efficiently the asymmetric cycloaddition of CO_2 with styrene oxide giving 94% yield of the resulting cyclic carbonate [120]. With other epoxides, specifically epichlorohydrin, the catalytic activity of ZSF-1 increases until to 99% of conversion thanks to the presence of electron-withdrawing Cl group, which promotes the nucleophilic attack of Br^- during the ring-opening process.

The chiral PCN-224-Mn(tart) (see Section 2.3) has been used in asymmetric CO_2 cycloaddition to styrene epoxide, its derivative showing conversions of 96% and 87%, respectively. The missing-linker defects in the Zr cluster and in the Mn center are Lewis acids inducing catalytic ability into the framework for CO_2 chemical fixation. In addition, the auxiliary chiral tartrate anions, and the co-catalyst (Bu_4NBr) act as nucleophiles generating a chiral epoxide, semi-intermediate, starting from prochiral styrene substrate. The CO_2 addition leads asymmetrically to cyclic carbonate with a high ee, and it is related to the interaction of the chiral centers and substrate pro R/S face. Moreover, catalytic reactions with PCN-224-Mn(tart) were performed at low energy and ambient pressure and temperature [71].

Table 5. MOF-based composites for cycloaddition reaction.

MOF	Substrate	Reaction Data T (°C) CO_2 P (atm) Time (h)			Cocatalys [a]	Conversion %	Cyclic Carbonate Selectivity%	Ref.
Hf-NU-1000	Styrene epoxide	r.t.	1	56	TBAB	100	100	[101]
	Propylene oxide	r.t.	1	26	TBAB	100	100	[101]
	Epoxide divinylbenzene dioxide	r.t.	1	19	TBAB	100	100	[101]
PCN-224-Mn(tart)	Styrene epoxide	60	1	15	TBAB	96 94 ee (S)	100	[71]
	(2,3-Epoxypropyl)benzene	60	1	15	TBAB	87 90 ee (S)	100	[71]
	Propylene oxide	60	1	15	TBAB	99 98 ee (S)	100	[71]
	1,2-Epoxybutane	60	1	15	TBAB	91 97 ee (S)	100	[71]
	1,2-Epoxyoctane	60	1	15	TBAB	78 96 ee	100	[71]
polyILs@MIL-101(Cr)	1-Butene oxide	45	1	48	-	94	100	[110]
	1,2-Epoxyhexane	70	1	24	-	89	100	[110]
	3-Hydroxy-1,2-epoxypropane	70	1	24	-	>99	100	[110]
	1,2-Epoxy-3-phenoxypropane	70	1	24	-	95	100	[110]
NH_2-MIL-101(Al)	Styrene oxide	120	18	6	TBAB	93.6	99	[115]
UiO-66-NH_2	Epoxy fatty acid methyl ester	120	30	12	TBAB	94	80	[117]
ILB@U6N	Epichlorohydrin	80	118	4	-	94	99	[109]
(I^-)Meim-UiO-66	Epichlorohydrin	120	1	24	-	100	93	[107]
VPI-100 (Ni)	Epichlorohydrin	90	10	6	TBAB	96	-	[112]
VPI-100 (Cu)	Epichlorohydrin	90	10	6	TBAB	94	-	[112]
$[Zn_4OL4_3]_n$	Epichlorohydrin	50	1	4	-	96	99	[107]
	2-Vinyloxirane	50	1	4	-	99	81	[107]
Hf-Bipy-UiO-67(Mn(OAc)$_2$	Epichlorohydrin	25	1	12	TBAB	83.2	99	[119]
Zn-MOF-184	Styrene oxide	80	1	6	TBAB	96	85	[102]
	Propylene oxide	80	1	6	TBAB	100	75	[102]
	Epichlorohydrin	80	1	6	TBAB	100	70	[102]
	Cyclohexene oxide	80	1	6	TBAB	69	85	[102]
SALI-4-Py-I-(Zr),	Styrene oxide	80	4	4	-	99	98	[111]
ILA@U6N	Epichlorohydrin	80	118	4	-	65	99	[109]
UiO-67-IL	Epichlorohydrin	90	1	3	TBAB	99	100	[108]
	Epichlorohydrin	90	1	3	-	99	96	[108]
{[Zn_2(TBIB)$_2$(HTCPB)$_2$]·9DMF·19H_2O}$_n$	Epichlorohydrin	r.t	1	24	TBAB	99	100	[105]
ZIF-95	Propylene oxide	120	118	24	TBAB	91	99	[106]
ZSF-1	Styrene oxide	100	1	20	TBAB	93	-	[120]
	Epichlorohydrin	100	1	20	TBAB	99	-	[120]
Zn(3,5-NH_2-Bpz)	Epichlorohydrin	120	5	24	-	98	50	[118]
Zn(BPZNH_2)	Epichlorohydrin	120	5	24	-	96	33	[118]

[a] TBAB = tetrabutylammonium bromide.

5. Conclusions

MOF-based catalysts are now a very promising class of compounds as they merge relevant characteristics of both homogeneous and heterogeneous catalysts. They can be easily modified by changing linkers substituents to increase affinity for reactants, or by growing the number of active catalytic sites.

In this review, we have explored the ability of MOFs, MOF nanocomposites and mixed metal species toward olefin epoxidation and carbon dioxide cycloaddition.

We have observed that the olefin conversion and the epoxide selectivity are strongly dependent on the metal nodes/clusters, Co and Cu species being the most efficient, in some cases as for the epoxidation of a-pinene by Co-MOF-150-2 a conversion and an epoxide selectivity close to 100% being found.

Mixed metal MOFs can be also successfully employed in styrene and cyclohexene epoxidation, the best results being obtained with Cu/Co, Mn/Cu, and Ni/V species.

Selected functional groups introduced in organic linkers can also act as catalytically active sites. Amino, pyridyl, amide and sulfonic acid groups, but also metalloporphyrins, vanadium and molybdenum acetylacetonate, tartaric acid, salen and analogous molecules can be inserted or deposited to obtain also greater selectivity. UiO-66, UiO67, and PCN-224, appropriately functionalized can induce a complete conversion and selectivity as in the case of the geraniol epoxidation.

MOF-based composites are often employed to increase the hydrostability of selected MOFs or to perform epoxidation also of specific substrates as norbornene or octadecatriene. Specifically, a porous dendrimer-like porous silica nanoparticles (DPSNs) used as a carrier to support Cu-BTC NPs overcame the poor hydrostability of $[Cu_3\text{-}BTC_2]$ MOF achieving high catalytic activity without by-products under mild reaction conditions.

Finally, MOFs and MOF-based composites show a great efficiency toward CO_2 cycloaddition to epoxides, conversion being generally in the range 70–100% and selectivity close to 100%. The use of chiral ligands and amine-functionalized ligands seems to be very promising. The CO_2 binding mode can in fact open new strategies for activation of CO_2 and its transformation.

However, the low reactivity and inert nature of CO_2 make its incorporation and activation into organic substrates still a challenge. Currently, the heterogeneous MOFs-based catalysts, as well as the technical system, remain at the laboratory scale and that makes the costs of productions of these materials extremely pricey. It is desirable that the improvement of MOFs-based catalysts might lead to technically viable efficiencies to industrial production to allow their large-scale application, in the next future. This review clearly shows that MOFs are now perspective materials and valid candidates for catalytic epoxidation and CO_2 cycloaddition reactions.

Author Contributions: Conceptualization, A.T. and C.P.; methodology, A.T. and C.P.; software, A.T. and C.P.; validation, A.T. and C.P.; data curation, A.T. and C.P.; writing—original draft preparation, A.T. and C.P.; writing—review and editing, A.T. and C.P.; funding acquisition, C.P. All authors have read and agreed to the published version of the manuscript.

Funding: This research was funded by University of Camerino.

Conflicts of Interest: The authors declare no conflict of interest.

Abbreviations

MOFs

Co6-MOF-3	$[(Co_6(OH)_6(TCA)_2(BPB)_3]_n$
Co-MOF-150-2	$[Co(BDC)]_n$
$Cu_{0.25}\text{-}Co_{0.75}$-MOF	$[(Cu_{0.25}\text{-}Co_{0.75})_3(BTC)_2]_n$
HKUST-1	$[Cu_3(BTC)_2]_n$
LaCoODa	$\{[La_2Co_3(ODA)_6(H_2O)_6]\cdot 12H_2O\}_n$
LaCuODA	$\{[La_2Cu_3(\mu\text{-}H_2O)(ODA)_6(H_2O)_3]\cdot 3H_2O\}_n$

MIL-100(Fe)	$[Fe_3O(OH)(H_2O)_2(BDC)_3]_n$
MIL-101	$[Cr_3O(H_2O)_2F(BDC)]_n$
$Mn_{0.1}Cu_{0.9}$-MOF	$[(Mn_{0.1}\text{-}Cu_{0.9})_3(BTC)_2]_n$
NENU-MV-1	$\{[Ni(4,4'\text{-bpy})_2]_2[V_7^{IV}V_9^VO_{38}Cl]\cdot(4,4'\text{-bpy})\cdot 6H_2O\}_n$
NH_2-MIL-101(Al)	$[Al_3O(OH)(H_2O)_2(BDCNH_2)_3]_n$
NH_2-MIL-101(Cr)	$Cr_3O(H_2O)_2F(NH_2\text{-BDC})$
NTUZ30	$\{[Co_3(\mu_3\text{-OH})(HBTC)(BTC)_2Co(HBTC)]\cdot(HTEA)_3\cdot H_2O\}_n$
PCN-222	$[Zr_6(\mu_3\text{-OH})_8(OH)_8\text{-}(TCPP)_2]_n$
PCN-224	$[Zr_6(\mu_3\text{-OH})_{12}(OH)_{16}\text{-}(TCPP)_{1.5}]_n$
TMU-16-NH_2	$\{[Zn_2(NH_2\text{-BDC})_2(4\text{-bpdh})]\cdot 3DMF\}_n$
UiO-66	$[Zr_6O_4(OH)_4(BDC)_6]_n$
UiO-66-NH_2	$[Zr_6O_4(OH)_4(NH_2\text{-BDC})_6]_n$
UiO-67	$[Zr_6(\mu_3\text{-O})_4(\mu_3\text{-OH})_4(BPDC)_6]_n$
UiO-68	$[Zr_6(\mu_3\text{-O})_4(\mu_3\text{-OH})_4(TPDC)_6]_n$
UiO-bpy	$[Zr_6O_4(OH)_4(bpy)_6]_n$
VPI-100(Cu)	$[Zr_6(\mu_3\text{-OH})_8(OH)_8(Cu\text{-L1})_4]_n$
VPI-100(Ni)	$[Zr_6(\mu_3\text{-OH})_8(OH)_8(Ni\text{-L1})_4]_n$
ZIF-67	$[Co(MeIm)_2]_n$
ZIF-8	$[Zn(MeIm)_2]_n$
ZIF-95	$[Zn(cbIm)_2]_n$
Zn-MOF-184	$[Zn_2(EDOB)]_n$
Zr-NU-1000	$([Zr_6(\mu_3\text{-O})_4(\mu_3\text{-OH})_4(OH)_4(H_2O)_4(TBAPy)_2]_n$
ZSF-1	$[Zr_6O_4(OH)_4(\text{metallosalen})_6]_n$
Hf-NU-1000	$[(Hf_6(\mu_3\text{-O})_4(\mu_3\text{-OH})_4(OH)_4(OH_2)_4(TBAPy)_2]_n$

Appendix A. Chart of the MOF Linkers Present in This Review and Their Relative Abbreviations

Structural Formula	Name	Abbreviation
	1-(4-Cyanobenzyl)-5-methyl-1H-imidazole	cbIm
	1,3,5-tri(1H-Benzo[d]imidazol-1-yl)benzene	TBIB
	1,3,5-tris(4'-Carboxy-phenyl-)benzene	H_3TCPB

Structural Formula	Name	Abbreviation
	1,3,6,8-(p-Benzoate)pyrene	H_4TBAPy
	1,3-bis(Imidazolyl)propane	H_2bip
	1,4-bis(Imidazol)butane	bib
	2,2′,6,6′-Tetramethoxy-4,4′-biphenyldicarboxylic acid	H_2L4
	2,2-Bipyridine-4,4′-dicarboxylic acid	H_2Bpy
	2,5-bis(4-Pyridyl)-3,4-diaza-2,4-hexadiene	4-bpdh
	2-Aminoterephthalic acid	H_2BDCNH$_2$
	2-Methylimidazole	HmeIm
	2-Picolinic acid	PCA
	3,5-Diamino-4,4′-bipyrazole	H_2-NH$_2$-Bpz

Structural Formula	Name	Abbreviation
	4-((1-Carboxy-2-(1H-imidazol-4-yl)ethylamino)methyl)benzoic acid	H$_2$L2
	4-(3,5-Dimethyl-4-H-1,2,4-triazol-4-yl) benzoate	Me$_2$trz-pba
	4,4′-Bipyridine	4,4′-bipy
	4,4′-(Ethyne-1,2-diyl)bis(2-hydroxybenzoic acid)	H$_4$EDOB
	4,4′,4″-Tricarboxyltriphenylamine	H$_3$TCA
	5-(3-Methyl-5-(pyridine-4-yl)-4H-1,2,4-triazol-4-yl) isophthalate	Me-4py-trz-ia
	5,10,15,20-Tetrakis(4-carboxyphenyl)porphyrin	H$_6$TCPP
	5-Dihydroxyterephthalic acid	H$_4$DHTA
	6,13-Dicarboxy-1,4,8,11-tetraazacyclotetradecane	L1

Structural Formula	Name	Abbreviation
	Benzene-1,3-5 tricarboxylic acid	H_3BTC
	Biphenyl-4,4′-dicarboxylic acid	H_2BPDC
	Imidazole	HIm
	Isonicotinate	Ina
	N,N′-bis(3-*tert*-Butyl-5-(carboxy)salicylide	H_4L5
	Oxydiacetic acid	H_2ODA
	Pyridine-2,5-dicarboxylic acid	H_2pdc
	Pyridine-2-aldehyde	PI
	Salicylaldehyde	SA
	Thiophene-2-carbaldehyde	TC

Structural Formula	Name	Abbreviation
	Triethylamine	TEA
	Tris(4′-carboxybiphenyl)amine	H_3L3

References

1. Batten, S.R.; Champness, N.R.; Chen, X.-M.; Garcia-Martinez, J.; Kitagawa, S.; Öhrström, L.; O'Keeffe, M.; Suh, M.P.; Reedijk, J. Terminology of metal–organic frameworks and coordination polymers (IUPAC Recommendations 2013). *Pure Appl. Chem.* **2013**, *85*, 1715–1724. [CrossRef]
2. Furukawa, H.; Cordova, K.E.; O'Keeffe, M.; Yaghi, O.M. The Chemistry and Applications of Metal-Organic Frameworks. *Science* **2013**, *341*, 1230444. [CrossRef]
3. Pettinari, C.; Marchetti, F.; Mosca, N.; Drozdov, A. Application of metal-organic frameworks. *Polym. Int.* **2017**, *66*, 93. [CrossRef]
4. Li, H.; Eddaoudi, M.; O'Keeffe, M.; Yaghi, O. Design and synthesis of an exceptionally stable and highly porous metal-organic framework. *Nat. Cell Biol.* **1999**, *402*, 276–279. [CrossRef]
5. Schaus, S.E.; Brandes, B.D.; Larrow, J.F.; Tokunaga, M.; Hansen, K.B.; Gould, A.E.; Furrow, M.E.; Jacobsen, E.N. Highly Selective Hydrolytic Kinetic Resolution of Terminal Epoxides Catalyzed by Chiral (salen)CoIII Complexes. Practical Synthesis of Enantioenriched Terminal Epoxides and 1,2-Diols. *J. Am. Chem. Soc.* **2002**, *124*, 1307. [CrossRef]
6. Stock, N.; Biswas, S. Synthesis of Metal-Organic Frameworks (MOFs): Routes to Various MOF Topologies, Morphologies, and Composites. *Chem. Rev.* **2011**, *112*, 933. [CrossRef]
7. Campagnol, N.; Souza, E.R.; De Vos, D.E.; Binnemans, K.; Fransaer, J. Luminescent terbium-containing metal–organic framework films: New approaches for the electrochemical synthesis and application as detectors for explosives. *Chem. Commun.* **2014**, *50*, 12545–12547. [CrossRef]
8. Zhang, Y.; Bo, X.; Nsabimana, A.; Han, C.; Li, M.; Guo, L. Electrocatalytically active cobalt-based metal–organic framework with incorporated macroporous carbon composite for electrochemical applications. *J. Mater. Chem. A* **2014**, *3*, 732. [CrossRef]
9. Masoomi, M.Y.; Morsali, A.; Junk, P.C. Rapid mechanochemical synthesis of two new Cd(ii)-based metal–organic frameworks with high removal efficiency of Congo red. *CrystEngComm* **2014**, *17*, 686. [CrossRef]
10. Wu, J.-Y.; Chao, T.-C.; Zhong, M.-S. Influence of Counteranions on the Structural Modulation of Silver–Di(3-pyridylmethyl)amine Coordination Polymers. *Cryst. Growth Des.* **2013**, *13*, 2953. [CrossRef]
11. Khan, N.A.; Jhung, S.H. Synthesis of metal-organic frameworks (MOFs) with microwave or ultrasound: Rapid reaction, phase-selectivity, and size reduction. *Coord. Chem. Rev.* **2015**, *285*, 11. [CrossRef]
12. Maurin, G.; Serre, C.; Cooper, A.; Férey, G. The new age of MOFs and of their porous-related solids. *Chem. Soc. Rev.* **2017**, *46*, 3104–3107. [CrossRef]
13. Murray, J.L.; Mircea, D.; Long, R.J. Hydrogen storage in metal–organic frameworks. *Chem. Soc. Rev.* **2009**, *38*, 1294–1314. [CrossRef]
14. Sun, Y.; Zheng, L.; Yang, Y.; Qian, X.; Fu, T.; Li, X.; Yang, Z.; Yan, H.; Cui, C.; Tan, W. Metal–Organic Framework Nanocarriers for Drug Delivery in Biomedical Applications. *Nano-Micro Lett.* **2020**, *12*, 103. [CrossRef]
15. Wang, H.-S. Metal–organic frameworks for biosensing and bioimaging applications. *Coord. Chem. Rev.* **2017**, *349*, 139–155. [CrossRef]
16. Gao, X.; Dong, Y.; Li, S.; Zhou, J.; Wang, L.; Wang, B. MOFs and COFs for Batteries and Supercapacitors. *Electrochem. Energy Rev.* **2019**, *3*, 81–126. [CrossRef]
17. Stavila, V.; Talin, A.A.; Allendorf, M.D. MOF-based electronic and opto-electronic devices. *Chem. Soc. Rev.* **2014**, *43*, 5994. [CrossRef]
18. Xu, W.; Yaghi, O.M. Metal–Organic Frameworks for Water Harvesting from Air, Anywhere, Anytime. *ACS Central Sci.* **2020**, *6*, 1348–1354. [CrossRef]
19. Zhang, L.; Li, H.; He, H.; Yang, Y.; Cui, Y.; Qian, G. Structural Variation and Switchable Nonlinear Optical Behavior of Metal–Organic Frameworks. *Small* **2021**, *17*, 2006649. [CrossRef]
20. Dhakshinamoorthy, A.; Opanasenko, M.; Čejka, J.; Garcia, H. Metal organic frameworks as heterogeneous catalysts for the production of fine chemicals. *Catal. Sci. Technol.* **2013**, *3*, 2509–2540. [CrossRef]

21. Vanesa, C.-C.; Rosa, M.M.-A. Advances in Metal-Organic Frameworks for Heterogeneous Catalysis. *Recent Pat. Chem. Eng.* **2011**, *4*, 1–16.
22. Farrusseng, D.; Aguado, S.; Pinel, C. Metal-Organic Frameworks: Opportunities for Catalysis. *Angew. Chem. Int. Ed.* **2009**, *48*, 7502–7513. [CrossRef] [PubMed]
23. Chen, Y.; Ma, S. Biomimetic catalysis of metal–organic frameworks. *Dalt. Trans.* **2016**, *45*, 9744–9753. [CrossRef]
24. Corma, A.; García, H. Lewis Acids as Catalysts in Oxidation Reactions: From Homogeneous to Heterogeneous Systems. *Chem. Rev.* **2002**, *102*, 3837–3892. [CrossRef]
25. Kravchenko, D.E.; Tyablikov, I.A.; Kots, P.A.; Kolozhvari, B.A.; Fedosov, D.A.; Ivanova, I.I. Olefin Epoxidation over Metal-Organic Frameworks Modified with Transition Metals. *Pet. Chem.* **2019**, *58*, 1255–1262. [CrossRef]
26. Arai, H.; Uehara, K.; Kinoshita, S.I.; Kunugi, T. Olefin Oxidation-Mercuric Salt-Active Charcoal Catalysis. *Ind. Eng. Chem. Prod. Res. Dev.* **1972**, *11*, 308–312. [CrossRef]
27. Liniger, M.; Liu, Y.; Stoltz, B.M. Sequential Ruthenium Catalysis for Olefin Isomerization and Oxidation: Application to the Synthesis of Unusual Amino Acids. *J. Am. Chem. Soc.* **2017**, *139*, 13944–13949. [CrossRef]
28. Gao, J.; Bai, L.; Zhang, Q.; Li, Y.; Rakesh, G.; Lee, J.-M.; Yang, Y.; Zhang, Q. $Co_6(\mu_3\text{-OH})_6$ cluster based coordination polymer as an effective heterogeneous catalyst for aerobic epoxidation of alkenes. *Dalton Trans.* **2014**, *43*, 2559–2565. [CrossRef] [PubMed]
29. Zhang, X.; Zhang, Y.-Z.; Jin, Y.-Q.; Geng, L.; Zhang, D.-S.; Hu, H.; Li, T.; Wang, B.; Li, J.-R. Pillar-Layered Metal–Organic Frameworks Based on a Hexaprismane $[Co_6(\mu_3\text{-OH})_6]$ Cluster: Structural Modulation and Catalytic Performance in Aerobic Oxidation Reaction. *Inorg. Chem.* **2020**, *59*, 11728–11735. [CrossRef]
30. Yu, F.; Xiong, X.; Huang, K.; Zhou, Y.; Li, B. 2D Co-based coordination polymer with a histidine derivative as an efficient heterogeneous catalyst for the oxidation of cyclohexene. *CrystEngComm* **2017**, *19*, 2126–2132. [CrossRef]
31. Zhang, H.; He, J.; Lu, X.; Yang, L.; Wang, C.; Yue, F.; Zhou, D.; Xia, Q. Fast-synthesis and catalytic property of heterogeneous Co-MOF catalysts for the epoxidation of α-pinene with air. *New J. Chem.* **2020**, *44*, 17413–17421. [CrossRef]
32. Lu, H.-S.; Bai, L.; Xiong, W.-W.; Li, P.; Ding, J.; Zhang, G.; Wu, T.; Zhao, Y.; Lee, J.-M.; Yang, Y.; et al. Surfactant Media To Grow New Crystalline Cobalt 1,3,5-Benzenetricarboxylate Metal–Organic Frameworks. *Inorg. Chem.* **2014**, *53*, 8529–8537. [CrossRef]
33. Sha, S.; Yang, H.; Li, J.; Zhuang, C.; Gao, S.; Liu, S. Co(II) coordinated metal-organic framework: An efficient catalyst for heterogeneous aerobic olefins epoxidation. *Catal. Commun.* **2014**, *43*, 146–150. [CrossRef]
34. Fu, Y.; Sun, D.; Qin, M.; Huang, R.; Li, Z. Cu(ii)- and Co(ii)-containing metal–organic frameworks (MOFs) as catalysts for cyclohexene oxidation with oxygen under solvent-free conditions. *RSC Adv.* **2012**, *2*, 3309–3314. [CrossRef]
35. Lashanizadegan, M.; Ashari, H.A.; Sarkheil, M.; Anafcheh, M.; Jahangiry, S. New Cu(II), Co(II) and Ni(II) azo-Schiff base complexes: Synthesis, characterization, catalytic oxidation of alkenes and DFT study. *Polyhedron* **2021**, *200*, 115148. [CrossRef]
36. Junghans, U.; Suttkus, C.; Lincke, J.; Lässig, D.; Krautscheid, H.; Gläser, R. Selective oxidation of cyclooctene over copper-containing metal-organic frameworks. *Microporous Mesoporous Mater.* **2015**, *216*, 151–160. [CrossRef]
37. Wang, B.; Jin, J.; Ding, B.; Han, X.; Han, A.; Liu, J. General Approach to Metal-Organic Framework Nanosheets With Controllable Thickness by Using Metal Hydroxides as Precursors. *Front. Mater.* **2020**, *7*, 37. [CrossRef]
38. Shi, D.; Ren, Y.; Jiang, H.; Cai, B.; Lu, J. Synthesis, Structures, and Properties of Two Three-Dimensional Metal–Organic Frameworks, Based on Concurrent Ligand Extension. *Inorg. Chem.* **2012**, *51*, 6498–6506. [CrossRef] [PubMed]
39. Li, J.; Ren, Y.; Qi, C.; Jiang, H. Fully meta-Substituted 4,4′-Biphenyldicarboxylate-Based Metal–Organic Frameworks: Synthesis, Structures, and Catalytic Activities. *Eur. J. Inorg. Chem.* **2017**, *11*, 1478–1487. [CrossRef]
40. Saha, D.; Gayen, S.; Koner, S. Cu(II)/Cu(II)-Mg(II) containing pyridine-2,5-dicarboxylate frameworks: Synthesis, structural diversity, inter-conversion and heterogeneous catalytic epoxidation. *Polyhedron* **2018**, *146*, 93–98. [CrossRef]
41. Dhakshinamoorthy, A.; Portillo, A.S.; Asiri, A.M.; Garcia, H. Engineering UiO-66 Metal Organic Framework for Heterogeneous Catalysis. *ChemCatChem* **2019**, *11*, 899–923. [CrossRef]
42. Vandichel, M.; Hajek, J.; Vermoortele, F.; Waroquier, M.; De Vos, D.E.; Van Speybroeck, V. Active site engineering in UiO-66 type metal–organic frameworks by intentional creation of defects: A theoretical rationalization. *CrystEngComm* **2014**, *17*, 395–406. [CrossRef]
43. Arrozi, U.S.; Wijaya, H.W.; Patah, A.; Permana, Y. Efficient acetalization of benzaldehydes using UiO-66 and UiO-67: Substrates accessibility or Lewis acidity of zirconium. *Appl. Catal. A Gen.* **2015**, *506*, 77–84. [CrossRef]
44. Zheng, H.-Q.; Zeng, Y.-N.; Chen, J.; Lin, R.-G.; Zhuang, W.-E.; Cao, R.; Lin, Z.-J. Zr-Based Metal–Organic Frameworks with Intrinsic Peroxidase-Like Activity for Ultradeep Oxidative Desulfurization: Mechanism of H_2O_2 Decomposition. *Inorg. Chem.* **2019**, *58*, 6983–6992. [CrossRef] [PubMed]
45. Liu, Y.; Klet, R.C.; Hupp, J.T.; Farha, O. Probing the correlations between the defects in metal–organic frameworks and their catalytic activity by an epoxide ring-opening reaction. *Chem. Commun.* **2016**, *52*, 7806–7809. [CrossRef] [PubMed]
46. Zalomaeva, O.V.; Evtushok, V.Y.; Ivanchikova, I.D.; Glazneva, T.S.; Chesalov, Y.A.; Larionov, K.P.; Skobelev, I.Y.; Kholdeeva, O.A. Nucleophilic versus Electrophilic Activation of Hydrogen Peroxide over Zr-Based Metal–Organic Frameworks. *Inorg. Chem.* **2020**, *59*, 10634–10649. [CrossRef] [PubMed]
47. Payne, G.B.; Deming, P.H.; Williams, P.H. Reactions of Hydrogen Peroxide. VII. Alkali-Catalyzed Epoxidation and Oxidation Using a Nitrile as Co-reactant. *J. Org. Chem.* **1961**, *26*, 659–663. [CrossRef]
48. Huang, K.; Yu, S.; Li, X.; Cai, Z. One-pot synthesis of bimetal MOFs as highly efficient catalysts for selective oxidation of styrene. *J. Chem. Sci.* **2020**, *132*, 139. [CrossRef]

49. Wang, F.; Meng, X.-G.; Wu, Y.-Y.; Huang, H.; Lv, J.; Yu, W.-W. A Highly Efficient Heterogeneous Catalyst of Bimetal-Organic Frameworks for the Epoxidation of Olefin with H_2O_2. *Molecules* **2020**, *25*, 2389. [CrossRef]
50. Santibáñez, L.; Escalona, N.; Torres, J.; Kremer, C.; Cancino, P.; Spodine, E. Cu^{II}- and Co^{II}-based MOFs: {[La_2Cu_3(μ-H_2O)(ODA)6(H_2O)$_3$]·$3H_2O$}n and {[La_2Co_3(ODA)6(H_2O)$_6$]·$12H_2O$}n. The Relevance of Physicochemical Properties on the Catalytic Aerobic Oxidation of Cyclohexene. *Catalysts* **2020**, *10*, 589. [CrossRef]
51. Tuci, G.; Giambastiani, G.; Kwon, S.; Stair, P.C.; Snurr, R.Q.; Rossin, A. Chiral Co(II) Metal–Organic Framework in the Heterogeneous Catalytic Oxidation of Alkenes under Aerobic and Anaerobic Conditions. *ACS Catal.* **2014**, *4*, 1032–1039. [CrossRef]
52. Otake, K.-I.; Ahn, S.; Knapp, J.; Hupp, J.T.; Notestein, J.M.; Farha, O.K. Vapor-Phase Cyclohexene Epoxidation by Single-Ion Fe(III) Sites in Metal–Organic Frameworks. *Inorg. Chem.* **2021**, *60*, 2457–2463. [CrossRef] [PubMed]
53. Ahn, S.; Thornburg, N.; Li, Z.; Wang, T.C.; Gallington, L.; Chapman, K.W.; Notestein, J.M.; Hupp, J.T.; Farha, O.K. Stable Metal–Organic Framework-Supported Niobium Catalysts. *Inorg. Chem.* **2016**, *55*, 11954–11961. [CrossRef]
54. Thornburg, N.E.; Nauert, S.L.; Thompson, A.B.; Notestein, J.M. Synthesis−Structure−Function Relationships of Silica-Supported Niobium(V) Catalysts for Alkene Epoxidation with H_2O_2. *ACS Catal.* **2016**, *6*, 6124–6134. [CrossRef]
55. Wang, S.; Liu, Y.; Zhang, Z.; Li, X.; Tian, H.; Yan, T.; Zhang, X.; Liu, S.; Sun, X.; Xu, L.; et al. One-Step Template-Free Fabrication of Ultrathin Mixed-Valence Polyoxovanadate-Incorporated Metal–Organic Framework Nanosheets for Highly Efficient Selective Oxidation Catalysis in Air. *ACS Appl. Mater. Interfaces* **2019**, *11*, 12786–12796. [CrossRef] [PubMed]
56. Hui, J.; Chu, H.; Zhang, W.; Shen, Y.; Chen, W.; Hu, Y.; Liu, W.; Gao, C.; Guo, S.; Xiao, G.; et al. Multicomponent metal–organic framework derivatives for optimizing the selective catalytic performance of styrene epoxidation reaction. *Nanoscale* **2018**, *10*, 8772–8778. [CrossRef] [PubMed]
57. Saedi, Z.; Safarifard, V.; Morsali, A. Dative and covalent-dative postsynthetic modification of a two-fold interpenetration pillared-layer MOF for heterogeneous catalysis: A comparison of catalytic activities and reusability. *Microporous Mesoporous Mater.* **2016**, *229*, 51–58. [CrossRef]
58. Berijani, K.; Morsali, A.; Hupp, J.T. An effective strategy for creating asymmetric MOFs for chirality induction: A chiral Zr-based MOF for enantioselective epoxidation. *Catal. Sci. Technol.* **2019**, *9*, 3388–3397. [CrossRef]
59. Noh, H.; Cui, Y.; Peters, A.W.; Pahls, D.; Ortuño, M.A.; Vermeulen, N.A.; Cramer, C.J.; Gagliardi, L.; Hupp, J.T.; Farha, O.K. An Exceptionally Stable Metal–Organic Framework Supported Molybdenum(VI) Oxide Catalyst for Cyclohexene Epoxidation. *J. Am. Chem. Soc.* **2016**, *138*, 14720–14726. [CrossRef]
60. Kardanpour, R.; Tangestaninejad, S.; Mirkhani, V.; Moghadam, M.; Mohammadpoor-Baltork, I.; Zadehahmadi, F. Efficient alkene epoxidation catalyzed by molybdenyl acetylacetonate supported on aminated UiO-66 metal−organic framework. *J. Solid State Chem.* **2015**, *226*, 262–272. [CrossRef]
61. Tang, J.; Dong, W.; Wang, G.; Yao, Y.; Cai, L.; Liu, Y.; Zhao, X.; Xu, J.; Tan, L. Efficient molybdenum(vi) modified Zr-MOF catalysts for epoxidation of olefins. *RSC Adv.* **2014**, *4*, 42977–42982. [CrossRef]
62. Liu, T.-F.; Vermeulen, N.A.; Howarth, A.J.; Li, P.; Sarjeant, A.A.; Hupp, J.T.; Farha, O.K. Adding to the Arsenal of Zirconium-Based Metal–Organic Frameworks: The Topology as a Platform for Solvent-Assisted Metal Incorporation. *Eur. J. Inorg. Chem.* **2016**, *27*, 4349. [CrossRef]
63. Pereira, C.; Biernacki, K.; Rebelo, S.L.; Magalhães, A.L.; Carvalho, A.P.; Pires, J.; Freire, C. Designing heterogeneous oxovanadium and copper acetylacetonate catalysts: Effect of covalent immobilisation in epoxidation and aziridination reactions. *J. Mol. Catal. A Chem.* **2009**, *312*, 53–64. [CrossRef]
64. Li, Z.; Wu, S.; Ding, H.; Lu, H.; Liu, J.; Huo, Q.; Guan, J.; Kan, Q. Oxovanadium(iv) and iron(iii) salen complexes immobilized on amino-functionalized graphene oxide for the aerobic epoxidation of styrene. *New J. Chem.* **2013**, *37*, 4220–4229. [CrossRef]
65. Ben Zid, T.; Khedher, I.; Ghorbel, A. Chiral vanadyl salen catalyst immobilized on mesoporous silica as support for asymmetric oxidation of sulfides to sulfoxides. *React. Kinet. Mech. Catal.* **2010**, *100*, 131–143. [CrossRef]
66. Zamanifar, E.; Farzaneh, F. Immobilized vanadium amino acid Schiff base complex on Al-MCM-41 as catalyst for the epoxidation of allyl alcohols. *React. Kinet. Mech. Catal.* **2011**, *104*, 197–209. [CrossRef]
67. Pourkhosravani, M.; Dehghanpour, S.; Farzaneh, F.; Sohrabi, S. Designing new catalytic nanoreactors for the regioselective epoxidation of geraniol by the post-synthetic immobilization of oxovanadium(IV) complexes on a ZrIV-based metal–organic framework. *React. Kinet. Mech. Catal.* **2017**, *122*, 961–981. [CrossRef]
68. Tan, C.; Han, X.; Li, Z.; Liu, Y.; Cui, Y. Controlled Exchange of Achiral Linkers with Chiral Linkers in Zr-Based UiO-68 Metal–Organic Framework. *J. Am. Chem. Soc.* **2018**, *140*, 16229–16236. [CrossRef]
69. Huang, K.; Guo, L.L.; Wu, D.F. Synthesis of Metal Salen@MOFs and Their Catalytic Performance for Styrene Oxidation. *Ind. Eng. Chem. Res.* **2019**, *58*, 4744–4754. [CrossRef]
70. Xia, Q.; Li, Z.; Tan, C.; Liu, Y.; Gong, W.; Cui, Y. Multivariate Metal–Organic Frameworks as Multifunctional Heterogeneous Asymmetric Catalysts for Sequential Reactions. *J. Am. Chem. Soc.* **2017**, *139*, 8259–8266. [CrossRef]
71. Berijani, K.; Morsali, A. Construction of an Asymmetric Porphyrinic Zirconium Metal–Organic Framework through Ionic Postchiral Modification. *Inorg. Chem.* **2021**, *60*, 206–218. [CrossRef] [PubMed]
72. Martínez, H.; Cáceres, M.F.; Martínez, F.; Páez-Mozo, E.A.; Valange, S.; Castellanos, N.J.; Molina, D.; Barrault, J.; Arzoumanian, H. Photo-epoxidation of cyclohexene, cyclooctene and 1-octene with molecular oxygen catalyzed by dichloro dioxo-(4,4′-dicarboxylato-2,2′-bipyridine) molybdenum(VI) grafted on mesoporous TiO_2. *J. Mol. Catal. A Chem.* **2016**, *423*, 248–255. [CrossRef]

73. Portillo, A.S.; Navalón, S.; Cirujano, F.G.; I Xamena, F.X.L.; Alvaro, M.; Garcia, H. MIL-101 as Reusable Solid Catalyst for Autoxidation of Benzylic Hydrocarbons in the Absence of Additional Oxidizing Reagents. *ACS Catal.* **2015**, *5*, 3216–3224. [CrossRef]
74. Kou, J.; Sun, L.-B. Fabrication of Metal–Organic Frameworks inside Silica Nanopores with Significantly Enhanced Hydrostability and Catalytic Activity. *ACS Appl. Mater. Interfaces* **2018**, *10*, 12051–12059. [CrossRef] [PubMed]
75. Zhou, Z.; Li, X.; Wang, Y.; Luan, Y.; Li, X.; Du, X. Growth of Cu-BTC MOFs on dendrimer-like porous silica nanospheres for the catalytic aerobic epoxidation of olefins. *New J. Chem.* **2020**, *44*, 14350–14357. [CrossRef]
76. Zhao, J.; Wang, W.; Tang, H.; Ramella, D.; Luan, Y. Modification of Cu^{2+} into Zr-based metal–organic framework (MOF) with carboxylic units as an efficient heterogeneous catalyst for aerobic epoxidation of olefins. *Mol. Catal.* **2018**, *456*, 57–64. [CrossRef]
77. Wu, Y.; Wang, H.; Guo, S.; Zeng, Y.; Ding, M. MOFs-induced high-amphiphilicity in hierarchical 3D reduced graphene oxide-based hydrogel. *Appl. Surf. Sci.* **2021**, *540*, 148303. [CrossRef]
78. Canioni, R.; Roch-Marchal, C.; Sécheresse, F.; Horcajada, P.; Serre, C.; Hardi-Dan, M.; Férey, G.; Grenèche, J.-M.; Lefebvre, F.; Chang, J.-S.; et al. Stable polyoxometalate insertion within the mesoporous metal organic framework MIL-100(Fe). *J. Mater. Chem.* **2011**, *21*, 1226–1233. [CrossRef]
79. Wang, S.-S.; Yang, G.-Y. Recent Advances in Polyoxometalate-Catalyzed Reactions. *Chem. Rev.* **2015**, *115*, 4893–4962. [CrossRef]
80. Song, X.; Hu, D.; Yang, X.; Zhang, H.; Zhang, W.; Li, J.; Jia, M.; Yu, J. Polyoxomolybdic Cobalt Encapsulated within Zr-Based Metal–Organic Frameworks as Efficient Heterogeneous Catalysts for Olefins Epoxidation. *ACS Sustain. Chem. Eng.* **2019**, *7*, 3624–3631. [CrossRef]
81. Villabrille, P.; Romanelli, G.; Gassa, L.; Vazquez, P.; Caceres, C. Synthesis and characterization of Fe- and Cu-doped molybdovanadophosphoric acids and their application in catalytic oxidation. *Appl. Catal. A Gen.* **2007**, *324*, 69–76. [CrossRef]
82. Jin, M.; Niu, Q.; Liu, G.; Lv, Z.; Si, C.; Guo, H. Encapsulation of ionic liquids into POMs-based metal–organic frameworks: Screening of POMs-ILs@MOF catalysts for efficient cycloolefins epoxidation. *J. Mater. Sci.* **2020**, *55*, 8199–8210. [CrossRef]
83. Ke, F.; Guo, F.; Yu, J.; Yang, Y.; He, Y.; Chang, L.; Wan, X. Highly Site-Selective Epoxidation of Polyene Catalyzed by Metal–Organic Frameworks Assisted by Polyoxometalate. *J. Inorg. Organomet. Polym. Mater.* **2017**, *27*, 843–849. [CrossRef]
84. Li, N.; Mu, B.; Lv, L.; Huang, R. Assembly of new polyoxometalate-templated metal–organic frameworks based on flexible ligands. *J. Solid State Chem.* **2015**, *226*, 88–93. [CrossRef]
85. Zhang, Y.; Li, Y.-X.; Liu, L.; Han, Z.-B. Palladium nanoparticles supported on UiO-66-NH_2 as heterogeneous catalyst for epoxidation of styrene. *Inorg. Chem. Commun.* **2019**, *100*, 51–55. [CrossRef]
86. Limvorapitux, R.; Chou, L.-Y.; Young, A.P.; Tsung, C.-K.; Nguyen, S.T. Coupling Molecular and Nanoparticle Catalysts on Single Metal–Organic Framework Microcrystals for the Tandem Reaction of H_2O_2 Generation and Selective Alkene Oxidation. *ACS Catal.* **2017**, *7*, 6691–6698. [CrossRef]
87. Jin, T.; Yang, Q.; Meng, C.; Xu, J.; Liu, H.; Hu, J.; Ling, H. Promoting desulfurization capacity and separation efficiency simultaneously by the novel magnetic Fe_3O_4@PAA@MOF-199. *RSC Adv.* **2014**, *4*, 41902–41909. [CrossRef]
88. Li, J.; Gao, H.; Tan, L.; Luan, Y.; Yang, M. Superparamagnetic Core-Shell Metal-Organic Framework Fe_3O_4/Cu_3(btc)$_2$ Microspheres and Their Catalytic Activity in the Aerobic Oxidation of Alcohols and Olefins. *Eur. J. Inorg. Chem.* **2016**, *2016*, 4906–4912. [CrossRef]
89. Hou, J.; Luan, Y.; Yu, J.; Qi, Y.; Wang, G.; Lu, Y. Fabrication of hierarchical composite microspheres of copper-doped Fe_3O_4@P4VP@ZIF-8 and their application in aerobic oxidation. *New J. Chem.* **2016**, *40*, 10127–10135. [CrossRef]
90. Qi, Y.; Luan, Y.; Yu, J.; Peng, X.; Wang, G. Nanoscaled Copper Metal-Organic Framework (MOF) Based on Carboxylate Ligands as an Efficient Heterogeneous Catalyst for Aerobic Epoxidation of Olefins and Oxidation of Benzylic and Allylic Alcohols. *Chem. A Eur. J.* **2015**, *21*, 1589–1597. [CrossRef]
91. Yang, X.; Yuan, S.; Zou, L.; Drake, H.; Zhang, Y.; Qin, J.; Alsalme, A.; Zhou, H.C. One-Step Synthesis of Hybrid Core–Shell Metal–Organic Frameworks. *Angew. Chem.* **2018**, *130*, 3991–3996. [CrossRef]
92. Aresta, M.; Dibenedetto, A.; Angelini, A. Catalysis for the Valorization of Exhaust Carbon: From CO_2 to Chemicals, Materials, and Fuels. Technological Use of CO_2. *Chem. Rev.* **2013**, *114*, 1709–1742. [CrossRef]
93. North, M.; Pasquale, R. Mechanism of Cyclic Carbonate Synthesis from Epoxides and CO_2. *Angew. Chem. Int. Ed.* **2009**, *48*, 2946–2948. [CrossRef]
94. Shi, L.; Xu, S.; Zhang, Q.; Liu, T.; Wei, B.; Zhao, Y.; Meng, L.; Li, J. Ionic Liquid/Quaternary Ammonium Salt Integrated Heterogeneous Catalytic System for the Efficient Coupling of Carbon Dioxide with Epoxides. *Ind. Eng. Chem. Res.* **2018**, *57*, 15319–15328. [CrossRef]
95. Shi, X.-L.; Chen, Y.; Duan, P.; Zhang, W.; Hu, Q. Conversion of CO_2 into Organic Carbonates over a Fiber-Supported Ionic Liquid Catalyst in Impellers of the Agitation System. *ACS Sustain. Chem. Eng.* **2018**, *6*, 7119–7127. [CrossRef]
96. Ziaee, M.A.; Tang, Y.; Zhong, H.; Tian, D.; Wang, R. Urea-Functionalized Imidazolium-Based Ionic Polymer for Chemical Conversion of CO2 into Organic Carbonates. *ACS Sustain. Chem. Eng.* **2018**, *7*, 2380–2387. [CrossRef]
97. Yang, Z.; Yu, B.; Zhang, H.; Zhao, Y.; Chen, Y.; Ma, Z.; Ji, G.; Gao, X.; Han, B.; Liu, Z. Metalated Mesoporous Poly(triphenylphosphine) with Azo Functionality: Efficient Catalysts for CO_2 Conversion. *ACS Catal.* **2016**, *6*, 1268–1273. [CrossRef]
98. Sun, Q.; Aguila, B.; Perman, J.A.; Nguyen, N.T.-K.; Ma, S. Flexibility Matters: Cooperative Active Sites in Covalent Organic Framework and Threaded Ionic Polymer. *J. Am. Chem. Soc.* **2016**, *138*, 15790–15796. [CrossRef]

99. Beyzavi, M.H.; Stephenson, C.J.; Liu, Y.; Karagiaridi, O.; Hupp, J.T.; Farha, O.K. Metal-organic framework-based catalysts: Chemical fixation of CO_2 with epoxides leading to cyclic organic carbonates. *Front. Energy Res.* **2015**, *3*, 63. [CrossRef]
100. Bon, V.; Senkovskyy, V.; Senkovska, I.; Kaskel, S. Zr(IV) and Hf(IV) based metal–organic frameworks with reo-topology. *Chem. Commun.* **2012**, *48*, 8407–8409. [CrossRef]
101. Beyzavi, M.H.; Klet, R.C.; Tussupbayev, S.; Borycz, J.; Vermeulen, N.A.; Cramer, C.J.; Stoddart, J.F.; Hupp, J.T.; Farha, O.K. A Hafnium-Based Metal–Organic Framework as an Efficient and Multifunctional Catalyst for Facile CO_2 Fixation and Regioselective and Enantioretentive Epoxide Activation. *J. Am. Chem. Soc.* **2014**, *136*, 15861–15864. [CrossRef]
102. Tran, Y.B.N.; Nguyen, P.T.K.; Luong, Q.T.; Nguyen, K.D. Series of M-MOF-184 (M = Mg, Co, Ni, Zn, Cu, Fe) Metal-Organic Frameworks for Catalysis Cycloaddition of CO_2. *Inorg. Chem.* **2020**, *59*, 16747–16759. [CrossRef]
103. Agarwal, R.A.; Gupta, A.K.; De, D. Flexible Zn-MOF Exhibiting Selective CO_2 Adsorption and Efficient Lewis Acidic Catalytic Activity. *Cryst. Growth Des.* **2019**, *19*, 2010–2018. [CrossRef]
104. Hayashi, H.; Côté, A.P.; Furukawa, H.; O'Keeffe, M.; Yaghi, O.M. Zeolite A imidazolate frameworks. *Nat. Mater.* **2007**, *6*, 501–507. [CrossRef] [PubMed]
105. Huang, A.; Chen, Y.; Wang, N.; Hu, Z.; Jiang, J.; Caro, J. A highly permeable and selective zeolitic imidazolate framework ZIF-95 membrane for H_2/CO_2 separation. *Chem. Commun.* **2012**, *48*, 10981–10983. [CrossRef] [PubMed]
106. Bhin, K.M.; Tharun, J.; Roshan, K.R.; Kim, D.-W.; Chung, Y.; Park, D.-W. Catalytic performance of zeolitic imidazolate framework ZIF-95 for the solventless synthesis of cyclic carbonates from CO_2 and epoxides. *J. CO2 Util.* **2017**, *17*, 112–118. [CrossRef]
107. Liang, J.; Chen, R.-P.; Wang, X.-Y.; Liu, T.-T.; Wang, X.-S.; Huang, Y.-B.; Cao, R. Postsynthetic ionization of an imidazole-containing metal–organic framework for the cycloaddition of carbon dioxide and epoxides. *Chem. Sci.* **2017**, *8*, 1570–1575. [CrossRef] [PubMed]
108. Ding, L.-G.; Yao, B.-J.; Jiang, W.-L.; Li, J.-T.; Fu, Q.-J.; Li, Y.-A.; Liu, Z.-H.; Ma, J.-P.; Dong, Y.-B. Bifunctional Imidazolium-Based Ionic Liquid Decorated UiO-67 Type MOF for Selective CO_2 Adsorption and Catalytic Property for CO_2 Cycloaddition with Epoxides. *Inorg. Chem.* **2017**, *56*, 2337–2344. [CrossRef]
109. Kurisingal, J.F.; Rachuri, Y.; Pillai, R.S.; Gu, Y.; Choe, Y.; Park, D. Ionic-Liquid-Functionalized UiO-66 Framework: An Experimental and Theoretical Study on the Cycloaddition of CO_2 and Epoxides. *ChemSusChem* **2019**, *12*, 1033–1042. [CrossRef]
110. Ding, M.; Jiang, H.-L. Incorporation of Imidazolium-Based Poly(ionic liquid)s into a Metal–Organic Framework for CO_2 Capture and Conversion. *ACS Catal.* **2018**, *8*, 3194–3201. [CrossRef]
111. Pander, M.; Janeta, M.; Bury, W. Quest for an Efficient 2-in-1 MOF-Based Catalytic System for Cycloaddition of CO_2 to Epoxides under Mild Conditions. *ACS Appl. Mater. Interfaces* **2021**, *13*, 8344–8352. [CrossRef] [PubMed]
112. Zhu, J.; Usov, P.M.; Xu, W.; Celis-Salazar, P.J.; Lin, S.; Kessinger, M.C.; Landaverde-Alvarado, C.; Cai, M.; May, A.M.; Slebodnick, C.; et al. A New Class of Metal-Cyclam-Based Zirconium Metal–Organic Frameworks for CO_2 Adsorption and Chemical Fixation. *J. Am. Chem. Soc.* **2018**, *140*, 993–1003. [CrossRef] [PubMed]
113. Vitillo, J.G.; Savonnet, M.; Ricchiardi, G.; Bordiga, S. Tailoring Metal-Organic Frameworks for CO_2 Capture: The Amino Effect. *ChemSusChem* **2011**, *4*, 1281–1290. [CrossRef] [PubMed]
114. Torrisi, A.; Bell, R.G.; Mellot-Draznieks, C. Functionalized MOFs for Enhanced CO_2 Capture. *Cryst. Growth Des.* **2010**, *10*, 2839–2841. [CrossRef]
115. Senthilkumar, S.; Maru, M.S.; Somani, R.S.; Bajaj, H.C.; Neogi, S. Unprecedented NH_2-MIL-101(Al)/n-Bu_4NBr system as solvent-free heterogeneous catalyst for efficient synthesis of cyclic carbonates via CO_2 cycloaddition. *Dalton Trans.* **2018**, *47*, 418–428. [CrossRef]
116. Zhu, M.; Carreon, M.A. Porous crystals as active catalysts for the synthesis of cyclic carbonates. *J. Appl. Polym. Sci.* **2014**, *131*. [CrossRef]
117. Cai, T.; Liu, J.; Cao, H.; Cui, C. Synthesis of bio-based cyclic carbonate from vegetable oil methyl ester by CO_2 fixation with acid-base pair MOFs. *Ind. Crop. Prod.* **2020**, *145*, 112155. [CrossRef]
118. Mercuri, G.; Moroni, M.; Domasevitch, K.V.; Di Nicola, C.; Campitelli, P.; Pettinari, C.; Giambastiani, G.; Galli, S.; Rossin, A. Carbon Dioxide Capture and Utilization with Isomeric Forms of Bis(amino)-Tagged Zinc Bipyrazolate Metal–Organic Frameworks. *Chem. A Eur. J.* **2021**, *27*, 4746–4754. [CrossRef]
119. Norouzi, F.; Khavasi, H.R. Diversity-Oriented Metal Decoration on UiO-Type Metal-Organic Frameworks: An Efficient Approach to Increase CO_2 Uptake and Catalytic Conversion to Cyclic Carbonates. *ACS Omega* **2019**, *4*, 19037–19045. [CrossRef]
120. Li, J.; Ren, Y.; Yue, C.; Fan, Y.; Qi, C.; Jiang, H. Highly Stable Chiral Zirconium–Metallosalen Frameworks for CO_2 Conversion and Asymmetric C–H Azidation. *ACS Appl. Mater. Interfaces* **2018**, *10*, 36047–36057. [CrossRef]

MDPI
St. Alban-Anlage 66
4052 Basel
Switzerland
Tel. +41 61 683 77 34
Fax +41 61 302 89 18
www.mdpi.com

Inorganics Editorial Office
E-mail: inorganics@mdpi.com
www.mdpi.com/journal/inorganics

www.ingramcontent.com/pod-product-compliance
Lightning Source LLC
LaVergne TN
LVHW070222100526
838202LV00015B/2077